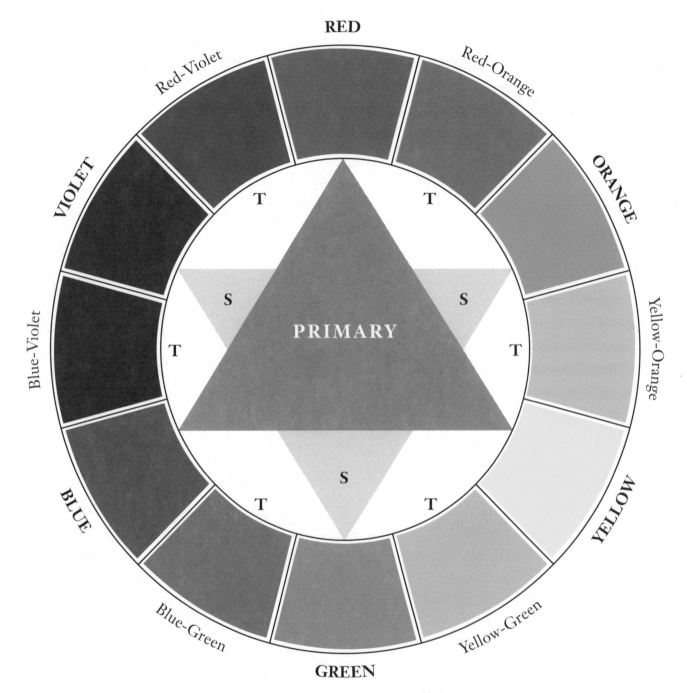

The color wheel shows an orderly arrangement of the basic colors—primary, secondary, and tertiary—and is helpful in selecting and mixing. The use of color in rendering, as well as the specifying of colors for other architectural applications, must start with a familiarity of the color wheel if pleasing combinations are to be achieved.

ARCHITECTURAL DRAWING
AND LIGHT CONSTRUCTION

FIFTH EDITION

ARCHITECTURAL
DRAWING
AND LIGHT
CONSTRUCTION

Edward J. Muller

James G. Fausett

Philip A. Grau III

PRENTICE HALL
Upper Saddle River, New Jersey   Columbus, Ohio

Cover photo: © Mason Morfit 1996/FPG International
Editor: Ed Francis
Production Editor: Christine M. Harrington
Design Coordinator: Karrie M. Converse
Text Designer: John Edeen
Cover Designer: Brian Deep
Production Manager: Patricia A. Tonneman
Marketing Manager: Danny Hoyt

This book was set in Jansen by York Graphic Services
and was printed and bound by Courier Kendallville, Inc.
The cover was printed by Phoenix Color Corp.

© 1999 by Prentice-Hall, Inc.
Simon & Schuster/A Viacom Company
Upper Saddle River, New Jersey 07458

Printed in the United States of America

10 9 8 7 6 5 4 3 2

ISBN: 0-13-520529-8

Prentice-Hall International (UK) Limited, *London*
Prentice-Hall of Australia Pty. Limited, *Sydney*
Prentice-Hall of Canada, Inc., *Toronto*
Prentice-Hall Hispanoamericana, S. A., *Mexico*
Prentice-Hall of India Private Limited, *New Delhi*
Prentice-Hall of Japan, Inc., *Tokyo*
Simon & Schuster Asia Pte. Ltd., *Singapore*
Editora Prentice-Hall do Brasil, Ltda., *Rio de Janeiro*

Library of Congress Cataloging-in-Publication Data
Muller, Edward John, 1916-1996
    Architectural drawing and light construction / Edward J.
Muller, James G. Fausett, Philip A. Grau. — 5th ed.
        p.   cm.
    Includes bibliographical references and index.
    ISBN 0-13-520529-8
    1. Architectural drawing.   2. Building—Equipment and
supplies—Drawings.   3. Computer-aided design.
I. Fausett, James G.   II. Grau, Philip A.   III. Title.
NA2700.M8   1999   98-26035
720′.28′4—dc21   CIP

This text has for many years been an excellent introduction to the fields of architectural drawing and construction. In recent years a number of new texts have been published purporting to cover the same information, but none provides a comparable introduction to the basics of architectural practice and design, or construction documentation and practices. The text was outdated in parts, however. My reason for working on this revision was to bring up to date those portions of the text that had not been revised in some years without losing the excellent, thorough treatment of beginning architecture and construction education.

Throughout the new edition we have combined the best of the existing manually drafted graphics with new computer-generated graphics. Most chapters have been updated to include information from the latest building codes and about new materials being introduced into residential and light commercial construction.

The order of the chapters in the text has been revised to more closely reflect how the subject matter is currently taught. Chapters 1–11 address drafting and drawing principles. Chapters 12–18 deal with light frame construction principles and the building design and documentation process. Chapter 19 looks at residential mechanical, electrical, and plumbing systems; and Chapter 20 reviews the design and construction of a small, energy-efficient commercial building.

Computers and Internet technology are introduced near the beginning of the text as components that are now integrated in the building design and construction process. There is updated information on hardware and software along with some discussion of what the future may hold in these rapidly changing areas.

The chapter on metric and modular drafting has been updated to reflect the current place of metric measure in construction. We have added an expanded appendix that includes typical metric sizes used in frame construction.

There is also significant expansion of the subject matter in Chapter 12 on light construction principles. New information addresses footing and foundation design, termite and radon control techniques, wood foundation design, glued engineered wood products, steel frame house construction, insulating foam concrete forms systems, and straw bale construction. We have also expanded the section on insulation and strategies to protect the whole house from heat loss, heat gain, and moisture problems, including weather barriers, moisture barriers, and damproofing.

Chapter 13 on structural member selection was updated to reflect current design standards for conventional wood frame techniques as well as for the newer systems such as steel and engineered wood introduced in Chapter 12. An expanded appendix includes design tables for a number of structural materials under various loading conditions.

Chapter 14 on typical architectural details has also been greatly expanded. Particularly significant is the section on new, innovative construction systems and materials that can be used as alternatives to wood framing and provide increased energy efficiency. One of the best current examples of this approach to design and construction is the 21st Century Townhouses built by the National Association of Homebuilders, with architects John T. Stovall of John Stovall and Associates and Daniel J. Ball, AIA, of Daniel Ball & Associates, which is used as a reference in the chapter.

Chapter 19 on mechanical, electrical, and plumbing systems has been updated to contain information relevant to current code requirements and various systems now available.

Chapter 20 on small commercial buildings has been significantly revised. Large-scale commercial building design and construction are topics beyond the scope of this text. Smaller commercial buildings are often designed and built using materials and construction systems very similar to those used today in residential construction. The Southface Energy and Environmental Resource Center, used as the primary example for this chapter, was developed as a state-of-the-art demonstration center for saving energy, water, and natural resources. It is built in a residential style in order to demonstrate the applicability of a wide variety of energy efficient features to both residential and small commercial construction.

I would like to extend a sincere thank you to all the individuals and organizations, too numerous to mention, who contributed material for this new edition. Special thanks to Dennis Creech of the Southface Energy Institute and to W. Allen Hoss, AIA, and Randy E. Pimsler, AIA, of Pimsler Hoss Architects of Atlanta, and their consultants, for their drawings and information about the Southface Resource Center. Also, I wish to thank the National Association of Home Builders and John T. Stovall of John T. Stovall Architects for the material on the 21st Century Townhouses.

I would like to recognize the incredible job done by Ed Francis, executive editor at Prentice Hall, who helped guide me through the complexities of putting together this manuscript, and to JoEllen Gohr, Steve Robb, Maggie Diehl, and Christine Harrington and all their coworkers of the Prentice Hall production editorial department for their hard work and patience in producing this excellent edition.

I would like to thank all of my former and present students, whose interest in architecture and construction has served to keep me motivated to stay abreast of current trends in these areas and to help them understand how these changes will affect their futures.

Finally, I would like to thank my wife Janet and my son Peter for their patience and understanding during this undertaking.

Philip A. Grau III
NCARB, AIA, CSI

CONTENTS

FRANK BETZ

"Man is a tool-using animal . . . without tools he is nothing, with tools he is all."
—*THOMAS CARLYLE*

The first consideration in the mastery of the mechanics of drafting is, naturally, the selection of the proper equipment. Many instruments are available at drafting supply stores—so many, in fact, that it would be bewildering for a beginning student to select appropriate and proper equipment without getting a few words of preliminary advice. Comparatively few pieces of equipment are necessary if carefully chosen. In selection of any of the instruments, the first advice is to buy the best quality you can afford. Second, purchase only equipment produced by reputable manufacturers; you will then be assured of accuracy and good design. Many instruments and pieces of equipment are continually being improved and redesigned to aid the drafter. Third, good equipment deserves proper care and treatment; do not abuse it or use it for any purpose other than the purpose for which it was designed.

To get the most from equipment, the aspiring architectural drafter must develop enough skill to have complete control of his or her instruments. Such skill comes with *practice*, deliberate and well-directed practice, which develops good work habits and eventually leads to the attainment of speed. Early training in drawing places particular emphasis on accuracy and skill until correct work habits are formed; speed will develop gradually with experience.

1.1

A LIST OF NECESSARY EQUIPMENT

The following list is the recommended minimum equipment for student use. Refer to Fig. 1–1, which illustrates each piece of basic equipment listed below. Although some of the equipment is identical to that used in engineering drawing, each piece is nevertheless specifically appropriate for architectural drawing:

1. Drawing board (24″ × 36″) (32″ × 48″)
2. T square or parallel bar (36″–48″)
3. 30°–60° plastic triangle (14″)
4. 45° plastic triangle (10″)
5. 12″ architect's scale, triangular
6. Two mechanical pencils, several 3H, 2H, H, F, and non-print leads
7. Pencil pointer, either sandpaper or mechanical pointer
8. Erasers, 1 Pink Pearl, 1 artgum, 1 white plastic, 1 kneaded
9. Roll of drafting tape
10. Erasing shield
11. Irregular curve or French curve
12. Adjustable triangle (a small size is sufficient)
13. Small set of case instruments
14. Lettering guide
15. Architectural template (¼″ = 1′-0″) (select one with circles)
16. Dusting brush
17. Tracing paper

The following list of supplemental equipment, shown in Fig. 1–2, will be useful for advanced students or drafters working in architectural offices:

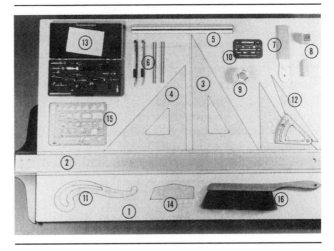

Figure 1–1 Basic drafting equipment.

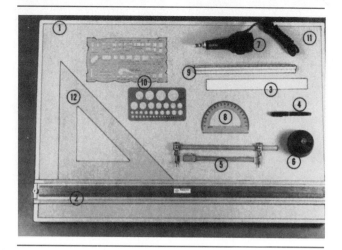

Figure 1–2 Supplementary drafting equipment.

1. Large drawing table (38″ × 72″), drafting stool (30″)
2. Parallel bar (60″)
3. Flat architect's scale, 4-bevel
4. Technical drafting pen, India ink
5. Beam compass
6. Mechanical pencil pointer
7. Electric eraser
8. Protractor, semicircular
9. Civil engineer's scale
10. Additional architectural templates
11. Vinyl plastic pad for drawing surface
12. Large 45° plastic triangle (14″)
13. Plastic drafting leads, E-1, E-2, and E-3
14. Mylar drafting film
15. Drafting lamp

Figure 1–3 Using the drafting equipment on the drawing table.

1.2

DRAWING BOARDS AND TABLES

The drawing board or table should be large enough to accommodate the anticipated drawing sheet sizes. Architectural drawings may be 24″ to 48″ long by 24″ to 36″ wide, thus requiring large boards; beginning work can usually be placed on a 24″ by 36″ board. The board should have a perfectly smooth surface and be rigid enough to prevent warping and bending. The edges that guide the T square must be straight and true. Students often rely on portable drawing boards for beginning work. Boards made of seasoned basswood or clear white pine, laminated or constructed to prevent warping, are relatively light and easily transported (Fig. 1–3). The drawing surface of a wood board should be protected with a paper or vinyl pad. An inclined board with a 10° or 15° slope from the horizontal is a desirable position for drafting. Portable boards with collapsible supports under the corners are available.

A flush, solid-core wood door 2′-6″ or 3′-0″ wide by 6′-8″ or 7′-0″ long supported on legs or cabinets is a popular choice for a drawing board (Fig. 1–4). Care should be taken to select a door that is not warped and that has one smooth surface. A 42″ to 60″ parallel bar and surface cover should be placed near the center of the door to allow space for instruments at the sides.

Although some drafters use a drafting chair and a board height of 28″ to 30″ at the lower edge, most prefer a drafting stool and a board height of 36″ to 40″ at the lower edge. It is desirable to select or construct a drafting board or table that can be adjusted in height and slope. The drafter occasionally moves from a seated position to a standing position to alleviate fatigue.

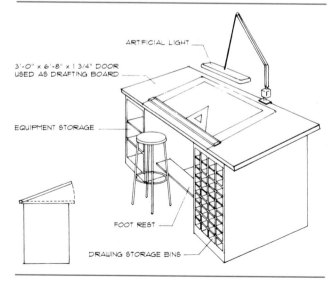

Figure 1–4 An economical table can be made from a flush door.

Many types of drafting boards and tables are marketed today. Some are constructed of melamine-covered particle board with lightweight tubular steel legs; these boards are more suitable for the occasional drafter. Elaborate tables are available with sturdy wood or vinyl-laminated metal tops. These tables have adjustable slopes and heights. Most have metal edges to prevent warping and provide an easy sliding edge for the T square.

A simple metal drafting stool with a wood or metal seat 27″ to 28″ high is frequently used by the beginning drafter. The swivel chair, 17″ to 18″ high, and the swivel stool, 27″ to 28″ high, on casters are used by the professional drafter. Most of these are designed with a foot rest and have upholstered seats and backs that are adjustable.

Drafting work stations are common in the professional workplace and in some schools. These work stations consist of a drafting table, a CADD monitor, a keyboard, a digitizer pad, a layout table, file drawers, equipment storage cabinets, shelving, and a drafting chair (Fig. 1–5).

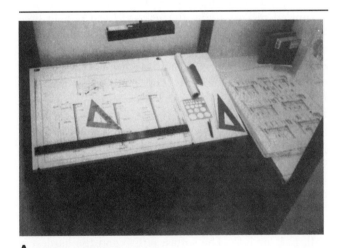

A

B

Figure 1–5 (A) Student's drafting space. (B) Professional's drafting work station.

BOARD COVERS AND DRAFTING SURFACES

The surface of the board is very important if good drafting techniques are to be attained. It is never desirable to draft on an unprotected wood drawing board unless the drafting medium is an illustration board or a very heavyweight paper. Sharp pencil leads etch the wood surface. Paper underlays are satisfactory if they have a slightly spongy effect; however, paper will usually retain grooves from linework, occasionally necessitating replacing the paper pads. Plastic-coated tinted paper with or without a ⅛″ vertical and horizontal grid is available for an underlay. The grid serves as guidelines for tracing paper overlays. The resilient sheet-vinyl pad is the most desirable board cover. Available in light tint, this product is durable and provides an excellent drafting surface.

All drawing board covers should be installed with care. Paper covers are usually wrapped around the edges of the drawing board and taped or tacked to the back surface. Vinyl pads simply lie on the drafting surface and can be stabilized with double-faced tape or tacks at the edges. Vinyls tend to expand and contract with temperature change. The use of glues and mastics to attach papers or vinyl pads should be avoided. Glues can dissolve the vinyl pad and produce irregular surfaces under paper and pad covers, making drafting difficult. Board covers should lie perfectly flat.

PARALLEL BARS AND DRAFTING MACHINES

Many drafters save time by using the parallel bar, which is always kept parallel by braided wires that are attached to the board at the corners and that run on pulleys through the bar (Fig. 1–2). Hardwood or hard plastic parallel bars with edges of clear plastic in length of 36″, 42″, 48″, 54″, 60″, 72″, 84″, and 96″ are standard. Some are adapted with small ball-bearing rollers underneath the bars to minimize direct contact with the drawing surface and to facilitate sliding the bar up and down the board. Parallel bars with a raised blade above the edges are desirable because the blade serves as a finger grip and allows the drafter to grasp the bar easily. A setscrew and a compression plate at the top of the drawing table held tight against the braided wire keep the bar parallel. The drafter can adjust the bar by loosening the setscrew, sliding the bar to the stops at the bottom edge of the drawing table, and then tightening the setscrew against the compression plate.

Drafting machines are more popular with the engineering drafter than with the architectural drafter. The arm-style drafting machine is a mechanical device that is attached to the top edge of the drafting table and has a pair of pinned arms with an elbow, a head with a pro-

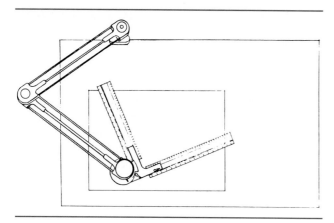

Figure 1–6 Arm-style drafting machine.

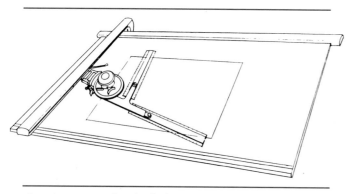

Figure 1–7 Track-style drafting machine.

tractor, and both vertical and horizontal scales (Fig. 1–6). The drafter can move the drafting machine horizontally, vertically, and diagonally across the drafting surface. It is a convenient device for drawing sloping lines. The protractor on the head can be set at any angle, and bars can be interchanged with different scales. The length of lines drawn at any one position is limited to the length of the scale, which is approximately 18″ long.

Track drafting machines are designed with a fixed horizontal track at the top edge of the drawing board and a movable vertical track attached to the horizontal track (Fig. 1–7). The protractor head is mounted on the vertical track. Two scales can be attached to the head. Track drafting machines offer versatility and accuracy. Both the arm-style and the track machines are fabricated from metals and hard plastics. Wide varieties of clear acrylic and metal scales ranging in length from 12″ to 18″ are available for the drafting machine.

1.5

PENCILS, LEADS, AND POINTERS

Traditional wooden pencils are hexagonal in shape and are available in 17 degrees of hardness, which are indicated by letters and numbers near the end of the pencil (Fig. 1–8).

Lead holders or mechanical pencils manufactured from metals and hard plastics hold sticks of lead and are popular tools of the drafter (Fig. 1–9). These pencils have a knurled surface at the finger area. Lead is held by a spring chuck and can be quickly released for sharpening or removal by pressing the cap. Lead holders accommodate 2-mm-diameter lead, and mechanical pencils hold 0.5-, 0.7-, and 0.9-mm sizes. Leads in varying degrees of hardness are available for both the lead holders and the mechanical pencils.

Graphite-clay leads are used predominantly with papers and illustration boards. The majority of architectural drafting can be accomplished with 2H, H, and F leads. Plastic leads in 6 degrees of hardness are available for drafting on film and are produced in 2-, 0.5-, 0.7-, and 0.9-mm sizes. The degree of hardness of plastic leads is designated by a letter followed by a number, such as E-0, E-1, or E-2. The softest is E-0 and the hardest is E-5. Most drafting on film is accomplished using E-1, E-2, and E-3 plastic leads. *Non-print* or *non-photo* pencils and leads are used for making guidelines and notes on drawings. Marks made by the blue-colored leads will not appear on most reproductions of the original drawing.

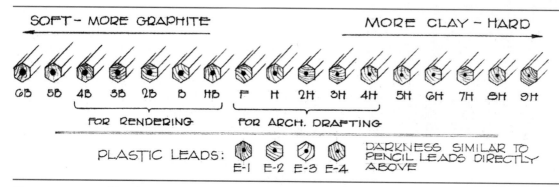

Figure 1–8 Pencil hardness chart.

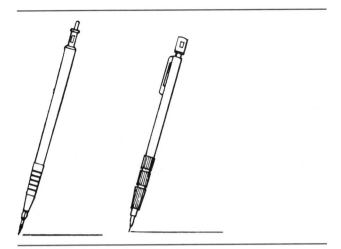

Figure 1-9 Lead holder and mechanical pencil.

A sharp lead point is essential to good line quality. Leads can be sharpened on sandpaper pads; however, most drafters use a mechanical lead pointer. Several models of mechanical pointers are available (Fig. 1-10). Some pointers are small and must be held in the hand and revolved to create a point. A weighted metal pointer that can be located at random on the drawing table uses a circular sandpaper insert to create a sharp point. Another style of lead pointer clips to the edge of the drafting table. Lead dust collects inside each pointer and must be removed periodically. Both graphite lead and plastic lead can be sharpened using the same style of pointer.

The main consideration in choosing a lead is the desired sharpness of lines and the opaqueness of the lead for reproduction. Softer leads wear faster and require re-pointing, whereas hard leads may groove or cut the paper. Different drafters produce various line weights with the same leads; experience will lead the individual to the correct lead to use for the desired lines. Remember that

the pressure used by the drafter, the texture of the paper, atmospheric conditions, the accuracy required, and the opaqueness needed for reproduction should all be considered in selection of a lead.

1.6

TRIANGLES

Relatively large triangles are needed for architectural drawings; a 10″, 45° triangle and a 14″, 30°–60° triangle are satisfactory for the work covered in this book (Fig. 1-1). Triangles should be made of heavy, transparent acrylic plastic about 0.06″ thick with true, accurate edges. Inside edges should be beveled to facilitate picking up the triangle with the fingernails. Lucite plastic triangles are available in tinted fluorescent colors that transmit light through their edges, thus reducing shadows near the edges of the instrument.

Triangles are used to draw vertical lines or diagonal lines at 30°, 45°, or 60°. First, position the head of the parallel bar against the edge of the board. Place the triangle with one of its legs against the upper edge of the parallel bar; slide the left hand to the lower part of the triangle and blade of the parallel bar. Maintain a slight pressure on the instruments until a line is drawn upward along the vertical edge of the triangle (Fig. 1-11).

The left-handed drafter should hold the parallel bar blade and the triangle with the right hand and draw vertical lines upward, away from the body. The vertical ruling edge of the triangle should be placed so that the vertical line will be drawn along the edge closest to the head of the parallel bar. Refrain from pulling the pencil down along the vertical side of the triangle. If long vertical or diagonal lines are required, stop the line short of the end of the triangle, and slide the triangle and parallel bar up so that the line can be continued (Fig. 1-12).

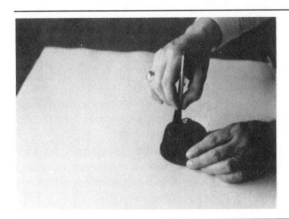

Figure 1-10 Pointing the lead with a mechanical lead pointer.

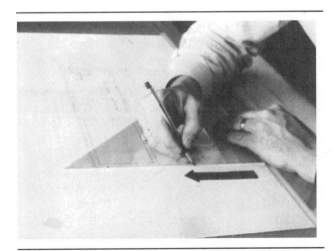

Figure 1-11 Drawing vertical lines with a triangle (reverse the position for left-handed drafters).

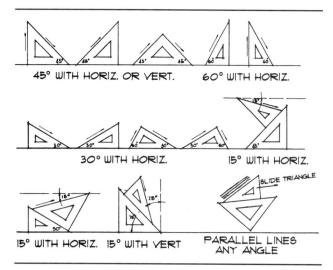

45° WITH HORIZ. OR VERT.　　60° WITH HORIZ.

30° WITH HORIZ.　　15° WITH HORIZ.

15° WITH HORIZ.　15° WITH VERT　PARALLEL LINES ANY ANGLE

Figure 1-12 Drawing various lines with the use of the triangle.

Figure 1-14 Erasing with a holder-type eraser.

1.7

ADJUSTABLE TRIANGLES

The adjustable triangle is used to draw diagonal lines at angles other than 30°, 45°, or 60°. It has a protractor scale adjustment from 0° to 45° from the horizontal or base line. The outer protractor scale indicates angles from 0° to 45° from the longer base, while the inner scale indicates angles from 45° to 90° from the shorter base (Fig. 1–13). Some adjustable triangles have scales that indicate roof slopes and stair risers. The triangle can be adjusted by loosening or tightening a setscrew, which also serves as a convenient lifting handle. A small adjustable triangle with a 6″ to 8″ leg is most useful.

1.8

ERASERS

Even the most advanced drafters must occasionally erase, but excessive erasing should be avoided. Selecting the proper eraser for a specific lead type that will not damage or discolor the surface of the paper or film is important. Block erasers such as artgum, soft pink, or white plastic are effective in removing graphic pencil or plastic lead from tracing paper and film. Paper-wrapped, pencil-shaped erasers can be reshaped by peeling the paper from the end of the eraser. Pencil-type eraser holders are available for soft pink or white plastic eraser sticks (Fig. 1–14). Block or stick erasers in pastel tints such as blue, yellow, or green and imbibed with an erasing fluid are suitable for erasing ink on film and some selected papers. A residue left from the erasing fluid should be removed from the film or paper with a standard soft pink or white plastic eraser before drafting commences.

1.9

ELECTRIC ERASING MACHINES

Mechanical erasers are popular with drafters; they are available in numerous sizes and styles (Fig. 1–15). Most have an electrical power cord, but some are cordless and operate with rechargeable batteries. All mechanical erasers use replaceable eraser sticks that are held securely on the end of the shaft with a screw-type collet. When using mechanical erasers, drafters must take care to prevent erasing holes in papers and films.

Drafters frequently clean a drawing during and after the drawing process. Some drafters use an artgum eraser for removing guidelines and smudges. Others prefer a dry cleaning pad, which is a small cloth bag filled with a granular eraser dust. Placing too much pressure on the eraser or dry cleaning pad can result in the removal of important lines.

Figure 1-13 Drawing lines with the adjustable triangle.

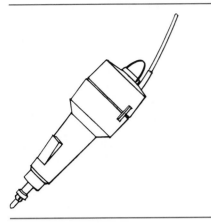

Figure 1–15 Electric erasing machine.

Figure 1–16 Using the erasing shield.

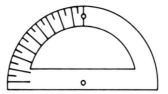

Figure 1–17 Protractor.

1.10
ERASING SHIELD

The erasing shield is a thin, metal instrument with holes of various shapes. To erase with the shield, select the opening that best fits the line or area to be eradicated and hold the erasing shield down firmly on the drawing. Then rub the line with the appropriate eraser (Fig. 1–16). Not only is it handy when eradicating small lines in confined areas without disturbing the lines, but it can also save time in other situations. For example, thin columns in front of a brick elevation view can be erased from the wall symbol linework much faster than starting and stopping the brickwork on each side of the column.

1.11
PROTRACTOR

Odd angles, other than those drawn with a 45° triangle, a 30°–60° triangle, or a combination of both, can be laid out with the protractor (Fig. 1–17). It is a semicircular, plastic instrument with degree calibrations marked off along its curved edge and a center or an apex identified in the center of its straight side. The two reversed scales along the curvature of the protractor allow angles to be laid out in either direction from the horizontal. More expensive protractors have ruling bars attached to the pivot point or apex for easier angle layout. Many drafters use the calibrated adjustable triangle instead of a protractor when odd angle layout is required.

1.12
IRREGULAR CURVES

Regular curves, that is, true circles and arcs, are drawn with a compass or a circle template, but often it is necessary to draw irregular curves—those with nonuniform curvature. Many clear acrylic curves are manufactured. Experience has shown that the most useful irregular curve is one that has a number of small curvatures, as well as a few long, slightly curved edges (Fig. 1–18).

To draw an irregular curved line, first accurately plot the points through which the curve should pass, and then select the edge on the irregular curve that will coincide with the most points on the paper. Intersecting lines on a curve should be continuous. If considerable curved work is involved and the curvatures are not too small, an adjustable curve will save time and result in neater lines. This instrument is made of a flexible metal core covered with a rubber or metal ruling edge (Fig. 1–19). It can be bent to the desired curvature before the line is drawn. Long, accurate curves often require the use of a flexible spline curve that is held in place on the drawing with metal weights. Ship curves or highway curves are convenient instruments for drawing long, smooth, uninterrupted lines.

Figure 1–18 Irregular curves.

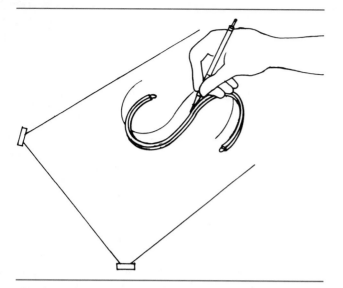

Figure 1-19 Adjustable curve.

Figure 1-21 Using the triangular architect's scale.

1.13

THE ARCHITECT'S SCALE

Most measurements on architectural drawings are made with the architect's scale. Scales are made of wood, plastic, or wood laminated with plastic. Sharp edges and distinct, machine-divided markings on the scale are necessary for accurate measurements. Scales are usually flat or triangular in shape and are available in 6″, 12″, and 18″ lengths, with the triangular-shaped 12″ length being the most popular. The edges of all scales are beveled (see Fig. 1–20). Scales are fully divided or open divided. Fully divided scales have each main unit throughout the entire scale fully subdivided. Open-divided scales have the main units undivided and a fully subdivided extra unit placed at the zero end of the scale; the architect's scale that is calibrated in feet and inches is open divided. Listed below are the 11 scales found on the triangular architect's scale:

Full scale—$^1/_{16}″$ graduations
$^1/_8″ = 1'-0″$
$^1/_4″ = 1'-0″$ (one-forty-eighth size)
$^3/_8″ = 1'-0″$
$^3/_4″ = 1'-0″$
$^1/_2″ = 1'-0″$ (one-twenty-fourth size)
$1″ = 1'-0″$ (one-twelfth size)
$1^1/_2″ = 1'-0″$ (one-eighth size)
$3″ = 1'-0″$ (one quarter size)
$^3/_{32}″ = 1'-0″$
$^3/_{16}″ = 1'-0″$

Two different scales are combined on each face except the full-size scale, which is fully divided into sixteenths. The combined scales are compatible; one is twice as large as the other, and their zero points and extra subdivided units are on opposite ends of the scale. The extra unit near the end of the scale is subdivided into twelfths of a foot or inches, as well as fractions of inches on the larger scales.

Accuracy in using the scale when preparing drawings is extremely important. To measure correctly with the scale, select the proper scaling edge and place it on the drawing with the edge parallel to and slightly below the line to be measured (Fig. 1–21). First, count from

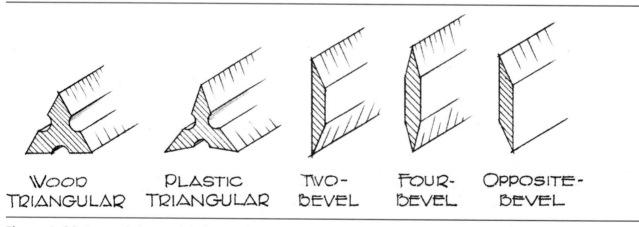

WOOD TRIANGULAR   PLASTIC TRIANGULAR   TWO-BEVEL   FOUR-BEVEL   OPPOSITE-BEVEL

Figure 1-20 Sectional shapes of drafting scales.

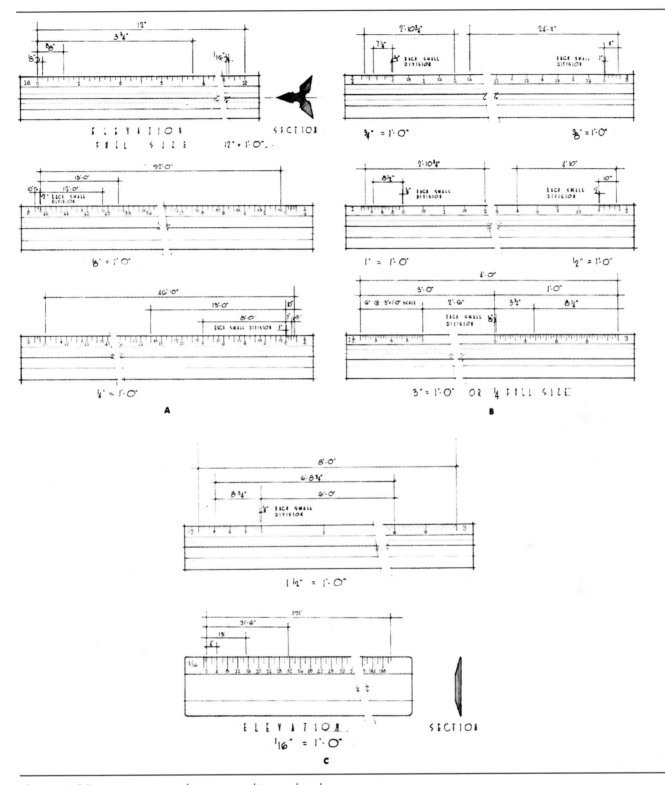

Figure 1–22 Measurements with various architectural scales.

zero the full number of feet in the dimension, and using a sharp pencil make a dash with a non-print pencil at right angles to the scale. Then count the inches (if the dimension has both feet and inches) from the same zero in the opposite direction on the scale, and mark the other limit of the dimension (Fig. 1–22). Laying out fractional parts of an inch at smaller scales will require interpolation between the calibrations. To avoid accumulative errors, a series of dimensions should be marked off, if possible, without moving the scale. Unequal dimensions should be added and then marked off consecutively from the original reference point. Always check the total length of the line to ensure that no accumulative error has resulted.

1.14

THE CIVIL ENGINEER'S SCALE

Occasionally, a civil engineer's scale (Fig. 1–23) is needed by architectural drafters for drawing plot plans, land measurement, and stress diagrams. Civil engineer's scales are always fully divided and can be obtained in any of the flat or triangular types (Fig. 1–20). Scales found on the triangular type are $1'' = 10'$, $1'' = 20'$, $1'' = 30'$, $1'' = 40'$, $1'' = 50'$, and $1'' = 60'$. Land measurement is always recorded in feet and decimal parts of a foot, rather than in inches. Therefore, the inch unit on the $1'' = 10'$ scale would have 10 divisions, and the inch unit on the $1'' = 20'$ scale would have 20 divisions, each representing feet throughout the other scales. If you needed a scale of $1'' = 100'$, you could use the $1'' = 10'$ scale by merely letting each inch represent $100'$ instead of $10'$. This method applies to other scales as well.

1.15

DRAFTING TEMPLATES

Templates are made of thin plastic with variously sized or scaled openings cut in the sheet. Many symbols used on architectural drawings can be found on templates. Circles, squares, rectangles, triangles, ovals, hexagons, arrows, door swings, furniture, plumbing fixtures, and electrical symbols are a few of the items that are available in template form. Architectural templates are available in scales of $1/8'' = 1'$, $1/4'' = 1'$, and $1/2'' = 1'$ (Fig. 1–24). In using the template, first select the proper-size symbol and hold it firmly over the drawing in the correct location. Then, using a conical point, draw a sharp, clear line around the opening in the template. To draw a circle with a template, lay out vertical and horizontal center lines on the drawing where the circle is to be drawn. Then superimpose the correct circle opening over the center line. To draw an arc, first lay out the tangent lines on the drawing. Then place the correct-size circle on the paper so that the quadrant lines coincide with the tangent lines. In placing the opening in the correct position on the drawing, you must make allowance for the width of the pencil line.

1.16

CASE INSTRUMENTS

The main items needed in architectural drafting are a compass, dividers, a ruling pen, a lead holder, needle points, and a screwdriver, all in a good-quality case. Quality instruments are carefully finished and have well-fitted components without play or tolerance in their joints or threaded parts (Fig. 1–25).

The compass is the most frequently used instrument and is available in three basic types of different sizes. The *friction* compass is adjustable at its hinged joints. The *center wheel* and *side wheel* compasses can be accurately set by revolving a thumb wheel on a threaded bar that passes through both legs of the instrument (Fig. 1–26). The *6" large bow compass* is the most versatile. When drawing a

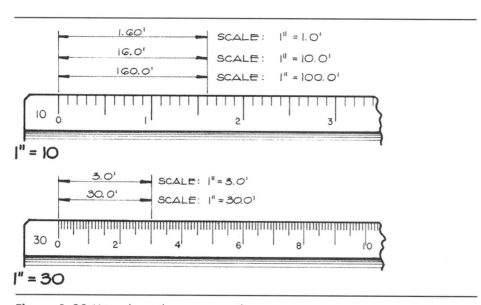

Figure 1–23 Using the civil engineer's scale.

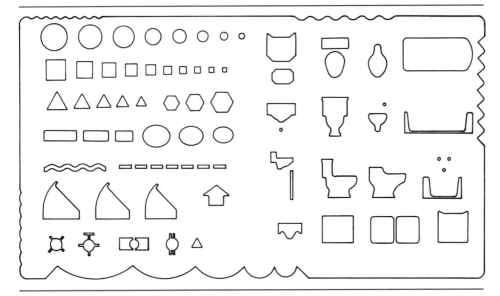

Figure 1-24 Architectural drafting templates.

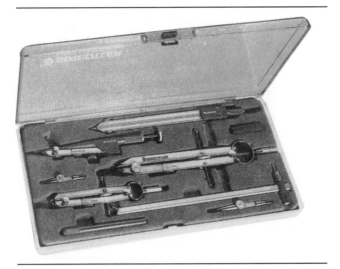

Figure 1-25 Case instruments.

circle with the bow compass, hold it with your thumb and forefinger, and lean it slightly in the direction of travel as you twirl it between your fingers. The needle end of the compass should be slightly longer than the pencil-lead-point end. Sharpen the compass lead point with the sandpaper pad by forming a single outside bevel on the lead (Fig. 1–27). A ruling pen attachment can be inserted in place of the pencil-lead attachment. By removing the pencil lead or inking attachment and inserting a needle point, you can use the compass as a divider. Some large bow compass sets are equipped with beam-lengthening bars, if circles up to 26″ diameter are needed. For still larger circles, the standard beam compass may be used with a rigid metal beam and beam extensions; these have adjustable needle points, pencil lead, and inking attachments.

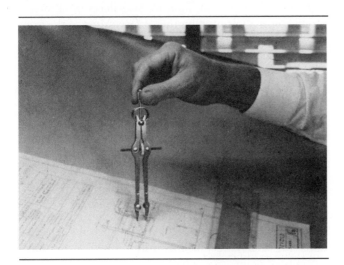

Figure 1-26 Drawing an arc with the compass.

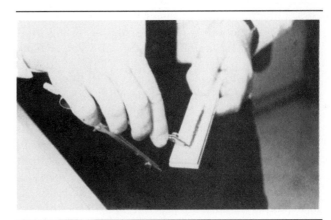

Figure 1-27 Sharpening the compass lead point with sandpaper pad.

1.17

DRAFTING (DUSTING) BRUSH

A drafting (dusting) brush (shown in Fig. 1–1) should be a part of every drafter's equipment. It will quickly remove eraser, dust, and graphite particles from the drawing without smudging the linework. Do not dust a drawing with your hands or blow on the drawing. Students can use an inexpensive 3″ paintbrush on small drawings as a substitute, but the regular drafting brush will save time on larger drawings. A clean cloth in the drafting kit is indispensable for wiping instruments and for other minor cleaning jobs.

1.18

DRAFTING TAPE AND DOTS

Drafting tape, ⅝″ or ¾″ wide, and drafting dots, ⅞″ in diameter, provide a convenient method for securing drawings to the board. Tape can be purchased in 10- to 60-yard rolls; some rolls come in packages with handy tear-off cutting edges. Drafting dots are attached to a coated paper tape and are boxed in units of 500. The round dots pop up from the paper tape as the tape is pulled from the storage box. Small pieces of drafting tape or dots are placed at the corners of the drawing to attach it to the board. The tape or dots can be removed without leaving marks or gum residue on the drawing.

1.19

DRAWING PAPERS AND FILM

Although architectural drafting work can be done on good-quality, opaque white paper, most work is done on tracing paper and films. A desirable paper or film should be durable, have a medium grain or tooth, and erase well. Many types of tracing paper are marketed. It is important to know the characteristics and quality of various papers. Thin, transparent tracing paper should be used for sketching and lettering; heavier grades of tracing paper should be used for drafting. Rolls of thin white or light-yellow tracing paper are available in widths of 10½″ to 48″ and 50-yard lengths. A similar, thin tracing paper is available in sketching pads. The drafter uses the thin paper for preliminary sketches and overlays.

When executing more permanent drawings that will be reproduced, you should use a high grade of tracing paper. The better-quality tracing papers are made from 100% rag stock. Vellums are rag stock tracing papers that have been made transparent with a synthetic resin. Before drafting on a tracing paper, hold the paper in front of a light and read the watermark on the paper. The drafting side of the paper is the side on which the watermark is legible.

Drafting films developed from polyesters have many advantages over tracing papers. Films are durable, transparent, and waterproof and resist deterioration. A single-matte, 3-mil-thick Mylar® is the most popular film for architectural drafting. The drawing side of this film has a special coating or matte that gives the surface a grain or tooth. A double-matte Mylar can be drawn on either side.

Graphite lead and ink are suitable for tracing papers. Plastic lead and ink are best for Mylars. Although ink may be difficult to erase on tracing paper, it can be easily erased on film using a white plastic eraser or an eraser imbibed with erasing fluid. The paper or film surface should be perfectly clean before inking. Oil from fingertips often inhibits ink from adhering to the drafting surface; it can be removed by using a soft eraser. No pounce or powders should be used on film, since they interfere with the chemical bond and clog the tip of the pen.

Heavier-quality tracing papers and films are available in rolls of widths of 24″, 30″, 36″, and 42″ in 20- and 60-yard lengths. These materials are also available in standard cut sheet sizes, which are listed below for both architects and engineers.

Sheet Designations	Actual Dimensions (″)	
	Architects	Engineers
A	9″ × 12″	8½″ × 11″
B	12″ × 18″	11″ × 17″
C	18″ × 24″	17″ × 22″
D	24″ × 36″	22″ × 34″
E	30″ × 42″	28″ × 40″

Drafting papers and films are available with a nonreproducible grid printed on the surface. Most of the grids are based on 1/16″, 1/8″, 1/4″, 1/2″, and 1″ divisions and are a light-blue or light-green color. The grid coordinates offer the drafter a convenient reference and eliminate use of the architect's scale for measuring every line.

1.20

TECHNICAL DRAFTING PENS

Most ink drafting is done with technical pens. These pens are fabricated from hard plastics and have standard steel, tungsten carbide, or jewel points. The tungsten carbide and jewel points last longer and are suitable for use on films (Fig. 1–28). Technical drafting pens are available in a variety of point sizes. The following chart indicates the designated sizes.

Point Size	Line Thickness (mm)	Point Size	Line Thickness (mm)
000000	0.13	2½	0.7
0000	0.18	3	0.8
000	0.25	3½	1.0
00	0.30	4	1.1
0	0.35	5	1.2
1	0.45	6	1.4
2	0.50	7	2.0

The care and the use of pens are of utmost importance. The pen has a clear, plastic reservoir which should be filled with black India ink or specialized inks from a squeeze bottle, dropper, or syringe. Use fresh ink; ink stored for long periods of time tends to separate. Avoid filling or shaking the pen over the drawing. Always have a cloth handy to wipe the pen after filling. Do not drop the pens, as the points are fragile and can be easily damaged. When not in use, pens must be sealed with their airtight caps to prevent the ink from drying in the pen. If the pens are not to be used for longer periods, ink should be removed and the pen flushed with water. When storing pens, remove all ink and clean the pen with a special cleaning fluid and syringe. No moisture or cleaning fluid should be left in the pen when it is stored. When ink has dried in the pen, it may be necessary to soak the pen in a pen cleaner or clean it with an electronic pen cleaner. The pen can be disassembled for cleaning and repairs, but since the inner parts damage easily, disassembly should be a last resort.

Most architectural drafting can be accomplished using the 00-, 0-, 1-, 2-, and 3-point sizes. These pens are avail-

able in a five-pen set. Pens can be purchased individually, and replacement parts can be obtained for most pens.

Technical drafting pens are a valuable asset to the drafter. Hold the pen against the T square, parallel bar, or triangle with a slight tilt, using little pressure so that the ink will not flow under the edges of the instruments. This takes practice. Some triangles and templates have special edges and are known as *inking instruments*. Elevating a standard triangle slightly above the drawing surface may help the beginning drafter.

Capillary cartridge pens with disposable ink cartridges are available and are similar to the technical drafting pens previously described. Available in a wide variety of point sizes, these pens eliminate the chore of manually filling the pen from the ink bottle.

Disposable drafting pens with fine to broad points are also available. These pens have felt-tip points and are discarded when the ink supply is depleted. Standard India ink is not used in the disposable drafting pens. The ink in these pens may vary in color or may not be suitable on some tracing papers or Mylar.

When properly cared for, technical drafting pens can provide years of good service. Purchase the correct pens for the medium on which the drafting is to be done.

1.21
OPTIONAL EQUIPMENT

Proportional dividers are an aid for reducing or enlarging a drawing. These dividers are a two-legged, metal instrument with needle points at each end, a ratio marked on the legs, and a sliding pivot. To use the dividers, move the legs together, set the sliding pivot to the desired ratio mark or setting on the leg, and tighten the screw (Fig. 1–29). When the legs are spread, the relative distance between the two needle points at one end and the two nee-

Figure 1–28 Technical drafting pen.

Figure 1–29 Using the proportional dividers.

dle points at the other end will always be in the same indication ratio. For example, when the dividers are set to a 3:1 ratio, the distance between the two long needles will always be three times greater than the distance between the two short points. This instrument can be used to divide a line or circle into equal parts, or it can be used to change meters to feet.

Drawings or tracings should be rolled, never folded. A fiberboard or plastic tube 3″ in diameter and 24″, 30″, or 36″ long with a slip-on or screw-on cap is available for storing or transporting work.

Paper shears or a knife with interchangeable blades should be used for cutting papers and films. Never cut on an unprotected drawing table surface or against the plastic edges of drawing instruments. Metal straightedges are available and should be used for cutting straight lines. Cutting against plastic instruments can make nicks and gouges, which will render them unusable for drafting.

Transparent mending tapes ¾″ wide by 5 to 36 yards long can be used to repair torn sheets. Mending kits are available for papers and films with transparent patches and heating tools that weld the patch.

LIGHTING AND LIGHT FIXTURES

Adequate lighting on the drafting table is very important. Natural north light or soft light of the proper intensity from the upper left is the most effective for right-handed drafters, and vice versa for left-handed drafters. Direct sunlight should never fall on the drafting board; it is too intense.

Most drafting boards are illuminated with artificial lighting from a fluorescent source ranging from 60 to 100 footcandles. Drafting lamps with an adjustable, floating arm holding either fluorescent or incandescent lamps, or a combination of both, can be clamped to the drafting board. Some models have illuminated magnifiers attached for small detail work. Daylight, bright white, and soft white lamps are available. Choose lamps or combine lamps that give soft illumination without glare or confusing shadows, and make sure that the level of intensity is proper.

REPRODUCTION OF DRAWINGS

The majority of all architectural working drawings are made on tracing paper, vellum, or film for the sole purpose of making economical reproductions. Drawings have little value unless they can be satisfactorily reproduced. Sets of the prints (often referred to as *blueline prints*) must be furnished to contract bidders, estimators, subcontractors, workers, and others who are concerned with the construction of the proposed building; the original drawings are re-

tained by the architectural or engineering office. Several types of reproduction equipment are in use for making prints; each type requires the original tracing to have opaque and distinct linework and lettering in order to produce legible copies. Some architectural offices have their own print-making equipment; others send their tracings to local blueprinting companies, which charge a nominal square-foot fee for the service.

1.23.1 Diazo Prints

Commonly known as *Ozalid* prints, diazo prints are produced on paper coated on one side with light-sensitive diazo chemicals. The sensitive paper, with the tracing above it in direct contact, is fed into a print machine (Fig. 1–30) and is exposed to ultraviolet light for a controlled amount of time. The pencil or ink lines of the tracing prevent exposure of the chemical directly below, while the translucent areas are completely exposed. After the exposure, only the sensitized paper is again fed into a dry developer, utilizing ammonia vapors which turn the background of the print white and the lines blue, black, or sepia, depending on which type of paper has been selected. Dry development neither shrinks the paper nor changes the scale of the drawing. The white background of prints also allows changes or notes to be added, if necessary, directly on the paper with ordinary pencils.

1.23.2 Photographic Reproduction

Various types of photographic reproduction are available today. Photostats can be made from an original drawing

Figure 1–30 A floor model diazo machine for reproducing working drawings. (GAF Corp.)

by use of a large, specially designed camera that produces enlargements or reductions from the original work. This direct print process delivers a negative with white lines and a dark background. When a photostat is made of the print, a positive image with dark lines and a light background results. A number of high-quality reproduction methods using a film negative made from the original drawing are available. Projection prints from a photographic negative can be reproduced on matte paper, glossy paper, vellum, and Mylar (Cronoflex™). These excellent-quality prints can be enlarged or reduced in scale with accuracy and are very durable. Microfilm negatives can be made from original drawings using a microfilm processing camera that prepares film ready for use within a few seconds. The negative is usually mounted on a standard-size aperture or data card and is systematically filed for future retrieval. A microfilm enlarger reader-printer with a display screen is used to review the images as well as reproduce print copies of various sizes from the negative. Microfilming is an excellent means of storing drawings, thus eliminating the need to retain cumbersome original tracings.

1.23.3 Contact Prints

The same size of the original drawing can be reproduced on autopositive paper or autopositive Mylar (Cronoflex). This process simply means that if the original is a positive, the print will also be positive. Prints made in a contact printer or a vacuum fame by exposure to high-intensity lights will have most of the background present in the original filtered out. The duplicate in most cases is sharper than the original.

1.23.4 Photocopy Reproductions

A number of high-speed electronic copiers are manufactured today that reproduce images from an original drawing onto paper, vellum, and various films. Some copiers have the capability of enlarging or reducing the size of the drawing from the original scale. Original work on paper, film, and illustration board can be reproduced on most photocopy machines. Photocopy reproductions are gaining in popularity.

In choosing the right reproduction process, you should know what the finished product will be. For architectural working drawings that are used for bidding and construction, diazo prints the same size as the original tracings are the most suitable and cost-effective reproductions. When only a few copies of a drawing are needed and enlargements or reductions are necessary, the photocopy process is desirable. Where high quality and durability are essential, photographic reproductions would be best. Cost and the sizes of prints vary greatly among the types of reproduction alternatives. It is always wise to consult professional printers who offer guidance in selecting the proper reproduction method for the desired results.

DRAWING LINES

The beginning drafter should study architectural drawings to become familiar with all types of lines, and he or she should also practice drawing lines of various types in order to improve drafting skills.

The various types of lines used in architectural drawings are as follows:

- The *primary lines* that define the major object in a drawing are heavyweight lines that are drawn dark and wide (Fig. 1–31).

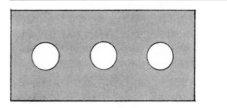

Figure 1–31 Object lines.

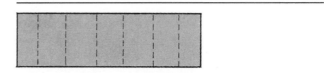

Figure 1–32 Hidden lines.

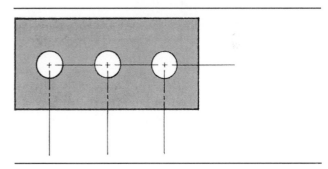

Figure 1–33 Center lines.

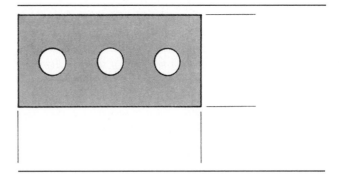

Figure 1–34 Extension lines.

- *Hidden lines* depicting features that are concealed from the viewer are dashed, medium-weight lines (Fig. 1–32).
- *Center lines* are lightweight lines drawn as alternating long and short dashes (Fig. 1–33).
- *Extension lines* that depict the extremities of a dimension are lightweight lines (Fig. 1–34).
- *Dimension lines* are lightweight lines terminated by an arrow, a dot, or a diagonal line (Fig. 1–35).

- *Cutting-plane lines* are very heavyweight lines with an arrow pointing in the direction in which the section is viewed (Fig. 1–36).
- *Break lines* are lightweight lines interrupted at intervals with zigzag lines (Fig. 1–37).
- *Leader lines* are lightweight lines with an arrow at one end connecting a note to a detail (Fig. 1–38).
- *Section lines* or *poché* are lightweight lines that symbolize a specific building material (Fig. 1–39).

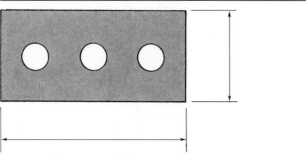

Figure 1–35 Dimension lines.

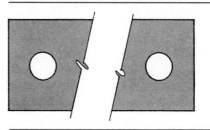

Figure 1–37 Break lines.

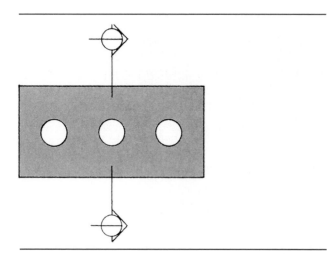

Figure 1–36 Cutting plane lines.

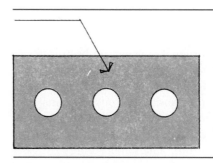

Figure 1–38 Leader lines.

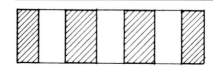

Figure 1–39 Section lines or poché.

EXERCISES
IN THE USE OF INSTRUMENTS

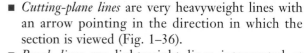

The following problems should be drawn accurately in pencil using your drawing instruments. They should be drawn lightly with a 3H pencil and then carefully retraced with an H pencil. Construction lines need not be erased if they are light and hardly visible. Keep the pencil sharp, and bear down slightly at the beginning and end of each line. Once you have mastered the use of the pencil and drawing instruments, you may trace the exercises using technical drafting pens and India ink.

1. *Parquet flooring* (Fig. 1–40). Draw a 3″ square. Divide the top and side into three equal parts, and draw in the nine squares. Complete the figure by drawing the horizontal and vertical lines as indicated in Fig. 1–40.

2. *Geometric shape* (Fig. 1–41). Using the scale 1″ = 1′-0″, draw a square with 1′-2″ sides, a circle 1′-6″ in diameter, a rectangle 8″ × 1′-9″, an equilateral triangle with 1′-5″ sides, a parallelogram 9″ × 1′-7″ using a 30°–60° triangle. Also, using the same scale, draw a trapezoid with the longest side 2′-3½″ and with a vertical height of 11″ using a 45° triangle.

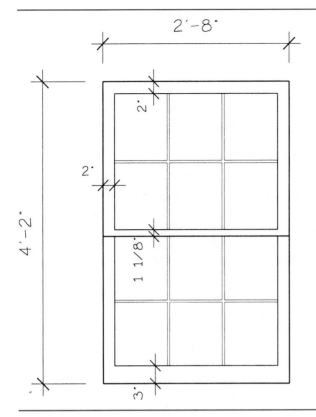

Figure 1–40

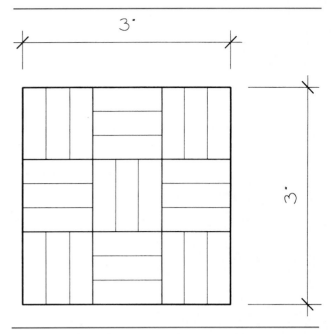

Figure 1–42

outside lines of the panels as shown by the dimensions in the figure; show the panel bevels 1½″ wide.

5. *Concrete block—top view* (Fig. 1–44). Using the scale 3″ = 1′-0″, draw the 8″ × 16″ rectangular shape. Divide the 16″ length into four equal parts, and draw vertical center lines for the cavities. Make the webs 1½″ thick at their thinnest points.

3. *Wood double-hung window* (Fig. 1–42). Using the scale 1″ = 1′-0″, draw the rectangle 2′-8″ × 4′-2″. Divide the height in two, and draw the 1⅛″ sash rail in the center of the height. Draw the inside of both sash according to the dimensions shown in Fig. 1–42. Complete the drawing by dividing the inside of the sash in half vertically and in thirds horizontally. Show the muntins with close, sharp lines.

4. *Six-panel colonial wood door* (Fig. 1–43). Using the scale 1″ = 1′-0″, draw the rectangle 3′-0″ × 6′-8″. Draw the

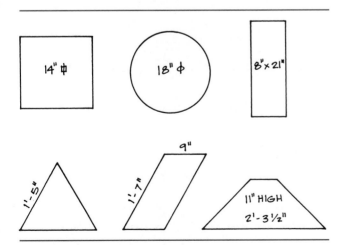

Figure 1–41

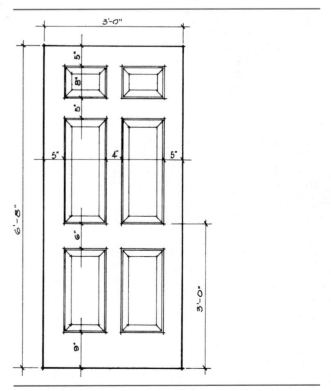

Figure 1–43

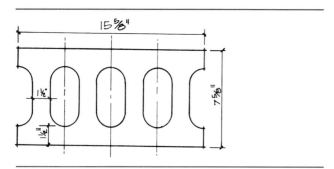

Figure 1-44

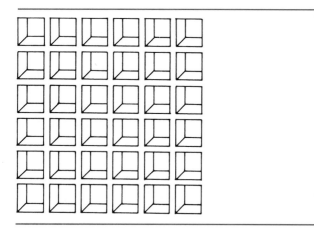

Figure 1-45

6. *Pigeonholes* (Fig. 1–45). Using the scale ³⁄₄″ = 1′-0″, draw 9″ squares with 2″ spaces between the squares; place a 6″ square in the upper right of each 9″ square. Connect the lower-left corners of each square using a 45° triangle.

7. *Wide-flange beam detail* (Fig. 1–46). Using the scale 3″ = 1′-0″, lay out the height of both views using the dimensions shown. Block in both views, and transfer sizes to the other with the T square. Make bolts symmetrical on the plate.

8. *Kitchen cabinet elevation* (Fig. 1–47). Using the scale 1″ = 1′-0″, draw the base rectangle 3′-0″ × 6′-0″. Add the 4″ backsplash. The pair of centered doors are 1′-4″ × 2′-5″, and a 1″ space surrounds all doors and drawers. Draw all lines very lightly; erase lines between features; darken lines where required; and indicate ¹⁄₄″ diameter × 3″ pulls placed 1¹⁄₂″ from edge of door and drawers. Dashed lines indicate door swings.

9. *Cubes* (Fig. 1–48). Using the scale 3″ = 1′-0″, draw a hexagon with 9″ sides; using a 30° angle, place three 4¹⁄₂″ cubes inside the hexagon shape.

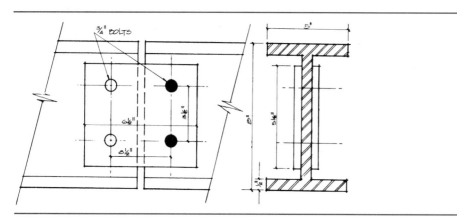

Figure 1-46

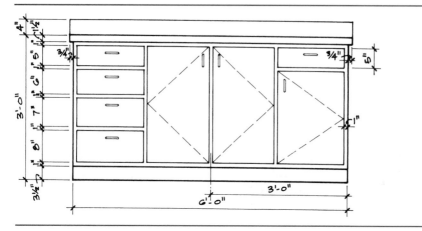

Figure 1-47

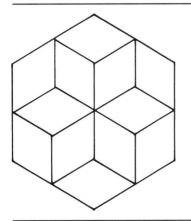

Figure 1–48

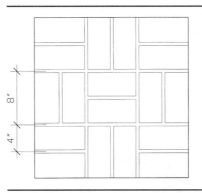

Figure 1–49

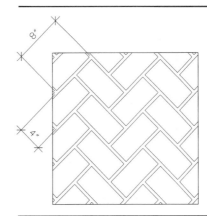

Figure 1–50

10. *Basket-weave brick pattern* (Fig. 1–49). Using the scale $1\frac{1}{2}'' = 1'\text{-}0''$, lay out a $2'\text{-}0'' \times 2'\text{-}0''$ square with light lines. Divide the height and width into thirds, and lightly draw the nine squares. Allow $\frac{1}{2}''$ mortar joints between the brick, and lay out the alternate brick as indicated. Erase the original layout lines, and darken the lines around each brick.

11. *Herringbone brick pattern* (Fig. 1–50). Using the scale $1\frac{1}{2}'' = 1'\text{-}0''$, lay out a $2'\text{-}0'' \times 2'\text{-}0''$ square with light lines. With the 45° triangle, draw the diagonals. From the diagonals lay out 4″ squares, and fill the entire space. Then lay out the $\frac{1}{2}''$ mortar joints $\frac{1}{4}''$ on each side of the 4″ square lines. Erase the proper construction lines, and darken the lines surrounding each brick as indicated.

12. *Arch and pediment* (Fig. 1–51). Using the scale $\frac{3}{8}'' = 1'\text{-}0''$, draw three 4″ risers and two 12″ treads. Four columns are 8″ diameter with $2\frac{1}{2}''$ high by 12″ diameter capitals and bases; the pediment has 45° slope.

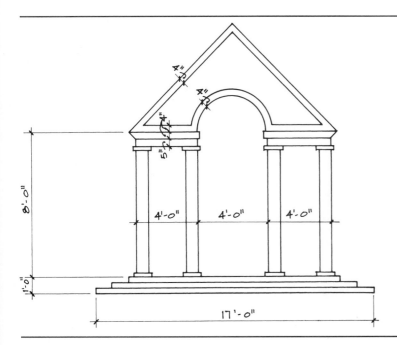

Figure 1–51

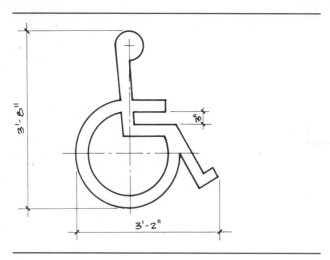

Figure 1-52

13. *Handicapped symbol* (Fig. 1–52). Using the scale 1″ = 1′-0″, draw a circle 2′-3″ in diameter for the wheel of the chair and a 7″ diameter circle for the head. The upper body is 10° off vertical; the leg is drawn with a 30°–60° triangle; and all lines are 3″ thick.

14. *Wood trim profiles* (Fig. 1–53). Using the 12″ scale, draw the figures as shown full size. Keep the pencil sharp, and let the intersection of lines run over at the corners. Draw the end-grain symbol of the wood freehand.

15. *Garage plan* (Fig. 1–54). Using the scale ¼″ = 1′-0″, draw the 4′-0″ × 4′-0″ grid lines lightly. Lay out the 6″ wide walls on the inside of the 12′-0″ and 20′-0″ grid lines. Center the doors, and make the small door and windows 2′-0″ wide. Fill the walls with a gray tone.

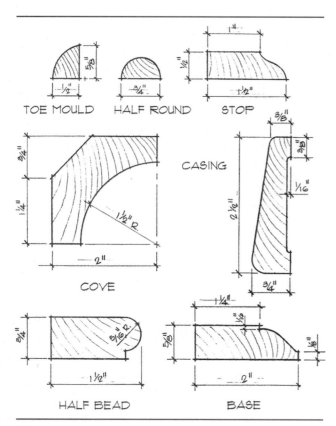

Figure 1-53

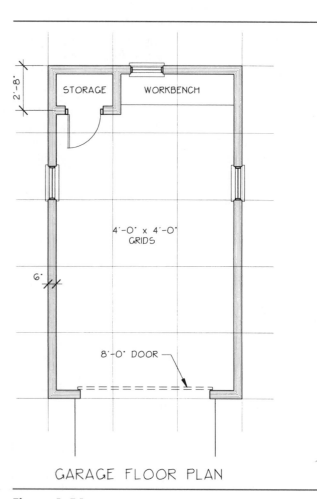

GARAGE FLOOR PLAN

Figure 1-54

1. Why are quality drafting instruments important for good work?

2. Why is the drawing board slightly inclined?

3. For what purpose is a pad used under tracing paper?

4. Do left-handed drafters place the triangles on the T square in the same manner as right-handed drafters? If so, why?

5. What is the advantage of transparent equipment such as triangles?

6. Why are drawing pencils often hexagonal in shape?

7. Is the 2H lead harder than the 2B?

8. For what purpose do you turn the pencil between your fingers when drawing a line?

9. Why is the adjustable triangle time-saving?

10. What is the difference between the architect's scale and the civil engineer's scale?

11. List the necessary qualities of suitable tracing paper.

12. What are the most ideal lighting conditions for drafting?

13. What is a parallel bar, and why is it preferred over a T square?

14. What is the most suitable type of lead for drafting on films?

15. What are templates, and where are they used?

16. How are drafting papers or films attached to the drawing surface?

17. What is an adjustable triangle, and how is it used?

18. What is the difference between the arm drafting machine and the track drafting machine?

19. Draw a line 3″ long and measure the line with the scale $3/4″ = 1′-0″$. What is the length of the line?

20. Draw a line $4\frac{1}{2}″$ long and measure the line with the scale $1/4″ = 1′-0″$. What is the length of the line?

"Would you tell me, please, which way I ought to go from here?"
"That depends a good deal on where you want to get to," said the Cat. "I don't much care where—" said Alice. "Then it doesn't matter which way you go," said the Cat. Alice and the Cheshire Cat in Alice's Adventures in Wonderland
—LEWIS CARROLL

COMPUTER-AIDED DRAFTING AND DESIGN 2

Although somewhat lighthearted, Lewis Carroll's point is important. If you don't know where you're going, then any road will get you there. This principle is important when you are learning about computers. Keep your needs and applications in mind so as not to be confused by the multitude of options available in hardware and software.

A basic understanding of graphic principles is also important before you learn how to operate computer-aided drafting and design (CADD) equipment and software. A beginner should already know the meaning and use of lines, the various types of drawings, and the graphic symbols used in technical drawing. It is also very useful to have an understanding of three-dimensional (3D) drawings such as axonometric and perspective drawings. CAD does not eliminate the need for knowledge of these principles. Sketching is still used by technicians and designers as a means of communicating preliminary ideas, basic design schemes, and rough drawings of details.

2.1

HISTORICAL OVERVIEW

Historians tend to divide history by materials (Stone Age, Bronze Age, Iron Age) and to mark progress by inventions. Starting with the development of language around 40,000 years ago, inventions that improved the collection and transmission of information—such as writing, printing, and the computer—have been the driving force behind society's evolution and have made progress in other areas possible.

The beginning of the Paper Age can be traced to the invention of the printing press 500 years ago. This invention marked the beginning of an information explosion that changed how society functioned. Over time, the main challenge to society became access to and transmission of this information. In the Information Age this challenge is being met by digital technology. Computers, initially designed as fast calculating machines, are now accepted as the primary tool to access, understand, and communicate large quantities of information.

Computers were introduced to the architecture and building professions as automated drafting machines. Their purpose was to make the most tedious and expensive part of traditional practice more efficient. Architects and builders are reinterpreting the computer as a tool for developing, processing, and communicating information about buildings. Included in this process is the development of animation, virtual reality scenes, interactive facilities management models, sun studies, real-time cost analyses, and working drawings. The goal in this Information Age is to increase the amount and change the nature of information available about a proposed or an existing building.

In the past 10 years, the design and construction field has realized that CADD (computer-aided drafting and design) can work and can be cost-effective. CADD can make a real difference in the way architecture is created, and it can improve the overall quality and value of design. Clients no longer see CADD as a novelty, and they have come to expect CADD work as a matter of course. CADD helps architects perfect building concepts and produce cost-effective construction drawings. Hours that would be spent hand drawing can be eliminated. Potential design mistakes can be caught early, before construction begins and corrections become expensive.

Computer-aided, design-based software such as AccuRender, 3D Studio, Virtus WalkThrough, QuickTime VR, AutoCAD, and ArchiCAD make it possible to move from designing on paper to designing on the computer. Once all the details have been entered in the computer and the computer-generated models have been created during a 16- to 20-hour development process, clients can be guided on a "flyby" or "walkthrough" adventure. They can explore their house or building on a predetermined path or a path of their own choosing. They can move around the house exterior for a 3D view of the roof and landscaping. This technology forms the basis for highly interactive computer-based presentations in which the designer and the client explore variations together and experiment with all aspects of a project.

CADD further enhances design by providing better

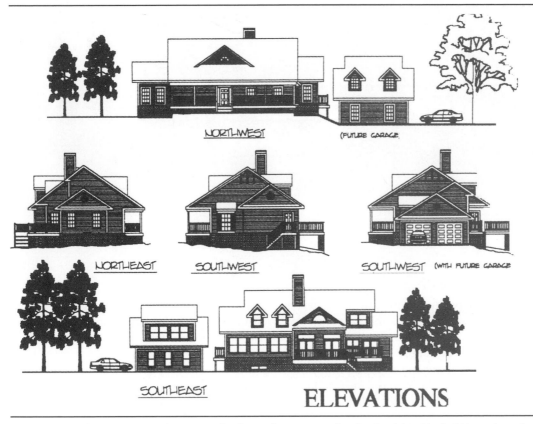

ELEVATIONS

Computer-aided presentation elevations of a home for a young family. (Paul M. Black AIA, architect)

communication of information among design professionals. A CADD drawing becomes an information resource readily accessible to all professionals involved in the design and construction process. Architects, project managers, engineers, and builders can work with "intelligent" building objects—walls, doors, pipes, ducts—that will have the same properties as those of the actual components. These objects automatically adjust themselves in relation to other drawing objects and accurately maintain these relationships throughout design changes. Building owners see the value of this shared electronic drawing information and increasingly require it for their own future building management and maintenance. Once completed, a CADD drawing can be modified and reused for facilities management with little effort.

Making the process even more realistic is the introduction of virtual reality headgear. The software company Virtus WalkThrough Pro recently introduced the first stereoscopic headgear to the architecture industry. This headgear provides a more realistic viewing of computer-digitized models using standard virtual reality modeling language and fitting two sets of images to each eye, thereby producing the actual feeling you have when walking around. This software can combine real photographs with computer-generated renderings. For example, photographs taken of a skyline and surrounding landscape can be scanned into the computer, and a com-

puter-generated building model can then be superimposed on the scanned background.

2.2

SYSTEM COMPONENTS

CADD systems are composed of several components; some involve hardware (electronic gear), and others involve software (programs). All of the components are used together to make the whole system work, but each must be in its proper place and interconnected with the other parts to function correctly. Figure 2–1 is helpful in picturing how each system works with the others to produce electronic drawings. Each component will be discussed in detail in the following sections.

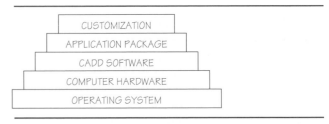

Figure 2–1 Components of a computer/CADD system

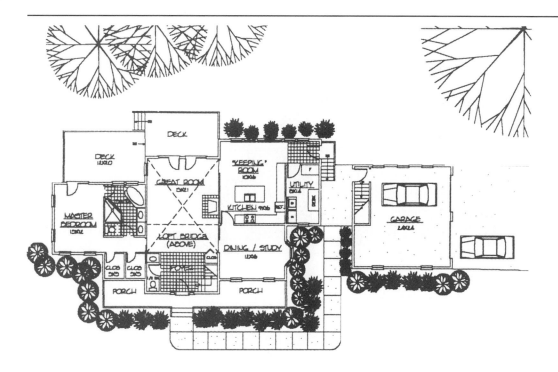

MAIN LEVEL

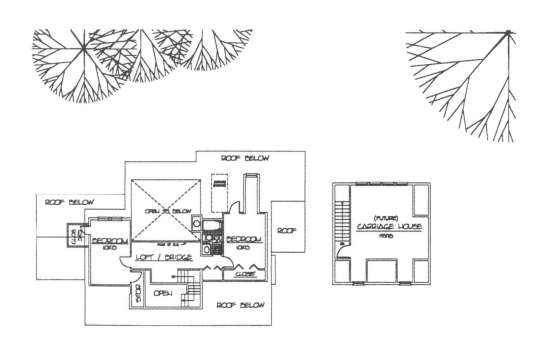

UPPER LEVEL

Computer-aided presentation floor plans of a home for a young family.

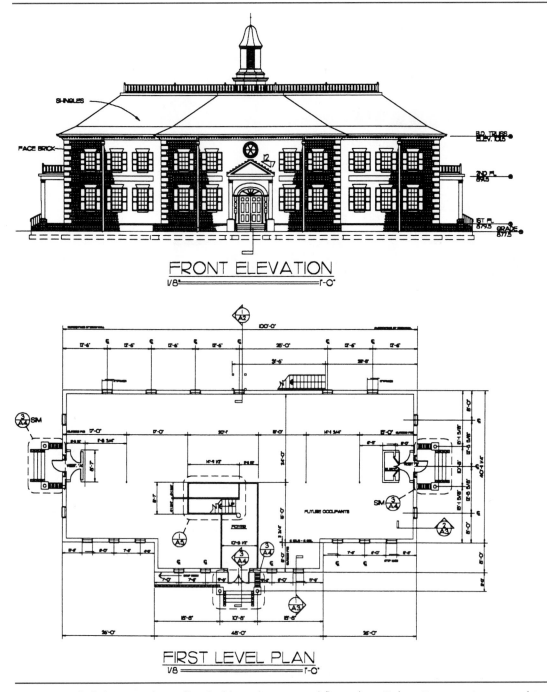

FRONT ELEVATION
1/8 = 1'-0"

FIRST LEVEL PLAN
1/8 = 1'-0"

Computer-aided drawing of an office building elevation and floor plan. (Robert Foreman Assoc., architect, Paul M. Black, AIA, project manager)

2.3

OPERATING SYSTEMS

An operating system (OS) is the software between the applications that a person uses and a computer's hardware such as memory, disk drives, and input/output devices. It is the foundation upon which all other programs are built. The performance and the usability of software applica-

tions are determined by the performance and the usability of the operating system.

Architectural software programs run on a variety of operating systems including

DOS

Windows 3.1

Windows 95

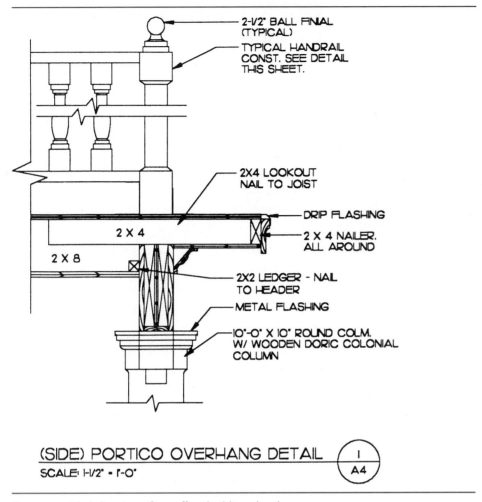

(SIDE) PORTICO OVERHANG DETAIL

SCALE: 1-1/2" = 1'-0"

1/A4

2-1/2" BALL FINIAL (TYPICAL)

TYPICAL HANDRAIL CONST. SEE DETAIL THIS SHEET.

2X4 LOOKOUT NAIL TO JOIST

DRIP FLASHING

2 X 4

2 X 4 NAILER. ALL AROUND

2 X 8

2X2 LEDGER - NAIL TO HEADER

METAL FLASHING

10"-0" X 10" ROUND COLM. W/ WOODEN DORIC COLONIAL COLUMN

Computer-aided drawing of an office building detail.

Windows NT

OS/2 Warp

MacOS

Today most architects use DOS/Windows computers, Macintosh computers, or a combination of both.

Some important features of current operating systems include multitasking, 32-bit memory addressing, and integrated networking:

- *Multitasking* refers to the ability of a computer to run more than one program at the same time. The term is misleading, however, because it gives the illusion that two or more things are happening at the same time. A multitasking system actively switches back and forth quickly between different programs.
- *Thirty-two bit memory addressing* means that the operating system can access up to 4 gigabytes (4 billion bytes) of memory at any given time.
- *Integrated networking* includes capabilities for built-in networking, e-mail, and access to on-line services.

2.3.1 DOS

DOS 7.0 entered beta testing in 1996, providing the 32-bit engine of Windows NT without the graphical user interface. Presently DOS can access only 640K of memory at any given time. DOS exists now primarily as a small portion of the Windows 95 operating system.

2.3.2 Windows 95

Windows 95 is a replacement for three operating systems: MS-DOS, Windows 3.1, and Windows for Workgroups 3.11. It supports plug-and-play hardware and includes an integrated information manager, built-in access to the Microsoft network, an on-line service, and a universal e-mail. Windows 95 also finally supports 32-bit memory addressing. It still supports DOS applications, although DOS is no longer required to run it. The most visible improvement of Windows 95 can be found in its user interface, which is often described as "Mac-like." Windows 95 is scheduled to be updated as Windows 98 and will provide integrated Internet access.

2.3.3 Windows NT

Windows NT is Microsoft's alternative to the performance and sophistication of UNIX. NT provides true multitasking, security, multiprocessing, and 32-bit software development tools. Windows NT has no DOS component and runs on five different kinds of multiprocessors. As a result, it is better suited for use as a network file server or a database server.

2.3.4 OS/2 Warp

OS/2 is a multitasking operating system that is used primarily for PC network servers and for custom applications developed within large corporations. OS/2 can simultaneously run OS/2, DOS, and Windows 3.x applications but not Windows 95 applications. IBM's right to use the Microsoft Windows source code within its products ended in September 1993; thus Windows' compatibility must be reverse-engineered by IBM, a costly and time-consuming process.

2.3.5 MacOS

The Macintosh represents only about 10 percent of all desktop computer sales, even though its operating system is generally recognized as more thoughtfully designed than Windows. The Mac's strength lies in the fields of desktop publishing, digital image editing, digital music editing, visualization, and multimedia authoring tools.

COMPUTER HARDWARE

Test driving different desktop computers using key software applications will always be the best way to evaluate their price and performance. Over the past several years, several key hardware technologies have become available and affordable to architects, including color scanners, digital cameras, color printers, and video capture hardware. Collectively they provide firms of all sizes—even the sole practitioner—access to publication and presentation techniques that only a few years ago were the exclusive domain of the largest architectural firms.

2.4.1 Desktop PC Components

Typical CADD hardware is shown in Fig. 2–2. The various components are described in the following paragraphs.

CPU   The CPU (central processing unit) is the microprocessor brain of the computer. Intel is the main CPU manufacturer for desktop computers with its 486, Pentium, and P6 CPUs. Other CPU manufacturers include Motorola (68030, 68040); Motorola/IBM/Apple (Power PC); Sun Microsystems (SPARC); Silicon Graphics (MIPS); and DEC (Alpha).

Some desktop computer buyers estimate the performance of a computer by the clock speed of the CPU,

Figure 2–2 Typical CADD hardware. Shown are the "box" containing the motherboard; a 3.5-inch diskette hard drive; and a CD-ROM drive. Also shown are a standard keyboard, a 17-inch monitor, speakers, a digitizing tablet with a four-button puck, a mouse, and a laser printer.

usually measured in megahertz (MHz). However, a computer clock speed is only one measure of performance and can be misleading. Consider the clock speed of a CPU as roughly equivalent to the rpm (revolutions per minute) of a car engine: The faster the rpm of an engine in a given gear, the faster the speed of the car. However, upshifting or downshifting the gears also affects the speed. This relationship between rpm and shifting is comparable to shifting between different brands of CPUs. A 75-MHz 486, for example, is roughly 50% faster than a 50-MHz 486, but a 60-MHz Pentium is faster than both.

Motherboard

The motherboard of a desktop computer is the printed circuit board that contains the CPU, the memory, the input/output (I/O) chips, and slots for expansion cards and other specialized circuitry for video sound or networking. Some components of a computer motherboard may operate at a slower clock speed than the CPU itself. For example, an Intel 100-MHz 486DX4 performs instructions at 100 MHz but fetches memory at only 33 MHz. This imbalance causes "wait states" in the CPU during which the CPU pauses to allow the motherboard to catch up.

Memory

RAM (random-access memory) is the fast, silicon memory accessed by the CPU; it is used to store programs and data for fast access. In general, the more memory, the faster the performance of the computer, depending on the software applications. Most RAM is packaged in what are known as SIMMs (standard in-line memory modules), which are small cards with memory chips that snap into memory slots on the motherboard. Memory capacity is measured in megabytes (MB, or about 1 million bytes of data). RAM is a volatile storage medium: It will not retain information when the computer is turned off.

Cache memory is a special, high-speed type of memory provided on most desktop computers. A cache is a temporary holding area for data and programs used by the CPU. Cache memory provides access to information much faster than ordinary RAM, but it is more expensive. Most desktop computers provide only a small amount of cache memory (128K or 256K).

Magnetic Disk Drives

Most desktop computers come with one 3.5-inch diskette drive as well as an internal hard disk drive that stores between 200 MB and 2 GB (1 gigabyte = 1000 megabytes) of data. Software and data stored on magnetic disks are retained when the computer is turned off. One measure of how fast the disk drive can access a track is called *seek time;* average seek time is 9 to 13 milliseconds.

Optical Disks

Optical disks are an excellent medium for publishing and distributing information and for archiving large amounts of digital data. A 5.25-inch CD-ROM (read-only) can store up to 650 MB of data, or tens of thousands of pages of text, thousands of graphics, or hundreds of photographs. CD-ROM drives are inexpensive and are standard equipment on computers. The quad-speed and six-speed drives are still substantially slower than hard disk drives. Read/write optical disk drives capable of storing 128 to 256 KB of data (3.25-inch), to 500 MB to 4 GB (5.25-inch) of data, are also available; and at 4 cents per megabyte their costs compare favorably with costs of other storage media.

Display Monitor/Graphics Card

The display monitor is where the user and the computer meet eye to eye. For CAD, rendering, image editing, desktop publishing, or word processing applications that require high graphics resolution, a 17-inch monitor is necessary. A 19- to 21-inch monitor is a nice option but is often too expensive.

Dot pitch refers to the resolution of the screen and is an important criterion for determining screen sharpness. A 0.28-mm dot pitch is a typical value for most desktop computer displays.

The quality of a digital image on the computer display depends on the dimensional size of the monitor, the resolution and color depth of the image, and the amount of display memory.

The display memory available and the number of bits used to store the color value of each pixel of the image determine the number of colors available and the maximum resolution of the screen. The CGA display standard of early IBM-compatible PCs was a 4-bit color display (16 colors) with a resolution of 320×200. Storing a CGA image in memory required $320 \times 200 \times 4$ bits (32,000 bytes). Computers equipped with standard VGA displays ($640 \times 480 \times 8$ bits) required 307,200 bytes of display memory. The requirements of present-day, 24-bit, high-resolution graphics displays are much higher. The average Windows PC includes a graphics card with 2 MB of display memory, and many high-end systems are equipped with graphics cards with 4 MB of display memory.

A 2-MB graphics card is the minimum requirement if you intend to use high-resolution, color image editing tools. It can support 24-bit color up to 800×600 display resolution, and 16-bit color up to 1024×768 resolution. A 4-MB graphics card is better, especially with 19-inch monitors or larger. It can display true-color, 24-bit images at 1280×1024 resolution.

Keyboard

Keyboards should be slightly below the work surface to minimize bending of the wrists. Ergonomic keyboards which look as if they have been sawn in half are designed to minimize the amount of user wrist rotation. They are available from Apple and Microsoft, among others (Fig. 2–3).

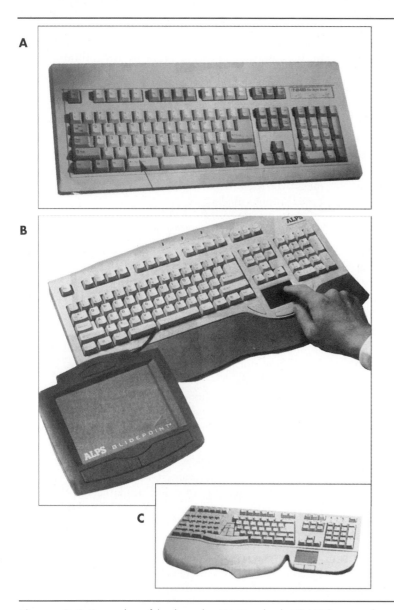

Figure 2–3 Examples of keyboards. (A) Standard. (B) With a touchpad. (C) Ergonomic with a touchpad. (GLOBAL Computer Supplies catalog)

Pointing Device The mouse is the most common pointing device used with desktop computers (Fig. 2–4). A mouse is a *relative* pointing device; that is, the cursor movement depends on the relative movement of the mouse.

Digitizing tablets are *absolute* pointing devices. They contain a grid of wires just below their surface which detect the location of a puck moved over it. An area of the tablet represents the screen drawing area (Fig. 2–5). Digitizing tablets are typically used in CAD applications.

Modems The word *modem* is an acronym for *modulator-demodulator.* A modem is the hardware component responsible for converting the digital signal of a computer into an analog signal which can be transmitted over telephone lines. Your computer must have a modem to access on-line services such as AIA-ONLine, CompuServe, or the Internet. Modems support different speeds. Accessing the Internet requires a minimum 28.8-K modem.

Tape Drives A tape drive is necessary for backing up and storing data; 500-MB and 1-GB disk drives are standard equipment on most computers. Most tape backup systems use 4-mm tapes of varying length. Two types of drives are available: (1) low-capacity, low-cost, 120- to 250-MB tape drives and (2) high-capacity, high-cost, 2- to 8-GB tape drives. The major limitation of backup tape drives is their sequential linear format. Accessing a file that is stored at the end of a tape takes time if the tape must be forwarded and rewound.

Figure 2-4 Mouse. The two versions available are (A) the two-button Microsoft mouse and (B) a three-button, programmable mouse from Micro. (GLOBAL Computer Supplies catalog)

Digital Cameras Digital cameras can be used to capture images without needing to store the images on film (Fig. 2–6). Some cameras, such as Apple's QuickTake cameras and Kodak's Digital Camera, combine a scanner with the optics of a traditional camera. The major limitations of these cameras is the resolution of the images they capture: approximately 640 × 480, similar to the resolutions of video. Kodak's professional digital cameras provide resolutions matching those of 35mm film, but they are still very expensive.

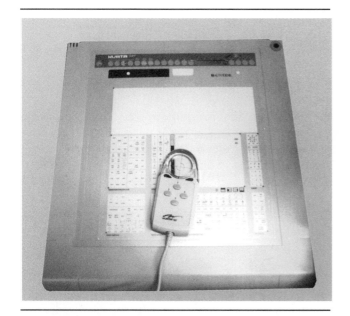

Figure 2-5 Digitizing tablet. Shown here is the Kurta XGT tablet with a four-button puck.

Figure 2-6 Digital cameras. Two relatively inexpensive cameras are (A) the Casio QV-10 and (B) the Epson PhotoPC. (*PC* magazine, February 1996)

Photographic service bureaus provide Kodak PhotoCD scanning services for slides or film negatives at a cost of $1.00 to $2.00 per image. Kodak PhotoCD provides a cost-effective way to convert existing slide libraries into digital images. Kodak's low-cost viewing software, PhotoEdge, can be used to browse CD-ROMs containing PhotoCD images.

Kodak, Epson, and Casio have all recently introduced relatively inexpensive digital cameras. The Casio QV-10 is being advertised for under $400, and the Kodak DS-20 sells for under $350. The latter camera can store between 8 and 16 color pictures, depending on the image resolution selected, and can transfer them to a Macintosh or a Windows-based computer. The camera comes with a software program, Kai's Power Goo from Meta Tools Inc., that allows the user to manipulate the image on the computer in various ways.

2.4.2 Peripheral Devices

Over the last 15 years, a variety of plotting technologies have been introduced. Pen, thermal, electrostatic, inkjet, and electrophotographic (LED) plotters have been introduced with great fanfare. Today only the LED and inkjet plotters continue to hold significant market share. The number of pen plotters and electrostatic plotters being sold to the AEC market are small. Many users are trading their older plotters in favor of new inkjet and LED plotters. The resolution of most printers and plotters (except for pen/pencil plotters) is typically measured in dots per inch (dpi).

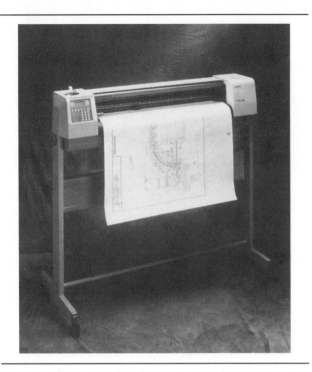

Figure 2–7 Pen/pencil plotter, the Mutoh XP-150 series. (*CADALYST*, July 1993)

Pen/Pencil Plotter   With the XP-500 series, starting at about $3000, Mutoh America offers the industry's fastest vector plotters, featuring a maximum plotting speed of 50 inches per second (ips) (Fig. 2–7). All versions of the XP-500 series feature the unique "fuzzy logic" capability of vector sorting. Fuzzy logic enables the computer to simulate the drawing data into more efficient and smoother output operations.

These plotters feature automatic pencil lead feeders, which can hold as many as 720 0.2-mm pencil leads, 480 0.3-mm pencil leads, 280 0.4-mm pencil leads, 200 0.5-mm pencil leads, 120 0.7-mm pencil leads, or a combination of lead sizes and hardnesses. As lead runs out, each holder automatically feeds another lead. The plotter carousel can hold up to 8 drawing devices and can mix pen and pencil devices within the same drawing.

Inkjet Plotter   Inkjet plotters range in price from $2000 to $7500 and are available from companies such as Hewlett Packard, CalComp, Encad, Summagraphics, Mutoh, and Selex (Fig. 2–8).

Inkjet has become the choice for many firms. Inkjet plotters use standard ink cartridges, the kind used in inexpensive desktop inkjet printers. They provide very good print quality with little or no operator intervention. All that is required is the loading of a new ink cartridge when the old one runs out. Most inkjet plotters are capable of creating a D-size, draft-quality plot in less than 3 minutes.

The standard language used by inkjet plotters is HP-GL/2, which is Hewlett Packard's graphic language. Although most manufacturers offer AutoCAD and Windows-specific driver software for their inkjet plotters, most will run quite acceptably with the built-in software drivers which ship with these products.

Inkjet plotters are available in both monochrome and color models. Higher-end color models, particularly the Encad Novajet® III, are suitable for printing photoreal-

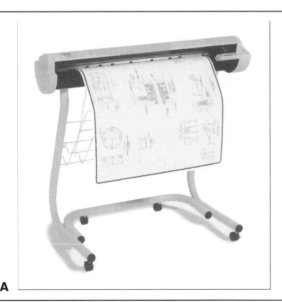

Figure 2–8 Inkjet plotters. (A) ENCAD® Novajet. (B) CADJET™. (Images courtesy of GAViota Graphics and the Autodesk Image Library; from *Cadence*, April 1996)

istic images when loaded with specially coated media, while lower-end color inkjet units, such as the Summa-Jet 2C and the Encad Cadjet™, are more suitable for spot color on CAD drawings.

A few limitations are associated with inkjet plotting. One is that inkjet plotting involves spraying ink on paper, and filling large black areas of paper is costly. Also, the ink used in inkjet plotters is water based, which creates questions about the permanence of the image. Some of the newer inkjet cartridges use inks containing a polymer-based pigment, and these inks are more resistant to fading and more archival than earlier inks. Media manufacturers offer specially coated inkjet media which are designed to be stable for long-term storage.

Monochrome D-size plotters have reached retail prices of $2000 or less. The Hewlett Packard DesignJet 230 monochrome sheet-fed plotter often sells in this range.

LED Plotter

The high end of the plotter market is dominated by electrophotographic (LED) plotters (Fig. 2–9). Prices for these units start at around $20,000 and range upward to $110,000. They are available from Xerox, Cal-Comp, Océ, JDL, Shacoh, and Mutoh. LED plotters have a resolution of 400 dpi and are designed for unattended operation. They are configured for roll-feed media, automatic cutting, and automatic data format recognition.

Top-end LED plotters are generally available as part of hybrid digital copying systems combining both a plotter and a scanner. Xerox, Océ, and Shacoh offer units that can copy or plot equally well. These units can scan a number of drawings and then create multiple collated sets without having to rescan the drawings. LED printed sets are fully collated, have no errors, and are of original quality. Large-format LED plotters are economical when used for high-volume production. The fastest LED plotter, currently the Océ 9800, can create up to 8 E-size copies per minute.

Desktop Printers

The preferred resolution for desktop, small-format printers (Fig. 2–10) is 600 dpi; 300 dpi is a minimum. One of the most important recent developments in printing technology for architects is the affordable 600-dpi, 11 × 17-inch, black-and-white laser printer. The Hewlett Packard 4MV printer is fast (16 pages per minute) and is an excellent printer for CAD checkplots. The 11 × 17-inch format, at 600 dpi, is suitable for a wide range of small projects and is excellent for reduced-size printing of large projects.

Scanners

Document scanning is a technology in which an existing paper drawing is captured as electronic data. Once converted, the "virtual" drawing is available for use in any number of CAD applications or for transmittal and reprinting.

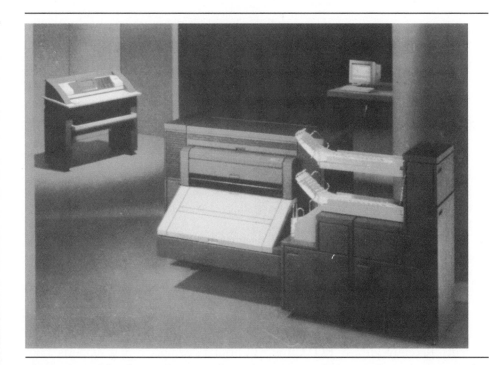

Figure 2–9 LED plotters. The Océ 9800 plotter/copier is designed to maximize productivity and ensure quality in digital, hard copy, or hybrid environments. (*A/E/C Systems*, Sept./Oct. 1995)

Figure 2–10 Desktop laser printers, the Hewlett Packard 4V and 4MV. (Hewlett Packard Product Information brochure)

Document scanners range from desktop, A-sized units to large scanners capable of reproducing full E-sized and larger drawings. They can record a range of images from simple lines to full, continuous color tones. They enable the operator to clean up badly damaged documents so that subsequent prints are better than the original.

The scanning process consists of taking a snapshot of a drawing with a digital camera. This photographic process uses one or more cameras that record a narrow portion of a drawing as it is moved on rollers past the camera. The digitized data are then passed to a computer which processes the information.

Scanner capability is based on image resolution and can range from 150 up to 1000 dpi. This resolution is directly related to what you intend to do with the data. The higher the resolution, the finer the detail. If the drawings are simple and relatively clean, then a lower resolution may be acceptable. Older drawings or drawings that are faded or damaged should be scanned at a higher resolution.

Several very high speed scanners are now available. At the very high end is the Océ 9800 system: a full-function, hybrid scanning/plotting/copying system featuring a 400-dpi scanner. The Océ 9800 is expensive. A complete system, including printer, starts at $162,000. At the economy end of the wide-format scanning market there have been some interesting developments. Recently Vidar introduced their TruScan 800 (400-dpi) scanner at $14,500. ANAtech has introduced a $13,000, desktop, wide-format scanner.

Color Scanners   Color desktop scanners which support resolutions up to 600 dpi are available for under $600 (Fig. 2–11). These are excellent for scanning small-format sketches or photographic prints.

2.4.3 Notebook Computers

Most portable computers today are sold with optional docking stations which connect the monitor, keyboard, modem, mouse, and network through a single adapter (Fig. 2–12). Connecting a portable computer to any of these resources simply requires plugging the notebook into the port. Notebook displays are starting to rival those of desktop computers, with 1024×768 resolution and true, 24-bit color at 640×480 resolution.

2.4.4 Hardware Recommendations

The best way to decide what kind of computer you should purchase is to talk to other people who have recently purchased hardware. Current high-end PC systems are running Intel Pentium II (300 MHz), Mac PowerPC chips, or Cyrix 686 processors. Systems should have a minimum of 16 MB RAM for Mac or Windows, 32 MB for CAD, or 64 MB for 3D work. Hard drives should be 4.0 GB or larger, with seek times under 13 ms. CD-ROM drives should be 16 speed minimum. If your computer is not networked, buy a tape backup unit or a recordable CD-ROM. You will need a fax/modem that supports 33.6-baud (bit-per-second) rate or a 56-K-capable $\times$ 2 modem. Finally, a 17-inch monitor with 0.28-mm dot pitch is a minimum for CAD work.

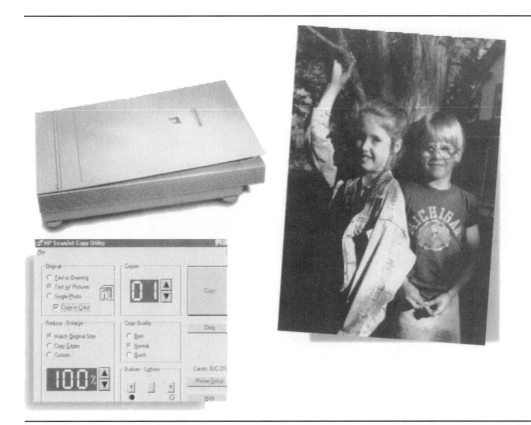

Figure 2-11 Color desktop scanner. Hewlett Packard ScanJet 4c is a 30-bit, 600-dpi color scanner. It comes with Copy Utility, which lets you turn the scanner and the color printer into a color copier. (*PC* magazine, December 1996)

SOFTWARE

The types of software applications used by most architects have expanded dramatically. In addition to CADD, software programs used by architects include word processing, spreadsheets, databases, contract management, project scheduling, desktop publishing, multimedia presentation, digital image editing, e-mail, and on-line services. Three-dimensional model building, visualization, animation, and virtual reality tools are becoming part of the design and communication process for many firms.

2.5.1 Computer Drafting Software

Many graphics software packages are available in the marketplace. A full-featured architecture and engineering software package provides a vector format for the storage of data and the means to create, change (edit), combine, and print drawings. It contains features that allow for the drawing of 3D objects (solids modeling); the addition of shade and color to the drawings; and the creation of moving pictures of the objects described in the

database. AutoCAD is a widely used, general-purpose package that can be combined with many special-purpose programs to accomplish a variety of tasks. MicroStation is another general-purpose package that can also be combined with special-purpose programs. DataCAD, written primarily for architects, features commands and procedures commonly used by architects in manual drafting. Macintosh-based CADD software includes such programs as MiniCAD, ArchiCAD, and Power Draw.

CADD software, especially 3D model building and visualization, is one of the most complicated and difficult tools that architects learn to use. This complexity stems partly from the practice of software developers of continually adding features so that the scope of CADD programs expands to become an all-in-one modeling and documentation tool.

Object-Oriented Software   Most software programs are organized into small modules that perform a specialized function on a set of data stored in files outside the program. This program structure is followed by most of the software currently used by architects. Each is a fairly large, complex program that provides hundreds of commands or operations that can be performed on a set of ex-

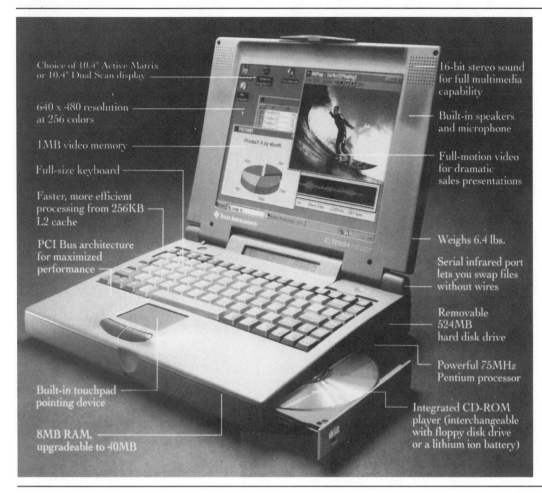

Choice of 10.4" Active Matrix
or 10.4" Dual Scan display

640 x 480 resolution
at 256 colors

1MB video memory

Full-size keyboard

Faster, more efficient
processing from 256KB
L2 cache

PCI Bus architecture
for maximized
performance

Built-in touchpad
pointing device

8MB RAM,
upgradeable to 40MB

16-bit stereo sound
for full multimedia
capability

Built-in speakers
and microphone

Full-motion video
for dramatic
sales presentations

Weighs 6.4 lbs.

Serial infrared port
lets you swap files
without wires

Removable
524MB
hard disk drive

Powerful 75MHz
Pentium processor

Integrated CD-ROM
player (interchangeable
with floppy disk drive
or a lithium ion battery)

Figure 2–12 Notebook computer, the Texas Instruments TI Extensa 550CD. (*PC* magazine, April 1996)

ternal data such as in a CADD drawing, a publication, an image, or a written document.

Unfortunately, most software programs arrange data in a very specialized way that makes it very difficult to share information between or among different programs. It is important, however, to architects, engineers, and clients for their electronic drawings to be compatible so that they can check and consult with one another. The DXF and IGES standards have been developed to allow a drawing created in one package to be displayed in another, but they are not always accurate.

CADD is best understood as a design information management system, and it should be well integrated as a general-purpose tool with other software applications. In many architectural firms the documents that describe the total building process are being developed using not a single, monolithic, CADD program but rather a variety of smaller, general-purpose software applications that share information with each other.

The basic idea of object-oriented software is to combine software and data into the same object. Object-oriented concepts are applied today in software applications that support Microsoft's OLE (object linking and em-

bedding). Using OLE, developers build applications that allow information from other applications to be combined into compound documents. Users can drag and drop objects from one document to another and create a link between the two. OLE is implemented on both Windows-based and Macintosh computers. Most major CAD software developers, including AutoDesk and Bentley Systems, are working on new object-oriented versions of their software.

Over time, traditional CADD software packages have become systems that supply all of the features needed within a self-contained environment. Object-oriented CADD software will provide the opportunity to use Windows systems resources more effectively. Whereas the traditional CADD file was an all-inclusive monster document, the new software will be a compact set of Windows-based files with intelligent links. The linked technologies will make it possible to experiment with changes wherever it is most convenient to do so. Making changes in the design module will also update the spreadsheet, database, and word processing documents. Making changes in the spreadsheet or database program will update the drawing and the word processing document. The new software will make it easier for architects and builders to balance all considerations when planning and designing buildings.

2.5.2 Resources

In the near future, architects will no longer maintain shelves of printed product catalogs. Instead, they will use the Internet and other private on-line resources to access more current information. For most product manufacturers and distributors, digital publishing will be less expensive than producing catalogs and distributing updates. However, external on-line resources do not yet provide high enough throughput. Downloading CAD drawings or high-resolution, full-color images can take minutes, and sometimes hours, even with fast modems.

CD-ROM resources, on the other hand, do provide fast access. Quad-speed and six-speed CD-ROM drives, now standard equipment for most desktop computers, provide throughput and performance that are substantially faster than what can be achieved on-line at present. The January 1996 issue of *The Construction Specifier* has a listing of more than 250 construction industry companies that offer manuals, specifications, construction details, and other references on computer disk and CD-ROM. The most recent edition of *Architectural Graphic Standards* is now available on CD-ROM as well (Fig. 2–13).

SweetSource, developed by Sweet's Group, is an interactive product selection tool that operates within Microsoft Windows (Fig. 2–14). SweetSource allows you to find information quickly on a specific building product or manufacturer. With SweetSource you can collect and organize the products you select, create your own slide show presentations, and conduct side-by-side product comparisons. Information on a specific product may contain text describing the technical features, CAD drawings, tables of product data, and photographs. You can copy or export these elements to other software products, including word processors, graphics applications, and CAD programs. Product information may also be a multiscreen presentation, like a slide show, which can be viewed, printed, and organized into customized client presentations.

The *Buyer's Guide on CD-ROM*, developed by Hanley-Wood, Inc., contains information on all of the companies and products typically listed in *Builder* magazine's *Builder's Yearly Buyers Guide* issue. Some of the companies include videos highlighting the features of their products, and some show their entire catalog—from color images to specification sheets (Fig. 2–15).

A fairly recent development allows builders and buyers to peruse house plans and elevations in a 3D format. The presentation shows 20 to 30 photorealistic views of a simple floor plan from both inside and out. The viewer can tour the house from room to room, stopping or panning to see a closer view or a different angle of a room. This product was developed through a collaboration among architect Larry Garnett, Houston; software engineering firm Macromedia Technologies Inc., Minneapolis; Pella Corporation, Pella, Iowa; and *Professional Builder* magazine. Additional CD-ROM house plans are also

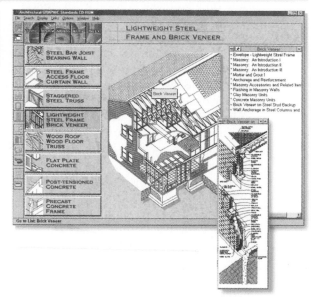

Figure 2-13 Graphics standards on CD-ROM. Shown is one option to locate data organized by systems and components for several common building technologies. (John Wiley & Sons, Inc., Product Information brochure)

A

B

Figure 2-14 SweetSource. This CD-ROM contains product information developed by Sweet's Group, a division of Mc-Graw-Hill. It combines text and graphics to describe materials and their applications.

available from HomeDesigns Multimedia Encyclopedia (HomeStyles Interactive), AbbiSoft's Home Plan Finder, Design Search, and Turning Point Publishing.

2.6

THE INTERNET

Imagine potential clients looking up your office listing in a directory and finding a history of your work, including drawings and photos. Also imagine these clients quickly looking up the testimonials of other clients and the credentials of your staff and consultants, finding a map to your office, or receiving an invitation to call or correspond by e-mail to review information and services that might be of value to them. A home page and a site on the World Wide Web can offer these services and can provide a chance for prospective clients to see what you have to offer, to see what sets you apart from others, and to see what you offer that most closely matches their needs.

The Internet, born in 1995, is a term used to describe an interconnection of worldwide computer networks. Operating on the Internet are a variety of computer services such as e-mail, UseNet newsgroups, and the World Wide Web. The Internet is a massive global network of networks connecting literally millions of computers.

Virtually every possible potential design client will be reachable on the Internet. Some will be accessible directly through their own Web sites, whereas others will be available through their membership in professional and trade associations.

Designing and deploying a Web site requires a notable commitment of time and resources, but the benefits can be quite impressive. The Internet ends traditional limitations of state and international borders. It is especially exciting for those whose specialized services have had limited local markets or those with small practices who could not see themselves offering services in other locales. Clients, consultants, and employees are accessible from all over the world.

The Internet is becoming a growing source of information. Research data from many sources, satellite images, and reference information from standards organizations such as ASTM are all available. The Internet will be both a global library and a universal communications medium.

Progressive design firms are already using the Internet. E-mail is displacing the fax as the medium of preference in many firms, and the global e-mail capabilities of the Internet make it possible for project managers to operate "virtual offices" anywhere. Some firms are already upgrading their brochures using 3D, sound, animation, and walkthroughs in their Web pages. Firms search for new personnel by reviewing portfolios and resumés on the Internet. Many students have already created personal portfolio Web sites.

Once security issues have been addressed, architects and their consultants can work together on data maintained on Web servers. There will be on-line Project Management Information Groups that not only house drawings and meeting minutes but also allow project managers to view project information wherever and whenever they wish. Drawing sketches and specifications will be sent back and forth among architects, engineers, clients, associates, consultants, contractors, and job-site representatives.

Design resources are also a growing presence on the Internet. In addition to a few CSI (Construction Specification Institute) -based product services, a number of vendors have current product information available at their Web sites. BUILDER Online (http://www.

CD-ROM

Features over 300 new products. Includes videos and product catalogs.

This electronic desktop reference is the most comprehensive guide to building products available.

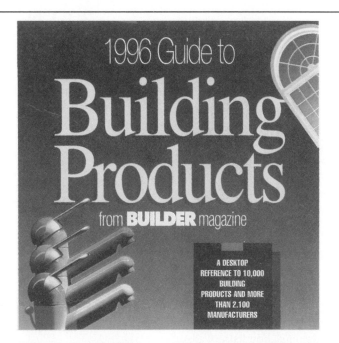

1996 Guide to **Building Products** from **BUILDER** magazine

A DESKTOP REFERENCE TO 10,000 BUILDING PRODUCTS AND MORE THAN 2,100 MANUFACTURERS

From the main menu, click on any category to see an alphabetical list of manufacturers. Or use the menu buttons to go directly to the index, product catalog or video.

Manufacturer catalogs are included, with color photographs and product specifications. Click on any photo to read product specification sheets (when offered).

Manufacturers are listed in alphabetical order. Read across for products each company makes. Click on a company name to get company's address and phone number to appear. Distribution information is also listed.

Figure 2–15 *Buyer's Guide to Building Products* on CD-ROM. (*Builder* magazine, May 1996)

builderonline.com), from *Builder* magazine, provides, among other things, a house plan section containing all of the magazine's house plans, instant price quotes, and on-line ordering (Fig. 2–16). Products made for direct transfer to your computer are the wave of the future.

The Internet, using a technique called *virtual reality modeling language* (VRML), will provide a means of manipulating images in 3D space. This technique will apply to a variety of activities, including construction staging on difficult sites, interior design simulations, signage visibility studies, and space planning.

VRML is meant to be a way to interact with 3D models over the Internet. You see the 3D object and interact with it: Move closer to it, walk around it, or jump to other linked models. VRML lets project data and building design and construction personnel be located anywhere in the world with an Internet connection.

Figure 2–16 *Builder* Online. This Web site gathers in one place all of the information that home builders need.

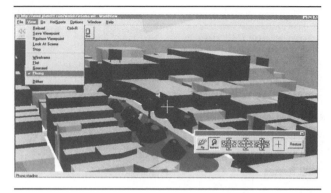

Figure 2–17 VRML SOMA. Shown is a south view of the Market Area in San Francisco. (Developed by WorldView, from InterVista, the first VRML browser for Windows-based computers; from *Cadence,* April 1996)

TriSpectives (3D-Eye), WalkThrough VRML (Virtus), StudioPro (Strata), and Fountain (Caligari) are 3D drawing packages that output VRML directly. However, you do not have to create any VRML worlds because WRL files abound on the Internet. You can walk along a street in Poland as it appears in local time; witness the collision of two black holes; fly over downtown San Francisco's SOMA district; and design a custom kitchen—all over the Internet via VRML (Fig. 2–17).

A relatively new option for Internet users is the Web TV, often referred to as the "empty computer." The concept is that the Internet, and through it the World Wide Web, will contain much of the computing horsepower needed and that the desktop unit will simply provide access to the network. You are connected to powerful servers all over the world through a low-cost modem in order to access data and execute programs. Senior executives from SunMicrosystems, Oracle, and IBM are among the people pushing this idea. Low-cost systems are available now, although their long-term viability is unknown.

2.7

FUTURE DEVELOPMENTS

Converting from manual drafting to computer-aided design and drafting was a small step compared to the potential of today's technology. The next major advancement will be the linking of digital data and remote construction activities by affordable telecommunications. Tomorrow's industry will be an interactive environment of virtual buildings constructed from binary bits, design and construction teams connected by keyboards and video eyeballs, and a new set of uniform drawing standards to help organize the entire design-build process.

2.7.1 Building Design and the Construction Process

Some of the endless possibilities in store for the design professional of the twenty-first century include code research, virtual design, live construction documents, interactive plan checking, on-line bidding, and a construction process using 3D models.

In the area of code research, current updates could be downloaded soon after agency approval. Notification of changes could be made through code compliance forums whereby design professionals communicate with code enforcement officials and code development agencies to clarify unusual design situations.

Another aspect of this virtual model is that a project will be represented digitally by objects in space rather than by lines, arcs, and so forth. These objects will represent familiar construction components such as doors, windows, stairs, and structural items, and they will be able to respond to criteria established for a project. If the interior changes, the object will adjust to what is appropriate for the new conditions. For example, a roof structure designed for heavy snow loading in one part of the country will automatically recalculate component member sizes if the loading conditions change, such as in moving the building to a climate with no snow. Linked to these objects will be information that can be used in the generation of schedules and project specifications.

Interactive plan check will allow plan checkers access to the 3D model database. Comments will be added to a plan check database for review by the design professional as the project proceeds.

Bidders will have read-only remote access to the project's graphic and written databases. A bidders' forum will allow questions in the form of messages, and the responses will be available to all bidders electronically. During construction, the drawing data will remain accessible to users by remote access, and the contractor will update a copy of the drawing data to produce project record drawings. Progress photos can be taken with digital cameras and incorporated as graphic files.

2.7.2 Building Form and Design

Changes in technology will influence building form and design as well. For most of history, architects and builders have been concerned with the body's immediate sensory environment. Their goal has been to provide shelter, warmth, and safety. Buildings are distinguished from one another by their different uses, and their physical plan organization reflects these differences. The floor plan of a library, school, bank, office, or home clearly shows how it works. The various activities that are housed together are integrated with a circulation system of doors and passageways.

In the future, activities and the buildings that house them will be transformed in response to the emergence of digital technologies. Buildings will become computer interfaces, and computer interfaces will become buildings. The task of designing will be fundamentally redefined: It will no longer be one of laying out and constructing a building with storage and circulation areas. It will become one of designing and programming the computer tools.

Traditional home builders may well become extinct. Their role will most certainly be different. Existing technologies will give clients control over decisions on residential building projects. Clients will have the ability to design a new home from a computer, with every house becoming a customized, "one-of-a-kind" home.

Designers will have to consider the "cyber home." Computer-controlled displays of various sizes will be built into houses. Wires to connect components will be installed during construction, and thought will have to be given to the placement of displays in relation to windows in order to minimize reflection and glare. When information appliances are connected, there will be less need for many items such as reference books, stereo receivers, compact disks, fax machines, file drawers, and storage boxes for records and receipts.

The living room might contain an information appliance connected to display a device such as a large-screen television (Fig. 2–18). This display will become a powerful organizer of space and activities. Provisions will have to be made for groups of people to sit around them in living rooms, view them from a distance of 8 to 10 feet, and probably control them with hand-held remote devices. Appliances such as this are already available. For example, the Destination Big Screen PC developed by Gateway 2000 is described as a "group-computing, TV-viewing, multime-

dia-blasting, Internet-cruising mothership." It has a 31-inch monitor, a wireless keyboard, and a remote control.

KEY GRAPHIC PRINCIPLES FOR CAD

2.8.1 Increased Accuracy and Precision

Computers can store information about graphics in two ways. In the most precise, known as *vector format*, lines and curves are stored according to their mathematical coordinates. In the other, known as *rastor format*, graphic images are stored as a picture composed of a series of dots. Most full-featured CAD software packages use the vector format, but drawings are occasionally converted to rastor format during scanning, plotting, or printing.

As drawings are created in CAD, each item is given an exact mathematical description; therefore, it is more precise. In architectural drawing, the size of a component drawn can be approximated only to as close as the scale of the drawing will permit. For example, in a ¼″ = 1′-0″ scale drawing, using a drafting scale, nothing smaller than 1″ can be accurately drawn. With CAD, since every item is entered by its exact size, it is drawn exactly to size.

2.8.2 Cartesian Coordinate System

In a full-featured microcad software package, all objects are stored by their mathematical coordinates in a coordinate system. Long the basis for analytic geometry, the Cartesian coordinate system is the basis for describing objects in CAD. In its two-dimensional form, it contains X and Y axes projecting vertically and horizontally from a point known as the *origin* (Fig. 2–19). Positive numbers are indicated on the top and right quadrants. Negative

Figure 2–18 The appliance of the future. Shown is the Destination Big Screen PC by Gateway 2000; it has a 31-inch monitor, a wireless keyboard, and a remote control. (*Windows Sources*, May 1996)

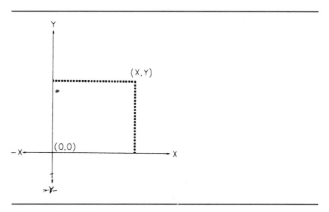

Figure 2–19 Cartesian coordinate system.

numbers are indicated in the left and bottom quadrants. Motion to the left or down is indicated by a negative number, and motion to the right or up is indicated by a positive number.

Each point in the description of an object is indicated by an X and Y value, usually displayed as (X, Y). Because a line is defined by two points, lines in microcad are defined by two points, each with an X and Y coordinate on the coordinate system. "Displacements," in CAD terminology, are the distance one wishes to move along an axis (X, Y, Z) described in terms of points on the coordinate system.

2.8.3 Linear Displacement

Linear displacement is motion along an axis from point to point (Fig. 2–20). It is the most fundamental method of drawing. At the keyboard you would draw from point (X, Y) to point (new X, new Y). There are two types of linear displacements. One is by absolute coordinates, which measure the location of all points from the origin. With other types, relative coordinates allow the user to reset the origin temporarily to the last point entered so that the new distances can be entered relative to the last point. Relative coordinates are frequently used because they are quicker.

2.8.4 Angular Displacement

Angular displacement (Fig. 2–21) is motion from a known (X, Y) point with a distance and an angle. This type of drawing is called *drawing with polar coordinates*. Angles in the electronic drawing area in CAD are generally measured from east. North, then, represents 90° west; 180°; and so on. Angles measured counterclockwise have a positive number; measured clockwise, they have a negative number.

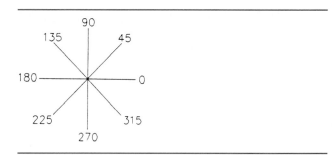

Figure 2–21 Angular displacement.

2.8.5 Database

All of the points and lines that define the objects in the drawings, taken together, represent the database of that drawing. Each time the drawing is added to or otherwise changed, the database is updated. As you might imagine, it takes many numbers to define the complex objects often depicted in CAD drawings; therefore, CAD requires special equipment and software and large amounts of computer memory.

2.8.6 Three-Dimensional Space (*X, Y, Z* Coordinates)

Even though we have discussed only the X and Y axes so far, the third coordinate, Z, has been there all along (Fig. 2–22). The third axis helps to describe three-dimensional objects. In three-dimensional CAD, a 3D model is being created, not just a drawing. Therefore, the model can be rotated and viewed from any direction. Most CAD packages display 3D drawings as axonometric drawings, and most can create perspective views that make attractive presentations. For example, once the model is created, it can be shaded. Moving pictures or "animations" can be made to simulate a walk-through or fly-over, making truly dynamic presentations.

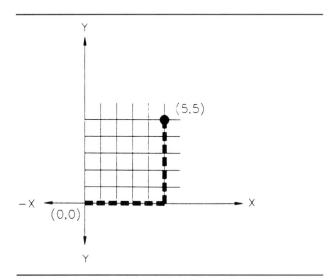

Figure 2–20 Linear point displacement.

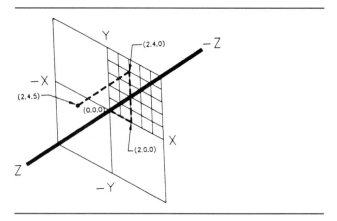

Figure 2–22 Three-dimensional coordinate system.

BASIC COMPUTER-AIDED DRAFTING

Once loaded onto a computer, most CAD packages can be started with a short keyword or by choosing an application icon from the screen menu. Once the program has begun, there are often several ways to drive it, including typing from the keyboard, pointing with a mouse or a digitizing tablet, and interacting with items as they appear on the screen.

2.9.1 Menus

When the drawing editor is active, a *screen menu* can be displayed on the edge of the graphics screen (Fig. 2–23). This menu lets you enter a command by simply pointing to the command on the screen with a pointing device or by using the keyboard's arrow keys.

- *Pull-down menus*, which you pull down from the menu bar at the top of the screen, may be available (Fig. 2–24).
- *Tablet menus* (Fig. 2–25) are stiff cards attached to

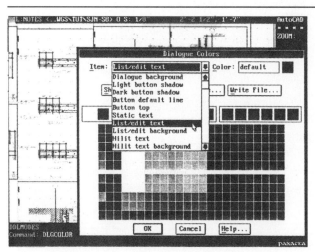

A

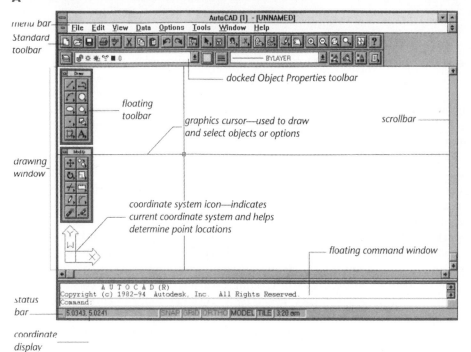

B

Figure 2–23 AutoCAD Release 12 screen menus for (A) DOS and (B) Release 13 for Windows. (*AutoCAD Power Tools*)

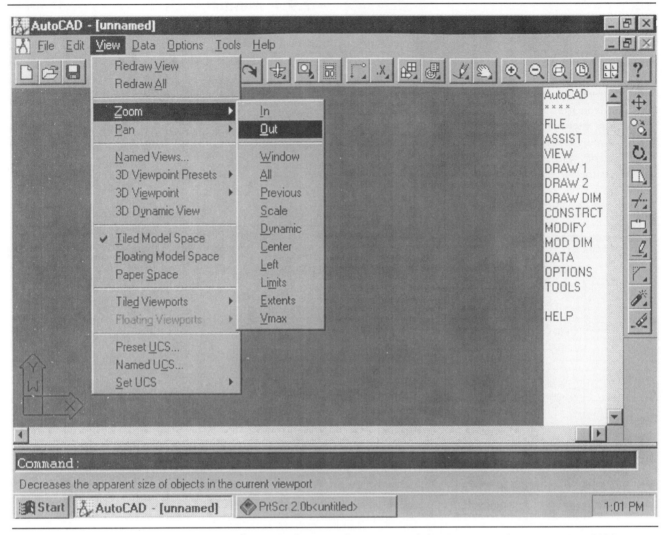

Figure 2-24 AutoCAD pull-down menus. Shown are the View, the Zoom, and the Out menus. (*Mastering AutoCAD*)

the digitizing tablet. They contain a pointing area that equates to the screen on the monitor for drawing, and they have illustrations of the commands that allow the user to select an operation quickly with the cursor while drawing.

- *Dialog boxes* (Fig. 2–26) are screens that appear when you need to make changes in values and it would be advantageous to review the current settings of those values. These boxes provide a quick visual means of viewing the current setup and making changes.

2.9.2 Commands

The work in CAD is created primarily by a series of operations known as *commands*. Each allows for the creation or change of entities within the drawing. Once started, a command may have several steps to define what is to be done before it can be executed. Commands can be invoked from the keyboard, from screen menus, from pull-down menus, or from the digitizing tablet with a tablet menu.

Draw Commands These commands take the user's input and create the geometric entities intended to be shown in CAD. Most systems have commands that will create points, lines, circles, polygons, ellipses, and the basic geometric shapes. Each shape is constructed by establishing first its location and then its size in accordance with its geometric properties. For example, a point must be defined by two coordinates, a line by two points, a circle by its center point and radius, and a polygon by its number of sides. Beyond simple geometric entities, Draw commands allow for more complex operations. Sketching consists of creating freehand lines with the cursor or arrow keys. Hatch patterns can be added to enclosed geometric shapes showing brick, concrete, steel, and much more. The annotating (lettering) of drawings is simplified with CAD. Select a typeface or font, and the size and location, and then type the text from the keyboard. Notes can be easily changed and moved.

Edit Commands These commands are used to revise entities that have already been created within the program.

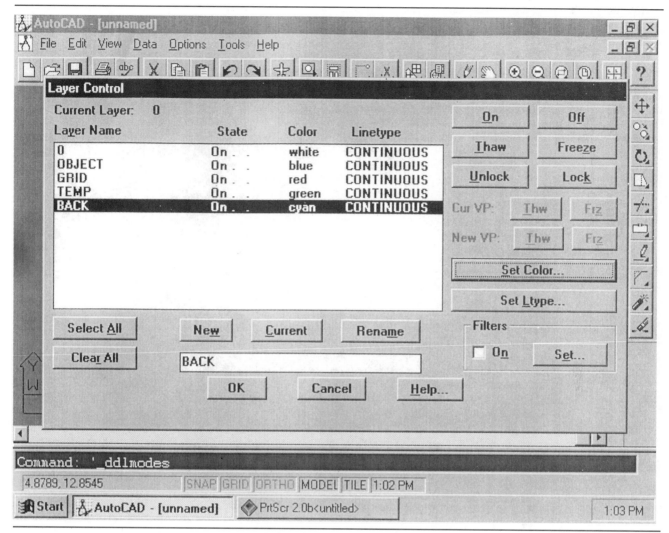

Figure 2-25 AutoCAD Tablet menus. (*Mastering AutoCAD*)

The power of CAD becomes obvious with the Edit commands. Entities can be erased and then restored if necessary. Lines can be cut or trimmed to meet others. Items can be easily moved from place to place on the drawing. They can be enlarged and reduced in size and rotated a full (or any part of) 360°. Entities can be stretched or enlarged in any direction. Through a process known as *mirroring*, upside-down or reverse copies can be made, yet the writing on the drawing can stay "right reading." Probably the most powerful edit command, Array, allows for multiple copies of an entity in a circle or a rectangular matrix.

Inquiry Commands   These commands enable the user to obtain details from the database and to obtain the status of the drawing environment. They can provide a listing of the size and a geometric description of the entities in the drawing, and they can calculate the perimeter and area of shapes. You can use inquiry commands to identify the location in the Cartesian coordinate system of points, lines, and shapes.

2.9.3 The Drawing Environment

Recall the Cheshire Cat. Most CAD software packages offer so much flexibility that the user can easily become confused. Many CAD packages can be used for a wide variety of applications. The drawing environment is extremely flexible in CAD, and when this flexibility is understood, it becomes another of its many advantages.

As our prime concern here is architectural drawing, think of the decisions that must be made before starting to draw with a pencil. How big a drawing table is necessary? What units will be used to draw? Architectural? Decimal-feet? Will a paper with a grid be used? Will the drawing be full size or scaled?

CAD programs have *settings* to control all of these factors and more. Settings control the size of the area in which the object will be drawn, the units in which it will be measured, and the relative size or scale at which it will be depicted. Furthermore, drawing aids are available that

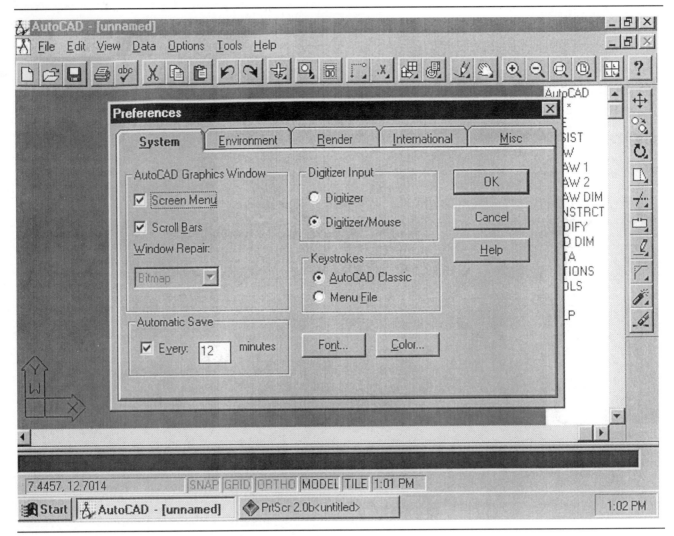

Figure 2–26 AutoCAD dialog box. Shown are different versions of the dialog boxes available. (*Mastering AutoCAD*)

can set up a reference grid in the background to help keep your place. Other drafting aids, impossible in manual drafting, allow you to draw only straight lines, draw only from grid point to grid point, and draw from one particular part of an object to another.

Layers in CAD drawings help to sort and organize data. They are derived from a drafting technique known as *overlay drafting*. In overlay drafting, a series of translucent sheets were used to create a composite drawing. All of the drawings were held in registration with a pin bar attached to the drawing board. The walls and doors were drawn on one layer, and the electrical plan and HVAC plans on others. A composite print could be made by placing the drawings together on a flat-bed diazo machine and making a blueline print.

CAD software packages offer the same capability and more. Using layers in a single "drawing," you can clearly organize the work and make it much easier to revise later. Each layer can have a separate name, color, and type of line (dashed, dotted, continuous, and so on). Each layer can be turned off, making it invisible on the screen, and then on again. Layers can also be temporarily excluded from the

database (frozen) to save time. Items can be moved from layer to layer as desired. Only one layer can be edited at a time. Therefore, only the active layer can be changed.

The Display command is used to enlarge a portion of a drawing. If you are creating an object as large as an Olympic-sized (50-meter) swimming pool in CAD and looking through a 15-inch monitor, it will probably appear too small to work on, especially when drawing the pattern of slots in the grid on the drain at the bottom. Although this example illustrates another advantage of CAD, the ability to draw infinite detail, there must be a way to move closer and farther away from your subject; and there is. A group of *Display commands* helps to control the point of view. The primary command for moving in or out on an object is Zoom. With this command, you can move in or out at specified increments. The primary way to move from side to side on an object is Pan. A linear displacement is required for Pan. It can be entered either from the keyboard or graphically on the screen. When zooming and panning, you should remember that the object always remains the same; only the viewpoint changes.

2.9.4 Polylines

A complex CAD drawing can have a large database. It is not uncommon for a single architectural drawing to require more than one diskette. It is important, then, that CAD drawings be kept to a manageable size. A series of detailed architectural elevations, for example, would best be stored as a series of separate drawings, known as *polylines*. They can be combined just prior to plotting, or each can be plotted separately on the same sheet. CAD packages, however, are designed to minimize the database whenever possible. One way to accomplish this goal is to put related objects together as polyline groups.

2.9.5 Combining Drawings

Architectural drawings are often composed of symbols that are transferred from drawing to drawing. Sometimes they contain actual parts from other drawings. Drafters of the past laboriously copied drawings from one set to another or, more recently, copied, cut, and pasted them into new drawings. Just the ability to copy an entity in a single drawing would relieve this tedium, but there is even more.

The Block feature allows the reuse of drawings and symbols within a drawing or from drawing to drawing. In some CAD systems these blocks are called *parts*. Whatever the name, it is one of the most powerful features of CAD. When the Block command is invoked, the user is prompted to select the items to be included. After selection, the items are saved as a part of the drawing file. They can be recalled into the drawing with the Insert command. If the user wishes to use those items in another drawing, then the block is converted into a drawing file with the WBlock (for Write Block) command. The items contained in the block are now contained in a drawing file. When the items are needed again *in any drawing*, they can be recalled with the Insert command. The beauty of these commands is that any drawing can be combined with any other drawing in CAD.

The power of blocks has two primary implications. First, drawing symbols that are frequently used, such as doors, windows, plumbing fixtures, and furniture, can be created just one time and reused. Symbol libraries are composed of these items. Second, parts of drawings can be combined with other drawings. For example, a title block can be created for a 24 × 36 sheet one time and then called into other drawings as required. With Attributes, another powerful CAD feature associated with Blocks, the user can be prompted to input the sheet number and date. Finally, details commonly used in an office can easily be reused on other projects.

2.9.6 Customization

The Prototype Drawing   CAD packages are very flexible, but they can also be customized for a particular use or user. Blocks are the first step in customization. The next step is automating the procedure by which the drawing environment is created. Normally, while booting up, the CAD software looks for a seed drawing or a *prototype*. Much as a word processing system starts with an 8.5 × 11 sheet, CAD packages must have something from which to start; therefore, the manufacturer usually provides one with the package. The exact information contained in the prototype drawing is normally duplicated in a new drawing. Therefore, you can duplicate the drawing environment of previous drawings with ease simply by changing the prototype drawing.

In some applications, *other prototype drawing*s are required, for example, title blocks for 18 × 24 sheets, 24 × 36 sheets, and 30 × 42 sheets. In this case, not only the previous settings but also the previous drawing (the title block) should be duplicated. It is possible to set any one of these drawings as the pattern (prototype) for a new drawing, thus saving the time required to create again.

Macro is short for "macro command," meaning a large or long command. A macro is a series of commands attached together so as to perform a task. Menus actually consist of prewritten macros. An advanced user can rewrite these menus or add new ones to string together series of frequently used commands. Keyboard characters have special meanings in the menu, almost like a primitive programming language. The result is a customized menu. In the case of the tablet menu, blank spaces or cells are available that can be assigned to custom macros so that they can be activated with the cursor.[1] Macros can be written for the buttons on a cursor and the screen menu. An example of a simple two-step macro would be created to enter the command Zoom Window in one step, thereby minimizing input from the user. Zoom Window could then be included on and activated from a screen menu.[2]

LISP is a programming language that was derived from XLISP. AutoLISP[3] is a dialect of LISP and coexists with AutoCAD itself within the AutoCAD program.[4] *LISP* is short for "list processing." It can be used to create programs within AutoCAD using more than just the resident commands. "An example of an AutoLISP program would be one that would expedite creation of a staircase in a building. If properly written, the program would prompt you for the distance between the upper and lower floors, ask you for the size and/or number of steps (risers and treads), and automatically draw the detailed staircase for you."[5]

2.9.7 Third-Party Development

Many software packages are general purpose, that is, intended for many different applications. Unfortunately, what is gained in flexibility is lost in utility. In other

---

[1] D. Raker and Harbort Rice, *Inside AutoCAD*, 5th ed. (Thousand Oaks, CA:New Riders Publishing, 1989) 18–1 through 18–8.

[2] Terry T. Wohlers, *Applying AutoCAD, Step by Step* (Mission Hills, CA: Glencoe Publishing, 1989), 350.

[3] AutoLISP™ is a registered trademark of Autodesk, Inc.

[4] Raker and Rice, *Inside AutoCAD*, 20–1.

[5] Wohlers, *Applying AutoCAD*, 380.

words, the basic package can be used for several different disciplines, but each discipline must adapt it for its particular way of drawing.

We have just had a brief overview of customization, in which new menu features and LISP routines can be added to increase speed and efficiency. Many users not proficient in customization and programming prefer to use customized features developed by others and incorporate them into their system. These include special type fonts, hatch patterns, text editors, macros, and LISP routines. Users sometimes share these add-ons through user groups, bulletin boards, and personal contacts, taking care not to violate copyright law. Some items are available directly from the developer, such as text fonts. Macros and LISP routines are regularly published in customization books and in CAD industry journals.

Some construction product manufacturers prepare details of their products in CAD; write custom programs that enable the user to select one of their products, including model, size, and features; and combine the product with the working drawings. Such programs are now available from a number of manufacturers.

Furthermore, packages that contain entire detail files are available that work within a CAD software package. Within these packages, the designer can choose among thousands of predrawn details, verify that they apply to the current project, change them if necessary, and then incorporate them into the working drawings. Currently, third-party software developers and construction product manufacturers are working together to combine their details into massive databases.

2.9.8 Applications Packages

A special type of software known as an *applications package* is available. Each is for use with specific CAD packages. In some cases they are developed by the primary CAD software company, but frequently they are produced by another company, a third party. They run within the primary CAD software and are a coordinated set of customized features, such as a drawing setup routine, layering schemes, drawing settings, menus, symbols, and macros.

Most applications packages are developed in modular coordinated sets. They have the advantage of being integrated so that design professionals can easily exchange files and information. Softdesk, Inc. (recently purchased by AutoDesk), manufactures an integrated line of products to run within AutoCAD: Civil/Surveying, Architectural, Facilities, and Structural. Within the architectural package are programs such as Auto-Architect, HVAC, and Architectural Electric and Plumbing. Auto-Architect includes routines to draw double lines for walls; draw stairs, doors, and windows; and perform space planning. It also contains an extensive symbol library.

HOME DESIGN OF THE FUTURE: COMPUTER INTEGRATION

With available software, the following scenario is not only possible today but will be commonplace in the near future.

You have purchased a piece of land on which you want to build your "dream" house and have just sold your current house. You determine that you have a $160,000 budget for your new house and a deadline of September 1 to move out of your present house.

You have three factors to consider: design, cost, and time. Cost

---

Text and figures are from the article "The 'New CADD' Trades Brawn for Smarts" by Barbara Goode, Softdesk, Inc., in the May 1996 issue of *Design/Build* magazine.

must be your primary consideration because it is inflexible. You do not qualify for more than a $160,000 mortgage. Timing is next. You have a deadline of when you have to be out of your current house. You could put your things in storage for a month, but this entails additional expenses. Design comes last. You have definite ideas of what you want in a house, but it is the one place where you have some flexibility.

Keeping these criteria in mind, you sit down at your computer and start designing. Once you have the basics down and are within budget and on time, you can start playing with options. You decide that you would like to put an arched window above the front door. You add it to your design and, thanks to a link to the spreadsheet, find that it keeps you within budget. Unfortunately, you also find out that it will take 8 weeks to order the model you have selected. Another arched window is available now, but it costs much more than your original choice and puts you over budget. Subsequently you find that all of the arched windows available are either too expensive or not available soon enough. You decide to choose a standard window instead. You have compromised on the design somewhat but have done so knowing all of your options. There has been no guesswork.

FIG. 1: *Highlighted text is hot-linked to your drafting and spreadsheet files.*

The "New CADD" software has intelligent links with other applications such as word processors and spreadsheets. This layout links with standard Microsoft Word and Excel programs, so that changes you make—in the letter (Fig. 1), invoice (Fig. 2), spreadsheet (Fig. 3) or layout (Fig. 4)—are updated in all four. This means you can accomplish work wherever it's easier for you.

FIG. 2: *The view from the invoice.*

FIG. 3: *The spreadsheet changes as you change the design—or vice-versa!*

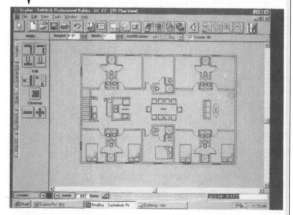

FIG. 4: *It may look like your typical CADD software, but it's more flexible.*

REVIEW QUESTIONS

1. Why doesn't learning to draw with a computer eliminate the need to learn to draw by hand?

2. What is meant by the term *model* in computer-aided design?

3. In what ways can computer-aided design aid creativity?

4. Which two major technological developments with computer hardware have fueled the rapid rise in popularity of microcomputers?

5. What is the proper name of the main printed circuit board that contains the microprocessor, the numeric coprocessor, and expansion slots?

6. How many pages of text can be stored on a typical CD-ROM disk?

7. Why is it important that CAD users develop good typing skills?

8. Discuss the advantages and disadvantages of pen and electrostatic plotters.

9. What device (other than a fax) is used to transmit files electronically over telephone lines?

10. What are the major purposes of an operating system?

11. Using CAD conventions, give an example of each of the following:
    a. linear displacement by absolute coordinates
    b. linear displacement by relative coordinates
    c. angular displacement by polar coordinates

12. What are four different techniques to issue commands using CAD?

13. In a typical CAD drawing session, which type of commands are used more frequently, Draw or Edit?

14. Why does a square 1 inch long on each side require the same amount of computer memory as a square 1 mile long on each side?

15. Points, lines, and circles are examples of *entities*. What is the name of a compact group of entities in CAD?

16. Polylines can greatly increase the *speed* of CAD work, but they can hinder *detailed work*. Why?

17. What are the advantages of combining drawings?

18. Explain how symbol libraries are created in CAD.

19. Explain the difference between graphic images stored in vector format and raster format.

20. When objects are drawn at full scale in CAD, how are scale drawings generally produced?

BIBLIOGRAPHY

Autodesk, Inc., *AutoCAD User's Guide Release 14*, 1997.

Conner, Frank L., *The Student Edition of AutoSketch, User's Manual*. Reading, MA: Addison-Wesley Publishing and Benjamin/Cummings Publishing, 1991.

Hart, Roger. Plotters, Teaming Up Your Compaq with the Right One. *PAQ Review* (Summer 1988).

Hoskins, Jim. *IBM Personal System/2, A Business Perspective*. New York: John Wiley & Sons, 1987.

Jeffris, Alan, and David Madsen. *Architectural Drafting and Design*. 2d ed. Albany, NY: Delmar Publishers, 1991.

Lloyd, Bill. *Technical Notes, Note 102, Electrostatic Technology*. Versatec, 1989.

Popular Electronics. "World's First Minicomputer Kit to Rival Commercial Models . . . 'Altair 8800.' " January 1975.

Raker, Daniel, and Harbort Rice. *Inside AutoCAD*. 5th ed. Thousand Oaks, CA: New Riders Publishing Co., 1989.

Sheldon, Thomas. *Hard Disk Management in the PC & MS DOS Environment*. New York: McGraw-Hill Book Company, 1988.

Wohlers, Terry T. *Applying AutoCAD, Step by Step*. Mission Hills, CA: Glencoe Publishing, Division of Macmillan, Inc., 1989.

REFERENCES

Books

Gates, Bill. *The Road Ahead*. New York: Viking Penguin, 1995.

Guidelines Publishing. *The Internet for A/E's*. Orinda, CA, 1995.

Mitchell, William. *City of Bits*. Cambridge, MA: MIT Press, 1995.

Omura, George. *Mastering AutoCAD Release 12*. Alameda, CA: Sybex, Inc., 1992.

Sanders, Ken, AIA. *The Digital Architect*. New York: John Wiley & Sons, 1996.

Smith, Bud, Jake Richter, and Mark Middlebrook. *AutoCAD Power Tools*. New York: Random House, 1993.

Periodicals

A/E/C System. Sept./Oct. 1995; Nov./Dec. 1995; May/June, 1996.

Architecture. June 1995; August 1995; May 1996.

Builder. April 1994; July 1995; May 1996; June 1996.

Cadalyst. July 1993.

Cadence. April, 1996.

The Construction Specifier. January 1996.

Design/Build Business. May 1996.

PC Magazine. December 1995; February 1996; April 1996.

PC World. February 1995.

Professional Builder. December 1994; Mid-January 1995.

Progressive Architecture. September 1995.

Sun Coast. May 1995.

Windows Sources. May 1996.

PUBLISHERS' ADDRESSES

Books

Guidelines Publishing
P.O. Box 456
Orinda, CA 94563

MIT Press
5 Cambridge Center
Cambridge, MA 02142

Random House, Inc.
201 East 50th Street
New York, NY 10022

Sybex, Inc.
2021 Challenger Drive
Alameda, CA 94501

Viking
A Division of Penguin USA
375 Hudson Street
New York, NY 10014

John Wiley & Sons, Inc.
605 Third Avenue
New York, NY 10158

Periodicals

A/E/C Systems Computer Solutions
A/E/C Systems, Inc.
P.O. Box 310318
Newington, CT 06131

Architecture
BPI Communications
1515 Broadway
New York, NY 10036

Builder
Builder Magazine
655 15th Street N.W., Suite 475
Washington, DC 20005

Cadalyst
Advanstar Communications, Inc.
859 Williamette Street
Eugene, OR 97401

Cadence
Miller Freeman, Inc.
600 Harrison Street
San Francisco, CA 94107

The Construction Specifier
Construction Specifications Institute
601 Madison Street
Alexandria, VA 22314

Design/Build Business
McKellar Publications, Inc.
333 E. Glenoaks Boulevard, Suite 204
Glendale, CA 91207

PC Magazine
Ziff-Davis Publishing Company, L.P.
One Park Avenue
New York, NY 10016

PC World
PC World Communications, Inc.
501 Second Street #600
San Francisco, CA 94107

Professional Builder
Professional Builder
1350 E. Touhy Avenue
P.O. Box 5080
Des Plaines, IL 60017-5080

Progressive Architecture
Penton Publishing
1100 Superior Avenue
Cleveland, OH 44114

Sun Coast
McKellar Publications, Inc.
333 E. Glenoaks Boulevard, Suite 204
Glendale, CA 92107

Windows Sources
Ziff-Davis Publishing Company, L.P.
One Park Avenue
New York, NY 10016

W. D. Farmer, AIBD

"Practice is the best of all instructors."
—PUBLILIUS SYRUS

Legible lettering on a drawing fulfills an important requirement. Information that cannot be revealed by graphic shapes and lines alone must be included in the form of notes, titles, dimensions, and identifications to make the drawing informative and complete. The lettering can enhance the drawing by making it simple to interpret and pleasant to look at or it can ruin an otherwise good drawing by making it difficult to read and unsightly in appearance.

3.1
LETTERING SKILL

Skill in freehand lettering adds style and individuality to a drafter's work. Even though the various mechanical devices available aid in producing the uniform technical lettering required in some offices, freehand lettering is generally preferred on architectural drawings and should be an important phase of training. In fact, many employers evaluate drafting skill in direct relation to lettering skill. If the student learns the basic letter forms (Fig. 3–1) and gives deliberate application to their mastery during early practice, he or she will soon be able to do a creditable job of lettering.

Anyone who can write can learn to letter. After considerable experience, every drafter's freehand lettering acquires an individuality—much the same as his or her handwriting—which is soon identifiable and which contributes favorably to good technique in architectural drawing.

Lettering is also occasionally incorporated into the design of structures. Inscriptions, plaques, signs, and other applications comprising well-designed letter forms in the architectural aspects of the building often fall upon drafters for their execution. Therefore, the basic styles of alphabets and their various characteristics should be of particular interest to you.

3.2
ALPHABET CLASSIFICATIONS

By closely observing the alphabets and typefaces in use by the modern commercial world, you will notice that each fits one of the following basic letter classifications (Fig. 3–1):

Figure 3–1 Basic alphabet types. All lettering can be classified under one of these types.

■ Roman Perfected by the Greeks and Romans and modernized during the eighteenth century by typefounders, this alphabet imparts a feeling of grace and dignity and is considered our most beautiful typeface family. The classic example can be found on the column of Trajan in Rome. Thick- and thin-width strokes and serifs are the distinguishing features. *Serifs* are the spurs or "boots" forming the ends of the strokes, which can be used on gothic letters as well as on roman. Although the lettering found on old Roman stonework is lightface, modern roman lettering has a bolder characteristic. Extensive modifications of this beautiful alphabet are in wide use today.

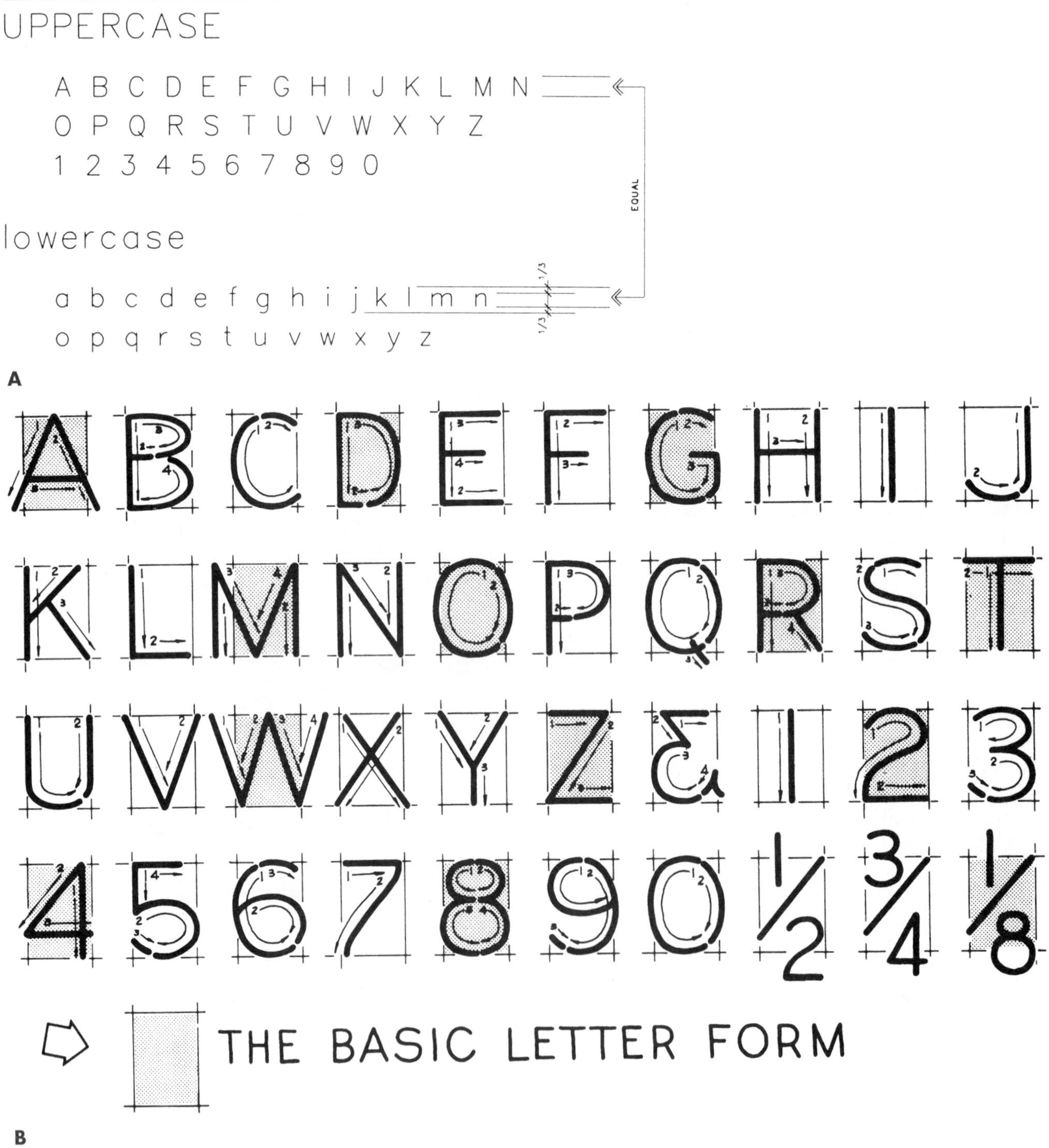

UPPERCASE

A B C D E F G H I J K L M N
O P Q R S T U V W X Y Z
1 2 3 4 5 6 7 8 9 0

lowercase

a b c d e f g h i j k l m n
o p q r s t u v w x y z

A

THE BASIC LETTER FORM

B

Figure 3–2 (A) Uppercase and lowercase letters. (B) The vertical, single-stroke technical alphabet, showing proportion and sequence of strokes. Master this basic alphabet before trying the architectural variations.

- **Gothic** This basic alphabet has been in use for many years as a commercial, block-type letter, which is comparatively simple to execute and easy to read. Its distinguishing characteristic is the uniformity in width of all of the strokes. Its modifications include inclined, the addition of serifs, rounded, squared, boldface, and lightface. The gothic alphabet is the base from which our single-stroke technical lettering has evolved.
- **Script** This completely different basic alphabet is cursive in character and resembles handwriting. Interconnected lowercase letters are used within words; capital letters, of course, are used only at word or sentence beginnings, a practice not required of roman or gothic letters. The free-flowing, continuous strokes produce a delicate and personal temperament, making this alphabet popular for many commercial trademarks. Modifications of either vertical or inclined scripts range from very thin lines to bold or thick-and-thin varieties of many modern interpretations.
- **Text** Sometimes referred to as Old English, this alphabet was first used by central European monks for recording religious manuscripts before the advent of the printing press. The strokes of different width are due to the fact that a flat quill pen was employed. It is similar to script in that lowercase letters must be used throughout the words, with the use of capitals restricted to word beginnings only. Text is difficult to read and draw.

The term *italics* is generally given to typefaces having inclined, lightface, and curved characteristics. It is not given a separate grouping because all of the above-mentioned letter types can be made in the italic form.

3.3

UPPERCASE AND LOWERCASE LETTERS

Nearly all alphabets have both uppercase (capitals) and smaller, more rounded, lowercase letters. The height of the lowercase body is usually two-thirds the height of its capitals, with ascenders extending to the capital line and descenders dropping the same amount below [Fig. 3–2A]. If a great deal of copy or an unusually large note is necessary on a drawing, lowercase letters will be found to require less space, and they are usually more readable. Yet capitals are used for the majority of the lettering done on architectural drawings. Because of the universal acceptance and popularity of capitals, the material in this chapter will consider mainly their various characteristics.

3.4

BASIC LETTERING PRACTICE

The vertical, single-stroke gothic alphabet shown in Fig. 3–2B is the basic style used by drafters for most of the lettering found on technical drawings. Because of its legibility and simplicity of execution, it has been universally accepted as a standardized alphabet. Lettering on architectural drawings is known for its freedom of style and individuality, yet you can develop no style that is easily read unless you thoroughly learn the basic alphabet. You should not attempt the architectural variations until you have accomplished the rudiments of letter construction. (For a beginning lettering exercise, it is suggested that you fasten a sheet of tracing paper or Mylar® film over Fig. 3–2B and make a direct duplication of the basic characters.)

Select an H or a 2H drafting lead or an E-1 or E-2 plastic lead for lettering, and keep the point medium sharp at all times. Figure 3–3 shows the proper method of holding the pencil. A comfortable, natural posture is the most desirable. Be sure that your entire forearm rests on the flat table surface; this position allows for more stability than trying to letter over instruments or equipment on the drawing table. Rotate the pencil continually between strokes to maintain a uniform pencil point. A point that is too sharp is difficult to control—it will either break or create grooves in the paper if a desirable, firm pressure is applied. A pencil that becomes too dull, on the other hand, produces ragged and awkward strokes (Fig. 3–4). As a preliminary exercise, try a few of the practice strokes shown in Fig. 3–5 before starting the actual letter forms.

Figure 3–3 Holding the pencil for lettering.

AVOID LETTING THE PENCIL GET **TOO DULL**

Figure 3–4 Maintain a medium-sharp point to avoid variations in stroke width.

||| ☰ //\\ (C)) ≋ ≋ SS

Figure 3–5 Preliminary practice strokes.

GUIDELINES

All lettering is done with the aid of penciled guidelines. Even experienced drafters carefully draw horizontal guidelines for their lettering. It is nearly impossible to maintain straight lettering without guidelines—use them as the experts do. See Fig. 3–6 for the method of drawing uniform guidelines with one type of plastic lettering guide. Until you have developed skill in *vertical* lettering strokes, use vertical guidelines also; after you have attained confidence, dispense with the vertical guidelines—but never with the horizontal ones.

In pencil lettering, the guidelines are always left on the drawing. Therefore, you should take care in placing the lines neatly on the paper with a sharp pencil or a nonprint pencil. Architectural drafters give more prominence to their horizontal guidelines and make no effort to minimize them; in fact, they usually draw guidelines well be-

yond the intended extremities of the actual letters. Only on inked lettering are the lines erased (Fig. 3–7).

One of the available lettering guides or triangles (Braddock, Ames) will aid in maintaining uniform letter heights. *To draw lines with the guide*, insert a sharp pencil into the proper countersunk hole, and pull the guide along the top edge of the parallel bar. Then insert the pencil into a lower hole to form the height of the letters, and return it along the parallel bar blade. A series of uniform guidelines can be easily drawn in this manner. Holes in the guide are grouped for capitals and lowercase; the numbers below each group indicate the height of the caps in thirty-seconds of an inch. For example, holes in the No. 4 row would lay out lines for capitals $^4/_{32}$″ or $^1/_8$″ high.

HEIGHT OF LETTERS ON WORKING DRAWINGS

Lettering on construction documents should be clear, legible, and consistent. In order to achieve a level of consistency, you should have uniform lettering of appropriate size and with adequate spacing between letters and lines of lettering. Typically, lettering on architectural drawings should all be uppercase, and the size of the letters should be approximately twice as high as the space between the lines of lettering (Fig. 3–8).

General dimensions and notations are $^1/_8$″ high, with common titles $^1/_4$″ high. Special titles and labels should be lettered $^1/_2$″ high. (Some firms prefer using $^3/_{32}$″ and $^3/_{16}$″ in place of the $^1/_8$″ and $^1/_4$″.) When lettering notes that require more than one line, keep the space between

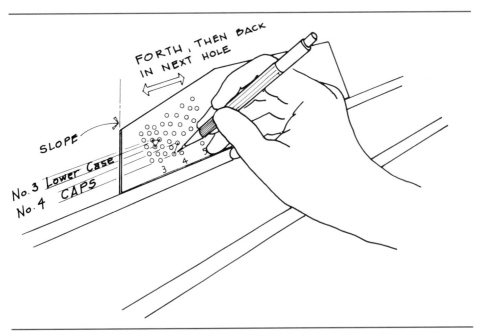

Figure 3–6 Using the lettering guide.

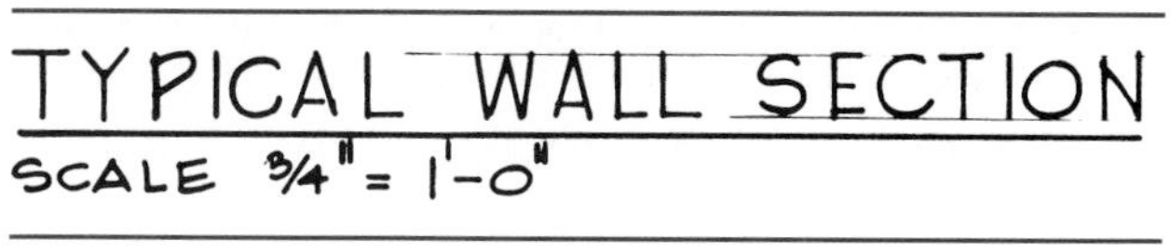

TYPICAL WALL SECTION
SCALE 3/4" = 1'-0"

Figure 3-7 Using guidelines for lettering.

the lines a minimum of $\frac{1}{2}$ the height of the letters and never more than equal to the height of the letters. If possible, make the space between lines equal to the height of the letters if the letters are $\frac{1}{8}''$ or smaller. If drawings are to be plotted at a reduced size or microfilmed for storage, it is preferable to use $\frac{1}{8}''$ lettering as a minimum size. Generally, small lettering carefully fitted into open spaces on a drawing is more legible than larger letters crowded into insufficient spaces. For the sake of readability, never allow lettering to touch any of the drawing lines.

Beginning lettering exercises should be done with letter heights of $\frac{1}{4}''$ or $\frac{3}{8}''$ until correct letter forms have been mastered. The distance between sets of guidelines for long notes should be spaced the same as the height of the letters; that is, if the letters are $\frac{3}{32}''$ high, the vertical space between each line should be $\frac{3}{32}''$, which will prevent a crowded appearance. A short note having only several lines of lettering can be made with the rows of letters slightly closer together and still be satisfactory.

3.7
LETTER PROPORTION

After studying the basic letter forms in Fig. 3–2B with the suggested sequence and direction of strokes, you will notice evidence of different letter widths relative to the basic rectangular shape. These definite letter

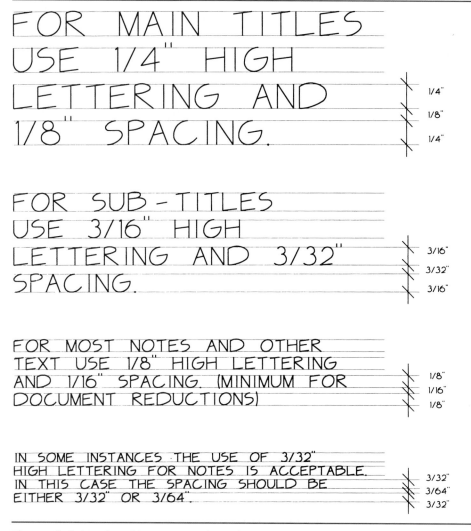

Figure 3-8 Heights of letters on working drawings.

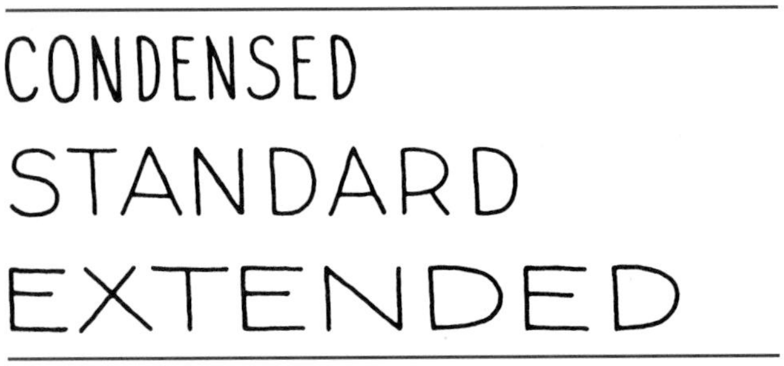

CONDENSED
STANDARD
EXTENDED

Figure 3–9 Condensed and extended lettering.

widths are identified with each character and should be maintained if the lettering is to be legible and harmonious. Not only do widths vary from the single line of the letter I to the widest character, the W, but all alphabets as a whole can be distorted, if necessary, in general width (Fig. 3–9). The narrower distortion is called *condensed* lettering, and the wider is called *extended*. In either method the letters vary in width, yet they retain the relative basic characteristics. Condensed lettering occasionally becomes necessary in confined areas on drawings, yet it is more difficult to read than standard or extended letters. Extended lettering is appropriate when a word or a title must identify a large area or a group of drawings. Remember that visual unity is attained when condensing or extending has been consistent and the recognizable silhouette or design of the individual letter has not been altered.

The effect of visual gravity is evident in correct letter construction. If the horizontal center bar of the letters B, E, F, and H is placed at midheight, it will appear to be below center. This effect is called *visual gravity*, and all lines, especially horizontal, are influenced by it when you are observing a graphic figure. Lines on a page have optical weight. Notice that to counteract visual gravity, the central horizontal stroke of the above-mentioned letters is placed slightly above center. Even the upper portions of the B, S, 3, and 8 are reduced in size to gain stability.

Each character is a well-designed form in itself, and you should not attempt to improve or alter an accepted letter form. Put more grace in the letter M by extending the center "V" portion to the bottom guide line, rather than terminating it above the line. Keep the W vertical by making the third stroke parallel with the first and the fourth stroke parallel with the second. Both "Vs" in the W are narrower than the regular letter A or V. Make circular letters full height, or even slightly larger than the space between the guidelines, with smooth, curving strokes. If vertical lettering is used, make the vertical strokes perfectly vertical. Check your work occasionally with a triangle—strokes that may appear to be perfectly vertical may nevertheless be slightly inclined.

3.8
LETTERING TECHNIQUE

The beginning and the ending of each stroke are important—emphasize them with a slight pressure of the pencil to bring the strokes to sharp and clean-cut terminations. Eliminate careless gaps in lettering by carefully intersecting the strokes. Make each stroke definite and firm; going over a stroke twice ruins the appearance of the letter. To gain positive pencil control when lettering, pull strokes toward you rather than away from you. Develop perfection of the letters before attempting speed. *There is no substitute for diligent practice.* Concentrate on letter forms before you concern yourself with spacing.

3.9
SPACING OF LETTERS AND WORDS

Because each letter has a different profile and width, the spacing of characters within each word becomes a visual problem rather than one of mechanical measurement between letters. The spaces between letters in a word should be nearly identical in *area* if the word is to appear uniform in *tone*. This effect can be achieved only by eye or by optical judgment (Fig. 3–10). Notice that the letters of the lower line are incorrectly spaced when they are equidistant from each other. Letters with vertical adjacent sides or edges must be kept well apart to allow a similar amount of space between very irregular, adjacent characters. The relatively wide space between the I and the N can be seen in the top row of Fig. 3–10. On the other hand, an LT, an AT, or another similar offset combination may require the letters actually to overlap if a uniform background area is to be attained throughout. Adjacent circular shapes, such as OO or DC, usually must be kept comparatively close together. The exact area between each letter cannot always be made identical, yet the general area should appear somewhat uniform. To evaluate good spacing, squint your eyes and observe the gray tone throughout the lettering; if the tone appears spotty or varies too much, it is poorly spaced. Beginning students tend to compress individual letters and space

LETTER SPACING
VISUAL — *Good*
LETTER SPACING
MECHANICAL — *Poor*

Figure 3–10 Spacing of letters.

them too far apart. Poor spacing often results in awkward appearance or poor legibility.

Spaces between words should be adequate to make word divisions obvious. A good rule of thumb is to make the space between words about twice the area between letters within the words, since spacing is a variable problem. If letters are tightly spaced, words can be put rather close together. If a group of words must be spread over a wide space, the space between each word must be sufficient to give the words identity and make the note readable. Too much space between words, however, produces a disconnected, spastic effect.

3.10
INCLINED LETTERING

Occasionally, to produce contrast or variety on drawings, inclined lettering can be used. Although vertical lettering is used primarily, inclined letters, if used appropriately for minor notes, will often add more expression to the lettering. Figure 3–23 shows an architectural, inclined alphabet; the sequence and the direction of strokes are similar to those of the vertical alphabet. To keep the incline uniform, lightly draw random guidelines, using the 67½° inclined edge of the lettering guide. This is the American Standard slope of 2 to 5 for inclined letters.

3.11
LETTERING WITH A TRIANGLE

Some experienced drafters who have developed unusual speed in lettering use their triangle for drawing all vertical strokes (Fig. 3–11). They simply slide the triangle along the T square with their left hand as they letter. When a vertical stroke is needed, they quickly set the triangle into place and draw perfectly vertical strokes with the edge. All of the other strokes are made freehand, and a personalized lettering style results. The system is used mainly with the alphabet shown in Fig. 3–24. Using the triangle to make vertical strokes is especially helpful with letters ⅜″ high or higher.

3.12
LETTERING TITLES

Designing titles with larger lettering requires more skill in composition and spacing. Defects that may go unnoticed in the small notes of a drawing frequently become obvious in larger titles. The size of a title should be consistent with the importance that it carries—titles of similar magnitude should be similar in size. Select the guideline heights after establishing title magnitudes so that the titles will have the correct importance throughout the drawing. Often titles can be given more importance by the use of extended lettering or by the use of a thicker stroke with the pencil. Underlining also provides a means of emphasis (Fig. 3–12).

Another important consideration is the finished silhouette formed by the title. If a number of words are used, care should be given to the general shape of the completed title, especially if the words must be put on two or three lines (Fig. 3–13). Allow sufficient surrounding space to avoid a crowded appearance. Be sure that the title is close enough to the view or detail that it identifies to ensure positive association. Titles can be easily misinterpreted if they are placed midway between two views. If a title must identify two views, position it to indicate an obvious relationship to both views. Extended

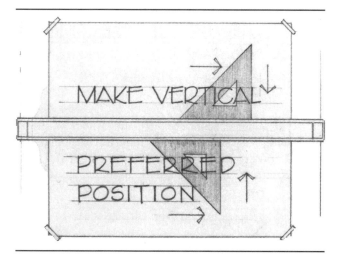

Figure 3–11 Making vertical lettering with the use of a triangle.

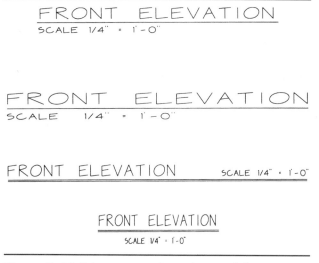

Figure 3–12 Title variations for architectural drawings.

Figure 3–13 Arrangements of lettering blocks.

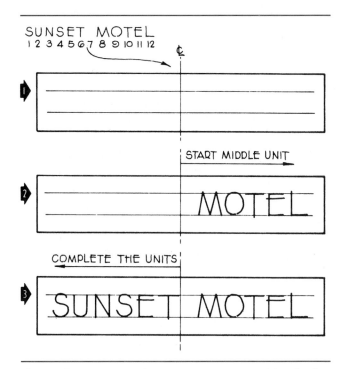

Figure 3–14 Steps in laying out a symmetrical (justified) title within a title block.

lettering with spread spacing must be used if a title is to ramble over a wide area. If possible, place a title directly below, rather than above or at the side of, the view or detail. When reading prints, a reader will search below a drawing for its title, and consistent relationships between title sizes and their placement will make drawings easier to interpret.

3.13

SYMMETRICAL TITLES

Occasionally, a title must be *justified*, or centered within a given space. This requirement presents the problem of spacing and layout to make the end of a line fall at a predetermined point. For example, lettering that must be put into title blocks appears more pleasing if it is centrally located (Fig. 3–21).

Several methods can be used to determine the title's length and to control its symmetry. One method is to count the number of letters (include the space between words as one letter width) to find its exact center. Then, after the guidelines and a vertical center line have been drawn in the desired space, you can put the title in by starting with the middle letter, or space, on the center line. Complete the last half of the line; then, working from the center letter toward the beginning, complete the first half of the title (Fig. 3–14). Differences in letter widths are usually compensated for by similar-size letters on both sides of the center. If more wide letters, especially W or M, happen to fall on one side, adjustments can be made to allow for them.

Another method that may be preferred by some drafters is first to lay out the required title on a separate

slip of paper. The length is measured, and its center is indicated with a mark. This mark establishes symmetry and is centered directly above the space to be titled. The final title can be duplicated below with the spacing taken from the preliminary layout.

These methods are only suggestions for fitting titles into given spaces; each individual may find various methods for arriving at the desired result.

3.14

LAYING OUT LARGE TITLES

To be graceful in appearance, large letters usually require the use of instruments for making both straight and curved strokes. Unless you are very accomplished in lettering art, freehand letters over an inch or so in height are difficult to make. After guidelines of the proper height are laid out, begin by constructing the letter forms lightly in pencil first, working over the construction until spacing and proportion have been perfected (from an established alphabet). Then, using a triangle and a T square for straight lines and a compass and an irregular curve for curved lines, carefully finish each letter in a mechanical technique. The freehand construction can be erased as soon as the settings for the instruments have been established. Width of lines should be uniform, and tangencies between curves and straight lines should be neat and unbroken. When straight lines are tangent to an arc (as in the U or the B), the arc should be drawn first and

ABCDEFGHIJKLMN
OPQRSTUVWXYZ&
1234567890
abcdefghijklm
nopqrstuvwxy

Figure 3–15 Helvetica Light alphabet.

the tangent lines attached to it. This method is much simpler than joining an arc to existing straight lines. If roman lettering is used, be sure that the correct inclined strokes are thick and the established thickness is uniform on both straight and curved strokes.

Avoid mixing freehand and mechanical lines on the same lettering. If instruments are used, make all of the lines with instruments.

If a freehand technique is desired on a large title (often on display drawings), construct all letters lightly with instruments after they have been spaced and proportioned by sketching. Then go over the mechanical lines freehand, using neat, deliberate pencil lines. The mechanical guides will maintain straightness and form, and the freehand finish will soften the appearance.

The Helvetica typeface (Figs. 3–15 and 3–25) is now the most widely used for architectural applications throughout this country and Europe. The simple, clean-cut lines of the alphabet make it easy to read and universally adaptable. Study the alphabet carefully, and consider it when beauty and ease of reading are important.

3.15

LETTERING WITH INK

The student should devote some practice to ink lettering, since ink is often the medium used for presentation drawings. Also, if the lettering on working drawings is expected to reproduce very clearly, drafters may do the lettering in ink. Small, freehand lettering and titles (Fig. 3–2B) can be effectively done with the use of a technical drafting pen. Many point sizes are available, and regardless of the pressure applied, the width of the line will always remain uniform. Usually, lightface freehand ink lettering is more readable than boldface. Observe the comparison in Fig. 3–16. Care should be taken in selecting the proper-size point according to the height of the letters.

LIGHTFACE
BOLDFACE

Figure 3–16 Comparison of narrow and wide ink point sizes used in lettering.

ABCDEFG
HIJKLMN
OPQRSTU
VWXYZ &
123456789
abcdefg
hijklmn
opqrstu
vwxyz.!

Figure 3–17 Boldface alphabets. Various sizes and styles on pressure-sensitive paper are available for students and drafters, and the lettering is easily applied to drawings.

An example of one of the many pressure-sensitive alphabets suitable for major titles is shown in Fig. 3–17. Many sizes and styles are available at drafting and art supply stores. If a title is to be put on tracing paper, a grid sheet placed below the tracing will facilitate its layout.

3.16

LETTERING MACHINES

Lettering machines are available that produce consistent letters on the drawing or on a medium suitable for transfer to the drawing. These machines vary in type, and each functions differently.

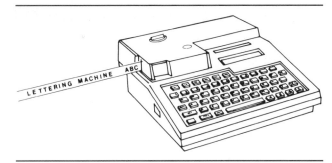

Figure 3–18 Lettering machine.

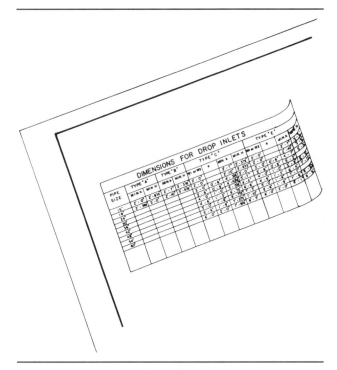

Figure 3–19 Sticky-back transfer.

The *Kroy*® machine allows the drafter to select a size and style of letter from a series of fonts. The drafter then manually operates the machine, which prints out letters on a plastic, adhesive-backed tape. The tape is applied to the drawing (Fig. 3–18).

Word processors are also used as lettering machines. Lettering is typed on the keyboard and monitored on a screen. A laser printer transfers the lettering to a plastic sheet known as a *sticky-back*. When you remove the thin film from the rear of the sticky-back, the sheet becomes ready for mounting on the original drawing. This method of lettering is desirable when notes are extensive and speed is essential (Fig. 3–19).

3.17

NOTES

Most of the hand lettering done by the drafter is in the form of general notes on drawings. These notes play an important role in depicting specific information about a detail. Notes must be clearly lettered, concise, and readable. Since architectural drawings become legal instruments in construction contracts, the importance of accurate lettering cannot be overemphasized.

Uniformly spaced guidelines should be drawn with a hard lead or a non-print pencil before notes are lettered. Notes are blocked or justified at the left. Most notes are brief and can be expressed on one or two lines of lettering. Leader lines with arrows relate the note to a certain item shown on the drawing (Fig. 14–3).

3.18

TITLE BLOCKS

Each sheet of a set of drawings requires identification, usually in the form of a title block placed in the lower-right corner of the sheet. Various types and sizes of blocks can be used. School projects need to have only simple title blocks; architectural offices often require more complex ones. Several examples are shown in Figs. 3–20 and 3–21. Many offices have sheets with borders and title blocks preprinted, ready for drafting use; others have title block stamps, called *hand stamps*, made up that can be quickly imprinted on each sheet before the drafting is started; others use *appliqué*, or preprinted title blocks of transparent plastic that will adhere to tracing paper and reproduce effectively. The use of any such type conserves drafting time.

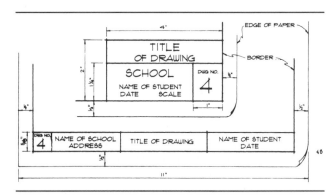

Figure 3–20 Title blocks satisfactory for school drawings of small size.

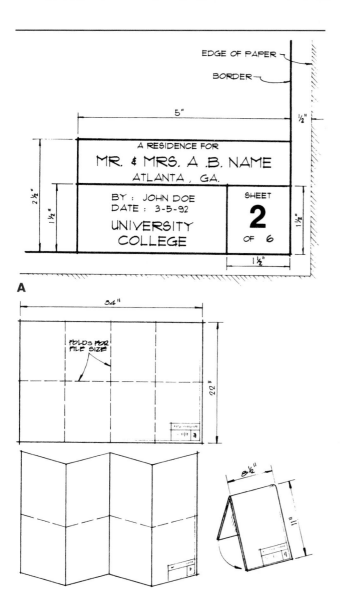

Figure 3–21 (A) Title block for a set of architectural working drawings. (B) Folding a print to file size so that the title block is exposed.

Information in title blocks varies, but a completed title block usually contains the following data:

1. Name and location of the structure
2. Name and location of the owner or client
3. Name and address of the architect
4. Date the drawing was made
5. Date(s) of revisions of the drawing, if any
6. Initials of the drafter
7. Initials or signature of the checker
8. Architect's or engineer's seal
9. Identification number of the sheet

LAYING OUT SIMILAR TITLE BLOCKS ON CONTINUOUS SHEETS

A very convenient method for laying out similar title blocks on a number of sheets in a set is as follows: First, carefully lay out the required block on a small piece of paper. Then simply slip the layout under each sheet of tracing paper as it is needed, and duplicate the block. Most of the lettering can be accurately spaced on the original layout, and considerable time will be saved in duplicating the linework. Keep the lettering symmetrical within the block for the most pleasing appearance. To aid in quick observation, sheet numbers are usually made very bold and placed in the lower right of the block.

3.20

DEVELOPING AN ARCHITECTURAL STYLE

The architectural alphabets (Figs. 3–22, 3–23, and 3–24) are included here as suggested letter styles for the student. Each sample has been taken from competent drafters' work and represents accepted lettering currently being used on working drawings. It is recommended that you duplicate the most appropriate styles carefully and try them experimentally on several drawings. A few trials will indicate whether they satisfy the personal taste of the student and whether they can be drawn easily and rapidly. You should study in detail the one style that seems to be the most satisfactory and practice it until you can do every character automatically. Uniformity in appearance should be attained, even in different sizes. Refinement should be a continual process until you develop a polished style that can be proudly put on drawings.

If the need arises, one or two of the other alphabets can be used as alternates, for emphasis or de-emphasis. Regardless of the alphabet selected, avoid extreme modifications—flairs with no meaning or awkward curves result only in faulty letter structure. Straightforward and uncomplicated lettering is the most acceptable.

3.20.1 No. 1 Extended

Number 1 extended (Fig. 3–22) is an extended style with very little space between each letter—the opposite of No. 1 condensed (not shown). It is widely used and represents the practice of extending each letter. Each word should appear as a tight unit; otherwise, use of the alphabet would require an extensive amount of space on the drawing. It is very readable, even in small sizes. If the low middle bar is used on E and F, similar low strokes should be used on H, K, R, and so on.

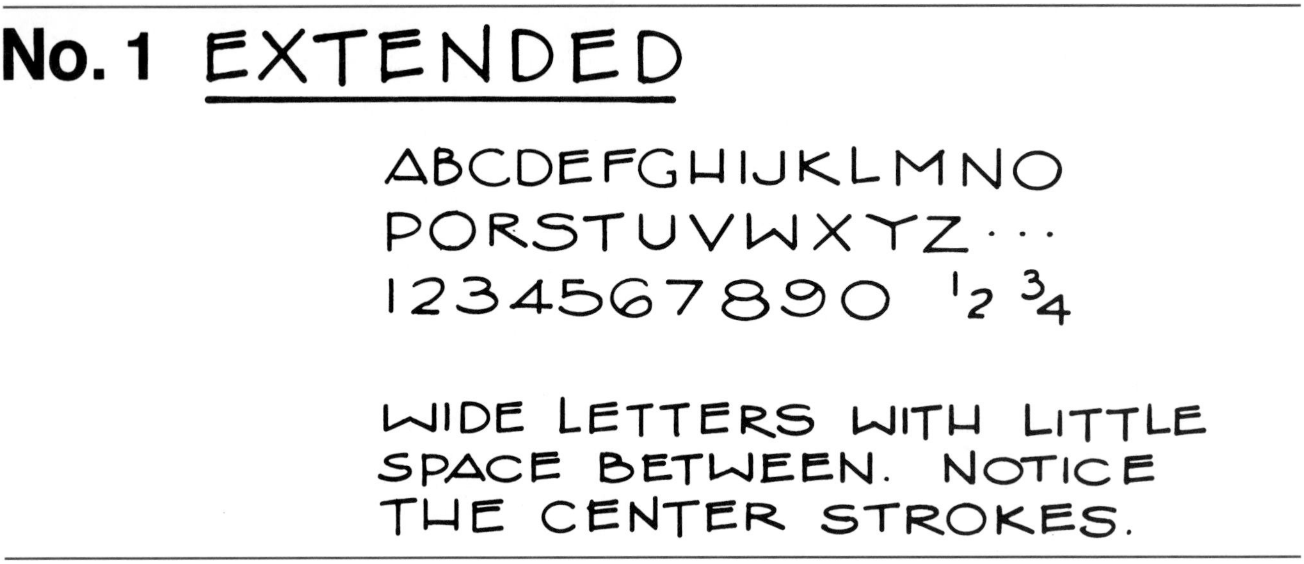

No. 1 EXTENDED

ABCDEFGHIJKLMNO
PORSTUVWXYZ···
1234567890 $\frac{1}{2}$ $\frac{3}{4}$

WIDE LETTERS WITH LITTLE
SPACE BETWEEN. NOTICE
THE CENTER STROKES.

Figure 3–22 Extended lettering.

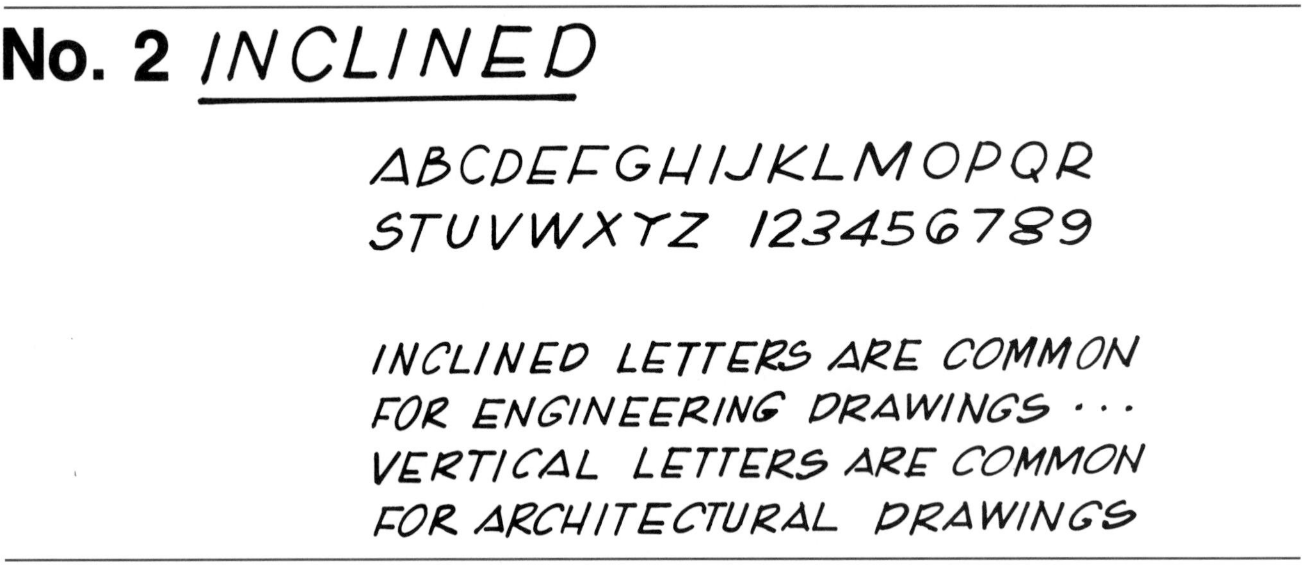

No. 2 *INCLINED*

ABCDEFGHIJKLMOPQR
STUVWXYZ 123456789

INCLINED LETTERS ARE COMMON
FOR ENGINEERING DRAWINGS···
VERTICAL LETTERS ARE COMMON
FOR ARCHITECTURAL DRAWINGS

Figure 3–23 Inclined lettering.

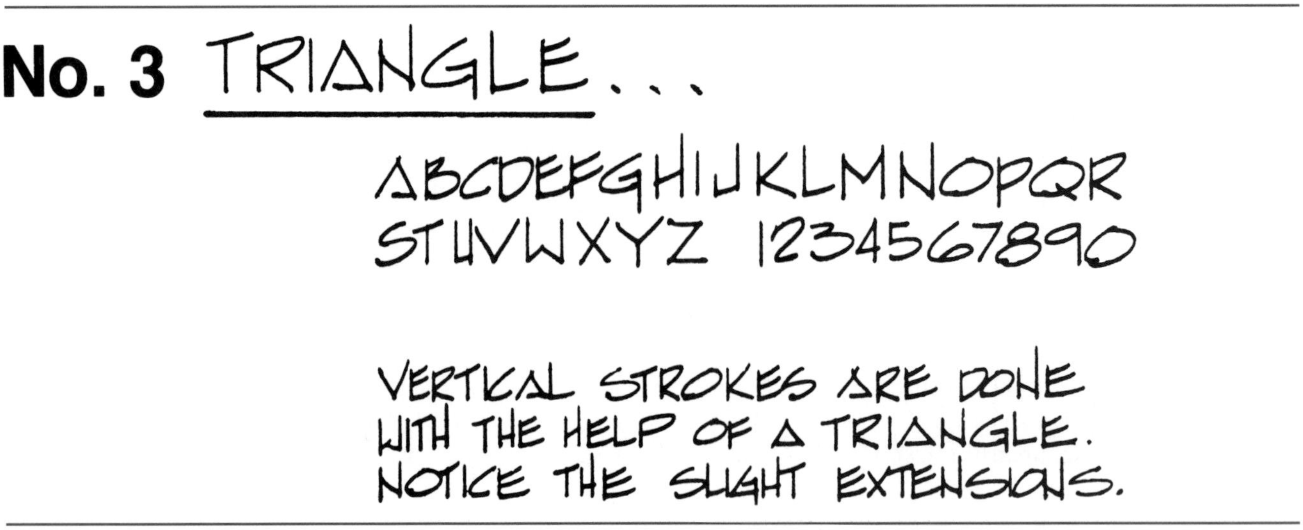

No. 3 TRIANGLE...

ABCDEFGHIJKLMNOPQR
STUVWXYZ 1234567890

VERTICAL STROKES ARE DONE
WITH THE HELP OF A TRIANGLE.
NOTICE THE SLIGHT EXTENSIONS.

Figure 3–24 Triangle lettering.

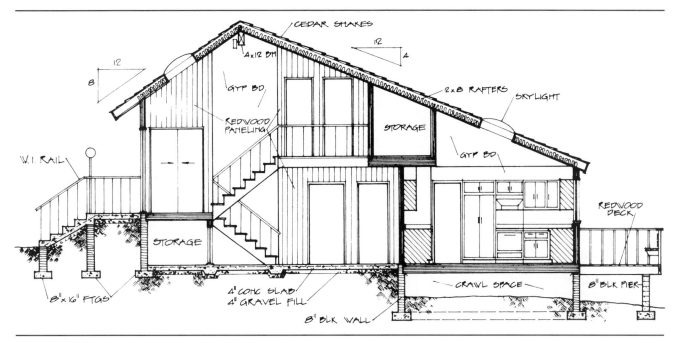

Figure 4–2 Line weight variation on section drawings.

be given emphasis at their ends with slight pencil pressure, and the dashes should not be spaced too far apart; that is, they should have more line than openings. Make the dashes uniform in length and spacing. If several features are represented with broken lines, use a different dash length for each representation. Each will then have an obvious meaning. For example, on floor plans, broken lines are used to represent roof outlines, electric wiring to an outlet and switch connections, and so on. When selecting a dash technique for each, you could

make the roof outline have ¼″ dashes, the electric wiring to an outlet and switch connections ½″ dashes, and so on. Should they cross each other or appear near one another, there will be an obvious difference, and no mistake will be made in their identification (Fig. 4–5A).

If possible, complete each type of broken line on the drawing before starting other types so that individual techniques will be maintained. When a broken line changes in direction at a corner, make dashes intersect at the actual corner, rather than at an opening with no corner point (Fig. 4–5B).

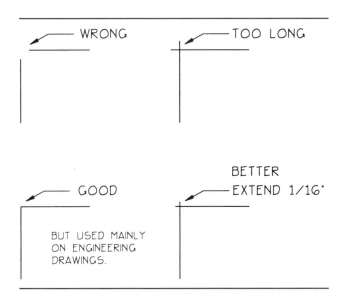

Figure 4–3 Line intersections on architectural drawings.

MAKE THE DIMENSIONS EASY TO READ

The usefulness of a working drawing depends to a large extent on the correctness and the manner of placement of its dimensions. If the dimensions are orderly and have been determined with the builder's needs in mind, information is readily obtained from the drawing. Numerous expensive mistakes are often the result of faulty dimensions.

Regardless of the scale of a drawing, numerical dimensions always indicate the actual size of the structure or feature. For example, if the side of a carport is represented by a line 5″ long on a plan drawn to the scale of ¼″ = 1′-0″, the numerical dimension shown on the drawing should be 20′-0″, which is the dimension workers must use to build the wall.

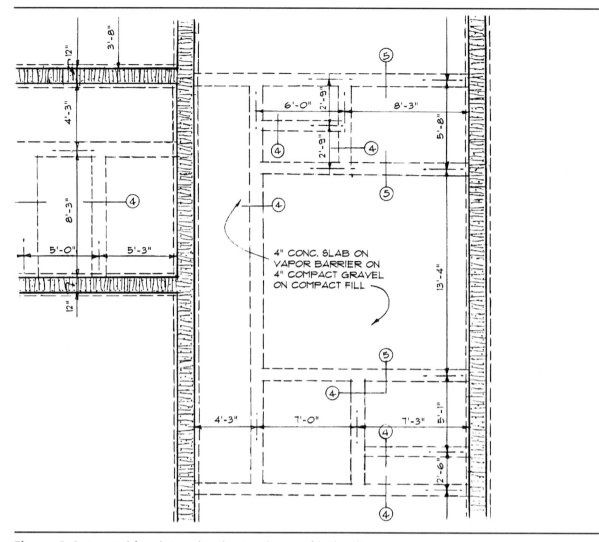

Figure 4–4 A partial foundation plan showing the use of broken lines.

4.3.1 Extension Lines

Extension lines are light lines drawn from the extremities of a feature requiring a dimension (Fig. 4–5C). Extension lines should not touch the feature, but rather they should begin $1/32''$ from the object line and extend about $1/8''$ beyond the terminator of the dimension line. If features within a drawing are best dimensioned outside the view, extension lines pass through the object lines and are unbroken.

4.3.2 Dimension Lines

Dimension lines are also light lines and have slashes on each end, indicating the exact extremities of the dimension (Fig. 4–5C). On architectural and structural drawings, dimension lines are made continuous, and the numerical dimension is placed just above the line, about midway between the slashing (Fig. 4–5C). When placing a numerical dimension above adjacent parallel dimension lines, be careful to place it closer to the line it identifies.

On shop or engineering drawings, the dimension line is broken near its center, and the numerical dimension is centered in the opening.

4.3.3 Terminators

Terminators are drawn at the ends of dimension lines. Various types are seen on drawings (Fig. 4–5D), but generally dimension limits that are indicated with 45° slashes are cleaner. Dots should be used to indicate dimensions to center lines of structural members; note leaders can end with arrows or dots.

Dimensions should be neatly lettered, and they should be made to read from the bottom and right side of the drawing—never from the top or left side. Dimensions under 12″ are given in inches, such as 6″, $3^5/8''$, or $11^1/2''$. Over 12″ long they are usually given in feet and inches, such as 3′-4″, 25′-7½″, or 12′-0″. However, on detail drawings, some features that are well over 12″ in size are commonly expressed in inches only. For example, joist

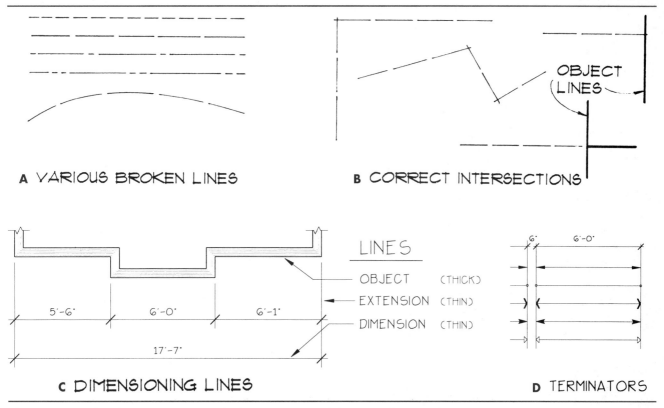

A VARIOUS BROKEN LINES

B CORRECT INTERSECTIONS

LINES

OBJECT (THICK)

EXTENSION (THIN)

DIMENSION (THIN)

5'-6"   6'-0"   6'-1"

17'-7"

C DIMENSIONING LINES

D TERMINATORS

Figure 4–5 Broken lines; intersections; object, extension, and dimension lines; and terminators.

spacing is usually indicated as 16″ on center, kitchen base cabinets as 36″ high, and minimum crawl space as 18″. When lettering dimensions, neither place the numbers directly on linework nor allow the numbers to touch lines on the drawing. To do so often mutilates the figure, and after a print is made of the drawing, the dimension is usually misread.

4.4

PLACE DIMENSIONS FOR OBVIOUS ASSOCIATION WITH THEIR FEATURES

Dimension lines should be placed outside the view or drawing, if possible. On large plans, however, features such as partitions that are well inside the exterior walls must be dimensioned inside the view for obvious association. Dimensions that are too remote from their features are difficult to find, and the excessively long extension lines that are then necessary make the drawing confusing. The placement of dimensions within a drawing should be studied carefully to be sure that the dimensions do not interfere with other lines on the view. Avoid a dimension line that coincides with an object line at the arrowhead; offset the dimension slightly. Judgment must also be used in placing dimensions around the pe-

riphery of a plan or detail. If the dimensions are too close, they interfere with the outline of the view, and the numerical dimensions often become crowded. If they are too far away, they give the impression of pertaining to features other than their own.

4.5

LINE UP DIMENSIONS IN SERIES FOR SIMPLICITY

If possible, a series of dimensions should be lined up so that the dimensions can be quickly drawn with one setting of the ruling edge, even if the view has offset edges (Fig. 4–6). With slashes at each extension line intersection, the effect of the continuous dimension line simplifies the appearance of the drawing. The smallest dimensions are placed nearest to, and the larger ones farther away from, the view. On a floor plan, for example, which often has many features requiring complex dimensions, similar features should be brought to each series for unity. The location of window and door center lines can be dimensioned along one series, closest to the plan; the offsets in the exterior wall and abutting partitions can be dimensioned along another series about ½″ farther out; and the overall length of the building can be shown as a single dimension about the same distance out again. Almost

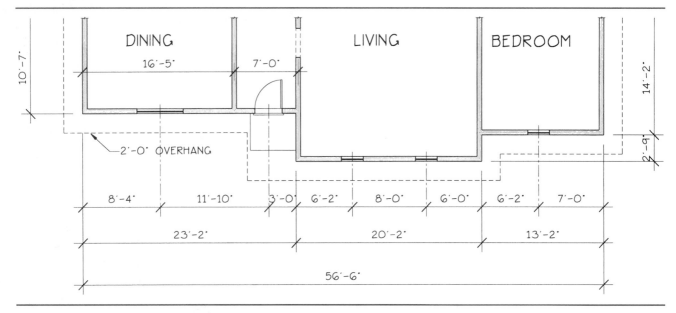

Figure 4–6 In-line dimensions on a plan view.

each dimensioning situation requires a different solution. Be sure to check the total of a series, arithmetically, to verify the overall dimension. Occasionally, it becomes necessary to cross extension lines, but avoid crossing dimension lines.

DO NOT DUPLICATE DIMENSIONS

Basically, dimensions are either *size* dimensions or *location* dimensions. Only those that contribute to the construction of the building or feature should be shown—needless repetition only clutters. Often the method of construction dictates the correct choice. For example, in frame wall construction, the window and door openings are located to their center lines on the plan, whereas in solid masonry construction, they are dimensioned to their edges or rough openings (Fig. 4–7). In frame partitions, the locations of doors are often superfluous, inasmuch as their symbol indicates an obvious placement. Correct dimensioning eliminates the need for workers to make unnecessary addition or subtraction calculations on the job.

USE LEADERS TO AVOID OVERCROWDING

Leaders are straight or curved lines leading from a numerical dimension or note to the applying feature (Fig. 4–8). Usually, an arrowhead or a dot is drawn on the fea-

ture end of the leader. Various leader techniques have been devised by drafters, but the most effective leaders are those that identify themselves as leaders and thus do not become confused with object lines or other lines on the drawing. Leaders should appear as connectors only and should not be misinterpreted. This is the purpose for using angular or offset-curve leaders.

It is usual practice to add leaders to the drawing after dimension lines have been completed. Then you can find satisfactory, uncrowded spaces for the leader dimension or note without placing this information too far from the feature. If a series of leaders must identify a number of materials or features, on section views, for example, be sure that the notes are arranged in a logical sequence so that the leaders do not cross (Fig. 4–8A). For readability, bring the leader to the beginning of a note if the note is located to the right of the object. Start the leader at the end of the note if the note is located to the left of the object. Leaders with circles or balloons attached are effective for identifying parts of an assembly or similar drawing requiring a sequence of letters or numbers.

Use angular leaders also to label circular shapes and holes. Place the leader indicating a diameter dimension on the exterior of the circle, with the arrowhead touching the circumference and pointing toward its center. A regular, angular dimension line can be used to label large circles or to label dimension cylinders on their rectangular view. Draw the leaders indicating the radii of arcs from their radiating points to the arcs, with the arrowhead of the leader on the inner side of the curvature (Fig. 4–8B) and the letter R following the numerical radius.

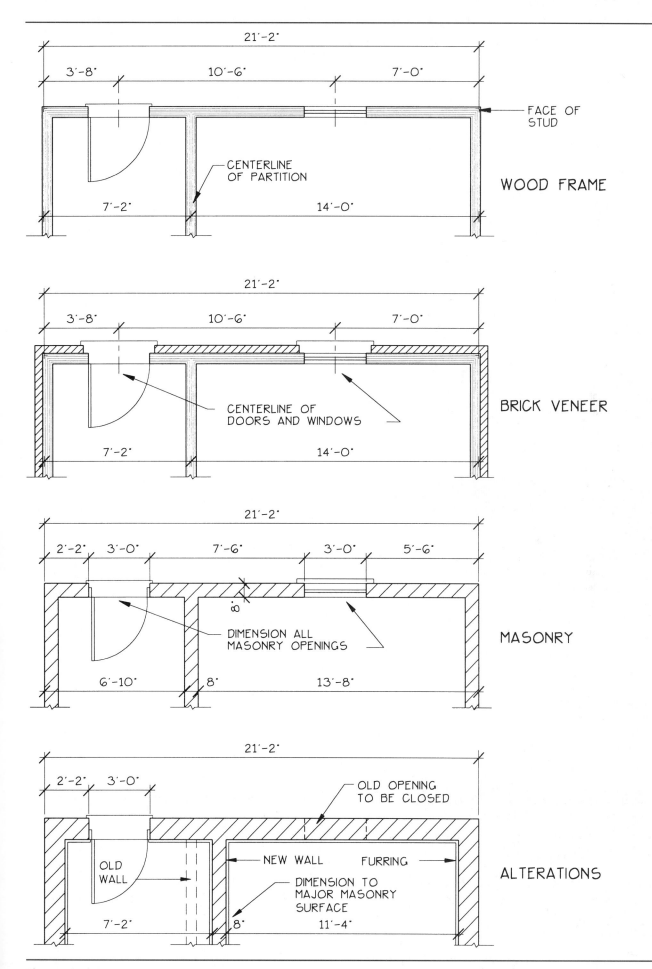

Figure 4-7 Dimensioning conventions.

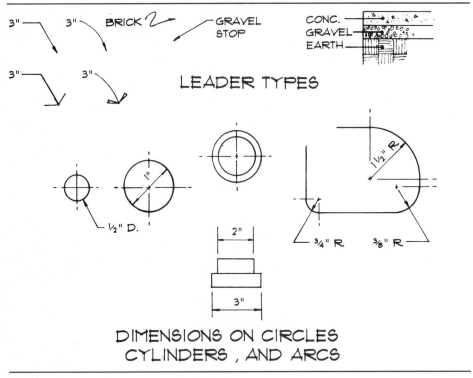

LEADER TYPES

DIMENSIONS ON CIRCLES
CYLINDERS , AND ARCS

Figure 4–8 (A) Various types of leaders. (B) Dimensions for circles and arcs.

4.8

USE NOTES SPARINGLY ON A DRAWING

Unless a note clarifies the graphic representation, it can probably be better included in the written specifications and thereby save the drafter's time. If a note is needed, however, use small lettering and make it clear and concise. If a note, such as "See detail A–A," is used, be sure there is a detail A–A to which the reader can refer. Check spelling carefully; even drafters need a dictionary. Extensive notes should be organized in neat blocks away from linework. These lettered areas of unusually long notes (if they definitely must be put on the drawing) should be considered in the sheet layout planning. Allow breathing space around a note, regardless of size, so that the reader can understand it.

4.9

ABBREVIATIONS SAVE TIME, BUT THEY ARE OFTEN CONFUSING

Even though the American Standards Association has established abbreviation standards for the construction industry, many variations still exist throughout the country. Abbreviations are used on architectural drawings to save space, reduce production time, and accurately convey a message to the reader. The abbreviations and terms used on architectural drawings should be defined at the front of each set of drawings.

Some architectural offices insist on abbreviating every word possible on their drawings, mainly to economize; other offices, knowing the inconsistencies and chances of misinterpretation, will not tolerate the use of abbreviations. The drafter must use discretion, however, and abbreviate only those words that have commonly identified abbreviations, and he or she must use them in such a way that they cannot possibly be misconstrued. Capital lettering is usually employed, and, as a rule, the period at the end is omitted. If it becomes necessary to use unusual abbreviations throughout a set of drawings, indicate their meanings on a legend. A list of common construction abbreviations is included in Appendix A.

4.10

INDICATE MATERIALS AND FEATURES WITH SYMBOLS

Because most buildings are a complexity of different materials and components, a simplified method of representation is necessary on a drawing—the smaller the scale, the simpler the symbol. Many of the conventional architectural symbols are shown in Fig. 4–9. Study these

MATERIAL SYMBOLS - SECTION

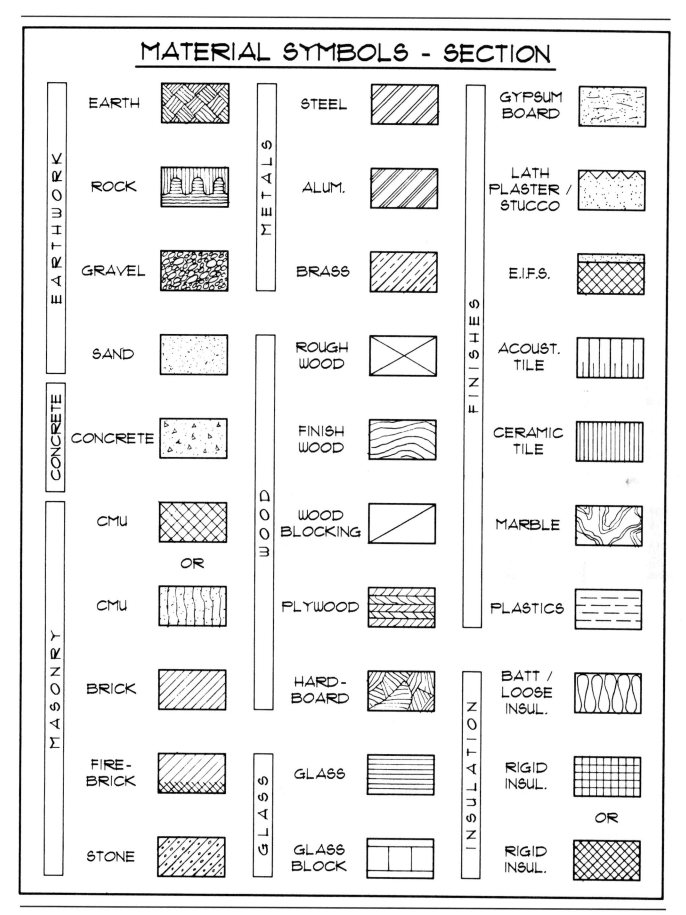

Figure 4–9 Material symbols used on working drawings.

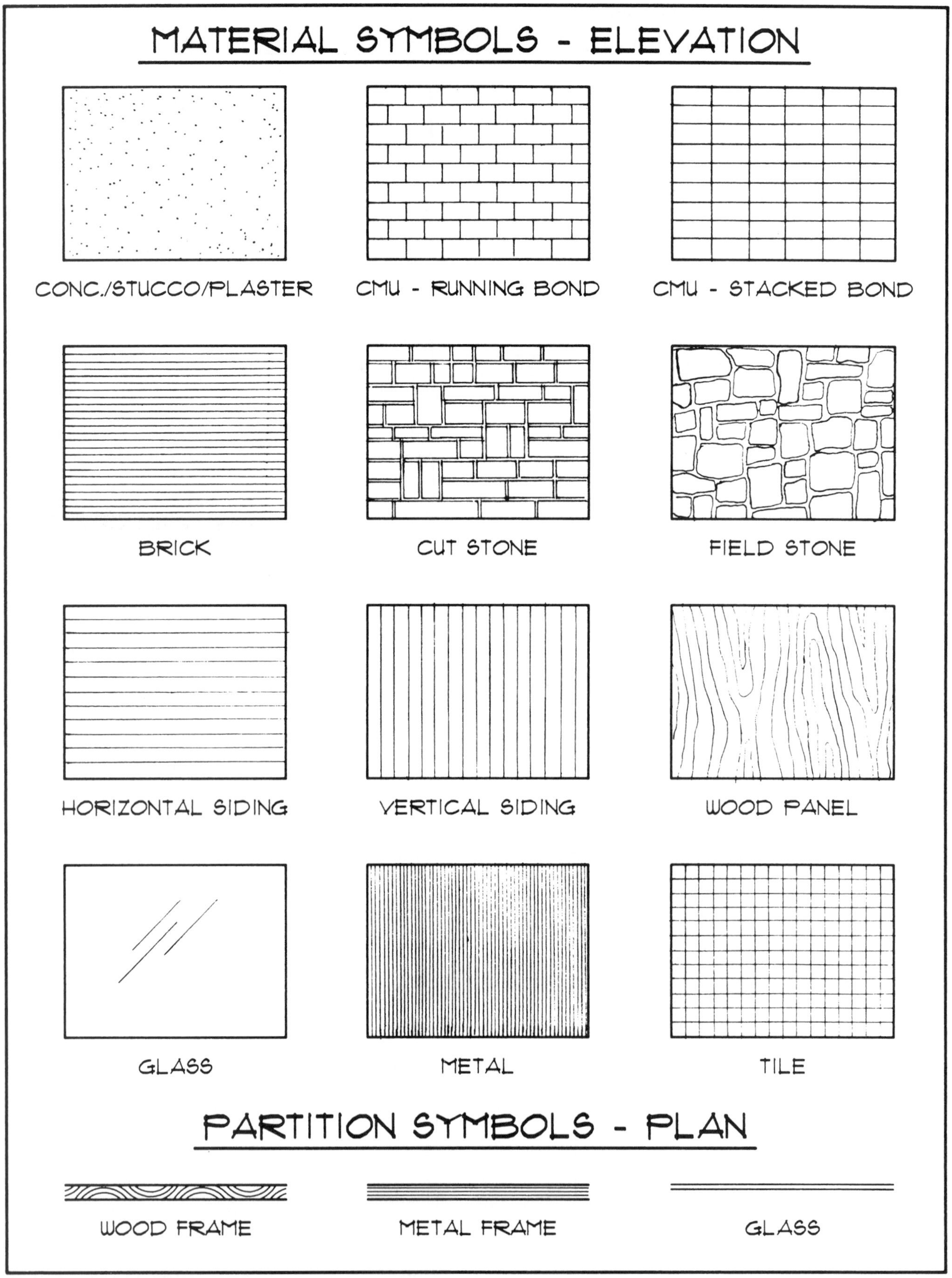

MATERIAL SYMBOLS - ELEVATION

CONC./STUCCO/PLASTER

CMU - RUNNING BOND

CMU - STACKED BOND

BRICK

CUT STONE

FIELD STONE

HORIZONTAL SIDING

VERTICAL SIDING

WOOD PANEL

GLASS

METAL

TILE

PARTITION SYMBOLS - PLAN

WOOD FRAME

METAL FRAME

GLASS

Figure 4-9 *(continued)*

symbols carefully, and commit to memory the important, most used ones: brick, concrete, stone, earth, wood, concrete masonry unit, insulation, glass, gravel, doors, and windows. Notice that most symbols on elevations are different from those in section or on plans.

Like other aspects of architectural drawing, symbols also vary slightly in different areas and in different offices. If nonstandard symbols are used on drawings, a symbol legend can be drawn on one of the sheets. Although simply drawn, all symbols should be indicative of the material they represent; if a symbol is misleading or gives the impression of the wrong material, clarify it with a simple note. Material symbols or hatching should not be made with heavy line weights. Their representation in rather large areas will appear more pleasing if the linework fades out near the center, thus resembling a highlight on the surface. The linework should be given more emphasis near the edges (Fig. 4–10). Make the highlights nongeometric in shape to eliminate any possibility of their appearing as construction features on the surface. The edges of hatched areas should be distinct.

Contrast of adjacent materials is especially important on section views. To aid expression, hatch adjacent pieces on section views in different directions. Otherwise, use the same tone and technique for a material or a member reappearing in different places throughout the section. If an overall gray pencil tone is used to indicate a predominant material, usually it is more tidy to apply the shading on the reverse side of the tracing. Wood structural members in section are indicated merely with diagonals (Fig. 4–9); finish lumber is indicated with an end-grain symbol done freehand.

If notes or dimensions must go into hatched areas, leave out enough of the linework so that the number or the note can be inserted and can be read without difficulty.

In drawing wood frame construction, you can represent the exterior walls of the floor plan of a home with parallel lines, scaled 6″ apart, which is usually close to the actual thickness of the combined studs, inside material, and outside wall coverings. Parallel lines are often the only symbol needed to represent the wood walls. But by the addition of either straight or wavy lines (resembling wood grain) within the parallel lines, a more prominent outline of the plan is attained (Fig. 4–11). The tone of the symbol makes the plan more readable.

This principle of giving prominence to important features by shading, either freehand or mechanical, is adaptable to many situations in drafting. The symbols for windows and doors on plans drawn at small scales are shown in Fig. 4–7. Several methods are shown for indicating doors. On elevation views, draw windows and doors accurately to reveal their characteristics.

4.11
AVOID UNNECESSARY REPETITION OF SYMBOL HATCHING

When drawing elevation views of a symmetrical nature, you can often omit the material symbol on one side of the center line if there is no special reason for hatching the entire elevation (Fig. 4–12). In design drawings, however,

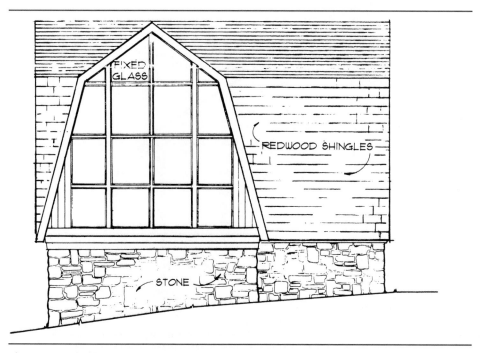

Figure 4–10 Elevation showing application of material symbols.

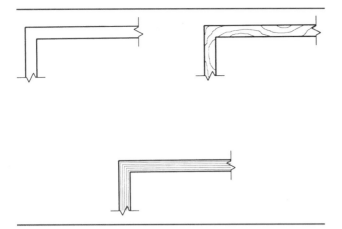

Figure 4–11 Representation of wood frame walls on plan views.

completion of the various material tones may be necessary in analyzing the esthetic aspects of the building.

Another method of suggesting materials on an elevation without taking the time to hatch the entire drawing is shown in Fig. 4–13. A light diagonal line that does not conflict with other important lines is drawn on the elevation to limit the hatching to a typical area. Sometimes merely a note is sufficient to indicate the material if it does not require careful delineation.

4.12

USE A DRAFTING TEMPLATE FOR DIFFICULT SYMBOLS

Many symbols—such as bathroom fixtures, electrical symbols, kitchen equipment, circles, and irregular shapes—can be quickly drawn with appropriate architectural templates (see Fig. 1–1). Refer to Section 1.15 for the correct use of drafting templates. Drafters may develop their own templates or symbols for items such as north arrows (Figs. 4–14 and 4–15).

4.13

AVOID SMUDGED PENCIL LINEWORK

Slight smudging of pencil linework with triangles and instruments cannot be entirely avoided, but care should be taken to refrain from unnecessary rubbing or touching the finished pencil lines. Beginning students, especially, are confronted with the problem of smudged drawings. Listed below are a number of suggestions for keeping a pencil drawing clean while on the drawing board:

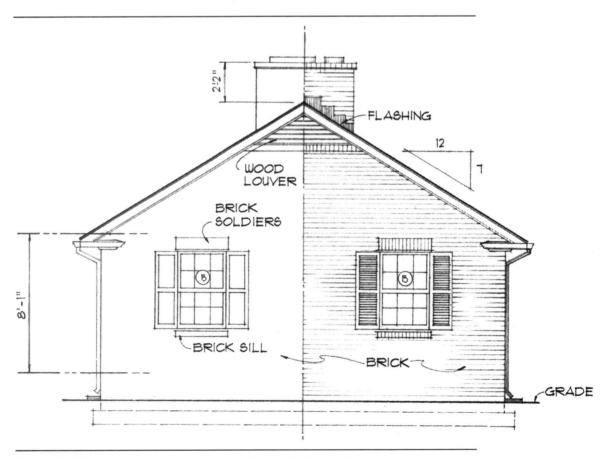

Figure 4–12 A symmetrical elevation view showing half-symboling.

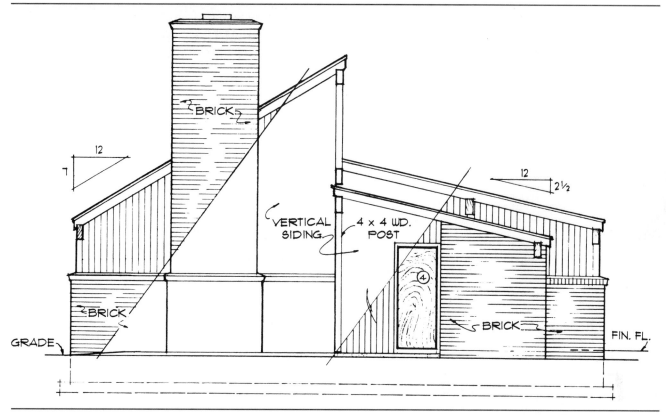

Figure 4-13 Partial symboling with the use of a diagonal line.

1. Use a systematic and organized procedure in the sequence of drawing lines so that instruments will not have to be placed over existing lines any more than necessary. For example, start from the top of the layout, and draw as many major, horizontal lines as possible, proceeding to the bottom of the layout. For vertical lines, start from the left of the layout, and draw as many vertical

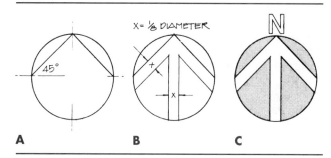

Figure 4-15 Steps in drawing a standardized north arrow.

lines as convenient, proceeding toward the right of the drawing (if left-handed, proceed toward the left). In other words, draw lines so that the triangle can be moved away from the lines already completed. This method will become a time-saving habit if conscientiously applied during early training.

2. Protect completed parts of a drawing by covering them with extra sheets of paper. Do so also at the end of the day or the work period.
3. Small triangles will reduce smudging if extensive work must be done on small details.
4. Form the habit of raising the parallel bar and the triangle blade slightly when moving them up or down on the drawing board. You can do so by merely bearing down on the head as it is moved.

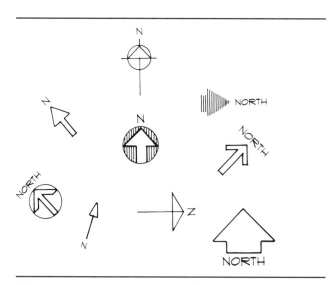

Figure 4-14 Suggestions for north-point arrows.

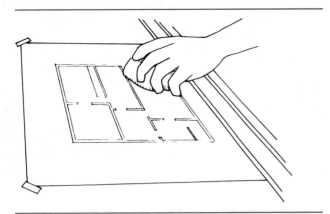

Figure 4–16 Using a dry cleaning pad to prevent smudged linework.

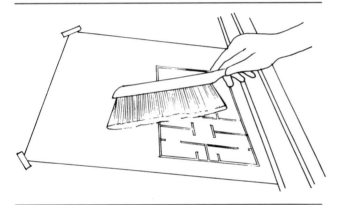

Figure 4–17 Sweeping loose graphite from drawings.

5. Keep the instruments clean by wiping them with a damp cloth regularly. Avoid soap and water on wooden parts. The fine graphite from pencil leads has the habit of lodging in the surface pores of almost any material. Always wipe the pencil point after sharpening. Do not store the pencil pointer in direct contact with other drafting equipment, either on the drawing board or in portable kits. Keep it in an isolated container.

6. Use the dry cleaning pad sparingly to prevent smudged linework (Fig. 4–16). Sweep the drawing occasionally with the drafting brush to reduce the amount of loose graphite on the surface. Avoid rubbing the drawing with your hand or any hard objects, as well as with sleeves or cuffs (Fig. 4–17).

7. Be sure that your hands are clean and dry. If necessary, use a little talcum powder on them to reduce perspiration.

4.14

USE SCHEDULES TO ORGANIZE EXTENSIVE INFORMATION

A schedule is an organized arrangement of notes or information usually lettered within a ruled enclosure and conveniently placed on the drawing, although it is becoming increasingly common to find schedules bound in with the project manual. Figure 4–18 is an example of this type of schedule. Extensive information placed in a schedule, rather than lettered in various places on the drawing, makes the information easily accessible to estimators and builders, who appreciate the organized arrangement.

When drawing a schedule enclosure, make the horizontal ruled lines $\frac{1}{4}''$ or $\frac{3}{8}''$ apart so that there will be sufficient space between the lines for the lettering (which needs to be only $\frac{1}{8}''$ high) to avoid having it touch the ruled linework. The title should be slightly larger, and a top-horizontal and a left-vertical column for identification and headings should be isolated from the remaining spaces with heavy lines. Draw a heavy border line around the entire schedule.

Many different types of schedules are used in construction drawings and documents. Schedules organize information in either a matrix or a picture format. *Room finish* and *window schedules* are examples of matrix formats (Figs. 4–19 and 4–20). A *door schedule* is an example of a picture format (Fig. 4–21). Door schedules in picture format are combined with matrix-type schedules (Fig. 4–22) in order to provide more extensive information than can be communicated in the picture-type format alone.

Room-finish information generally will not fit within a room on a floor plan. A *room finish schedule* provides a condensed format for presenting a large amount of finish-related information. The schedule, however, does not show where one finish material stops and another starts. This information should be indicated on floor plans or on interior elevations.

It is important to develop a numbering system to assign finishes clearly to specific rooms within a building. Many different systems can be used; but regardless of the system chosen, room numbers should be unique, and each space in the building should have its own number. Spaces that are part of a larger space often are not assigned a new number but rather are given a suffix (A, B, C, etc.) which identifies their relation to the main room.

Room numbers and door numbers are located on the floor plans. Window types can be identified on either the plan or the elevations, whichever will be clearer—but not on both.

Identify windows with letters (A, B, C, D) both in the schedule and near the windows on the plan. Rough-opening sizes for windows installed in frame walls are important and should be included in the schedule, along with

MILWAUKEE AREA TECHNICAL COLLEGE KS #93030

Rm.#	Room Name	Floor	Base	N. Wall	E. Wall	S. Wall	W. Wall	Ceiling	General Notes
100	BUILDING LOBBY	ET–1	VBC--2	PT–1	PT–1	PT–1	PT–1	ACT–1	NOTE 1, 11, 57
100A	VESTIBULE	ET–1/MAT	BRK	PT–1	PT–1	PT–1	PT–1	ACT–1	NOTE 1, 11
101	ELEVATOR EQUIP. ROOM	VCT–1	VBC–1	PT–1	PT–1	PT–1	PT–1	ACT–1	
102	STUDENT WORK AREA	VCT–1	VBC–1	PT–1	PT–1	PT–1	PT–1	ACT–1	NOTE 3, 66
103	ELEVATOR	VCT–1	—	—	—	—	—	—	NOTE 60
104	ELECTRICAL SWITCHGEAR	VCT–1	VBC–1	PT–1	PT–1	PT–1	PT–1	ACT–1	
104A	ELEV. SUMP.	VCT–1	VBC–1	PT–1	PT–1	PT–1	PT–1	ACT–1	
105	MEN'S TOILET ROOM	CT–1,2, 3,4	CB–2	CTW– 1,2,3	CTW– 1,2,3	CTW– 1,2,3	CTW– 1,2,3	ACT–2	NOTE 4,5
105A	VESTIBULE	CT–1,2, 3,4	CB–2	CTW– 1,2,3	CTW– 1,2,3	CTW– 1,2,3	CTW– 1,2,3	ACT-2	NOTE 4,5
106	MECHANICAL RM.	VCT–1	VBC–1	PT–1	PT–1	PT–1	PT–1	ACT–2	
105A	EMERG. GEN. ROOM	VCT–1	VBC–1	PT–1	PT–1	PT–1	PT–1	ACT–2	NOTE 66
107	JANITOR CLOSET	VCT–1	VBC–1	PT–1	PT–1	PT–1	PT–1	ACT–2	
108	GEN. PURPOSE CONFERENCE RM.	CPT–1	VBS–1	VWC–2	VWC–2	VWC–2	VWC–2	ACT–1	NOTE 11
109	WOMEN'S TOILET ROOM	CT–1,2, 3,4	CB–2	CTW– 1,2,3	CTW– 1,2,3	CTW– 1,2,3	CTW– 1,2,3	ACT–2	NOTE 4, 5

Figure 4–18 Schedule bound in project manual.

ROOM FINISH SCHEDULE

ROOM NO.	ROOM NAME	FLOOR	WALLS				CEILING		NOTES
			N	S	E	W	MATL	HEIGHT	

Figure 4-19 Room finish schedule. (© Copyright 1997, Construction Specifications Institute, 601 Madison Street, Alexandria, VA 22314-1791)

WINDOW SCHEDULE

MARK	SIZE		TYPE	MATERIAL	NOTES
	WIDTH	HEIGHT			

Figure 4–20 Window schedule. (© Copyright 1997, Construction Specifications Institute, 601 Madison Street, Alexandria, VA 22314-1791)

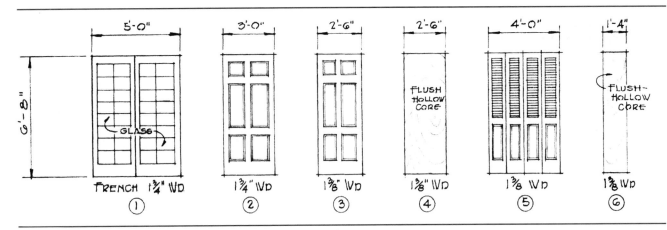

Figure 4–21 Door schedule in picture format.

DOOR SCHEDULE														
	DOORS						FRAMES							
				SIZE						DETAILS			Fire Rating	Hdw. Group	
Door #	TYPE	MAT.	FIN.	WIDTH	HEIGHT	THICK	TYPE	MAT.	FIN.	JAMB	HEAD	SILL			REMARKS

Figure 4–22 Door schedule in matrix format.

window size, type, and material. Often the name of the manufacturer and the item's exact catalog number are also provided.

You can also show door information by drawing small elevation views of each door type. Sometimes the small elevations are included as part of a lettered schedule. If the doors are similar stock heights, they become simplest to draw if arranged in a horizontal row (Fig. 4–21). Dimensions and material information should both be included in this type of graphic schedule.

DRAFTING EXPRESSION

Do the following exercises on 8 1/2" × 11" drawing paper with a 1/2" border, using drawing instruments.

1. Using guidelines 1/4" high and 1/2" between each set, to the left of the paper letter the titles of the following types of lines, and to the right graphically illustrate the lines: object lines, hidden lines, extension lines, dimension lines, center lines, broken lines, leader lines, section lining (poché), and section cut lines. Show the proper weight for each line.

2. Draw sixteen 1/2" high by 1" wide rectangles equally spaced on the paper; 1/4" below each rectangle draw guidelines 3/16" high. Letter the titles of the following material symbols, and poché the proper symbol in the rectangle above: compact fill, gravel, rock, concrete, C.M.U., brick, firebrick, field stone, steel, finish wood, rough wood, plywood, hardboard, batt insulation, rigid insulation, and glass.

3. Draw four floor plans of a small storage facility 8'-0" wide by 12'-0" long (outside dimensions) at a scale of 1/4" = 1'-0". Center a six-panel 3'-0" × 6'-8" door on the 12'-0" wall and a 4'-6" × 5'-2" double casement window on the 8'-0" wall. Four different materials will be used for wall construction. Draw one plan with 4" wood stud walls; draw the second plan with 6" C.M.U. walls; draw the third plan with 8" concrete walls; and draw the fourth plan with 6" wood stud walls and 4" brick veneer. Place exterior and interior dimensions on each plan.

4. Draw elevation views and plan views of the following doors at a scale of 3/8" = 1'-0": six-panel wood doors, 3'-0" × 6'-8"; flush wood door, 2'-8" × 6'-8"; full-louvered wood door, 2'-4" × 6'-8"; two-panel bi-fold doors, 6'-0" × 6'-8"; and two-panel, nine-light 3'-0" × 6'-8" door. All doors are 1 3/4" thick. Add dimensions to and label each door.

5. Using two of the floor plans developed in Exercise 3 at a scale of 1/4" = 1'-0", draw the side and end elevation views of the door and window walls. Use the 6" C.M.U. plan with 12" board-and-batten siding at the triangular gable ends of the roof. Use the brick veneer plan with a hip roof. Floor-to-ceiling heights are 8'-0"; roof pitches are 7:12 with fiberglass shingles; overhangs are 1'-4" with a 1' × 6' fascia; and buildings are slab on grade. Poché and label all materials.

REVIEW QUESTIONS

1. What do we mean by *contrast* in linework?

2. How may broken lines or dashed lines be varied to represent various features on a drawing?

3. What is a typical architectural characteristic of line intersections at corners?

4. From what directions on a drawing are dimensions made to read?

5. Why is a series of dimensions best placed along a continuous line on a floor plan?

6. What is the conventional method of expressing architectural measurement with numbers?

7. Why are the basic structural features of a plan the first dimensions to be given consideration?

8. Why is it preferable to use only those abbreviations that have been standardized and commonly accepted?

9. How can pencil smudging be removed from tracings?

10. Describe three techniques used by drafters to keep drawings neat while drafting.

11. Draw the proper symbol for expressing a 5:12 roof pitch on a working drawing.

12. What is another term used for symbol hatching?

13. On floor plans of commercial buildings, dimensions are shown from wall to wall. How are the corresponding dimensions shown on residential floor plans?

14. Why is it necessary to draw a north arrow on floor plans?

15. An arrowhead is used at the ends of dimension lines. What other symbol may be used at the ends of dimension lines?

16. Identify two types of schedules used on working drawings, and describe how they are organized.

17. What is the advantage of using a schedule when considerable information is involved?

"Now, this is not the end. It is not even
the beginning of the end. But it is,
perhaps, the end of the beginning."
—WINSTON CHURCHILL

MODULAR AND METRIC DRAFTING 5

To meet the rising cost of construction, the building industry has developed the *modular method* for simplifying the details of structures so that building components, materials, and equipment can be assembled easily and with a minimum of alterations, such as cutting, fitting, and other minor modifications. Experts feel that by reducing hand labor for such job-site alterations, the application of this principle effects the only definite economies yet to be realized in present-day construction, and that this fuller utilization of our technical advances in the building industry can be made without sacrificing any of the architectural qualities of the design.

In 1988, federal law mandated the *metric system* as the preferred system of measurement in the United States. This law required that metric measurement be used in all federal procurement, grants, and business-related activities, to the extent feasible, by September 30, 1992.

This chapter discusses the use of the modular method and the metric system of measurement.

5.1
THE PRINCIPLES OF MODULAR DRAFTING

To realize the benefits of the modular method, the building industry recommends *modular coordination*, which involves these four aspects:

1. thinking modular during all stages of the design process;
2. drawing the working drawings so that modular concepts are represented simply, yet with some flexibility;
3. sizing building products so that they will conform to modular assembly; and
4. devising construction techniques that will economically utilize coordinated materials with a minimum of modification and waste.

Modular drafting concerns itself with these concepts. Some architectural offices use modular coordination and dimensioning on their working drawings; others, because of the occasional difficulty in obtaining a full range of modular products and local difficulty in obtaining construction personnel who fully understand the system, are reluctant to convert to the modular method. However, there is merit in coordination; and, as with any advance in an industry involving large numbers of people in various areas and trades, time is required before the method is universally understood and accepted.

Beginning students should first learn the conventional methods for representing and dimensioning working drawings. Then, after students have acquired an intimate acquaintance with construction methods and building products, the benefits of modular methods will become apparent to them. Well-thought-out conventional drawings often result in modular characteristics, whether or not the drafter realizes it. The first step is "thinking modular," both in conceiving the building structurally and esthetically and in continuing this thinking throughout the working drawings. Modular drawings are not necessarily restricted to the use of modular materials only.

The following points are given briefly to serve as an introduction to the fundamentals of modular drafting.

5.1.1 Use a Modular Grid in Laying Out Preliminary Design Sketches

From the very outset, design trial sketches must be developed over a modular grid, which has only one restriction—it must be sized in 4″ multiples: 16″, 24″, 4′-0″, 10′-8″, 24′-0″, and so on. The selection of the basic grid may be almost any convenient multiple size (Fig. 5–1). If plywood or other sheet materials are to be a dominant material in the construction, a 4′-0″ × 4′-0″ grid size might be a logical selection. If the spans of structural members seem to satisfy a 12′-0″ spacing, a grid size of

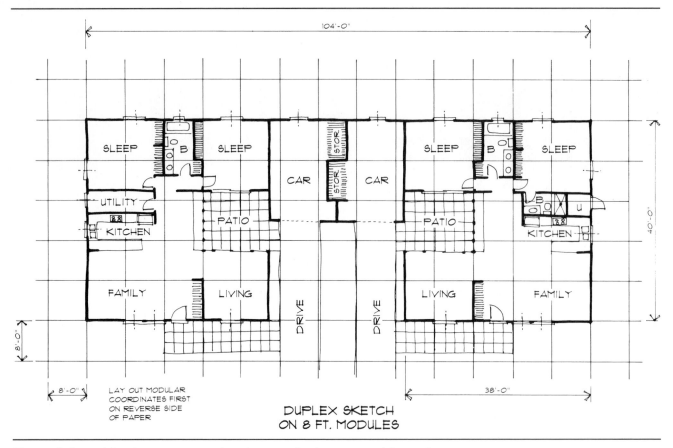

Figure 5-1 Modular drawings start with preliminary sketches over modular grids.

12′ may be appropriate. If concrete masonry units are to be a major material, a 16″ grid size could be used (it may be a bit small for design sketches, however). In selecting a design grid size, you should consider the size of the building or the repeating spaces within the building, just so that the selected size of the grid can be divided by 4.

After careful study, the basic 4″ module was adopted by the American Standards Association as being the most convenient unit from which major materials could be sized and construction features could be dimensioned. The module is three-dimensional, resembling a 4″ cube (Fig. 5–2), yet it appears in all three principal building

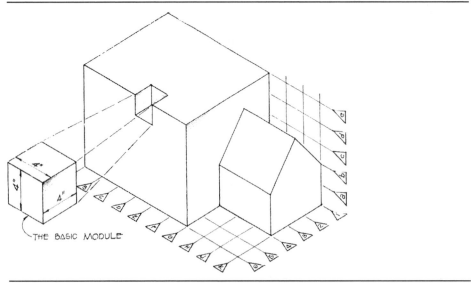

Figure 5-2 The basic module of an architectural volume.

planes as a 4″ grid. Some drawings do not show the grids; but because the building was created over a grid, important points will fit, even if the grid lines are imaginary. The basic module is also compatible with sizing of products from countries using the metric system; 4″ is nearly identical with 10 cm.

When the preliminary sketches are converted to large-scale drawings, multiples of the 4″ module become easy to dimension, and all parts of the building can be conveniently related. The design grid can be laid out directly on the sketch paper, or it can be inserted below tracing paper as an underlay. If the structural system and the planning layout can be combined into unified modules, both the working drawings and the construction will be simplified. Grid-disciplined design sketches are the foundation for modular coordination, making possible the benefits of modular drawings.

5.1.2 Develop the Details of the Building on Grid Lines

In developing the details of a building, you must draw the 4″ grid lines first, and they must become a part of the drawing. The grids can be drawn on the reverse side of the tracing paper, or they can be done in ink so that they cannot be erased if changes must be made. Draw the grids only where each detail will appear, but the grids must be made to show clearly on the reproduction prints. Because of their small size, 4″ grid lines are not put on a small-scale drawing. Any scale smaller than 3/4″ = 1′-0″ is considered a small-scale drawing. On complex plans and elevations it may be necessary to use reference grid lines to relate the details to their correct placement in the building. Some of the grids in the details may coincide with the reference grids shown on the plans and elevations and others may not. Regardless, the continuity of the 4″ module is related with modular dimensions at key points throughout the building.

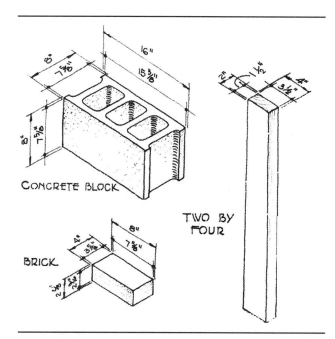

Figure 5–3 Actual and nominal sizes of common building materials.

The accompanying details show where units and surfaces usually fit best within or on grid lines. Each unit of a detail does not necessarily have to fall on grid lines; nor does the modular system restrict the use of materials to only modular. However, key points have been worked out in conventional construction to utilize standard materials best in the system. Notice that generally the nominal sizes and surfaces fit the grid lines (Figs. 5–3 and 5–4). The actual size is usually a fraction of an inch smaller than the nominal size (an exception is the height of a wide-flange beam). For example, the nominal cross-sectional size of a 2″ × 4″ wooden lumber actually measures only 1½″ × 3½″ when the lumber is dressed and dry (19% moisture con-

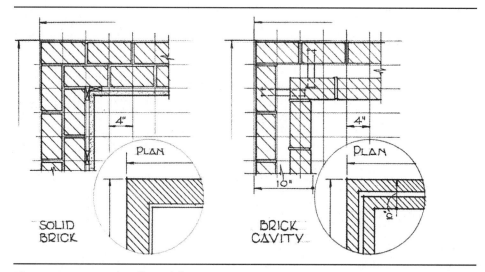

Figure 5–4 Typical walls and their treatment.

tent or less); the actual size of an 8″ × 8″ × 16″ concrete block measures 7⅝″ × 7⅝″ × 15⅝″. To eliminate the need for excessive fractions, the grid line conveniently falls within the joint in masonry and just outside the edge of a 2″ × 4″ member, allowing some flexibility. Because ⅜″ is usually the width of a mortar joint between masonry units, 3/16″ is a constant dimensional relationship of unit surfaces to the controlling grid.

It becomes convenient for grid lines to fall on the center lines of columns, window openings, partitions, and the like. When grid lines indicate normal sizes, it is important that the drafter as well as the construction contractor be fully aware of this fact. Working drawings can also be based on actual sizes, but a consistent method must be used throughout the drawings. Comparatively few fractions are required in modular drafting, which should make the system of particular interest to drafters.

5.1.3 Use an Arrowhead to Indicate a Grid Dimension; a Dot to Indicate a Nongrid Dimension

The dots and the arrowheads at the ends of dimension lines have a definite significance on modular drawings. To distinguish between grid and off-the-grid dimensions, an arrowhead is always used to indicate a dimension on a grid line, and a dot to indicate a dimension off the grid lines (Fig. 5–5). Dimensions that have an arrowhead at each end of the dimension line will be in 4″ multiples. Dimensions that have dots on each end may or may not be modular, but their extremities do not fall on grid lines. If a dimension has an arrowhead on one end and a dot on the other, it cannot be a 4″ multiple, and it therefore is not a modular dimension.

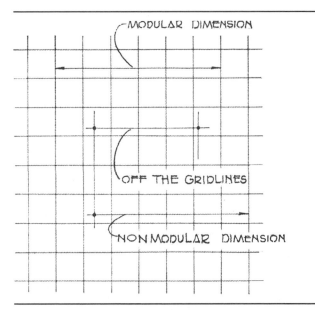

Figure 5–5 The use of arrowheads and dots on modular drawings.

Even though grid lines do not actually appear on a modular drawing, the designer or drafter imagines them to be there, and their presence is always implied by the 4″ multiple dimensions. As we mentioned, reference grids with heavy triangles at their ends, enclosing a consecutive numbering system (Fig. 5–2), can be used to give additional coordination between details and the plans and elevations. Any point within the total volume of the building can be located with the reference grids.

5.1.4 Coordinate Vertical Dimensions with Modular Floor Heights

Usually the nominal finished floor is placed on a grid line; the actual finished floor surface is located ⅛″ below a grid line (Fig. 5–6). In wood frame construction, however, the top of the subfloor or the surface of a slab on grade foundation is placed on a grid line (Fig. 5–7). Three brick heights as well as one 8″ concrete masonry unit height coincide with two grid heights.

A difficult point of modular coordination concerns the use of stock doors. If the finished floor coincides with a grid line, then the 6′-8″ or 7′-0″ stock door heights will not allow enough space for the head jamb to make the rough opening modular. If special doors and frames are not available to fit the modular heights, consideration must be given to this point in the details. Other important vertical heights are windowsills, ceiling heights, and table heights.

5.1.5 Use a Note to Indicate That the Drawings Are Modular

Because a set of drawings drawn and dimensioned to modular measure appears very much like a set of conventional drawings, a note stating that the drawings are modular should be used as a guideline for contractors, bidders, estimators, and workers. The note, usually placed on the first sheet of the drawings, should be simple and clear; if it is accompanied by several simple drawings illustrating the method of dimensioning employed, as well as by any special representations found on the drawings, all of the advantages of modular measure will be readily understood.

To standardize the modular note, an easily applied, translucent appliqué[1] has been devised by the AIA modular coordinator and is recommended by the Modular Measure Committee of the American Standards Association (Fig. 5–8). It can be affixed by a pressure-sensitive adhesive to the tracing itself and will clearly show on all prints. Some architectural offices design their own modular note and merely letter it, along with any illustrative drawings, directly on the first sheet of their drawings.

---

[1] Available from Standpat Co., 150–42 12th Road, Whitestone, NY 11357.

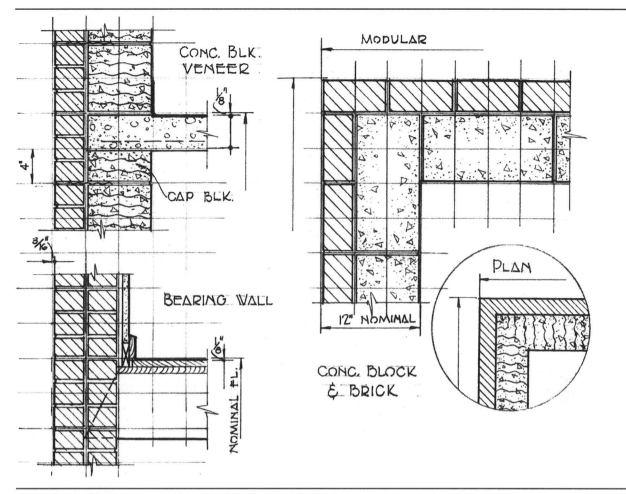

Figure 5–6 The placement of floor levels within modular grids.

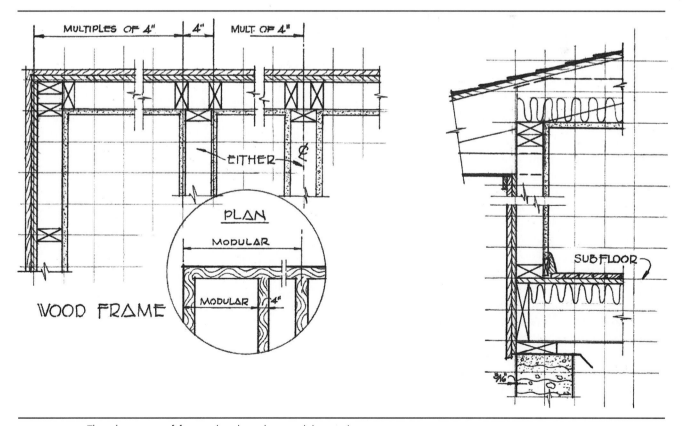

Figure 5–7 The placement of frame details within modular grids.

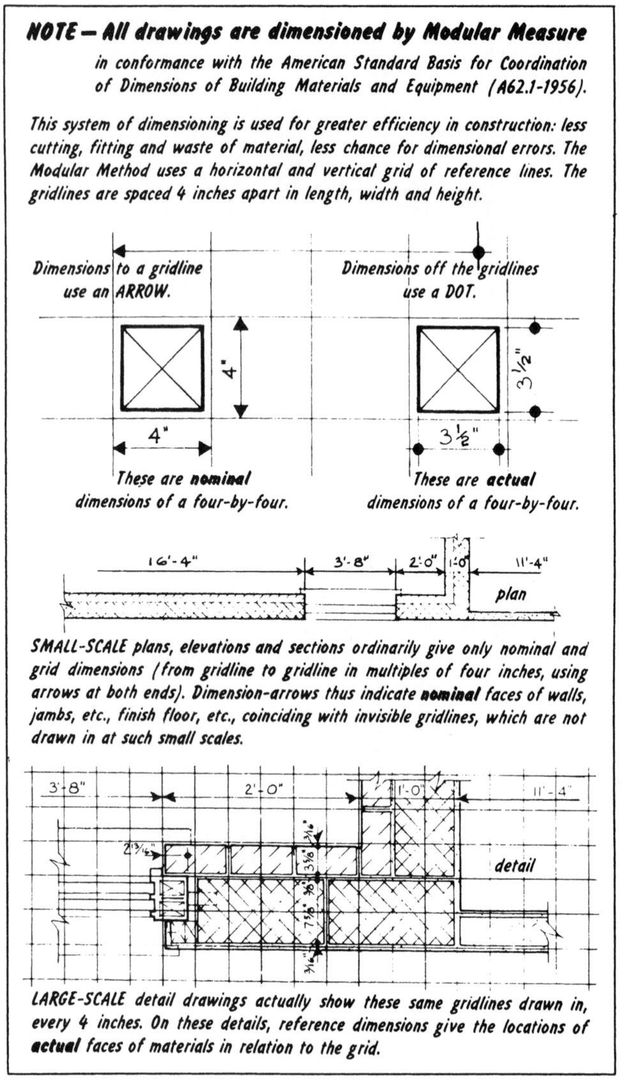

NOTE — All drawings are dimensioned by Modular Measure
in conformance with the American Standard Basis for Coordination
of Dimensions of Building Materials and Equipment (A62.1-1956).

This system of dimensioning is used for greater efficiency in construction: less
cutting, fitting and waste of material, less chance for dimensional errors. The
Modular Method uses a horizontal and vertical grid of reference lines. The
gridlines are spaced 4 inches apart in length, width and height.

Dimensions to a gridline use an ARROW.

Dimensions off the gridlines use a DOT.

These are nominal dimensions of a four-by-four.

These are actual dimensions of a four-by-four.

SMALL-SCALE plans, elevations and sections ordinarily give only nominal and
grid dimensions (from gridline to gridline in multiples of four inches, using
arrows at both ends). Dimension-arrows thus indicate nominal faces of walls,
jambs, etc., finish floor, etc., coinciding with invisible gridlines, which are not
drawn in at such small scales.

LARGE-SCALE detail drawings actually show these same gridlines drawn in,
every 4 inches. On these details, reference dimensions give the locations of
actual faces of materials in relation to the grid.

Figure 5–8 A preprinted appliqué available for attachment to working drawings to indicate that they are modular.

Figures 5–9 and 5–10 show a residential floor plan and various details drawn and dimensioned modular.

A type of modular construction for small buildings is shown in Fig. 5–11 (MOD 24, devised jointly by the major U.S. wood products associations). The framing is placed on 24″ centers so that standard 4′ × 8′ plywood and sheeting materials can be used with a minimum of waste and cutting. Many factory-manufactured small homes are now produced using a 4″ module for sizing.

5.2

THE METRIC SYSTEM OF MEASUREMENT

As mentioned in the introduction to this chapter, a 1988 federal law mandated the metric system as the preferred system of measurement in the United States. This law required that metric measure be used in all federal procurement, grants, and business-related activities, to the extent feasible, by September 30, 1992. The intent of this law is to make the United States more competitive in international trade by bringing its measurement system into line with that of the rest of the world. The United States remains the only industrialized country in the world that still employs the English system of measurement.[2]

5.2.1 Economics

The driving force behind metric conversion efforts in the United States today is economics. Almost all foreign markets, including Canada and Mexico, use metric measure. Thus, products made in the United States must be designed and manufactured to metric standards in order to compete. The European Community (EC) construction industry is ready to take advantage of opportunities opening in Eastern Europe, and the Pacific Rim countries, particularly Japan, are well trained and ready to invest in the vast markets of China.

5.2.2 Federal Construction

Today, virtually all major federal construction programs have been converted to the metric system, with about $15 billion in metric projects presently under construction or already completed. By the year 2000, total government metric work could amount to as much as $100 billion. Such large expenditures will expose a significant portion of the U.S. construction industry to metric. No one will want to work with two different systems of measurement for very long. As a result, U.S. construction will probably convert to predominantly metric within the next five to ten years.[3]

Architects and engineers are expected to use metric units in their work. Contractors and tradespersons need to understand and bid on metric contract documents. They also need to prepare shop drawings and perform on-site work in metric. Product manufacturers have been advised to include metric units in their product literature catalogs and advertising and to design new products in rounded, rational metric sizes.

During the transition to metric, about 5% of the products currently used in building construction in the United States will require a physical change in composition or size and configuration standards. We have been using the metric system for quite a while in many areas. For example, flu and measles shots are measured in cc's, or cubic centimeters; the speedometer on most cars is marked in both kilometers and miles per hour; beverages are sold

---

[2] *Metric in Construction*, May 1992.

[3] *Metric in Construction*, September 1992.

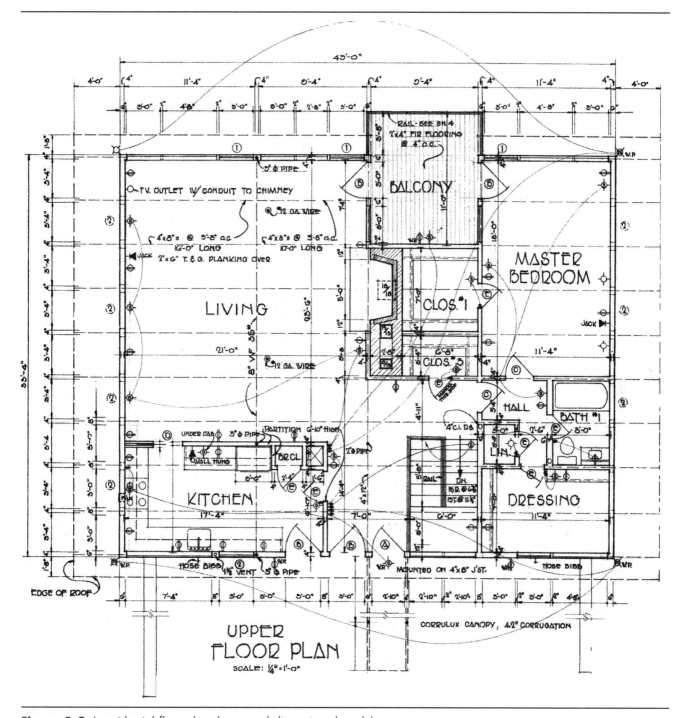

Figure 5–9 A residential floor plan drawn and dimensioned modular.

in liter quantities; and the glass in windows is manufactured in thicknesses measured in millimeters.

The National Council of Architectural Registration Boards has announced that it will implement the metric system of measurement in its Architectural Licensing Examination and it will phase out inch–pound units by the year 2004. The tenth edition of AIA's *Architectural Graphic Standards* (1998) is completely metric.

SYSTÈME INTERNATIONAL

The system of measure commonly referred to as *metric* is the International System, or SI, the abbreviation for the French term *Système International*. The main unit of length in the metric system is the meter, which equals 39.370079 inches exactly. The symbol for meter is m.

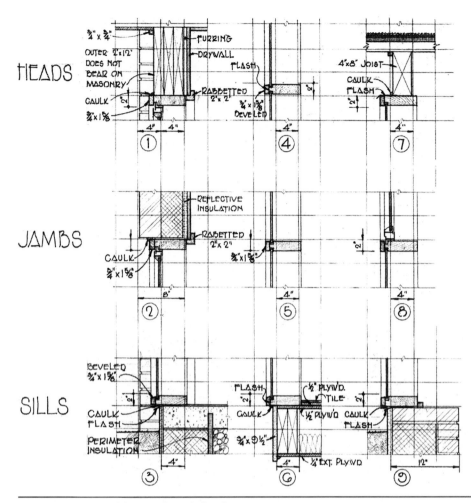

Figure 5–10 Modular residential details.

5.3.1 Measurement Coordination

The metric system coordinates the measurements of length, area, volume, and mass into a simple decimal system. The metric system requires no conversion from unit to unit as does the English inch–pound system. Conversions from inches to feet, or from ounces to pounds, are not required in the metric system.

For example, to calculate the amount of concrete needed for a floor 200 feet long, 180 feet wide, and 5.5 inches thick, you must first convert the thickness to feet by dividing by 12. Next you must multiply that value by the floor length and width to obtain cubic feet and then divide the number of cubic feet by 27. This conversion requires three steps to determine that you need 611 cubic yards of concrete. To find out how much concrete is needed for a floor 61 m long, 55 m wide, and 140 mm thick, you need only change 140 mm to 0.14 m and multiply the floor length, width, and thickness to obtain 470 m.[4]

---

[4] *Metric in Construction,* Nov./Dec. 1994.

5.3.2 Building Construction Units

Six basic metric units are used in building construction:

meter for distance

kilogram for mass

second for time

ampere for electrical current

kelvin for temperature

candela for illumination

The metric units preferred for linear construction measurements are the millimeter (mm), the meter (m), and the kilometer (km). The millimeter is used for small dimensions, the meter for large dimensions, and the kilometer for longer distances. Items that used to be measured in fractions of inches will be measured in millimeters. Items that used to be measured in yards will be measured in meters. Items that used to be measured in miles will be measured in kilometers (Fig. 5–12).

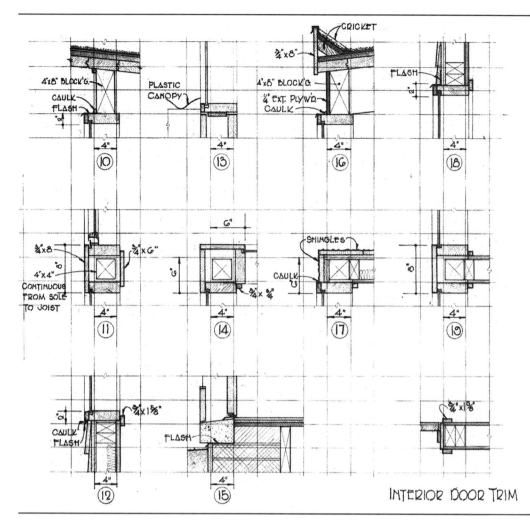

Figure 5–10 *(continued)*

5.3.3 Metric Conversion

Metrication will be introduced to the construction industry in two phases: *soft*, which means relabeling existing dimensions in SI, and *hard*, which will require changes in modular components to be in rounded-off metric units.

Soft Conversion   Soft conversion from the inch–pound system to the metric system will be necessary until all existing documents have been converted. Soft conversion in the metric sense implies no dimensional change in a product and expresses the dimension to the nearest practical metric decimal place. When soft conversions are made, a rational equivalent would be more appropriate. For example, 12 inches (exactly 304.8 mm) is not a clean, rational number; it should be rounded to 300 mm. Other examples include the following:

$$1'' = 25 \text{ mm}$$
$$4'' = 100 \text{ mm}$$
$$6'' = 150 \text{ mm}$$
$$12'' = 300 \text{ mm}$$
$$16'' = 400 \text{ mm}$$
$$24'' = 600 \text{ mm}$$

Recommendations for rationalizing linear dimensions can be found in Table 5–1.

Hard Conversion   Hard conversion implies making dimensional changes in a product that will provide more easily used numbers in the metric system. For example, 16″ becomes 400 mm, and a 16″ stud spacing becomes a 400-mm stud spacing. Only products that must fit together in a modular grid are logical candidates for hard metric conversion—that is, an actual dimensional change. Such products include dry wall, plywood and other wood-based panels, suspended ceiling components, raised flooring, tile, brick, and block.

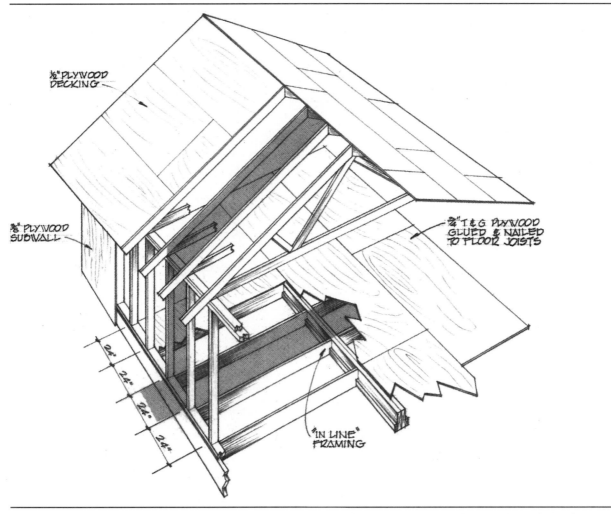

Figure 5–11 MOD 24, an economical modular framing system for light construction developed jointly by the American Plywood Association, Western Wood Products Association, and Southern Forest Products Association.

DIMENSIONS ON DRAWINGS

On architectural drawings, dimensions are to be expressed in millimeters, and areas in square meters. Whole numbers indicate millimeters, and decimal numbers, meters. Because only the millimeter and the meter are used for building dimensions on drawings, there is no need to use the unit symbol after each dimension, although they should be included in lettered notes when a dimension is given. For the present it is recommended that each drawing carry the following note:

ALL DIMENSIONS ARE GIVEN IN MILLIMETERS
UNLESS OTHERWISE INDICATED.

MODULAR COORDINATION[5]

With the change to a metric-based measurement system will come a strong push to adopt dimensional coordination in construction and design.

---

10 mm = 1 cm
100 cm = 1 m
1000 m = 1 km

Figure 5–12 Metric conversions.

---

[5] This section is adapted from Hans J. Milton, FRAIA, "Dimensional Coordination in Building," *AIA Metric Building and Construction Guide, 19XX.*

Table 5-1 Rationalizing of converted metric values. In building construction, fractions of the millimeter are rarely required, but where such use is unavoidable (thickness of sheet metal, paint coatings, and so on), the decimal fraction should be kept to one place, for example, 1.6-mm-thick sheet metal. If small fractions are needed, use micrometer (μm).

Order of Preference	Very Small Dimensions (Up to 50 mm)	Small Dimensions (Up to 500 mm)	Medium Dimensions (Up to 5000 mm)	Large Dimensions (For Example, Site Layouts)
1st preference	Multiples of 10 mm	Multiples of 100 mm	Multiples of 500 mm	Multiples of 10 m
2nd preference	Multiples of 5 mm	Multiples of 50 mm	Multiples of 300 mm	Multiples of 5 m
3rd preference	Multiples of 2 mm	Multiples of 25 mm	Multiples of 100 mm	Multiples of 1 m
4th preference	Multiples of 1 mm	Multiples of 10 mm	Multiples of 50 mm	Multiples of 0.5 m
5th preference			Multiples of 25 mm	Multiples of 0.1 m
6th preference			Multiples of 10 mm	

Source: William J. Hornung, "Tables Showing Metric Equivalent Sizes for Some Building Materials," *Metric Architectural Construction Drafting and Design Fundamentals* (Englewood Cliffs, NJ: Prentice-Hall, 1981).

Dimensional coordination is based on the use of an accepted fundamental unit of size, called the *basic module*, and a set of selected multiples and submultiples as preferred dimensions in building design, production, and construction. Modular coordination is a particular form of dimensional coordination, based on the internationally accepted basic module of 100 mm.

When Great Britain, Canada, and the other English-speaking countries converted to the SI, they introduced dimensional coordination and the use of the 100-mm module in conjunction with the change in the building industry. Since 1973 the International Organization for Standards (ISO) has issued a range of international standards on modular coordination in building based on the basic module of 100 mm.

5.5.1 Modules

Basic Module   The basic module of 100 mm replaces the customary module of 4". Any multiples of the basic metric module are immediately visible in the dimension;

for example, 3200 mm represents 32 modular units (Fig. 5–13).

Multimodules   Multimodules are whole multiples of the basic module. They are used for structural and planning grids and for determining ranges of sizes for building elements and components (Fig. 5–14). There are two types of multimodules:

- **Horizontal multimodules**   Preferred horizontal multimodules for construction are 300, 600, 1200, 3000, and 6000 mm. Additional multimodules for special uses are 900 and 1500 mm. In small buildings built with masonry units, a horizontal multimodule of 200 mm may be used. The large, 6000-mm multimodule provides a useful systems module for the overall coordination of buildings. It is a whole multiple of 200-, 300-, 400-, 500-, 600-, 1000-, 1200-, 1500-, 2000-, and 3000-mm multimodules.

- **Vertical multimodules**   Preferred vertical multimodules are 200, 300, and 600 mm.

100 mm

Figure 5-13 The basic module for the construction industry is 100 mm, an internationally accepted value. The basic module should apply to building components as well as entire buildings. (Milton, "Dimensional Coordination")

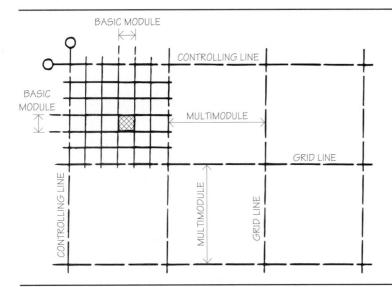

Figure 5-14 Multimodules, if carefully selected, can coordinate with the controlling dimensions for a building and can limit component sizes to a minimum. (Milton, "Dimensional Coordination")

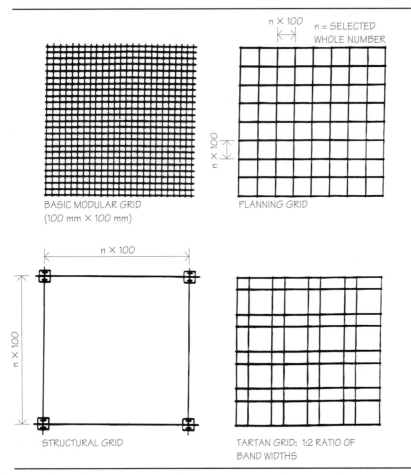

Figure 5-15 Grids for use with dimensional coordination. (Milton, "Dimensional Coordination")

Table 5–2 Preferred sizes for building components and assemblies.

Category	Typical Examples of Components and Assemblies	Dimensional Preference (mm)		
		First	Second	
A. Small components (under 500 mm)	Bricks, blocks, tiles, paving units	100 200 300 400	25 50 75 150 250	
B. Medium-size components and assemblies (under 1500 mm)	Sheets, panels, partition units, doorsets, windows, slabs	600 800 900 1200	500 700 1000 1400 See Note a.	
C. Large-size components and assemblies (up to 3600 mm)	Precast floor units, precast wall units, panels, door assemblies, window assemblies, precast stairs, precast ducts	1800 2400 3000 3600	($n \times 300$) 1500 2100 2700 3300	($n \times 200$) 1600 2000 2200 2600 2800 3200 3400 See Note b.
D. Very large components and assemblies (over 3600 mm)	Prefabricated building elements, precast floor and roof sections	4800 6000 7200 8400 9600 10,800 12,000	($n \times 600$) 4200 6600 7800 9000 10200 11400	($n \times 1500$) 4500 7500 10500 See Note c.

Source: Milton, "Dimensional Coordination."

[a]For the purposes of rationalization, those multiples of 100 mm which are above 1000 mm and which are prime numbers (for example, 1100, 1300, 1700, 1900, 2300, 2900) constitute a lower order of preference and should be considered only when special requirements exist.

[b]Alternative second preferences are shown; for vertical dimensions the use of multiples of 200 mm may sometimes be more appropriate than the use of multiples of 300 mm, particularly in conjunction with masonry materials.

[c]Alternative second preferences are shown; for some projects it will be more appropriate to size large components or assemblies in multiples of 1500 mm.

Nonmodular Nonmodular dimensions occur in practice and need to be accommodated in any system of dimensional coordination. Quite a few component dimensions will be nonmodular, especially thickness.

5.5.2 Grids

Reference grids provide a constant space reference system from the inception to the completion of a project (Fig. 5–15). Grids assist designers in the coordination of building dimensions and in the expression of how the coordination among structure, elements, and systems is to be achieved.

Basic Grid The basic modular grid is a grid with consecutive parallel lines spaced 100 mm apart horizontally and vertically. Its greatest application is in large-scale drawings (1:1 to 1:20) where it clarifies component locations.

Multimodular Grid The multimodular grid is used for determining the layout of a building and the location of the main structural features. The grid interval is normally chosen from preferred horizontal and/or vertical multi-modules. This grid is most useful in planning and in structural positioning.

Special Grids In certain circumstances a tartan grid may be used. A typical tartan grid has alternate bands of 100 mm and 200 mm in plan.

5.6
PREFERRED SIZES FOR BUILDING COMPONENTS

Preferred sizes for building components and assemblies are an integral part of a dimensioning system. These coordinating dimensions define the space occupied by a component or an assembly, including all necessary allowances for tolerances and joints. The coordinating dimensions can also be the distances between the center lines of joints or, in the case of overlapping components, the distance representing the overlap (Table 5–2).

REVIEW QUESTIONS

1. What is the main unit of length in the metric system?
2. Define *soft* and *hard* conversion in relation to the use of metrics in construction.
3. What is the size of the *basic module* used in metric construction?
4. Describe modular coordination.
5. What is the primary unit of measurement for modular dimensioning?

REFERENCES

Books

AIA Metric Building and Construction Guide. Braybrooke, Susan, ed. New York: John Wiley & Sons, 1980.

Metric Architectural Construction Drafting and Design Fundamentals. Hornung, William J. Englewood Cliffs, NJ: Prentice-Hall, 1981.

Periodicals

Architecture. September 1992.

The Construction Specifier. March, July, August, September, November, December 1992; January, February, March, September 1993.

Metric in Construction. May, July/Aug., Sept./Oct. 1992; Sept./Oct. 1993; May/June 1994; Sept./Oct., Nov./Dec. 1994; Jan./Feb., March/April, July/August 1995; Jan./Feb., March/April 1996.

Progressive Architecture. April 1992.

PUBLISHERS' ADDRESSES

Books

Prentice Hall
A Division of Simon & Schuster
240 Frisch Court
Paramus, NJ 07652

John Wiley & Sons, Inc.
605 Third Avenue
New York, NY 10158

Periodicals

Architecture
  BPI Communications
  1515 Broadway
  New York, NY 10036

The Construction Specifier
  Construction Specifications Institute
  601 Madison Street
  Alexandria, VA 22314

Metric in Construction
Construction Metrication Council
National Institute of Building Sciences
1201 L Street, N.W., Suite 400
Washington, DC 20005

Progressive Architecture
Penton Publishing
1100 Superior Avenue
Cleveland, OH 44114

FRANK BETZ

"Geometry . . . is the only science that it
hath pleased God to bestow upon
mankind."
—THOMAS HOBBES

A solid foundation in architectural drawing requires knowledge of multiview drawings or orthographic projections, sectioning, auxiliary views, and geometric construction. Each of these fundamentals is of equal importance; each is discussed here for the benefit of those students who may not possess the necessary background, and for those who may need an accelerated review. Further study of architectural drawing assumes that these essential concepts have been mastered.

Knowledge of *multiview* or *orthographic* drawings serves as a conventional and universally accepted method of representing three-dimensional objects in a two-dimensional way on architectural drawings. *Sections* are necessary to show information about construction that is normally hidden on exterior views. *Auxiliary projections* illustrate how views that lie on other orthogonal planes are drawn. *Applied geometric construction* provides methods for arriving at general graphic lines and shapes that form most elements of architectural design.

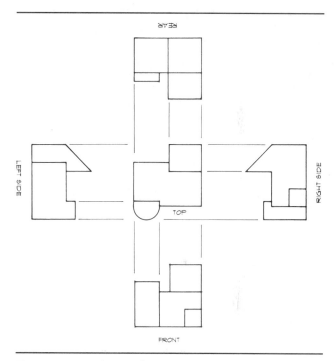

Figure 6–1 Multiview drawings.

6.1

MULTIVIEW TWO-DIMENSIONAL DRAWINGS

We live in a three-dimensional world, but in most instances we use two-dimensional drawings to represent the three-dimensional objects around us. We are familiar with the terms *top*, *bottom*, and *side* as we use them to describe a three-dimensional object. In essence, we are using *multiviews* to communicate in a three-dimensional language. We cannot visualize an object in three dimensions if only the top is graphically represented; the sides and possibly the bottom must also be drawn (Fig. 6–1). Thus a number of multiview drawings must be assembled to create the total effect that could be accomplished by one single-view drawing (Fig. 6–2).

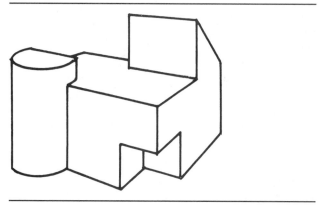

Figure 6–2 Single-view drawing.

Figure 6-3 Pictorial single-view drawing of a building.

The representation of a three-dimensional object such as a building on a flat piece of drawing paper is a fundamental and important consideration of the drafter. A pictorial, single-view representation would be the most realistic in appearance (Fig. 6–3); yet from the standpoint of an accurate graphic description for technical use, multiview drawings known as *orthographic* projections are commonly used. (*Ortho* means "perpendicular" or "at right angles.") Over the years orthographic projections have proved to be the clearest method for showing true shapes, relationship of features, and necessary dimensions of an object (Fig. 6–4).

A multiview, or orthographic, drawing requires as many related views as necessary to reveal all necessary information about an object. Some objects, such as simple gaskets, require only one view, since the slight thickness can be shown better by a note than by graphic methods. On the other hand, if a more complex object, such as a building, is drawn, often four or five views might be needed. Graphic experience, together with knowledge of

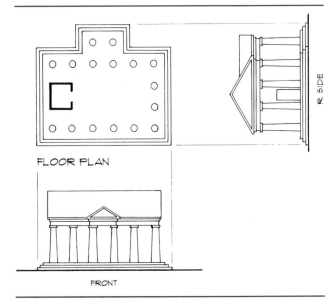

Figure 6-4 Orthographic projections.

construction, provides the architectural drafter with the means to determine the number and the type of views necessary to represent accurately a structure on paper.

6.1.1 Principles of Projection: The Glass Box

For the simplest method of understanding this graphic language, let us think of an object, such as the basic house shown in Fig. 6–5A, as being enclosed by transparent picture planes, or actually a glass box. Notice that the sides of the house are parallel to the glass planes, and if we viewed each side through each transparent plane, a true-shape image of the side would result.

The inclined surfaces of the roof are not parallel to the upper horizontal glass plane; therefore, they do not appear in their true shape. The ridges of the roof, on the

other hand, appear as full-length lines on the top view because they are parallel to the top picture plane. The profile of the gable roof, which is not evident on the top view, appears in true shape on the end views. Theoretically, the surfaces of the house that are perpendicular to the transparent planes and lines formed by the intersections of inclined surfaces are brought to the projection planes with parallel projectors. Also, the images created by these projectors on the planes comprise what is known as an *orthographic projection*.

6.1.2 The Relationship of Views

If the glass box is then opened, using theoretical hinges as shown in Fig. 6–5B, an orthographic, multiview drawing is formed on a flat plane. If required, six views of the house could be extended to the planes of the box

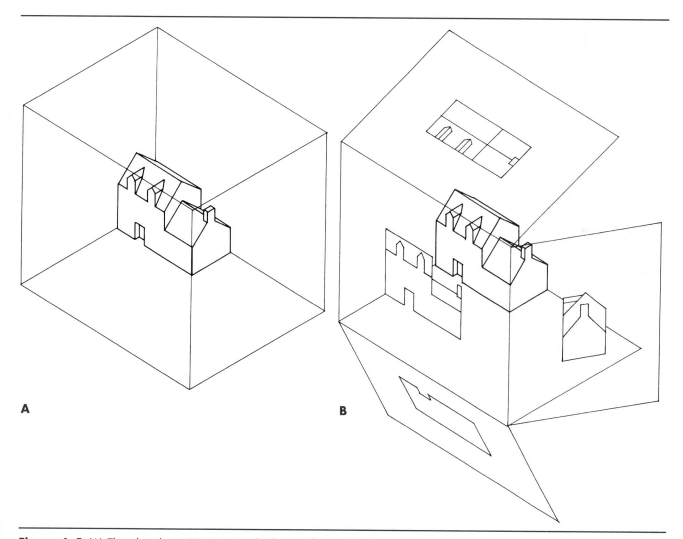

A   **B**

Figure 6–5 (A) The glass box. (B) Opening the box to form a flat plane.

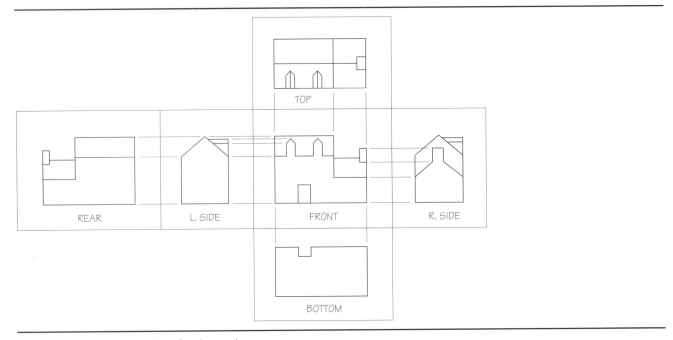

Figure 6–6 The relationship of orthographic views.

(Fig. 6–6); however, six views would seldom be necessary. After opening the box, you will notice that the top view is directly over the front view; the right-side view is directly to the right of the front; the left-side view is directly to the left of the front; and so on. This alignment of views allows features of the house to be clarified by cross-observation of more than one view. Only two dimensions (height and width) can be shown on each orthographic view. Thus, by aligning the views, the drafter can obtain an accurate, three-dimensional description from the combined views.

6.1.3 Third-Angle Projection

The convention for most engineering drawings in this country is to use the front view (the principal view), the top view, and the right-side view. Usually, these three views will give a complete graphic description of the object. (If important information is located on the left side, this view is substituted for the right-side view.) Ordinarily, the left-end view of the house would only duplicate the information shown on the right end, and the left-end view also has the disadvantage of showing part of the roof as hidden surfaces. The same is generally true of the rear or bottom views; they only duplicate information and are therefore unnecessary. As long as all information is shown, a minimum number of views are used. To produce the three conventional views (front, top, and right side), place the object within the third octant of the transparent plane assembly shown in Fig. 6–7. The top view is projected to the horizontal plane, the front view is projected to the frontal plane, and the right-side view is projected to the profile plane. This method of projection and

of showing relationship of views is called *third-angle projection* and is universally used for technical drawings throughout the United States and Canada.

6.1.4 Hidden Surfaces

Often surfaces and features on orthographic views are hidden from the observer. If these features contribute necessary information, they are represented by dashed (hidden) lines. In Fig. 6–6, the hidden roof surfaces are shown as dashed lines on the left-side view since they are behind the higher portion of the object. The drafter should make dashed lines carefully. Uniform dashes should be used; very little space should be left between each dash; and the ends of each dash should be made prominent (see Fig. 4–4).

6.1.5 Architectural Views

In architectural drawings, orthographic views that are projected to vertical planes, such as the front, side, and rear views, are called *elevations*. Views that are projected to horizontal planes and observed from the top (or bottom) are called *plans*. Hereafter, we will use this terminology when referring to the different architectural views.

Another variation in architectural work is the placement of the different elevations and plans on paper. As previously mentioned, when drawing the orthographic views of a small object, the drafter must relate the views for proper cross-reference. Because the elevations and the plans of a building are often comparatively large, it becomes difficult to place them all on the paper in definite orthographic relationship. For convenience, the plan and

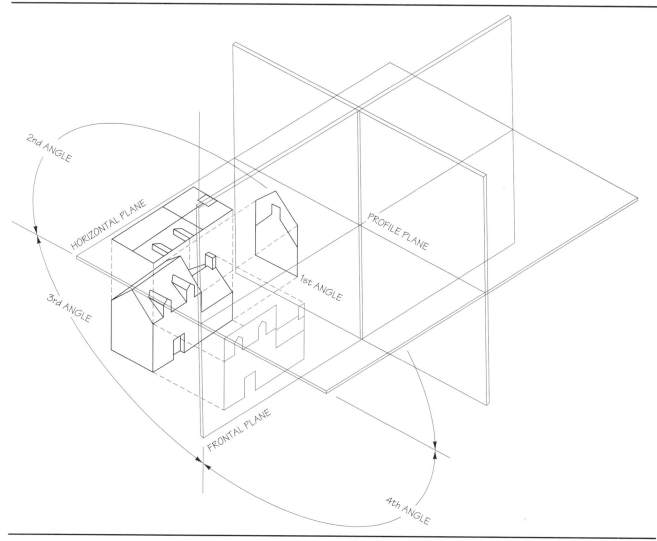

Figure 6-7 Third-angle projection planes.

the elevations are frequently put on separate sheets. Such an arrangement allows the drafter to draw each view large enough to facilitate reading and interpretation. However, each plan and elevation must be properly identified so that all views can be related in the reader's mind. Elevations should be shown horizontally, and, if space permits, the right elevation should be to the right of the front, and the left elevation should be to the left of the front. Avoid an elevation "on end" of the paper—it would not appear in a natural position. Otherwise, the elevations are true orthographic projections with few, if any, hidden surfaces or features shown.

Top views of buildings are limited mainly to roof framing plans showing the layout of complex roof structural members. Occasionally, a roof plan is necessary to show skylight, drains, and other important information on commercial buildings. Top views of houses would be largely for the purpose of developing different roof intersections and overhang treatments during planning stages. It is seldom necessary to draw a top view of a house showing only the roof surface.

The main drawing of a set of working drawings, the *floor plan*, is actually a horizontal section view of the building as seen from above. The cutting plane (Fig. 6–8) is theoretical and passes through important vertical features, such as doors, windows, walls, stairs, and fireplaces, to show their location and size within the building (Fig. 6–9). Thus the cut must often have offsets to pass through features that might be located high or low on the walls, yet the offsets are not shown on the plan. You will see this conventional method of floor plan representation after studying a number of the floor plans shown throughout this book. Floor plans will be considered further in Chapter 18, "Working Drawings of Small Homes."

Other architectural views of only portions of a building, such as structural members and small sections, come under the broad, general heading of *details* and will be discussed throughout later chapters.

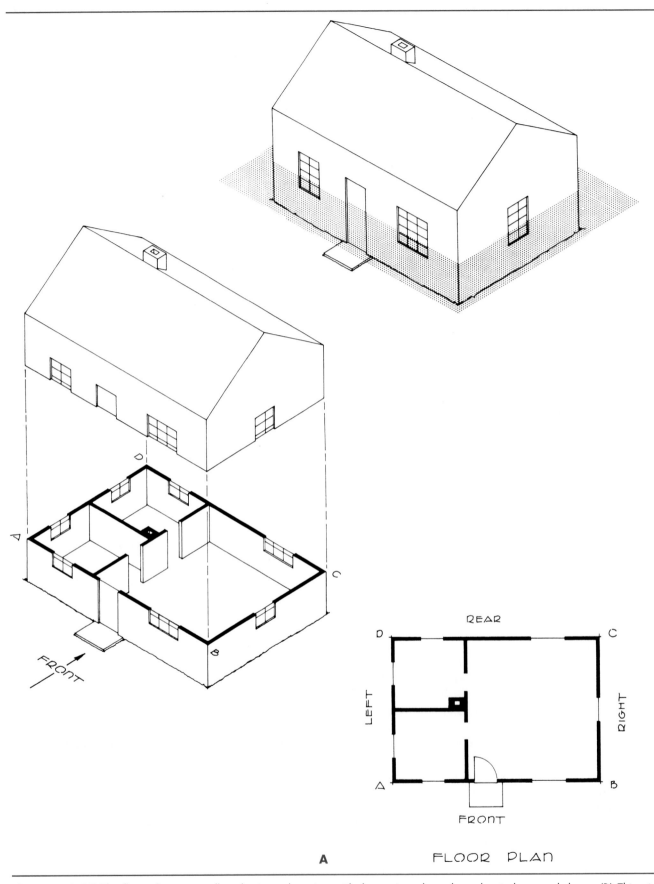

REAR

D   C

LEFT   RIGHT

A   B

FRONT

A   FLOOR PLAN

Figure 6–8 (A) The floor plan is actually a horizontal section with the cutting plane through windows and doors. (B) This pictorial plan of a house shows offsets in the cutting plane for the purpose of including important features. (C) In the plan view, the offsets are not shown.

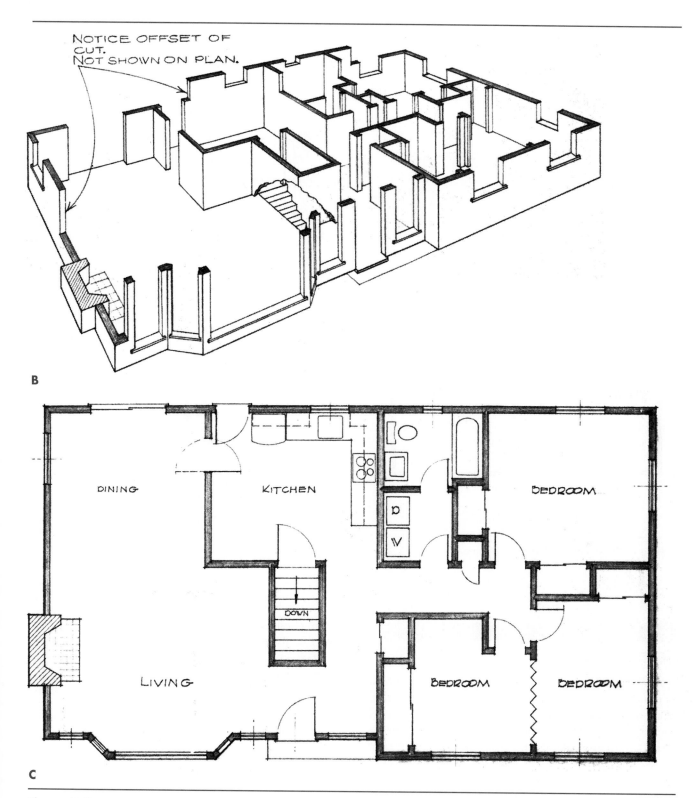

B

NOTICE OFFSET OF
CUT.
NOT SHOWN ON PLAN.

DINING

KITCHEN

BEDROOM

DOWN

LIVING

BEDROOM

BEDROOM

C

Figure 6–8 (*continued*)

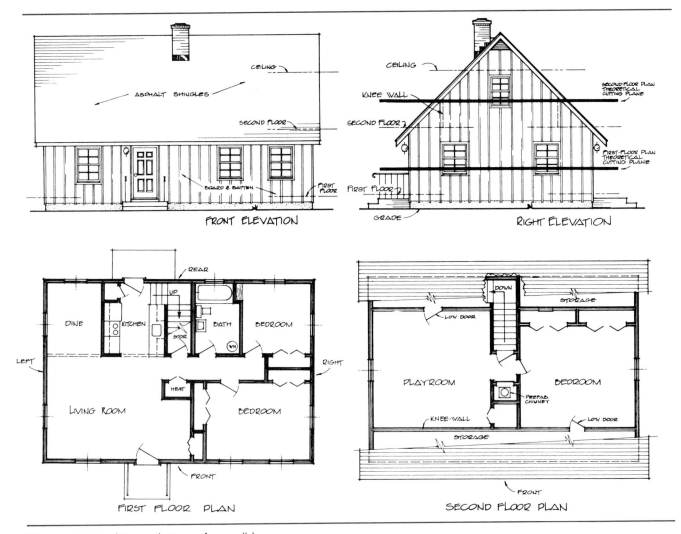

Figure 6–9 Architectural views of a small house.

6.2

SECTIONING

A *section* is defined as an imaginary cut made through an object, or a combination of objects, with the material in front of the cut removed. A view of this cut or plane, then, becomes a direct method of revealing important internal information about the object. Usually, sections are drawn through walls, floors, windows, doors, structural members, cornices, fireplaces, stairs, footings, trim moldings, cabinet work, special construction, and the like.

6.2.1 The Importance of Architectural Sections

When someone is looking at the elevation of a building, only the exterior materials and features are visible. If dashed lines were used to indicate the many hidden materials and their shapes and interrelationships, the re-

sulting hodgepodge of lines would make the drawing impossible to read (Fig. 6–10). A simple method must be used, therefore, to show clearly each important feature of interior construction on working drawings for the benefit of builders and workers. Sections are drawn mainly for this structural purpose, although occasionally a simplified section is used on design drawings.

On typical architectural working drawings, drafters seldom employ the use of hidden lines to the extent that they are employed on many engineering drawings. Rather, the architectural drafter uses a series of sections taken through strategic points of the construction throughout the building and arranges them for easy reference to the other views. Different types of sections are used, and each has an appropriate application, determined by the drafter, to show necessary information simply and comprehensibly. Often, for clarity in dimensioning, sections are drawn at a larger scale than the view from which they are taken.

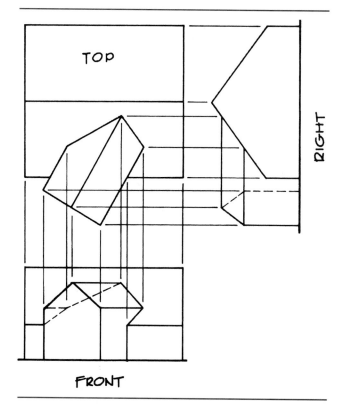

Figure 6–10 Projecting roof intersections on architectural views.

6.2.2 Indicating the Cutting Plane

Unless the plane of the section is obvious, a *cutting-plane line* should be used to identify the imaginary cut on the related drawing (Fig. 6–11). Cutting planes are usually represented by a heavy dash and double-dot line; arrows are usually placed at the ends of the line to indicate the direction of observation revealed by the section view. If the section is identical when viewed in either direction, the arrows are unnecessary.

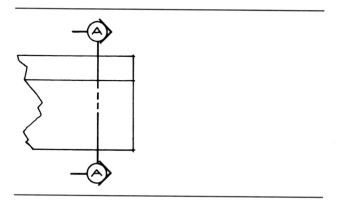

Figure 6–11 Cutting-plane line identification.

As previously mentioned, floor plans are actually horizontal sections, and no cutting plane is conventionally needed to identify them (Fig. 6–8). To understand this concept, think of the floor plan as a means of showing information about only one floor level; as long as the theoretical cut is somewhere through that floor level, there is no reason for showing a cutting-plane line on an elevation view (Fig. 6–9). Even if the cut is offset in some parts of the building to encompass features found at different heights in the walls, no cutting plane is needed since the elevation views show the correct heights of such features.

Also, typical wall sections, showing uniform construction throughout major exterior walls, require no cutting planes for identification.

On drawings having two or more cutting planes and section views, each section must be related to the proper cutting plane; usually letters, such as A–A or B–B, are used for this identification (Fig. 6–11). Complex drawings having section views on several sheets must have letter identifications as well as sheet identification to relate the view correctly. Figure 14–6 shows a satisfactory method for relating numerous sections to their cutting planes when views must be arranged on different sheets. The small circles can be quickly drawn with a circle template (Fig. 6–11).

6.2.3 Sectioning Conventions

The section must distinctly show the outline of all solid materials that have been cut and also the voids or spaces that may fall between the materials. Occasionally, features lying beyond the cutting plane must also be shown for clarity (Fig. 6–12). These features are brought to the plane of the section by perpendicular projectors, very much as the features of an object are brought to the projection plane of an orthographic projection drawing. Usually, hidden features are not shown on section views; dashed lines only confuse sectioned details. On some views, however, it is necessary to violate true projection for added clarity; features such as holes, ribs, and spokes, which do not fall on the cutting plane of a symmetrical object, are revolved to the cutting plane and are then projected to the section view (Fig. 6–13). It is also conventional practice not to section bolts, screws, rivets, shafts, rods, and the like, which may fall in the plane of the section. They are more clearly shown unsectioned (Fig. 6–14).

6.2.4 Crosshatching Sections

Other than the above-mentioned conventions, the materials that are cut must be crosshatched (symbol line tone) with the proper material symbol. (See Fig. 4–9 for the various standard sectioning symbols used on construction drawings.) Adjacent materials are crosshatched in different directions for easier identification; the 45° triangle

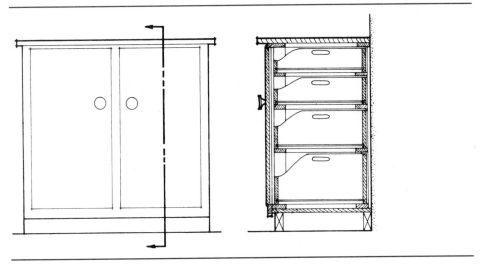

Figure 6-12 Section showing features beyond cutting plane.

can be used for the majority of the linework; and to conserve time, the spacing of crosshatch lines should be done by eye—experience will develop uniformity of tone. Small, sectioned pieces should have darker tones; larger surfaces can be lighter. Wood structural members are indicated merely with diagonals drawn with instruments; others, such as finish lumber and concrete, are executed freehand. Thin members, such as sheet metal, building paper, or gaskets, are filled-in solid black—usually only a heavy line. If the same member is revealed in several places in a section view or in different sections, the crosshatch technique should always be identical. Important profile lines of a section should be made prominent.

As mentioned previously, many architects use a technique called *poché* to make sectioned areas of a drawing appear more prominent, especially on display drawings. The term means to shade, color, or otherwise darken the value of an area to provide contrast.

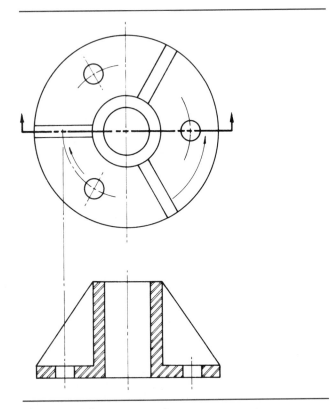

Figure 6-13 Conventional symmetry in sections.

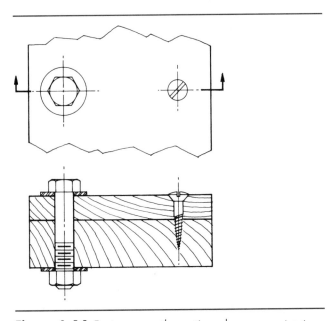

Figure 6-14 Fasteners on the cutting plane are not cut.

6.2.5 Full Sections

A full section shows an object or a building *cut entirely across* so that the resulting view is completely "in section." The cutting plane can be a straight plane (Fig. 6–12), or it can be offset to pass through features that it may have missed if it were a straight, nonoffset plane (Fig. 6–13). An offset does not show on the section view, as the change in plane is purely imaginary and is meant to gather as many features as necessary to make the section inclusive and complete. Sections can be either horizontal or vertical. Although drawings showing horizontal section views through a building are in reality one of the types of sections, it is customary to call these *plans*, whereas drawings showing vertical section views are commonly referred to as *section details*. Section details are the usual method of showing cuts through walls and are the most predominant way of showing sections in architectural drawings. A vertical section, if completely taken through the narrower width of a building, is called a *transverse section;* taken through the long dimension of a building, it is called a *longitudinal section.* Either type is usually informative and often reveals construction detail that is difficult to show with other drawing procedures.

6.2.6 Half-Sections

Instead of passing entirely through the object, as in the full section, the cutting plane in a half-section goes only to the center and gives the appearance of having the front quadrant removed (Fig. 6–15). On symmetrically shaped objects, half-sections conveniently show both the interior and exterior in only one view. Because of symmetry, some objects do not warrant a full section. When drawing a half-section, you do not need to show the hidden features on the uncut half of a symmetrical object with dashed lines, since the lines would merely duplicate the features of the section. Notice that a center line separates the interior and exterior portions of the half-section view.

6.2.7 Broken-Out Sections

A broken-out section is a convenient method of showing important interior features that can be given sufficient description in a small area of an exterior view. Usually, the removal of only a small portion of exterior material will not affect the appearance of a view, and the flexibility of a broken-out section allows it to be placed where important exterior features need not be removed (Fig. 6–15). Irregular break lines indicative of the material removed are used to show the limits of the section area.

6.2.8 Revolved Sections

The cross-sectional shapes of I beams, wood moldings, irregularly shaped structural members, and so on, can be effectively shown by the use of revolved sections. To draw each section, assume a cutting plane perpendicular to the axis of the member, as shown in Fig. 6–15, which has

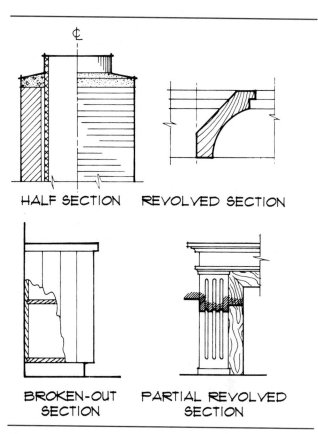

Figure 6–15 Various types of minor sections used on drawings.

been revolved through a 90° arc to the frontal plane and superimposed directly on the elevation view of the member. Lines of the elevation view that may underlie the section should be removed before the crosshatch lines are applied.

6.2.9 Revolved Partial Sections

Occasionally, elevations of traditional fireplaces, exterior door trim, molded woodwork, and so on, requiring only information about their exterior profiles, can conveniently employ a revolved partial section (Fig. 6–15). Using a bold line, the drafter draws only the profile directly over the elevation. (Notice that the offsets in the profile coincide with the lines on the elevation that represent the offsets). Short crosshatch lines may be drawn along the material side of the heavy profile line for more emphasis. This method of sectioning clearly shows the projection and the contour of the elevation surface without the need for a separate view (Fig. 6–16).

6.2.10 Removed Sections

Removed sections are similar to revolved sections; instead of being superimposed on the exterior view, they are removed to a convenient position near the view from which

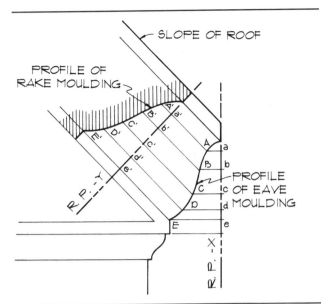

Figure 6–16 Plotting the profile of a rake molding (partial revolved section):

1. Lay out the profile of the horizontal molding and the slope of the roof with a side view.
2. Draw the reference plane X (RP–X) vertically and the reference plan Y (RP–Y) perpendicularly to the roof slope.
3. Establish arbitrary points A, B, C, and D. Transfer distances A–a, B–b, and so on, to RP–Y on the projection lines.
4. Connect points A′,B′, C′, and so on, with an irregular curve to form the profile of the rake mold.

they are taken, and a cutting-plane line is used to show their origin. Identification letters (Section 6.3.2) relate the cutting planes to their sections; several methods can be used (refer to Fig. 10–6). The advantage of the removed section over the revolved section is the possibility of enlargement, if necessary. Small details can be shown and dimensioned more easily at larger scales. If a series of sections relating to a main view is drawn, each section should be arranged in a logical manner and should be located, if possible, near its cutting plane. Whenever workable, a removed section should be drawn in its natural projected position. Sections difficult to orient will create complications in interpretation of a drawing.

6.2.11 Pictorial Sections

Sometimes a pictorial section becomes the most effective method of showing structural assemblies or other construction features that are awkward to explain with orthographic views alone (see Figs. 7–19 through 7–21). Pictorial sections are discussed in Chapter 7.

6.3

AUXILIARY VIEWS

As previously mentioned in the discussion of orthographic projection, the principal surfaces of a rectangular object, such as our miniature building in Fig. 6–15, are shown in true shape when their features are observed from the surface in a perpendicular manner. That is, the surfaces, or elevations, appear in true shape if their projection planes are parallel to each surface. However, when the sloping roof appears on the front and rear elevations, the view of the roof is not a true shape because the surfaces are inclined to the front and rear projection planes (Fig. 6–6). The top view also shows the roof surfaces distorted because of their inclination to the horizontal projection plane (Fig. 6–17).

Usually, inclined roof surfaces do not require true-shape representation; therefore, principal elevations are adequate. However, suppose that we have a building with a wing, or part of an elevation, extending in a nonparallel direction from the other elevations. In contemporary architecture, many buildings have oblique wings or nonparallel sides; cabinets are designed at angles other than right angles; and structural members have elevations and surfaces that are sometimes inclined to the principal planes. These inclined surfaces would appear distorted on the principal views or elevations. Occasionally, several important surfaces are inclined and often must be shown in true shape on working drawings (Fig. 6–18). To reveal the true shape of an inclined surface, an *auxiliary view* is necessary.

To draw an auxiliary view, assume an auxiliary plane *parallel* to the inclined surface; then project the features of the surface to the auxiliary plane with parallel projectors that are perpendicular to the surface. The direction of observation also becomes perpendicular to the inclined

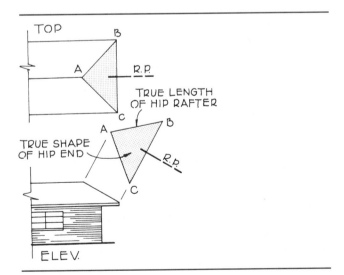

Figure 6–17 Drawing the true length of a hip rafter.

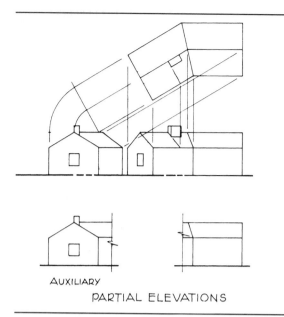

Figure 6-18 Auxiliary elevation views.

surface and the auxiliary plane. In most respects, an auxiliary view is similar to an orthographic view, except that the plane to which it is projected is *inclined* from the principal orthographic planes of projection. It is also convenient to think of the auxiliary plane as being hinged to the (profile) plane from which it is developed. *Reference planes*, one of which appears as a line on a principal view and the other as a line parallel to the auxiliary projection plane, are assumed to facilitate the construction of the auxiliary view. An auxiliary view, then, is a view that has been projected to any plane *other than* one of the six orthographic planes and that has been projected from a surface that is parallel to the auxiliary plane.

6.3.1 An Auxiliary View of an Irregular Object

The step-by-step construction shown in Fig. 6–19A and B is meant to simplify the procedure for developing a typical auxiliary view. The steps are as follows:

Step 1: Draw two views of the object, one showing the profile of the auxiliary surface and one showing the top and depth characteristics. Indicate the direction from which the auxiliary is to be taken.

Step 2: Assume a reference line on the top view, from which depths of the auxiliary are taken. A reference plane is established in back of this object for convenience, and it will show as a line on both the top view and the auxiliary view. (On symmetrical objects, the reference plane is best established on the center line.) Draw the reference line (a heavy dash and double-dot line) on the top view.

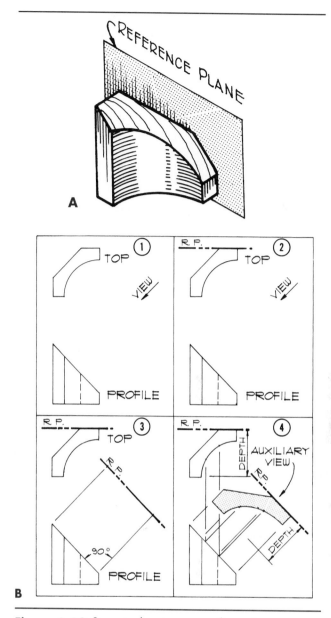

Figure 6-19 Steps in drawing an auxiliary view.

Step 3: Draw the auxiliary reference line parallel to the inclined surface of the front view. Be sure that it is located far enough away from the front view to allow enough space for the auxiliary view.

Step 4: Draw projection lines from the front view perpendicular to the reference line, as shown in Fig. 6–19B. Transfer the measurements of the various necessary points of the top view to the corresponding features of the auxiliary view, with the dividers or by measurement. Notice that the auxiliary view will form between the auxiliary reference plane and the front view, because the reference plane has been assumed to be in back of the object on the top view. You can plot irregular or curved surfaces accurately by projecting as many points as needed along the curved surface of the top view to the inclined surface of the profile view and then to the auxiliary reference line as shown. Depth measurements are transferred in the same way that other features are transferred. Use an irregular plastic curve to connect the plotted points with a smooth line.

Step 5: Connect all of the proper points on the auxiliary surface with lines, noting the visibility of each. Strengthen the lines. Usually, only a partial auxiliary is necessary, but the complete object as viewed from the auxiliary plane can be drawn for graphic experience.

6.3.2 Partial Auxiliary Views

The drafter will encounter auxiliary views mainly in drawing elevations of buildings whose exterior walls are nonparallel to the principal planes (Fig. 6–18). If these elevations have important information and need to be shown in true shape, a partial auxiliary elevation is drawn. If the complete elevation is drawn on the auxiliary plane, a portion of the elevation will appear distorted and will therefore have little pictorial value. Generally, this distorted portion is not drawn; the same holds true for the inclined surface in the principal elevation. The termination between the true shape and the distorted shape in either elevation is shown with a conventional break line.

To draw a *partial auxiliary elevation*, assume an observation direction perpendicular to the auxiliary surface. (The projection plane then becomes parallel to the auxiliary surface, as previously mentioned.) It is preferable to draw the auxiliary view in a natural, horizontal position, rather than in its projection position. Show all features, such as windows and doors, in true shape and size and all horizontal dimensions in true length. These fea-

tures can be projected from the plan view, which has been fastened above the drawing and inverted in a position to make the auxiliary wall horizontal. Take the heights from the other main elevations, or assume a horizontal reference plane at the ground line of the other elevations, and transfer the heights with the dividers or by measurement.

6.3.3 Inclined Oblique Auxiliary Views (Secondary Auxiliary Views)

If an inclined surface cannot be seen as a line in any of the six principal views and is not perpendicular to any of the six principal planes, its true shape must be constructed by the use of a secondary auxiliary, rather than a direct-projection method. Such views are rarely required, yet the drafter should be able to develop inclined oblique surfaces, principally to determine the true length of structural members that may have to be placed in an inclined oblique position within a building.

A convenient method is shown in Fig. 6–20. *To construct the true shape* of the shaded portion of the roof, first draw a primary auxiliary (1) projected from the top view. This auxiliary view is needed first because it shows the *edge view* of the shaded roof; the required surface is represented as a line (X–Y). Remember that auxiliary views of inclined surfaces are projected from views that show the inclined surface as a line. From the primary auxiliary view, construct parallel projectors that are perpendicular to the edge view line X–Y; these projectors form the true width of the shaded surface. For convenience, place the reference plane on the end of the oblique wing of the top view, and on the secondary auxiliary view draw it parallel to line X–Y of the primary auxiliary. Transfer the true lengths of the shaded surface on the top view of the second auxiliary view (2).

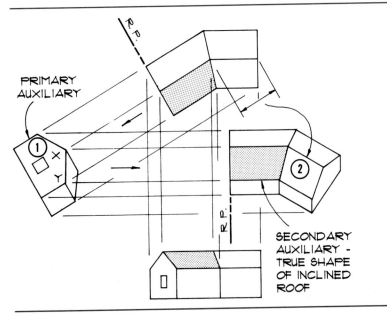

Figure 6–20 Drawing the true shape of an inclined oblique plane (secondary auxiliary).

Connect all visible points of intersection. The entire house can be drawn by projecting all points from the primary auxiliary view and by transferring lengths of all points from the reference line of the top view to the corresponding projection lines of the second auxiliary.

GEOMETRIC CONSTRUCTION

For centuries architects have relied on the principles of plane and solid geometry to design two- and three-dimensional elements in the built environment. Most structures are composed of squares, rectangles, triangles, circles, arcs, polygons, pyramids, cones, cylinders, and so on (see Fig. 8–30). It is essential that the drafter become proficient in geometric constructions. Such knowledge will enable the drafter to bisect lines and angles, divide lines into equal parts, draw arcs and circles, construct regular and irregular polygons, and so on. Once these skills have been mastered, the drafter should be able to develop plans, elevations, sections, details, and pictorial drawings with greater ease.

The parallel bar, triangles, compass, and irregular curve are the tools required for beginning geometric construction exercises. To gain accuracy in drafting, use a well-sharpened 3H or 4H pencil. Often the locations of center points from which arcs or circles are drawn become the important steps in the construction. Place these carefully.

6.4.1 How to Draw a Line Through Two Points Fig. 6–21)

After the two points have been established, hold the pencil firmly on either point, and pivot the triangle edge against the anchored pencil until the other point is aligned. Draw a line between the two points with the pencil. Hold the triangle firmly during this exercise.

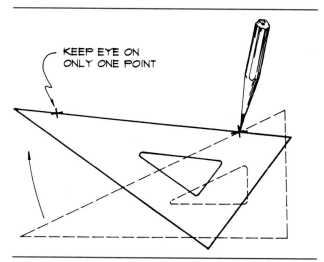

Figure 6–21 Convenient method of drawing a line through two points.

6.4.2 How to Divide a Line into Equal Divisions

First Method   Given line *AB* to be divided into, for example, five equal parts, draw line *BC* at any convenient angle, and, along it, measure or step off five equal divisions (Fig. 6–22A). Any suitable increment can be measured with the scale or stepped off with the dividers. Connect point *A* with the fifth point on line *BC*, and transfer the remaining points to line *AB* with parallel connectors.

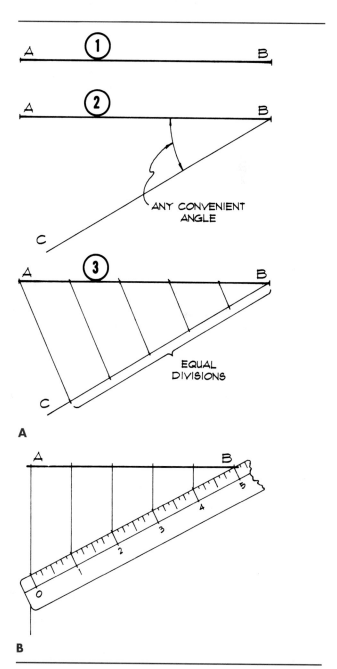

Figure 6–22 (A) Dividing a line into equal divisions (first method). (B) Dividing a line into equal divisions (alternate method).

Alternate Method   Draw a perpendicular from line *AB* as shown in Fig. 6–22B, and place the scale between point *B* and the perpendicular. If five divisions are required, select five divisions on the scale, and pivot it from point *B* and the perpendicular until the end of the five divisions coincides with point *B* and the other with the perpendicular. (Slightly larger divisions on the scale must be used than the intended divisions on line *AB*.) Indicate the remaining divisions along the scale, and transfer them with perpendiculars to line *AB*. Any number of divisions can be made, and considerable latitude in choice of increment size on the scale can be employed. However, the smaller the angle between the line to be divided and the edge of the scale, the more accurate the divisions will be.

Division by Trial and Error   Many drafters merely step off the divisions with the use of dividers, if the number of divisions is not excessive. Several trials usually result in an accurate division.

6.4.3 How to Bisect a Line with a Perpendicular Bisector

With the Compass   Given line *AB*, with the compass set larger than half the length of the line, draw the same arc from both ends (Fig. 6–23A). A line passing through both intersections of the arcs will divide line *AB* with a perpendicular bisector.

With the T Square and a Triangle   Given line *AB*, from both ends draw equal angles, as shown in Fig. 6–23B, with the triangle. A perpendicular from line *AB* passing through the intersection of angular lines will bisect line *AB* with a perpendicular bisector.

6.4.4 How to Draw an Arc Tangent to the Lines of a Right Angle (Fig. 6–24)

Given the right angle *CAB*, with its center at corner *A*, draw the arc with given radius *R*, intersecting *AB* at *S* and *AC* at S_1. Then, using points *S* and S_1 as centers, construct arcs of the same radius. Their intersection at *P* forms the center for the given radius arc *R*, which will be tangent to the lines of the right angle.

6.4.5 How to Draw an Arc Tangent to Two Straight Lines That Are Not Perpendicular (Fig. 6–25)

Given lines *AB* and *CD* forming an acute angle as shown in Fig. 6–25 (use the same procedure for lines forming an obtuse angle), set the compass for the given radius, and draw arcs *R* at random points on each line. Locate the intersection *O* by drawing light lines through the extremities of the arcs. The lines will be parallel to the given lines *AB* and *CD* and will be equidistant.

6.4.6 How to Draw an Arc of Any Radius *R* Tangent to a Given Circle and a Straight Line (Fig. 6–26)

Let R_1 be the radius of the given circle or arc and *XY* be the given line. Draw line *ST* parallel to *XY* at *R* distance away as shown in Fig. 6–26. Using *P* as the center and the sum $R_1 + R$ as the radius, construct an arc intersecting line *ST* at point *O*. The arc of *R* radius drawn from point *O* will be tangent to the given circle and line *XY*.

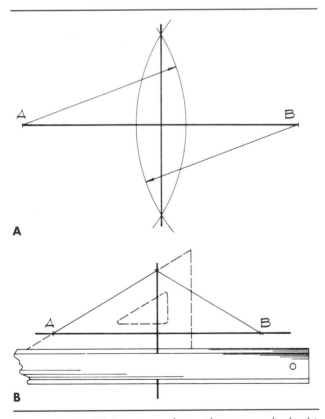

Figure 6–23 (A) Bisecting a line with a perpendicular bisector. (B) Bisecting a line with a triangle.

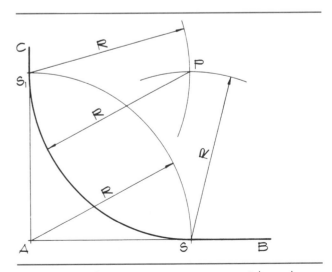

Figure 6–24 Drawing an arc tangent to a right angle.

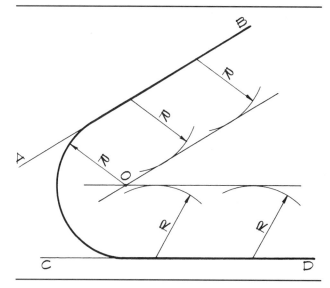

Figure 6–25 Drawing an arc tangent to two straight lines (the arc center must be equidistant from both lines).

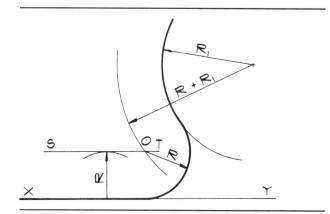

Figure 6–26 Drawing an arc tangent to a circle and a straight line.

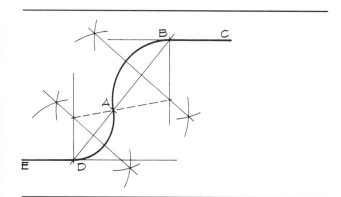

Figure 6–27 Drawing an ogee curve.

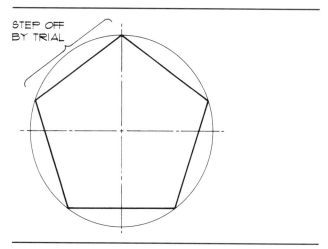

Figure 6–28 Drawing a regular pentagon.

6.4.7 How to Draw an Ogee or a Reverse Curve (Fig. 6–27)

Given two parallel lines *BC* and *ED*, connect *B* and *D* with a line, and upon it indicate the reversal point *A*. (The placement of point *A* is arbitrary, depending on the desired nature of the ogee curve.) Erect perpendiculars from line *BC* at *B* and from line *ED* at *D*. Construct perpendicular bisectors *BA* and *AD*. The intersections of the bisectors and the previous perpendiculars are the centers for the required arcs. Set the compass at the required radii, established graphically by the perpendiculars. A line intersecting both centers of the arcs will pass through point *A*, revealing the accuracy of the construction and the tangent point of the two arcs.

6.4.8 How to Draw a Regular Pentagon (Fig. 6–28)

Given the size of the circle within which the pentagon is to be inscribed, draw the circle, divide its circumference into five equal parts by trials with the dividers, and connect the points. The points on the circle can also be located with the protractor: Step off five 72° angles (360° ÷ 5).

6.4.9 How to Draw a Regular Hexagon

When the Distance Across Flats Is Given   The flats are the parallel sides (Fig. 6–29A). A hexagon can be dimensioned either across its parallel sides or across its corners. Construct a circle with a diameter of the given distance across flats. Circumscribe a hexagon around the circle that is tangent to the center of each side using the 30°–60° triangle and a T square.

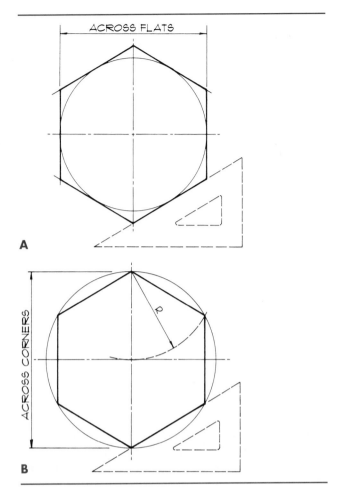

A

B

Figure 6–29 (A) Drawing a hexagon with the distance across flats given. (B) Drawing a hexagon with the distance across corners given.

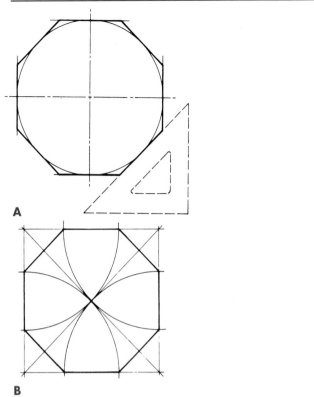

A

B

Figure 6–30 (A) Drawing an octagon with a circle. (B) Drawing an octagon with a square.

When the Distance Across Corners Is Given   Construct center lines, and draw a circle of the diameter required across corners of the hexagon (Fig. 6–29B). Using the 30°–60° triangle and a T square, inscribe the hexagon within the circle starting at an intersection of one center line and the circle. The radius of the given circle will be the length of each side.

6.4.10 How to Draw a Regular Octagon

Construction Around a Circle   Given the distance across flats (between parallel sides), draw center lines, and construct a circle of the given diameter (Fig. 6–30A). Using a 45° triangle and a T square, draw the sides of the octagon tangent to the circle as indicated.

Construction Within a Square   Draw a square with sides equal to the required distance across flats (Fig. 6–30B). Draw diagonals, and, using half the length of the diagonal as a radius and the corners of the square as centers, draw the arcs as shown. Connect the intersections of the arcs and sides to form the octagon.

6.4.11 How to Construct a Regular Polygon with Any Number of Sides (Fig. 6–31)

Given the length of one side, let, for example, seven sides be required. From any point *A*, draw the semicircle *CDB* using *AB* as the radius. By trial with the dividers, divide the semicircle into seven equal parts; *AD* will become the second side. Bisect side *AB* and *AD* with perpendicular bisectors to locate point *O*, which will be the center of the polygon. Draw the circle through points *A*, *B*, and *D*. To locate the remaining vertices, step off chord *AB* along the remainder of the circle, or extend lines *A*3, *A*4, and so on, to the circle.

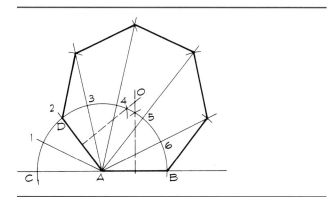

Figure 6-31 Drawing a polygon.

6.4.12 How to Draw an Ellipse by the Concentric-Circle Method (Fig. 6-32)

Given the major axis AB and the perpendicular minor axis CD, draw concentric circles of both diameters from point O. Divide each quadrant into an arbitrary number of angular dimensions; the example in Fig. 6-32 shows four divisions in each quadrant that have been conveniently laid out with the triangles—more accuracy can be obtained by more divisions. From the intersections of the division lines and the outer circumference, draw vertical lines parallel to CD. From the intersections of the angular division lines and the inner circle, draw horizontal

lines parallel to AB as indicated. The intersections of these perpendicular construction lines produce points from which the ellipse can be constructed. To complete the drawing, smoothly connect the points with the use of the irregular curve. If all of the points in one quadrant of the ellipse can be aligned with one setting of the irregular curve, you can complete the other quadrants by merely marking the curve and tipping it over for the remaining curved lines. Elliptical arches can be quickly drawn with the foregoing construction.

6.4.13 How to Draw an Ellipse by the Parallelogram Method (Fig. 6-33)

Given the major axis AB and the minor axis DE, from the axes draw a rectangle or a parallelogram with sides equal to the axes. Divide AO and AC into an equal number of parts. Draw light lines from point E through the AO divisions, and from point D through the AC divisions. Intersections of similar numbered lines will create points along the ellipse. Complete the remaining three quadrants by the same procedure, and draw the ellipse with an irregular curve.

6.4.14 How to Draw an Approximate Ellipse by the Four-Center Method (Fig. 6-34)

Given the axes AB and CD, connect points A and D with a diagonal line. With AO as the radius, draw the arc AA_1 using O as the center. With DA_1 as the radius, draw the arc FA_1 using D as the center. Bisect line AF with a perpendicular bisector as shown in Fig. 6-34; this locates points G and H, which become the centers for the required arcs. Points G_1 and H_1 are located equidistant from O as points G and H, respectively. The distance GO

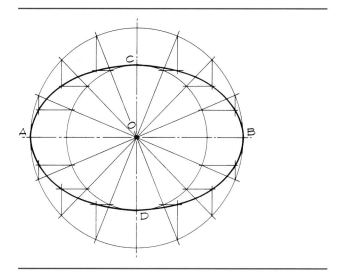

Figure 6-32 Drawing an ellipse by the concentric-circle method.

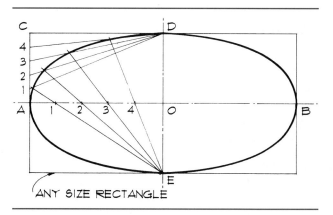

Figure 6-33 Drawing a parallelogram ellipse.

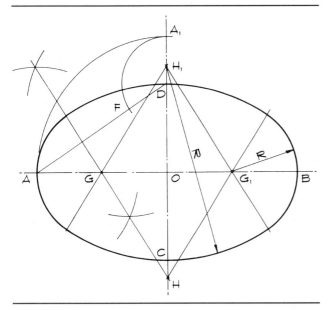

Figure 6–34 Drawing a four-center approximate ellipse.

equals OG_1, and HO equals OH_1. Draw the radius AG arc from points G and G_1 and the radius HD arc from points H and H_1 to complete the ellipse. Notice that an ellipse drawn by the four-center method is not as graceful as the previous ellipses.

6.4.15 How to Draw a Parabola Within a Rectangle (Fig. 6–35)

Given a rectangle $CDEF$ as shown in Fig. 6–35, the dimensions shown are the usual ones, indicating the width and the depth (or span and rise) of the needed parabola. Divide AC and CD into the same number of parts as shown. The intersections of like-numbered lines become

points on the parabola. Notice that one set of lines is parallel and the other radiating. Connect the points with the use of an irregular curve. The parabola is finding use in various contemporary structures.

6.4.16 How to Enlarge or Reduce an Irregular Drawing (Fig. 6–36)

Enlargement   If a drawing is to be enlarged, first lay out a uniform grid field over the figure as shown in Fig. 6–36A. Use a grid size convenient for enlargement. For example, if a drawing is to be enlarged five times or 500%, then it would be appropriate to select a spacing for the grid lines of the smaller field that could be conveniently increased five times for the larger grid. Next, lay out a larger, compatible grid field, using the same number of grid lines as in the smaller field. For convenience, number the corresponding grid spaces in both fields as shown. The detail within each small grid space can then be enlarged within the corresponding larger grid block. Complete one block at a time until the entire figure is enlarged.

Reduction   To reduce the size of a drawing, use the opposite procedure. Take the detail from the large grid field, and reduce it to the corresponding smaller one.

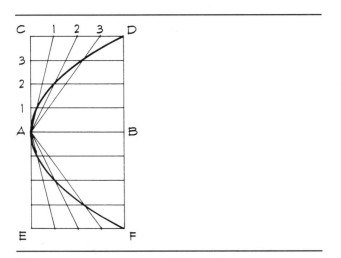

Figure 6–35 Drawing a parabola within a rectangle using equal increments.

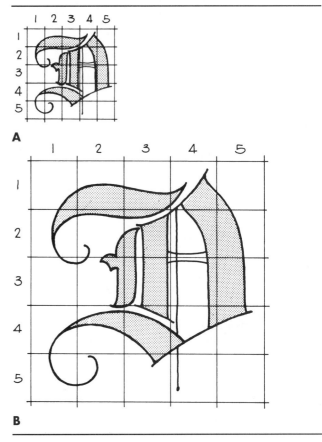

Figure 6–36 Enlarging or reducing a figure.

6.4.17 Esthetic Proportions Through Geometry

A number of interesting geometric proportioning methods have appeared from time to time in the history of architecture and art. Some have withstood the test of time and should be accepted for their esthetic worth, even by modern standards. Not only do they provide the student of drawing with valuable construction exercise, but they also indicate an application of geometry to the esthetics of structures.

The Golden Section This section was originally a Greek method of dividing lines and form into pleasing proportions (Fig. 6–37A). Mathematically, its ratio may be expressed as follows: $(\sqrt{5} + 1)/2 = 1.618$. The ratio of one division to the other, then, can be stated as $1:1.618$. Given line AE, to be divided according to the Golden Section, first bisect the line at point C. Erect a perpendicular to line AE at point E, equal in length to CE. On the hypotenuse (AD) of the triangle, locate point B so that BD equals DE. On the original line AE, locate point F so that AF equals AB. F divides AE according to the proportions of the Golden Section.

The Golden Rectangle The Golden Section provides the stable proportions for the Golden Rectangle (Fig. 6–37B). The long side of the rectangle is in proportion to the short side as AF is to FE in Fig. 6–37A. For convenience in drawing a predetermined size of the rectangle, the numerical proportion $76:47$ units will provide a satisfactory method of scaling. As a matter of interest, the short side of the rectangle can be laid out within it to form a square, and the remaining space will always be another Golden Rectangle (see the dashed line).

Construction of the Basic Triangle for a Classic Pediment (Vignola) Given the width of the base AB, bisect line AB with a center line intersecting it at C (Fig. 6–37C). With AC (one-half of AB) as the radius, construct the semicircle AEB. With the radius EB, construct the arc AGB from point E. Connect points AG and GB.

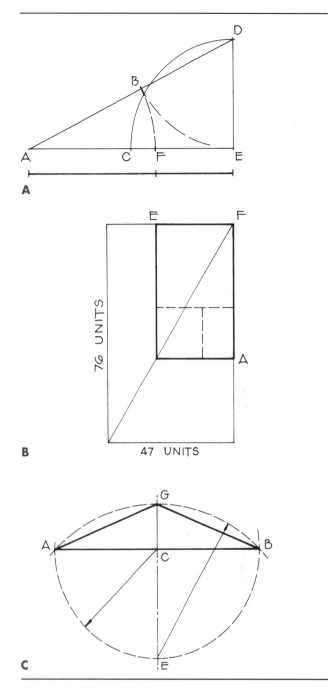

Figure 6–37 (A) The Golden Section. (B) The Golden Rectangle, made from the Golden Section proportions. (C) The classic pediment triangle (Vignola).

EXERCISES
GEOMETRIC CONSTRUCTION

On $8\frac{1}{2}'' \times 11''$ paper, lay out a $\frac{1}{2}''$ border, and divide the area into nine equal spaces similar to those shown in Fig. 6–38. With the following information, complete the exercises as shown, matching the numbers in the figure to the exercise numbers.

1. Construct a $1''$ arc tangent to the given right angle.

2. Find the center of the given circle using chord perpendicular bisectors.

3. Construct a $\frac{3}{4}''$ arc tangent to the given arc and straight line.

4. Divide line AB into five equal divisions.

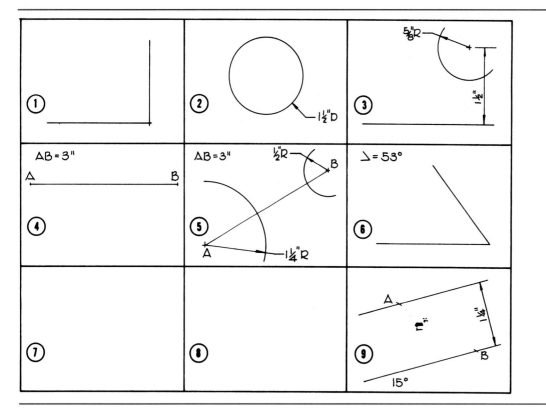

Figure 6–38 Exercises in geometric construction.

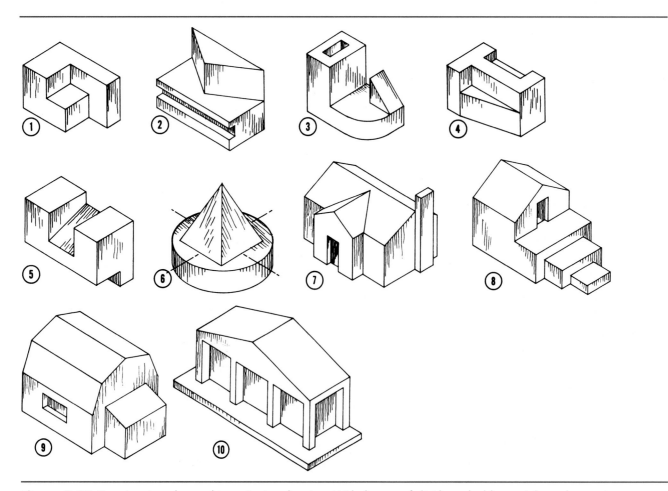

Figure 6–39 Exercises in orthographic projection drawing. With the use of dividers, double or triple each exercise.

5. Construct a ⁷/₈″ arc tangent to the given circle arcs.

6. With the compass, bisect the given angle.

7. Construct a hexagon 1¹/₂″ across corners.

8. Construct a 1³/₈″ octagon.

9. Construct an ogee curve from point *A* to point *B*, passing through point *E* (*EB = 2AE*). The line passing through *A* is parallel to the line passing through *B*.

ORTHOGRAPHIC PROJECTION

Study the pictorial drawings in Fig. 6–39, and develop the orthographic views of each problem. The exercises are meant to provide you with experience in understanding projection principles. Upon their completion, your instructor should examine them to ensure proper interpretation.

AUXILIARY VIEWS

1. Construct the true shape of the end of the hip roof shown in Fig. 6–17 (surface *ABC*). Select a convenient scale to fit your paper.

2. Construct the true shape of 45° miter cut from wood profiles (Fig. 6–40):
   a. 1¹/₄″ dowel
   b. 2″ × 2″ molding
   c. 3″ × 3″ ogee molding

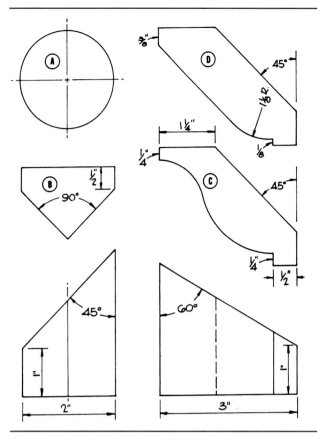

Figure 6–40 Exercises in auxiliary projection.

REVIEW QUESTIONS

1. Describe the term *orthographic projection*.

2. What is a *frontal* plane?

3. Why do drawings of many objects require more than one orthographic view?

4. Which orthographic view is considered the major view?

5. How are hidden surfaces shown on an orthographic view?

6. Why are orthographic views of a building often drawn in an unrelated position on working drawings?

7. What is the *cutting plane* of a section?

8. Why are cutting planes sometimes offset rather than straight planes?

9. Why are hidden lines seldom used in section views?

10. What is the advantage of a partial section?

11. Why are auxiliary surfaces projected to planes that are parallel to the surfaces?

12. How are numerous sections identified as to their position on the major view?

13. Describe a *transverse section*.

14. Describe a *longitudinal section*.

15. Define the term *poché*.

16. How many orthographic views can be made from a cube?

17. What two views of a house could be revealed if a horizontal cutting plane passed midway between the floor and the ceiling and completely bisected the house?

18. Why is it important to draw a vertical wall section from the footing to the roof of a house?

19. Why are pictorial sections used?

20. What is the *Golden Section?*

W. D. FARMER

"Step after step the ladder is ascended."
—GEORGE HERBERT

Pictorial or *paraline* drawings enable the drafter to create single-view images that represent three-dimensional space. Since we view the objects and the space that surround us in a three-dimensional way, it becomes easier for us to comprehend our three-dimensional environment when *pictorial* drawings are used (Fig. 7–1.) The drafter has the option to choose the three-dimensional view that he or she wishes to draw from a family of pictorial techniques.

Axonometric and oblique drawings are pictorial in nature, yet they are compatible with orthographic drawings in that three sides are shown and the principal lines of each side or plane are measured directly. As mentioned previously, orthographic views have certain advantages, but they also have limitations. For example, two or three isolated views are required to present the information clearly (Fig. 7–2A). This information must be shown only on the orthographic planes; visualizing their features makes training in technical graphics a prerequisite. Even trained persons find that orthographic views are not always entirely adequate for representing construction in the best way. On the other hand, the average person finds a pictorial drawing easy to interpret because three surfaces are combined into one realistic view.

To represent an object pictorially so that the height, width, and depth can be observed in one composite representation, the drafter turns the object to make three sides visible (*axonometric*); or the drafter shows the top and side by projecting oblique lines from a frontal orthographic view (*oblique*). An example of each is shown in Fig. 7–2B. The sides of an object on either type of drawing are represented with parallel lines, conveniently executed with parallel bar and triangles, and sometimes collectively termed *paraline* drawings.

Figure 7–1 Pictorial drawing.

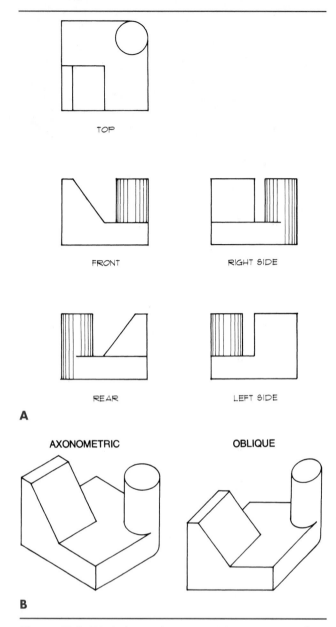

Figure 7–2 (A) Orthographic views. (B) Comparison of axonometric and oblique pictorial drawings.

Because these pictorial views are easy to draw with instruments, they become suitable for use on working drawings and are satisfactory for occasional design drawings. They are also handy for quick freehand sketches (see Chapter 8)—for analyzing construction details or for graphic communication between coworkers. Therefore, drafters should be able to draw and interpret the conventional pictorial drawings covered in this chapter. Other types of technical-pictorial drawings have been devised; however, they are more time-consuming, and they do not find application in architectural work. Axonometric and oblique drawings, used intelligently, will be found entirely adequate for presenting supplementary construction information whenever needed.

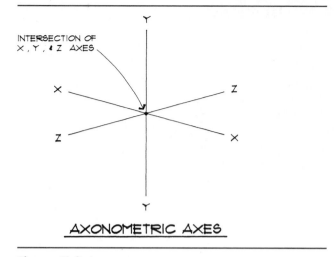

Figure 7–3 Axonometric axes.

<h3>7.1</h3>

AXONOMETRIC DRAWINGS

An *axonometric* is an accurately scaled drawing that depicts an object that has been rotated on its axes and inclined from a regular parallel position to give it a three-dimensional appearance. The word is derived from the Greek term *axon*, meaning "axis," and the French term *metrique*, meaning "pertaining to measurement." Three systems of the family of axonometrics are commonly used: isometric, dimetric, and trimetric. (Some other variations of axonometrics will be described later in this chapter.) Graphically, the three axes X, Y, and Z are used to create the illusion of three-dimensional space on a two-dimensional plane (Fig. 7–3).

A characteristic of axonometric axes (assuming that the Y axis is vertical) is that neither the angle formed between the X and Y axes nor the angle formed between the Y and Z axes may be drawn at 90°, although each of these referenced angles could represent 90° in space (Fig. 7–4).

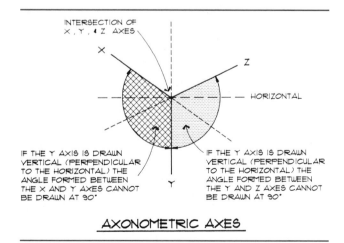

Figure 7–4 Axonometric drawing.

7.1.1 Isometric

When the rotation and the inclination of the planes result in a drawing with three equally divided angles about a center point, the principle of construction is known as *isometric* (the prefix *iso*, from the Greek word *isos*, means "equal"). The axes may be turned in several different positions, but the most pleasing and generally the most convenient position shows one axis vertical and the other two 30° above the horizontal. The three angles created by the intersection of the *X*, *Y*, and *Z* axes are all equal, measuring 120° each (Fig. 7–5). In Fig. 7–6, the concrete masonry unit (C.M.U.) with equal exterior dimensions clearly shows the uniformity of the angles and sides. Regardless of the position of the axes, the three faces of the C.M.U. are equally foreshortened; intersecting surfaces produce 120° angles (Fig. 7–7), and similar intersections of the surfaces produce parallel lines. Because the lines that recede from the viewer must be foreshortened in a true perspective, making them difficult to measure, the drafter merely uses scaled lengths for convenience and speed. For example, if the actual size of the C.M.U. in Fig. 7–8 is 8″ × 8″ × 8″, then the lines representing the surfaces are scaled to the same measurement. No allowance is made for foreshortening, but all measurements must be made on the axes lines. The front, top, and side views of the block are given equal importance, and they can be easily drawn with the 30°–60° triangle. If it becomes necessary to show the underside of an object (so that it appears to be observed from below), the drafter employs a *reverse axis* (Fig. 7–9). This axis position is more appropriate when drawing a high cornice view or an overhead structural feature. The object appears to have been tilted backward instead of forward, as in the conventional position. Care must be taken to keep the major planes of the reverse axis clearly in mind.

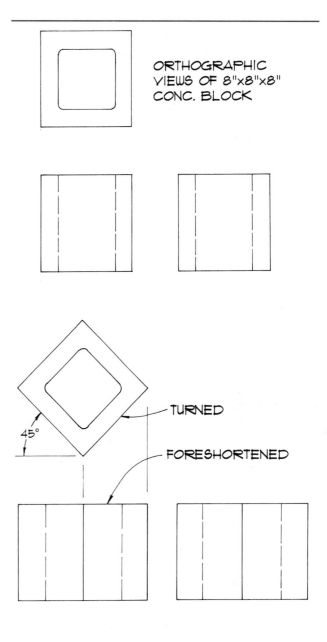

ORTHOGRAPHIC VIEWS OF 8″x8″x8″ CONC. BLOCK

TURNED

FORESHORTENED

45°

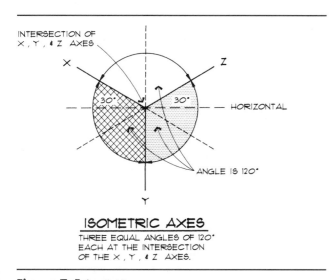

INTERSECTION OF X, Y, & Z AXES

X

Z

30°   30°   HORIZONTAL

ANGLE IS 120°

Y

ISOMETRIC AXES

THREE EQUAL ANGLES OF 120° EACH AT THE INTERSECTION OF THE X, Y, & Z AXES.

Figure 7–5 Isometric axes.

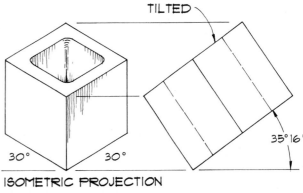

TILTED

30°   30°

35° 16′

ISOMETRIC PROJECTION

Figure 7-6 Method of developing a true isometric projection, which is time-consuming for the drafter.

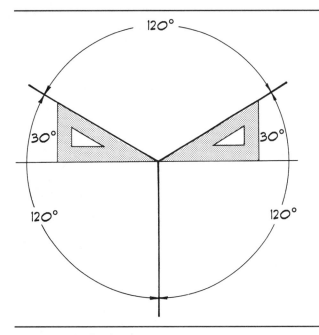

Figure 7-7 The three isometric axes

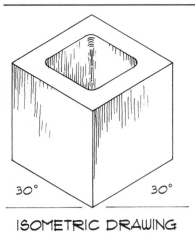

ISOMETRIC DRAWING

ALL LINES ON AXIS
ARE TRUE LENGTH

Figure 7-8 The same block drawn with lines on the axis the same length as in the orthographic views. An isometric drawing is about 22% larger than an isometric projection, yet it has equal pictorial value and is simpler to draw.

7.1.2 Dimetric

When two equally divided angles are created that are not 90° or 120° at the intersection of the *X*, *Y*, and *Z* axes, the principle of projection is known as *dimetric* (the prefix *di*, in Greek, means "twice" or "double"). The most popular adaptation of this principle makes the angle between the *X* and *Y* axes and the angle between the *Y* and *Z* axes equal and greater than 90° but less than 120° (Fig.

Figure 7-9 An isometric drawing exposing the underside.

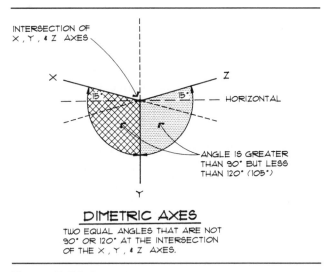

DIMETRIC AXES
TWO EQUAL ANGLES THAT ARE NOT
90° OR 120° AT THE INTERSECTION
OF THE X, Y, & Z AXES.

Figure 7-10 Dimetric axes.

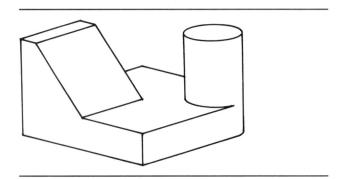

Figure 7-11 Dimetric drawing.

7–10). Important surfaces can be effectively illustrated by this principle. The drawing in Fig. 7–11 demonstrates this rotation. The two surfaces in the foreground take prominence over the top surface.

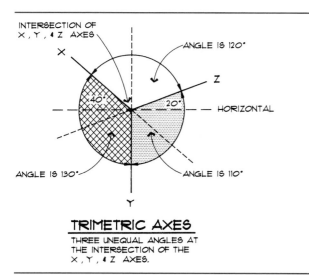

Figure 7-12 Trimetric axes.

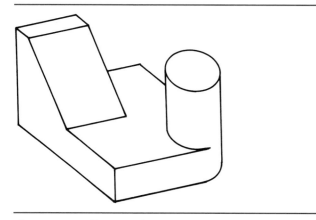

Figure 7-13 Trimetric drawing.

7.1.3 Trimetric

When the intersecting *X, Y,* and *Z* axes form three unequal angles, the system of construction is known as a *trimetric* (the prefix *tri* comes from Greek and means "three"). No two angles can be 90° angles, and the sum of all three angles must equal 360°. The wide choice of angle values gives the drafter flexibility and control of the pictorial view. Greater emphasis can be placed on a single plane in a trimetric view (Fig. 7–12). The right foreground plane of the object in Fig. 7–13 takes precedence over the left side and the top.

7.2
CONSTRUCTING A PICTORIAL DRAWING

To construct a pictorial drawing, first consider the most advantageous angle from which to view the object. Then adopt the view that will show the important information on the three planes, even if it requires turning the object around. Avoid the need for hidden lines; they are usually omitted on paraline drawings. Then look for the major

geometric forms. Box these forms in by drawing the correct axes lines, measuring the lines, and completing the boxed-in effect by drawing the surrounding parallel lines (Fig. 7–14B). This becomes a framework (very much like

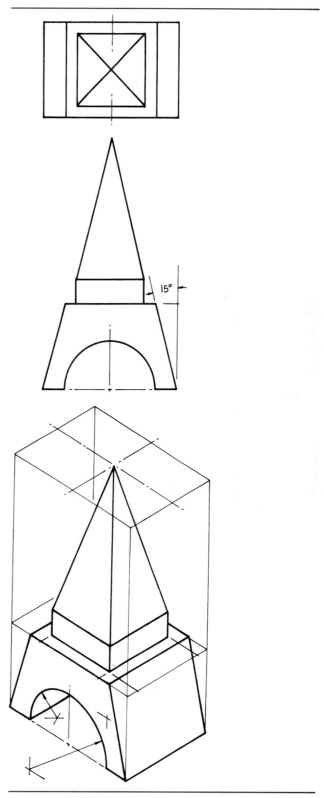

Figure 7-14 Constructing an isometric drawing within a box framework with axes distances taken from the orthographic views.

a cage) from which or to which minor shapes or features can be taken away or added. The preliminary box construction aids in placement of the drawing: If the view seems to be developing off to one side, the construction lines can be changed quickly, and little time is lost. The basic enclosure also acts as reference surfaces from which offset features can be measured during the development of the drawing.

If an object similar to that shown in Fig. 7–15 is to be drawn, a block-on-block method of construction may be simpler. The drafter adds each basic shape to another, much as blocks are piled together, rather than drawing a complete cage around the entire area of the object. After drawing the base block using this method, the drafter should take care to locate correctly the position of the succeeding block. Notice that the position of each new block is first laid out on the surface of the previous block. If the adjacency of the blocks falls on a hidden surface, often the features of the hidden surface must be constructed, even though they will not show and must be erased later. The block-on-block method usually requires

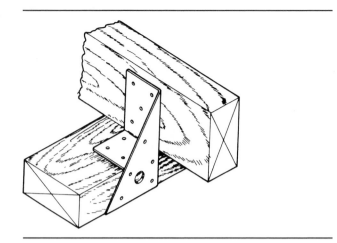

Figure 7–16 Isometric drawing of a metal framing anchor connection.

less construction linework than the boxed-in method (Fig. 7–16).

7.2.1 Angular Shapes

Angles specified in degrees do not appear in their shape on a three-dimensional drawing. The construction of an angular shape generally requires the use of coordinates from an orthographic view to determine the angular extremities (Fig. 7–14A). The coordinate lengths are then transferred to the corresponding three-dimensional planes.

7.2.2 Irregular Curved Shapes

In three-dimensional drawings, irregular curved shapes are executed with the use of offsets (Fig. 7–17). The drafter takes points from an orthographic view and transfers them to the framework of the three-dimensional projection. Profile sections can be used, or after the points are located on the drawing, they can be connected with curved lines to produce the pictorial feature. Coordinate distances must be obtained from orthographic views of the same scale, and they are projected on the three-dimensional planes.

7.2.3 Circles

Circles also appear distorted on three-dimensional drawings. A true circle becomes an ellipse. Other than construction with an elliptical template, satisfactory circles are generally constructed by the *four-center method* (Fig. 7–18). The following procedure is used for drawing a three-dimensional circle:

1. Construct center lines to locate it.
2. Construct a square (rhombus) the size of the required diameter.
3. From corners *A* and *B*, construct lines to the midpoint of the opposite sides of the square, which become perpendiculars of the sides.

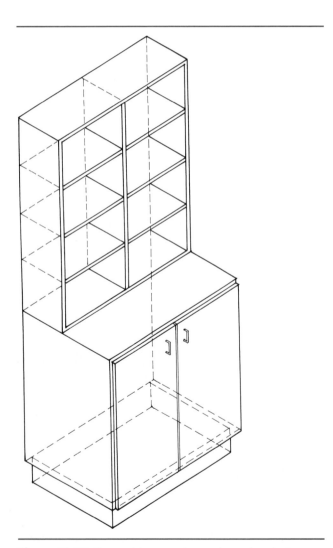

Figure 7–15 Some objects are better drawn by placing one block on the other.

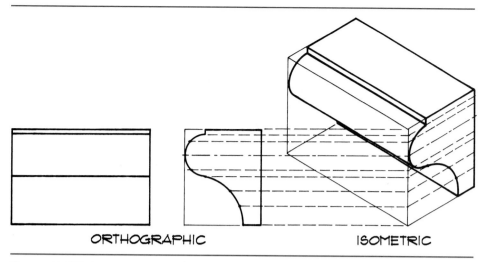

ORTHOGRAPHIC   ISOMETRIC

Figure 7-17 Constructing irregular curved shapes from orthographic views.

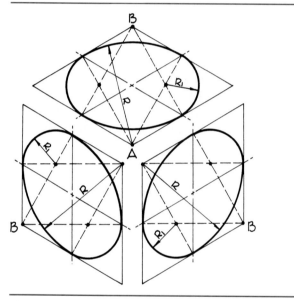

Figure 7-18 The four-center method of drawing isometric circles on the three axes.

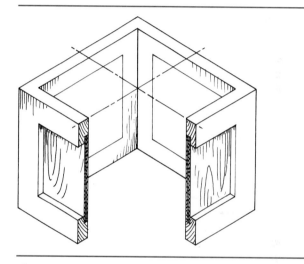

Figure 7-19 Isometric half-section.

4. The intersections of these construction lines become the centers of the small arcs; draw the arcs using radius R_1.

5. Points A and B become centers of the large arcs; draw both arcs of R radius, taking the size of R from the construction.

All of the construction is done with the use of a parallel bar, a 30°–60° triangle, and a compass. If an elliptical template is available, first the center lines are drawn; then the template is lined up accurately on the center lines. Circles and arcs on the three planes can be drawn by either method. If the circles fall on nonaxonometric planes, they can be plotted by the coordinate method from an orthographic view (Fig. 7–17).

PICTORIAL SECTION VIEWS

Because hidden lines are seldom used in pictorial drawings, sectioned surfaces showing interior assembly or construction often become necessary. The object is cut on either of the three-dimensional planes. If it is symmetrical, it is cut on center-line planes (Fig. 7–19), and principles similar to orthographic sectioning are followed. On small, symmetrical objects, generally a *half-section* is used (Fig. 7–19). First, the major shapes are drawn lightly; then the profiles of the sectioned surfaces are established, and the remaining lines are completed. The front quadrant of the object is removed to reveal the

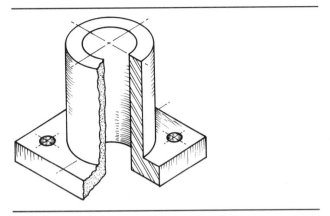

Figure 7-20 Isometric section, partially broken.

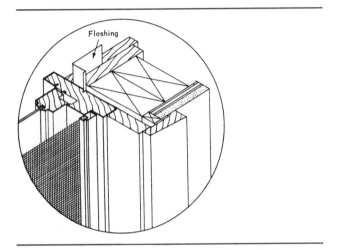

Figure 7-21 Isometric section of a window detail.

interior best. Usually, more of the exterior of a symmetrical object can be shown if a *broken-out* pictorial section is used (Fig. 7–20). The axis cut is made on only one center-line plane, and only enough of the frontal material is removed to reveal the profile of the sectioned surface. The other broken edge of the opening is drawn with a freehand line. Crosshatch lines indicating the material are made less prominent, and their angle should not run parallel to the outline of the section. Diagonals are sufficient to indicate wood structural members (See Figs. 7–21 and 4–9 for building material symbols).

When strengthening the finish lines of pictorial drawings, start with the nearest features, and continue toward the back of the object. This procedure saves time because it minimizes erasing; the result will be a neater piece of work.

7.4

OBLIQUE DRAWINGS

An oblique drawing shows an object with one set of planes in true orthographic projection on the frontal surface. The receding perpendicular sides are represented with parallel lines at various angles, which are projected from the frontal, true shape (Fig. 7–22).

The drawings in Fig. 7–22 resemble isometric drawings in that three sides are shown, and the three sides are represented with parallel lines upon which a drafter can make true measurements. The concrete block (C.M.U.) shown in the illustration has a true square, similar to an orthographic view, as its front surface. The receding top and right sides appear nearly axonometric.

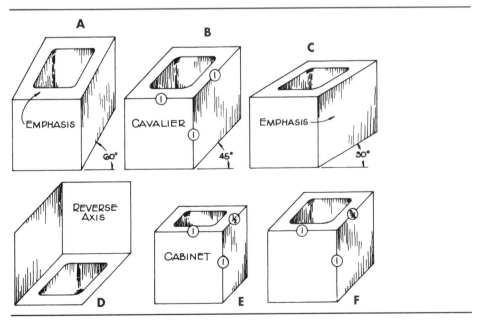

Figure 7-22 Pictorial appearance of various types of oblique drawings.

7.4.1 Various Pictorial Effects

One advantage of oblique drawings is their versatility in creating pictorial effects. By varying the angles of the receding axes, the drafter can emphasize either one of the receding sides (Fig. 7–22A, B, and C). Because drafting triangles conveniently provide angles of 30°, 45°, and 60°, these angles are most commonly used for the oblique axes. A 30° axis from the horizontal adds side-view emphasis; a 45° axis gives equal emphasis to both of the receding surfaces (commonly called a *Cavalier drawing*); and a 60° axis results in emphasis on the top view. If important information falls on one of the receding planes, the angle should be given careful consideration.

The axis can be projected to either the right or the left, depending on which side better describes the object. Also, the axis can be projected below the horizontal if the bottom surface yields more information than the top (Fig. 7–22D).

On conventional oblique drawings, as we mentioned, the receding lines are measured true length, producing a distorted effect. This method applies especially to objects of considerable depth. The heavy appearance on receding surfaces often makes an oblique rendition objectionable. However, you can usually achieve more pleasing proportions and a more realistic appearance by reducing the depth by ¼, ⅓, or ½. The lines on the frontal planes remain full scale, giving the oblique lines proportional ratios of 1 : ¾, 1 : ⅔, or 1 : ½ (Fig. 7–22E and F). A drawing with the ratio of 1 : ½ is referred to as a *Cabinet drawing* and generally has an appearance that is too thin for most objects other than shallow cabinets.

One other expediency of oblique drawings is their convenience for showing circles or irregular shapes on the frontal planes. Circles can be drawn easily with a compass, and the other shapes do not have to be distorted (Fig. 7–23). This is a definite advantage in giving technical information. Therefore, when selecting the layout of an oblique drawing, if possible, place the circular and irregular features on the frontal plane. Also, if a long object is drawn, place the long dimension on the frontal plane; the drawing will appear shallower in depth. Actually, objects of little depth are better suited to oblique drawing because the distortion on their receding surfaces will hardly be noticeable (Fig. 7–24).

If *circular shapes* fall on one of the oblique planes, they are made elliptical, similar to axonometric circles. Because of the construction involved, circles on oblique axes should be avoided; but if they become necessary, a rhombus similar to an axonometric drawing is constructed around the intended circle, and the approximate, four-center method of ellipse construction is used. Different angular axes necessarily require differently shaped rhombus construction; the appropriate method can be taken from Fig. 7–18. Notice that the large arcs are drawn from the intersection of the two perpendicular bisectors of the two sides that form the larger rhombus angle; the smaller arcs are drawn from the intersection of the perpendicular bisectors of the two sides that form the smaller angle. The circle arcs must be tangent to the midpoints of the oblique-square sides. On the oblique axes, arcs representing rounded corners (and the like) are drawn by the same method, even though they may require only part of the construction.

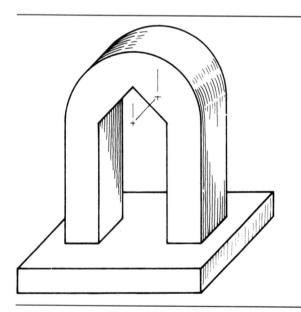

Figure 7–23 Circles and irregular shapes should be placed on the frontal plane of oblique drawings.

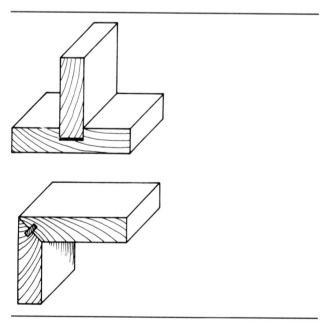

Figure 7–24 Sectioned orthographic details can be given a pictorial appearance by the addition of oblique receding surfaces.

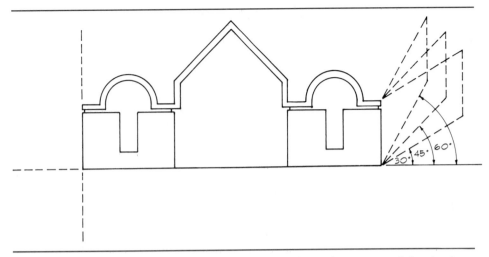

Figure 7–25 To draw an elevation oblique, turn a true-shape elevation parallel to the drawing surface.

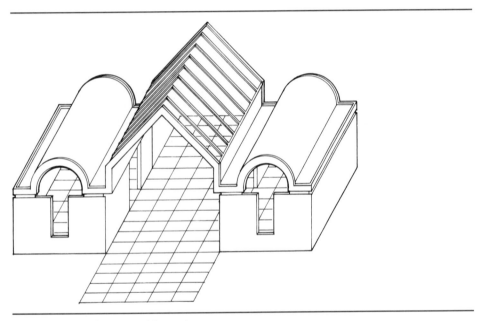

Figure 7–26 Elevation oblique drawing.

7.5

ELEVATION OBLIQUE DRAWINGS

When the façade of a building or an important elevation of an object needs emphasis, the *elevation oblique* pictorial drawing is appropriate. To draw this type of pictorial, turn a true-shape elevation parallel to the drawing surface. A side and the top of the object will usually be drawn at a 30°, 45°, or 60° angle oblique to the elevation. This variation of the oblique pictorial is easy to construct and is appropriate when a slightly distorted depth is not objectionable (Fig. 7–25). A print of an orthographic pro-

jection is placed under the drafting paper with the elevation located parallel to the X and Y axes. Triangles are used to draft the side and top views. Considerable emphasis is placed on the elevation or façade in this type of drawing (Fig. 7–26).

7.6

PLAN OBLIQUE DRAWINGS

A useful variation of the conventional oblique pictorial drawing is the *plan oblique* (referred to as an *axonometric* by many drafters). To create a plan oblique, place a true-

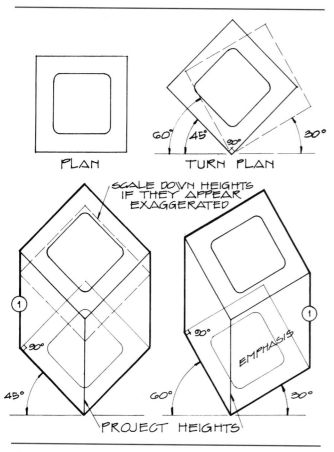

Figure 7–27 Plan oblique with vertical projection.

shape plan beneath the tracing surface. Then project the heights or vertical planes from the plan, and draw them to scale.

The plan oblique has two subclasses: the *plan oblique with vertical projection* and the *plan oblique with angular projection.*

7.6.1 Plan Oblique with Vertical Projection

In the execution of this pictorial drawing, a true-shape plan is commonly rotated at a 45° or 30°–60° angle from the horizontal. Using a triangle, the drafter draws the vertical lines and planes perpendicular to the horizontal (Fig. 7–27). This projection method is popular because of the ease with which it can be drawn. The plan oblique

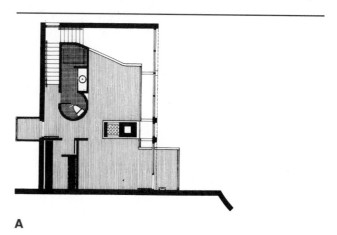

A

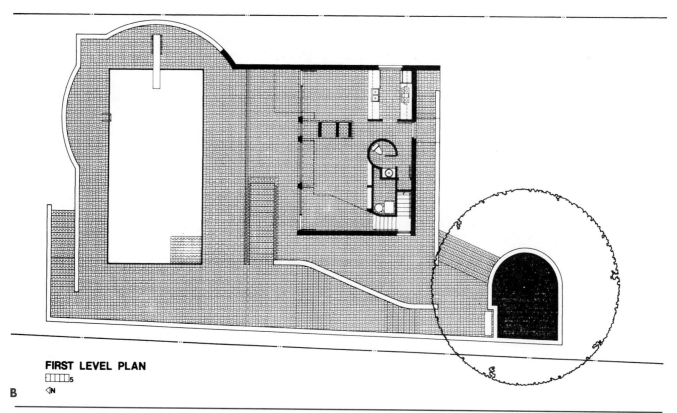

FIRST LEVEL PLAN

B

Figure 7–28 Turning a plan and projecting heights to draw a plan oblique drawing. Floor plans (A and B) and pictorial views (C) of a small house. (Tony Ames, architect)

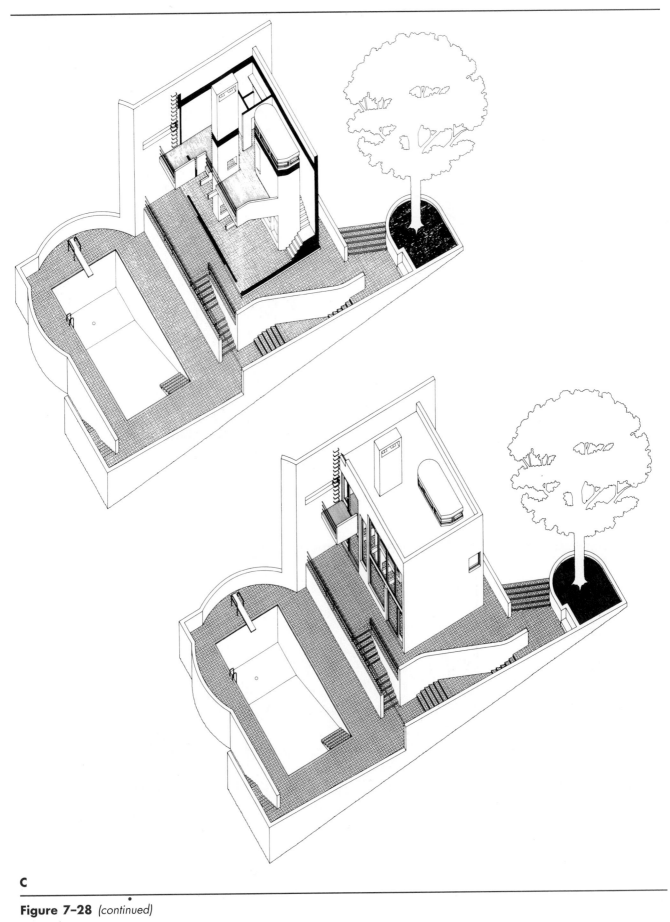

C

Figure 7-28 (continued)

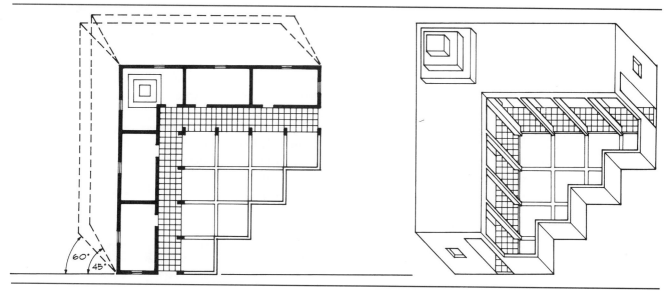

Figure 7–29 Plan oblique with angular projection.

7.6.2 Plan Oblique with Angular Projection

When a true-shape plan is placed parallel to the picture plane, and the vertical planes are projected upward with vertical projection can convey infinite detail about interior space and structural elements, as well as exterior mass and form (Fig. 7–28).

at an angle, usually 45° to 60°, and are scaled, a plan oblique with angular projection drawing is created. This form of projection places the viewer almost directly above the object and reveals a bird's-eye view (Fig. 7–29). Although not as widely used as the vertical projection method due to the appearance of greater distortion, the angular projection method is easily constructed (Fig. 7–30).

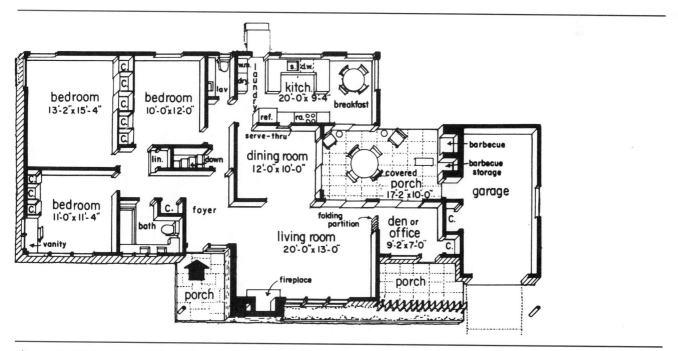

Figure 7–30 A residential floor plan given a pictorial nature with the addition of oblique lines. Cabinetwork and furniture may also be represented.

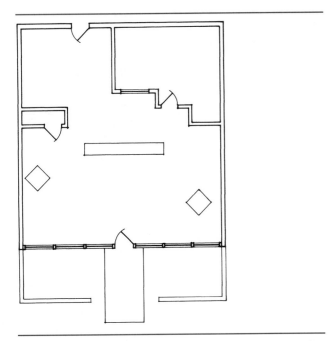

Figure 7–31 A true-shape plan is positioned with one of its sides parallel to the picture plane.

7.7

PLANOMETRIC DRAWINGS

Also referred to as an *axonometric* projection by some drafters, the *planometric* is probably the least used of all pictorial drawings. A true-shape plan is positioned with one of its sides parallel to the picture plane. A triangle is used to extrude vertically the receding planes that are perpendicular to the picture plane. The planometric negates all right-angled side planes (Fig. 7–31). Elements that are skew to the horizontal and vertical are best defined by this technique. Rectilinear distortion is common in planometric drawings (Fig. 7–32).

7.8

PICTORIAL DIMENSIONING

Although many of the rules for dimensioning orthographic drawings are also valid for pictorial drawings, the rules have slight variations due to the pictorial nature of the drawings. The following rules should be observed:

1. Keep the dimension number midway between the arrowheads, and use a logical system to place the smallest dimension closest to the view and the successively larger farther out (Fig. 7–33).
2. If possible, dimension *visible* surfaces and features only.

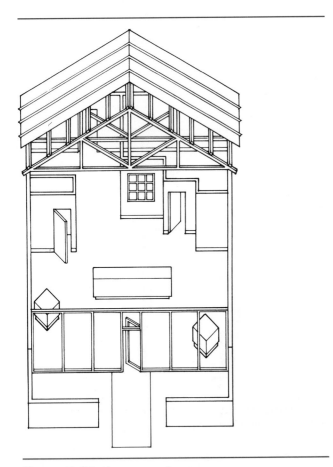

Figure 7–32 Planometric drawing.

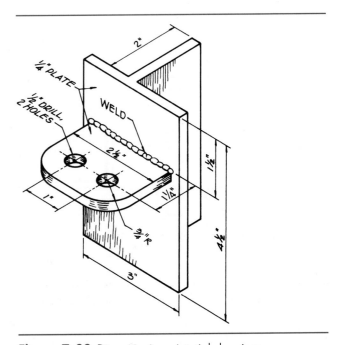

Figure 7–33 Dimensioning pictorial drawings.

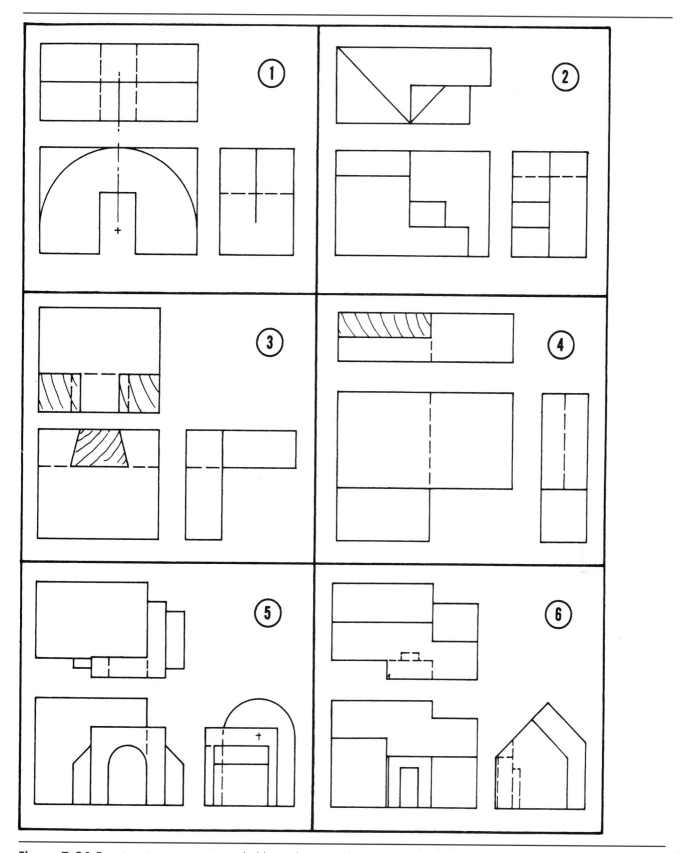

Figure 7–34 Exercises in axonometric and oblique drawing. The exercises should be doubled.

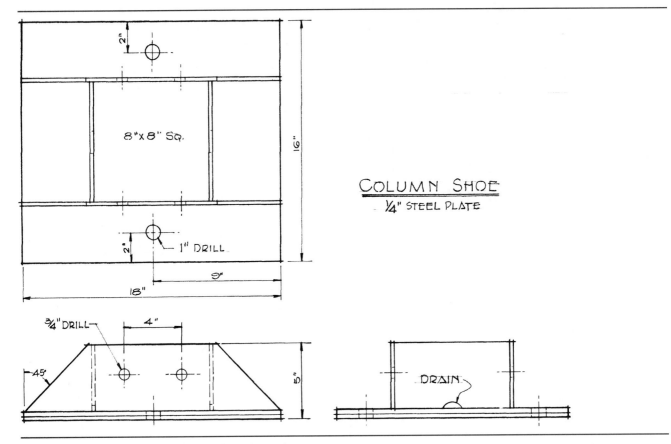

COLUMN SHOE
¼" STEEL PLATE

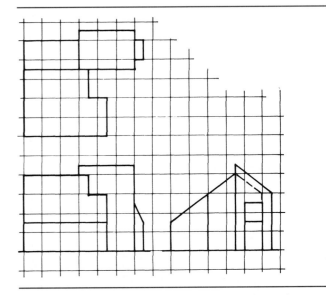

Figure 7-35 Exercise in axonometric or oblique drawing.

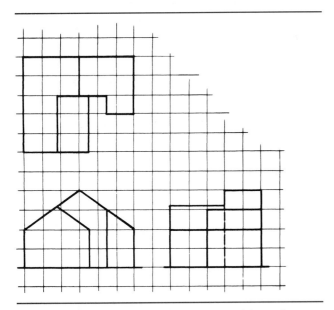

Figure 7-36 Exercise in axonometric or oblique drawing. Enlarge the grids to a convenient size.

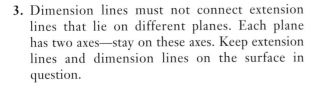

Figure 7-37 Exercise in axonometric or oblique drawing. Enlarge the grids to a convenient size.

3. Dimension lines must not connect extension lines that lie on different planes. Each plane has two axes—stay on these axes. Keep extension lines and dimension lines on the surface in question.

4. Occasionally, dimensions can be placed directly on the object, provided that clarity is maintained and there is sufficient space. This method will usually eliminate unnecessary extension lines running through the object. Otherwise, place dimensions

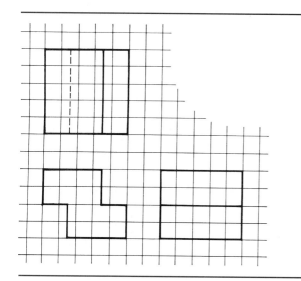

Figure 7–38 Exercise in dimetric and trimetric drawing. Double the grids.

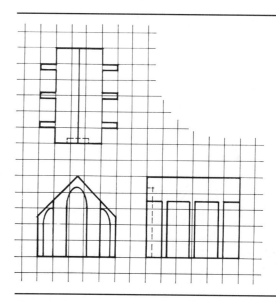

Figure 7–40 Exercise in elevation oblique drawing. Double the grids.

out beyond the extremities of the view so that they do not seem to crowd the exterior profile of the drawing.

5. Leaders should not be made parallel with the axes of the drawing. Leaders that dimension circular shapes should just touch the arc, and the arrowhead should point toward the center point. The other end of the leader should extend out to an uncrowded space and have a ¼″ line attached, which should lie on the same axis as the circle center lines (Fig. 7–33). Arc leaders dimensioning radii should originate at the center point of the arc, and the arrowhead should touch the *inner side* of the arc line.

6. Notes may be lettered on either of the pictorial planes or in a horizontal position. Whichever

method is adopted should be maintained throughout the drawing.

7. On construction drawings, dimension lines should be made continuous and the dimension number placed just above the line (Fig. 7–33). Letter the number so that it appears to lie on the same plane as the surface being dimensioned. You achieve this effect by using vertical lettering yet keeping the vertical axis of the lettering parallel with the extension lines of the dimension. To make the lettering appear pictorial, distort each figure very much as a square is distorted into a rhombus when it is placed on a pictorial plane.

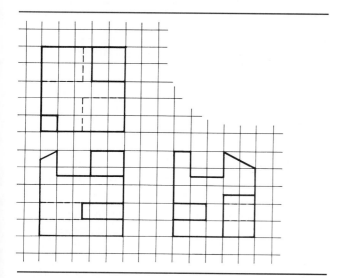

Figure 7–39 Exercise in dimetric and trimetric drawing. Double the grids.

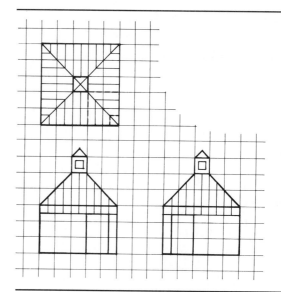

Figure 7–41 Exercise in elevation oblique drawing. Double the grids.

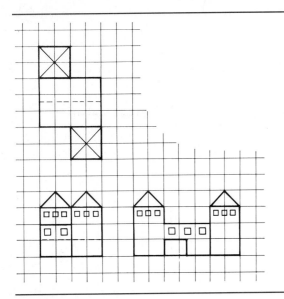

Figure 7–42 Exercise in plan oblique with vertical projection. Double the grids.

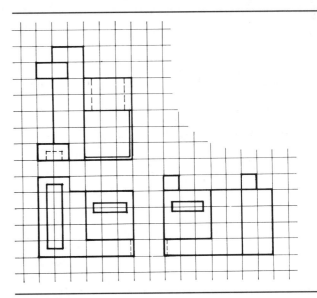

Figure 7–44 Exercise in plan oblique with angular projection. Double the grids.

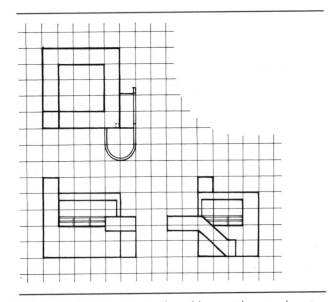

Figure 7–43 Exercise in plan oblique with vertical projection. Double the grids.

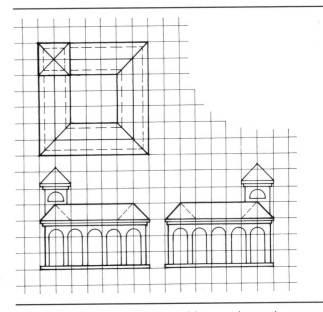

Figure 7–45 Exercise in plan oblique with angular projection. Double the grids.

EXERCISES

PICTORIAL DRAWING

For this set of exercises, draw the pictorials in Figs. 7–34 through 7–37 with instruments on small (8 1/2″ × 11″) sheets of drawing paper. All problems are shown in orthographic projection so that you can convert them into both isometric and

oblique drawings. Double the size of each problem shown in Fig. 7–34; dimensions can be transferred with dividers. When drawing Figs. 7–36 and 7–37, merely let the grids equal any convenient size to fit your paper.

Now draw the pictorials in Figs. 7–38 through 7–46 with

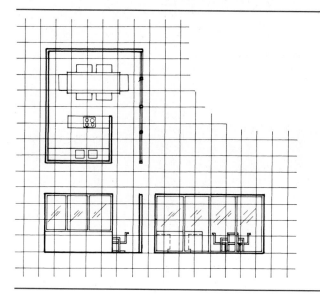

instruments on 11" × 17" sheets of drawing paper. These problems are shown in orthographic projection. You should double them in size using proportional dividers when converting them to the required three-dimensional drawings. Do the drawings in Figs. 7–38 and 7–39 in dimetric and trimetric views. Do the drawings in Figs. 7–40 and 7–41 in elevation oblique views. Do the drawings in Figs. 7–42 and 7–43 in plan oblique with vertical projection views. Do the drawings in Figs. 7–44 and 7–45 in plan oblique with angular projection. Do the drawing in Fig. 7–46 in a planometric view.

Figure 7–46 Exercise in planometric drawing. Double the grids.

REVIEW QUESTIONS

1. What is the advantage of a mechanical pictorial drawing over an orthographic drawing?

2. What are the applications of axonometric and oblique drawings on working drawings?

3. Why does the oblique type of drawing have more versatility than the axonometric?

4. What angle from the horizontal gives emphasis to the top view in an oblique drawing?

5. Describe a *Cabinet drawing*.

6. Why is it advantageous to place the irregular shapes of an object on the frontal plane when making an oblique drawing?

7. How may a plan view of a building be quickly made into a pictorial drawing?

8. In dimensioning pictorial drawings, why do you draw leaders at angles other than those of the principal axes?

9. How are dimension numbers made to appear as though they lie on the planes of the drawing?

10. Are the longest dimensions placed closest to the view or farthest from the view, as compared with the shorter dimensions?

11. What is the difference between an axonometric drawing and a perspective drawing?

12. What is the proper title of a pictorial drawing that has X, Y, and Z axes with two of the angles between adjacent axes equal?

13. What view of an object is revealed by a reverse axes drawing?

14. List the types of pictorial drawings that require a true-shape plan view that can be traced as a basis for the drawing.

15. Why are 30°, 45°, and 60° angles used in preparing most pictorial drawings?

16. What is the proper title of the pictorial drawing that has equal angles between the X, Y, and Z axes?

17. Define *paraline drawings*.

18. List the types of pictorial drawings that require true measurements for all lines making up the drawing.

19. What is the difference between a plan oblique with vertical projection and a plan oblique with angular projection?

20. What characteristic is lost in the planometric drawing that is evident in the plan oblique drawing?

FRANK BETZ

*"Practice and thought might gradually
forge many an art."*
—VIRGIL

FREEHAND SKETCHING 8

The often overlooked *freehand sketch* is a vital means of technical expression, not only for generating ideas, but also for making concepts workable realities. Drafters frequently find themselves in the position of needing to do a freehand sketch. Thus, without sketching ability, they are handicapped in many day-to-day drafting situations.

Generally, every structure, no matter what size, and every graphic problem that confronts the drafter or designer originates with a pencil sketch. Whether crudely or meticulously done, the sketch becomes an important exploratory instrument, as well as a means whereby technically trained people can converse with one another. Whether you are concerned with the esthetic aspects of a building or with its structural properties, the ability to make spontaneous sketches is a considerable asset for architectural drafters, and therefore development of this ability should be an important part of your training.

To gain proficiency in this mode of expression, the student must invite situations entailing sketching at every opportunity. Graphic problems requiring solutions should be analyzed by preliminary sketches—not one, but as many as are needed to develop the concept before it is put in final form on the drawing. Therein lies one of the many advantages of sketching: Graphic ideas can be quickly developed on sketch paper before being drafted to scale on the board.

In addition, a sketch is necessary in planning and organizing intelligently the sheet layout of a complete set of drawings, which often includes many views and details. Numerous layout sketches are necessary to arrange drawings in proper sequence and to utilize the paper efficiently.

However, students are occasionally reluctant to resort to sketches when experimenting with their initial concepts. They seem to feel that their early attempts will look unprofessional, but appearance is inconsequential at the outset. Rather, the desire to develop the concepts graphically should be the incentive for using sketches—appearance will improve with experience. Even the most proficient drafters had to start with crude beginnings.

The secret of learning to sketch, especially for those who are doubtful of their aptitude, is to begin by drawing simple things, such as lines and rectangular views, circles and circular shapes, and complete orthographic sketches. Then advance to axonometric and oblique pictorial sketches, and finally to realistic perspective drawings. This sequence, which leads the student to progressively more difficult work, also provides experience in all of the necessary types of drawings that drafters encounter.

So that the beginner will not be burdened with rules and restrictions, the following material is suggested as a series of guides for making freehand sketching as simple and easy as possible and as a means for eventually making it an enjoyable and rewarding accomplishment.

8.1
SKETCHING MATERIALS

Few materials are needed for sketching: pencils, paper, pencil pointer, sandpaper pad, and an eraser. The majority of linework can be done with an H or a B pencil. Occasionally a 2B will be handy for darker accents, and a 2H pencil, if one is available, can be used for preliminary construction layout. Points on the pencils can be conical, or if both wide and thin lines are required (on floor plan sketches), a chisel-type point (Fig. 8–1) can be dressed on a sandpaper pad to allow variation with the same point. Dressing the pencil point becomes very important in later shading exercises to represent various architectural textures, but for beginning work the medium-sharp point is satisfactory.

The pencil should be held in a natural way; avoid clutching it too close to the point (Fig. 8–2). Some drafters prefer to hold the pencil in a horizontal position when laying in preliminary construction lines.

The most useful type of paper (for the student as well as the professional) is inexpensive tracing paper or an inexpensive sketch pad. As mentioned in Chapter 1, either

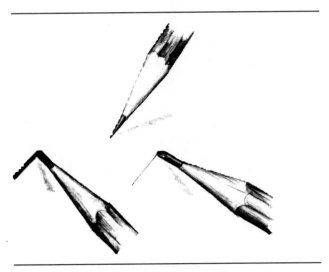

Figure 8-1 Pencil points used in sketching.

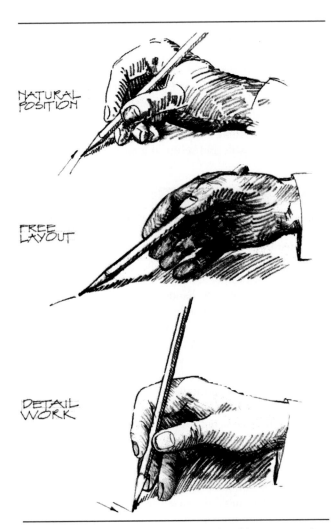

Figure 8-2 Hold the pencil in a comfortable position when sketching.

a 12″ roll of tracing paper or an inexpensive pad (11″ × 17″) is convenient. The advantage of sketching on tracing paper is the ease with which you can modify or redevelop sketches simply by placing the transparent paper over previously done sketches to minimize repetitious construction. Also, by turning the paper over and observing the sketch in reverse, you can find flaws that might possibly go unnoticed. An eraser on the end of the pencil will be handy, but generally most erasing will be unnecessary when sketches can be redrawn more quickly than mistakes can be erased.

Have a pointed felt-tip pen handy with which to practice basic exercises after trial work with the pencil.

8.2

BEGINNING LINE EXERCISES

Start your sketching practice with simple line exercises, drawn horizontally and vertically and inclined in several directions (Fig. 8–3). Make the short lines with finger and wrist movements, and make longer lines by attaching a series of shorter lines of more comfortable length to each other (Fig. 8–4). After each stroke, move the ball of your hand to a new position along the path of the long line. Usually, a very light trial line will help to establish the

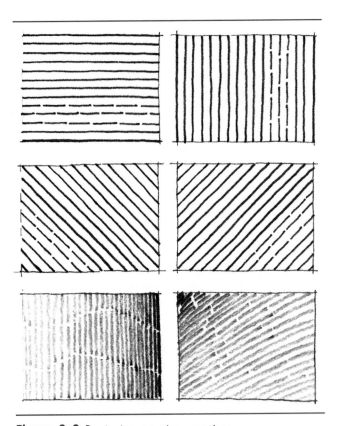

Figure 8-3 Beginning practice exercises.

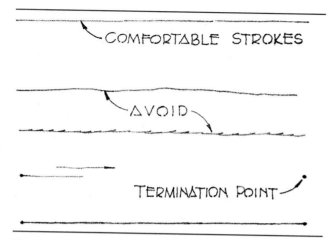

Figure 8–4 Sketching long lines freehand.

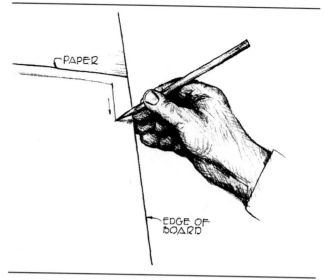

Figure 8–6 Drawing semi-freehand lines with a convenient edge.

general direction of a long line. Or, before starting the long line, indicate a termination point: Concentrate on the point as the line progresses, and almost without exception the line will end at the point. Establish *parallel lines* by placing a few points along the original line, equidistant from it, and sketch the parallel line through the points, as in Fig. 8–5.

Right-handed persons should sketch horizontal lines from left to right (the reverse if left-handed) and should draw vertical lines toward themselves. If using a small sheet of paper, you can turn the paper in a favorable position when sketching lines; usually it becomes easier to sketch a line in a horizontal direction. For practice, sketch straight lines in all directions without turning the paper; this improves coordination of the hand and eye.

If a board or a pad is used when sketching, a semi-freehand method of drawing long lines, shown in Fig. 8–6, can be employed. The edge of the board or pad is used as a guide, and the free hand holds the paper in a position parallel to the edge of the board. This method is especially helpful in establishing important center lines or major horizontal roof lines on an elevation, from which other lines on the sketch can be made parallel. Borders are also easily made with this method.

DIVIDING LINES AND AREAS EQUALLY

The ability to divide lines and areas into equal spaces is frequently necessary in arriving at many of the common geometric forms found in sketches. A line can be quickly divided if the pencil itself is used as a measuring stick. With the aid of the pencil, lay off approximate half-lengths of the line from each end; if the approximation is not exact, the remaining small space is easily bisected by eye to produce an accurate division of the entire line.

Another method of bisecting lines is by *visual comparison*. Observe an entire line and weigh it optically in determining its fulcrum, or point of balance. Compare each half visually before placing the bisecting point. This procedure can be repeated any number of times to divide a line into any number of equal divisions merely by dividing and redividing its line segments (Fig. 8–7). It is especially effective in maintaining symmetry of sketches when lines as well as spaces must be kept uniform.

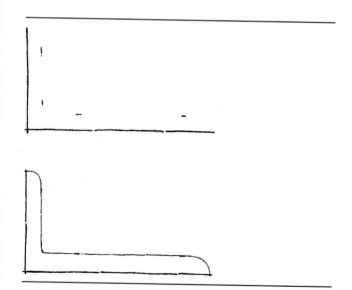

Figure 8–5 Laying out parallel lines.

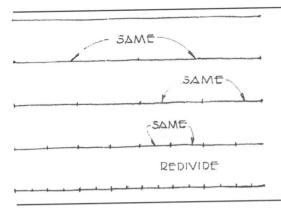

Figure 8–7 Bisecting lines by visual comparison.

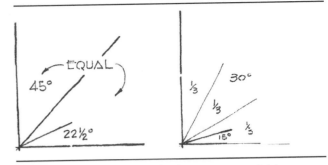

Figure 8–9 Sketching angles by convenient divisions.

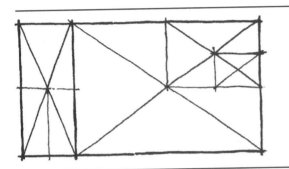

Figure 8–8 Finding centers by sketching diagonals.

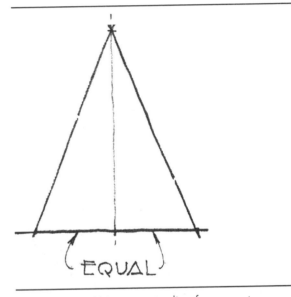

Figure 8–10 Using a center line for symmetry.

You can find centers of rectangular figures by drawing their diagonals. If necessary, the halves can be divided with diagonals for smaller divisions (Fig. 8–8).

 8.4

SKETCHING ANGLES

The 90° angle is a predominant feature in the majority of construction sketches. Thus, the ability to sketch right angles accurately becomes important, even if it entails checking them with the triangle occasionally. Frequently, the perpendicular edges of the drawing paper can serve as a guide for comparison. It is also helpful to turn the sketch upside down; nonperpendicular tendencies of horizontal and vertical lines will become evident. Correct shape of the right angles gives the sketch stability, without which effectiveness is lost.

To make a 45° angle, divide a right angle equally by visual comparison. To make a 30° or 60° angle, divide the right angle into three equal parts, which is slightly more difficult. Always start with the right angle for the most accurate estimation of angle shape (Figs. 8–9 and 8–10).

8.5

SKETCHING CIRCLES AND ARCS

Small circles are usually sketched without construction other than center lines, but larger circles require some preliminary construction to help maintain their symmetry and also to locate them properly on the sketch. Start larger circles (over ¹/₂″) with perpendicular center lines, around which you lay out a square, and carefully check each quadrant of the square as it develops for unity of size and shape (Figs. 8–11 and 8–12). If the square is not true, the finished circle will not come out round. Then, starting at the intersection of the top of the square and the center line, sketch each half of the circle with full, curved strokes that touch each center-line intersection within the square. Practice sketching circles of different sizes until they become balanced and well rounded. Be careful to maintain a "blown-up," full character on all circles, rather than a flattened or elliptical appearance.

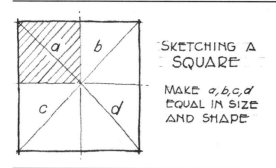

Figure 8-11 Sketch an accurate square by starting with center lines.

If several circles are required on a sketch, the *semimechanical* technique shown in Fig. 8–13 will produce them accurately and save considerable time. Try a few of different sizes for exercise. Hold two equally long pencils rigidly in your hand so that they act very much as a compass. One pencil becomes the pivot; the other scribes the circle when you revolve the paper under the pencils with your other hand.

Freehand arcs are drawn with similar construction as mentioned for circles (Figs. 8–14 and 8–15). Often it is advisable to construct an entire square and an entire circle in order to maintain the character of a regular curve on arcs. The construction linework can be quickly erased.

8.6
DEVELOPING GOOD PROPORTION

True proportioning is an aspect of the art of sketching which is difficult to master. Without a feeling of proper proportion, the sketch can be misleading and therefore of little value. Continually strive for correct proportion by astute observation and analysis of your sketching subject. Drawing in proportion is largely a matter of discerning how one thing relates to another—for instance, how length relates to width, height to length, the size of one feature to another, and one space to another. The trained drafter automatically sees these relationships and checks for accuracy at frequent intervals while sketching.

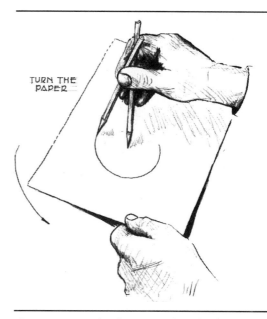

Figure 8-13 Sketching a circle with the use of two pencils.

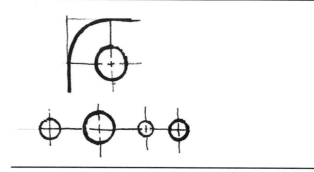

Figure 8-14 Sketching small circles.

One of the simplest methods of arriving at correct proportion is seeing through incidental details of the subject and recognizing the *basic geometric forms* inherent in it (Figs. 8–16 and 8–17), whether the subject is orthographic or pictorial. If these basic forms can be recognized, an important step in proportion has been accomplished (Figs. 8–18 and 8–19). At this point we are concerned mainly with plane figures; pictorial sketches are discussed later.

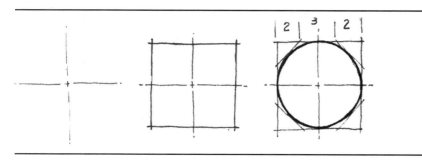

Figure 8-12 Sketching a circle.

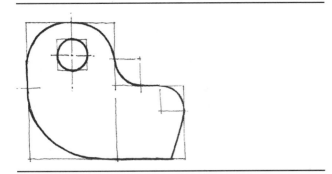

Figure 8–15 Sketching arcs with construction.

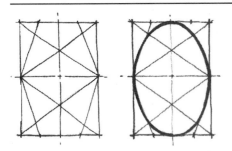

Figure 8–16 Construction for sketching an ellipse.

When sketching from a dimensional drawing, be aware that the dimensions themselves provide a definite basis for proportion—if you carefully observe them. To set the general size of a sketch, establish a major line or feature first; then make the remaining lines proportional to that original line according to the given dimensions. For example, if a 20′ × 40′ floor plan is to be sketched, the 20′ is represented with a line of any convenient length, depending on the size of the paper (preliminary sketches should be small); the line for the 40′ should then be drawn twice this length. Similarly, if a partition within the plan is, say, 10′ long, the line representing it should be only one-half as long as the 20′ line. This procedure introduces adequate proportion throughout the sketch.

However, suppose that we are sketching an object or a view that has neither given dimensions for measurement nor any other tangible means for proportion. This presents more of a problem to an easily discouraged beginner, and some system must be developed to help create the necessary sense of relationship. Use of the simple *square*, as a shape for comparing basic proportion, is the most effective and appropriate method. Although not particularly attractive in itself, the square is the ideal geometric form to look for when arriving at proportion of height and width. It has uniform sides, regardless of its size, and can be accurately sketched by comparison of these sides. For example, if you were analyzing the proportion of the vase in Fig. 8–20, you would observe that the body of the vase will fit neatly into a square. If you were sketching the lamp in Fig. 8–21, you would note that its shade has the gen-

eral shape of a square and that its height is a little less than twice the height of the square. The width of the base is one-third the width of the square. Also, note that two squares quickly give you the basic proportion of the piece of furniture in Fig. 8–22. If you were sketching the exterior elevation of the house in Fig. 8–23, a square would organize its general shape, and smaller squares would give you the proportions of minor details.

The square, therefore, makes a suitable unit for simplifying a subject under observation and for quickly duplicating the units in the sketch. When you become "square conscious," you will automatically see squares of different sizes in your subjects; when you use these same squares as guides in your sketch, you will have arrived at a handy method of proportioning plans, elevations, and section sketches of buildings as well as other subjects.

It is also important to be familiar with the general proportions of commonly used building materials, such as lumber, brick, concrete block, stock moldings, and stock doors (Figs. 8–24 and 8–25). Once you have committed the proportions of these materials to memory, other aspects of the sketch will assume relative proportions.

ACQUIRING GOOD LINE TECHNIQUE

At the start of a sketch, make the lines light and indefinite while the forms are still in a plastic state of development. These exploratory and trial lines should be retained until satisfactory shapes emerge and the general nature of the sketch takes on a realistic proportion. Next, perfect and strengthen important lines so that the corrected shapes predominate the trial lines; work on minor details after major shapes seem right. In this way, the entire sketch will mature simultaneously. Light trial lines can be erased as the sketch progresses, although if they are left in, the light lines throughout the drawing seem to soften it and give it "sketchy" quality.

The finished lines should have weight variations according to their importance. Object lines and key profile lines are made bold, whereas less important lines are made accordingly less prominent. Line variations provide the sparkle and depth so important in a successful sketch. Occasionally, a freehand line is broken at uneven intervals rather than made continuous, and the beginning and the end of each stroke are made definite instead of appearing feathered out. One interesting quality of a freehand line is its slight waviness as compared with the sharp straightness of a mechanical line. This quality adds personality to the finished sketch, and there is no reason for trying to avoid it. On the other hand, a choppy, hesitating line technique, shown in Fig. 8–4C, is neither forthright nor pleasing in appearance. Progress in sketching is made only with *practice*; each successive exercise should be a deliberate effort to improve.

Figure 8–17 A page from the sketchbook of a thirteenth-century French architect. Villard D'Honnecourt, showing the use of geometric forms in sketching.

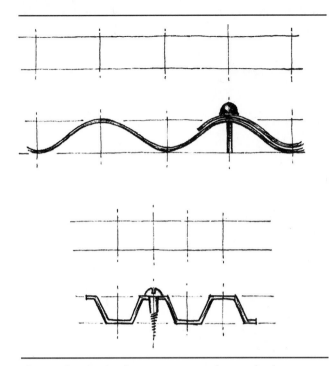

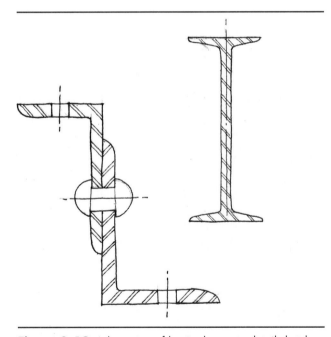

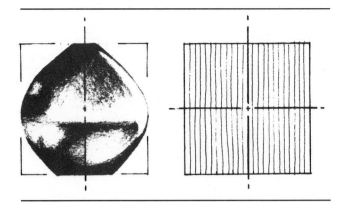

Figure 8–18 Sketching construction for simple shapes.

Figure 8–20 Using a simple square for arriving at proportion.

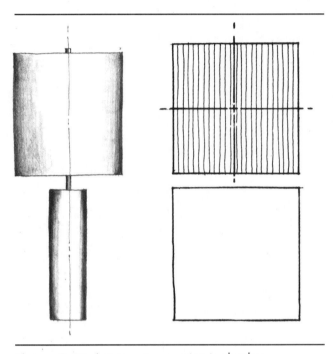

Figure 8–19 Adaptation of basic shapes to detail sketches.

Figure 8–21 Arriving at proportion in sketches.

8.8

SKETCHING PLANS AND ELEVATIONS

The floor plan sketch is the logical start for depicting the architectural idea. After preliminary planning has produced a few general concepts of the structure, but definite limitations are not yet established, a *schematic sketch*

showing possible room locations is drawn. Use only circular shapes for these trial layouts (Fig. 8–26). Make similar sketches if necessary for second-floor plans as well. From these schematics, a more developed plan is made.

The best procedure for working out preliminary studies of tentative floor plans and elevations is the use of $\frac{1}{8}''$ coordinate paper. Either work directly on $8\frac{1}{2}'' \times 11''$ sheets, or tape your sketching tracing paper over a coordinate underlay pad which has been fastened to the drawing board as described in Chapter 1. The coordinate paper facilitates sketching horizontal and vertical lines and provides a preliminary sense of scale. Each coordinate space represents 1 sq ft, and the sketch will have a convenient scale of $\frac{1}{8}'' = 1'\text{-}0''$. This scale is satisfactory for early studies. Frame walls on coordinate paper can be made a half-space wide (6''); large buildings can be put

Figure 8–22 Many sketching subjects can be fitted into squares.

on a small sheet; and, more important, the small scale allows more command of the development. Even the scale of $1/16'' = 1'-0''$ will be found adaptable to beginning sketches on coordinate paper. Large sketches are clumsy to handle during the planning stage.

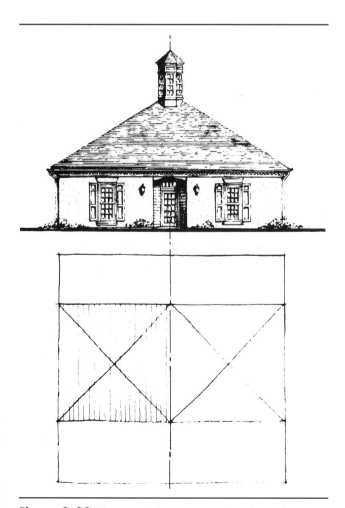

Figure 8–23 The use of the square in sketching elevations.

8.8.1 How to Sketch a Plan

To sketch a plan, lightly draw the exterior walls with a single, continuous line; do not show windows and doors yet. Lay out all partitions with a fine line, and let the lines continue through doors and other openings. Establish all clothes closets with a minimum of 2′ in depth, and fill in with close lines parallel to the 2′ dimension. Indicate the placement and width of all windows and doors on the wall lines. With the pencil dressed to a chisel point, widen the exterior walls and partitions with a broad line; stop at the window and door openings. Next, complete the window and door symbols. (Fig. 8–27) with a sharp point. On design sketches, door indications can merely be left as openings. Sketch the approximate locations of stairs; indicate risers with parallel lines (actual size about 11″ apart), and indicate the direction of travel with an arrow and the note UP or DOWN, as the case may be. Draw in fireplaces, entrance platforms, terraces, walks, and other minor features. Kitchen cabinets and bathroom fixtures are blocked-in freehand. Label the rooms and add as many dimensions as needed; usually only interior room sizes are indicated at this stage. Show the plan with the front side *down* on the paper.

If a second-floor plan is necessary for a 2-story house, develop the second-floor plan on tracing paper placed over the completed first-floor plan. Place bearing walls and major partitions above first-floor walls, if possible, for sounder construction. Stairwells and chimneys must coincide, and plumbing wet-walls are best aligned above each other. The vent stack of a first-floor bath must pass through a second-floor partition. Draw an end elevation of 1½-story houses before attempting the second-floor plan. Locations of half-walls and dormer windows become simpler on elevation views showing the roof profile.

Before the elevations can be sketched, a preliminary roof layout must be made on the plan view. The character of the roof plays an important role in the total design of a house, not only for the sake of appearance, but also from the standpoint of accommodating the outline of the

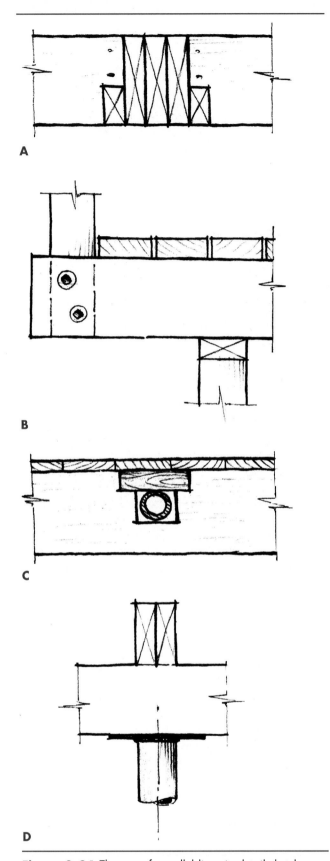

A

B

C

D

Figure 8–24 The use of parallel lines in detail sketches.

plan. Consequently, several sketches may be necessary to find the most desirable type of roof. For economy, keep the roof simple; on the other hand, if an unassuming single-gable or hip roof is used, its austere silhouette usually appears monotonous and contributes little architectural character to the structure. (Refer to Fig. 12–160 for illustration of roof types.) Uniform roof overhangs are easier to frame than various extensions, but protection from the weather over entrances, walks, and terraces makes the roof more serviceable. Show the roof overhang with dashed lines, and indicate ridges, valleys, hips, and roof offsets. If the roof slopes are uniform, the intersections of gables and hips are always shown with 45° lines.

8.8.2 How to Sketch an Elevation

To sketch an elevation, place the floor plan sketch under a fresh sheet of tracing paper, take the longitudinal features from the plan, and merely transfer features from each exterior wall (Fig. 8–27). You can establish the ridge and eave lines of sloping roofs by first sketching the end elevations if necessary. Then transfer them to the other elevations after satisfactory roof slopes have been developed. Stock door heights are shown about 6′-8″ above the estimated finished floor level, and the tops of windows are usually aligned with door tops in residential construction. Important details showing the character of the windows and the doors should be added, as well as other significant features such as chimneys, columns, exterior steps, and symbols of exterior materials. Sketching some foliage near the ends of the elevation will project an impression of unity with the landscape (Fig. 8–28).

8.8.3 How to Sketch Section Details

Sketching section details requires definite information about the construction of the tentative structure; nevertheless, exploratory section sketches must be made in conjunction with plan and elevation development.

Start details by blocking in important structural members, around which subordinate materials will be added. If masonry is involved in the section, establish its shape before applying wood framing or other details. Add exterior and interior wall coverings, moldings, finish flooring, and the like, after orienting major structural pieces correctly. Care must be taken to relate each member according to its structural position; proportion is important to reveal workability of the detail.

Because section details are composed mainly of parallel and perpendicular lines, the right-angle nature of the lines deserves attention. Strengthen the outlines of important elements, and make the symbols of the different sectioned materials lighter. In large areas of sectioned surface, except for the periphery of the surface, much of the symbol linework can be omitted (Fig. 8–25).

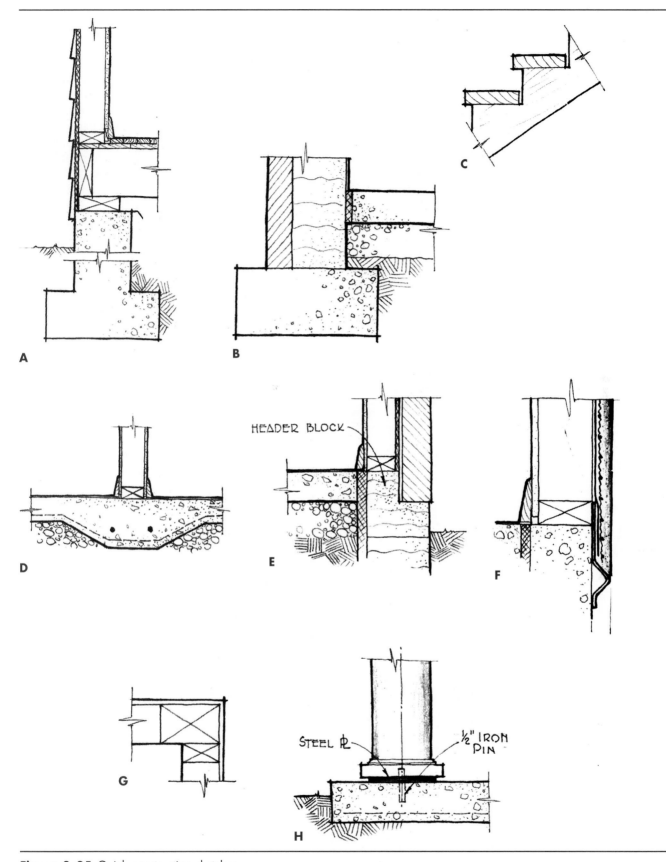

Labels within the figure:

C

HEADER BLOCK

STEEL ℙ   ½" IRON PIN

A   B   D   E   F   G   H

Figure 8–25 Quick construction sketches.

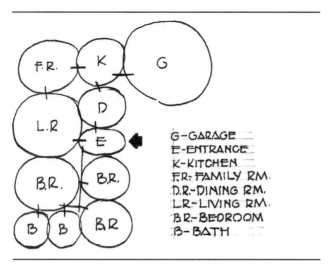

Figure 8-26 A schematic sketch for starting a floor plan.

Coordinate paper with 12 × 12 divisions within each 1″ square can be used to sketch accurate details in the following conventional scales: ¼″ = 1′-0″, ½″ = 1′-0″, 1″ = 1′-0″, and 3″ = 1′-0″ (Fig. 8–29). Especially when using the 1″ = 1′-0″ scale, you will find that section details can be quickly sketched over this paper, if scaled sketches are needed. In fact, many drafters find that details sketched over 12 × 12 per inch coordinate paper are perfectly satisfactory for developing working drawings. Using the different scales with the paper, each small square is equal to the following number of inches:

¼″ = 1′-0″	4″
½″ = 1′-0″	2″
1″ = 1′-0″	1″
3″ = 1′-0″	(3 small squares = 1″)

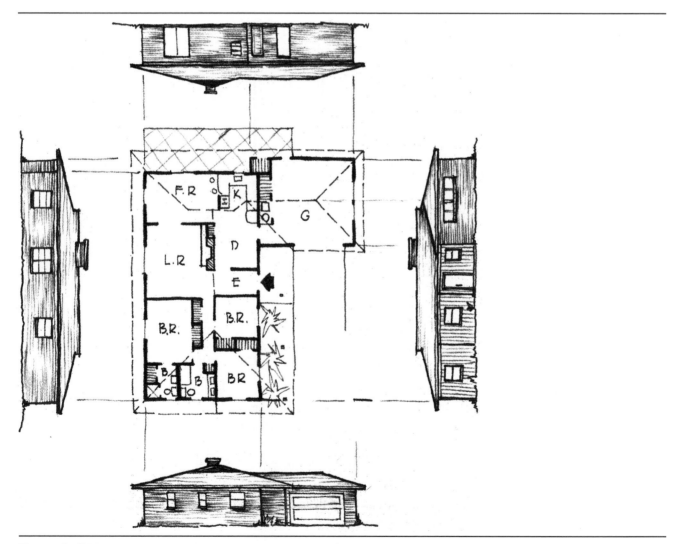

Figure 8-27 Plan development sketch and relating elevations.

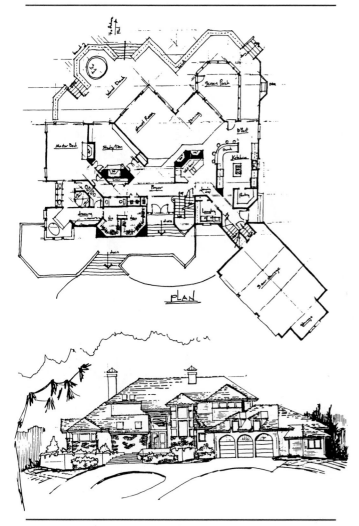

Figure 8–28 A professional preliminary sketch of a residence.

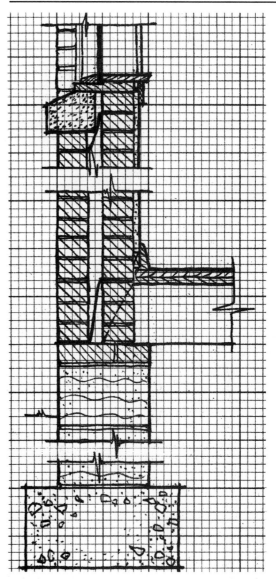

Figure 8–29 A scaled detail sketch on grid coordinate paper.

AXONOMETRIC, ISOMETRIC, AND OBLIQUE SKETCHING

Not only are axonometric, isometric, and oblique sketches the next step in the logical development toward perspective sketching, but they also provide a quick method of examining tentative construction details pictorially if the need arises.

The principles of pictorial and orthographic sketching are similar, except that now we will deal with volumes rather than flat planes. Instead of the square being a convenient unit of proportion as in orthographic views, the *cube* becomes the ideal unit of pictorial proportion. In addition, all objects are simplified to their basic volumetric or platonic forms, and these forms are the first consideration in the pictorial sketch (Fig. 8–30). *Block out the basic forms of the object before attempting the minor details.* Since this simple procedure is effective for experienced technical illustrators, it should serve the student equally well. Study the forms shown in Figs. 8–30 and 8–31; careful analysis of any object being sketched reveals that these basic shapes are inherent forms, although sometimes only parts of these shapes or several of them in combination are perceived.

First, practice sketching the basic geometric shapes, using the construction principles indicated in Chapter 7. Then look for these shapes in the object that you are about to sketch, and concentrate on the basic form representation. The object can be enclosed in a cube or a box (see Fig. 7–14), or it can be built up by a series of differently shaped blocks, one on top of the other (see Fig. 7–15), depending on the nature of the object. Details are added or "carved" from these main blocks after shape and proportion are established.

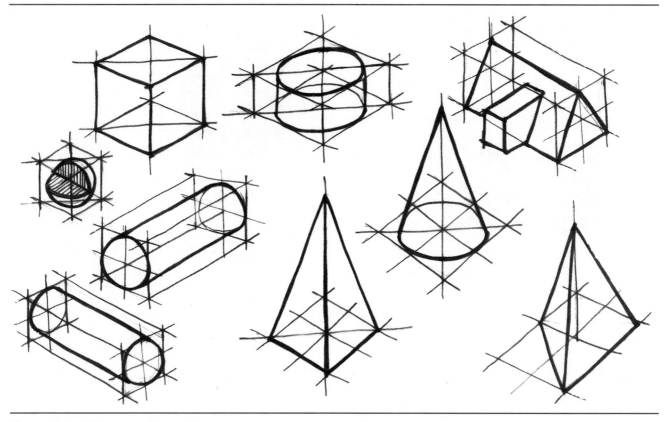

Figure 8–30 Volumetric platonic forms.

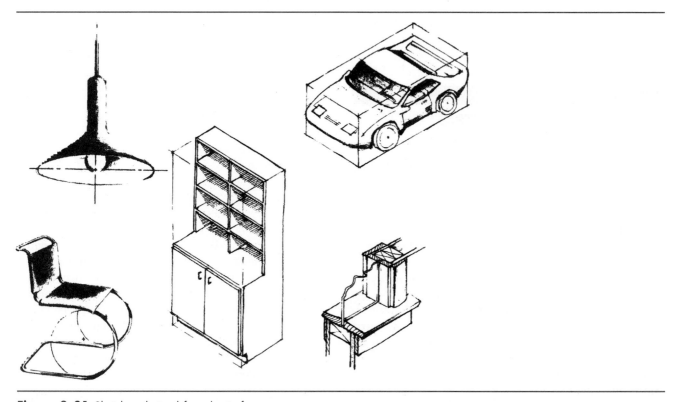

Figure 8–31 Sketches derived from basic forms.

If you are sketching from orthographic plans or elevations, the lines of each view can often be laid out on their perspective pictorial planes for easier construction. Occasionally, it becomes advisable to combine a pictorial sketch with an orthographic section sketch. Partial pictorial sections are also revealing sketches. Appropriate exercises should be devoted to sketching from actual models, drawing equipment, or even books—or from anything else at hand.

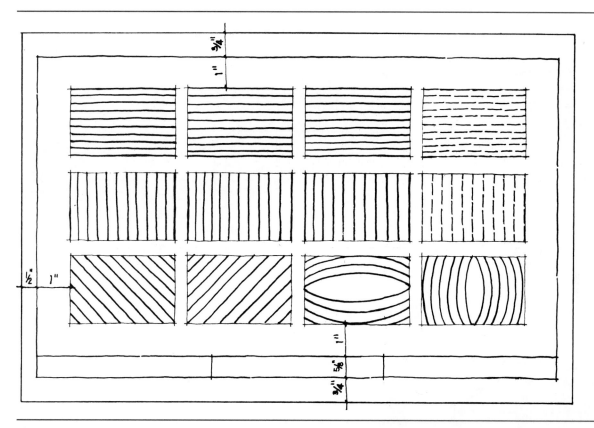

Figure 8–32 Layout of beginning sketching exercise.

EXERCISES
SKETCHING

1. On 8¹⁄₂″ × 11″ drawing paper, lay out a border, and divide the paper into 12 equal rectangles freehand, as shown in Fig. 8–32. With an HB pencil, carefully fill the rectangles with horizontal, vertical, and inclined lines as shown.

2. Sketch a 3″ square and bisect each side by eye. Connect the points to form an inverted square within the original, as shown in Fig. 8–33. Continue drawing squares within squares until the space is filled.

3. Sketch Fig. 8–19, showing the angle-iron assembly.

4. Sketch a section view of an I beam as shown in Fig. 8–19.

5. Sketch four 1¹⁄₄″ squares and draw circles within each (Fig. 8–12).

6. Sketch an ellipse as shown in Fig. 8–16 with major axis 3″ and minor axis 2″.

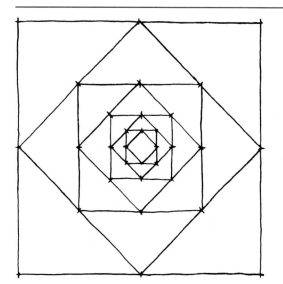

Figure 8–33 Sketching exercise.

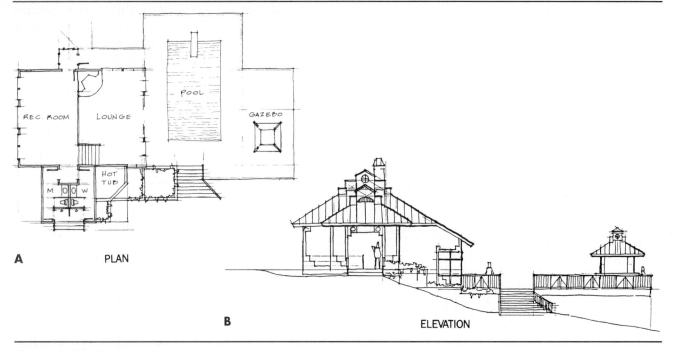

A   PLAN

B   ELEVATION

Figure 8–34 Sketching exercise.

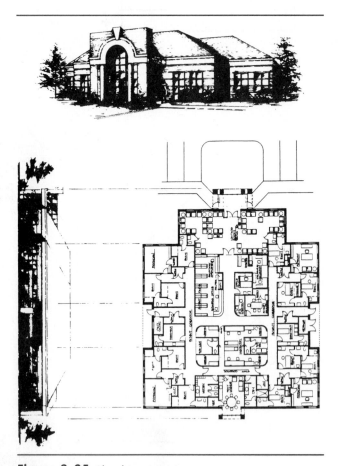

Figure 8–35 Sketching exercise.

7. Sketch the corrugated metal sections in Fig. 8–18.

8. Sketch the lamp in Fig. 8–21.

9. Sketch the box sill section in Fig. 8–25A.

10. On coordinate paper, sketch the floor plan shown in Fig. 8–34A. Sketch your own grid coordinates over the drawing in the book.

Figure 8–36 Sketching exercise.

Figure 8–37 Sketching exercise.

11. On coordinate paper, sketch the front elevation shown in Fig. 8–34B. Sketch your own grid coordinates over the drawing in the book.

12. On 8½″ × 11″ drawing paper, sketch the basic isometric shapes in Fig. 8–30.

13. Sketch the building in an oblique view in Fig. 8–35.

14. Make an isometric sketch of the lamp in Fig. 8–21.

15. Sketch the sphere in Fig. 8–36.

16. Sketch the glass block in Fig. 8–37.

REVIEW QUESTIONS

1. Why does ability in sketching require a strong sense of observation?

2. Explain the *structural* or *skeleton* method of developing a sketch.

3. In organizing a sketch, why do you reduce the sketching subject to its simplest geometric forms and draw the details later?

4. What type of practice sketches should precede pictorial sketches?

5. Which simple geometric form aids in establishing proportion?

6. Why do architectural drafters prefer to do their sketching on inexpensive tracing paper?

7. For what reason is sketching a necessary part of drafting training?

8. What is meant by a *schematic* sketch of a floor plan?

9. Why is the use of coordinate paper underlays especially helpful in sketching floor plans?

10. In perspective sketching, why may it be necessary to try several basic-block forms before arriving at suitable vanishing points when drawing a building?

11. Why is it essential for the beginner to practice sketching basic geometric shapes?

12. Describe the method used for drawing a freehand circle using two pencils.

13. How is good line technique acquired?

14. What method is recommended for sketching symmetrical objects?

15. Why is it important for the drafter to have the ability to make freehand sketches?

"Thus shadow owes its birth to light."
—JOHN GAY

PERSPECTIVE DRAWINGS

It is of great advantage for the architectural drafter to be able to draw objects as they appear to the casual observer. A realistic representation is still the most effective way of showing a client, who may have a minimal knowledge of graphics, the appearance of a proposed structure. Through the years, architects have used perspectives in presentation drawings and also for preliminary planning sketches. Unquestionably, the value of perspective drawings lies in the fact that architectural designs are shown in the most natural way. Drafters with a working knowledge of perspective will be better prepared for presentation work and, equally important, will find themselves with a keener sense of three-dimensional space visualization. To the beginner, perspective may seem difficult, but after careful study of the principles, even the novice with comparatively little experience will be able to do surprisingly well.

This chapter is not an exhaustive study of the theory of perspective; rather, it is meant to show the beginner the fundamentals and the methods commonly employed in drawing perspectives with the least difficulty and in the most practical way. We will concern ourselves mainly with the how rather than the why of perspective drawing by the use of step-by-step illustrated instructions.

9.1

THEORY AND NOMENCLATURE

Perspective drawing is a pictorial method of representing a building or an object, very much as the lens of a camera records an image on film. We can say that a perspective is the projection of an object onto a fixed plane as seen from a fixed point. Remember, a view that we observe with our eyes is actually two coordinated views from two points, usually not fixed as we move our heads about, producing three-dimensional realism difficult to obtain on a drawing. The drawing can show the image only as

it will appear from one point of view, and that point is fixed. This limitation is inherent in every perspective drawing. Occasionally, even with the most accurate projection, an entirely unrealistic representation will result. For that reason, slight modifications of points and geometric arrangement in drawing may be necessary before you arrive at a satisfactory picture. The student must be willing to make several trials, if needed, for the sake of appearance and true architectural proportions.

The basic theory of perspective drawing assumes that the image is produced on a transparent vertical plane called the *picture plane* (Fig. 9–1), very much as orthographic views (discussed in Chapter 6) are formed on transparent planes—the only difference being that the orthographic views have parallel projectors, whereas the perspective views have projectors radiating from a single point, similar to the visual lines of sight from a person's eye (Fig. 9–1). The outline of these projection points, as they pierce the picture plane, forms the perspective drawing. For comparison, we could produce a perspective on a window glass if we were looking through it at a building across the street and drew the exact outlines as we saw them through the glass. On our paper, however, we will plot all points of the perspective with projection lines from the orthographic plan and elevation view, thereby eliminating any guesswork. A pictorial representation of the planes and points necessary to draw a typical perspective is shown in Fig. 9–2. Later, we will discuss how each is laid out on paper to produce the finished perspective.

Down through the years, many artists and drafters have contributed to the system of perspective projections as we know it today. The woodcut shown in Fig. 9–3, *Demonstration of Perspective* by Albrecht Dürer, is from the artist's treatise on geometry, written in 1525. It illustrates a rather crude yet effective early attempt to prove the principle of how projectors form a true image as they pierce a vertical plane.

To interpret both the drawings and the written material, we must have an acquaintance with perspective

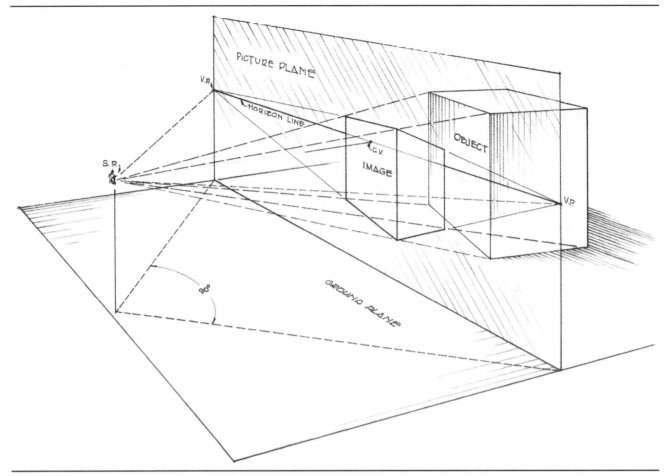

Figure 9–1 In perspective drawing, the image of an object is produced on a theoretical, transparent plane, called the *picture plane,* as we look through the plane at the object.

terms. At this point you should study the drawings in this chapter and observe the significance of each term defined in the following subsections.

9.1.1 Picture Plane (P.P.)

As we mentioned, the *picture plane* is conveniently thought of as a transparent, vertical plane upon which the perspective is drawn. The lower edge of the plane intersects the ground plane. On the plan view of the layout, it appears as a simple, straight line parallel to the ground plane, and is usually placed between the station point and the object (Fig. 9–2). All horizontal measurements are marked on the picture-plane line and are then projected to the perspective. Any part of an object touching the picture plane will have true-height characteristics, and these heights can be projected directly from an elevation of the object or can be measured directly as long as the feature touches the picture plane. You will notice in the different drawings that the farther back from the picture plane the features fall, the smaller they will appear on the perspective. On the lower portion of the perspective construction layout of Fig. 9–2, the surface of the paper becomes the

picture plane. For convenience, we will label the picture-plane line P.P. on the drawings; the other terms will also be designated by their initial capital letters.

9.1.2 Ground Plane (G.P.)

The *ground plane* must be horizontal and is represented with a line on the elevation portion of the layout in Fig. 9–2. If the object touches the picture plane on the plan, it must also touch the ground line in a similar manner on the perspective (Fig. 9–2). If the object is placed in back of the picture plane in plan, it will appear above and in back of the ground line on the elevation. The ground line is always parallel to the horizon line and represents the intersection of the picture plane and the ground.

9.1.3 Station Point (S.P.)

The *station point* is the origination of the observer's lines of sight as the object is seen through the picture plane. It will appear as a point on both the plan and the elevation construction. However, it is usually not obvious in

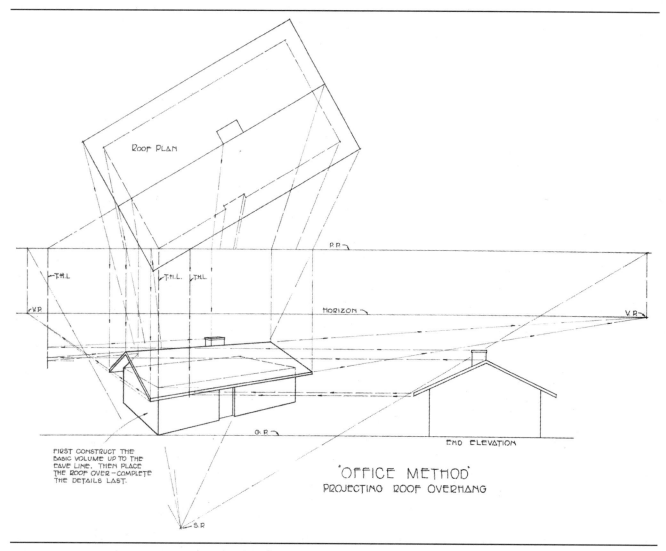

Roof Plan

P.P.

T.H.L. T.H.L. T.H.L.

V.P. HORIZON V.P.

G.R.

END ELEVATION

FIRST CONSTRUCT THE
BASIC VOLUME UP TO THE
EAVE LINE. THEN PLACE
THE ROOF OVER—COMPLETE
THE DETAILS LAST.

'OFFICE METHOD'
PROJECTING ROOF OVERHANG

S.P.

Figure 9–2 A 2-point perspective done by the office method.

Figure 9–3 A woodcut by Albrecht Dürer in 1525 verifying the principles of perspective. The string passing through the picture plane locates points to form the image.

elevation, because it falls on the horizon line. The placement of the point on the plan view obviously determines the view of the building, and this choice of placement would be very much like actually walking around the building to determine the most favorable position for observation. (Generally, it is most expedient to choose the front view showing the entrance.) It also becomes evident that the distance between the station point and the picture plane affects the sense of distance in the perspective (Fig. 9–7).

In early attempts, students often make the mistake of placing the station point too near the picture plane. In general, the station point should be placed about twice as far away from the picture plane as the length of the building being drawn. Another method of determining the proper distance for a pleasing perspective is with the use of a 30° triangle. Place the triangle so that its sides enclose the extremities of the plan (Fig. 9–4), and the apex will locate a satisfactory station point. Usually, the point

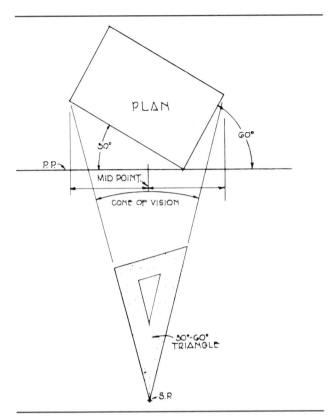

Figure 9–4 A convenient method for locating the station point.

is located in the center of vision (Fig. 9–7B), rather than off to one side. The most desirable placement of station points comes only after experience is gained in dealing with the correct appearances of the many architectural forms and features that are generally encountered.

9.1.4 Horizon Line (H.L.)

The *horizon line* represents the height of the observer's eye and therefore is represented only on the elevation portion of the drawing. Usually, the horizon line is placed above the ground line; the amount above determines the height of observation, since the horizon plane is always at eye level. If a view from 30′ high were desired, for example, the 30′ would be scaled from the ground line to locate the horizon line.

9.1.5 Vanishing Points (V.P.s)

Vanishing points for horizontal lines always fall on the horizon line. Receding, horizontal lines that are not parallel to the picture plane vanish at these points (Fig. 9–5). On the plan portion, perpendicular lines parallel to the sides of the building are projected from the station point to locate the vanishing points on the picture plane; to bring the points to the elevation portion of the setup, they are projected down to the horizon line (Fig. 9–2). This method is used for a 2-point, angular perspective. In a 1-point perspective, the vanishing point is simply placed in the most favorable position on the horizon line.

In drawing perspectives, note that the vanishing points make convenient terminations for the horizontal receding lines, and as long as the proper lines vanish at the correct point, the picture will develop with little trouble. *Sloping* (nonhorizontal) *surfaces*, which will be discussed later, have vanishing points lying on a vertical trace through the original vanishing points. Unless a 3-point perspective is drawn, all vertical lines are drawn vertically. Very few architectural delineators have found use for the 3-point type of perspective. In reality, the sides of tall buildings would converge vertically as we look up at them. However, this distortion does not lend itself to accurate presentation; therefore, we will not concern ourselves with 3-point perspective in this material.

Occasionally, when working on large drawings, you will find that the vanishing point falls a considerable distance from the paper. If a large board is not available to overcome this problem, some method must be employed to vanish lines at the distant points. Often an adjacent table can be used, and a thumbtack in the vanishing point will aid in aligning an especially long straightedge. Another method is shown in Fig. 9–6 with the use of an offset-head T square and a curved template fastened to the board.

9.1.6 True-Height Line (T.H.L.)

If a vertical line of the object touches the picture plane in plan view, the line will appear at the same scale on the finished perspective, thus providing a convenient method for projecting true heights—from an elevation view or by measurement directly on the *true-height line*, if necessary (Fig. 9–2). If difficulty is encountered in establishing

Figure 9–5 Photograph of a house with major lines extended to locate the vanishing points.

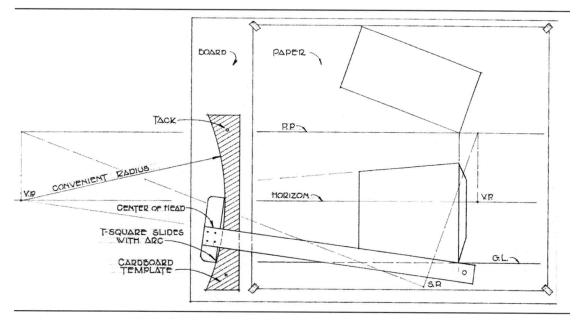

Figure 9–6 Handy method of drawing lines when their vanishng points fall off the drawing board.

heights for a certain feature of the floor plan with the use of only one true-height line, you can solve the problem by projecting the feature to the picture plane. Wherever the projection intersects the picture plane, a new true-height line can be drawn for measuring the height of that feature only. Remember that height measurements can be made only on those features that touch the picture plane.

9.2

PERSPECTIVE VARIABLES

Briefly, the variables in perspective construction, other than the actual scale change of the orthographic views, are the relationships between the station point, the picture plane, and the object (Fig. 9–7). Naturally, there can be an infinite number of relationships, and the drafter should know the various ways in which these variables can be manipulated for the most desirable pictorial appearance.

9.2.1 Relationship of the Object to the Picture Plane

First, a decision must be made as to which sides of the building should appear on the perspective (Fig. 9–7). Ordinarily, the front is shown and is given the most emphasis; occasionally, an interesting feature in the rear will call for a view from that side. The drafter attains emphasis by placing the important side at a small angle from the picture plane—the larger the angle, the less the emphasis. Usually, the 30°–60° angles are convenient for laying out the plan in relation to the picture plane, with the 30° angle given to the more important side. A 45° angle produces equal emphasis on the two observable sides of

a building. The perspective view may be more interesting if a 30°–60° angle is used.

It is worth remarking that without changing the scale of the orthographic views, you can control the size of the finished perspective to a certain extent by merely changing the relationship of the plan to the picture plane (Fig. 9–7G, H, and I). Usually, the front corner of the plan is placed on the picture plane, but by bring the plan *down*, with more of it (or all of it) in front of the picture plane, you can achieve a larger perspective. Conversely, if a smaller perspective is desired, the plan can be placed in back of the picture plane.

9.2.2 Distance from the Station Point to the Picture Plane

After observing the drawings, note that the closer the station point is to the picture plane, the smaller the perspective becomes. Also, close station points produce images with sharp angles on their forecorners, resulting in distorted and displeasing perspectives. On the other hand, if the station point is placed too far from the picture plane, it will usually fall off the paper and therefore become troublesome. Under ordinary conditions, a station point placed to produce a 30° angle of vision with respect to the extremities of the plan produces satisfactory images (Fig. 9–4). The cone of vision should not be more than 45° in width.

The station point can also be moved to the right or the left of the center of vision (Fig. 9–7A, B, and C), but placing it too far either way will produce distortion. You can gain similar effects by changing the angle of the plan in relation to the picture plane, as previously mentioned. The latter method is advisable since it keeps the station point and the center of vision in a perpendicular rela-

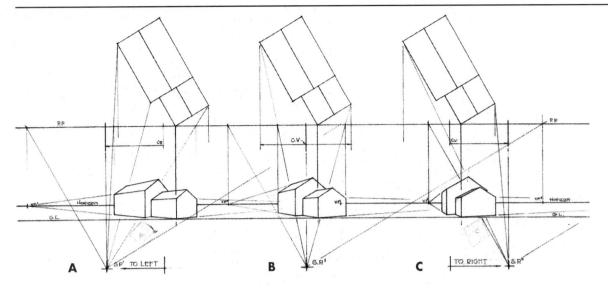

EFFECT OF CHANGING S.P. POSITION

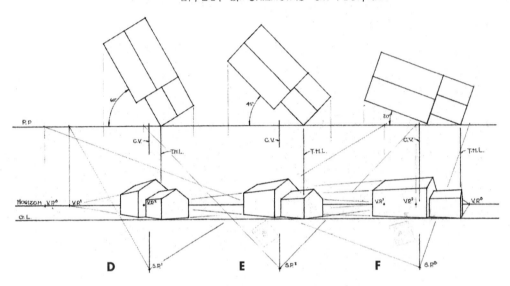

EFFECT OF CHANGING PLAN ANGLE

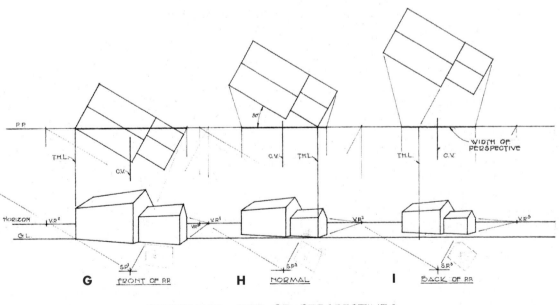

CONTROLLING SIZE OF PERSPECTIVES

Figure 9–7 Perspective variables.

tionship to the picture plane, eliminating unnecessary distortion.

9.2.3 Height of the Horizon from the Ground Plane

The placement of the horizon line with respect to the ground line determines the eye-level height in observing the building (Fig. 9–8). The horizon represents eye level. If it is placed above the height of the roof, a bird's-eye view will result; if it is placed below the foundation of the building, the perspective will give the viewer the impression of looking up at the building from a low position, such as a valley. The normal position of the horizon is 5'-6" or 6'-0" above the ground line; this distance represents the eye-level height of the average person standing on level ground. Care should be taken not to place the horizon line at the same height as a dominant horizontal feature on the building, such as a strong roof line. The feature will then coincide with the horizon line and thereby lose much of its interest and importance. Low buildings, such as houses with flat roofs, are usually given more interest if the horizon line is placed 25' or 30' high. Although this placement gives prominence to the roof, it nevertheless reduces strong, nearly horizontal roof lines (Figs. 9–2 and 9–9).

Note that the variables that exist in the setting up of a perspective layout make it possible for the drafter to adapt a mechanical projection method to a variety of perspective situations. With experience, the modifications can be made to give variety to the perspectives, which will be limited only by a drafter's reluctance to experiment and improve the quality of her or his work. Remember, however, that interest is important on a drawing, but not at the expense of misrepresenting true architectural conditions.

TYPES OF PERSPECTIVES

All linear perspectives (those defining outlines) used by delineators can be classified as either 2-point or 1-point perspectives (as we stated, 3-point perspective is not effective in architectural presentation). The 2-point angular perspective (Fig. 9–2) is the most popular type for showing the exteriors of buildings. Two sides of the building are seen, and the angular nature of these sides reveals the important information without excessive distortion. Two methods of construction have been developed:

1. The common or office method (direct projection)
2. The perspective-plan method

The *office method* is of particular importance to the beginner; it is widely used and most often the simplest method for orientation. Although more complex, the *perspective-plan method* has more versatility, and drawings can be completed more quickly once the principles have been mastered. Knowledge of these two methods will be sufficient for any perspective work encountered, and both will be discussed in what follows.

The 1-point perspective (Fig. 9–25) depicts a building or an interior with one side parallel to the picture plane. Note that the horizontal lines of the parallel side are drawn horizontally, producing a true orthographic shape of the side. The receding, parallel sides are formed by lines converging to a single point, the *vanishing point* (V.P.), usually placed within the view. Interior views of rooms are often drawn with the 1-point method; it presents an accurate description of the facing wall, combined with observation of both receding side walls.

Another typical application for 1-point perspective is a street flanked by buildings. As the viewer is looking di-

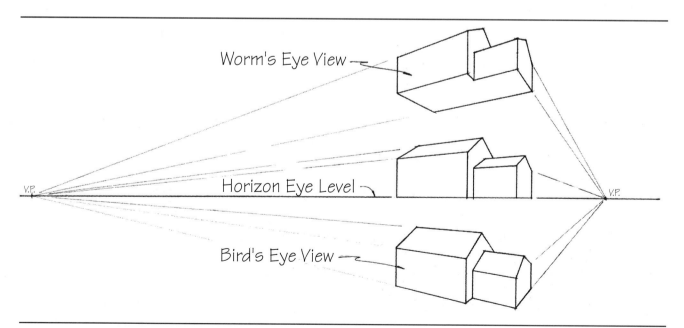

Figure 9–8 Effect of viewing a building from different heights.

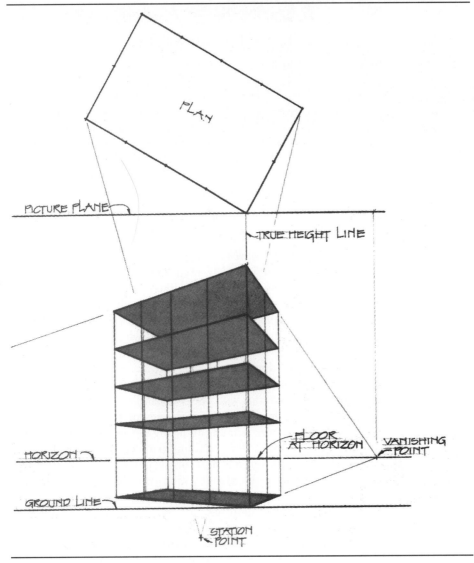

Figure 9-9 Appearance of floor levels at various heights.

rectly down the street, the vanishing point falls at the end of the street. The receding building and street lines are then conveniently drawn to the one vanishing point (Fig. 9–26). Other dramatic applications can be found for the 1-point method, especially when formal architectural arrangements are involved. Many of the principles of 2-point perspective apply equally to 1-point, with only minor variations in setting up the change in perspective.

9.4

SKETCHING A QUICK PERSPECTIVE

The realistic appearance of a perspective sketch is often helpful, not only in analyzing many of the esthetic aspects of the structure before the working drawings are

completed, but also in presenting preliminary ideas to a client. Exterior features that appear quite satisfactory on elevation views may not always be pleasing when observed from an oblique angle. Axonometric and oblique drawings usually are not entirely reliable for scrutinizing these design aspects, whereas perspective sketches are especially adaptable to critical examination of architectural ideas during their planning stages.

Briefly, a perspective drawing or sketch is based on the fact that all horizontal lines that recede from the observer appear to converge at a distant point (that is, if one side of the building is parallel to the picture plane). The receding lines will appear to vanish at two distant points if the building is observed from an angular position (Fig. 9–10). Also, these points of convergence, called *vanishing points*, as mentioned previously, fall on a horizontal line or a plane that is theoretically at the same level as the observer's eye and that appears on the drawing very similarly to the way that the actual horizon appears on a

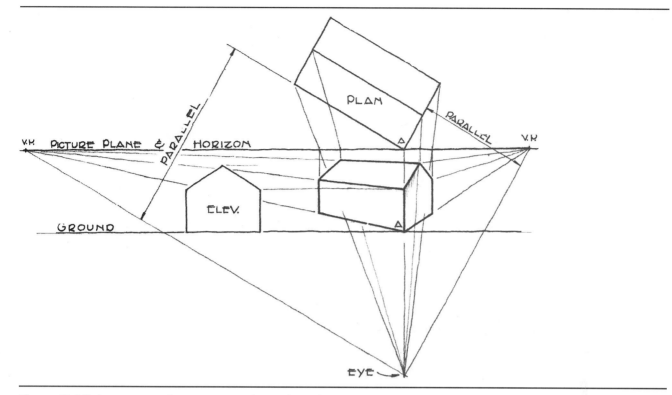

Figure 9–10 A perspective layout using a plan and an elevation view.

landscape. Depending on the placement of the horizon line on the drawing, a structure will appear to have been viewed from either a raised or a lowered position. Usually, the sketch is made with the natural height of observation in mind: about 6′ high, the height of a person's eye above the ground.

In sketching a perspective, the drafter usually arrives at vanishing points and proportion by eye; in fact, the points are commonly omitted except in early exercise sketches because the eye soon becomes trained to establish their correct position and horizon level mentally. Often the points do not fall within the confines of the paper.

Early exercises should consist of simple block studies as shown in Figs. 9–11 and 9–12. Basic building forms and shapes with a minimum of detail should be sketched in various positions until they look realistic; then more complex shapes can be attempted. Generally, the angular-type perspective, as shown in Fig. 9–13, is appropriate for small, planning-stage sketches. For sketches of interiors, the 1-point parallel perspective, shown in Fig. 9–15, is adequate for most situations.

In general, follow these simple steps in sketching perspectives of architectural subjects:

Step 1: Draw a vertical line to represent the closest vertical corner of the building. On this line, assume a sense of vertical scale by stepping off equal divisions of, say, 5′ to any height needed for the building. The divisions can be actually about ½″, 1″, or any similar length, depending on the desired size of the finished sketch.

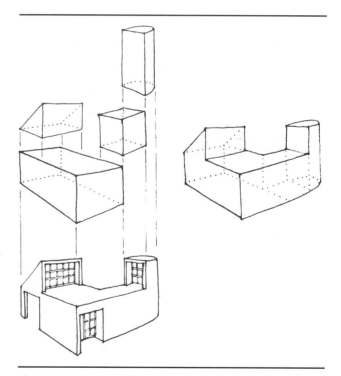

Figure 9–11 The basic geometric forms in a perspective sketch.

Step 2: At a point nearly 6′ high (approximately the height of the average person) on the vertical line, sketch a horizontal line across the paper to represent the horizon (Fig. 9–13). Occasionally, you can improve the

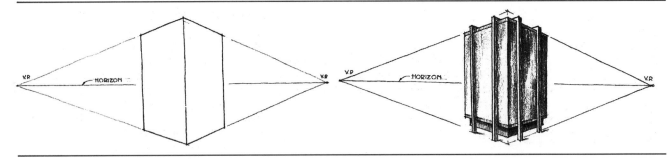

Figure 9–12 Blocking in forms to get proportion.

Figure 9–13 Blocking in residential forms.

appearance of the building by raising the horizon line to more than 6′; however, avoid placing it at a level coinciding with major roof lines or other important horizontal features of the building.

Step 3: Establish vanishing points on the horizon line by trial (Fig. 9–14). Place the points on the horizon line well out beyond the area for the intended building so that the receding lines from the vertical forecorner will result in realistic forms.

Step 4: From the vertical line, sketch a *trial block* of the general size and proportion of the building. The scale of the plan can be taken from the vertical line, but depth proportions become smaller as they recede from the viewer. If the basic block does not appear to have the desired shape, move the vanishing points to new locations on the horizon line. Even the horizon can be moved to create more desirable appearances. A few trials will produce the correct construction for the building. Wall lines of the block must converge at their respective vanishing points, and care should be taken to avoid a sharp angle formed by the receding walls at the bottom of the forecorner (Fig. 9–14). If the building is a house

with a gable or a hip roof, block in the entire height of the roof at the ridge height; use the ground line as the bottom of the block.

Step 5: Develop the irregularities and offsets of the building, using the basic block as a cage or framework. Symmetry can be maintained with the proper use of center lines. Remember that similar features will appear wider near the observer than they will farther away. Depth proportions are established by eye so that the entire sketch maintains a satisfactory appearance.

To simplify beginning perspective sketches, use grid sheets that can be placed under tracing paper when sketching. These sheets or charts are laid out for various angles of view, and the grids eliminate the need for establishing vanishing points. Adopt a suitable scale to establish the basic block of the sketch; after this scale appears to be correct, add the details, with grid lines serving as a guide for receding lines.

*Interior sketch*es are started with an orthographic, rectangular view of the facing wall (Fig. 9–15), inasmuch as we will be looking directly into the wall if we sketch a parallel perspective sketch. The height of the rectangle on the ver-

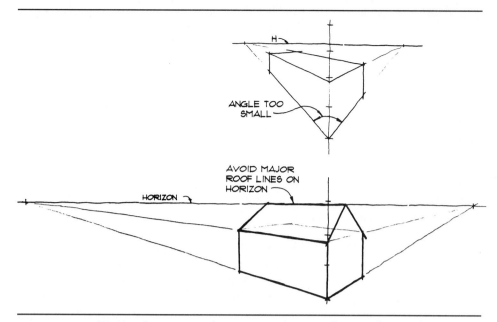

Figure 9–14 (A) Vanishing points that are too close together produce a distorted appearance. (B) Vanishing points placed further apart produce a natural appearance.

Figure 9–15 Interior sketch using 1-point perspective.

tical plane can be considered to be the standard 8′ ceiling height, for convenience, and the length is made accordingly. For example, if the wall is 16′ long, it is drawn twice as long as whatever height we have made the 8′ ceiling. Place a vanishing point about 6′ high near the center of the rectangle. Side-wall lines are drawn from the corners of the rectangle outward. Notice that the side-wall lines, if extended into the rectangle, converge at the vanishing point. All lines that are perpendicular to the picture plane converge at this point. Heights of relative features are horizontal on the facing wall and can be carried along the side walls as long as they vanish toward the vanishing point.

This construction forms the basis for the sketch; all features such as doors, windows, beams, and furniture are added in proportion after completion of the basic room interior.

DRAWING A 2-POINT ANGULAR PERSPECTIVE OF AN EXTERIOR USING THE OFFICE METHOD

Figures 9–16, 9–17, and 9–18 show the step-by-step sequence usually employed in this method. The steps are described in the following paragraphs.

Step A: The Plan View

1. Draw the floor plan, or roof plan as shown in the figures, on a 30°–60° relationship with the horizontal (Fig. 9–16). Or you can tape a separate plan down in a similar position.
2. Draw the horizontal picture-plane line touching the lower corner of the plan. (Other relationships can be used later, if desired.)
3. Locate the center of vision (C.V.) midway on the horizontal width of the inclined plan.

4. Establish the station point on the center of vision far enough from the picture plane to produce a 30° cone of vision (Fig. 9–4).
5. From the station point, draw projectors, parallel to the sides of the plan, to the picture plane. These points represent the vanishing points as seen in the plan view and are often called *distance points*.

Step B: The Perspective Setup

6. Draw a horizontal ground line a convenient distance below the plan view far enough from the picture plane to allow sufficient space for the perspective layout (Fig. 9–17).
7. Draw the elevation view on the ground line. Place it off to the side of the perspective area; even if projection lines run through the elevation view, no harm will result. This view supplies the heights; therefore, it must be drawn to the same scale as the plan. Usually, the end elevation is sufficient if the major heights are shown. (If a perspective is being drawn from a separate set of plans, the elevation,

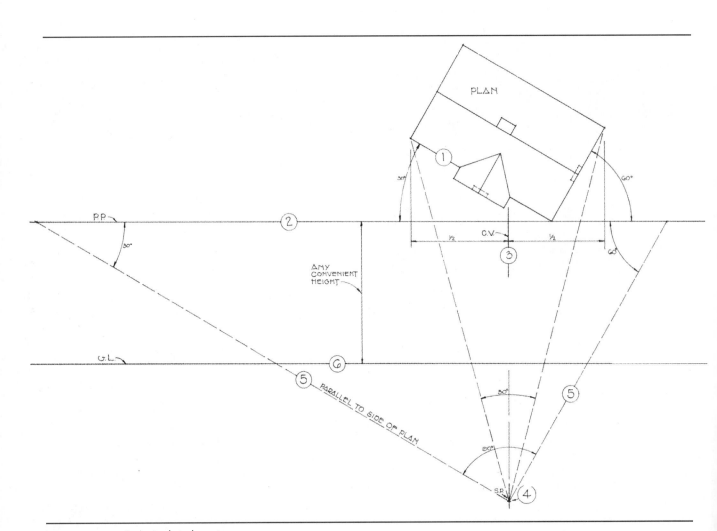

Figure 9–16 Step A: the plan view.

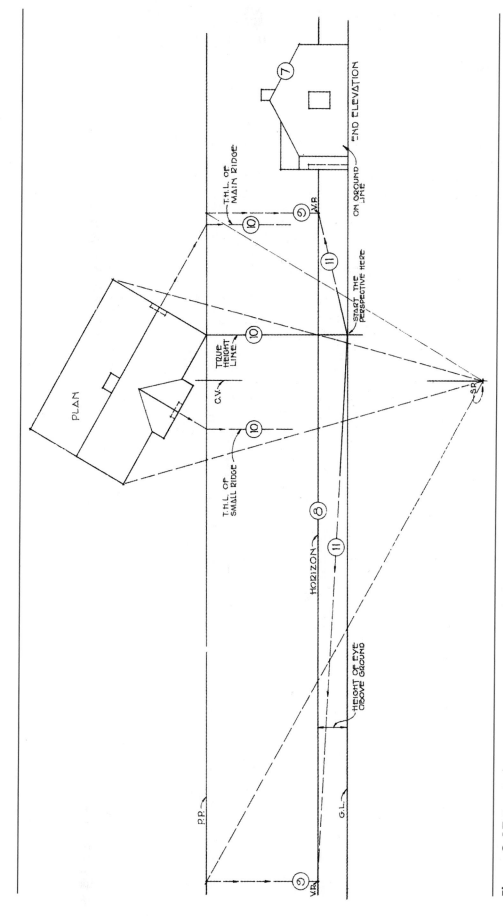

Figure 9-17 Step B: the perspective setup.

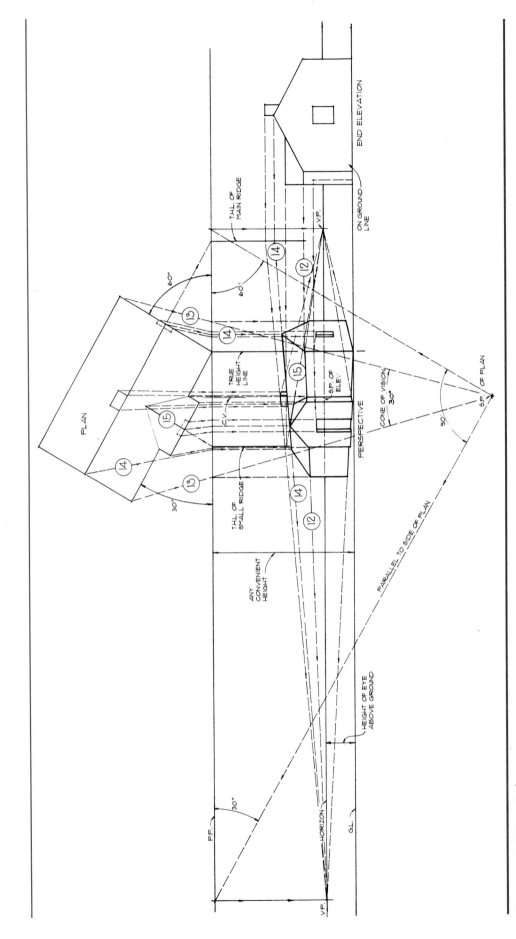

Figure 9-18 Step C: the perspective view.

like the plan, can be merely taped on the ground plane in a convenient position.)

8. Draw the horizon line as shown in Fig. 9–17. The heights of the elevation view will aid you in determining the most effective eye-level height. Usually, if a level view is desired, the horizon line is scaled 6′-0″ above the ground-plane line. This distance is optional.

9. Drop vertical projectors from the picture-plane distance points in the plan view (found in Step A, Number 5) to the horizon line. These points on the horizon line are the vanishing points of the perspective and should be made prominent to avoid mistaken identity.

10. Draw vertical true-height lines from the corner of the plan that touches the picture plane and the extension of the two roof ridges as they intersect the picture plane. Unless they are boxed in, ridge lines of gable roofs should be brought to the picture plane, where a true-height line can be established. Usually, this is the simplest method of plotting their heights; from the true-height line, the ridge height is vanished to the proper vanishing point.

Step C: The Perspective View

11. Now we are ready to start the perspective itself. From the intersection of the main true-height line at the corner of the building and the ground plane, construct the bottom of the building by projecting the point toward both vanishing points (Fig. 9–18). All perspectives should start at this point.

12. Next, continue developing the main block mass of the building. Mark the height of the basic block by projecting it from the elevation view to the true-height line. Again, project this point (on the true-height line) to both vanishing points.

13. To find the width of the basic block, we must go to the plan. With a straightedge, project both extreme corners of the plan toward the station point. Where these projectors pierce the picture plane, drop verticals to the perspective, establishing the basic-block width. You can locate the back corner, if desired, by vanishing the outer corners to the correct vanishing points.

14. Next, plot the main ridge so that the roof shape can be completed. Project the height of the ridge from the elevation view to the main ridge true-height line. Vanish this point to the left vanishing point. The ends of the ridge must be taken again from the plan view. Project both ends of the ridge on the plan toward the station point; where the projectors pierce the picture plane, drop verticals to the vanished ridge line. This defines the main ridge, and the ridges of the roof can then be drawn to the corners of the main block.

15. The small ridge of the front gable roof can be established by the same method as above. Because this ridge is perpendicular to the main ridge, the small ridge is vanished to the right vanishing point. The remaining corners and features of the gable extension in front of the main block can be taken from the plan by the method previously mentioned and vanished to the correct vanishing point.

16. Continue plotting the remaining lines and features on the perspective by locating each from the plan as usual and projecting their heights from the elevation view. After heights are brought to the true-height line, they must usually be projected around the walls of the building to bring them to their position. Remember that true heights are first established *on the picture plane* and are then vanished along the walls of the building to where they are needed. Drawing a horizontal circle in 2-point perspective by the office method is shown in Fig. 9–19.

9.6

DRAWING A 2-POINT ANGULAR PERSPECTIVE OF AN INTERIOR USING THE OFFICE METHOD

Figure 9–20 illustrates the method of drawing an interior view with 2-point perspective. The principles are the same as for exterior views. However, notice that only a partial plan is drawn, and the rectangular shape of the interior touching the picture plane is drawn on the perspective. The view forms within this rectangle; later, the rectangle can be removed if a feathered-out drawing is desired.

A pole has been placed in the room to indicate the method of plotting any point in space; other points can be located in a similar manner. Heights of features on the walls are projected to the true-height line and are carried along the walls to their correct position, which is located from the plan. When setting up the perspective, keep the station point about twice as far from the picture plane as the greatest width of the plan being drawn; this effects a desirable appearance. Coordinates can be laid off on the floor, resembling square tile, if odd locations or shapes are required within the room.

The picture plane can be placed in positions other than those shown on the figure; regardless, projectors locating the features must be brought to the picture plane before they are dropped to the perspective.

9.7

THE PERSPECTIVE-PLAN METHOD AND MEASURING LINES

Comparison will show that the perspective-plan method requires less space on the drawing board, has more versatility, and is obviously more sophisticated than the of-

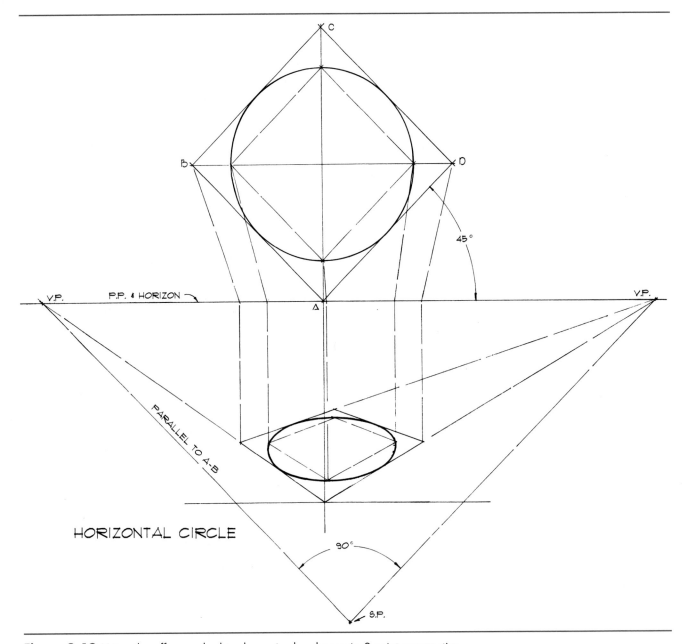

Figure 9–19 Using the office method to draw circular shapes in 2-point perspective.

fice method. Many professional architectural delineators use the perspective-plan method exclusively. Although several new variations in procedure are encountered, the basic principles of the office method are still applicable.

The plan method requires no orthographic plan from which projections are taken. Rather, the perspective plan is drawn from measurements laid off on a horizontal measuring line. From this plan, vertical projectors establish widths and feature locations on the finished perspective. Heights are measured on a true-height line rather than projected.

It is usual practice to draw a perspective from a set of working drawings. The site plan serves as a guide for correctly orienting the station point. The plan and elevation views furnish all of the measurements for drawing the perspective. Here lies one of the advantages of this method: When transferring the dimensions from the working drawings to the perspective layout, the drafter can control the size of the perspective by changing the scale of the dimensions during transfer. Also, the method allows trials of various horizon heights without a major amount of reconstruction. With the use of tracing paper over the original perspective plan, experimentation becomes a simple matter.

Figures 9–21, 9–22, and 9–23 illustrate the three major steps necessary in completing a simple perspective by the plan method. The given conditions and dimensions, similar to those of a typical problem, are shown in Fig. 9–24. The steps are described in the following list.

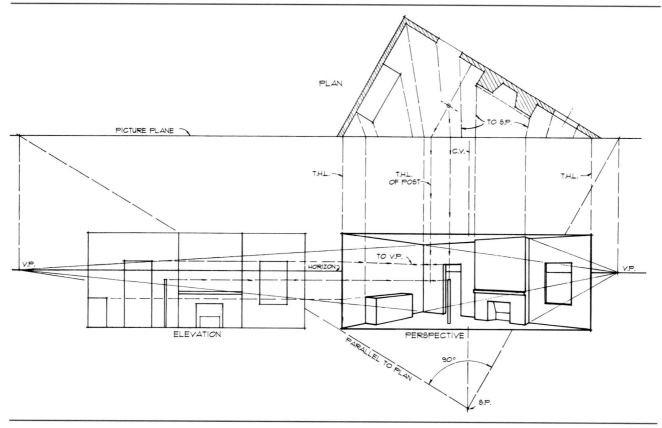

Figure 9-20 Interior 2-point angular interior perspective (office method).

Step A: Locating the Preliminary Points

1. Start with the horizon line and draw it near the upper part of the paper (Fig. 9–21). For convenience, this line also may be used as the picture plane in plan.

2. Establish the station point at the given distance on the center of vision (to scale), and draw the 30°–60° projectors to the picture plane. This locates the left and right vanishing points.

3. Construct the left and right measuring points (M.P.) on the picture plane. To locate the left measuring point, bring the distance between the left vanishing point and the station point to the picture plane with the use of an arc swung from the left vanishing point. From the right vanishing point, scribe the radius (the right vanishing point to the station point) to the picture plane; this point becomes the right measuring point. These measuring points will be vanishing points for the horizontal measurements that we will use in our next step.

4. Draw the ground-plane line 6'-0" (scaled) below the horizon line. This establishes our eye level.

The orthographic floor plan shown with dashed lines in Fig. 9–21 is unnecessary in an actual layout; it is added merely to give the beginner a visualization of the plan and picture-plane relationship, which, to the more expe-

rienced, would be indicated by the points just established on the horizon line.

Step B: Drawing the Perspective Plan

5. At an arbitrary location below the horizon, draw a horizontal measuring line (H.M.L.) (Fig. 9–22). The plan in perspective will develop from this line, making it actually a ground line for the plan only, as well as a line upon which horizontal measurements of the building are laid off. It is helpful to know that projections from the plan to the finished perspective will be more accurate if the measuring line is placed well below the horizon. The exaggerated shape of the resulting plan will not adversely affect the perspective, and sufficient space will be gained for the development of the picture. Transfer the corner of the building touching the picture plane to the measuring line (point *A*).

6. From point *A*, draw lines to both the left and the right vanishing points. These lines are the left and right edges of the plan and are referred to as *base lines*. Measurements laid off on the measuring line, when projected to the measuring point, will terminate at these base lines. To lay out the dimensions of the plan on the measuring line, start at point *A*. All dimensions of the front side of the building (*A–B*), starting at the forecorner, are

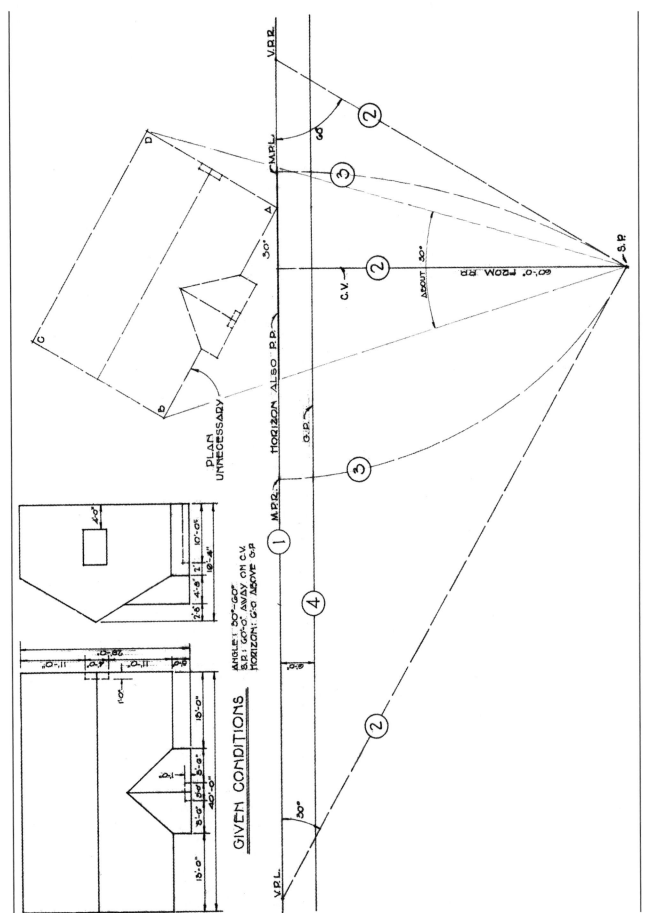

Figure 9-21 Step A: locating the preliminary points.

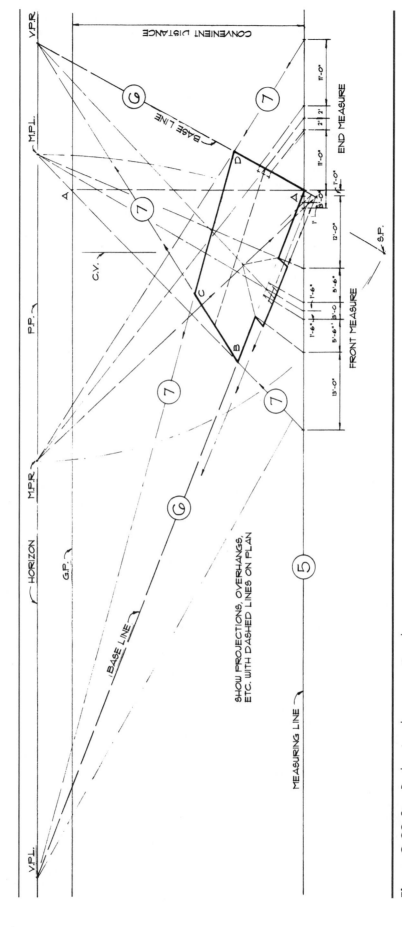

Figure 9-22 Step B: drawing the perspective plan.

stepped off to the left of point *A*, and those for the right end of the building (*A–D*) are stepped off to the right of point *A*. The depth of the small front entrance projection would necessarily extend in front of point *A*, and therefore its depth measurement would be laid off *to the left of point A* instead of to the right.

When part of a plan falls in front of the measuring line (Fig. 9–23D), the base line must continue through point *A* below the measuring line; the measurements must also be laid off in continuity through point *A*. If the left side is in reference, you should lay off the measurements of the features extending in front of the measuring line *to the right of point A*, and you should extend the projection line to the left measuring point below the measuring line to locate the features on the base line. The opposite construction would be needed for similar right-side measurements.

7. To complete the plan, draw lines toward the correct vanishing point from the foreshortened measurements on the base lines.

Step C: Completing the Finished Perspective

8. Lay out a true-height line, projected from point *A* on the picture plane (Fig. 9–23). Establish the bottom of the line on the ground plane, which has been scaled 6′-0″ below the horizon. This line represents the corner of the building on the picture plane, and all scale heights are measured on it.
9. Transfer corners and features from the plan to the perspective with vertical projectors.
10. Project heights of the building to the vanishing points as discussed in the office-method construction. Complete the perspective as shown in Fig. 9–23 by first blocking in the major forms and then adding the projections, openings, and other minor features after the general shape is found to be satisfactory.

When a number of similarly sloped features are needed on a perspective and the amount of slope or pitch is given, it may be advisable to locate the vanishing points of the sloping planes (Fig. 9–24). Slope vanishing points must lie on traces (vertical lines) that pass through the vanishing points located for horizontal surfaces, V.P.L. and V.P.R. Notice that the V.P. for the 30° inclined roof plane, labeled No. 1, is located on the right trace *above* V.P.R. If a V.P. for a horizontal surface is located on the horizon line, then the V.P. for an inclined surface must be located *above* or *below* the horizon.

To locate the V.P. for slope No. 1, start at M.P.R., and lay out the given slope (30° in this case) from the horizontal, as shown by the shaded area in Fig. 9–24, and extend the line to the right trace. This locates the V.P. of the inclined roof. The V.P. for the declining roof plane on the back part of the house is located from the same M.P.R., and the slope is laid out *below* the horizon and extended to the right trace *below* the V.P.R.

Opposite-slope V.P.s are located in a similar manner as shown by the shaded areas Nos. 3 and 4 and their related projection lines.

9.8

DRAWING A 1-POINT PARALLEL PERSPECTIVE USING THE OFFICE METHOD

The 1-point perspective has several typical applications, such as interiors, street scenes, and exterior details of entrances or other special features. Sometimes the 1-point method is the only effective way to represent buildings of unusual shape. For example, a U-shaped house being observed toward the U is most faithfully represented with a 1-point perspective. This method is usually easier to draw than the 2-point, angular perspective, and it is the only type that reveals three wall planes. All receding horizontal lines converge at only one vanishing point, and lines parallel to the picture plane in plan are parallel to the horizon in the perspective. Usually, less board space is needed.

Figure 9–25 shows a simple room interior drawn with the 1-point method. Notice that many of the perspective principles previously discussed are applicable. All of the variations with respect to the placement of the picture plane, station point, and horizon equally affect the finished 1-point perspective drawing.

9.8.1 How to Construct a 1-Point Parallel Perspective

To construct a 1-point parallel perspective, start by drawing the plan and elevation views as shown in Fig. 9–25. Establish the picture plane at the lower part of the plan (it may be placed in front, in back, or at an intermediate area of the plan). On the center of vision below the plan, locate the station point at approximately the width of the plan away from the picture plane; or a 60° maximum cone of vision in this case will satisfactorily locate the station point. The elevation view can be placed on either side of the area reserved for the perspective drawing. From the plan and elevation views, project the frame of the perspective representing the picture plane in elevation. Locate the one vanishing point within the frame at the desired distance above the ground line (the bottom of the frame). No horizon line is needed.

If other than a room interior is to be drawn, start the perspective by drawing the features touching the picture plane; project their lines from both the plan and the elevation views. Project interior wall lines toward the vanishing point. Locate the horizontal spacing of points and vertical lines by projecting the features from the plan toward the station point; at the intersection of the projectors and the picture plane, drop verticals to the perspective in the same manner as described in 2-point per-

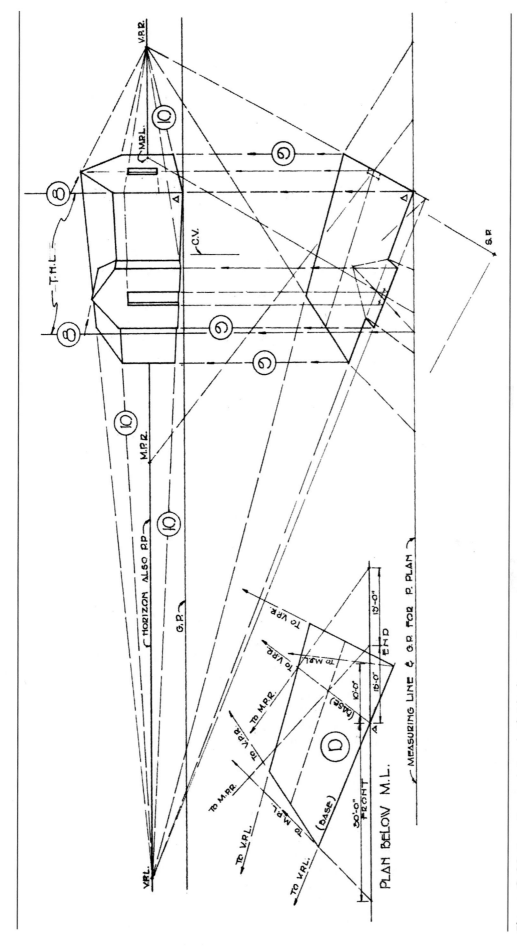

Figure 9-23 Step C: completing the finished perspective.

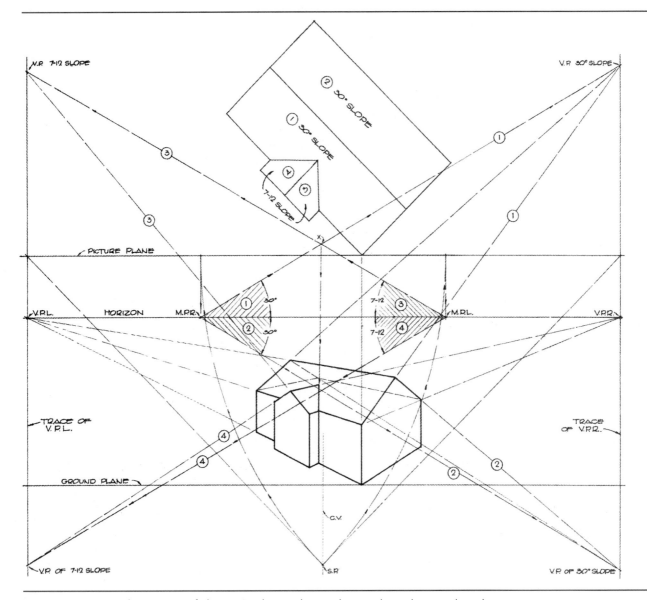

Figure 9–24 Vanishing points of slopes. Similar numbers indicate relative lines and angles.

spective (Section 9.5, Step C, Number 13). Project heights from the elevation view to the true-height line. Notice that heights are carried along the walls, floors, or ceilings to where they are needed. You can locate heights for objects away from the walls by first establishing their heights on the nearest wall. Then, after projecting the objects horizontally to the same wall on the plan view, you can easily bring their heights and locations down to the perspective view. This procedure is indicated by the arrowheads on projectors from the tall box in the room in Fig. 9–25.

The perspective-plan method can also be adapted to 1-point perspective construction, yet the office method is usually less time-consuming.

Figure 9–26 shows the construction of a simple *exterior perspective* using the 1-point method. Notice that the street lines converge at the vanishing point in the center of the drawing and that the buildings have one wall parallel to the frontal picture plane.

When drawing 1-point perspectives of room interiors (frequently used by interior designers), the student often finds it difficult to place furniture in its desired position within the floor area. One method that will simplify the location of objects is to use grid lines (Fig. 9–27). Notice that a scaled orthographic plan is first needed with the furniture laid out in its correct position. Convenient grid lines, similar in appearance to large, square floor tile, are lightly drawn on the plan and are numbered consecutively if necessary both ways. The same grid lines are then drawn in perspective on the perspective drawing floor area.

Locate all furniture outlines in perspective from the plan diagram using the correct grids for placement. Next, build up the heights of the furniture with measurements as previously mentioned to have correctly blocked-in forms. To complete the actual appearance of the furniture, add the details last.

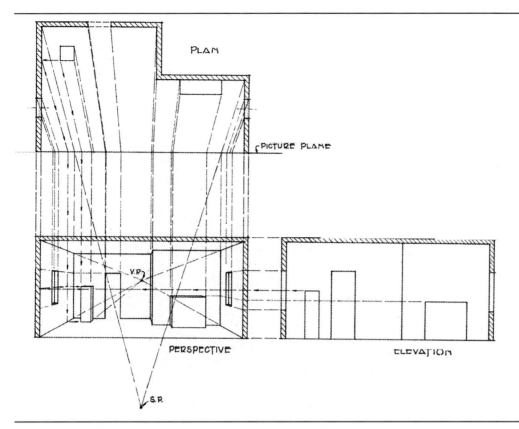

Figure 9–25 One-point parallel perspective (office method).

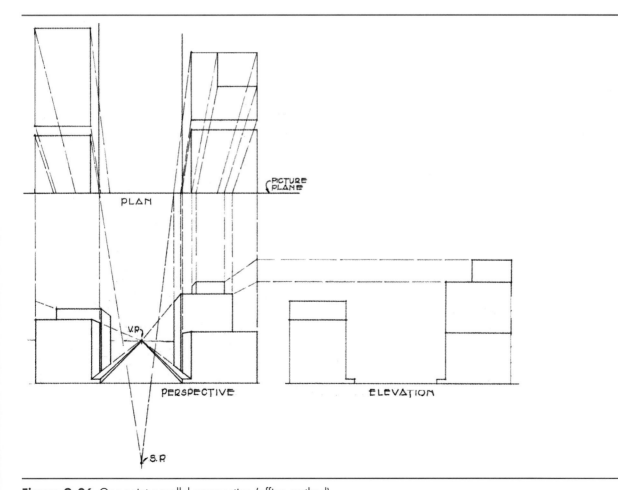

Figure 9–26 One-point parallel perspective (office method).

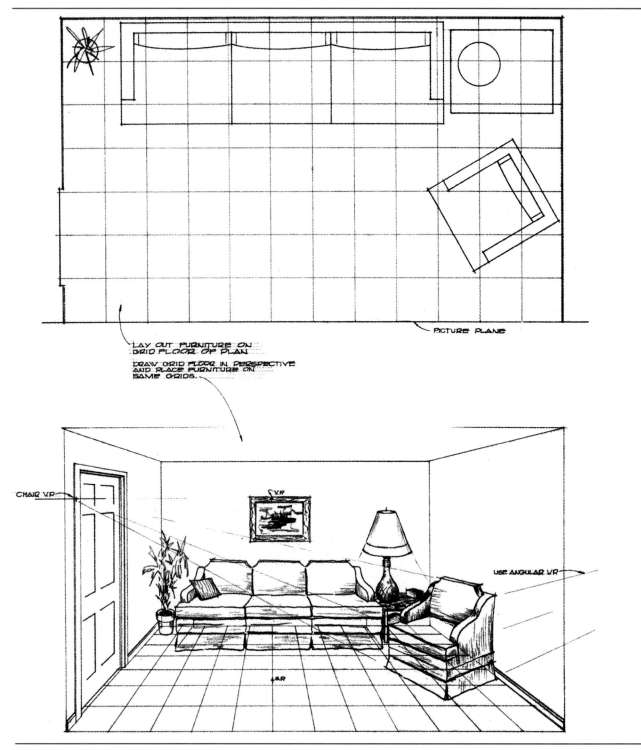

Figure 9–27 Using a floor grid field on a 1-point interior perspective to position furniture correctly.

A PROFESSIONAL METHOD OF DRAWING PERSPECTIVES

As we have already seen, setting up the perspective construction for the average building is time-consuming for the drafter. The professional delineator, who is continu-

ally concerned with architectural perspective, must adopt a rapid yet versatile system that consistently produces satisfying and faithful drawings. One method having these characteristics combines the perspective-plan and measuring-point principles with a simple way of modifying and controlling the setup during construction. It begins with a pictorial plan of very small scale, drawn by the perspective-plan method, mainly for the purpose of early

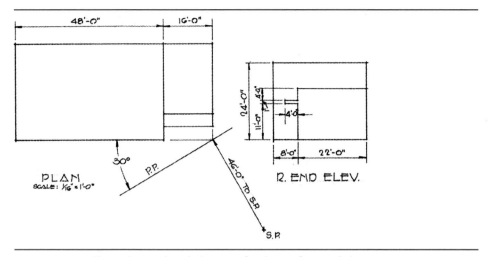

Figure 9–28 Floor plan and end elevation for the professional diagram perspective.

study before the finish perspective is started. This small, preliminary layout, called a *diagram*, is the secret of good perspective without unnecessary, large-scale, trial-and-error construction.

After a small diagram has been perfected as to angle of observance, distance to station point, height of horizon, and the like, only the necessary lines and points are enlarged to scale for the final perspective. Several drawings employing this system will convince the student that it is as effective in the classroom as it is in an architectural office. Follow the sequence of the numbers shown in the accompanying drawings and in the written instructions that follow.

Before beginning to draw, study the site plan, if available, or the proposed site arrangement. Determine which angle of observation will show the important elements of the building. Draw a line on the site plan indicating the line of sight you have chosen. On this line, establish the station point by laying a 30° triangle on the line so that the angle represents the angle of vision (Fig. 9–28). The apex of the 30° angle, as previously indicated, locates a satisfactory station point when the length across the building forms the length of the side opposite the 30° angle of the triangle. This point will help you visualize the tentative picture and will orient you to the problem at hand. Study the elevation views, which will then be visible, and concern yourself only with them; if you are using a set of working drawings, lay the other drawings aside.

Step A (Fig. 9–29)

1. Draw the horizon line as shown.
2. Construct a vertical center-of-vision line. It will also serve as a true-height line later.
3. Establish the station point on the center of vision by scaling the distance that you indicated on the site plan. Use a small scale. The civil engineer's scale can be used for enlargements of scale, if desired; it provides convenient multiples of 10.

4. Through the station point, draw a horizontal line; this will be the picture plane in plan upon which all horizontal measurements of the building can be made. Occasionally, it may be necessary to construct an auxiliary picture plane for measuring; this construction will be discussed later.

Step B (Fig. 9–29)

5. Locate the right and left vanishing points on the horizon line by projecting from the station point as shown (if a 30°–60° angle of observance is satisfactory). Any angle can be used as long as the included angle between the two projectors is 90°. (These projectors can serve as base lines of the perspective plan if the drawing is small and space is not critical.)
6. Next, locate the left and right measuring points on the horizon line. With a compass, swing the distance between the station point and the right vanishing point to the horizon line, using the right vanishing point as a center, to locate the right measuring point. This point will be the vanishing point of all parallel measuring lines laid off to the right station point. Follow the same procedure for bringing the distance between the station point and the left vanishing point to the horizon line, and locate the left measuring point.
7. Establish the desired eye level of the perspective by measuring down (in scale) from the horizon on the center of vision to locate the corner of the building on grade. From this point, construct lines to both the left and the right vanishing points. These lines become the base legs of our basic rectangle in perspective.

In the following step (Step C), it may be advisable to construct a new picture plane in order to keep the drawing lower on the paper when working with larger buildings and scales. On small preliminary diagrams, the picture plane indicated in Step B will usually serve the

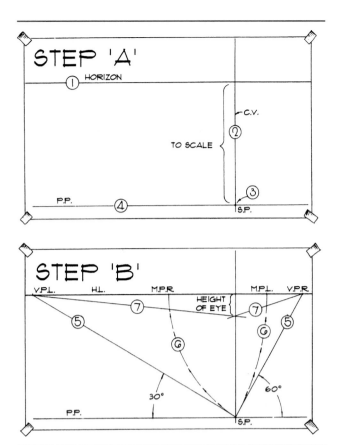

Figure 9–29 Preliminary steps (A and B) for the professional diagram perspective.

center of vision (see Section 9.7, Step B). Project right-side measurements to the right measuring point, and left-side measurements to the left measuring point. *When they intersect the base line,* vanish them toward the corresponding vanishing point.

Step D (Fig. 9–30)

11. Lay off measurements and construct all projections, overhangs, and roof lines if necessary. Intersections of sloping roofs will be needed to complete the perspective on gable and hip roofs. To avoid confusion later, when projecting the plan up to eye level, draw roof intersections, overhangs, exterior stoops, and the like, as broken lines on the plan.

12. Project all visible corners from the plan up to the perspective. This final step is generally done on a separate tracing paper *overlay;* only the perspective will then be on the clean sheet. All construction is made from points under the overlay. Of course, preliminary diagrams are done on the original sheet for study when only general shapes are necessary.

13. Take heights from the working drawings and convert them to scale. Starting at ground level, mark off the heights on the center of vision. The center of vision is on the elevation picture plane and therefore can be used for true-height measurements. If necessary,

purpose without being cumbersome. It will be found satisfactory for laying off measurements and constructing the perspective plan. However, if a large building is drawn, requiring a large piece of paper and considerable space on the board, when enlarging the diagram, you will find an auxiliary picture plane more convenient. This new picture plane will replace the original for measurements and is arbitrarily placed between the station point and the ground corner of the perspective. After the vanishing points and the measuring points have been located on the horizon line, the station point is no longer necessary and can be removed.

Step C (Fig. 9–30)

8. Draw the auxiliary picture plane, as shown in Fig. 9–30.

9. Draw the new base lines from the intersection of the picture plane and the center of vision to both vanishing points. Using the original vanishing points, extend the lines below the auxiliary picture plane to take care of overhangs and offsets of the plan that occasionally fall outside the basic rectangle.

10. Draw the basic rectangular shape of the plan as shown. Notice that measurements are taken from the working drawings, converted to the working scale, and laid out on the auxiliary picture plane, on both sides from the

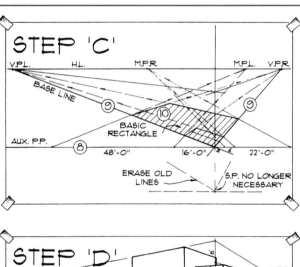

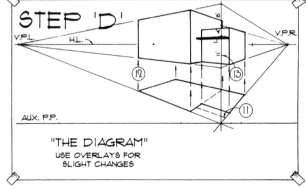

Figure 9–30 Steps C and D in drawing the diagram.

project heights around the walls of the building to where they are needed, as explained in previous methods. Block in basic shapes first, and then make measurements and project details last to avoid confusion resulting from numerous lines.

Step E Enlarging to Desired Size (Fig. 9–31)

14. After several diagrams are studied and one is found to be satisfactory, a larger perspective can easily be constructed at the desired size. Lay out the horizon line, and transfer the center of vision and all points from the diagram to a new scale. The size of the finished perspective can be controlled by the scale selected. For example, if the finished perspective is to be four times the size of the diagram, use a scale four times as large as the diagram for the new measurements along the horizon line, center of vision, and auxiliary picture plane.

15. At larger scales, horizontal measurements for constructing the plan will occasionally fall beyond the paper. To overcome this difficulty, follow the procedure shown in Fig. 9–32. Draw a horizontal line to the edge of the paper from a point on the base line where the longest measurement has been made. Make the additional measurement from the center of vision, and, instead of projecting it toward the measuring point, project it to the vanishing point on the same side of the center of vision until it intersects the horizontal line. From this point it is treated as previously shown in Step C. The construction merely brings the

dimension back in perspective to the point of maximum measurement and lays it off on that plane, rather than on the original forward picture plane.

TIME-SAVING SUGGESTIONS

After making several drawings, you may find slight variations of the procedure that save time and overcome minor difficulties, should they arise. The following list offers some suggestions for saving drafting time:

- **Similar perspectives** If a number of similar perspectives are to be drawn, use perspective grid charts as an underlay. Various charts are available at drafting supply stores (Fig. 9–33).
- **Diagonals** Use diagonals of rectangular areas for quickly locating centers and for checking construction of the perspective as it develops (Fig. 9–34).
- **Subdividing vertical heights** When a vertical height must be divided into equal subdivisions, a scale can be positioned at a convenient point for measuring (Fig. 9–34).
- **Reflections** When showing the reflections of buildings in water or on other shiny surfaces, draw the reflections to the same vanishing point as the building. The water's edge is the dividing line between the reflections and the true images. Locate a reflected point as far below the shiny surface as the point above (Fig. 9–35).

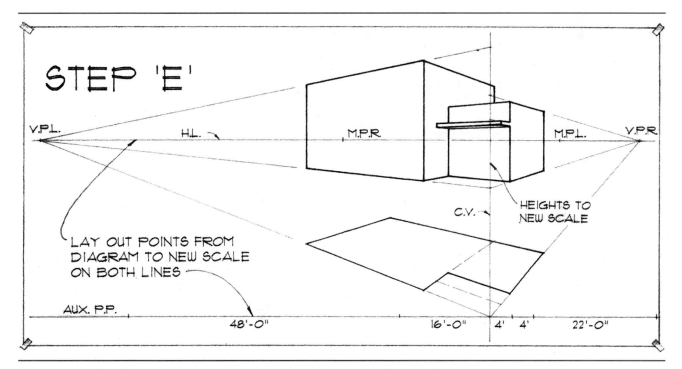

Figure 9–31 Step E: laying out the perspective at a convenient scale from the diagram.

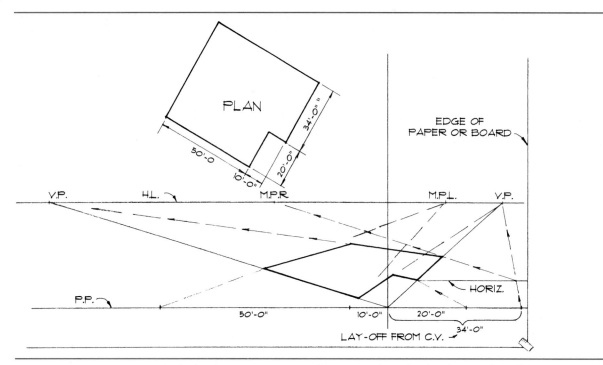

Figure 9–32 Step E: a method of making measurements that are too long for the drawing board.

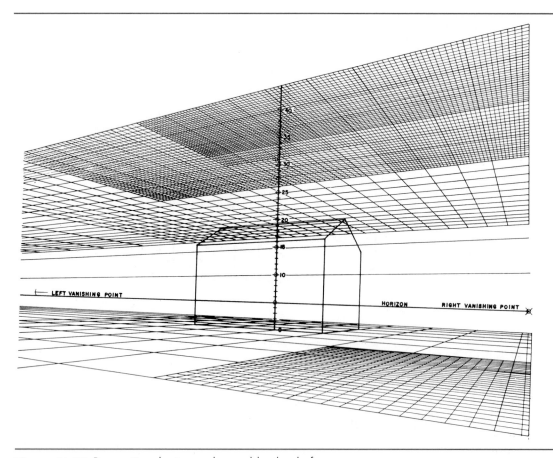

Figure 9–33 Perspective charts may be used by the drafter.

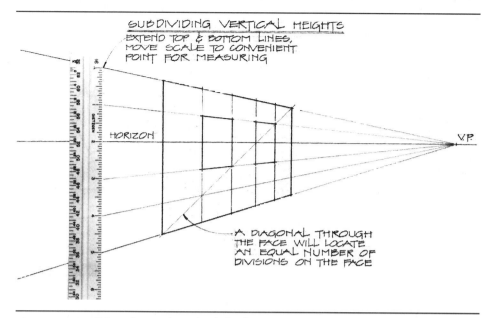

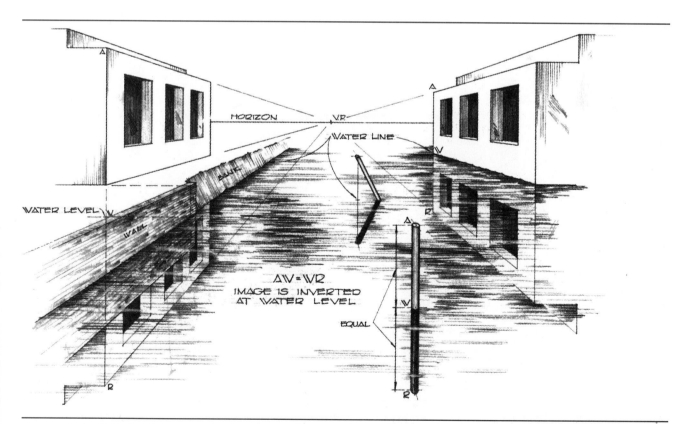

SUBDIVIDING VERTICAL HEIGHTS
EXTEND TOP & BOTTOM LINES,
MOVE SCALE TO CONVENIENT
POINT FOR MEASURING

HORIZON

V.P.

A DIAGONAL THROUGH
THE FACE WILL LOCATE
AN EQUAL NUMBER OF
DIVISIONS ON THE FACE

Figure 9–34 Using diagonals and subdividing perspective spaces.

HORIZON

VR

WATER LINE

WATER LEVEL

WALL

AV = VR
IMAGE IS INVERTED
AT WATER LEVEL

EQUAL

Figure 9–35 Reflections on water or other shiny surfaces can be quickly projected with the use of points as shown.

EXERCISES

1. Make a perspective sketch of the piece of furniture in Fig. 8–22 (Chapter 8).

2. Make a perspective sketch of the small pool house in Fig. 8–34 (Chapter 8).

3. Make a perspective sketch of a small home from a magazine.

4. Using the office method, draw an angular perspective of Fig. 9–36A, B, and C.

5. Using the office method, draw an angular perspective of the interior views of Figs. 9–37 and 9–38.

6. Using the perspective-plan method, draw an angular perspective of Figs. 9–39 and 9–40.

7. Using the 1-point perspective method, draw a parallel perspective of Fig. 9–37.

8. Using the office method, draw an angular perspec-

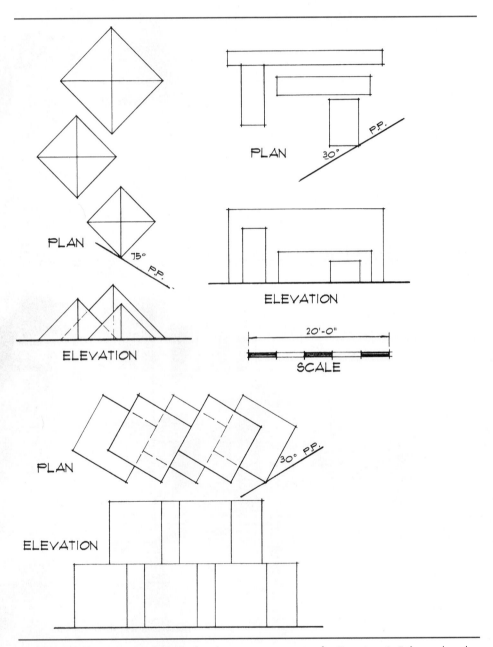

Figure 9–36 Geometric forms to be drawn in perspective for Exercise 4. Enlarge the plan and the elevation to suit your paper.

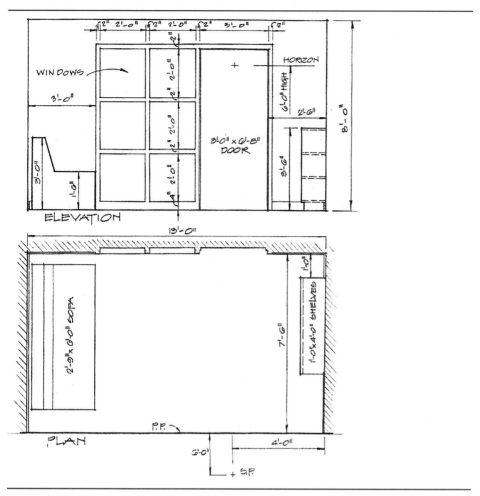

Figure 9–37 Problem for 1-point and 2-point interior perspective (Exercises 5 and 7).

tive of Fig. 9–41 on a 24″ × 36″ sheet with the 36″ dimension placed vertically on the drawing table. Draw the structure four times the size shown in the book. Locate the picture plane 9″ below the top edge

of the sheet. Place the station point 9″ below the picture plane. Locate the ground line 11″ below the picture plane, and the horizon line 3″ above the ground line. Center line *A* on the sheet.

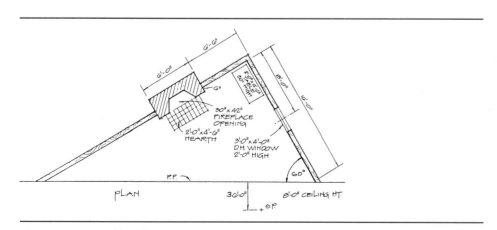

Figure 9–38 Problem for 2-point interior perspective (Exercise 5).

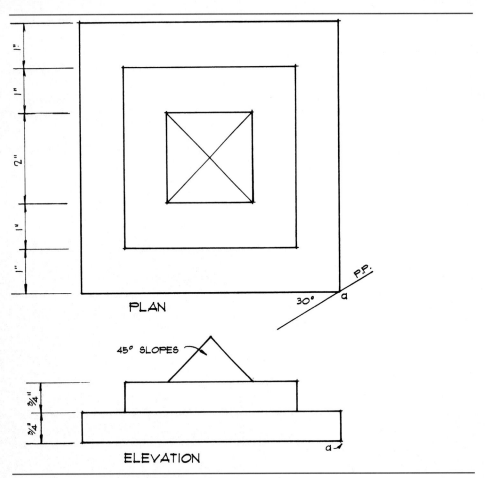

PLAN

P.P.

30° a

45° SLOPES

ELEVATION

a

Figure 9–39 Problem for drawing a perspective using the perspective-plan method. Place the station point 6″ below picture plane and 3″ to the right of point *a*. Place ground line (1) 3″ below the picture plane and ground line (2) 11″ below the picture plane.

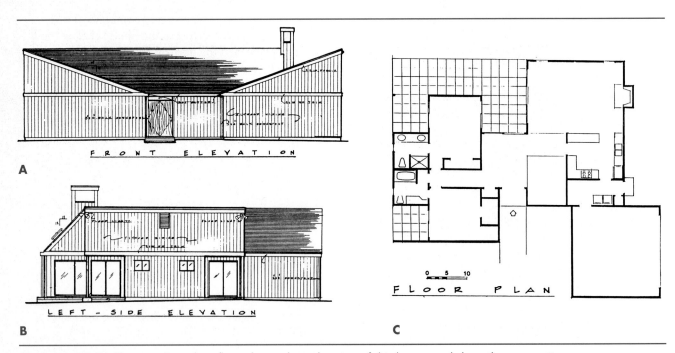

FRONT ELEVATION

A

LEFT - SIDE ELEVATION

B

FLOOR PLAN

C

Figure 9–40 For Exercise 6, scale a floor plan and an elevation of this house, and draw the perspective.

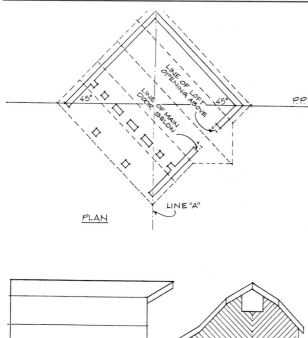

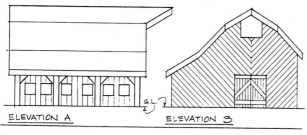

Figure 9–41

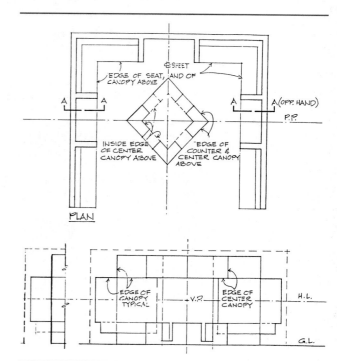

Figure 9–42

9. Draw a 1-point perspective of the interior of the room in Fig. 9–42. Use a 12″ × 24″ sheet with the 24″ dimension placed vertically on the drawing table. Draw the plan and the elevations four times the size shown in the book. Locate the picture plane 6″ from the top edge of the sheet. Place the station point 14″ below the picture plane, centered on the sheet. Locate the ground line 12″ below the picture plane, and the horizon line 2″ above the ground line. Place the vanishing point on the horizon line, centered in the room.

REVIEW QUESTIONS

1. What is meant by *linear perspective?*
2. How does the projection of a perspective view to the picture plane differ from the projection of an orthographic view?
3. What is the point called that approximates the eye when a person draws an object in perspective?
4. What effect is produced on a perspective when the horizon line is raised in relation to the ground line?
5. What effect is produced on a perspective when the object is placed farther away from the picture plane?
6. What type of perspective view is established when the ground line and the horizon line are placed close to each other?
7. Describe the step-by-step procedure used in setting up a freehand, 2-point perspective.
8. Describe in detail the differences between a 1-point perspective and a 2-point perspective.
9. Why must true heights of an object be established only on the picture plane?
10. If an object is placed in an oblique position in relation to the picture plane, why are two vanishing points necessary?
11. What are the advantages of using the perspective-plan method for drawing a perspective?

12. In a 1-point parallel perspective view of an interior, is the single vanishing point used to draw a piece of furniture that is in a diagonal position?

13. When would it be appropriate to use a 3-point perspective?

14. Why are most 2-point perspectives drawn with the horizon line located about 6′ above the ground plane?

15. Why is it important to limit the cone of vision from the station point to the picture plane to an angle of 30° to 45°?

16. When would it be appropriate to use a bird's-eye view perspective?

17. Describe a convenient method for drawing a circle in a 2-point perspective.

18. Why must true heights of an object be established only on the picture plane?

"Our eyes are made to see forms in light;
light and shade reveal these forms . . ."
—LeCORBUSIER

The geometric forms of light and shade produced by the action of the sun on architectural subjects are of particular interest to the architect and the drafter. Good architectural forms have the property of producing pleasing shadows regardless of the sun's position. To the observer, shadows are an integral part of an architectural composition, and their representation becomes almost as important as the building itself. This chapter discusses the roles of shades and shadows in architectural drawings.

10.1

SHADES AND SHADOWS

Linear perspective, as we mentioned earlier, produces only the outlines of objects. Realism is attained not by outlines, but by the sensitive selection of values of light and shade as well as texture to represent various surfaces. The effect of light on surfaces and materials produces the true image; often, outlines are almost entirely obscure. The study of shades and shadows is a further step in creating graphic realism. First the student must understand the action of light; then he or she must define it geometrically as it creates various patterns. Then the student must give these areas or patterns the correct value or tone, in keeping with composition, contrast, and visual interest, to produce the desired pictorial effect.

On actual renderings, shadows can be overdone; if they are too mechanical and hard, much of their three-dimensional expression is lost. The uniformity of the shades and shadows on the illustrations in this material is for the introduction of principles only and should not be taken as the correct representation of shadow values. Other finished perspectives, such as shown in the chapter opening illustration on the opposite page, should be observed for this quality. Note that shades and shadows of finished work seldom have uniform tones throughout; in fact, the interplay of reflected light usually produces a gradation of tone. Shadows on architectural subjects are generally most prominent close to the observer, near the center of interest, and those farther away from the center of interest become more neutral and indefinite as they recede. Contrast and intensity of shadows near the observer, then, should be given the most consideration by the drafter.

10.1.1 Light Source

Usually, on elevation view shadowing, the light source is considered to be coming from the upper left. The conventional method of illustrating light ray direction is to show it passing through a cube diagonally, from upper left to lower right, as shown in Fig. 10–1. Notice that it appears as a 45° line on both the elevations and the plan, and it can be easily drawn with the triangle. Another advantage of using the conventional 45° light source is that it conveniently reveals depth–dimensional characteristics. Shadows will fall to the right of and below the object.

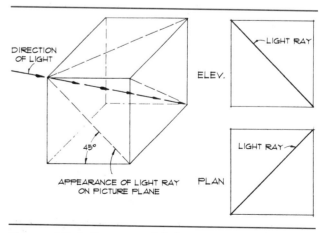

Figure 10–1 Conventional light source on orthographic views.

The shadow of a point will be the same distance below as it is to the right of the original point. Therefore, the shadow clearly indicates the depth of recessed features.

By its convenience for transferring distances from the horizontal to the vertical, and vice versa, the 45° triangle actually serves as a handy tool for measuring when plotting shadows. However, if a different shadow effect is desired, the 30°–60° triangle can also be used. In the casting of shadows, light rays are assumed to be parallel.

10.1.2 Orthographic Shadows

As an introduction to the characteristics of shades and shadows, it would seem logical to begin with shadows pro- duced on orthographic views. You might ask, "Why learn to put shadows on orthographic views, which are com- monly used only for working drawings?" It is true that or- thographic views are mainly for working drawings, and they show depth information by association with related orthographic views. Yet many architectural offices have found that front elevations of buildings (as well as other elevations), with skillfully applied texture indications and shadows, make very adequate and often very attractive pre- sentation drawings (Figs. 10–2 and 10–3). Such drawings are used to show clients the tentative appearance of a build- ing. The greatest advantage of shadowed elevations over perspectives is the tremendous saving in preparation time, and time and cost are usually important.

Figure 10–2 An elevation rendering showing the use of shadows.

Figure 10–3 An elevation view rendering of a small commercial building showing the use of shadows.

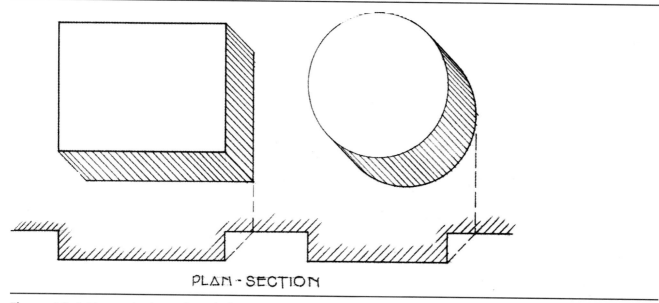

PLAN - SECTION

Figure 10–4 Plotting shadows of simple forms in orthographic views.

At the very outset, we can say that the casting of shadows is affected by three conditions:

1. direction of the light source,
2. shape of the object, and
3. manner and shape of the surface on which the shadow falls.

In analyzing the action of light, you are encouraged to observe the shadows of buildings and different objects found in everyday life, even those of models in an artificial light source; the importance of astute observation of actual shadows cannot be overemphasized. After observing actual shadows and studying the accompanying shadow drawings, you will note a number of obvious *consistencies*. A few general ones are listed below and should be remembered:

1. Only an object in light casts a shadow.
2. A shadow is revealed only when it falls on a lighted surface.
3. The shadow of a point must lie on the light ray through that point.
4. On parallel surfaces, a shadow is parallel to the line that casts it.
5. The shadow of a plane figure will be identical to the outline of the figure if the shadow falls on a plane parallel to the outline of the figure.
6. The shadow of a line perpendicular to the picture plane will be inclined if it falls on a surface parallel to the picture plane.

In the plotting of orthographic shadows, usually two views are necessary for the projection (Figs. 10–4 through 10–9). Sometimes it may be a plan and an elevation; other times it may be two elevations. The important view is the

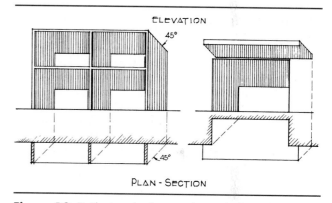

Figure 10–5 Plotting shadows to show relief in orthographic views.

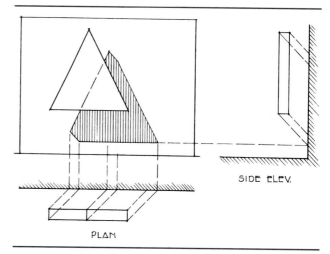

Figure 10–6 Plotting shadows on removed surfaces in orthographic views.

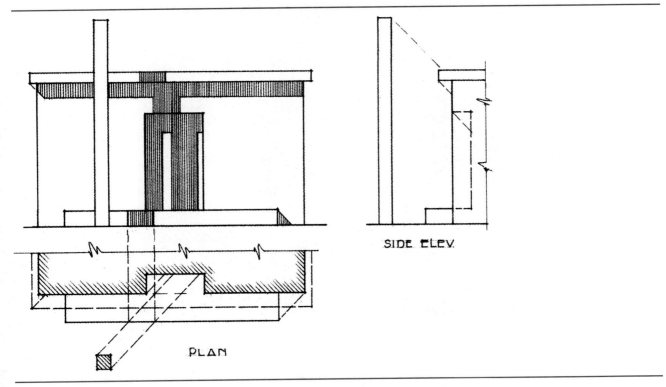

SIDE ELEV.

PLAN

Figure 10–7 Plotting shadows on architectural features in orthographic views.

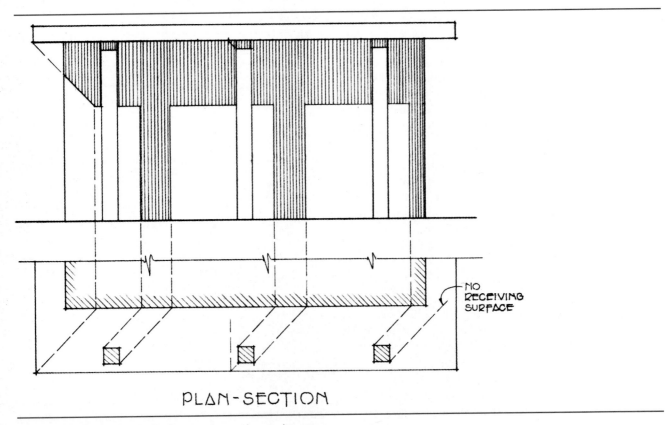

NO RECEIVING SURFACE

PLAN-SECTION

Figure 10–8 Shadows of columns in an orthographic view.

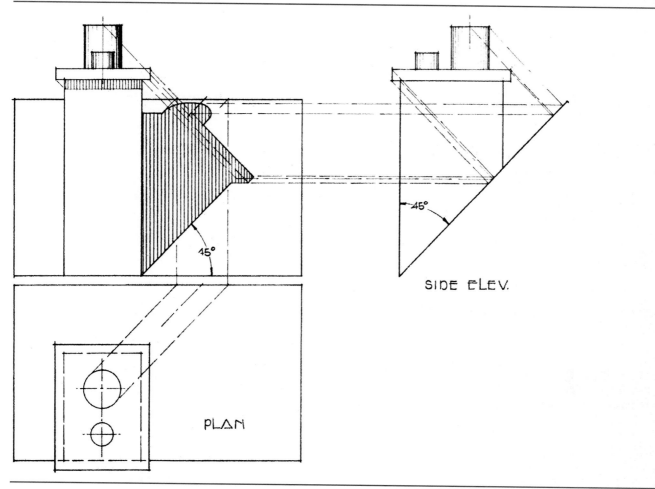

Figure 10–9 Plotting shadows falling on inclined planes in an orthographic view.

one having the surface that receives the majority of shadows. Plot each point or corner casting a shadow, and complete the one shadow profile before going to the next. Check the resulting shadow to be sure that each point is accounted for. If the result does not appear logical to the eye, the construction is usually faulty.

10.1.3 Perspective Shades and Shadows

In plotting shadows of pictorial subjects, you will encounter principles that are similar to those we found in orthographic shadow casting. On perspective drawings, often entire surfaces are on the opposite side from the light source and therefore receive no light. These surfaces must be shown in darker tones; yet they are not shadows. We refer to the darker surfaces of the object not receiving light as *shades* (Fig. 10–10). Determining the outlines of both shades and shadows (as well as occasional highlights) plays an important part in giving realism to perspective drawings.

Notice that the same vanishing points are used for both the shadows and the horizontal lines of the perspective itself (Fig. 10–10). If the light source is parallel to the picture plane, the shadows of horizontal lines will vanish at the same point as the object lines themselves. Also, the shadows of vertical lines will appear as horizontal shadows if they fall on a horizontal surface. Plotting shadows with a light source parallel to the picture plane, of course, limits shadow casting to either the right or the left of the object, never in an oblique manner. In angular perspective, then, one exposed wall will be in light and one will be in shade. You can obtain various shadow characteristics by using different angles of the light source. A high angle of light produces a narrow shadow on a horizontal plane such as the ground, whereas a low angle produces a wide shadow. Usually, 45°, 60°, or 30° angles are most convenient because of their construction with drafting triangles (Fig. 10–11). For the student, casting shadows on 2-point, angular perspectives with *the light source parallel to the picture plane* will produce adequate realism for most situations. For that reason, we will concern ourselves mainly with this method of shadow construction.

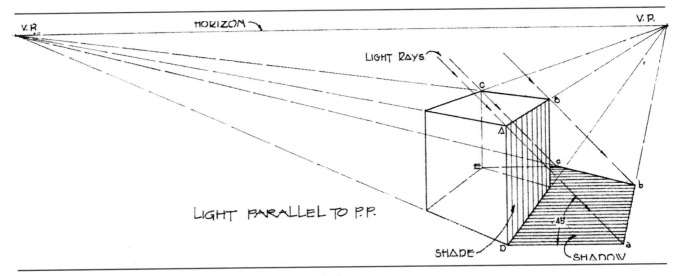

Figure 10–10 Perspective shadows falling on a horizontal plane.

Actually, light striking an object such as a building that is drawn in an angular position produces rather interesting and revealing shadows when the light source is parallel to the picture plane. The shadows from overhangs, offsets, and other features can be made to contribute effective composition elements to the finished drawing.

In Fig. 10–10, a 45° light source produces the shade and shadow of a perspective cube as shown. Point *A* casts its shadow at point *a*, point *B* casts its shadow at *b*, and point *C* casts its shadow at *c*. The shadow of line *A–D* is drawn horizontal, inasmuch as the shadow falls on a horizontal plane. Line *A–B* creates the shadow line *a–b*, which must vanish at the same right vanishing point as line *A–B*. Line *b–c* is the shadow of *B–C* and therefore must vanish at the same left vanishing point. By plotting points and then the lines connecting these points, you can complete the entire shadow outline. Notice that the

shadow of the hidden corner *C–E* is plotted on the figure merely to show the horizontal relationship of *E* to *c*. From Fig. 10–10, we see the following:

1. The shadow cast by a vertical line on a horizontal plane is horizontal.
2. On parallel surfaces, a shadow is parallel to the line that cast it and therefore vanishes at the same vanishing point.

Figure 10–12 shows the shadow of a vertical line being interrupted by a vertical wall plane. The shadow of point *A* cannot be established until the horizontal shadow line from point *E* is projected to the receiving wall. The remaining diagonal line above point *a* is a part of the shadow of line *A–D* and is completed after the horizontal shadow of line *A–D* has been projected on the top of the small block (line *x–y*). Line *a–x–y* is the shadow of *A–D* falling on perpendicular surfaces. From Fig. 10–12, we see the following:

1. The shadow of a vertical line is vertical if it falls on a vertical surface.
2. The shadow of a horizontal line is inclined if it falls on a vertical surface.

Figure 10–13 illustrates the effects of a horizontal shadow cast on various levels of a simple stairs. The shadow is located on each level and is vanished to the right vanishing point, and the shadows on the vertical risers merely connect the shadows falling on the treads. Notice the convenient points used to establish the width of each horizontal shadow.

Shadows falling on inclined surfaces (Fig. 10–14) present interesting projection problems. To find the shadow of the chimney, project the ridge at point *A* to point *B* on the forepart of the chimney. Point *B* is projected horizontally back to the ridge at point *C*. Line *C–D* will then be the shadow line of corner *D–E*; the 45° projection

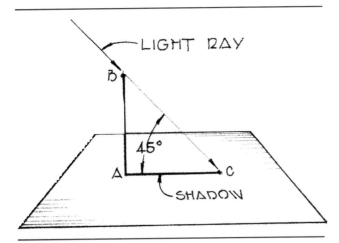

Figure 10–11 On perspective drawings, a light source parallel to the picture plane is convenient for casting shadows.

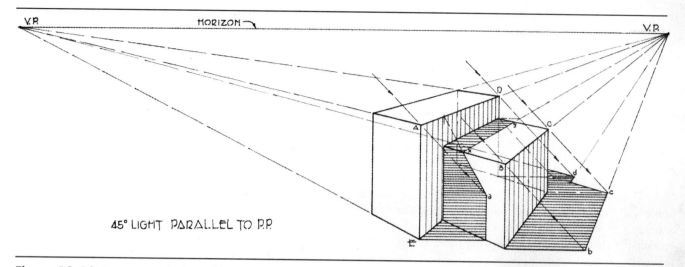

Figure 10–12 Perspective shadows falling on a vertical plane.

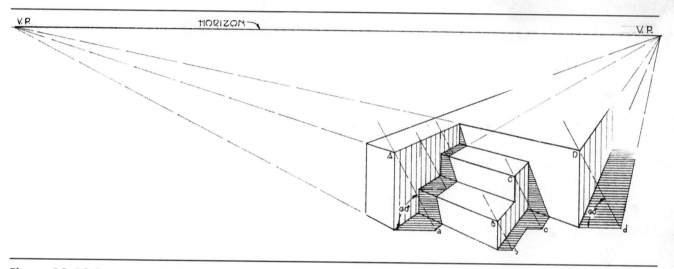

Figure 10–13 Perspective shadows on stairs.

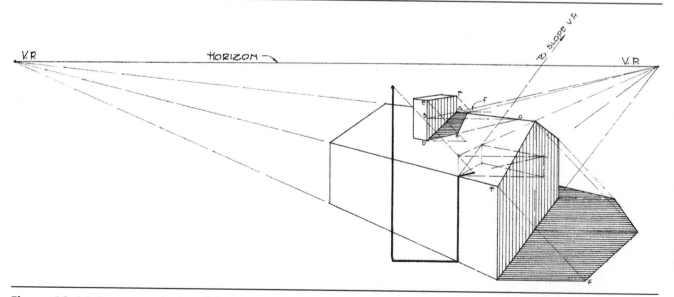

Figure 10–14 Perspective shadows falling on inclined surfaces.

from corner *E* to *e* describes its length. To find the shadow of point *F*, we can consider a theoretical horizontal plane extending back from the ridge height. The shadow of *F* will fall at *f* on the imaginary plane; by projecting *f* to the ridge, we have the shadow of line *E–F* on the inclined roof.

A similar procedure is needed to plot the shadow of the flagpole in Fig. 10–14 after it reaches the incline of the roof. Notice that a line extending up the incline from the shadow at the eave produces a condition similar to the one that the chimney provided. The shadow of the flagpole top is brought down to the vertical plane of the wall (just above the eave on a vertical line); from that point a theoretical horizontal plane is assumed that will intersect the roof, and a horizontal plane is assumed at the eave level. The diagonal connecting both planes will be the shadow of a vertical, such as the flagpole, as it falls on the inclined roof. This diagonal is plotted in the same manner as the shadow of vertical line *D–E* of the chimney.

From Fig. 10–14, it can be deduced then that the shadows of vertical lines are inclined if they fall on inclined surfaces. From Fig. 10–15, note that shadows from surfaces parallel to an incline will be parallel to the inclines.

Perspective shadows with the light source *oblique* to the picture plane can be projected if an actual exterior light condition is desired (Fig. 10–16). Notice that two shadow vanishing points are needed. They can be located at random or by actual bearing and altitude angles of the light source. The light-bearing V.P. is on the horizon, and the altitude V.P. is on a trace through L.H.V.P.

Shadows from a single source such as a light fixture in an interior can be plotted as shown in Fig. 10–17.

10.2

SHADING PICTORIAL SKETCHES

A limited amount of shading on pictorial sketches adds to the effect of realism, although only lines may define the limits of a three-dimensional object. Shading also makes interpretation much simpler for the untrained person, and therefore its use is justified. A few major points seem appropriate here for the purpose of improving sketching technique.

The observation of light as it affects the images of objects is especially helpful to students when learning to add shading to their sketches. Without light we would see nothing; light, therefore, must be considered if we are to represent a three-dimensional object realistically.

The *light source* is generally regarded as coming from an upper-left position, slightly in front of the object. We can think of the source as coming over our left shoulder as we work on the sketch. The top of our object, then, will be the lightest in value or tone, since it receives the most intense light; the front side will have a lesser value, and the right side will appear darkest (Fig. 10–18). Also, reflected light plays on the sides of an object and must be shown. For that reason, surfaces are usually not shaded with a monotone over their entire area; generally, a graduating shade is more effective.

Highlights, which appear entirely white, are the result of direct reflection of the light source from a surface into our eyes. These white areas play an important part in representing the character of each surface and, therefore, should be handled carefully (see Fig. 10–18, which shows

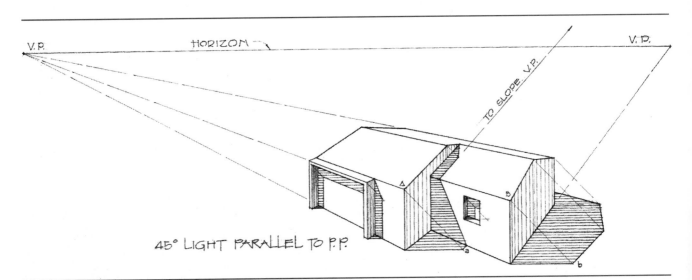

Figure 10–15 Various shadows showing the characteristics of buildings.

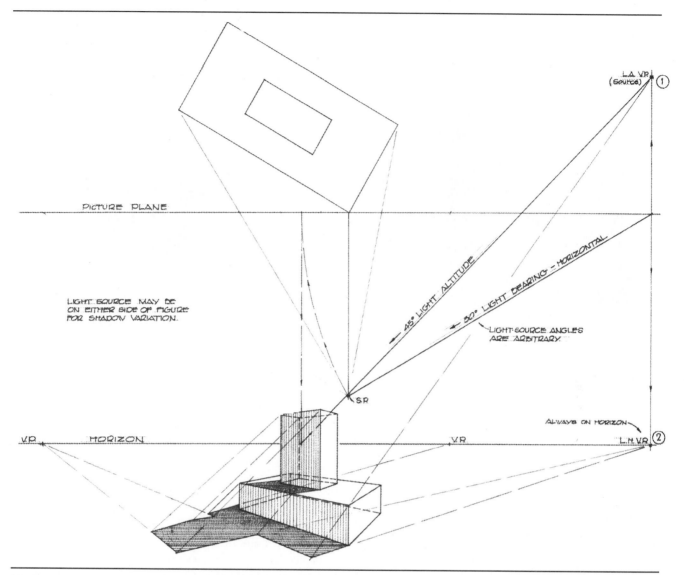

Figure 10–16 Perspective shadows with the light oblique to the picture plane. Notice that two shadow vanishing points are needed. They can be located at random or by bearing and altitude angles of a light source. The light-bearing V.P. is on the horizon, and the altitude V.P. is on a trace through L.H.V.P.

the nature of highlights on basic forms). Usually, it is advisable to omit the highlight in small areas until the shading is complete; then the exact shape of the highlight is rubbed out with an eraser. Overhangs and undercuts, as well as holes and recesses, require the use of shadows to reveal their relief and give them emphasis. Cylindrical surfaces must be made to appear round by application of a parallel highlight slightly to the side of the light source and darker tones near their receding edges.

Light patterns created by strong sunlight on outdoor subjects require more contrast at the shadows than interior subjects.

Contrast can be a useful tool for emphasizing important features on a sketch. When a white area and an extremely dark tone are placed adjacent to each other, the contrast provides a definite emphasis at their junction and gives the impression of bringing the white area forward. This quality is helpful in obtaining three-dimensional effects, especially when adjoining surfaces form abrupt intersections. Remember that shading indications merely supplement the outlines of the sketch and should not be overdone; nor should they produce conflicting geometric shapes that interfere with the important features of the drawing.

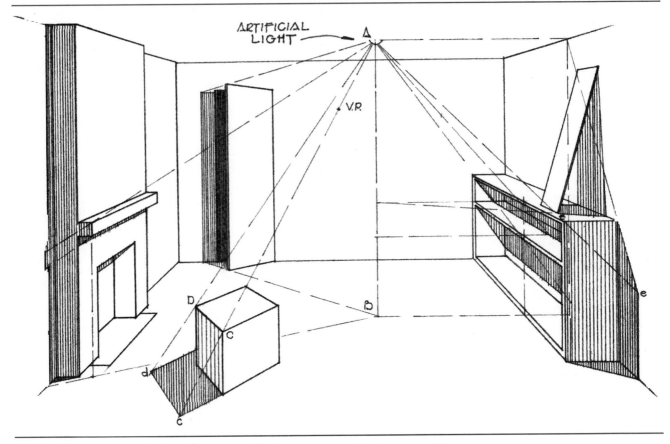

Figure 10–17 Plotting shadows from a single light source on an interior perspective.

An example of good shading technique, made with careful pencil strokes, is shown in Fig. 10–19. Darker tones of the shade are made with wide strokes spaced closely together and with considerable pressure on the pencil. Lighter tones are produced with fine lines spaced farther apart, and the lines are usually feathered out at their ends. Shade lines should reveal the basic form of the surface they represent—straight lines on flat planes and curved lines on curved surfaces (Fig. 10–19). Other shading techniques, such as a series of dots or smudging, can be employed to depict definite surface textures; smudge tones give the impression of a smooth satiny surface, whereas dots appear to represent a rough texture. See Section 11.4 for further information on pencil rendering.

Figure 10–18 Shading of basic forms and masses.

Figure 10–19 A quick perspective sketch.

EXERCISES

1. Construct the shadows on the orthographic views of Figs. 10–20B and C, 10–21, and 10–22.

2. Draw an angular perspective of Fig. 10–20A–C, and complete their shades and shadows as indicated. Use a light source parallel to the picture plane. Make the shadows a darker tone than the shades.

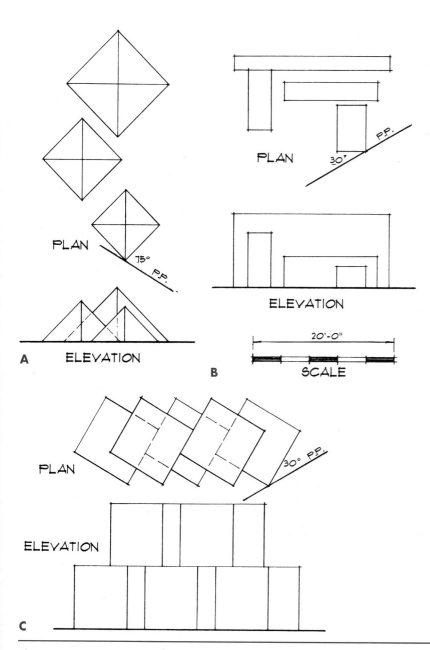

Figure 10–20 Geometric forms to be drawn in perspective. Enlarge the plan and elevation to suit your paper.

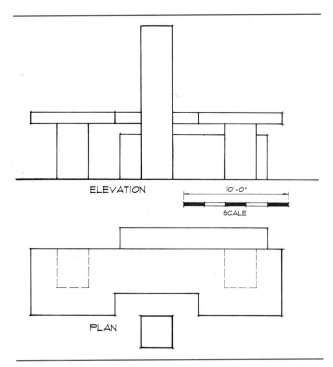

ELEVATION

10'-0"

SCALE

PLAN

Figure 10–21 A problem for plotting orthographic shadows or for drawing a perspective.

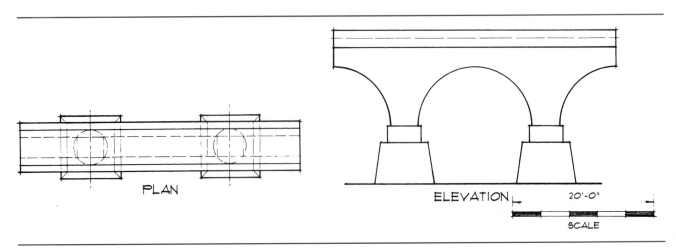

PLAN

ELEVATION

20'-0"

SCALE

Figure 10–22 A problem for plotting orthographic shadows or for drawing a perspective.

3. Using the perspective-plan method, draw an angular perspective of Fig. 10–23, including shades and shadows, on a 24″ × 36″ sheet with the 36″ dimensions placed horizontally on the drawing table. Locate the picture plane 2″ below the top edge of the sheet. Place the station point 10″ below the picture plane and 10″ from the left edge of the sheet. Place point *A* 10″ from the left edge of the sheet. Locate the ground line 13″ below the picture plane. Place the horizontal measuring line 5″ below the ground line. Locate the horizon line 3″ above the ground line. Draw the structure four times the size shown in the book. The light source is parallel to the picture plane and at 45° to the ground, shining down from the left.

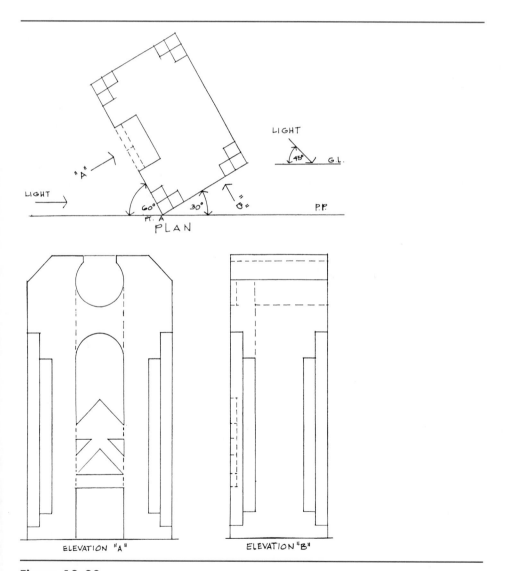

Figure 10–23

4. Using the office method, draw an angular perspective including shades, shadows, and reflections in the pond of Fig. 10–24, on a 24″ × 36″ sheet with the 36″ dimensions placed vertically on the sheet. Locate the picture plane 15″ below the top edge of the sheet. Place the station point 10″ below the picture plane and centered on the sheet. The ground line shall be 9″ below the picture plane. Locate the horizon line ³/₄″ above the ground line. Point *B* shall be directly above the station point and centered on the sheet. Draw the structure four times the size shown in the book. The light source is parallel to the picture plane and at 60° to the ground, shining down from the left.

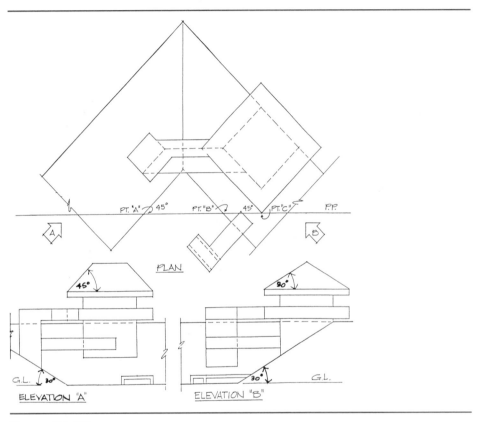

Figure 10–24

REVIEW QUESTIONS

1. Why is the action of light on architectural subjects important to the designer?

2. Why is it convenient to use a 45° light source direction?

3. Explain the difference between a *shade* and a *shadow*.

4. What purpose do orthographic shadows serve?

5. When the light source is parallel to the picture plane,

is the shadow of a vertical corner drawn vertically when it falls on a wall?

6. When you are drawing lines to represent shadows falling on a ground plane, would it be best to draw the lines vertically or horizontally?

7. Why are shades and shadows essential elements of the sketch?

8. What advantage does the felt-tip-pen sketch have over the pencil sketch when reproductions of the sketch is important?

"Good painting is like good cooking: it
can be tasted, but not explained."
—VALAMINCK

PRESENTATION DRAWINGS AND RENDERINGS

Before working drawings can be started, it is generally necessary to prepare a presentation drawing of the tentative structure. This drawing is done for two reasons. First, it combines the efforts of preliminary planning into a tangible proposal for a client's approval, making the presentation an actual marketing instrument. Second, it offers a means whereby all concerned parties can do a careful study of the structure's appearance, upon which improvements or changes for final development can be based.

The presentation is the designer's graphic concept of a building, in its natural setting, made to represent the structure honestly, realistically, artistically, and in a manner easily understood by the layperson. A great deal of unnecessary technical information is avoided to make the general concept more pronounced. Rendered perspective or elevation views showing the realism of light and shade, landscaping, and textures are usually the most effective elements of successful presentation drawings. However, considerable latitude in the selection of views is possible, depending on the nature of the proposal and the time allowed for its completion. Student projects usually include a perspective, a floor plan, several elevations, and a site plan. The duplication of similar information on the various selected views should be avoided so that an attractive yet uncluttered graphic proposal results.

11.1
TYPES OF PRESENTATION DRAWINGS

Many presentations consist of a perspective only. In fact, an architectural office will often give presentation work of complex buildings to professional delineators, specialists in this type of work, for elaborate perspective renderings—frequently in color. Many such presentations have sufficient artistic merit to be used for promotional work, in brochures, and even in national publications.

If time is limited, effective presentations can be made with only a floor plan and a rendered elevation. Others may include a transverse section view or an interior perspective, if such features seem worthy of special consideration. Residential presentations often include a site plan to show the proposed orientation of the home and grounds. In other situations, a well-executed freehand sketch may be adequate in showing a client sufficient information about a structure. We see, then, that on presentation drawings, drafters can give their creative abilities free rein in depicting the most interesting qualities of buildings.

Successful renderings and presentations are done in almost all of the different art media; the most frequently used are pencil, pen and ink, colored pencil, watercolor, tempera, acrylic, pastel, and charcoal pencil (Figs. 11–1 through 11–8). Usually, the mastery of one medium or another is the deciding factor in making the choice. After experimentation the drafter will find that each medium has advantages as well as limitations. For example, pencil and pen and ink, capable of executing sharp, fine lines, can be used to reveal small details and various textures, but they are time-consuming to use when you are covering large areas. The pencil is comparatively easy to control, yet it should be selected as the medium for rather small drawings, as should pen and ink. The combining of pencil or ink linework with transparent watercolor wash for large areas overcomes this limitation and is often an intelligent choice of media for many architectural subjects. Although color commands attention and excites the viewer, it requires considerable skill in handling, especially in representing finer detail.

Beginning students would be wise in developing their ability in pencil work first before attempting other media. After some degree of skill is acquired in representing values, textures, light and shade, trees, and shrubbery, and after command of composition and balance is attained, facility in other media becomes only a matter of mechanics.

A

B

Figure 11–1 (A) Pencil-and-ink sketch. (B) Pencil rendering.

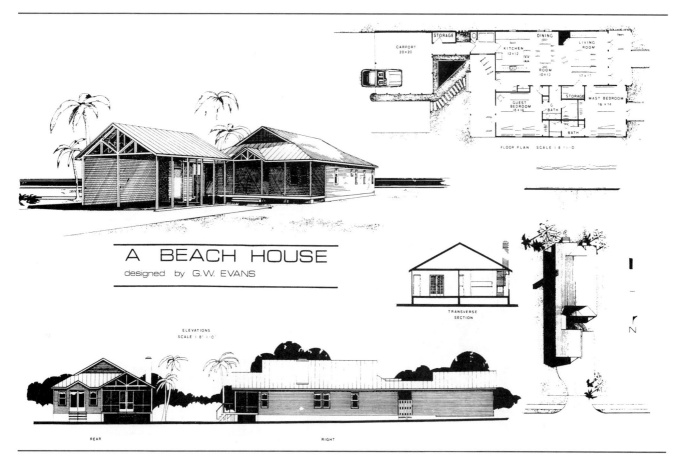

Figure 11–2 Residential presentation in ink.

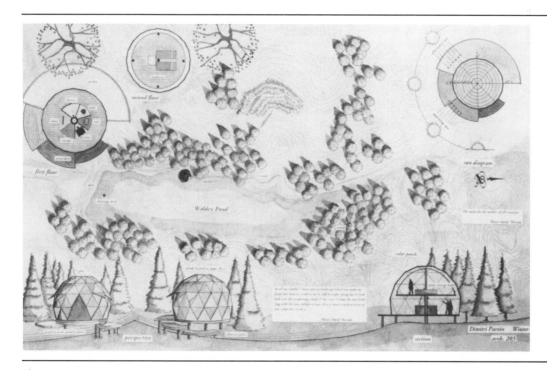

Figure 11–3 Presentation in watercolor.

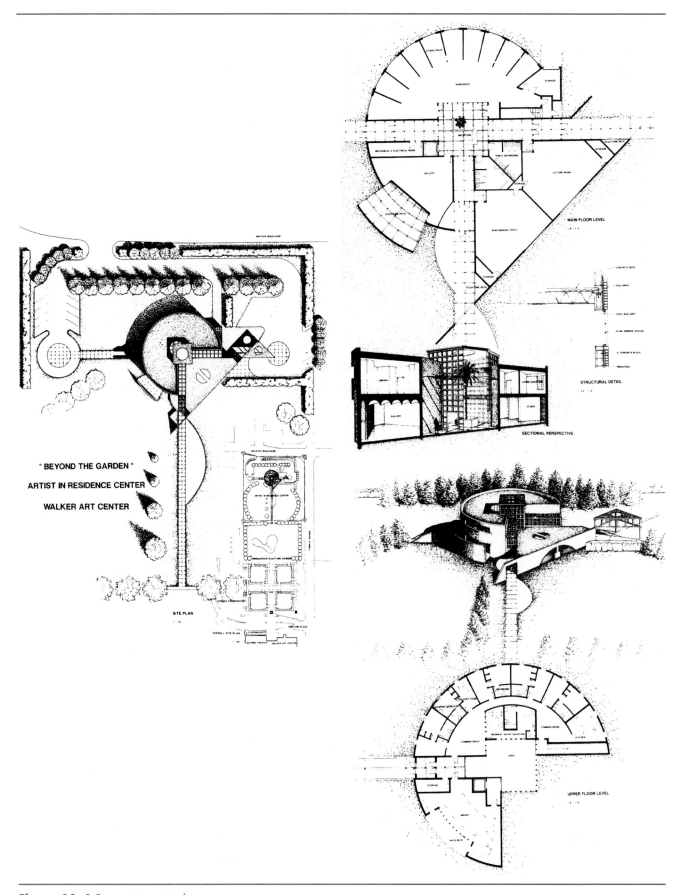

Figure 11-4 Presentation in ink.

Figure 11–5 Perspective rendering in ink.

PRESENTATION PAPER OR BOARD

Pencil drawings may be done on a good grade of tracing paper, for convenient reproduction, or they may be done on illustration board stock for more rigid display drawings. As mentioned earlier, tracing paper, because of its transparency, aids trial-and-error composition with little wasted effort. We can trace elements from previous sketches, rearrange them to suit the composition with the final tracing as an overlay, and even improve on them when we do the finish work. Also, weak points of the composition, when viewed from the reverse side, can be quickly observed.

Frequently it is necessary to display a tracing paper drawing or a paper print in an upright position. Since tracing paper and print paper have little stiffness, it is desirable to attach drawings prepared on such media to a firm backing. These drawings may be merely taped to a suitable mounting board and, if desired, surrounded with a neat frame cut from mat board stock (Fig. 11–37). If permanent mounting is required, drawings on tracing paper or print paper may be mounted on illustration board or polystyrene core board that is commonly known as *foam core board*. Use a contact cement glue to attach the paper to the board backing; take care to eliminate air pockets between the drawing and the backing. Many print shops offer vacuum frame mounting. This professional mounting process usually guarantees a perfectly smooth surface and permanent bonding.

When preparing a presentation on heavy illustration board, first draw each drawing on tracing paper. This is especially important when a number of views or drawings must be arranged into a pleasing composition. Cut out all views so that they are individual units. Next, arrange the views on the board for balance and interest. Do not hesitate to try two or three completely different arrangements before you are satisfied that you have the best possible one. To transfer drawings from tracing paper to the board, scrub each drawing on its reverse side with a soft pencil to act as a carbon, making sure that there is sufficient graphite in back of each line. Then tape each view in place, and go over the linework with a 2H pencil to transfer the drawings to the board. The illustration board is then ready for final rendering.

COMPOSITION

Composition is the art of arranging lines, values, spaces, masses, and other elements into a thing of beauty. Whether you are rendering a single perspective or a pre-

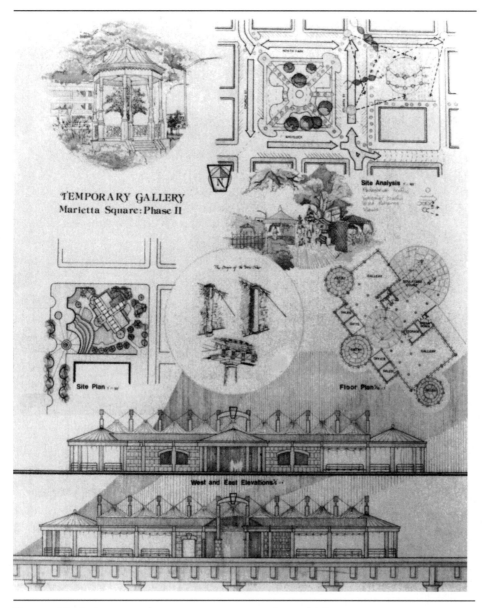

Figure 11–6 Commercial presentation in watercolor and ink.

sentation drawing having a number of views, the principles of composition are equally important. Often a well-designed building will appear rather mediocre if the presentation is poorly planned. Thorough study of professional work will indicate the careful attention given to composition, which to the layperson is not readily apparent. To a fortunate few, pleasing composition comes easily and may require little more than a chance to experiment. To others, it means acquiring a thorough knowledge of composition by repeated applications in an effort to improve. Even if beginning work seems disap-

pointing, composition can be mastered, and the following suggestions may be of help.

11.3.1 Begin with Pleasing Forms

Study the basic geometric shapes involved in the layout. Be sure that the simple forms, both in themselves and in the surrounding spaces, are harmonious and attractive. There must be variety. Avoid mechanical and uniform patterns, unless minor repetition is necessary for harmony. Avoid strong lines near and parallel to the edges

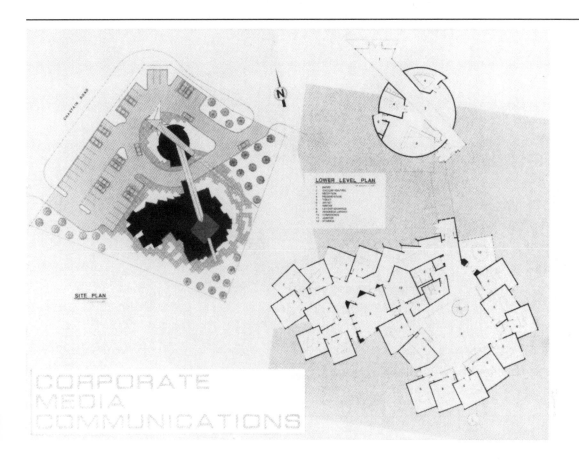

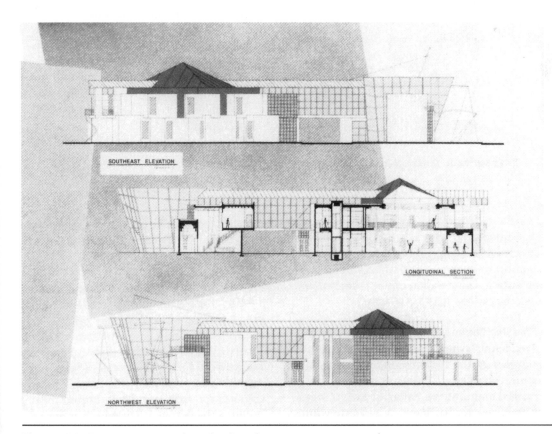

Figure 11-7 Presentation made with ink and pressure-sensitive transfers.

Figure 11–8 Elevation drawing in ink.

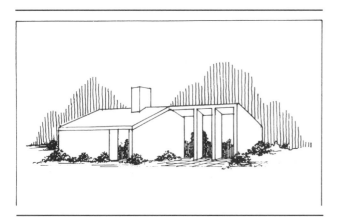

Figure 11–9 Avoid uniform foliage shape around the building.

of the paper; an outline of distant trees should not parallel a strong roof line (Fig. 11–9). Avoid two or three strong elements in a straight line.

11.3.2 Provide Optical Balance in the Composition

Each element is affected by a feeling of gravitational pull and therefore has optical weight (Fig. 11–10). Larger elements have more weight than smaller; darker elements have more weight than lighter; and intense colors have more effect than pale colors. These graphic weights must balance, and they must be placed to prevent the composition from appearing bottom-heavy. Make the building the dominating element or one view the dominating view; the other elements should be subordinate to it. Provide a center of interest, usually an entrance on a perspective, which acts as an optical fulcrum from which the various other elements are balanced. Avoid a static balance, yet combine features and spaces to appear restful and to retain attention without carrying the eye off the paper. Avoid placing the building or other prominent element in the dead center of the paper.

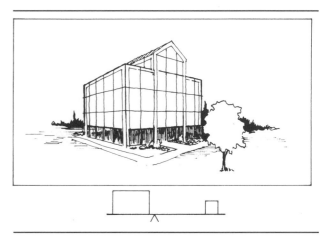

Figure 11–10 Keep the composition balanced.

11.3.3 Use Contrast for Producing Emphasis

To achieve contrast in a composition, adjoin entirely different elements. The elements may be contrasting in value, size, shape, or texture, or a combination of any of these attributes. For example, you can emphasize an important architectural feature shown in a very light tone by surrounding it with a dark background of tree forms or a dark shadow nearby. Conversely, a light background can be made to emphasize a dark feature. Usually, the building is shown bathed in warm sunlight, and subordinate areas, whether shadows, foliage, sky, or similar elements, are shown in various darker tones to lend emphasis as well as relief to the building (Fig. 11–11). Also, occasional strong contrasts create interest. *Remember that carefully drawn shadows help explain the surfaces on which they fall.* In your planning of contrasts and values, one of the very first decisions that you must make is the direction of light. This decision is very important. Even though you have certain liberties in this choice, adopt the most advantageous direction from the standpoint of revealing the major architectural features as well as producing the most interesting shadow patterns.

In general, features in the foreground can be made darkest; features, in mid-distance, intermediate; and distant features, lightest. Other value arrangements may also be found satisfactory. Usually, a dark shadow falling across the foreground in front of a building is effective in creating a feeling of depth. Make sure that the structure appears firmly attached to the ground and has space around, behind, and in front of it. Temper hard architectural lines with planting. Planting is also helpful in emphasizing an important entrance, by leading the eye to it with a hedgerow or a similar strong contrast along a walkway. The strong outline created by extreme contrasts, however, must be interesting.

When using color, we may achieve emphasis by making important features bright and subordinate elements subdued or grayed. On line drawings, emphasis is produced by line weights. Heavy lines become more important than thin, light lines. Give the center of interest the strongest lines and the various subordinate features weaker lines, often actually fading out near the edge of the composition. Variety holds the attention of the viewer.

11.3.4 Hold the Composition Together

Avoid strong, isolated elements that tend to fall away from the layout. To tie elements together, indicate trees, shrubs, or grass, but do not place a large tree in the foreground; it will detract from or hide important information about the building. This problem can be eliminated, if need be, by making the tree in the foreground appear almost transparent, or several branches with foliage might be made to look as though they were part of a tree behind the viewer.

Tree forms and planting indications must also be typical of the area where the building is to be built; palm trees around a New England residence would be ludicrous. Avoid strong indications of clouds or sky, birds, or smoke coming from a chimney; they are rarely needed on an architectural presentation.

11.3.5 Do Not Overburden the Composition with Too Much Detail

Pick out essentials and leave some of the detail to the imagination. Carefully detail important features of the structure for interest, and avoid details on subordinate elements. To create rhythm, however, some elements must recur throughout the composition. Too much of any one surface or texture representation will be found monotonous. On pencil drawings, you will find the eraser an important tool in composing. Erasing areas of superfluous detail can often add considerable interest to the finished drawing.

11.3.6 Keep All Elements of the Composition in Proportion

To maintain reality, do not overscale a tree or other feature. Occasionally, a simple human figure or group of figures, if drawn the correct size, contributes an easily associated relative scale. The size of the building is then easily compared to the size of the figure. Sometimes a related object such as a vehicle can be shown to suggest the activity for which the building is designed. Keep these suggestions simple (Figs. 11–1 through 11–7) and in the correct scale; avoid laborious, diligent studies. Also, consider the building itself in proportion to the size of the paper; the effect will not be pleasing if the building appears to be lost in the space or is so big as to appear overpowering. Beginning students often make the mistake of placing important elements of a composition too close to the outside edges of the paper or board. The finished drawing must give the impression of being tailor-made for the sheet on which it is placed.

Figure 11–11 Use contrast for emphasis.

11.3.7 Learn to Represent the Basic Architectural Textures

Correct indications of the materials used in construction should be evident to the observer, especially in the foreground. In the distance, the various materials usually will appear only as values. The viewer should be able to distinguish among stone, brick, concrete masonry, wood, glass, roofing materials, and the like. Notice that some appear rough and others smooth, some dark and others light. In bright sunlight, even a rough texture is occasionally lost and therefore is left completely white.

11.3.8 Study the Silhouettes and the Character of Trees and Shrubs

Get to know their forms, their trunk and branch structures, their foliage textures, and the manner in which they appear in groups. Notice the action of light and shade on the different types of foliage and on the bark, giving individuality to the representation of each species. Many drafters simplify and stylize their representations of trees and planting; this is advisable for students as long as they retain the basic natural characteristics of the subject. Study the planting on professional renderings and the way it relates to the total development. Sketch trees and outdoor scenes from nature to learn firsthand this necessary element of architectural presentation. Effective tree indications can be shown with only trunk and branch structures (as they would appear in winter) as long as their skeletal character is faithfully maintained.

11.4

RENDERING IN PENCIL

For pencil rendering we recommend that you have available two each of the H, HB, 2B, 4B, and 6B drawing pencils. Sharpen one of each grade to a cone point (see Fig. 8–1) for outlines and fine detail work. Remove about ³⁄₈″ of the wood on the other set, and sharpen or wear down to a chisel point (see Fig. 8–1) for laying in various tones, often referred to as *pencil painting*. In beginning work, the pencil will be more expressive if both sharp and broad points are employed. Later, your individual style, developed after much experimentation and practice, dictates the use of either fine-line or broad-stroke linework.

Hold the pencil in a natural position, similar to writing, with the hand resting comfortably on the paper. (Always place a small sheet of paper under your hand to eliminate smudging the finished drawing.) Work in a restful, uncramped position. If you are working on thin paper, be sure that you have several sheets of underlay paper underneath; this makes the drawing paper more responsive to pencil stroking. One of the secrets of velvety pencil tones is continual *extreme pressure* on the pencil, even on light tones. It makes the strokes more definite and minimizes grainy pencil work. Use harder pencils for lighter tones and softer ones for darker tones. If a tone seems to be too dark when pressure is applied, pick up a harder pencil, but always maintain firm pressure on the pencils if you want to develop interesting textures. Give each stroke identity, especially at its beginning and ending. Notice that occasional white areas left out between strokes contribute sparkle and variety.

Artists develop individuality by going about a drawing in their own way; hence, students must feel free to develop drawings in their own way, as long as the results are satisfactory. Each person's work will differ, much like handwriting. Students will find it interesting and also very helpful to observe how professionals treat various features. In addition, it will be good experience—and many accomplished artists have profited by it—to reproduce some of the drawings of the old masters found in art museums.

One point that students must remember is that they must occasionally exaggerate features such as contrast between light and dark areas in a composition rather than try to make the rendering resemble a photograph (Fig. 11–12). Plan a well-balanced tone arrangement (with values ranging from extreme darks to white, as shown in Figs. 11–13 and 11–14) before you start rendering. Have the main outlines carefully laid out with light lines, even major shadows, so that you will avoid mistakes and not lose control of your drawing. Fine detail such as window muntins can be omitted in the preliminary layout. Landscaping and tree studies are first done on tracing paper overlays. When their composition and general effect are satisfactory, transfer them to the final drawing with light lines. While most of the strokes representing natural foliage and tone work are done freehand, strong roof lines and other architectural edges may be completed with the help of the straightedge.

11.4.1 Practice Strokes

With a grade B pencil sharpened to a conical point, practice the fine line strokes shown in Fig. 11–15A, B, E, F, and G. Try all of these strokes several times to develop discipline in pencil control. The exercises are mainly for showing outline information and suggestions of surfaces and textures. Keep the pencil uniformly dressed throughout the exercises. With a softer pencil (2B or 3B) sharpened to a chisel point, and, using its flat surface, practice the broad-stroke exercise shown in Fig. 11–13C, D, and N. These strokes become useful for pencil washes and various background tones throughout a drawing. Notice that definite pressure is required at the ends and beginnings of strokes. Try graduated tone exercises, which will be continually useful in areas varying from a dark to a light value. Try the strokes in the full range of values, from darks to light. Experiment with other methods of your own for creating other tones, since there are unlimited ways of representing them with a pencil. Before applying pencil tones to a final rendering, evaluate their suitability first on practice paper.

For filling large areas, a broad sketching pencil, resembling a carpenter's pencil and available at art supply stores, can be useful, and a graphite stick may also be

Figure 11–12 A pencil presentation and a photograph of the finished building.

found appropriate for showing sharp edges and gradual, uniform tones. The wide pencil will save time in laying in larger backgrounds and in representing many other features. In Figs. 11–12 and 11–21A, observe the manner in which E. A. Moulthrop has so ably given a charming and casual quality to typical architectural subjects with the use of a broad sketching pencil. These drawings were originally done on tracing paper.

11.4.2 Pencil Textures

Draw Brickwork Differently at Different Scales The need for rendering brick walls arises frequently in architectural subjects. For this, a pencil is especially suitable because it is a medium that lends itself well to soft, tonal variations. Care must be taken to plan the general tone of all brick surfaces with respect to light and shade in the com-

position. In full sun, for instance, the tone value of brick will usually be very light, shown with generous areas of white, devoid of detail. In shade, the values become darker tones with more of the brick detail in evidence. Corners, where contrast is intense, must be carefully done to reveal the rough nature of the brick coursing joints.

When drawing brick, first establish brick background values with the broad stroke; put in shade and shadow areas where they seem necessary. Change the intensity of some of the strokes to avoid monotony. Usually, shadow lines are suggested and allowances made for reflected light during this step. The representation of the actual brick is then applied over the tonework.

For drawing brick at a small scale, use a series of parallel wavy lines, broken occasionally, to produce a convincing texture (Fig. 11–16A). It is important that the lines be in perspective and that their width and spacing be scaled properly. Remember that the height of one brick course with a mortar joint is only about 2½″. Although the linework is broken at irregular intervals, vertical joints of the brick are ignored. Do not try to indicate each brick. As suggested previously, let a bit of pure white show in a few areas. Be sure that shadows expressing the relief of corners, offsets, and reveals at window and door openings are carefully handled. Brick walls can often be a dark value at corners opposite the light, and as they recede, they can be made gradually lighter, to where extremely dark tree backgrounds provide sharp contrast.

Brick surfaces near the viewer must be larger in scale and must therefore be given more individual brick detail (Fig. 11–16B). For best results, use a pencil dressed to the proper brick width, and show (with various short strokes) the brick and mortar joints between. Go over occasional strokes to bring out variations in color. Make bold, sharp-ending strokes. Fine-line shadows along the lower edge and side of the individual brick, here and there, help emphasize the texture. Give brick in shadow a very dark value. Do not forget the shadows below brick windowsills. Shadows from nearby foliage falling on a wall will help create interest; they are drawn with vertical strokes briskly applied over the finished brick.

Stonework Can Appear Smooth or Rough Much of the artistry of stonework is provided by the arrangement of the various unit sizes and the mortar joint pattern throughout the wall. The appearance of stone varies greatly, depending on coursing, unit sizes, and roughness. Cut stone usually has uniform coursing and a rather smooth texture, whereas field stone may have various unit sizes, and shadows indicate that the stones are rough and irregular. The rendering should capture the intended texture. The bold treatment of the stonework contributes to the rustic appearance of a building.

Begin stonework by lightly indicating the unit pattern throughout the wall, regardless of the desired texture. As a rule, patterns showing a dominance of horizontal courses are the most pleasing (Fig. 11–16F). Place larger stones at

Figure 11-13 Rendering done in pencil.

Figure 11-14 Rendering showing various textures in pencil.

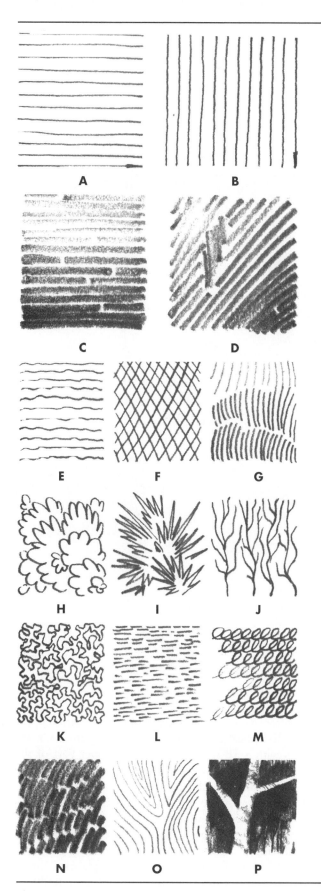

Figure 11-15 Practice strokes in pencil.

corners for stability, and vary the sizes and shapes of the smaller ones between to avoid regularity. Next, shade each stone with parallel, broad strokes, making sure that the stones vary in value for interest and that the strokes change in direction occasionally. Leave generous amounts of white areas, especially if you plan dark features in back of the stone. Possibly, the suggestion of shadows, drawn in fine lines emanating from the lower edge of a few stones, may be all that needs to be shown in these areas. Let the mortar joints remain light and not very wide. As in brickwork, treat stone corners carefully, particularly where strong light and shade contrasts appear and where stone is bordered by sky. Emphasize stone textures near the center of interest where they should be darkest.

Use a dotted technique to represent stucco. A similar method can be used for concrete block by the addition of a few fine lines to suggest coursing (Fig. 11–16E).

Wood Siding Is Drawn with Parallel Lines The general tone of wood siding must be considered with respect to composition and light and shade—and with respect to color, if painted. If the siding is painted white, only the shaded areas may require a background value. If, on the other hand, the wood is to be stained, more of the values will appear darker. However, use white highlights even on the darker wood surfaces to prevent monotony. Apply background values with the broad stroke mainly in the direction of the wood grain.

Show clapboard siding with fine horizontal lines (occasionally a few heavy lines may be needed), which represent the shadow below each board (Fig. 11–16D). Spacing of the linework depends on the width of the boards exposed to the weather. Vary the weight of the lines, but in areas of a wall where light intensity is the strongest, omit them altogether. Be sure that the lines are in perspective. Take care to represent the sawtooth profile at corners or where the clapboards abut vertical wood trim. Show wood louvers in a similar way—with parallel lines and a sawtooth shade at the shadow side of the louver strips.

Represent board-and-batten siding by carefully drawing the edges of each vertical batten strip with light lines. Give the boards between the strips a light-gray value, and leave the strips white, except in shade. Accent the shadow side of the batten strips with a fine, dark line. In shade, darken the boards rather than the battens (Fig. 11–17A). Shadows falling across the wall must give evidence of the batten offsets.

To render wood shingle siding, first draw the horizontal coursing with freehand, wavy lines to indicate the shadow at the butts. Break the lines frequently and vary their darkness to give the wall a weathered, rustic effect. Draw staggered vertical joints between shingles only here and there with the fine point (Fig. 11–16C). Then apply broad strokes in slightly varying tones vertically throughout (to simulate the wood grain); the tone value of these strokes should create a pleasing overall value composition. Indicate shadows with dark, bold strokes.

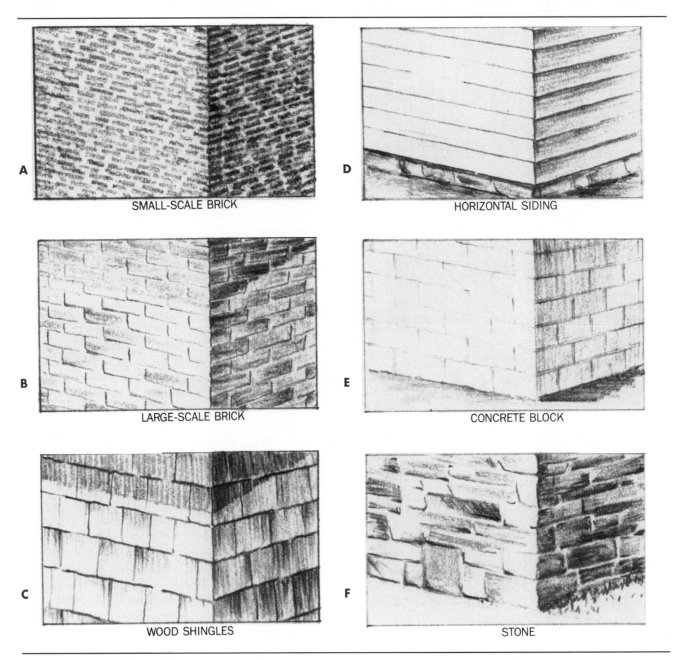

Figure 11–16 Architectural textures in pencil.

Roofing shingles must be given an identification value in the composition; usually, it will vary throughout the roof surface. Coursing is drawn with wavy, horizontal lines, and very few vertical strokes are shown (Fig. 12–17). Background tones are drawn down the slope of the roof to indicate a slightly weathered effect. In some roofing materials the shingle coursing is not pronounced; the drawing must make this clear. Built-up roofing may be indicated with a dot technique in shadow and at important corners only.

Glass Is Usually Dark

In general, glass appears dark unless there are drapes, blinds, or shades directly in back

of it to receive the sunlight. Highlights on glass appear white. To indicate the dark areas, make definite broad strokes with a 3B or 4B pencil. The upper part of the shaded area is made the darkest. For highlighting, leave white openings that slope in the direction of the light; these will suggest the smoothness of the glass. Reflections of foliage or of surrounding features will add interest (Fig. 11–12). In direct sunlight or on planes facing the viewer, windows are especially dark and more reflective on their shaded sides or on planes slanting away from the viewer. Leave muntins and mullions white if they are in direct sunlight. Large glassed areas may require suggestions of furniture within, since an ex-

A

B

Figure 11–17 (A) Pencil rendering showing reflections in glass. (B) Ink rendering showing reflections in glass.

panse of glass admits a considerable amount of light to the interior.

The surface of glass is generally recessed several inches or more from the surface of the wall; thus, reveals and surrounding trim must be carefully drawn. When drawing window units such as double-hung or casement (or even entire glass walls), show drapery or curtains along the inside edges where they would ordinarily be seen. Indicate the folds of drapery with vertical light and dark areas, and show shadows from muntin bars as wavy lines across the drapery (Fig. 11–17A). Curtains or other window treatment should look realistic. Remember that muntins and division bars, although white in sunlight, appear darker on the shadow side.

Trees Provide a Natural Setting When drawn in correct proportion to a building, trees establish the rel-

ative scale in the viewer's mind and add a note of softness and charm to the rendering. Often they are drawn in natural groupings as background, or a few may be shown in the foreground to achieve good composition.

Earlier in this chapter we stressed the importance of direct observation from nature. You will find such observation particularly helpful in becoming familiar with the individual characteristics of different tree species, and, before long, you will be able to draw trees from memory. Their silhouette, trunk and branch structure, bark texture, appearance at the base of the trunk, and foliage clusters and leaf forms all play a part in giving tree species identity.

Before drawing a tree, study its silhouette (Fig. 11–18) so that you get to know the character and form of the trunk and the angle of the branches in relation to it. Then lightly lay out the trunk and branch structure, which be-

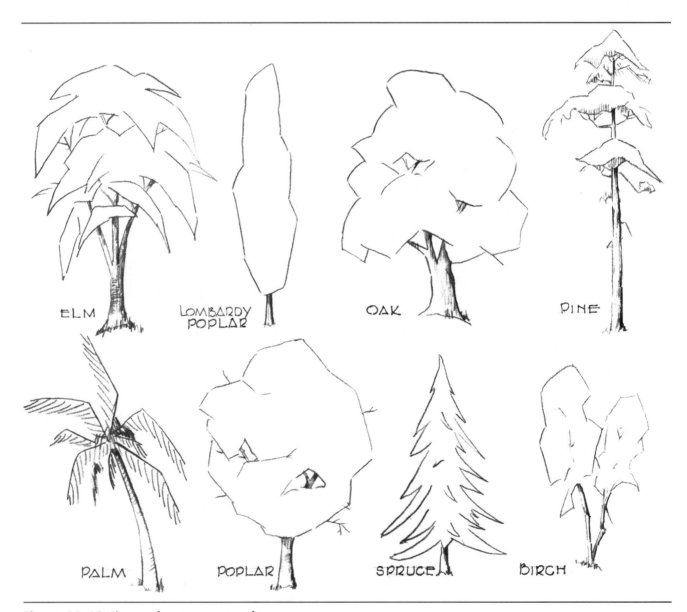

Figure 11–18 Shapes of various species of trees.

comes the skeleton of the tree. Give the structure personality, but do not overly distort it—it will only detract from the building. Reserve slight distortions for needed interest in an otherwise weak composition. Notice that branches radiate in all directions (in front and back of the trunk, as well as the sides) and become consistently smaller toward the top. The pine tree, however, has a slightly different trunk system. The thickness of the trunk remains similar for most of its height, and only near the top does it gradually diminish, where small branches radiate almost horizontally with needle clusters of various sizes. No two trees, even of the same species, are exactly alike. The drawing should show slight variations, yet the identifying characteristics should be present.

After the basic structure has been established, draw the clusters of foliage lightly (Fig. 11–19B) with some darker areas to indicate shade; strive to make the general pattern a pleasing one. In the foreground, more detail must be shown on tree representations. Using a 4B pencil and starting with the darkest values, render the foliage with a broad stroke. Have various tones throughout the foliage, using a fanlike series of continuous pencil strokes. Some foliage may be better represented by slightly different stoking, but accent the bottom edges of foliage clusters. Leave branches or parts of the trunk white where dark foliage falls in back of them. Remember as you work that highlights on leaf clusters appear white, similar to the highlight of a sphere, whereas shade appears dark below clusters. Treat the trunk as a cylindrical form to give it volume. Then draw coarse bark, if the species requires

it, with most of the bold dashes near the edges of the trunk to give it a rough, uneven appearance. In direct sunlight, leave some of the trunk entirely white. Soften any harsh vertical lines at the base of the tree by indicating grass or shadows. Undulating tree shadows falling on grass provide a convenient way of showing rise and fall of the terrain.

Show Shrubbery near the Building Evergreen shrubs used for foundation planting and as accents help to tie building and ground together and eliminate hard, vertical architectural lines, especially at corners. Pleasing groups of various indigenous species contribute a landscaped appearance and help to emphasize the center of interest, usually the front door. Shrubs can be shown in their natural forms or trimmed into continuous geometric forms, such as a hedge along a walkway (Fig. 11–20).

In rendering shrubbery, outline the groupings first before applying the broad stroke patterns shown in Fig. 11–15. Again, leave white areas. On some, show indications of the branch structure, and be sure that the darkest values are near the ground where shadows appear. Shrubs in front of a building should be rendered before you draw the linework on the building. This eliminates needless erasing of wall textures and lines.

Tie Elements Together with Grass Indications Grass may be represented with broad horizontal lines, drawn with a T square if necessary, and accented along edges of drives and walks. Make the indications lighter

A SKELETON   B FOLIAGE   C VALUES

Figure 11–19 Steps in drawing a tree.

Figure 11–20 Shrubbery suggestions.

in sun than in shade. You can suggest the presence of grass by drawing it only where shadows fall along the ground, leaving lighted areas entirely white; too much grass might de-emphasize the center of interest. Short, vertical strokes made with the fine point can then be shown along borders and edges. Some delineators use a series of looping strokes drawn horizontally and close together to represent grass (Fig. 11–15M); others use rows of short, vertical strokes, varying in value, throughout the shaded areas. Be sure that the edge of grass indications, extending to the edge of the rendering, forms an interesting outline (Fig. 11–21A).

Clouds and sky are usually not indicated on pencil renderings. If they seem necessary for the composition, make clouds rather subdued, and remember that sky appears lightest near the horizon and gradually darker above.

As we previously mentioned, trees and shrubbery on a rendering should be merely suggestive rather than too complex—that is, they should be subordinate to the architecture. For that reason, many renderings are shown with stylized indications. In this technique, trees and planting are simplified, often showing only outlines or branch structures; others show only flat tones of foliage masses. The treatment is effective if the natural forms and tree characteristics have been retained. Nature must be the source for ideas in arriving at authentic stylized treatments. Even the representation of people can be simplified (Fig. 11–22).

Use a Value Scale to Plan Values and Contrast

Nearly everything we see is made up of an almost unlimited number of shadings of colors or grays ranging from black to white. The variation of lightness or darkness of color or grays is known as *value*. To the beginner, most subjects appear to be muted in values that are difficult to distinguish. Early trials by students often result in dull-gray renderings that are composed of nearly identical values, but this dullness must be avoided at all costs.

Contrast is defined as the striking difference in adjacent elements of a composition; contrast is achieved by the use of varying color, value, texture, outline, or treatment. We know that the appearance of contrast is relative (Fig. 11–23). A black value next to a gray value will not appear as dark as a black value adjacent to a white area. Likewise, a relatively loud sound in a quiet bedroom

may be very disturbing, whereas the same sound in a busy street may go unnoticed.

Before starting a drawing, develop a value scale that ranges from black to white much as the one shown in Fig. 11–24 does. A value scale to the artist is similar to a score to the musician. Each brings a sense of order to the creation—the musician with sound value, the artist with color value. White and black with three or four intermediate grays between are satisfactory for the range of most drawings. Plan your drawings with the use of the value scale so that important features are accented with extreme contrasts (whites against blacks or blacks against whites), and make less important features or outlines with lesser amounts of contrast. Plan a definite arrangement of the values; such arrangement is necessary even if reality is exaggerated to some degree. However, if you want sparkle and interest in your pencil drawings, have light and dark values provide contrast where the dominance counts most.

The size of the white or black areas also plays a part in developing brightness. Position the values in the composition so that they have variety yet balance (Fig. 11–25). Use dramatic profiles where the highest contrast occurs. This *shape accent* should be reserved for the dominating features only, since too many accents tend to cancel each other. The pattern or repeating quality that holds the whole together is known as *matrix*.

11.5

RENDERING IN INK

The following materials are recommended for producing ink renderings:

1. ***Black waterproof India ink:*** The standard brands can be diluted with water for gray washes or gray ink linework.
2. ***Technical drafting pen, points No. 0 and No. 2:*** This pen is used for lettering and other uniform linework.
3. ***Technical drafting pens:*** These pens are used for various-width, ruled lines.
4. ***Hunt's No. 102 Crow Quill pen:*** This pen has a fine, flexible point for delicate linework.

A

B

C

Figure 11–21 Representing grass in foreground.

Figure 11–22 Indications of people must be drawn to the correct scale.

5. *Hunt's No. 99 pen:* This pen has a nibbed point for general work.
6. *Speedball No. B6:* This pen has a round nib for bold linework.
7. *Small pointed watercolor brush.*
8. *Small bottle of white tempera color.*

Select a strong paper or board that will take ink well. A smooth surface is preferable, because a rough surface tends to interfere with fine pen stroking. Good grades of tracing paper and illustration board are satisfactory; plate-finish illustration board, if available, will be found very desirable.

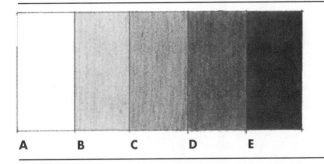

Figure 11–24 Use a value scale to help you plan the light and dark areas.

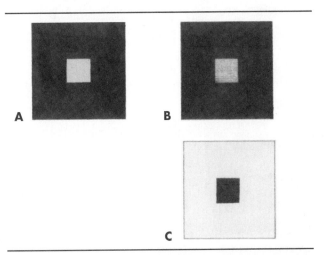

Figure 11–23 Contrast of values.

Pen-and-ink drawing requires more care in planning than pencil. Pencil can be erased or darkened by repeated strokes until a desired value is attained, but ink work is difficult to remove or modify once it is put on paper. Successful pen renderings also require a carefully balanced composition of white, black, and intermediate values. Various line or dot techniques must be found to represent the intermediate values since the ink remains black unless, of course, it is diluted. Elements of the composition must therefore be reduced to their simplest expression to give the rendering a crisp yet delicate appearance.

Before you begin, a word of advice: Do not make the mistake of using the same-width pen point throughout your entire drawing. Unless you are experienced in pen-and-ink technique, the drawing will probably look monotonous and uninteresting. Use two or three points of different widths to give the lines character. If you want

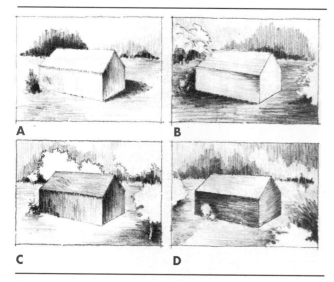

Figure 11–25 Planning value layouts.

to soften the appearance of the ink rendering, do all of the linework freehand; it is usually more appealing. To obtain a quality of perspective depth, draw tree and planting indications in the far background with a fine pen, the mid-distance trees with an intermediate point, and the foreground features with a broader pen.

Begin by practicing the various pen strokes shown in Fig. 11–26. Acquaint yourself with the capabilities of each pen; try each tone shown in the figure. Experiment with the tone representations before you use them on your finished rendering. By all means, have the outlines and the edges of all values laid out lightly in pencil (as well as other major linework), using care not to groove the paper, before you do any inking. Pencil lines can be easily erased when the ink is thoroughly dry. If you are not pressed for time, make a *value study* of the composition. After the perspective is completed in pencil, place a sheet of tracing paper over it and quickly establish the general value patterns rather roughly, using a soft pencil or charcoal. Avoid details on the study; instead, strive for pleasing value patterns that can be created by contrasts of light and shade and from the landscaping composition. When finished, this becomes your guide for choosing line technique, values, and landscaping on the final ink drawing (Figs. 11–27 and 11–28).

To avoid smearing ink lines, start at the top of the drawing and work down. Remember that ink work cannot be rushed. To save time in filling large black areas, outline them with the pen, and fill in the remainder with a small watercolor brush. Be sure that there are sufficient areas left white to give the impression of strong light, as we mentioned earlier. If a minor mistake should occur in the linework, white tempera applied with a brush will cover it. Check the value study from time to time to see that the total desired effect is developing. The same basic principles of composition that were discussed in connection with pencil rendering apply equally to work done in ink.

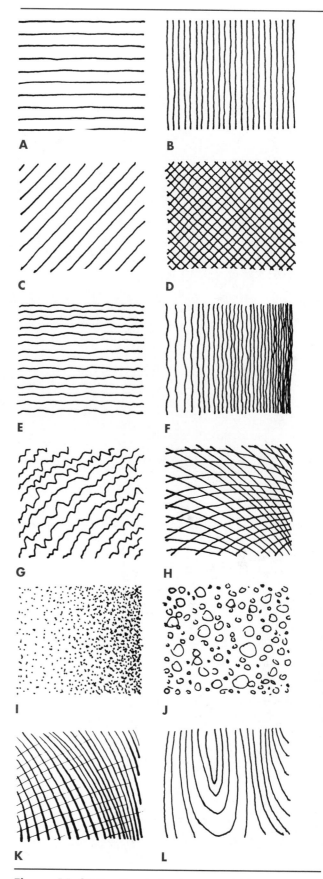

Figure 11–26 Practice strokes in pen and ink.

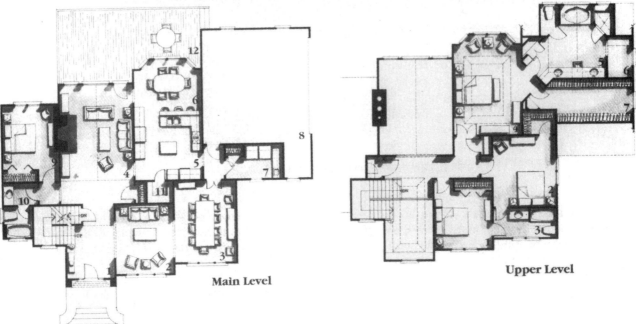

Figure 11-27 Residential presentation in ink and felt-tip markers.

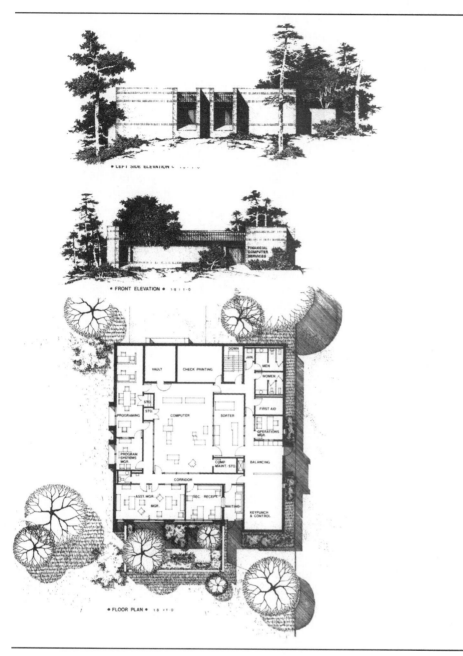

Figure 11-28 Presentation made with ink and watercolor.

RENDERING WITH COLOR

After you have made a number of pencil and pen-and-ink renderings, you may have the urge to try working in color. There is something exciting about color work. It is dramatic and commands attention, and there is no question about its compelling qualities on architectural presentations if the color has been carefully handled. If, on the other hand, the coloring appears hard, unharmonious, and detracting, it would have been better if the presentation had been done in black and white.

Color is a completely different graphic dimension that requires mental application not required in other work. The study of color is rather extensive, entailing scientific and psychological aspects, and there are many good books covering the subject. Here we can discuss only the major points, primarily to introduce the beginning student to the basic principles of color and to clear up any misconceptions he or she may have.

Let us start our study with an examination of the color wheel (inside the front cover) showing the interrelationship of basic colors. The color wheel, as you will notice, places colors in an organized position around the circle, relative to each other, for convenient reference by the student or the color artist, and the wheel is the result of extensive study about color characteristics. It is handy for selecting actual color combinations or as an aid in mixing pigments to arrive at a desired color on the palette. Many of the rules governing the effective use of color are based on the relationship of the colors on the wheel. It does not solve all of the problems concerning color, yet it can guide the bewildered novice in his or her early work. Later, these color characteristics will come automatically.

The following list defines the most important and frequently encountered color terms:

- **Primary colors** Red, yellow, and blue, from which, theoretically, all other colors can be made.
- **Secondary colors** Orange, green, and purple, made by mixing adjacent primary colors on the color wheel.
- **Tertiary colors** Colors located between primary and secondary colors on the color wheel and made by the mixing of adjacent colors.
- **Hue** The term used to designate the name of a color. Red, blue-green, orange, blue, and so on, are different hues.
- **Value** The lightness or darkness of the same hue, such as light red or dark red.
- **Chroma** The intensity or purity of a color, that is, the degree to which it has been diluted or neutralized.
- **Shade** A darkened value of a color.
- **Tint** A lightened value of a color, usually by the addition of white.
- **Complements** Colors located opposite each other on the color wheel and considered harmonious. Complements mixed together will produce a gray if the result ends midway between the two colors. A color can be darkened by the addition of a slight amount of its complement. Complements of equal chroma and value are usually not pleasing, but by varying their values or chromas in a subtle manner, for example, a dark green and a light red, you can achieve a more pleasing color combination.
- **Monochromatic** A color scheme using values of only one color. For instance, sepia produces a very artistic monochromatic color scheme.
- **Analogous** A color scheme employing two or three adjacent colors on the color wheel. Usually, more interest is created if one of the colors predominates the scheme, for example, yellow, yellow-green, and green, with the yellow predominating.
- **Warm colors** Reds, oranges, and yellows, which seem to advance toward the viewer.
- **Cold colors** Blues, greens, or violets, which appear cool and seem to recede when seen from the same distance as warm colors. Gray can be either cold or warm.

Pure pigments, made from plants or minerals, reflect particular colored waves, while absorbing others. Pigments do this in varying degrees, so experimentation in mixing various colors is necessary; arriving at the desired colors is a matter of trial and error. When mixing colors, you will notice that certain pigments have a greater affinity for one color than for another.

Colors on renderings, as we mentioned, should be kept soft and subdued, often grayed. Greens for foliage, for instance, should be yellowish-green hues, rather than hard, high-intensity greens. If you look about at the surrounding landscape, you will notice that nature's colors are restful to the eye. Few abrupt, hard colors exist unless put there by people. Even the coloring on buildings should utilize soft, natural colorings of building materials to complement the landscape. Try to have three basic harmonious colors in your composition: one in the foreground, one in the mid-distance, and one in the background. In general, you can treat these three distinct planes in a monochromatic color scheme by using various values of one color for light and shade effects. Stay with simple, uncomplicated color combinations; usually, a few colors will be sufficient and more appropriate and also easier to handle than many colors. The choice of colors is personal, and you will find with experience that impressions created by color are affected by many things—light, adjacent colors, paper, patterns of the drawing, and so on. Make it a habit to "see" the colors around you and learn to identify them, and you have taken the first step in trying to master this troublesome technique.

11.6.1 Colored Pencils

One of the simplest methods of coloring a rendering is with colored pencils. Although many colors are available, they lack subtlety because a direct application is required. Some pencils, however, have water-soluble colors; the lines made by this type can be blended in with a wet brush to create a watercolor effect.

A rather effective color pencil rendering can be done on a dark black line diazo print of an elevation view. The pencil or ink tracing is run through the machine so that the background of the print becomes gray rather than white, giving it an intermediate tone throughout the paper. Light pencil colors are used to bring out the architecture and to contrast the gray background. Major colored areas are filled in with close, parallel lines made with instruments, and the foliage indications are added in an abstract manner. A rendering, for example, can be done very quickly in this manner (Fig. 11–29).

11.6.2 Watercolor

Down through the years, watercolor has been an important medium in architectural presentation. The two general types of watercolor work are transparent and opaque. As its name implies, *transparent* watercolor uses

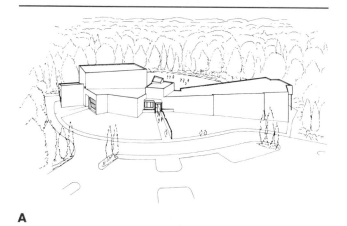

A

B

Figure 11–29 (A) Blackline diazo print. (B) Rendering with colored pencils.

A

B

C

Figure 11–30 (A) Rendering in transparent watercolor. (B) Rendering in acrylics. (C) Rendering in tempera.

transparent-type watercolor pigments (Fig. 11–30A). Thin washes are spread over special paper stretched carefully on the drawing board or over commercially prepared watercolor board. In some cases the washes may have to be superimposed over each other to arrive at desired values and effects. *Opaque* watercolors—tempera, acrylic, or casein—are slightly different in that they have the capacity to cover different colors beneath. They are applied in a creamy consistency rather than as a thin wash. Although the techniques of transparent and opaque watercolors are basically different, beautiful work can result by using both media, separately or in combination (Fig. 11–30B).

Because tempera does not necessarily have to be diluted, and because of its opaqueness, it is the ideal medium for a student's early work. Tempera can be put on regular, inexpensive illustration board; and even though the paint is water soluble, it can be easily applied in successive layers, making it ideal for the beginner. You will gain self-confidence when you find that mistakes can be covered without much trouble or carefully sponged off with a damp sponge. However, tempera colors possess strong, vivid qualities that require toning down,

which is done by mixing with white or gray or with their complements. In mixing, as well as in method of application, tempera paints handle much as artist's oil colors (Fig. 11–30C).

11.6.3 Colored Markers

Felt-tip markers in a wide spectrum of colors and tip sizes are used over pencil and ink drawings. Different techniques can be accomplished with markers depending on the method of application. Colors may be dabbed or washed. Care must be taken to select less vivid colors. Colored markers are unforgiving; however, warm or cool tone grays can be placed over bright colors to subdue them. Practice on blueline or blackline prints before applying color on the original. Most media, such as illustration board, vellum, and a variety of prints, are suitable for colored marker renderings (Fig. 11–31).

Purchase the materials shown in the material list of Table 11–1 at your art or drafting supply store. You will find that the more expensive brands are more enjoyable to work with. Brushes must be cleaned in water before the tempera dries; never allow brushes to rest on their bristles, since they will lose their shape and become worthless.

11.6.4 Painting a Color Wheel

To acquaint yourself with the tempera pigments, make your first color exercise the duplication of a color wheel. Not only will it be a helpful exercise, but it will also become a serviceable reference when you are doing other color work.

Lay out the wheel on illustration board using a 6″ outer and a 4″ inner diameter. Divide it into 12 equal 30° segments as shown on the inside front cover of this book. Allow a small amount of white space between each segment so that the colors do not run together. Paint the primary colors first, making sure that you have enough remaining pigment to mix the other colors. Use cadmium yellow (pale) for the yellow, alizarin crimson with a bit of cadmium red for the red, and Prussian blue mixed with a little ultramarine blue for the blue. In mixing the secondary colors, see if you can get them from the primaries; otherwise, you may have to use the other tube pigments.

Figure 11–31 Rendering in felt-tip markers.

Table 11–1 Materials for tempera, caesin, or acrylic rendering.

Colors (in tubes)	Brushes	General
Chinese white Alizarin crimson Cadmium red (medium) Cadmium yellow (pale) Yellow ochre Prussian blue Ultramarine blue Viridian Ivory black Burnt sienna Burnt umber	One No. 4 red sable, pointed watercolor brush One No. 1 and one No. 4 red sable "Brights" flat-end brush One No. 5 flat-bristle color brush One regular painter's nylon-bristle brush (1″)	Muffin tin Pint jar (for water) Old dinner plate (for a palette) Baby sponge

Viridian requires a touch of yellow for the proper green. Then mix and paint in the tertiary colors to complete the color wheel. Label all of the colors correctly as shown on the inside front cover.

You can gain further experience by mixing various pigments and noting the results. See what happens when white is mixed with other colors. Try mixing a slight amount of black with the various colors as well. Try mixing complements. Be sure that you feed small amounts of a strong color into a weaker color when mixing; otherwise, you may end up with more color than you need. All of this experimentation will be helpful before you attempt an actual rendering. Other exercises should include architectural textures such as stone, brick, and wood and the painting of trees, shrubs, and foliage. Study professional tempera renderings for ways of handling these features.

11.6.5 Helpful Hints

Some artists cover the entire rendering with a thin sepia or pale-yellow transparent wash before applying the tempera.

Do not overstroke the paint; let some of the brushwork give character to the forms.

Blending can be done with a finger while the paint is wet. Blends can also be accomplished with a *drybrush* technique.

Avoid using too much white for light value tints. The overuse of white pigment results in a chalky painting, often appearing cold and lifeless.

Values in color rendering are just as important as they are in black-and-white renderings. Here, too, careful value studies are necessary in planning color renderings (Figs. 11–30 and 11–31).

To create harmonious color combinations, the renderer does not have to stay with the colors found in nature—green grass, blue sky, and the like. Slight modifications of natural colors, as long as they are plausible and not extreme, create interest and a note of abstraction.

If the tempera painting is to occupy only a small rectangular area of a larger presentation, surround the area with masking tape before starting to paint so that the paint will stay within bounds.

Start the painting with flat tones in the major areas; then add the texture indications and the detail strokes in a systematic manner over the backgrounds.

When painting the landscaping, apply indications in the distance first, usually with flat tones along the horizon. Then add the mid-distance foliage, and finally the trees and shrubs in the foreground. Paint any figures last.

If necessary, tempera can be reduced to a thin wash and brushed over underlying colors to soften their appearance and yet retain their identity.

Paint mullions over the general glass areas with a pointed brush or a ruling pen, or use a small (No. 1) flat-end brush if the mullions have more width. The flat brush is easier to control.

For painting straight lines on a building, use a bridge such as shown in Fig. 11–32 that can be made from an

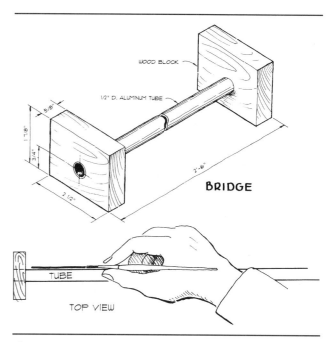

Figure 11–32 Convenient handrest for painting.

aluminum tube or a wooden dowel (about 30″ long) with blocks of wood at each end. Have about ½″ clearance between the tube and the working surface. When using the bridge, rest several fingers and your thumb comfortably along the tube, with the brush held between your thumb and forefinger as you move the brush uniformly along the surface. For more versatility, make the clearance at the top of the blocks ⅞″ so that it can be turned over and used with larger brushes.

A more economical bridge can be made with an ordinary brass-edged ruler having artgum erasers (cubes) glued to each end (Fig. 11–33).

11.7

SKETCHING WITH THE FELT-TIP PEN

Many architects use a felt-tip pen on tracing paper to produce quick sketches that can be reproduced if necessary. The pens are convenient and responsive, lending themselves to many architectural subjects. They are economical and easily available in most stationery and drafting supply stores. Many stores carry a full line of colors as well as black. The only drawback for the student is that the lines are positive and cannot be erased as pencil lines can. For that reason, use a felt tip only after you have gained confidence with the pencil, or lay out the principal lines of the sketch with pencil first before completing the drawing with the pen.

Several types of tips are available, but the most popular are the *cone* and the *chisel* points (Fig. 11–34). The cone point is best for outlines and finer detail requiring uniform-width lines, whereas the chisel is more versatile

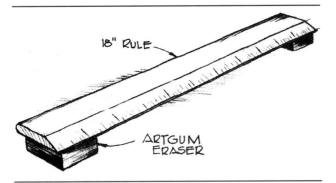

Figure 11–33 To make an inexpensive bridge, glue artgum erasers to a brass-edged rule.

Figure 11–34 Many types of felt-tip pens with chisel (left) or cone (right) tips are available for sketching.

and allows different-width lines with the use of the corner, the edge, or the broad side for flat sweeps when larger areas need to be covered. Various pressures also produce different widths with either point. Experiment with the chisel point by holding it in different ways to see the many free-flowing lines that are possible.

Avoid using the felt-tip pen on porous paper since the ink will blot and spread throughout the paper. On some

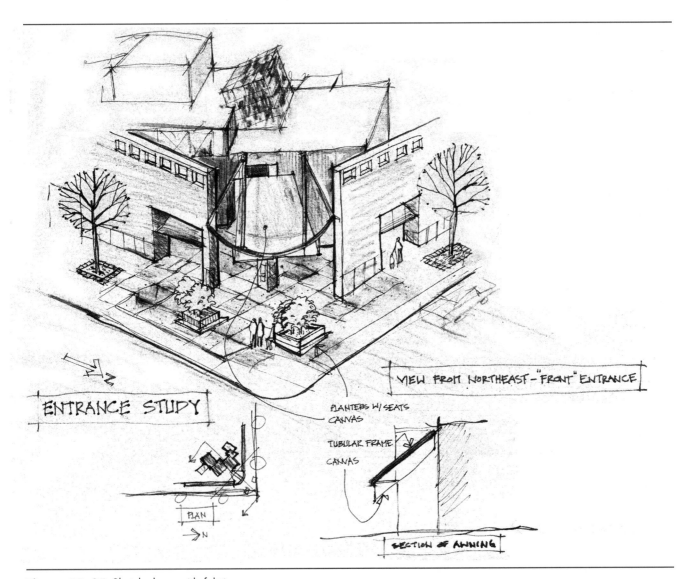

Figure 11–35 Sketch done with felt-tip pen.

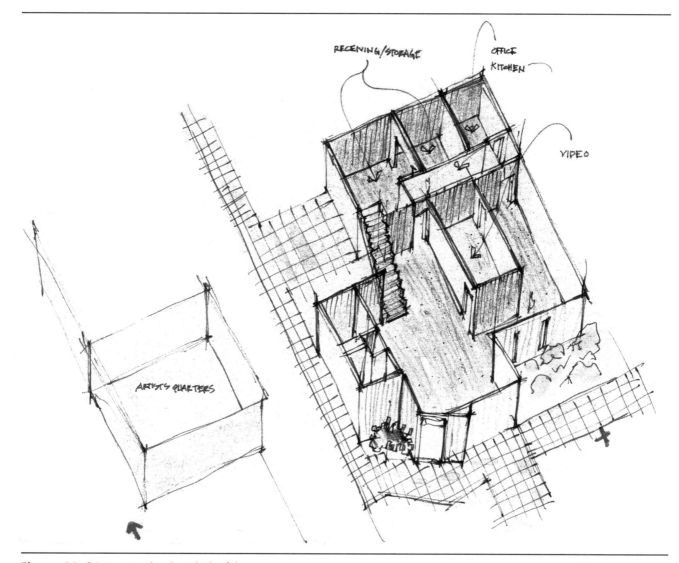

Figure 11–36 Interior sketch with the felt-tip pen.

thin papers, the ink may soak completely through and even stain paper below. A sized paper should be used; as previously mentioned, tracing paper is very satisfactory.

Keep the pens tightly capped when not in use, or they will soon dry out. Rather than trying to revive a dried-out point, it is usually better to discard it. Points, however, can be re-dressed with a razor blade should they become worn and blunt from repeated use.

After you have tried a few felt-tip sketches, it will be evident that they take on their own definite character, different from that of other media. Black linework combined with bold areas of soft gray or very light tan pro-

duces interesting architectural combinations. Bright colors should be avoided. With the fine point, often masses of foliage can be stroked with bold sweeps, and the edges of the masses can be given repeating character lines, to represent landscaping quickly. The pen is also well adapted to masses of light and shadow on buildings (Figs. 11–35 and 11–36).

Blackline sketches on tracing paper can be easily mounted over colored board so that the tint shows through. Then they can be framed with a white mat to dramatize or display the sketch if necessary (Fig. 11–37).

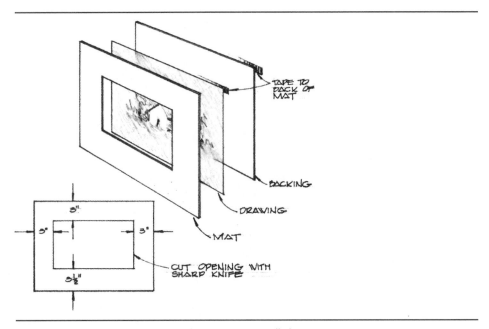

Figure 11–37 Making a mat and mounting a small drawing.

EXERCISES

1. On $8\frac{1}{2}'' \times 11''$ white drawing paper, lay out six equal rectangles with $\frac{1}{2}''$ space between as shown in Fig. 11–16. Using a soft pencil, render the following material textures in the rectangles: brick, stone, horizontal siding, concrete block, board-and-batten wood siding, and wood shingle siding.

2. Draw a realistic deciduous tree, a coniferous tree, and a group of evergreen shrubs—all in pencil.

3. Double the size of the plans, and draw the elevation view shown in Fig. 11–27. Render it in pencil, and add appropriate landscaping. Assume a vertical scale proportional to the horizontal scale and the perspective.

4. On $8\frac{1}{2}'' \times 11''$ white drawing paper, duplicate the ink textures shown in Fig. 11–26.

5. Using the scale $\frac{1}{4}'' = 1'-0''$, make an ink presentation drawing of a lakeside lodge. Use $20'' \times 30''$ illustration board, and include a floor plan, two elevations, and a perspective view.

6. Make an ink presentation drawing of an original home design using the scale $\frac{1}{8}'' = 1'-0''$. Use $20'' \times 30''$ illustration board, and include a site plan, a floor plan, two elevations, and a perspective view.

7. Make an ink presentation drawing of a small commercial bank building. Include a site plan, a floor plan, and four elevations using the scale $\frac{1}{4}'' = 1'-0''$.

8. With ink and colored pencils, render a perspective of a playground showing buildings in the background. Use either $20'' \times 30''$ illustration board or a blackline diazo print.

9. On $20'' \times 30''$ dark-tinted illustration boards, render a small, two-story commercial building in white pencil. Show the site plan, floor plan, elevations, and perspective view.

10. Make an ink-and-felt-tip-marker presentation drawing of a site plan. Draw a lake, an amphitheater, a parking lot, and a playground on the site plan. Use $20'' \times 30''$ white illustration board, or apply colored marker to a diazo print that is reproduced from an ink tracing.

1. What are the advantages, for client and architect, of preparing a presentation drawing of an architectural proposal?

2. For what reason are perspective views generally effective in presentation work?

3. What is meant by *optical balance?*

4. Why are plumbing fixtures and furniture usually shown on presentation plan views?

5. What media are appropriate for rendering small drawings?

6. On perspective renderings, why should the details nearest the observer be made more prominent than those in the distance?

7. What methods can the renderer employ to produce emphasis?

8. What do landscaping indications contribute to a perspective? What do indications of people contribute?

9. What is meant by a *value study?*

10. How are analogous colors related to each other on the color wheel?

11. When you are rendering sky, should the sky appear darker near the building or near the top edge of the rendering?

12. What is the most effective method for rendering glass?

13. What is the first step you should take in drawing a tree?

14. List three types of water-based paints frequently used in rendering.

15. Why is it important to draw people on a rendering of a building?

16. What drawing medium would be appropriate if a rendering must be prepared very quickly?

17. What is the best way to achieve contrast in a rendering?

18. Why are colored pencils a good medium for the beginner?

19. Why is waterproof ink essential when it is combined with a wet medium?

20. What would be the advantage of preparing a pencil or an ink rendering on vellum or Mylar® instead of illustration board?

"Ah, to build, to build. That is the noblest
art of all the arts."

—HENRY WADSWORTH LONGFELLOW

PRINCIPLES OF LIGHT CONSTRUCTION 12

This chapter introduces and explains many of the accepted principles and methods of light construction found in residences and small commercial buildings. The written material is supplemented with typical sections taken through walls, floors, roofs, and structural members. Pictorial drawings are also used when necessary to present this information as simply as possible. Some of the methods shown are adaptable to one area of the country; others are common construction practice in other areas, depending on building codes and local conditions. Some methods are conventional practice, having been in use for years; others are comparatively new. Together, they form details widely used and generally found satisfactory. More exhaustive treatments of construction details can be found in specialized texts to supplement this material. The actual drawings of details is treated later, but you should not attempt design drawing until you have made a thorough study of construction methods. Tables and charts from the Uniform Building Code, referenced usually as, for example, "Table D–1," are found in Appendix D.

12.1

DEVELOPING A BACKGROUND IN LIGHT CONSTRUCTION METHODS

At this point in their training, students should acquire as much information and background relative to construction as they can find from various sources. Whether it is pursued by people who create and design the most elaborate structures or by others called on to draw the various working drawings of the simplest buildings, architecture requires a searching interest and enthusiasm about the methods of building. To be successful in this field not only requires skill in knowing *how to draw*, but, equally important, requires the acquisition of background

information in knowing *what to draw*. Unless drafters continually strive to increase their knowledge of construction details, they can expect to become no more than junior drafters or merely tracers.

Technical courses in architectural print reading, building materials, and construction methods, which in technical schools often precede or are prerequisites to architectural drawing, provide this needed background. Students and drafters should continually read monthly periodicals and trade magazines containing current information. Literature from the many building construction trade organizations will be found valuable. Various governmental agencies concerned with housing and standardized construction practice, including the U.S. Government Printing Office, offer inexpensive manuals that should not be overlooked. Drafters must become familiar with *Sweet's Architectural Catalog File* (F. W. Dodge Corp., New York) for convenient reference on many construction materials. Many good books on construction are available in libraries.

Part-time employment with builders or contractors, as an adjunct to drafting training, is desirable for learning construction methods. In fact, actual on-the-job observation of a building in the process of construction is time well spent by drafters. Not only is it a thrilling experience, but it offers the opportunity of seeing many of the building materials in use and the way that they are put together within the building—mainly the structural members that are often concealed once the building is completed. Students who have familiarized themselves with the methods and techniques of construction will certainly be better equipped to draw more intelligently, progress more rapidly, and grow in stature as architectural drafters.

This text makes no attempt to treat special situations requiring structural engineering design to solve the problems. Such information, which is beyond the scope of this book, is left to specialized texts.

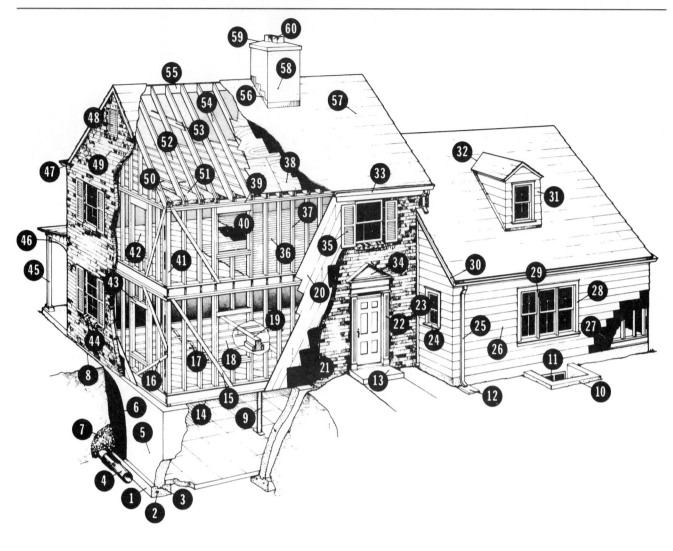

1. Footing
2. Reinforcing rod
3. Keyway
4. Drain tile
5. Foundation wall
6. Waterproofing
7. Gravel fill
8. Grade line
9. Metal column
10. Areaway wall
11. Basement window
12. Splash block
13. Stoop
14. Sill plate
15. Corner brace
16. Knee brace
17. Bridging
18. Floor joist
19. Beam; girder
20. Sheathing
21. Building paper
22. Trim pilaster
23. Double-hung window
24. Window sill
25. Downspout; leader
26. Bevel siding
27. Fiberboard sheathing
28. Window trim
29. Mullion
30. Rake mold
31. Dormer
32. Valley
33. Gutter
34. Pediment door trim
35. Shutter
36. Finish flooring
37. Stud
38. Roof decking
39. Double top plate
40. Flooring paper
41. Corner post
42. Subfloor
43. Lintel; header
44. Brick sill
45. Porch post
46. Porch frieze board
47. Return cornice
48. Louver
49. Brick veneer; gable
50. End rafter
51. Insulation
52. Ceiling joist
53. Collar beam
54. Common rafter
55. Ridge board
56. Flashing
57. Shingles
58. Chimney
59. Cement wash; cap
60. Chimney flues; pots

Figure 12-1 Typical residential terms.

12.2

CONSTRUCTION TERMINOLOGY

Figure 12–1 is a guide to the terms found in light construction. You should memorize these terms and know where they apply within a typical building. Definitions of other terms, which may be unfamiliar, can be found in the Glossary of Construction Terms following the Appendices. You will occasionally encounter several different designations for similar features in construction terminology because workers frequently use terms that are indigenous to their particular region; elsewhere the terms might have quite different applications. Despite widespread national distribution of building construction products, terminology has not been entirely standardized, but as you continue to improve your vocabulary and construction background, the ambiguities will become less confusing.

Some of the present terms used in residential construction, especially in wood framing, have been handed down from New England colonial history, for example, *collar beam*, *soleplate*, and *ridge pole*. Some terms have come from the ancient Greek civilization, especially terms relative to classic building exteriors, such as *frieze*, *plinth*, and *dentils*. Other parts derive their names from their physical shapes. For instance, the *box sill*, which has a boxed-in characteristic, the *K brace*, which has the shape of the letter *K*, and the *T post*, which resembles the letter *T* when shown in cross section, are just a few of the terms that are so labeled because of their shape. Most of the terms, however, have been derived from the logical use or placement of the part in relation to the building, for example, *footing*, *foundation*, *baseboard*, *shoemold*, and *doorstop*.

Girders, *beams*, and *joists* are terms given to horizontal structural members used in buildings. Often the heavy, structural support for the floor framing in light construction is referred to as either a *girder* or a *beam*. In more complex construction, however, the heaviest members are referred to as *girders*. They support the beams, which are of intermediate weight and which support the joists, the lightest members of the floor or roof structure.

12.3

FOOTINGS AND FOUNDATIONS

Every properly constructed building must be supported by an appropriate foundation, which will support the weight of the building and sustain the structure throughout all weather conditions to which it may be subjected. The *footing*, the enlarged base of the foundation wall, must be massive enough, depending on local soil conditions, to distribute the weight of the building to the ground below, and it must be deep enough to prevent frost action below. Poured concrete is usually the best and most widely used material. Footings can be formed and poured where soils are stable enough to hold their shape; trenches are dug the proper depth, and concrete is poured directly into the trenches to form adequate footings.

Figure 12–2 illustrates the method of laying out footings and foundation walls. Leveled batter boards are used to lay out the foundation of a building. Wooden stakes are driven into the soil several feet outside the corners of the proposed building. The top of the batter board is positioned at the finished floor elevation of the building. The batter board is then securely nailed to the wooden stakes. A taut string line that locates the exterior wall of the building and foundation is run between batter boards, as shown in Fig. 12–2.

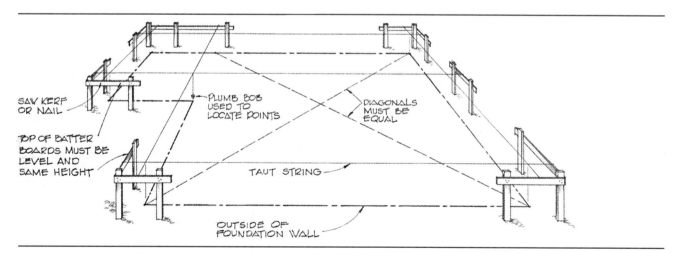

SAW KERF OR NAIL

PLUMB BOB USED TO LOCATE POINTS

DIAGONALS MUST BE EQUAL

TOP OF BATTER BOARDS MUST BE LEVEL AND SAME HEIGHT

TAUT STRING

OUTSIDE OF FOUNDATION WALL

Figure 12–2 Leveled batter boards are used to lay out the foundation of a building.

Frost action below the footing would raise and lower the building during freezing and thawing and would soon fracture the foundation, as well as disrupt other parts of the building. Naturally, frost lines vary throughout the country; local building codes usually indicate the depth to which footings must be placed below grade. Figure 12–3 shows average frost depths throughout the country, compiled by the U.S. Weather Service. The Uniform Building Code prescribes footings to be placed a minimum of 1'-0" below the frost line. Footings of buildings with basements are deep enough so that frost lines usually need not be considered. In northern areas, the cost of excavating for deep footings often makes the basement more economical than in southern areas, where footings are often merely trench footings several feet below grade.

Footings should also be placed on undisturbed soil. Occasionally, if a building must be put over a fill, the soil should be well compacted, and sufficient steel reinforcing should be in the footing to prevent cracking (Fig. 12–5). Local codes in some areas require steel reinforcing in the footings to counteract specific conditions (for example, the Florida Hurricane Code, and the California Seismic Code). In cold weather, care must be taken to ensure that foundations are neither placed on frozen soil nor poured in freezing weather. Footings should project a maximum of 6" beyond the face of the foundation wall. Table D–7 and Figure 12–4 contain guidelines for determining minimum footing depth and width based on building wall size and height.

The soil on which footings rest must be level and compact. In weak soils in which the load-bearing qualities are questionable, it is advisable to introduce steel reinforcing rods at intersecting points between chimney or fireplace footings and wall footings. In soft soils, this is advisable also at corners, at offsets, and at breaks in the direction or size of the footings. When footings are subjected to heavier loads, calculations should be made to determine their correct size in order to balance the various footings in relation to their various loads. The first consideration is the load-bearing qualities of the soil on which the footings rest.

Table 12–1 indicates typical supporting capacities of different types of soil conditions in tons and pounds per square foot. The soil capacities are only general, and we cannot rely on them entirely for calculating critical footing sizes. Soils vary considerably even in local areas; also, the moisture content of some soils changes its supporting qualities. Only a soil expert or a local testing laboratory after numerous tests will be able to analyze load-

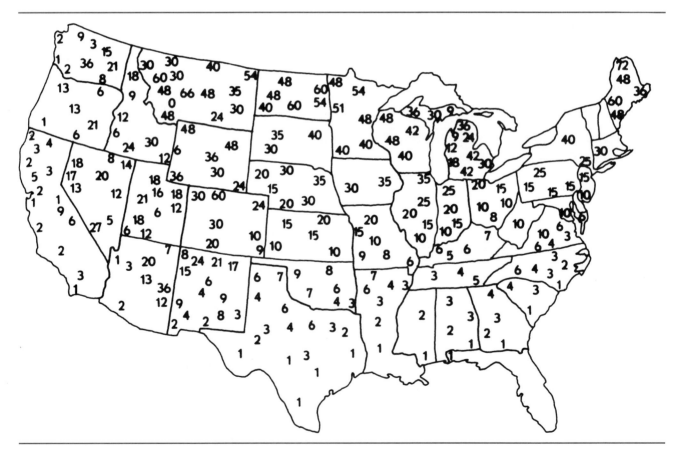

Figure 12–3 Average depth of frost penetration in inches for locations throughout the United States. (U.S. Department of Commerce Weather Bureau)

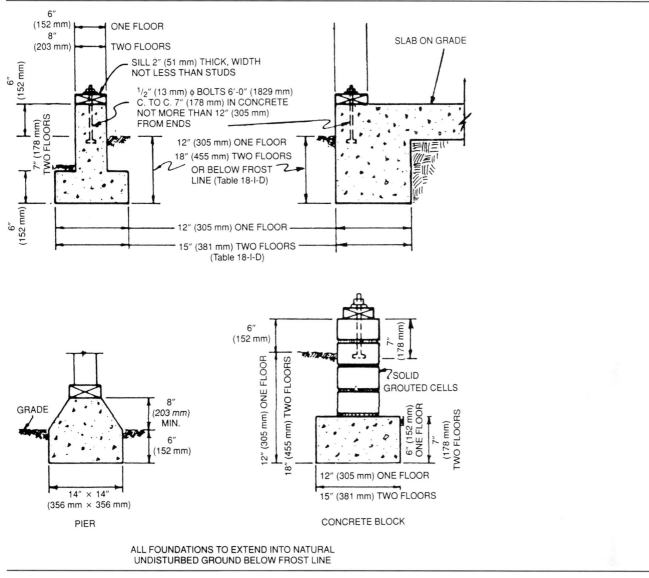

Figure 12–4 Footings: recommended depths and sizes. (Dwelling Construction Under the Uniform Building Code, 1994. International Conference of Building Officials.)

bearing characteristics of a soil for critical footing design. Local practices that have been found successful should be relied on for this purpose. Careful builders generally have the soil tested before footings are poured. If local codes require steel reinforcing rods, they should be placed near the bottom in wall footings and in column or isolated footings. Engineers have found that the steel is usually effective in case of uneven settlement in resisting tensile stresses (Fig. 12–5).

Because residences and frame buildings are comparatively light, they seldom require accurate footing design. However, if the soil is not trustworthy or if the building gives indication of being heavier than usual, the following method can be used to calculate the necessary size of the footings (Tables 12–2, 12–3, and 12–4). Both dead load and live load must be considered in the calculations. *Dead*

load refers to the weight of the structure as well as any stationary equipment fastened to it; *live load* refers to varying weights and forces to which parts of the building are subjected (people, furnishings, storage, wind, snow, and so on). Both weights are designated in pounds per square foot (lb/sq ft) (Fig. 12–6). Local codes generally require residential floors to be able to carry 40 lb/sq ft live load.

12.3.1 Footing and Foundation Design

Good foundation design and construction depend on a number of strategies. For example, appropriate structural design should be combined with insulation, and with moisture-, termite- and radon-control techniques where needed.

Table 12-1 Presumptive values of allowable bearing pressures for spread foundations.

Type of Bearing Material	Consistency in Place	Allowable Bearing Pressure (Tons/Sq Ft) Range	Recommended Value for Use
Massive, crystalline igneous and metamorphic rock: granite, diorite, basalt, gneiss, thoroughly cemented conglomerate (sound condition allows minor cracks).	Hard, sound rock	60 to 100	80.0
Foliated metamorphic rock: slate, schist (sound condition allows minor cracks).	Medium-hard, sound rock	30 to 40	35.0
Sedimentary rock: hard, cemented shales, siltstone, sandstone, limestone without cavities.	Medium-hard, sound rock	15 to 25	20.0
Weathered or broken bedrock of any kind except highly argillaceous rock (shale). RQD less than 25.	Soft rock	8 to 12	10.0
Compaction shale or other highly argillaceous rock in sound condition.	Soft rock	8 to 12	10.0
Well-graded mixture of fine- and coarse-grained soil: glacial till, hardpan, boulder clay (GW-GC, GC, SC).	Very compact	8 to 12	10.0
Gravel–gravel sand mixtures; boulder-gravel mixtures (SW, SP, SW, SP)	Very compact	6 to 10	7.0
	Medium to compact	4 to 7	5.0
	Loose	2 to 6	3.0
Coarse to medium sand; sand with little gravel (SW, SP)	Very compact	4 to 6	4.0
	Medium to compact	2 to 4	3.0
	Loose	1 to 3	1.5
Fine to medium sand; silty or clayey, medium to coarse sand (SW, SM, SC)	Very compact	3 to 5	3.0
	Medium to compact	2 to 4	2.5
	Loose	1 to 2	1.5
Homogeneous inorganic clay; sandy or silty clay (CL, CH)	Very stiff to hard	3 to 6	4.0
	Medium to stiff	1 to 3	2.0
	Soft	0.5 to 1	0.5
Inorganic silt; sandy or clayey silt; varved silt–clay–fine sand	Very stiff to hard	2 to 4	3.0
	Medium to stiff	1 to 3	1.5
	Soft	0.5 to 1	0.5

SOURCE: Foundations and Earth Structures, Design Manual 7.02, Naval Facilities Engineering Command.

NOTES:

1. Compacted fill, placed with control of moisture, density, and lift thickness, has allowable bearing pressure of equivalent natural soil.
2. Allowable bearing pressure on compressible, fine-grained soils is generally limited by consideration of overall settlement of the structure.
3. Allowable bearing pressure on organic soils or uncompacted fills is determined by investigation of the individual case.
4. If the tabulated recommended value for rock exceeds the unconfined compressive strength of the intact specimen, allowable pressure equals unconfined compressive strength.

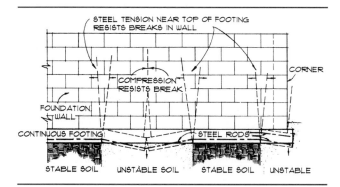

Figure 12–5 Reinforcing steel becomes most effective if placed in the lower part of a footing-supporting C.M.U.

The choices involved reflect the decision-making process used by a designer, builder, or homeowner dealing with foundation design development (Fig. 12–7). You must decide which type of foundation and which construction system will be used. If the foundation is a basement, you must decide whether the below-grade space will be heated and/or cooled. Second, you must choose the construction system: concrete, masonry, or wood. Third, the placement of insulation needs to be addressed. Be sure to check local codes to determine the minimum amount of insulation required for the type of foundation and construction system. Finally, once the amount and location of the insulation have been determined, the necessary construction details can be developed and construction documents can be finalized.

Radon   Controlling radon is a relatively new concern in the design and construction of basements and foundations. Radon is a colorless, odorless, tasteless gas found in soils and underground water at varying levels throughout the United States. It is potentially harmful if it decays while in the lungs. Because radon is a gas, it can travel through the soil and into a building through cracks, joints, and other openings in the foundation wall and floor (Fig. 12–8).

The initial step in addressing the radon problem is to determine to what degree it is present on the site. Various techniques to control radon levels can then be applied.

The Council of American Building Officials (CABO) has adopted the standards of the Environmental Protection Agency (EPA) for radon-resistant new residential construction. The methods for radon mitigation have been added as an appendix to the code. Many of the techniques that resist radon entry are already described in other sections of CABO's code because they effectively insulate buildings from temperature extremes and moisture. There are three basic approaches to dealing with radon:

1. the barrier approach,
2. soil gas interception, and
3. indoor air management.

Table 12–2 Typical weights of materials used in light construction.

Roof	lb/sq ft
Wood shingles	3
Asphalt shingles	3
Fiberglass shingles	2.5
Copper	2
Built-up roofing, 3-ply and gravel	5.5
Built-up roofing, 5-ply and gravel	6.5
Membrane roofing, without ballast	1
Slate, 1/4" thick	10
Mission tile	13
1" wood decking, with felt	2.5
2"× 4" rafters, 16" o.c.	2
2"× 6" rafters, 16" o.c.	2.5
2"× 8" rafters, 16" o.c.	3.5
1/2" plywood	1.5

Walls	
4" stud partition, plastered both sides	22
Window glass, DSB	2
2"× 4" studs, 1" sheathing, paper	4.5
Brick veneer, 4"	42
Stone veneer, 4"	50
Wood siding, 1" thickness	3
1/2" gypsum wallboard	2.5

Floors and Ceilings	
2"× 10" wood joists, 16" o.c.	4.5
2"× 12" wood joists, 16" o.c.	5
Oak flooring, 25/32" thick	4
Clay tile on 1" mortar base	23
4" concrete slab	48
Gypsum plaster, metal lath	10

Foundation Walls	
8" poured concrete, at 150 lb/cu ft	100
8" concrete block	55
12" concrete block	80
8" brick, at 120 lb/cu ft	80

The *barrier approach* is a set of techniques for constructing an airtight building foundation in order to prevent soil gas from entering. This approach is different for each foundation type.

Table 12–3 Calculation for exterior wall footings (Fig. 12–6).

Given: 1-story brick veneer frame, wood shingle roof, plaster interior walls, poured concrete foundation. Load-bearing value of soils = 2000 lb/sq ft.

Roof

Dead load
1. Live load, snow, wind (varies locally) — 30.0 lb/sq ft
2. Shingles, wood — 3.0
3. Wood deck, 1" thick, and felt — 2.5
4. Rafters, 2"× 6", 16" o.c. — 2.5

(Length of rafters measures 13'-0".) — 38.0 lb/sq ft × 13' = 494 lb

First-Floor Ceiling

Dead load
1. Live load (attic storage) — 20.0
2. Wood floor, 1" thick — 2.5
3. Wood joists, 2"× 8", 16" o.c. — 3.5
4. Ceiling plaster, metal lath — 10.0

(Half-span of ceiling is 5'-0".) — 36.0 lb/sq ft × 5' = 180 lb

Exterior Wall

Dead load
1. Brick veneer — 42.0
2. Sheathing, wood, 1" thick, and felt — 2.5
3. Studs, 2"× 4", 16" o.c. — 2.0
4. Plaster, metal lath, interior — 10.0

(Height of wall measures 9'-0".) — 56.5 lb/sq ft × 0 = 509 lb

Floor

Dead load
1. Live load (usual code requirements) — 40.0
2. Finished oak floor — 4.0
3. Subfloor, 1" thick — 2.5
4. 2"× 10" joists, 16" o.c. — 4.5

(Half-span of floor to center beam is 5'-0".) — 51.0 lb/sq ft × 5' = 225 lb

Foundation

Dead load
1. Concrete wall, 8" thick
(Height of foundation wall is 9'-0".) — 100.0 lb/sq ft × 0 = 900 lb
2. Concrete footing, 16" × 8" — 100.0 lb/sq ft × 1.33' = 133 lb

Total load on soil per lin ft of wall = 2471 lb

$$\text{Required area of footing} = \frac{\text{load}}{\text{soil-bearing capacity}} = \frac{2471}{2000} = 1.23 \text{ sq ft}$$

Total area of footing required per linear foot of wall = 1.23 sq ft

For convenience, use 1.25: 12" × 1.25 = **15" wide footing required.** To be on the safe side, it would be practical to make the footing 16" wide, which would be the width if calculated by rule of thumb.

Table 12–4 Calculation for column footing (Fig. 12–6).

1. Ceiling	180	lb/lin ft
2. Partitions	100	
3. First floor	255	
Total	535	lb/lin ft

The column supports 10 lin ft of beam:

$$535 \times 10 = 5350$$

Dead load of beam: 250

Total 5600 lb carried by column

$$\frac{\text{Total load}}{\text{Soil capacity}} = \frac{5600}{2000} = 2.8 \text{ sq ft footing required}$$

$$X = \sqrt{2.8} = 1.7 = \mathbf{1'\text{-}8''}$$

Required column footing size = 1'-8" × 1'-8"

Make footing 2'-0" × 2'-0" × 12" deep.

Soil gas interception uses vent pipes and fans to draw soil gas from a gravel layer beneath the foundation floor slab (Fig. 12–9). This approach can be used for basements and slab-on-grade foundations.

Air management techniques reduce the suction that a building applies to the surrounding soil gas by reducing the pressure differential across the building envelope. It is necessary to make the entire building envelope airtight and to control the amount of the incoming fresh air, the exhausted inside air, and the air supplied for combustion devices.

Many of these principles are essentially the same as those recommended for moisture vapor control and energy-efficient design. Figure 12–10 illustrates the CABO recommendations, and Figures 12–11, 12–12, and 12–13 show radon-mitigation strategies for the three basic foundation types.

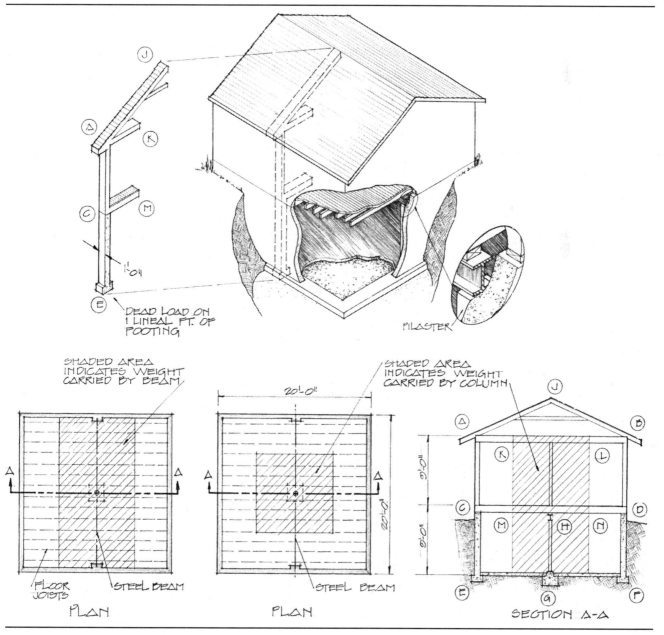

Figure 12–6 Footing loads.

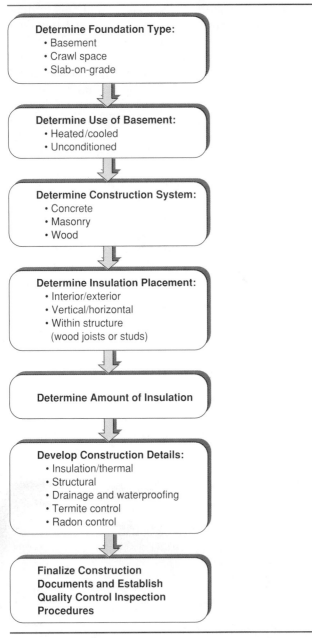

Figure 12–7 The decision-making process for foundation design.

The flowchart contains the following boxes, connected top to bottom by arrows:

Determine Foundation Type:
- Basement
- Crawl space
- Slab-on-grade

Determine Use of Basement:
- Heated/cooled
- Unconditioned

Determine Construction System:
- Concrete
- Masonry
- Wood

Determine Insulation Placement:
- Interior/exterior
- Vertical/horizontal
- Within structure (wood joists or studs)

Determine Amount of Insulation

Develop Construction Details:
- Insulation/thermal
- Structural
- Drainage and waterproofing
- Termite control
- Radon control

Finalize Construction Documents and Establish Quality Control Inspection Procedures

Termite Control Controlling the entry of termites through residential foundations is advisable in much of the United States. Be sure to consult with local building officials and codes for specific strategies that are acceptable.

The following general recommendations apply where termites are a potential problem:

1. Minimize soil moisture around the basement. Use gutters, downspouts, and runouts to remove roof water, or install a complete subdrainage system around the foundation.

2. Remove all roots, stumps, and scrap wood from the site.

3. Place a bond beam or a course of cap blocks at the top of all concrete masonry foundation walls; or fill all cores on the top course with mortar, and reinforce the mortar joint beneath the top course.

4. Be sure that the sill plate is at least 8″ above grade. It should be pressure-preservative treated to resist decay, and it should be visible from the inside for inspection. Termite shields are often damaged or installed incorrectly. They should be considered optional and should not be regarded as sufficient by themselves.

5. Be sure that the exterior wood siding and trim are at least 6″ above grade.

6. Construct porches and exterior slabs so that they slope away from the foundation wall and are at least 2″ below existing siding. They should also be separated from all wood members by a 2″ gap visible for inspection or by a continuous metal flashing soldered at all seams.

7. Form a termite barrier between the foundation wall and the slab floor by filling the joint with urethane caulk or coal-tar pitch.

8. Place wood posts on flashing or a concrete pedestal raised 1″ above the basement floor, or use pressure-preservative-treated wood posts on the basement floor slab.

9. Flash hollow steel columns at the top.

Plastic foam and mineral wood insulation materials have no food value to termites, but they can provide protective cover and easy tunneling. Exterior insulation system installations can be detailed to facilitate inspection, but they usually sacrifice thermal efficiency. Restrictions on widely used termiticides may make soil treatment unavailable or may cause the substitution of products that are more expensive and possibly less effective.

Figures 12–14, 12–15, and 12–16 show termite-control strategies for the three basic foundation types.

12.3.2 Foundation Types

The three basic types of foundations are full basement, crawl space, and slab-on-grade (Fig. 12–17). Each foundation type can be built using several construction systems. The most common systems, cast-in-place concrete and concrete block foundation walls, can be used for all foundation types. Other systems that can be used include pressure-preservative-treated wood foundations, precast concrete foundation walls, masonry or concrete piers, cast-in-place concrete sandwich panels, and various insulated masonry systems.

Slab-on-grade construction with an integral concrete grade beam at the slab edge is common in climates with shallow frost depths. In colder climates, deeper, cast-in-place concrete walls and concrete block walls are more

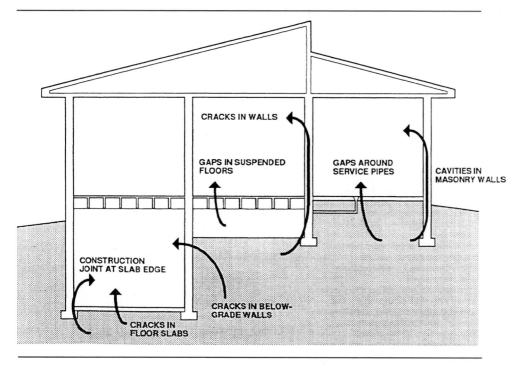

Figure 12–8 Points of radon entry into buildings.

prevalent. Site conditions, overall building design, climate, and local market preferences and costs can all affect the choice of foundation type and construction system.

Basement Foundations Basement walls must be designed to resist lateral loads from the soil and vertical loads from the structure above (Fig. 12–18). These walls are usually constructed of cast-in-place concrete, concrete masonry units (C.M.U.s), or pressure-preservative-treated wood.

Concrete spread footings provide support beneath basement concrete and masonry walls and columns. The footings must be of an adequate size in order to distribute the building loads to the soil. A compacted gravel bed serves as the footing under a wood foundation wall. Concrete slab-on-grade floors are generally designed to have sufficient strength to support floor loads without requiring reinforcing if they are placed on undisturbed or compacted fill.

In areas of expansive soils or high seismic activity, special foundation construction techniques may be necessary. In these cases, consultation with local building officials and a structural engineer is recommended.

Keeping water out of basements is a major concern in many areas of the United States. Water sources may include rainfall, snow melt, and sometimes surface irrigation (Fig. 12–19). There are three basic approaches to preventing water problems in basements:

1. surface drainage,
2. subsurface drainage, and
3. dampproofing or waterproofing on the wall surface.

Surface drainage is designed to keep water from surface sources away from the foundation by sloping the ground surface and by using gutters and downspouts for roof drainage.

Subsurface drainage is designed to intercept, collect, and carry away any water in the ground surrounding the basement. A subsurface drainage system can include placement of porous backfill, drainage mat materials or insulated drainage boards, and perforated drainpipes that lie in a gravel bed along the footing or beneath the slab and that drain to a sump or to daylight.

Waterproofing is intended to keep out water that finds its way to the wall of the structure. It is important to distinguish between the need for dampproofing versus waterproofing. A dampproofing coating covered by a 4-mil layer of polyethylene is recommended in most cases to reduce vapor and capillary draw transmission through the basement wall. However, a dampproofing coating is not effective in preventing water from entering through the wall. Waterproofing is recommended (1) on sites with anticipated water problems or poor drainage, (2) in a building for which a finished basement space is planned, or (3) on any foundation where intermittent hydrostatic pressure occurs against the basement wall.

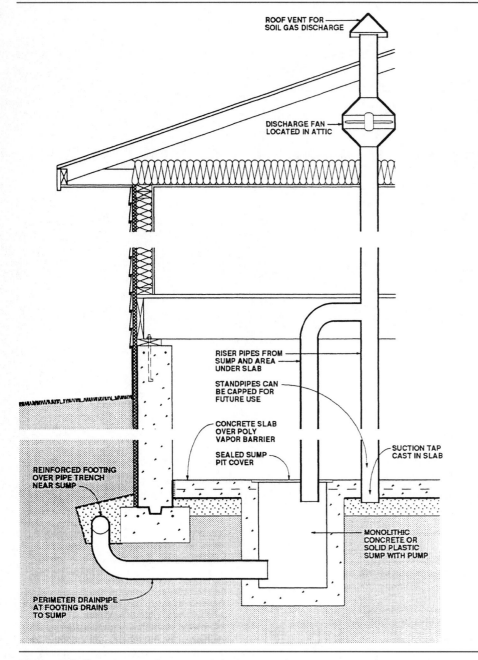

Figure 12–9 Soil gas collection and discharge techniques. (*AIA Architect.*)

Whether to place insulation inside or outside the basement walls is a key question in foundation design (see Appendix F). Placement of a rigid insulation on the exterior surface of a concrete or masonry basement has some advantages over locating the insulation on the inside:

It can provide continuous insulation with no thermal bridges.

It protects and maintains the waterproofing and structural wall at moderate temperatures.

It minimizes moisture condensation problems.

It will not reduce usable basement floor area.

Exterior insulation at the rim joists also leaves joists and sill plates open to inspection from the interior for termites and decay. On the other hand, exterior insula-

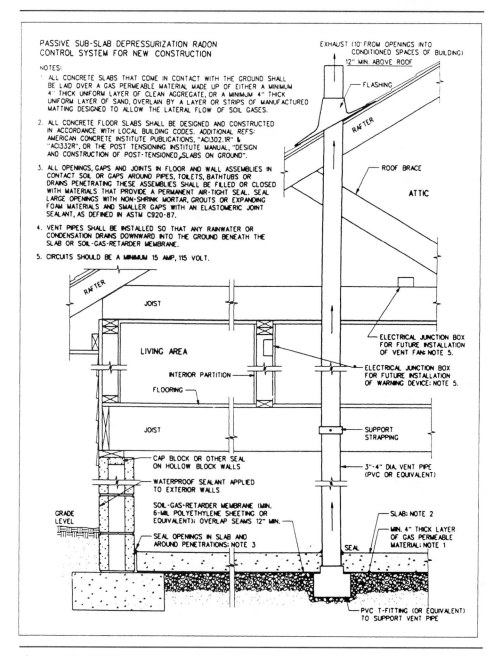

PASSIVE SUB-SLAB DEPRESSURIZATION RADON
CONTROL SYSTEM FOR NEW CONSTRUCTION

NOTES:

1. ALL CONCRETE SLABS THAT COME IN CONTACT WITH THE GROUND SHALL
   BE LAID OVER A GAS PERMEABLE MATERIAL MADE UP OF EITHER A MINIMUM
   4" THICK UNIFORM LAYER OF CLEAN AGGREGATE, OR A MINIMUM 4" THICK
   UNIFORM LAYER OF SAND, OVERLAIN BY A LAYER OR STRIPS OF MANUFACTURED
   MATTING DESIGNED TO ALLOW THE LATERAL FLOW OF SOIL GASES.

2. ALL CONCRETE FLOOR SLABS SHALL BE DESIGNED AND CONSTRUCTED
   IN ACCORDANCE WITH LOCAL BUILDING CODES. ADDITIONAL REFS:
   AMERICAN CONCRETE INSTITUTE PUBLICATIONS, "ACI302.1R" &
   "ACI332R", OR THE POST TENSIONING INSTITUTE MANUAL, "DESIGN
   AND CONSTRUCTION OF POST-TENSIONED SLABS ON GROUND".

3. ALL OPENINGS, GAPS AND JOINTS IN FLOOR AND WALL ASSEMBLIES IN
   CONTACT SOIL OR GAPS AROUND PIPES, TOILETS, BATHTUBS OR
   DRAINS PENETRATING THESE ASSEMBLIES SHALL BE FILLED OR CLOSED
   WITH MATERIALS THAT PROVIDE A PERMANENT AIR-TIGHT SEAL. SEAL
   LARGE OPENINGS WITH NON-SHRINK MORTAR, GROUTS OR EXPANDING
   FOAM MATERIALS AND SMALLER GAPS WITH AN ELASTOMERIC JOINT
   SEALANT, AS DEFINED IN ASTM C920-87.

4. VENT PIPES SHALL BE INSTALLED SO THAT ANY RAINWATER OR
   CONDENSATION DRAINS DOWNWARD INTO THE GROUND BENEATH THE
   SLAB OR SOIL-GAS-RETARDER MEMBRANE.

5. CIRCUITS SHOULD BE A MINIMUM 15 AMP, 115 VOLT.

EXHAUST (10' FROM OPENINGS INTO
CONDITIONED SPACES OF BUILDING)
12" MIN. ABOVE ROOF

FLASHING

RAFTER

ROOF BRACE

ATTIC

RAFTER

JOIST

LIVING AREA

INTERIOR PARTITION

FLOORING

JOIST

ELECTRICAL JUNCTION BOX
FOR FUTURE INSTALLATION
OF VENT FAN: NOTE 5.

ELECTRICAL JUNCTION BOX
FOR FUTURE INSTALLATION
OF WARNING DEVICE: NOTE 5.

SUPPORT
STRAPPING

CAP BLOCK OR OTHER SEAL
ON HOLLOW BLOCK WALLS

WATERPROOF SEALANT APPLIED
TO EXTERIOR WALLS

3"-4" DIA. VENT PIPE
(PVC OR EQUIVALENT)

GRADE
LEVEL

SOIL-GAS-RETARDER MEMBRANE (MIN.
6-MIL POLYETHYLENE SHEETING OR
EQUIVALENT); OVERLAP SEAMS 12" MIN.

SEAL OPENINGS IN SLAB AND
AROUND PENETRATIONS: NOTE 3

SLAB: NOTE 2

MIN. 4" THICK LAYER
OF GAS PERMEABLE
MATERIAL: NOTE 1

SEAL

PVC T-FITTING (OR EQUIVALENT)
TO SUPPORT VENT PIPE

Figure 12–10 CABO recommendations for radon mitigation. *(AIA Architect.)*

tion can provide a path for termites if it is not properly treated, and it can prevent inspection of the wall from the exterior.

Insulation placed on the inside is generally less expensive if the cost of interior finish materials is not included. It does not leave the wall with a finished, durable surface, however. Another concern is that energy savings may be reduced due to thermal bridges with some systems and details. Insulation can be placed on the inside of the rim joist, but it creates a greater risk of condensation problems on the rim joist. It also limits access to wood joists and sills for termite inspection from the interior.

Placing insulation in the ceiling of an unconditioned basement is another alternative. This approach should be used with caution in colder climates where pipes may freeze and structural damage may result from lowering the frost depth. With a wood foundation system, insulation is normally placed in the stud cavities.

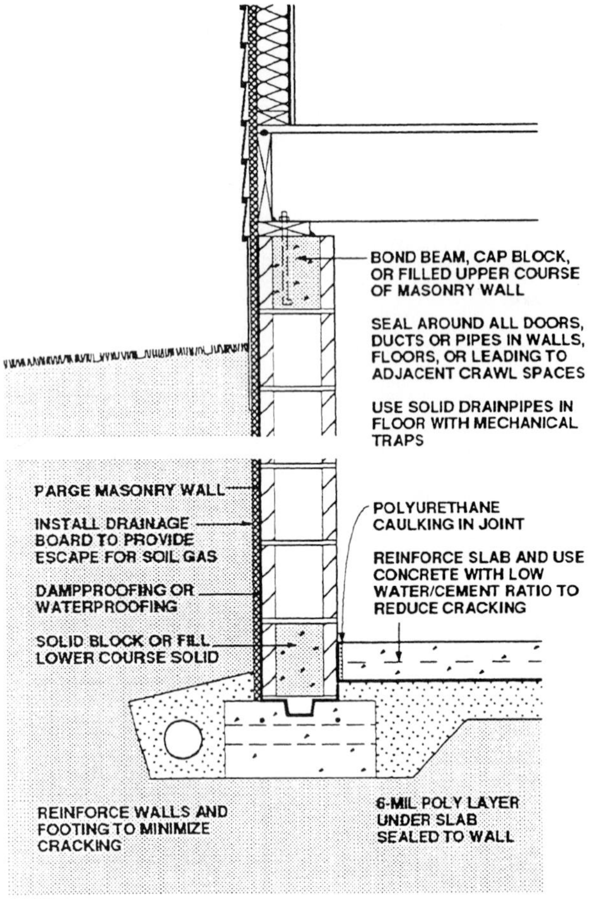
BOND BEAM, CAP BLOCK, OR FILLED UPPER COURSE OF MASONRY WALL
SEAL AROUND ALL DOORS, DUCTS OR PIPES IN WALLS, FLOORS, OR LEADING TO ADJACENT CRAWL SPACES
USE SOLID DRAINPIPES IN FLOOR WITH MECHANICAL TRAPS
PARGE MASONRY WALL
INSTALL DRAINAGE BOARD TO PROVIDE ESCAPE FOR SOIL GAS
DAMPPROOFING OR WATERPROOFING
SOLID BLOCK OR FILL LOWER COURSE SOLID
POLYURETHANE CAULKING IN JOINT
REINFORCE SLAB AND USE CONCRETE WITH LOW WATER/CEMENT RATIO TO REDUCE CRACKING
REINFORCE WALLS AND FOOTING TO MINIMIZE CRACKING
6-MIL POLY LAYER UNDER SLAB SEALED TO WALL

Figure 12–11 Radon-control techniques for basements.

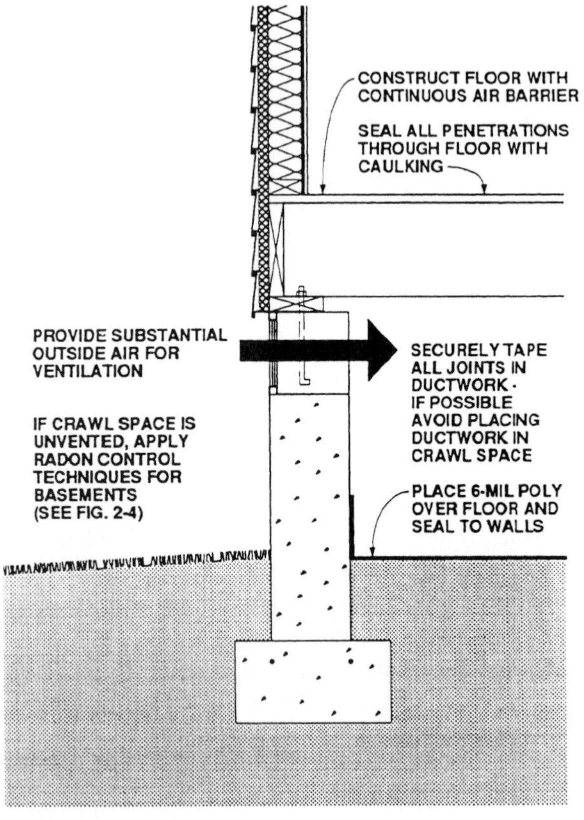
CONSTRUCT FLOOR WITH CONTINUOUS AIR BARRIER
SEAL ALL PENETRATIONS THROUGH FLOOR WITH CAULKING
PROVIDE SUBSTANTIAL OUTSIDE AIR FOR VENTILATION
SECURELY TAPE ALL JOINTS IN DUCTWORK - IF POSSIBLE AVOID PLACING DUCTWORK IN CRAWL SPACE
IF CRAWL SPACE IS UNVENTED, APPLY RADON CONTROL TECHNIQUES FOR BASEMENTS (SEE FIG. 2-4)
PLACE 6-MIL POLY OVER FLOOR AND SEAL TO WALLS

Figure 12–12 Radon-control techniques for crawl spaces.

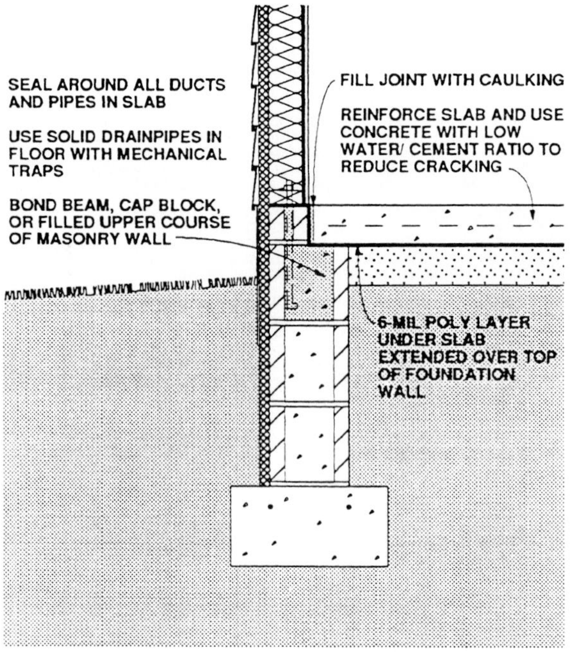
SEAL AROUND ALL DUCTS AND PIPES IN SLAB
USE SOLID DRAINPIPES IN FLOOR WITH MECHANICAL TRAPS
BOND BEAM, CAP BLOCK, OR FILLED UPPER COURSE OF MASONRY WALL
FILL JOINT WITH CAULKING
REINFORCE SLAB AND USE CONCRETE WITH LOW WATER/ CEMENT RATIO TO REDUCE CRACKING
6-MIL POLY LAYER UNDER SLAB EXTENDED OVER TOP OF FOUNDATION WALL

Figure 12–13 Radon-control techniques for slab-on-grade foundations.

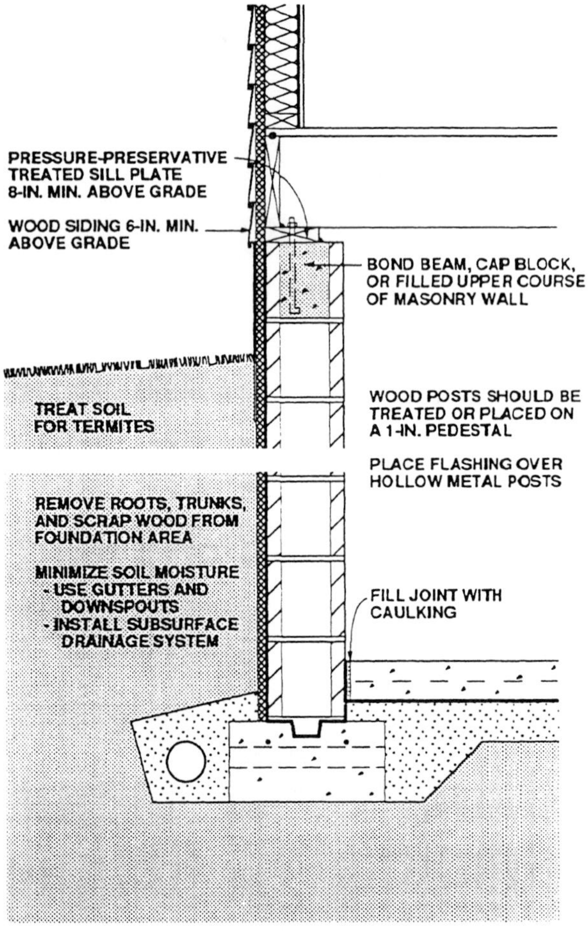
PRESSURE-PRESERVATIVE TREATED SILL PLATE 8-IN. MIN. ABOVE GRADE
WOOD SIDING 6-IN. MIN. ABOVE GRADE
BOND BEAM, CAP BLOCK, OR FILLED UPPER COURSE OF MASONRY WALL
TREAT SOIL FOR TERMITES
WOOD POSTS SHOULD BE TREATED OR PLACED ON A 1-IN. PEDESTAL
PLACE FLASHING OVER HOLLOW METAL POSTS
REMOVE ROOTS, TRUNKS, AND SCRAP WOOD FROM FOUNDATION AREA
MINIMIZE SOIL MOISTURE - USE GUTTERS AND DOWNSPOUTS - INSTALL SUBSURFACE DRAINAGE SYSTEM
FILL JOINT WITH CAULKING

Figure 12–14 Termite-control techniques for basements.

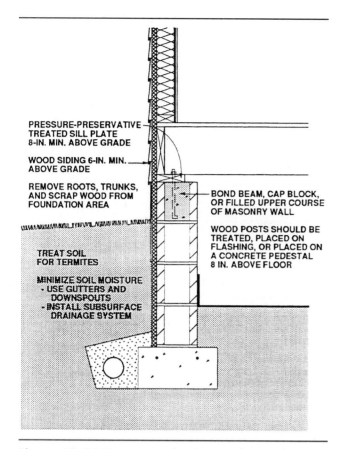

PRESSURE-PRESERVATIVE
TREATED SILL PLATE
8-IN. MIN. ABOVE GRADE

WOOD SIDING 6-IN. MIN.
ABOVE GRADE

REMOVE ROOTS, TRUNKS,
AND SCRAP WOOD FROM
FOUNDATION AREA

BOND BEAM, CAP BLOCK,
OR FILLED UPPER COURSE
OF MASONRY WALL

WOOD POSTS SHOULD BE
TREATED, PLACED ON
FLASHING, OR PLACED ON
A CONCRETE PEDESTAL
8 IN. ABOVE FLOOR

TREAT SOIL
FOR TERMITES

MINIMIZE SOIL MOISTURE
- USE GUTTERS AND
  DOWNSPOUTS
- INSTALL SUBSURFACE
  DRAINAGE SYSTEM

Figure 12–15 Termite-control techniques for crawl spaces.

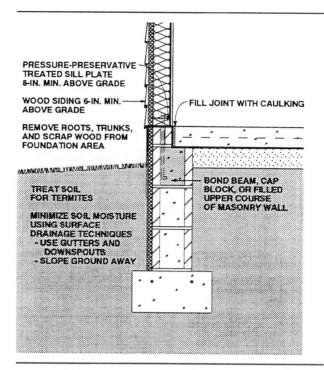

PRESSURE-PRESERVATIVE
TREATED SILL PLATE
8-IN. MIN. ABOVE GRADE

WOOD SIDING 6-IN. MIN.
ABOVE GRADE

FILL JOINT WITH CAULKING

REMOVE ROOTS, TRUNKS,
AND SCRAP WOOD FROM
FOUNDATION AREA

BOND BEAM, CAP
BLOCK, OR FILLED
UPPER COURSE
OF MASONRY WALL

TREAT SOIL
FOR TERMITES

MINIMIZE SOIL MOISTURE
USING SURFACE
DRAINAGE TECHNIQUES
- USE GUTTERS AND
  DOWNSPOUTS
- SLOPE GROUND AWAY

Figure 12–16 Termite-control techniques for slab-on-grade foundations.

In addition to more conventional interior or exterior placement, there are several systems that incorporate insulation into the construction of the concrete or masonry walls:

1. rigid foam plastic insulation cast within a concrete wall,

2. polystyrene beads or granular insulation materials poured into the cavities of conventional masonry walls,

3. systems of concrete blocks with insulating foam inserts,

4. framed interlocking rigid foam units that serve as permanent insulation forms for cast-in-place concrete, and

5. masonry blocks made with polystyrene beads instead of aggregate in the concrete mix.

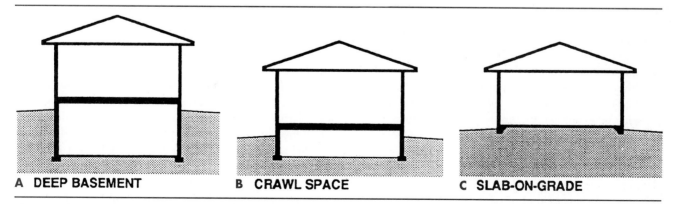

A DEEP BASEMENT   **B CRAWL SPACE**   **C SLAB-ON-GRADE**

Figure 12–17 Basic foundation types.

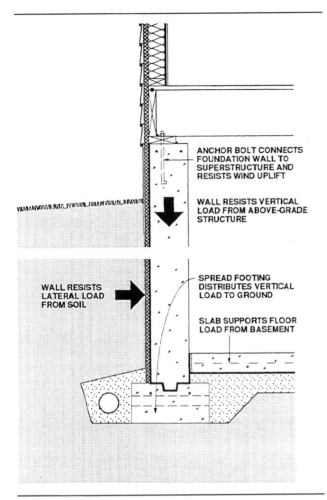

Figure 12-18 Components of a basement structural system.

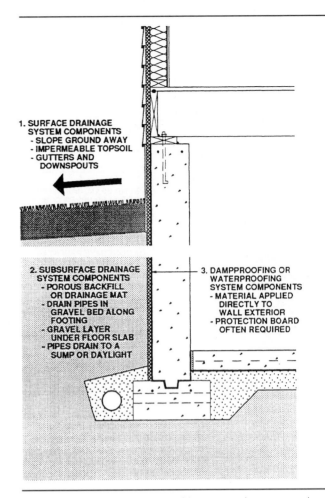

Figure 12-19 Components of basement drainage and waterproofing systems.

Crawl Spaces   Crawl spaces vary in their height and in their relationship to exterior grade. A standard crawl space has walls that are 2' high with only the upper 8" exposed above grade on the exterior side (Fig. 12–20). Support can be either (1) structural foundation walls with continuous spread footings or (2) piers or piles with beams between. The beams between piers support the structure above and transfer the loads back to the piers. Because the interior temperature of a vented crawl space may be below freezing in very cold climates, footings must be below the frost depth with respect to both interior and exterior grade.

Although a crawl space foundation is not as deep as a full basement, it is best to keep it dry. Good surface drainage is recommended, and subsurface drainage may be necessary (Figs. 12–21 and 12–22). Where the crawl space floor is at the same level or above the surrounding exterior grade, no subsurface drainage system is required. On sites with a high water table or poorly draining soils, keep the crawl space floor above or at the same level as the exterior grade. On sites with porous soil and no wa-

ter table near the surface, place the crawl space floor below the surface. If it is necessary to place the crawl space floor beneath the existing grade, and if the soil is nonporous, a subsurface perimeter drainage system similar to that used for a basement should be used. Waterproofing or dampproofing on the exterior foundation walls of crawl spaces is not considered necessary if there is adequate drainage.

The insulation in a vented crawl space is always located in the ceiling. Batt insulation is typically placed between the floor joists, leaving sill plates open for inspection for termites or decay.

With an unvented crawl space, you must decide whether to place insulation inside or outside the walls, similar to the situation for a full basement. Vertical exterior insulation on a crawl space wall can extend as deep as the top of the footing. You can supplement it by extending the insulation horizontally from the face of the foundation wall. Placing crawl space wall insulation inside is more common than placing it on the exterior, primarily because it is less expensive.

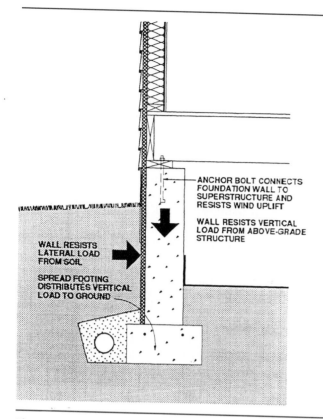

Figure 12-20 Components of a crawl space structural system.

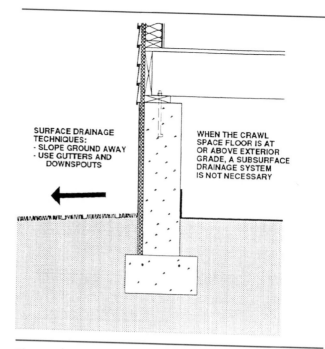

Figure 12-21 Crawl space drainage techniques.

Interior wall insulation may be less desirable than exterior insulation for the following reasons:

It increases the exposure of the wall to thermal stress and freezing.

It may increase the likelihood of condensation on sill plates, band joists, and joist ends.

It often results in some thermal bridges through framing members.

It may require installation of a flame-resistant cover.

Rigid insulation board is easier to apply to the interior wall than is batt insulation. It is continuous and may not need an additional vapor barrier.

Batt insulation is commonly placed inside the rim joist. This rim joist installation should be covered on the inside face with a polyethylene vapor retarder or a rigid foam insulation sealed around the edges to act as a vapor retarder.

With a pressure-preservative-treated wood foundation system, insulation is placed in the stud cavities, similarly to above-grade insulation in a wood frame wall.

Most major building codes require venting of crawl spaces in order to control moisture condensation within the crawl space. Generally, natural ventilation by reasonably distributed openings through foundation walls or exterior walls is required. A number of exceptions do allow for reducing or eliminating the openings. For example, when the ground surface within the crawl space is covered with an approved vapor barrier, the net vent area can be reduced to $1/10$ of that required. In this case, the

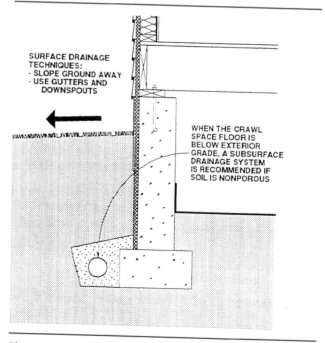

Figure 12-22 Crawl space drainage techniques.

ventilation openings can be equipped with operable louvers, allowing additional energy conservation benefits during the times of the year when moisture condensation is unlikely. Another exception allows the omission of ventilation openings on one side. This exception assumes that the remaining three sides will provide the necessary crossflow. The remaining three sides, however, must provide the total minimum net area as required. Under-floor spaces used as supply plenums for distribution of heated and cooled air do not have to be vented. Figure 12–23 shows the conditions used in calculating the required crawl space vent area for the following examples:

EXAMPLE 1: A house having a crawl space area of 1300 sq ft would require a total net clear area of openings not less than 8.7 sq ft (1300/150). With the 10 openings shown in Fig. 12–23, an opening size of 8″× 16″ (0.87 sq ft) would be acceptable.

EXAMPLE 2: If an approved vapor barrier were placed over the ground surface of the crawl space, the total area of openings to the exterior in Example 1 could be reduced to 0.87 sq ft (8.7 sq ft/10).

There are several advantages to designing crawl spaces as semiheated zones. Duct and pipe insulation can be reduced, and the foundation can be insulated at the perimeter instead of at the ceiling. As a result, less insulation is usually required, installation difficulties are simplified in some cases, and the design can be detailed to minimize condensation hazards.

Concrete Slab-on-Grade Construction In recent years, basements have been eliminated in many 1-story houses. Formerly, a basement was definitely necessary, for heating plants, fuel storage, and laundry or utility areas. With the advent of liquid and gas fuels, however, the need for bulky fuel and ash storage space has been eliminated. Both heating plants and laundry equipment have become more compact, thus requiring little space in the modern home. Basementless residences have become popular, especially in warmer climates where footings need not be placed so deep in the ground. With the elimination of the basement, one type of floor construction that has been found successful, if properly executed, is the concrete slab-on-ground construction (Fig. 12–24).

Precaution should be used in placing a slab on certain lots. Not only do sloping lots require considerable excavation, but they also present drainage problems. Low lots where moisture often accumulates would also be unfavorable. The ideal area would be a nearly level lot that requires a minimum of excavation and that is free of moisture problems. Finish floor should be at least 8″ above grade. Concrete floors have a tendency to feel cold and damp during winter and wet seasons; you can eliminate this coldness by introducing either radiant heating coils or perimeter heating ducts into the slab. Care must be taken to have an absolutely watertight membrane below the slab to prevent moisture from penetrating from the ground; and, particularly in northern climates, the periphery of the slab must be well insulated.

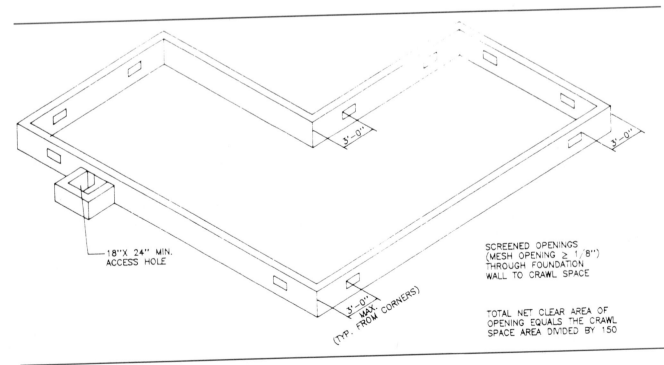

Figure 12–23 Crawl space ventilation.

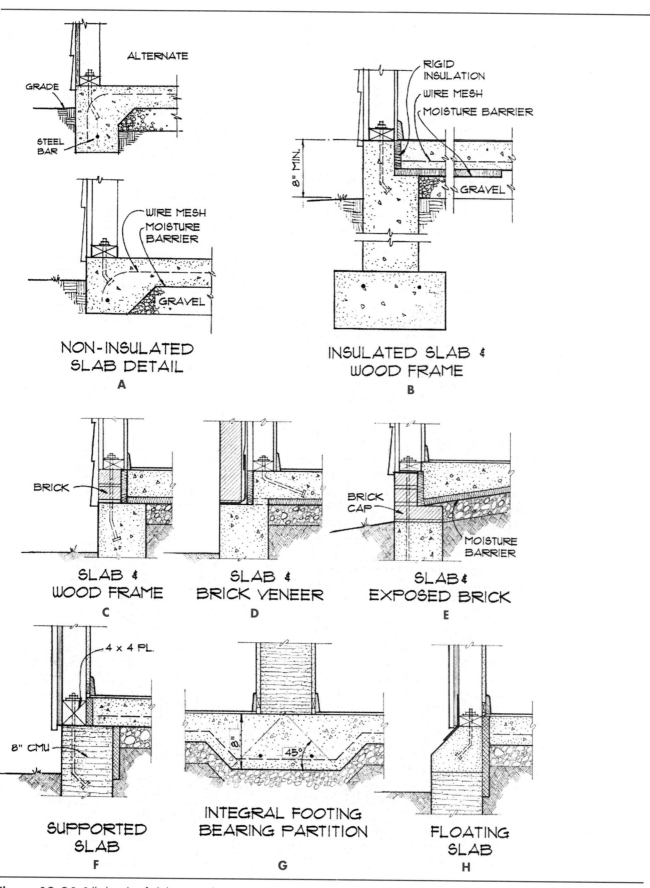

Figure 12–24 Sill details of slab-on-grade construction.

Concrete slabs can be finished with any of the many types of flooring materials: clay, ceramic, vinyl, or cork tile in mastic; carpeting over padding, or plywood block flooring in mastic; or hardwood strip flooring laid on sleepers over a concrete slab.

The following construction requirements should be met in using a concrete slab on ground:

1. Finish-floor level should be high enough above natural grade so that the finished grade will provide good drainage away from the building.
2. All debris, topsoil, and organic matter must be removed from below the slab. Loose soil must be compacted. Earth fill should be placed in 6″ layers.
3. All sewer, water, gas, and oil supply lines must be installed before the slab is poured. Gas lines must be placed in a pipe sleeve.
4. At least a 4″ layer of coarse gravel or crushed rock or coarse sand must be placed above the soil and must be well compacted.
5. A watertight vapor barrier must be placed over the crushed stone before the slab is poured to prevent moisture from seeping up into the slab from the soil.
6. A permanent, waterproof, nonabsorptive type of rigid insulation must be installed around the perimeter of the slab in accordance with the requirements of the climate. In very mild climates, insulation may not be required, but an expansion joint should be used.
7. The slab must be reinforced with wire mesh designated w 1.4 × w 1.4 fabricated in a 6″× 6″ grid from No. 10 gauge steel wire. The slab must be at least 4″ thick and must be troweled to a smooth, hard finish.
8. When ductwork is placed in the slab, it should be of noncorrodible, nonabsorbent material with not less than 2″ of concrete completely surrounding the ductwork. Heating coils and reinforcement, if used, must be covered with at least 1″ of concrete.

Slab-on-grade floors permit a lower silhouette in the overall appearance of the house. In areas where outdoor living spaces adjacent to the house are popular, the slab floor eliminates the need for four or five steps leading from interior floor levels to the surface of the outdoor terrace areas. Consequently, the terrace becomes more usable and more easily accessible from the interior of the house.

Slabs can be supported by the foundation wall (Fig. 12–24F), or they can be independent of the foundation and supported entirely by the soil below the slab (Fig. 12–24H). Either method is satisfactory if the soil below is well compacted. If settlement should occur below the slab, the floating-slab construction would be less likely to eventually crack or fracture. If considerable fill is necessary under a slab, it should be placed under the supervision of a soil engineer in accordance with acceptable en-

gineering practice. Such a slab should be supported by the foundation wall and should have reinforcing steel bars placed in the slab and in thickened portions of the slab throughout the slab area (Fig. 12–25). All load-bearing partitions and masonry nonbearing partitions should be supported on foundations bearing on natural ground independent of the slab, unless the slab is thickened and reinforced to distribute adequately the concentrated load (Fig. 12–24G).

In northern areas, where insulation from heat loss is of prime importance, several inches of Zonolite® or insulating concrete can be used below the regular concrete. Proper rigid insulation should also surround the entire slab (Fig. 12–24).

Structural Concrete Slabs These slabs are designed on the basis of a structural analysis of their intended load. They contain steel reinforcing to make them strong enough to withstand the load that they will bear, independent of any central supports below. For example, garages with basements underneath may require structural slab floors.

Integral Slab and Footing Where frost lines are very shallow, slabs and their footings can be poured at one time, thus providing an integral foundation and floor, economical for 1-story buildings (Fig. 12–24A).

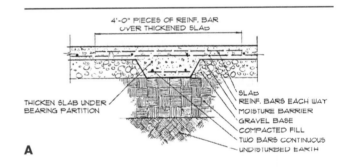

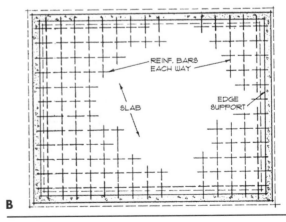

Figure 12–25 Use steel reinforcing in slabs placed on fill.

The footing is merely a flared thickening of the slab around its perimeter. For best results the bottom of the footing should be at least 1′ below the natural grade line and must be supported on solid, unfilled, and well-drained soil. Several No. 4 bars are placed near the bottom of the footing, and the wire mesh of the slab is rolled into the shape of the footing to tie the entire concrete mass together. Gravel fill and a waterproof membrane must be placed below the slab, similar to the other types of slabs. This type of slab is difficult to insulate, but construction time is gained by its use.

Insulation for Floor Slabs

In most areas of the United States, perimeter insulation must be installed around residential concrete slabs if the construction is to be satisfactory. Because of the density of concrete, it can be classified as a conductor of heat rather than an insulator (the reverse is true of wood and other cellular types of building materials).

Heating engineers maintain that practically all of the exterior heat loss from a slab on grade takes place along and near the edges of the slab. Very little heat is dissipated into the ground below the slab (Fig. 12–26). Therefore, careful consideration must be given to the insulation of the slab edges, especially in colder areas. Insulation should be rigid; resistive to dampness; immune to fungus and insect infestation; easy to cut and work; high in crushing strength; and, of course, resistive to transmission of heat.

Insulating qualities, referred to as *R* factors, are usually given for different insulation materials. The *R* factor for a material is the temperature difference in degrees Fahrenheit necessary to force 1 Btu/h through 1 sq ft of the material 1″ thick. Table 12–5 lists the resistance values, or *R* factors, that should be used in determining the minimum amount of insulation recommended for various design temperatures, with and without floor radiant heating. The table also indicates the minimum depth that the insulation should extend below grade.

Table 12–5 Resistance (R) values for determining slab-on-ground perimeter insulations.

Heating Design Temperature (°F)	Depth That Insulation Extends Below Grade	Resistance (R) Factor	
		No Floor Heating	Floor Heating
−20	2′-0″	2.00	3.00
−10	1′-6″	1.75	2.62
0	1′-0″	1.50	2.25
+10	1′-0″	1.25	1.87
+20	1′-0″	1.00	1.50

For example, if the table indicates that an *R* factor of 2.62 is needed, and if the insulation selected has an *R* factor of 1.50 per inch of thickness, then the total thickness needed would be calculated as follows:

$$\frac{2.62}{1.50} = 1.74'' \text{ of insulation}$$

For convenience and reliability, a 2″ thickness should be used, making the *R* factor $2 \times 1.50 = 3.00$

Cellular glass insulation board, asphalt-coated glass-fiber board, and synthetic plastic foam are favorable types of perimeter insulation for concrete slabs.

Stepped Footings

These footings are often required on sloping or steep lots where straight horizontal footings would be uneconomical (Fig. 12–27). The bottom of the footing is always placed on undisturbed soil and below the frost line, and horizontal runs should be level.

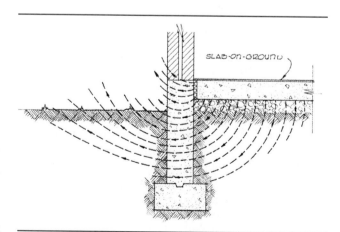

Figure 12–26 Heat loss of slab on ground.

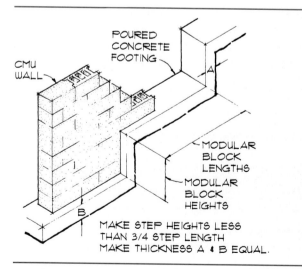

Figure 12–27 Stepped footings. Check local codes for various restrictions.

The vertical rises of the footings are usually the same thickness as the horizontal portion and are poured at the same time. The vertical distance between steps should not exceed 2'; if concrete masonry units (C.M.U.s) are used for the foundation wall, the vertical height of each step should be compatible with the unit height of the C.M.U. Horizontal lengths for the steps are usually not less than 3'.

Column or Post Footings   These footings are required to carry concentrated loads within the building. Usually, they are square with a pedestal on which the column or post will bear. A steel pin is usually inserted into the pedestal if a wooden column is to be used. The footing will vary in size, depending on the load-bearing capacity of the soil and the load that it will have to carry. Ordinary column footings for small residences might be 24" square by 12" deep, which would support 16,000 lb if the soil-bearing capacity were 4000 lb/sq ft (Table 12–6).

Footings for chimneys, fireplaces, and the like, should be poured at the same time as other footings, and they should be large enough to support the weights that they will have to carry. You can calculate total weights of different types of masonry construction by using Table 12–6.

Freestanding Pier Foundations   In warmer areas where floors need not be insulated, it is more economical to put 1-story houses on freestanding piers of masonry rather than on continuous foundations around the entire exterior wall (Fig. 12–28). The piers must be of solid masonry: brick, poured concrete, or concrete masonry units. Customarily, piers 18" or 24" high above grade are used, and they are never more than three times their least horizontal dimension unless reinforced. Usually 8"× 16" or 12"× 16" piers spaced not more than 8' apart are sufficient. Piers should be spanned with either wood or metal beams large enough to carry the weight of the exterior walls. The beams act as sills, and the floor is built above the beams, which are well anchored to the piers.

Brick veneer can be used with freestanding piers if the brick veneer is started on a poured footing below grade. The footings must be poured integrally with the pier footings, and each must be the proper size. Despite the

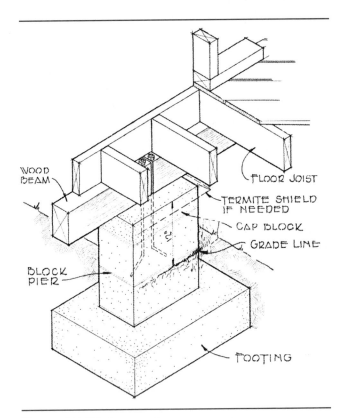

Figure 12–28 Freestanding masonry pier.

use of masonry piers, this brick-veneer curtain wall gives the exterior the appearance of a continuous foundation. When a brick-veneer curtain wall is used, sufficient ventilation and an access door must be provided as in continuous foundations (Fig. 12–29).

Piers   Piers are freestanding masonry posts. In basementless houses with wood floors, they are spaced 6' to 8' apart to support the girders. Usually, they need to be only several feet high and 16" square in section and capped with a 4" solid cap unit. The units should be laid in a lapped bond for strength, and if the piers support heavy loads, they should be made larger and filled with concrete.

Grade Beam Foundations   One-story frame buildings can be adequately supported on foundations made of a system of poured concrete piers with suitable footings, with the piers spanned by reinforced concrete beams that have been formed with trenches dug into the ground (Fig. 12–30). The concrete beams must be designed in accordance with good engineering practice to carry the weight of the building above. Unless the soil below the beam has been removed and replaced with coarse rock or gravel not susceptible to frost action, the bottom of the grade beam must be below the frost line. If structural analysis has not been made, the following minimum conditions (F.H.A.) or conditions according to local building

Table 12–6 Masonry weights.

Material	Weight (lb/cu ft)
Poured concrete	150
Concrete block (C.M.U.)	75
Solid brick	120
Stone	160

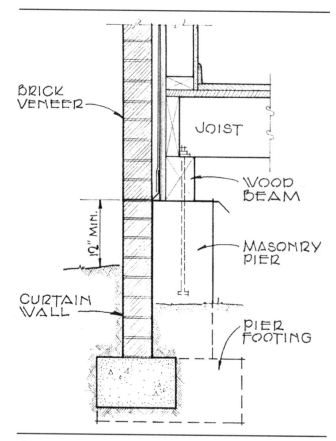

Figure 12-29 Economical pier construction with brick veneer and crawl space.

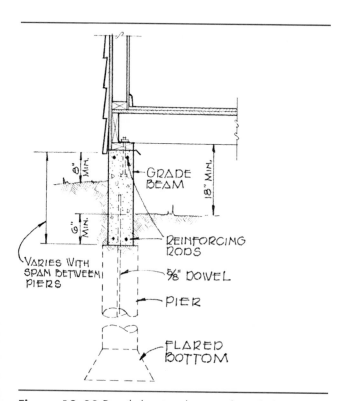

Figure 12-30 Detail showing the use of grade beam with frame construction.

codes should be met, based on a 1-story structure with average soil conditions:

1. *Piers:*
   (a) Maximum pier spacing, 8'-0" o.c.
   (b) Minimum size of piers, 10' diam. Pier should be reinforced with a No. 5 bar for the full length of the pier and extending into the beam.
   (c) Depth of pier should extend below the frost line and have a bearing area of at least 2 sq ft for average soils.

2. *Grade beam:*
   (a) Minimum width for frame buildings: 6". An 8" beam can be flared if covered by base trim. Masonry or masonry veneer should be 8" wide.
   (b) Minimum effective depth, 14". If grade beam supports a wood floor, the beam must be deep enough to provide a minimum of 18" crawl space below the wood floor.
   (c) Reinforce the beam with two top bars and two No. 4 bottom bars, if frame construction. If masonry or masonry veneer construction, reinforce with two top bars and two No. 5 bottom bars. If grade beam is flared at top, reinforce top with one No. 6 bar instead of two No. 4 bars.

Foundation Walls For light construction, foundation walls are generally built of poured concrete or C.M.U. Poured concrete requires time to set and harden. The integral, one-solid-piece concrete foundation will resist fracturing from settling and differential soil pressures. It lends itself to steel reinforcement at critical points, and dimensions are not restricted by unit sizes, as is the case with C.M.U., brick, or other materials. A good rule of thumb is to limit the height of unreinforced concrete foundation walls to *10 times their thickness*; otherwise, vertical reinforcing rods should be used.

Frequently, the thickness of the wall is determined by the thickness of the superstructure rather than the load that it is to carry. If masonry is used above, the foundation should be as thick as the masonry. If the brick or stone veneer frame construction is used, an 8" thick foundation will be sufficient for 1-story heights. The height of a 2-story building with a basement will require a 10" thick foundation wall.

Brick cavity walls, which are usually 10" thick, can be supported on an 8" foundation, provided that the 8" thickness is *corbeled* with solid masonry to the thickness of the cavity wall. Long foundation walls, over 25', should be increased in thickness or designed with integral pilasters at points in the wall where girders or beams rest. The pilasters should be extended ½ the thickness of the foundation wall (Fig. 12-31).

When concrete walls are poured separately from the footings, a key is used at their intersection to resist lateral ground pressures (Fig. 12-32).

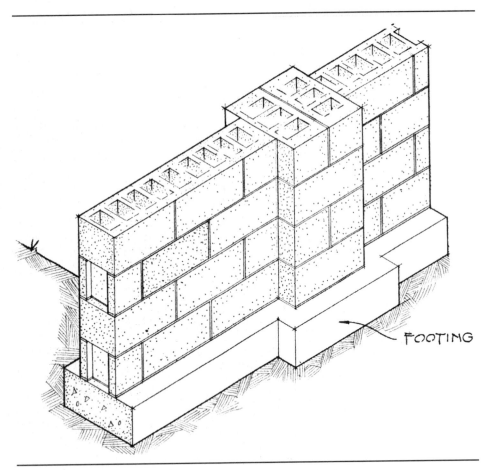

Figure 12–31 Block wall pilaster.

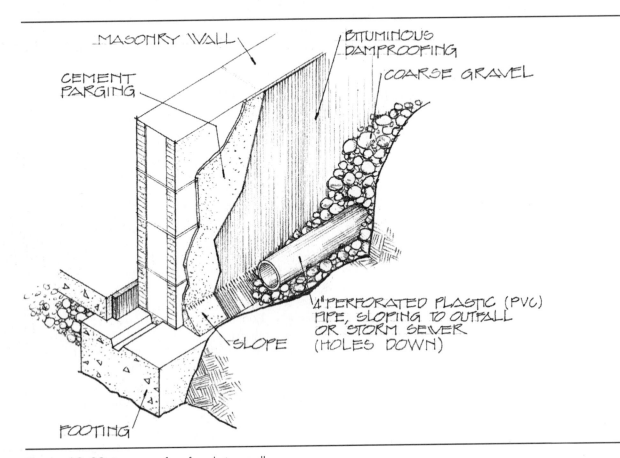

Figure 12–32 Dampproofing foundation walls.

Concrete Masonry Unit (Concrete Block) Because of their economy and speed of erection, these foundations are used for residences as well as other types of buildings. Concrete masonry units are acceptable when correct design has been used and good construction methods have been followed. The majority of foundation walls are made with the nominal 8″ × 8″ × 16″ and/or 8″ × 12″ × 16″ sizes. Their actual sizes are 7⅝″ × 7⅝″ × 15⅝″ and 7⅝″ × 11⅝″ × 15⅝″ (standard sizes vary slightly in different areas). The smaller height and length allow for vertical and horizontal mortar joints to attain modular dimensions. This feature can be a convenience to the designer or drafter. Common brick can be incorporated into C.M.U. walls if necessary—three brick courses are equivalent to the 8″ C.M.U. thickness; three brick wythes are the same as the 12″ C.M.U. thickness; and so on.

Concrete masonry unit foundation walls should be capped with 4″ high solid cap units. This cap provides a smooth, continuous bearing surface for wood sills and makes the walls more resistive to termite infestation. Corner, joist, jamb, header, and other special units are also made in the standard sizes (Fig. 12–33). Basement interior partitions can be built with 4″ or 6″ thick partition units. Very often, C.M.U. walls are veneered with face brick to produce economical yet pleasing masonry walls of different thicknesses.

Points on foundation walls of C.M.U. on which beams or girders rest should have the cores filled with 1:2:4 concrete from the footing to the bearing surface. If heavy loads are anticipated, vertical steel rods should also be introduced to the cores. In long C.M.U. walls, 8″ × 16″ vertical pilasters should be incorporated in the inside of the wall at a maximum of 10′-0″ intervals (or as local building codes require) and filled with concrete for rigidity (Fig. 12–31).

The load put on C.M.U. walls should be limited to 70 or 80 lb/sq in., which usually makes the 8″ or 12″ thick blocks adequate for foundations in the majority of light construction buildings. In most situations, the mortar joints are the weakest part of the wall. Codes usually require units to be laid in full beds of portland cement mortar. Fully loaded walls would still have a safety factor of approximately 4 against failure. Walls supporting wood frame construction should extend at least 8″ above the grade line. Entrance platform slabs, porch slabs, and areaways should be supported or anchored to foundation walls (Fig. 12–34). In straight walls, units should be laid in lap bond (Fig. 12–31) so that vertical joints in every other course are directly above each other. Intersecting block walls should also be bonded for rigid construction by interlocking or lapping of alternate courses.

For economy and the sake of good construction, care should be taken to design C.M.U. walls so that units will not have to be cut by the mason. Wall lengths should be in modules of 8″ if the 16″ long unit is used. Openings for doors and windows should be placed, if possible,

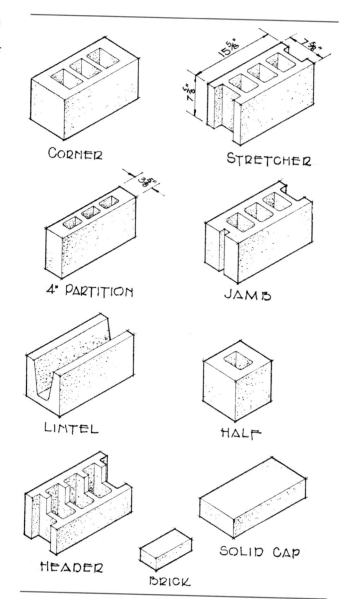

Figure 12-33 Typical concrete masonry units.

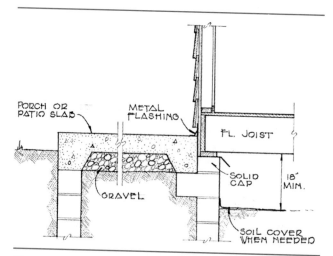

Figure 12-34 Anchoring exterior slabs to foundation walls.

where vertical joints occur; both the width and the placement of the opening in the wall must be considered. Similarly, heights of C.M.U. walls should be determined by the heights of the units used. If care is taken on the drawing board to provide wall lengths and openings compatible with the C.M.U. measurements used, sounder walls and neater construction will result.

Anchor bolts are required by codes in foundation walls. They are especially necessary to tie frame constructions such as open structures, carports, or garages to the foundations. Spaced 6'-0" apart with at least two bolts in each sill member, anchors will resist winds if they are long enough to extend down into the second course of C.M.U. construction. You can bend or secure the lower ends of the bolts by embedding a 4" square steel washer into the mortar joints and filling the cavities with concrete.

In areas subject to earthquakes, C.M.U. walls are not practical for foundations. To resist earthquakes, they must be reinforced with vertical rods in cores, and there must also be horizontal reinforcement in mortar beds.

Basement C.M.U. walls should be carefully laid with full, tight mortar joints. If the wall is to be waterproofed, the exterior surface below grade should be parged with a ½" layer of cement mortar and coated with hot tar or asphalt. In case of extreme moisture conditions, roofing felt is applied over the hot asphalt and is given another coat of hot asphalt. A foundation drainpipe set in gravel around the edge of the footings should also be used to carry off water that may build up around the wall (Fig. 12–32).

Lintels   Lintels over wall openings should be reinforced concrete rather than wood. Wood incorporated into C.M.U. walls for support often shrinks and causes cracks in the masonry work. Some codes require a reinforced bond beam capping the entire block wall. Precast concrete lintels are available that can be set in mortar above openings by the mason. Because of their weight in handling, split precast lintels are often used over wider openings. Lintel units can also be made into satisfactory lintels. The units are set on forms over the opening and are filled with concrete and several reinforcing rods (Fig. 12–35).

Areaways   Basements in residences require some method of ventilation and natural lighting. If the basement wall extends only 1' or 2' above grade, an areaway is needed so that basement windows that are on or below grade do not allow moisture to get into the basement (Fig. 12–36). The areaway walls must be tied to the basement walls, and the same masonry materials are usually used for the areaway as for the foundation.

Small areaways can be made of semicircular, corrugated, galvanized steel. The bottom of the areaway can have a masonry floor with a drain leading to the storm sewer, or it can be filled with sufficient crushed gravel or

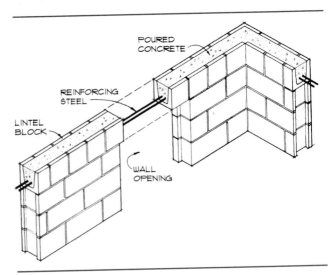

Figure 12-35 Lintel unit in masonry wall.

stone with drainage tile below to carry the water away from the foundation wall. The areaway walls should extend several inches above grade and at least 4" below the bottom of the basement window. If the areaway is over 24" deep, it should be provided with a grill above to prevent children or animals from falling into it.

Steel Reinforcement   Poured concrete used in foundations for lintels or structural beams must be reinforced with steel rods to carry the weight imposed on it. For structural use, concrete in itself possesses good compression resistance qualities but has relatively poor tensile strength. Mild steel and concrete have nearly the same coefficients of expansion and contraction; they are, therefore, compatible in that they will expand and contract practically the same amount during temperature changes and thus will not lose their bond when combined. When a force or weight is exerted downward on the top of a concrete beam, the tensile stresses will react outward on the lower part of the beam and cause it to break. For example, if a heavy weight were put on the center of a horizontal beam, supported on each end, the beam would crack open on its underside. To counteract this effect,

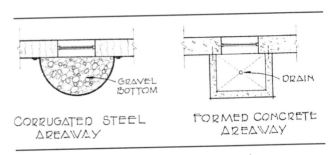

Figure 12-36 Areaways for basement windows.

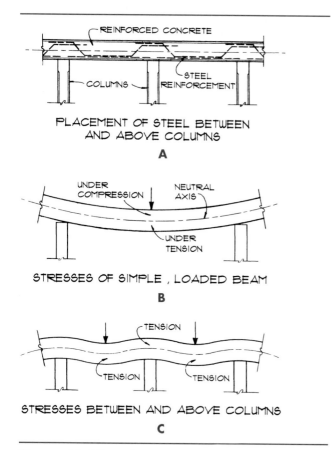

PLACEMENT OF STEEL BETWEEN AND ABOVE COLUMNS

A

STRESSES OF SIMPLE, LOADED BEAM

B

STRESSES BETWEEN AND ABOVE COLUMNS

C

Figure 12-37 Loading stresses of reinforced concrete beams. Similar stresses exist in wood beams.

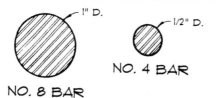

NO. 8 BAR   NO. 4 BAR

Figure 12-38 Reinforcing steel is sized according to its diameter.

steel, with its high tensile strength, should be placed in the lower part of the beam where the steel becomes the most effective. If the beam has a column or support below it at any point, the steel in the beam should continue over the column but be raised to the upper part of the beam directly over the column (Fig. 12–37A). The tensile stresses directly above the column are in the upper part of the beam.

Steel wire mesh, used in concrete slabs on grade, should be placed in the upper part of the slab for most effectiveness; in a 4″ slab the mesh should be placed 1″ from the top of the slab. Steel mesh, also known as *welded wire fabric*, is available in 5′ wide rolls, and the 6″ × 6″ mesh opening size is usually used in slab reinforcement unless structural slabs are required; w 1.4 × w 1.4 wire size is sufficient.

In brickwork, steel angles are usually used above openings, unless brick arches or other types of masonry lintels are used. Steel angle sizes, 3½″ × 3½″ × 5/16″ and 6″ × 4″ × 3/8″, are often used because they lend themselves to the width of the brick and the mortar joint thickness. However, the thickness and the size of the lintel depend on the load above and the span of the opening.

Steel reinforcement rods are available in sizes from No. 2 to No. 11, and they are labeled by their number or by their diameter (Fig. 12–38 and Table 12–7). The number labeling is preferable because it indicates how many eighths of an inch the rod is in diameter. A No. 5 rod, for example, would be 5/8″ in diameter; a No. 7 rod; 7/8″ in diameter; and so on. Deformed rods are rods that

Table 12-7 Standard sizes of reinforcing bars.

To determine the strength of a given size of rebar, multiply its yield stress (40,000 psi and 60,000 psi for grades 40 and 60) times its cross-sectional area.

Bar Size	Weight per Ft		Diameter		Cross-Sectional Area	
	lb	kg	in.	cm	in.	cm
#3	0.376	0.171	0.375	0.953	0.11	0.71
#4	0.668	0.303	0.500	1.270	0.20	1.29
#5	1.043	0.473	0.625	1.588	0.31	2.00
#6	1.502	0.681	0.750	1.905	0.44	2.84
#7	2.044	0.927	0.875	2.223	0.60	3.87
#8	2.670	1.211	1.000	2.540	0.79	5.10
#9	3.400	1.542	1.128	2.865	1.00	6.45
#10	4.303	1.952	1.270	3.226	1.27	8.19
#11	5.313	2.410	1.410	3.581	1.56	10.07
#14	7.650	3.470	1.693	4.300	2.25	14.52
#18	13.600	6.169	2.257	5.733	4.00	25.81

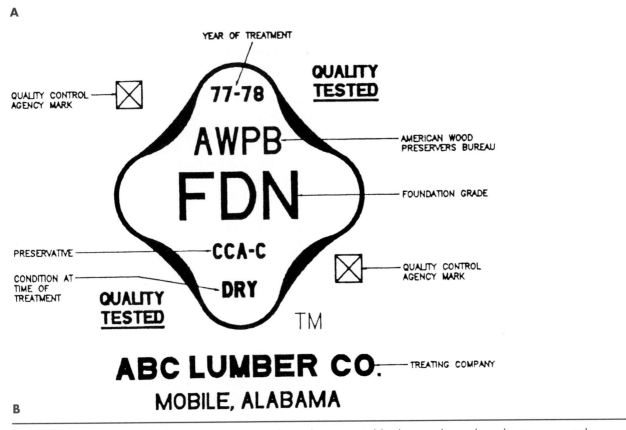

19 _ -19 _ ³
GROUND CONTACT ⁶

ABC ¹   .40 ⁵

AWPA — STDS ²
XXX ⁷   REPRESERVATIVE ⁴
KDAT ⁸
X-XX ⁹

1 - The identifying symbol, logo or name of the accredited agency.
2 - The applicable American Wood Preservers' Association (AWPA) commodity standard.
3 - The year of treatment if required by AWPA standard.
4 - The preservative used, which may be abbreviated.
5 - The preservative retention.
6 - The exposure category (e.g. Above Ground, Ground Contact, Etc.).
7 - The plant name and location; or plant name and number; or plant number.
8 - If applicable, moisture content after treatment.
9 - If applicable, length, and/or class.

A

YEAR OF TREATMENT

QUALITY TESTED

QUALITY CONTROL AGENCY MARK

77-78

AWPB
FDN

CCA-C

DRY

QUALITY TESTED

AMERICAN WOOD PRESERVERS BUREAU

FOUNDATION GRADE

QUALITY CONTROL AGENCY MARK

PRESERVATIVE

CONDITION AT TIME OF TREATMENT

TM

ABC LUMBER CO. —— TREATING COMPANY
MOBILE, ALABAMA

B

Figure 12–39 (A) Typical label for pressure-treated wood. (B) Treated lumber or plywood grade stamp example.

have ridges on their surfaces that provide a more mechanical bond between the steel and concrete. When drawing section details, you should indicate steel reinforcement and place it correctly.

Wood Foundations   The All-Weather Wood Foundation System is fabricated of pressure-treated lumber and plywood approved for below-grade use by an approved inspection agency (Fig. 12–39). All parts of the supporting element for the house structure are included in the foundation system. Foundation sections of nominal 2″ lumber framing and plywood sheathing may be fac-

tory fabricated or constructed at the job site. A good drainage system is an integral part of the wood foundation to keep the basement dry. Gravel is a key element, providing an unobstructed path for water to flow away from the foundation to a sump.

The construction is essentially a below-grade, load-bearing, wood frame system which serves both as the enclosure for basements and crawl spaces and as the structural foundation for the support of light frame structures. Footing plates for the foundation are nominal 2″ pressure treated wood planks resting on a 4″ or thicker leveled bed of gravel or crushed stone (Figs. 12–40 and 12–41).

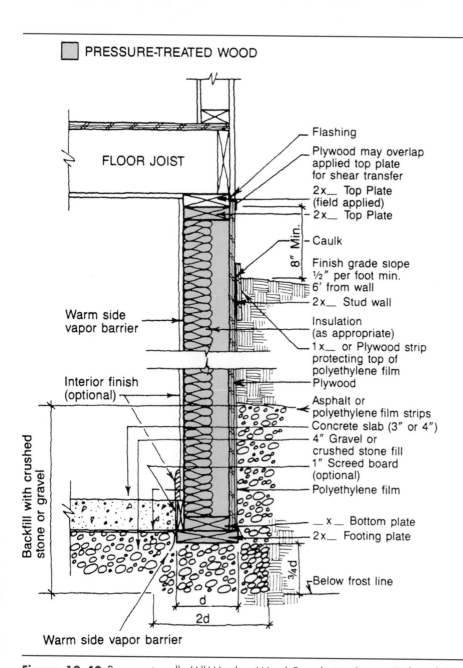

■ PRESSURE-TREATED WOOD

FLOOR JOIST

Flashing

Plywood may overlap applied top plate for shear transfer

2x__ Top Plate (field applied)

2x__ Top Plate

8" Min.

Caulk

Finish grade slope ½" per foot min. 6' from wall

2x__ Stud wall

Warm side vapor barrier

Insulation (as appropriate)

1x__ or Plywood strip protecting top of polyethylene film

Plywood

Asphalt or polyethylene film strips

Concrete slab (3" or 4")

4" Gravel or crushed stone fill

1" Screed board (optional)

Interior finish (optional)

Polyethylene film

Backfill with crushed stone or gravel

__x__ Bottom plate

2x__ Footing plate

¾d

d

2d

Below frost line

Warm side vapor barrier

Figure 12–40 Basement wall. (All-Weather Wood Foundation System Technical Report, 1972. American Forest and Paper Association.)

 PRESSURE-TREATED WOOD

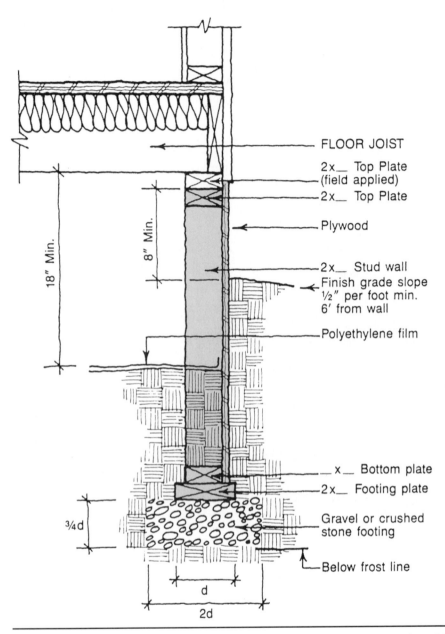

FLOOR JOIST

2x__ Top Plate
(field applied)

2x__ Top Plate

Plywood

2x__ Stud wall

Finish grade slope
½" per foot min.
6' from wall

Polyethylene film

__ x__ Bottom plate

2x__ Footing plate

Gravel or crushed
stone footing

Below frost line

18" Min.

8" Min.

¾d

d

2d

Figure 12–41 Crawl space wall. (All-Weather Wood Foundation System Technical Report, 1972. American Forest and Paper Association.)

Table 12-8 Pressure-treated wood foundation: minimum structural requirements for lumber framing.

Construction	House Width (ft)	Number of Stories	Height of Fill (in.)	25 lb/cu ft of Soil Pressure				30 lb/cu ft of Soil Pressure			
				Species and Grade of Lumber[a] Required	Stud and Plate Size (Nominal)	Stud Spacing (in.)	Size of Footing (Nominal)	Species and Grade of Lumber[a] Required	Stud and Plate Size (Nominal)	Stud Spacing (in.)	Size of Footing (Nominal)
Basement		1	24	C	2×4	16	2×6	C	2×4	16	2×6
	24 to 28		48	B	2×4	12	2×6	B	2×4	12	2×6
				C	2×6	16	2×8	C	2×6	16	2×8
			72	B	2×6	16	2×8	A	2×6	16	2×8
				C	2×6	12	2×8	B	2×6	12	2×8
			86	A	2×6	12	2×8	A	2×6	12	2×8
	29 to 32	1	24	B	2×4	16	2×8	B	2×4	16	2×8
				C	2×4	12	2×8	C	2×4	12	2×8
			48	B	2×4	12	2×8	C	2×6	16	2×8
				C	2×6	16	2×8				
			72	B	2×6	16	2×8	A	2×6	16	2×8
				C	2×6	12	2×8	B	2×6	12	2×8
			86	A	2×6	12	2×8	A	2×6	12	2×8
	24 to 32	2	24	C	2×6	16	2×8	C	2×6	16	2×8
			48	C	2×6	16	2×8	C	2×6	16	2×8
			72	B	2×6	16	2×8	A	2×6	16	2×8
				C	2×6	12	2×8	B	2×6	12	2×8
			86	A	2×6	12	2×8	A	2×6	12	2×8
Crawl space	24 to 28	1		B	2×4	16	2×6				
				C	2×6	16	2×8				
	29 to 32	1		B	2×4	16	2×8				
	24	2		C	2×6	16	2×8				
	25 to 32	2		B	2×6	16	2×8				
				C	2×6	12	2×8				

[a]Species and species groups having the following minimum properties as provided in National Design Specification (surfaced dry or surfaced green):

		A	B	C
F_b (repetitive member) psi:	2 × 6	1750	1450	1150
	2 × 4		1650	1300
F_c psi:	2 × 6	1250	1050	850
	2 × 4		1000	800
F_{c1} psi:		385	385	245
F_v psi:		90*	90	75
E psi:		1,800,000	1,600,000	1,400,000

*Length of end splits or checks at lower end of studs not to exceed width of piece.

Table 12-9 Pressure-treated wood foundation: structural plywood requirements.

Height of Fill (in.)	Stud Spacing (in.)	Minimum Plywood Grade and Thickness for Basement Construction[a]											
		Face Grain Parallel to Studs[b]						Face Grain Across Studs[b,c]					
		25 lb/cu ft Soil Pressure			30 lb/cu ft Soil Pressure			25 lb/cu ft Soil Pressure			30 lb/cu ft Soil Pressure		
		Grade[d,e]	Minimum Thickness	Identification Index	Grade[d,e]	Minimum Thickness	Identification Index	Grade[d]	Minimum Thickness	Identification Index	Grade[d]	Minimum Thickness	Identification Index
24	12	B	1/2	32/16	B	1/2	32/16	B	1/2	32/16	B	1/2	32/16
	16	B	1/2	32/16	B	1/2	32/16	B	1/2	32/16	B	1/2	32/16
48	12	B	1/2	32/16	B	1/2	32/16	B	1/2	32/16	B	1/2	32/16
	16	A / B	1/2 / 5/8	32/16 / 42/20	A / B	5/8 / 3/4	42/20 / 48/24	B	1/2	32/16	B	1/2	32/16
72	12	A / B	1/2 / 5/8	32/16 / 42/20	A / B	1/2 / 5/8	32/16 / 42/20	B	1/2	32/16	B	1/2	32/16
	16	A / B	5/8 / 3/4	42/20 / 48/24	B	3/4	48/24	B	1/2	32/16	A	1/2	32/16
86	12	A / B	1/2 / 5/8	32/16 / 42/20	A / B	5/8 / 3/4	42/20 / 48/24	B	1/2	32/16	B	1/2	32/16

[a]For crawl space construction, use grade and thickness required for 24″ fill depth.

[b]Panels that are continuous over fewer than three spans (across fewer than three stud spacings) require blocking 2′ above the bottom plate. Offset adjacent blocks and fasten through studs with two 16d, corrosion-resistant nails at each end.

[c]Blocking between studs is required at all horizontal panel joints fewer than 4′ from the bottom plate.

[d]Minimum grade: A—STRUCTURAL I C–D; B—STANDARD C–D (exterior glue). If a major portion of the wall is exposed above ground, a better appearance may be desired. In this case, the following exterior grades would be suitable: A—STRUCTURAL I A–C, B–C, or C–C (plugged); B—Exterior Group I A–C, B–C, or C–C (plugged).

[e]All panels shall be 5-ply minimum.

For most house designs and loading conditions, nominal 2×6 or 2×8 footings will be adequate. Brick veneer exterior construction requires the use of 2×10 or 2×12 footing plates to provide added width for supporting the veneer. The height of brick should not exceed 16'-0" unless the knee wall, footing plate, and gravel base are designed to support greater height (Tables 12–8 and 12–9).

The depth of the gravel bed under the footing plate for a continuous wall is ³/₄ of the required plate width. The bed extends from both edges of the plate a distance of ¹/₂ the plate width. For columns or posts, the depth and the width of the gravel bed should be greater to accommodate the concentrated load.

Lumber framing members in foundation walls enclosing a basement are designed to resist the lateral pressure of fill as well as vertical forces resulting from live and dead loads on the structure. The plywood sheathing is designed to resist maximum inward soil pressure occurring at the bottom of the wall. Foundation walls for crawl space construction can be designed to resist only vertical loads if the difference between the outside grade and the ground level in the crawl space is 12" or less.

The exterior of basement foundation walls is covered with a 6-mil polyethylene film. The film is bonded to the plywood, and the joints must be sealed and lapped 6" with a suitable construction adhesive. The top edge of the film is completely sealed to the plywood wall with adhesive. A polyethylene film 6 mils in thickness is applied over the gravel bed, and a concrete slab at least 3" thick is poured over the film. The slab should be high enough to provide at least 2 sq in. of bearing against the bottom of each stud to resist the lateral thrust at the bottom of the wall. Backfill should not be placed against the foundation walls until after the concrete slab is in place and has set and the top of the wall is adequately braced. Where the height of backfill exceeds 4', gravel should be used for the lower portion.

For habitable basement space, insulation can be installed between studs, and an interior vapor barrier and interior finish applied to wall framing. Permanent wood foundation basements are easy to finish: Nailable studs are already in place, no furring is needed to install paneling or insulation, and plumbing and wiring are simplified.

Cold weather precautions should be taken in the installation of the wood foundation system. The composite footing consisting of a wood plate supported on a bed of stone or sand fill should not be placed on frozen ground. Most important is the use of proper sealants during very cold weather. All manufacturers of sealants and bonding agents impose temperature restrictions on the use of their products. Only sealants and bonding agents specifically produced for cold weather conditions should be used.

Maxito® Foundation   The Maxito® foundation was designed for use with modular prefab houses.[1] The building is leveled on jacks above the foundation trenches; then the Maxito formwork panels are attached around the

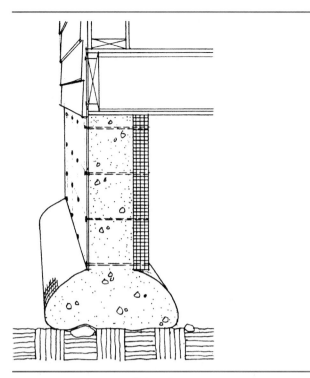

Figure 12–42 Maxito® foundation.

perimeter, extending down into the trenches. Each panel has a solid plastic exterior face and a foam plastic interior face joined by plastic form ties. A woven fabric bag is attached to the bottom of the panel. When concrete is poured into the form, the bag fills with concrete and molds itself to the contours of the bottom of the excavation. Soil height variations up to 8" are accommodated automatically (Fig. 12–42).

12.4

CONVENTIONAL WOOD FRAME CONSTRUCTION

The majority of houses built in North America employ wood frame construction. A frame structure will give many years of satisfactory service if careful planning has been done on the drawing board and if this careful planning is followed during construction. Correct details and good workmanship are two important factors in realizing durability from a frame building. However, the following methods will further ensure maximum service life from the use of wood:

---

[1] The Maxito foundation is manufactured by Maxito Industries, Ltd., Unit 224, 2570 King George Highway, South Surrey, British Columbia, V4P 1H5 Canada (1–604–535–7160).

Lumber grade. *Listed below are grades established under the National Grading Rules for Dimension Lumber and used by inspectors at mills throughout the country. Nationwide grading rules were first established in the 1920s through the cooperative efforts of mill operators, builders, architects and officials from the U. S. Department of Commerce.*

Abbreviation of grade name	Category	Dimension*	Use
CONST—construction STAND—standard UTIL—utility	Light framing	2 in. to 4 in. thick, 2 in. to 4 in. wide	This category is intended for use where especially high strength values are not required.
SEL STR—select structural #1 & BTR—#1 and better #1 #2 #3	Structural light framing	2 in. to 4 in. thick, 2 in. to 4 in. wide	These grades, typically used for trusses and tall concrete forms, are appropriate where higher strength is needed in light framing sizes.
SEL STR—select structural #1 #2 #3	Structural joists and planks	2 in. to 4 in. thick, 5 in. and wider	These grades commonly are used as joists and rafters.
STUD	Stud	2 in. to 4 in. thick, 2 in. and wider	A separate grade with lengths of 10 ft. or less. Relatively high strength and stiffness values make this grade suitable for use in load-bearing walls.

*Nominal

Mill. *The mill where lumber was sawn or manufactured, or the company that owns the mill, is identified by a name or a number. There are approximately 1,500 mills in the United States and about 500 in Canada. More than 95% of these mills belong to regional certifying agencies such as those listed in the box below.*

Reading a grade stamp

Lumber is graded to supply builders, architects, building officials and others with reliable information about its quality, characteristics and origin. Most grade stamps are composed of five elements: grade, species, moisture content, certifying agency and mill.

Species. *Species and species groups are identified by abbreviated symbols. Some of the more common symbols are shown below.*

 Douglas fir-larch

 Douglas fir-south

 Hemlock-fir

 Spruce-pine-fir (south)

Certification. *Listed below are a few of the agencies that supervise lumber grading at individual mills. There are ten such agencies in the United States and 15 in Canada that are accredited by the American Lumber Standard Committee. ALSC writes standards under which lumber is milled and graded.*

 Northeastern Lumber Manufacturers Association Inc.
272 Tuttle Road
P. O. Box 87A
Cumberland Center, Maine 04021
(207) 829-6901

 Western Wood Products Association
Yeon Building
522 SW Fifth Ave.
Portland, Ore. 97204-2122
(503) 224-3930

SPIB® Southern Pine Inspection Bureau
4709 Scenic Highway
Pensacola, Fla. 32504
(904) 434-2611

Moisture content. *Below are the abbreviations typically included in a grade stamp that provide information about the lumber's moisture content.*

Abbreviation	
S-GRN	Surfaced-green. This mark indicates that the moisture content of the lumber when it was planed was more than 19%.
S-DRY	Surfaced-dry. Indicates moisture content of the lumber when planed was 19% or less.
KD-19 **or KD**	Kiln-dried 19%. Indicates the lumber has been dried in a kiln to a moisture content of 19% or less.
MC-15 **or KD-15**	Moisture content 15%. Indicates the lumber has been dried to a moisture content of 15% or less.

Figure 12–43 Examples of grade stamps found on lumber.

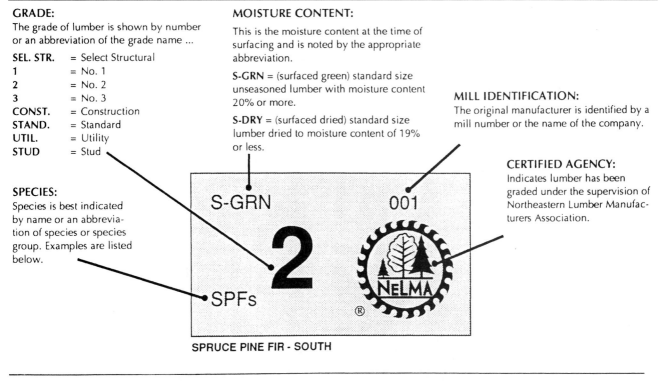

GRADE:
The grade of lumber is shown by number or an abbreviation of the grade name ...

SEL. STR.	= Select Structural
1	= No. 1
2	= No. 2
3	= No. 3
CONST.	= Construction
STAND.	= Standard
UTIL.	= Utility
STUD	= Stud

SPECIES:
Species is best indicated by name or an abbreviation of species or species group. Examples are listed below.

MOISTURE CONTENT:

This is the moisture content at the time of surfacing and is noted by the appropriate abbreviation.

S-GRN = (surfaced green) standard size unseasoned lumber with moisture content 20% or more.

S-DRY = (surfaced dried) standard size lumber dried to moisture content of 19% or less.

MILL IDENTIFICATION:
The original manufacturer is identified by a mill number or the name of the company.

CERTIFIED AGENCY:
Indicates lumber has been graded under the supervision of Northeastern Lumber Manufacturers Association.

SPRUCE PINE FIR - SOUTH

E HEMLOCK — B FIR · EASTERN SOFTWOODS · EASTERN WHITE PINE

Figure 12–44 Examples of grade stamps found on lumber.

1. Control the moisture content of the wood.
2. Provide effective termite and insect barriers throughout.
3. Use naturally durable or chemically treated wood in critical places in the structure.

To ensure that wood materials performing a load-carrying function conform to a minimum quality-control standard, all load-bearing lumber, plywood, and particleboard are required to be properly identified. In the case of load-bearing lumber, such information must be adequate to determine the bending strength and the modulus of elasticity for the purpose of using the allowable span tables found in the Uniform Building Code. For examples of grade marks, see Figs. 12–43 and 12–44.

To allow the widest range of lumber grades and species in floor construction, the code provides a series of tables permitting the selection of joists based upon the clear span of the floor joist, the modulus of elasticity, and the bending strength properties of the wood members (see Appendix D). Tables are organized to allow a different member spacing, and in codes where the floor joists are spaced not more than 24″ o.c., a repetitive member use bending value may be used.

12.4.1 Platform Frame Construction

Platform frame construction (also referred to as *western frame*) is identified by its story-level construction erected by the workers on each subfloor or platform (Fig. 12–45). This method of construction has become popular throughout the country in recent years, mainly for the following reasons:

1. Long studding pieces required in balloon framing are not always available and are usually more costly.
2. The platform built at each level is convenient for carpenters to work on (the construction builds its own interior scaffolding).

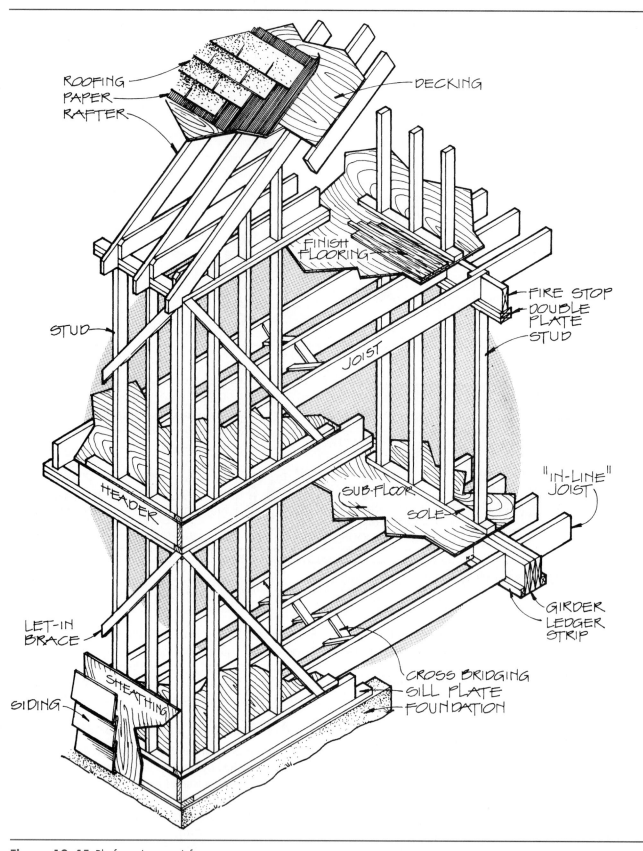

Figure 12–45 Platform (western) frame construction.

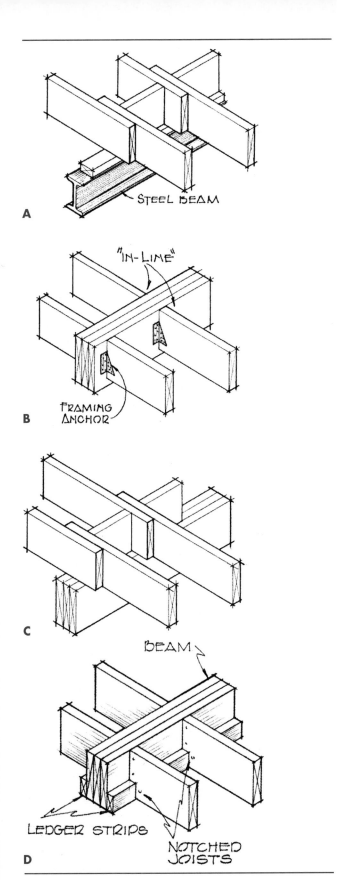

A — STEEL BEAM

B — "IN-LINE" / FRAMING ANCHOR

C

D — BEAM / LEDGER STRIPS / NOTCHED JOISTS

Figure 12–46 Joist and beam framing.

3. The construction prevents flue action within the walls in case of fire, thus eliminating the need for firestop material in critical places.
4. It lends itself to modern prefabrication methods.
5. It is quicker to erect; wall sections can be assembled quickly on the horizontal platform and tilted into place.

Beams or Girders Under wood floor framing, unless the joists are long enough to span between exterior walls, some type of heavy beam must be used to support the inner ends of the joists (Fig. 12–46). If a building is wider than 15′ or 16′, which is generally the case, it is necessary to use additional support under the floor joists to avoid using excessively heavy floor joists. For this, wood or metal beams are used, and they are supported by the exterior foundation walls and piers or columns (Fig. 12–47).

The beams under floors carry concentrated loads, and their sizes should be carefully considered. If steel is selected, S beams are usually used; their sizes for general residential construction can be selected from Table 13–6 (in Chapter 13). Steel has the advantage of not being subject to shrinkage. Wood beams can be either solid or built-up of nominal 2″ lumber, usually of the same-size lumber used for the floor framing joists. Solid wood beams contain more cross-sectional lumber than built-up beams of similar nominal size. For example, a 6″ × 8″ solid wood beam would have an actual size of 5½″ × 7¼″, whereas a built-up beam of 6″ × 8″ nominal dimension would be actually 4½″ × 7¼″. However, from a practical standpoint the built-up wood beams have several advantages:

1. Stock-size framing lumber is more readily available.
2. Smaller pieces are easier to handle.
3. They reduce splitting and checking.
4. They can be nailed to the frame more easily.

It is desirable to locate beams, if possible, under major interior partitions in order to eliminate the need for double joists under the partitions. Bathrooms and kitchens have heavier dead loads, and their weights should be considered in placement of the beams. If living space in basements is important, beam sizes are generally made larger to avoid numerous columns throughout the basement. Columns are usually spaced 8′ to 10′ o.c. in basements, depending on the beam sizes. Under basementless houses, piers are spaced no more than 7′ to 8′ o.c. Wood beams should have at least a 3″ bearing on masonry walls or pilasters, with a ½″ air space around the ends of the beams for adequate ventilation if they are surrounded with masonry (Fig. 12–48). If splices are necessary in wood built-up beams, they should be staggered and should be placed near the column for sound con-

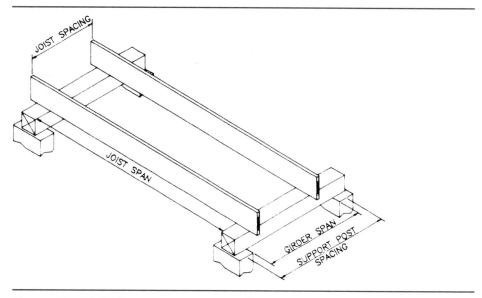

Figure 12-47 Girder span-spacing relationship.

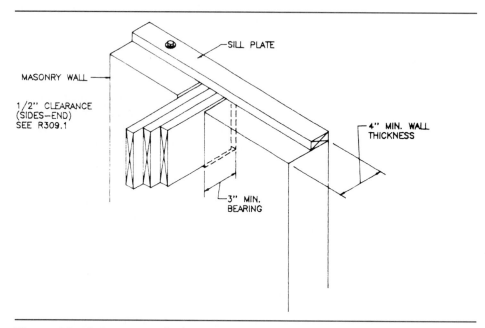

Figure 12-48 Beam or girder bearing on masonry.

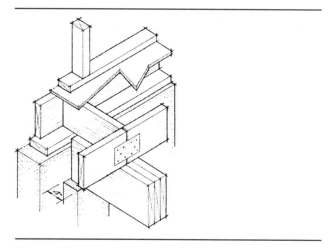

Figure 12-49 Box sill and girder support construction.

struction. The top edge of the beams must be the same height as the sill plates if joists are to rest on the beams (Fig. 12–49). If the joists are to butt into the beams, a ledger strip is nailed to the beam, and either notched or narrower joists are toe-nailed over the ledger strip to the beams (Fig. 12–45). Metal angle connectors or steel joist hangers can also be used for butt-fastening joists and beams.

Joists butting into beams allow more headroom in basements and facilitate the installation of ductwork and plumbing under the floor; basement ceilings can be finished without difficulty with unbroken surfaces; and plywood subfloors can be more easily put over in-line joists. However, the simpler method of floor framing is to use joists resting on the beams.

Beams carry half the weight imposed on all of the joists that rest on or are attached to the beams (Fig. 12–6). In 2-story houses, considerable weight is transferred from the upper story to the beams. For convenient selection of beam or girder sizes under wood floor construction, see Table 13–5 (in Chapter 13), which designates maximum spans for the usual beam sizes used. Beams supporting unusually heavy loads would have to be designed according to established engineering practice.

Columns   In residential basements, either wood or metal columns can be used under the beams. The columns are usually 6×6 wooden posts or, preferably,

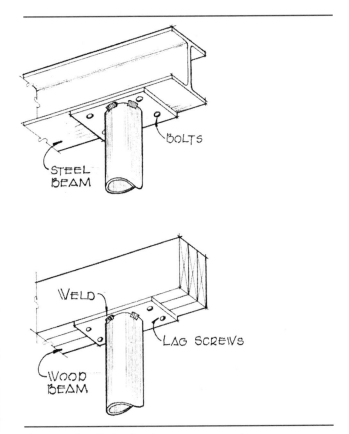

Figure 12-50 Steel columns supporting beams.

round steel columns of 3″ or 4″ diameter with cap plates welded to both ends (Fig. 12–50). Steel columns are available in stock lengths, or they can be fabricated in any lengths desired. Other types are adjustable. Select wood column sizes according to loads, unsupported heights, and lumber species (Table 13–3 in Chapter 13).

Sills   The box type of sill (Fig. 12–49) is generally used with platform frame construction. The sill plate, generally measuring 2×6 or 2×8, is placed over the foundation wall and is anchored with 1/2″ diameter anchor bolts, spaced according to local codes (usually 6′ o.c.). All plates, sills, and sleepers must be treated wood or foundation-grade redwood or cedar. Use of these woods reduces the hazard of termites.

Anchor bolts must be 1/2″ diameter bolts set 7″ deep in concrete or masonry foundations at least every 6′. There must be at least two bolts in every piece of plate material, and neither may be more than 12″ from the end. Occasionally, builders use a 4×6 or 4×8 treated sill plate in box sill construction. The wood sill must be at least 6″ to 8″ above finished grade. If exterior wood siding is used on the building, the outside surface of the framing should be placed within the outer edge of the foundation so that the outside surface of the sheathing becomes flush with the outer surface of the foundation wall (Fig. 12–45). This allows the finish siding to come down over the joint between the sill plate and the masonry wall to make a weathertight joint.

Floor Joists   Floor joists are the structural members of the wood floor. Joists rest on the sill plate, with a minimum of 1 1/2″ bearing surface (preferably 3″ on the plate); this should be worked out on the sill detail. Joists are selected for strength and rigidity. The strength is required to carry both dead and live floor loads; the rigidity is needed to prevent vibration and movement when live load concentrations are shifted over the floor. Joists are usually of 2″ (nominal) dimension thickness. The 6″, 8″, 10″, or 12″ (nominal) depths are used, depending on the load, span, spacing, and species and grade of lumber involved. Generally, joists are spaced 16″ o.c.; however, 12″ or 24″ spacing can be used when load concentrations vary.

Floor framing is simpler if the same-depth joists are used throughout the floor, and for greatest rigidity they should run in the direction of the shorter span. Builders can gain additional strength and stiffness in joists by using lumber finished on only two sides [the narrow edges (S2E)], although rough lumber will ignite more quickly in case of fire, and some codes prohibit its use. If joists lap over beams or supports, the lap should be no more than 6″ to 8″ and no less than 4″ for proper nailing of joists (Fig. 12–46A). If the lap is excessive, any sag or reflection at the center of the joists will cause the ends of the joists extending over the beams to raise the floor. If joists are supported by a steel beam, a 2″ × 4″ nailing strip

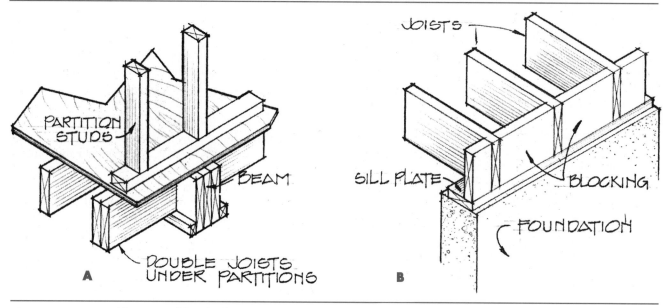

Figure 12–51 Double joist and header blocking.

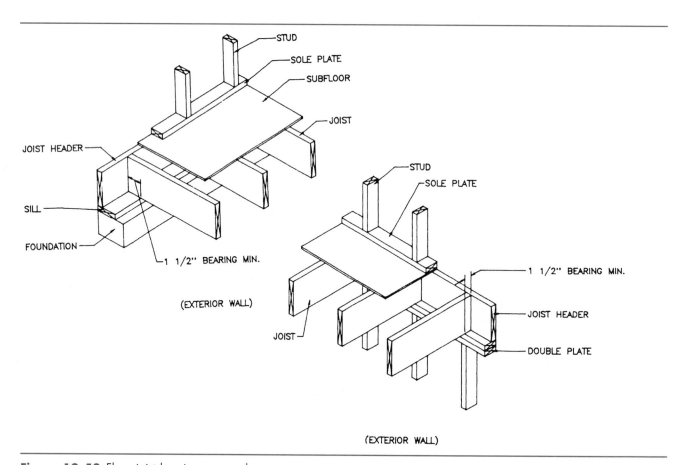

Figure 12–52 Floor joist bearing on wood.

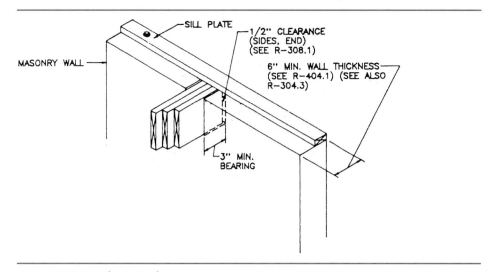

Figure 12–53 Floor joist bearing on masonry.

should be bolted to the beam and the joists fastened to it.

For economy, the lengths of the joists should be limited to 2′ increments. The header joists are end-nailed to each joist with 16-penny nails, and both the headers and the joists are toenailed to the sill plate or are attached with metal joist anchors. If the bearing surface at the ends of the joists is critical, blocks between the joists can be used as a header; the blocks will function very much as block bridging (Fig. 12–51B).

Local codes usually indicate the minimum joist sizes for various conditions. A rule of thumb for joist sizes using Douglas fir, southern yellow pine, western larch, or hemlock (J & P Grade) would be to make their spans in feet 1½ times their depth in inches, if the joists are spaced 16″ o.c. Thus, a 2″× 10″ joist of the species indicated will span 15′ if a 40-lb live load is required. Usually, lumber of a higher grade does not increase the allowable span, because rigidity is the controlling factor, which is not contingent on grades. For more accurate joist size selection according to spans required, refer to Chapter 13.

Bearing   To ensure the adequate transfer of floor loads to supporting elements, minimum lengths of bearing for several alternative support systems are specified by code. For joists, a minimum bearing of 1½″ on metal or wood must be provided (Fig. 12–52). When joists bear on masonry, a minimum bearing of 3″ is required (Fig. 12–53). Joists may be supported by a 1″× 4″ ribbon

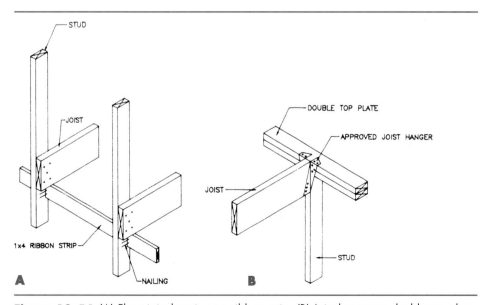

Figure 12–54 (A) Floor joist bearing on ribbon strip. (B) Joist hanger at double top plate.

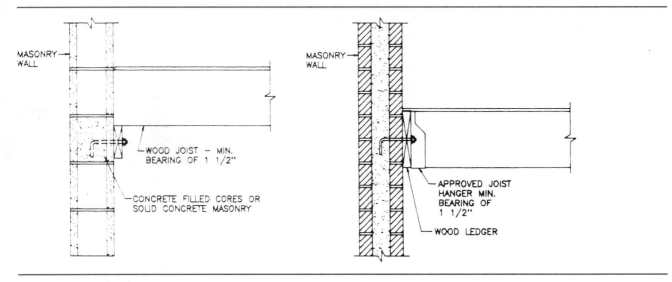

Figure 12–55 Joist-bearing minimums.

strip when they are nailed to adjacent studs, or they may be secured by the use of approved joist hangers (Figs. 12–54 and 12–55).

To ensure a reasonably concentric application of load from the joist to supporting beams or girders, joists framing from opposite sides of a beam or girder are required to lap at least 3″ or the opposing joists must be tied to-gether in an approved manner (Fig. 12–56). Joists framed into the side of a wood beam or girder must be supported by approved framing anchors or by ledger strips having a minimum nominal dimension of 2″ (Fig. 12–57).

Lumber associations have a unified standard for sizing lumber. Examples for a few of the more commonly used lumber sizes are shown in the following list:

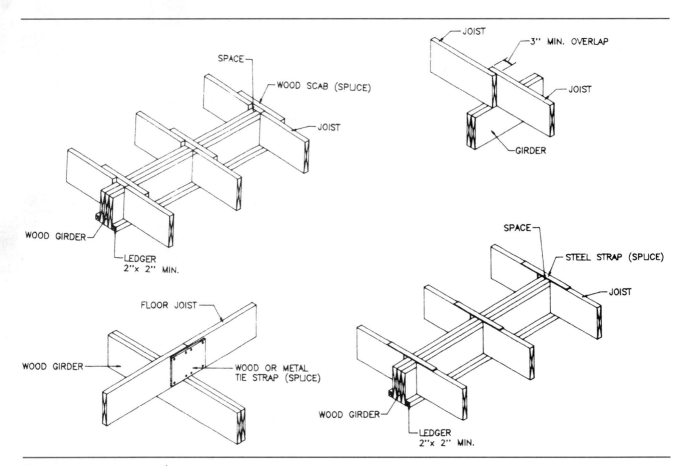

Figure 12–56 Joist at girder.

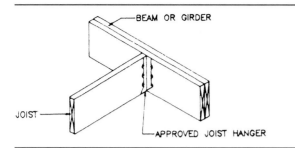

Figure 12-57 Joint hanger at girder.

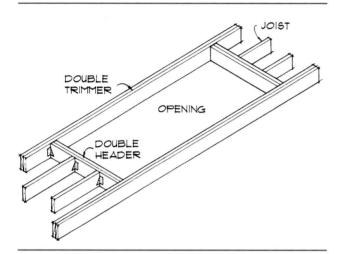

Figure 12-58 Framing around floor and ceiling openings.

Dressed size (S4S) in Inches

Nominal Size	Surfaced Dry	Surfaced Green
2 × 4	1½ × 3½	1⁹⁄₁₆ × 3⁹⁄₁₆
2 × 6	1½ × 5½	1⁹⁄₁₆ × 5⅝
2 × 8	1½ × 7¼	1⁹⁄₁₆ × 7½
2 × 10	1½ × 9¼	1⁹⁄₁₆ × 9½
2 × 12	1½ × 11¼	1⁹⁄₁₆ × 11½

Double joists should be used under all partitions running parallel with the joists, or some method of blocking must be employed so that partitions bear on framing rather than on the subfloor between the joists (Fig. 12–51A).

Trimmers and Headers Openings in floor framing for chimneys, stairways, and the like, must be surrounded by additional framing members for support. The joists that run parallel to the two sides of the opening are doubled and are called *trimmers*. The members that surround the other two sides of the opening are also doubled and are called *headers*; into these headers are framed the shortened joists called *header joists* or *tail joists* (Fig. 12–58).

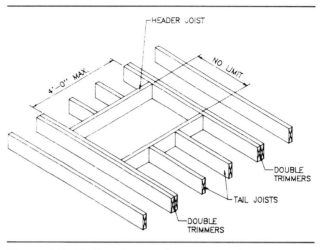

Figure 12-59 Floor framing for maximum 4' openings.

Header joists the same size as floor joists may be used for spans not exceeding 4'. Trimmers must be doubled to carry the additional load (Fig. 12–59). When the header joist span exceeds 4', the header joist must be either doubled or of sufficient size to support the floor joist framing (Fig. 12–60). Sometimes nailing is insufficient to transfer vertical loads. In these cases, positive connections must be employed. These hangers are required when the header joist span exceeds 6' (Fig. 12–61). Headers should not be over 10'-0" long; if the spans are short and loads around the opening are not heavy, the assembly can be spiked together; otherwise, joist hangers or additional support must be used. Framing should not be fastened into chimneys; fire codes require a 2" space around chimneys between the framed opening, which should be filled with incombustible insulation. If the chimney flues are well surrounded with masonry, corbel masonry may be extended from the chimney on which framing may rest. Tail joists over 6'-0" long must be connected to beams or girders.

Cutting Openings in Joists The cutting or drilling through of joists that have been set in place should be avoided if possible. However, plumbing or wiring must occasionally pass through floor joists. Small holes, if placed properly, are not too objectionable, but larger cuts often result in weakened or dangerous conditions (Fig. 12–62). A horizontal structural member, such as a floor joist, with a load imposed on it can be considered to be under compression throughout the upper half of its depth; it is under tension throughout its lower half, with a neutral axis in the center of its depth. The farther away the wood fiber is from the center of the depth of the joist, the more important the fiber is to the strength of the member. Therefore, the closer to the center axis a hole is cut, the less destructive it will be.

Joists may be notched at top or bottom if the notch does not exceed one-sixth of the depth of the joist, as long as the notch does not occur within the center one-

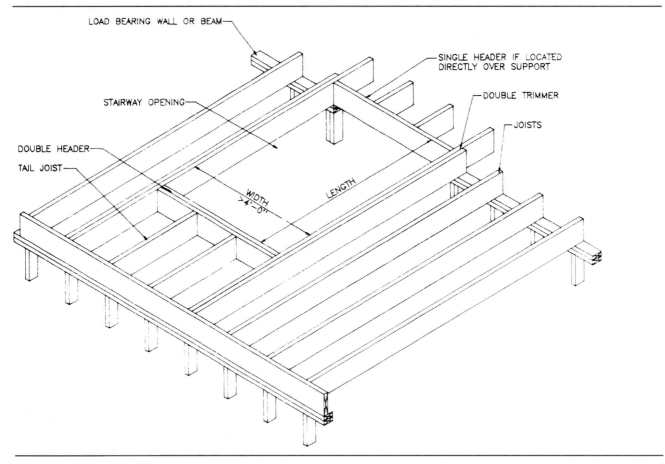

Figure 12-60 Floor framing for greater than 4' openings.

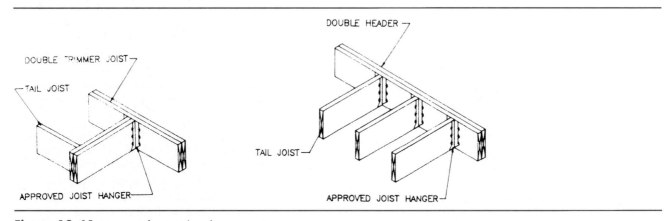

Figure 12-61 Hangers for joist-header connections.

third of the span. Holes with a maximum of 2″ diameter can be bored within 2″ of the edges of the joist without causing excessive damage.

Bridging   Bridging consists of rows of short members between joists to produce a firm, rigid floor. Bridging not only prevents the deep, narrow shape of the floor joists from buckling sideways, but it also distributes concentrated live loads to adjoining joists. Rows of bridging

should be uninterrupted throughout the floor. Three types of bridging are seen in floor framing:

1. 1 × 3 pieces of wood nailed diagonally between the joists,
2. blocks the same depth as the joists nailed in staggered fashion perpendicular to the joists, or
3. light metal bridging pieces nailed diagonally between the joists.

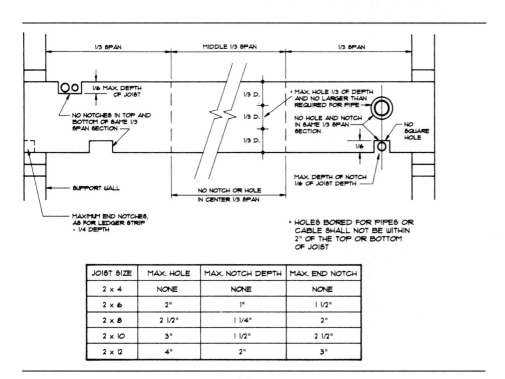

JOIST SIZE	MAX. HOLE	MAX. NOTCH DEPTH	MAX. END NOTCH
2 x 4	NONE	NONE	NONE
2 x 6	2"	1"	1 1/2"
2 x 8	2 1/2"	1 1/4"	2"
2 x 10	3"	1 1/2"	2 1/2"
2 x 12	4"	2"	3"

Figure 12–62 Cutting openings in wood floor joists.

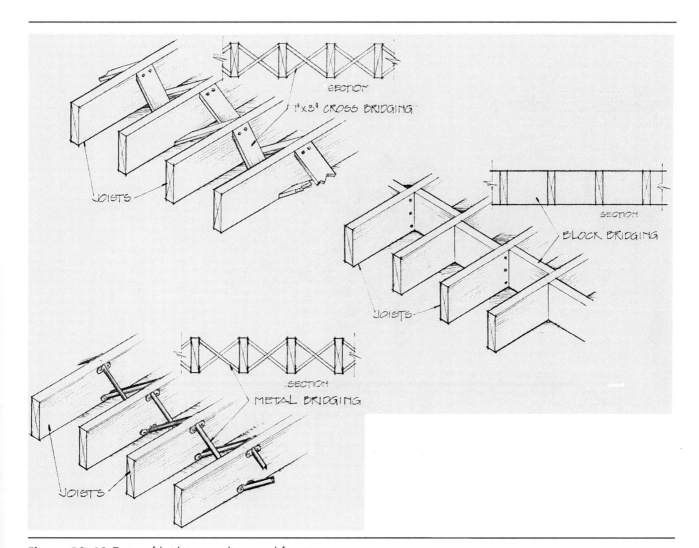

Figure 12–63 Types of bridging used in wood framing.

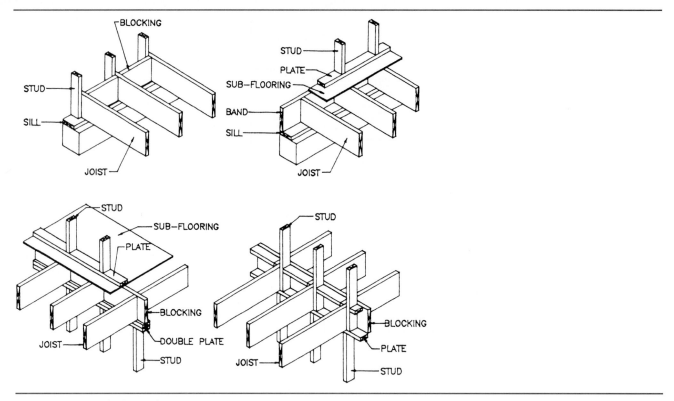

Figure 12–64 Blocking of joists.

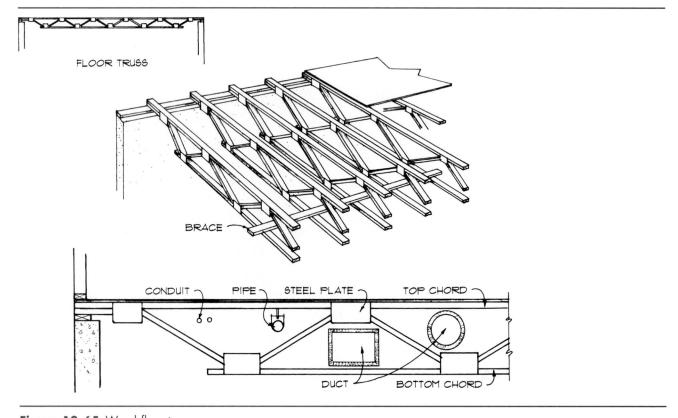

Figure 12–65 Wood floor trusses.

Crossbridging allows wiring and piping to pass through without cutting (Fig. 12–63). Joists with more than 8' of span should have one row of bridging, but rows of bridging should be no more than 7' or 8' apart in longer spans.

To ensure that joists are reasonably held in line and do not twist out of the plane of the applied load bridging, blocking or some other acceptable means of holding the joists in place is required. Lateral support at ends may be provided by full-depth solid blocking not less than 2" in thickness; or the ends of the joists may be nailed or bolted to a header band, rim joist, or an adjoining stud (Fig. 12–64).

In addition to the lateral support at the ends, joists may also be required to have intermediate lateral support at intervals not exceeding 10' when such members have a depth-to-thickness ratio exceeding 6:1 based upon nominal dimensions. Intermediate blocking is not required for joists 2"× 12" or smaller.

Floor Trusses An alternative to framing floors with conventional floor joists is to use floor trusses or truss joists (Fig. 12–65). These systems of floor framing are appropriate when long spans are required and the use of load-bearing walls or columns to support floor loads is undesirable. Floor trusses or truss joists are used for first-floor framing in residences where open space in basements is essential.

Floor trusses are usually factory-assembled units fabricated from 2 × 4 structural-grade lumber. Engineered as flat trusses with top chords, bottom chords, and web members connected by stamped metal plates, these structural members are capable of spanning greater distances than wood floor joists. Most floor trusses are spaced at 24" o.c., thus requiring a ³⁄₄" thick or thicker subfloor. Diagonal bridging and continuous lateral bracing are required between floor trusses. Wiring and ductwork can easily be routed through openings between web members of floor trusses.

Glued Engineered Wood Products Glued engineered wood products make up an increasing percentage of the wood used in all forms of residential and nonresidential construction. Glued engineered wood is classified by the APA/Engineered Wood Association into four general groups:

1. structural wood panels,
2. glued laminated timber (glulam),
3. structural composite lumber (SCL), and
4. wood I joists.

Structural adhesives are used to bond individual elements of wood in strands, veneers, or lumber to create structural shapes. Recent market studies project that the use of some of these products will more than double by the year 2000, partly because they use resources efficiently.

The most widely used of the glued engineered wood products are *structural wood panels*, which include plywood, oriented-strand board (OSB), and composite panels. *Plywood* is manufactured by bonding the veneers in a crossband orientation using structural adhesives. The crossbanding provides strength and dimensional stability. *OSB* is produced by bonding individual wood strands into a mat-formed product. The mat consists of multiple crossbanded layers. *Composite panels* consist of wood-face veneers and a reconstituted wood core. The face veneers are positioned parallel to the long dimension of the panel, with the wood-based core oriented 90° to the face veneers.

Glulam, introduced in the United States in 1935 after almost 40 years of use in Europe, is an engineered, stress-rated product created by adhesively bonding individual pieces of lumber having a net thickness no greater than 2". The individual pieces are end-joined to create long lengths referred to as *laminations*. These laminations are face-bonded to create the finished glulam product. The direction of the grain in all laminations is approximately parallel (Fig. 12–66).

Figure 12–66 Glue-laminated timber. (APA/Engineered Wood Association)

Figure 12–67 Laminated-veneer lumber. (Courtesy Arizona Public Service Co.)

Figure 12–68 Parallel-strand lumber. (Courtesy Arizona Public Service Co.)

Figure 12–69 Wood I joists. (Courtesy Arizona Public Service Co.)

Glulam, one of the most versatile engineered wood products, is used for a variety of structural applications, including floor and header beams in residential construction and beams in a variety of light commercial projects. Glulams come in several finish grades. Header grade is specified for concealed applications. Beam widths are the same as those for standard framing and provide added load capacity compared to faced glulam sizes. Widths are 3½″, 5½″, and 7¼″. Industrial-appearance-grade glulams are inexpensive timbers used in warehouses, garages, and other structures where appearance is not important. Architectural-appearance grade is for

projects where appearance is important. Premium-appearance grade and clear-appearance grade are used where tight control of surface defects is of prime concern.

Structural composite lumber (SCL) refers to a family of product including laminated-veneer lumber (LVL), parallel-strand lumber (PSL), and other composite lumber products (Figs. 12–67 and 12–68).

LVL, the most widely used of the SCL products, is produced by adhesively bonding thin wood veneers, not greater than 0.25″, such that the grain of all veneers is approximately parallel to the long direction of the mem-

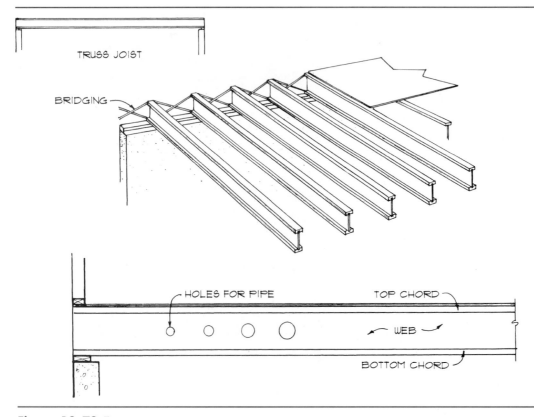

Figure 12–70 Truss joists.

ber. LVL members are typically used for dimension lumber in beams and headers. They are also used as flanges for wood I joists.

PSL is created by adhesively bonding relatively long strands of wood fiber such that the strands are oriented parallel to the long axis of the member. The most common sizes are comparable to those of dimension lumber, although members having cross sections up to $11'' \times 19''$ can be produced. PSL members are typically used in beams, headers, and columns.

Prefabricated wood I joists are probably the fastest-growing of the glued engineered wood products. Wood I joists are manufactured using either sawn lumber or LVL flanges and plywood or OSB webs (Fig. 12–69).

I joists are used extensively in residential floor construction. In general, nominal 10″ and 12″ deep I joists can be substituted on an equal-depth basis for most grades and species of 2×10 and 2×12 sawn lumbers. Special design verifications need to be made where I joists support concentrated loads or where they are cantilevered over supports. Because I joists can be supplied in long lengths, they are often used in multiple-span applications.

Diagonal bridging is essential between truss joists. Unlike with floor trusses, ductwork cannot be placed through the solid web; however, small openings can be cut through the webs to allow room for electrical wiring and plumbing pipes (Fig. 12–70). Such openings should be placed at strategic locations designated by the manufacturer of the truss joists.

Subfloor Nominal 1″ boards 4″, 6″, or 8″ wide, securely nailed to the joists, were traditionally used for subfloors. The boards were square edged, tongue and groove, or shiplap, and it was preferable to lay them diagonally (45° to the joists). Finish strip flooring was then laid in either direction over the subfloor. Joists in the subflooring fall in the center of the joists and are staggered. Square-edge boards laid up too tight produce squeaky floors; rather they should be nailed about 1/8″ apart. When subflooring is laid perpendicularly to the joists, the finish floor must be laid perpendicularly to the subfloor.

Because of the speed of erection, plywood is the most popular material for subflooring today. Plywood that is used in floor construction and that performs a load-carrying function must conform to a known quality-control standard. Compliance is indicated by a grade mark issued by an approved agency (Fig. 12–71).

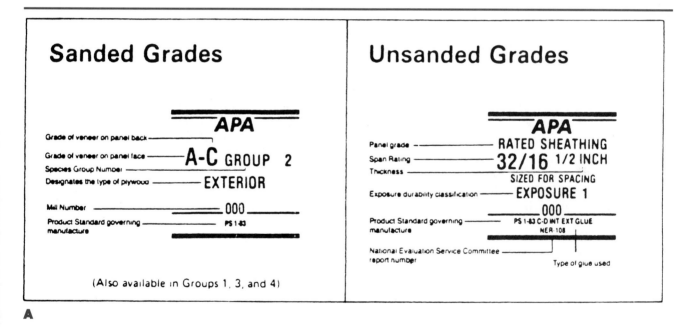

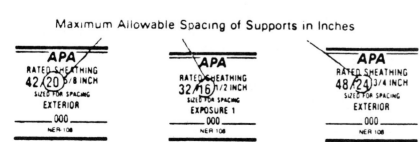

Figure 12–71 (A) Plywood-grade mark examples. (B) Identification of plywood subfloor span limitations. (APA/Engineered Wood Association)

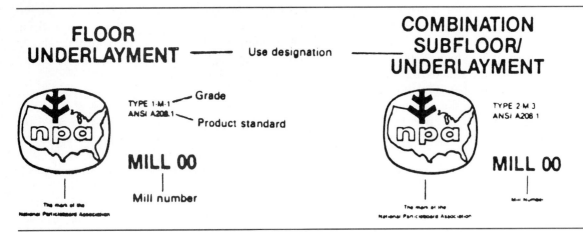

FLOOR UNDERLAYMENT ⟶ Use designation ⟶ COMBINATION SUBFLOOR/ UNDERLAYMENT

TYPE 1-M-1
ANSI A208.1 — Product standard
— Grade

MILL 00

Mill number

The mark of the
National Particleboard Association

TYPE 2 M 3
ANSI A208 1

MILL 00

Mill number

The mark of the
National Particleboard Association

Figure 12–72 Particleboard-grade mark examples. (NPA)

Allowable spans for plywood used as subflooring are shown in Table D–16. The span limitations are based upon the grade of plywood used. In the case of rated sheathing used as structural subflooring, the maximum span is easily identified through the use of the panel span rating stamp, where the denominator represents the allowable span of the plywood floor sheathing.

Particleboard used in floor construction that performs a load-carrying function must conform to known quality-control standards, as do plywood and lumber; see Fig. 12–72 for examples of approved marks. Plywood produces a smooth surface for block flooring, tile, carpeting, and other nonstructural flooring materials. Sheets of plywood, usually 1/2″, 5/8″, or 3/4″ thick for subfloors, should be laid with their face grain across the joists; the end joints of the sheets should be staggered, and should fall midway between adjoining sheets (Fig. 12–73). If flooring other than strip wood is used, the joints around the plywood should have blocking below, or tongue-and-groove edge plywood can be used. Plywood must be securely nailed around its edges as well as throughout central areas where backing occurs.

A hardboard or particleboard overlayment 1/4″, 3/8″, 1/2″, or 5/8″ thick should be placed over plywood subfloors 1/2″ or 5/8″ thick. Tongue-and-groove plywood 3/4″ thick or thicker does not require an overlayment when installed with screws and construction adhesive on floor joists spaced 16″ o.c. or less. Structural-oriented strand board or wafer board 7/16″ and 3/4″ thick is used as subflooring in locations where it is acceptable to building codes.

Subfloor Under Ceramic Tile

In bathrooms and other areas requiring ceramic tile floors, some method must be used to lower the subfloor so that a concrete base for the tile can be provided for. If the joists are adequate for the load, they can be dropped several inches, or the top of the joists can be chamfered and the subfloor dropped 3″ between the joists. If there is no objection to

the finished tile floor being an inch or so above the other finished floor surface, the regular subfloor could be left as is. In economical construction, ceramic tile can be laid on mastic, on top of a substantial wood subfloor, which neither requires a concrete subfloor nor entails dropping of the wood floor.

Soleplates

After the subfloor is in place, 2 × 4 plates, called *soles*, *shoe plates*, or *bottom plates*, are nailed directly over the subfloor to form the layout for all walls and partitions. Dimensions for this layout must be taken from the floor plan working drawing. Note that unless floor plans show dimensions from the outer edge of the exterior wall framing to the center line of partitions, accurate layout of soleplates by workers becomes rather difficult. The plates are generally run through door

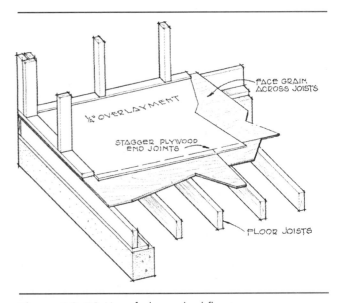

Figure 12–73 Use of plywood subflooring.

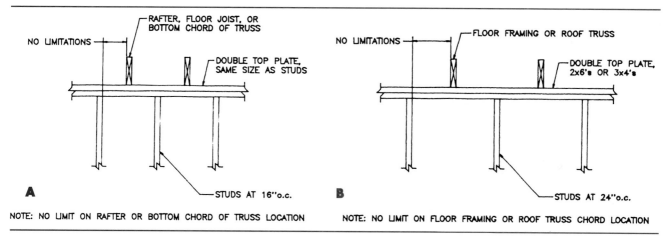

Figure 12-74 (A) Top plate with 16″ stud spacing. (B) Top plate with 24″ stud spacing.

openings and are then cut out after the framing and rough openings have been established. Plumbing walls must be built of 2 × 6 or 2 × 8 studs to enclose soil stacks; partitions between closets can be built of 2 × 4 studs sideways to increase usable space.

Studs   Studs, usually 2 × 4 (16″ o.c.), are toenailed to the soleplate and are capped with a double 2 × 4 cap plate. Double studs, which support proper-size lintels or headers above the openings, should be used around all door and window openings. In 1-story buildings, studs can be spaced 24″ o.c., unless limited by the wall covering.

The size, height, and spacing of studs shall be in accordance with Table D–15. Studs shall be placed with their wide dimension perpendicular to the wall. To ensure minimum load-carrying capabilities for conventional wood frame walls, a minimum grade of No. 3 standard- or stud-grade lumber is specified.

To allow the more economical use of lumber for bearing studs not supporting floors and for nonbearing studs, an exception is given permitting the use of utility-grade lumber. Utility-grade studs shall be spaced not more than 16″ o.c.; nor shall they support more than a roof and a ceiling or exceed 8′ in height for exterior walls and load-bearing walls or 10′ for interior, non-load-bearing walls.

Bearing and exterior wall studs shall be capped with double top plates installed to provide overlapping at corners and at intersections with other partitions. End joints in double top plates shall be offset at least 48″.

When stud spacing is 24″ o.c., the code allows for three options to account for the increased loading. The first option is an increase in the minimum top plate size (Fig. 12–74). The second option is a limitation on the location of the floor joists, floor trusses, or roof trusses being placed within 5″ of the supporting studs below (Fig. 12–75). The third option is solid blocking used as reinforcement for a double plate (Fig. 12–76).

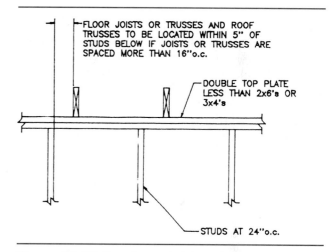

Figure 12-75 Top plate with 24″ stud spacing and bearing point limitations.

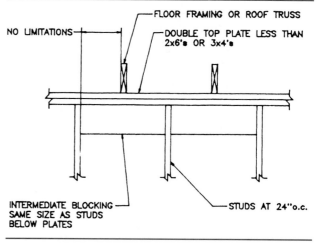

Figure 12-76 Blocked top plate with 24″ stud spacing.

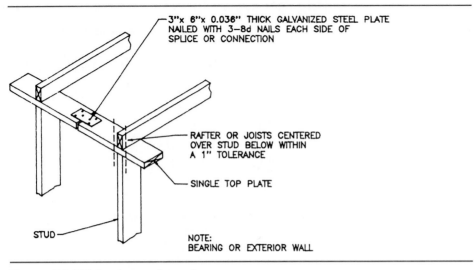

Figure 12–77 Single top plate splice.

With the advent of wider framing to accommodate increased thickness of insulation, a desire to save on material costs led to the allowance of a single top plate alternative. The use of a single top plate in bearing and exterior walls is permitted as long as adequate top plate ties are provided (Fig. 12–77).

If ceiling heights are to be 8′-0″ high, the studs are generally cut between 7′-8″ and 7′-9″ in length. Blocking, halfway up the height of the studs, stiffens the wall and prevents the studs from warping. If vertical paneling is to be used as interior wall covering, horizontal block-

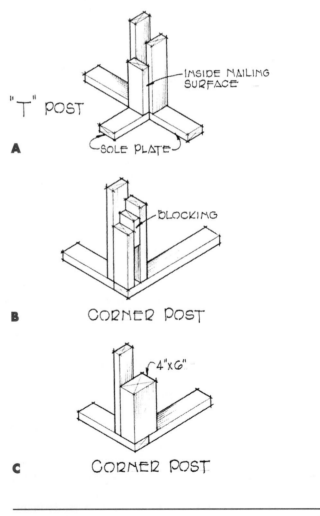

Figure 12–79 Stud framing at corners and intersections of walls.

Light Construction Principles

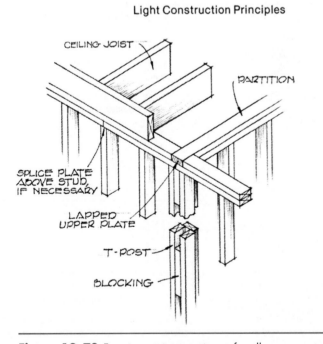

Figure 12–78 Framing at intersections of walls.

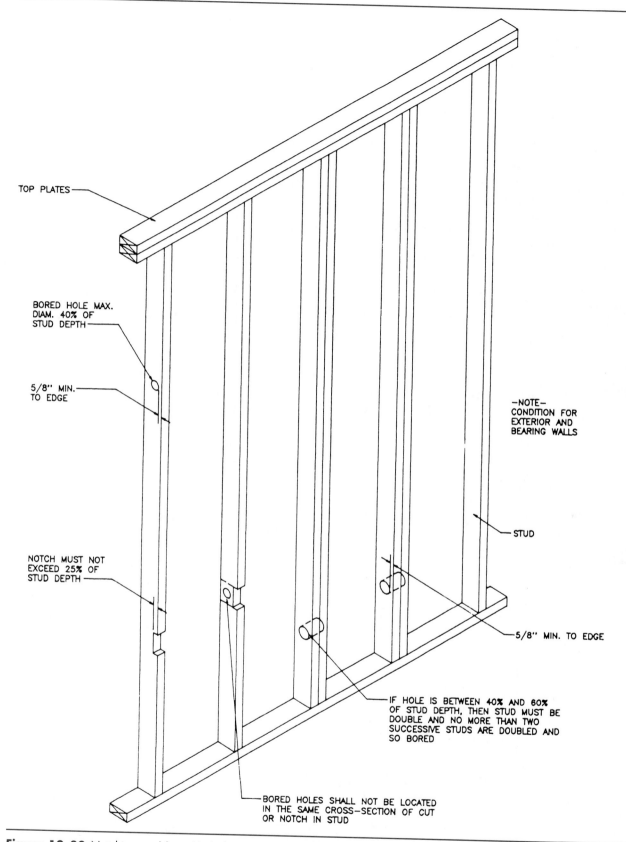

TOP PLATES

BORED HOLE MAX.
DIAM. 40% OF
STUD DEPTH

5/8" MIN.
TO EDGE

NOTCH MUST NOT
EXCEED 25% OF
STUD DEPTH

—NOTE—
CONDITION FOR
EXTERIOR AND
BEARING WALLS

STUD

5/8" MIN. TO EDGE

IF HOLE IS BETWEEN 40% AND 60%
OF STUD DEPTH, THEN STUD MUST BE
DOUBLE AND NO MORE THAN TWO
SUCCESSIVE STUDS ARE DOUBLED AND
SO BORED

BORED HOLES SHALL NOT BE LOCATED
IN THE SAME CROSS-SECTION OF CUT
OR NOTCH IN STUD

Figure 12–80 Notching and bored-hole limitations for exterior walls and bearing walls.

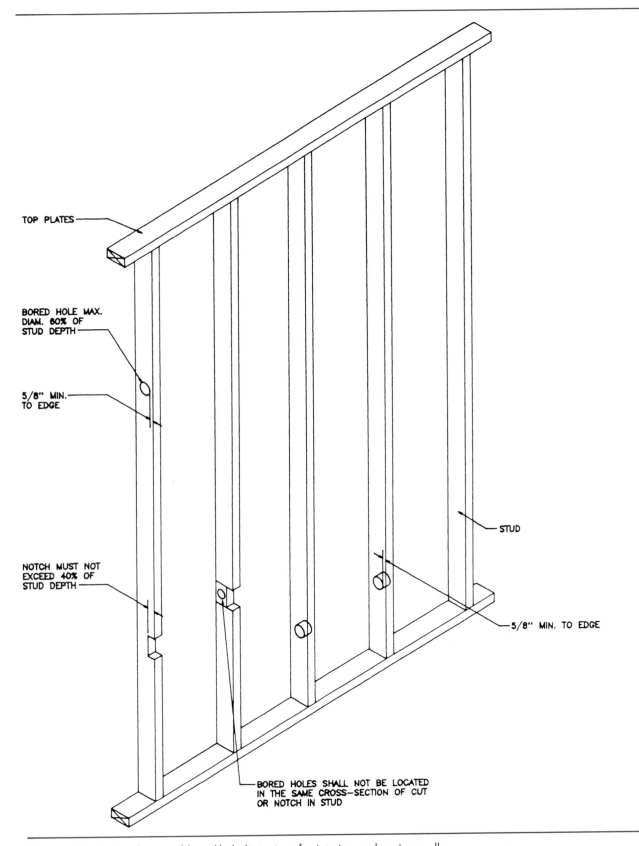

TOP PLATES

BORED HOLE MAX.
DIAM. 60% OF
STUD DEPTH

5/8" MIN.
TO EDGE

NOTCH MUST NOT
EXCEED 40% OF
STUD DEPTH

STUD

5/8" MIN. TO EDGE

BORED HOLES SHALL NOT BE LOCATED
IN THE SAME CROSS-SECTION OF CUT
OR NOTCH IN STUD

Figure 12–81 Notching and bored-hole limitations for interior nonbearing walls.

ing must be inserted between the studs in rows every 2'-0" of vertical height throughout the wall, or else 1 × 3 horizontal furring strips will have to be nailed to the studs to back up the vertical paneling.

With platform framing, it is convenient for workers to lay out each wall horizontally on the platform, nail the pieces together without the second cap plate, and then lift the wall into place where it can be temporarily braced. All of the walls and partitions are set in place and carefully trued. The top piece of the double cap plate is then attached so that all of the walls and partitions are well lap-jointed (Fig. 12–78). Corner posts, T posts, and posts forming intersections of partitions are built up of 2 × 4s, well-spiked together in a manner to provide for both inside and outside nailing surfaces at the corners (Fig. 12–79).

Drilling and Notching   Limitations on drilling and notching of studs used to frame partitions are based on ensuring the retention of structural or functional integrity (Figs. 12–80 and 12–81). In exterior walls and bearing partitions, any wood stud may be cut or notched to a depth not exceeding 25% of its width. Cutting or notching of studs to a depth not greater than 40% of the width of the stud is permitted in nonbearing partitions.

A hole not greater in diameter than 40% of the stud width may be bored in any wood stud. Bored holes not greater than 60% of the width of the stud are permitted in non-bearing partitions or in any wall where each bored stud is doubled, provided that no more than two successive double studs are bored. In no case shall the edge of the bored hole be nearer than ⅝" to the edge of the stud. Bored holes shall not be located at the same section of studs as a cut or a notch.

Fire Blocking   Fire blocking is required to restrict movement of flame and gases to other areas of the building through concealed combustible passages in building components such as floors, walls, and stairs. Such firestopping is required in order to cut off all vertical and horizontal concealed combustible spaces and to form a fire barrier between stories and between a top story and the roof space. The following locations shall be firestopped in wood frame construction:

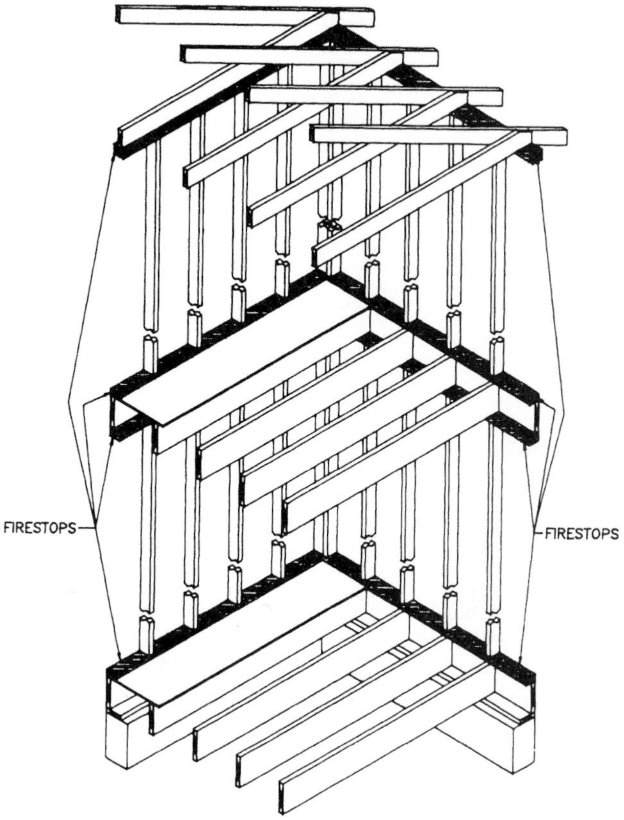

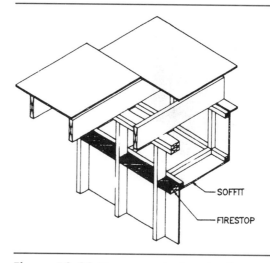

Figure 12–83 Firestopping by furred soffit.

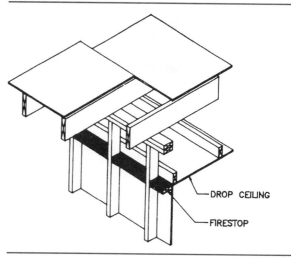

Figure 12–84 Firestopping by dropped ceiling.

Figure 12–82 Firestopping by platform framing.

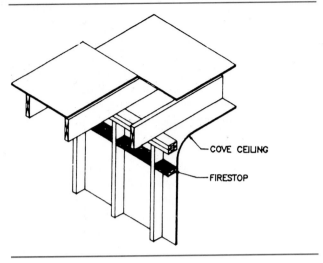

Figure 12–85 Firestopping by cove ceiling.

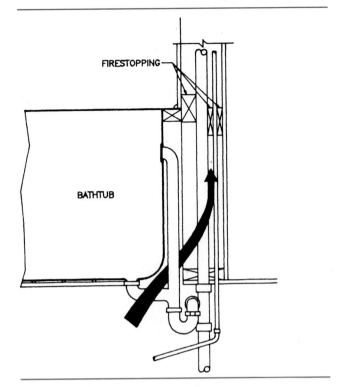

Figure 12–86 Firestopping at tub.

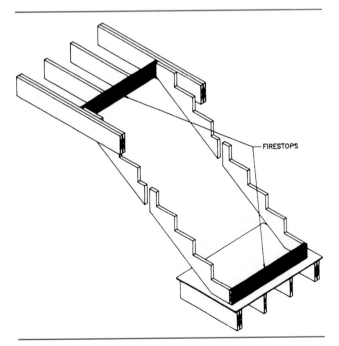

Figure 12–87 Firestopping at stairways.

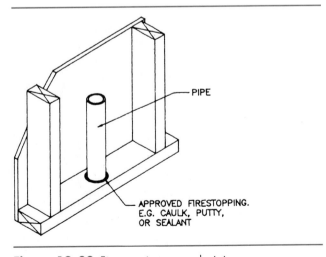

Figure 12–88 Firestopping around piping.

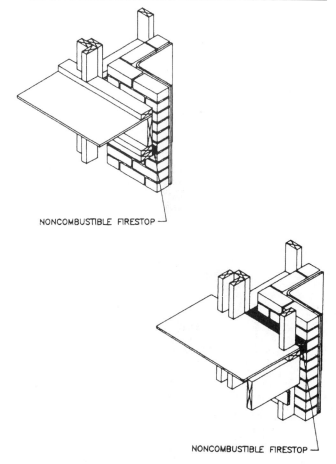

NONCOMBUSTIBLE FIRESTOP

NONCOMBUSTIBLE FIRESTOP

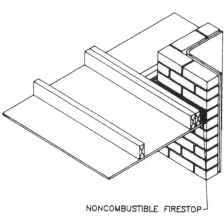

NONCOMBUSTIBLE FIRESTOP

Figure 12-89 Firestopping around chimneys and fireplaces.

1. In concealed spaces of stud walls and partitions, including furred spaces at the ceiling and floor levels and at 10′ intervals both vertically and horizontally (Fig. 12–82).
2. At all interconnections between concealed vertical and horizontal spaces, such as soffits, drop ceilings, cove ceilings, and bathtubs (Figs. 12–83, 12–84, 12–85, and 12–86).

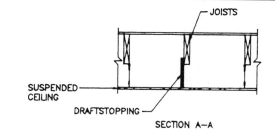

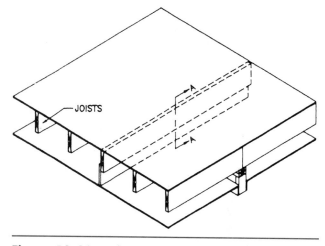

Figure 12-90 Draftstopping between floor joists and ceiling.

3. In concealed spaces between stair stringers at the top and bottom of the run (Fig. 12–87).
4. At openings around vent pipes, ducts, chimneys, and fireplaces at ceilings and floor levels with noncombustible materials (Figs. 12–88 and 12–89).

Materials approved by code for use as firestopping include the following:

2″ nominal lumber,

two thicknesses of 1″ nominal lumber with broken lap joints,

one thickness of $^{23}/_{32}$″ plywood with joints backed by $^{23}/_{32}$″ plywood, or

one thickness of $^{3}/_{4}$″ Type 2-M particleboard with joints backed by $^{3}/_{4}$″ Type 2-M particleboard.

Other approved materials that may be used include gypsum board, glass fiber mesh–reinforced backer board, or mineral fiber, glass fiber securely fastened in place.

Draftstopping Draftstopping is required to limit the spread of fire through combustible spaces in floor-ceiling assemblies when such spaces create a connected area beyond the normal joist cavity so that the area of the concealed space does not exceed 1000 sq ft. The concealed space must be divided into approximately equal ar-

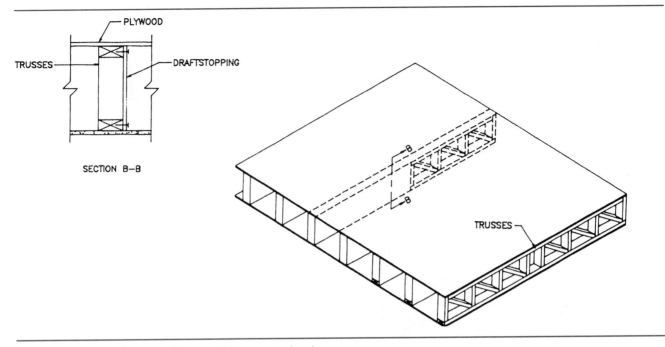

Figure 12–91 Draftstopping between floor truss and ceiling.

eas. The material must be placed parallel to the main framing members (Figs. 12–90 and 12–91).

Materials used to meet draftstopping requirements are ½″ gypsum board, ⅜″ plywood, or ⅜″ Type 2-M-W particleboard adequately attached to supporting members.

Bracing   Wall bracing is usually one of two types: *let-in bracing* or *structural sheathing*. Both types serve to carry or transmit forces from the upper portion of the structure to the foundation. When subjected to wind loads or seismic activity, the upper portion of the structure moves horizontally while the lower portion is restrained at the

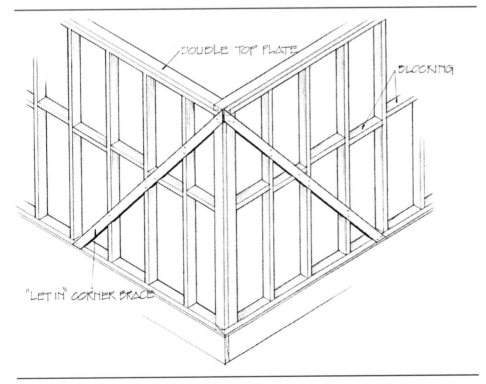

Figure 12–92 Diagonal bracing at corners.

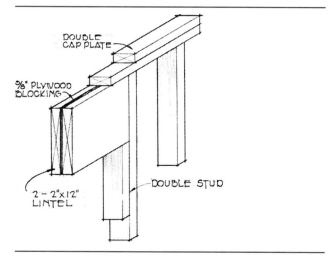

Figure 12-93 Lintels over typical wall openings.

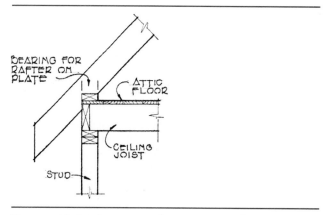

Figure 12-96 Fastening rafters to an attic floor.

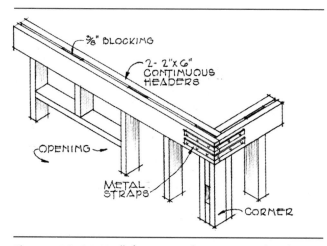

Figure 12-94 Wall framing with continuous headers instead of double cap plate.

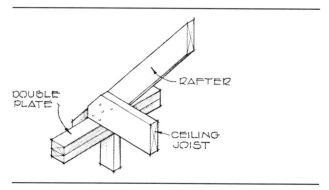

Figure 12-97 Fastening rafters and ceiling joists.

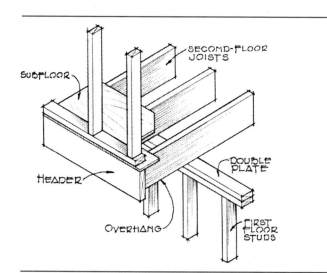

Figure 12-95 Framing a second-floor overhang.

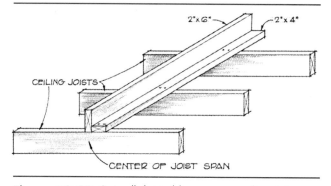

Figure 12-98 Catwalk lateral brace over ceiling joists.

ground level. Bracing resists this movement and thus prevents damage to the building.

Sufficient diagonal bracing must be used at the corners (Fig. 12–92) as well as in major partitions to counteract wind pressures and to prevent lateral movement of the framing. Let-in, 1 × 4 braces placed as close to 45° as possible and not over 60° should be used, and they should be tied into both the top and the bottom plates of the wall. If windows or openings are close to the corners, K braces can be used (Fig. 12–1). Diagonal bracing can be eliminated if structural sheathing is used and is nailed diagonally to the studs or if the frame is sheathed with plywood.

Lintels

The horizontal supports above openings are usually built up of two pieces of nominal 2 × 4, 2 × 6, 2 × 8, 2 × 10, or 2 × 12 lumber with the wide dimension vertical. Extra strength and the exact 3½″ stud width can be gained by sandwiching a ³⁄₈″ plywood panel between the two vertical lintel pieces (Fig. 12–93).

Headers and lintels over openings 4′ wide or less can be made of double 2 × 4s on edge. For each 2′ of opening over 4′, increase the lumber size 2″. Use 2 × 4s on edge over a 4′ opening, 2 × 6s over a 6′ opening, 2 × 8s over 8′, and so on.

All headers and lintels must have at least 2″ solid bearing at each end to the floor or the bottom plate unless you use other approved framing methods or joist brackets. Many builders use double 2 × 12 lintels above all wall openings, thereby eliminating the need for cripples or vertical blocking between the lintel and the top plate (Fig. 12–93). For volume construction, some builders use a double 2 × 6 cap plate, set vertically, over all bearing walls, thus eliminating lintels. Such buildings generally have openings no wider than 4′-0″ (Fig. 12–94).

Second-Floor Framing

Each story in platform framing is a separate unit; if two stories are planned, the second-floor joists and headers are attached to the top plate of the first floor. The second-story walls are built on the platform or subfloor of the second floor. Second-story floor joists then become the first-floor ceiling joists. If possible, second-story partitions should fall directly above partitions below; otherwise, additional floor joists or blocking is required. Occasionally, second stories are cantilevered out beyond exterior walls below. Joists below cantilevered walls should be perpendicular to the walls and well anchored (Fig. 12–95).

Ceiling Joists

Ceiling joists generally rest on the top wall plate if the roof is put directly above. If an attic floor is required (Fig. 12–96), headers and joists can be built up to the outside surface of the walls, with the attic floor nailed to the joists and 2″× 4″ soleplate nailed over the

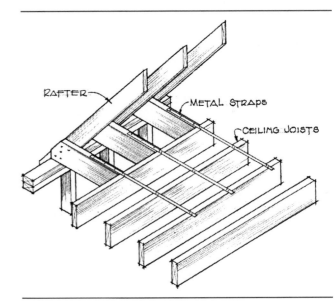

Figure 12–99 Framing with ceiling joists at right angles to rafters.

floor around its perimeter on which the rafters bear. In ceiling framing, if a gable-type is used, it is more desirable to put the ceiling joists parallel to the roof rafters, making the ceiling joists actually bottom chords for the truss shape of the roof rafters. Ceiling joists are nailed to both the top plate and the roof rafters (Fig. 12–97). A lateral brace or a catwalk brace for ceiling framing is shown in Fig. 12–98.

When ceiling joists must be placed at right angles to the rafters, short joists are run perpendicularly to the long joists over the plate and are nailed to both the plate and the rafters. The inner ends of the short joists are butted and anchored to a double-long joist. Metal straps are used to tie the tops of the short joists across at least three of the long joists if no wood flooring is used (Fig. 12–99).

Gable-Roof Rafters

Wind pressures against gable roofs (and hip roofs as well) not only exert downward forces on the windward side, but also exert upward pressures on the leeward side, especially if the roof has a considerable overhang. Therefore, rafters must be well anchored at their lower ends or bearing cuts. The triangular notch cut into the rafters for horizontal bearing is called a *bird's-mouth* (see Fig. 12–100 for roof-framing terminology). It should be deep enough so that the rafter bears on the full width of the 2 × 4 plate. The bird's-mouth is toenailed to the plate, and the rafter is also nailed to the ceiling joist; metal anchors are sometimes employed for more positive anchorage. Rafters are placed directly opposite each other at the ridge so that they brace against each other. Rafter lengths and cuts are carefully laid out on a pattern piece to ensure uniformity of all common rafters and to simplify their cutting. The ridgeboard can

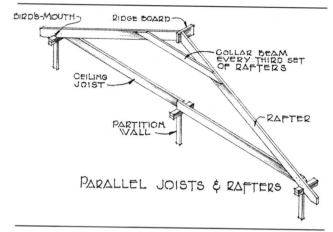

Figure 12–100 Arrangement of opposite rafters and ceiling joists.

be either 1″ or 2″ nominal lumber, but it should be 2″ wider than the rafters so that the entire beveled ridge cut will make contact with the ridgeboard.

Valley Rafters   Valleys are formed by the intersection of two sloping roofs; the rafter directly below the valley is called the *valley rafter* (Fig. 12–101). Valley rafters carry additional loads and should be doubled as well as widened to allow full contact with the diagonal cuts of the jack rafters.

Jack Rafters   Jack rafters are shorter than common rafters and run from hip or valley rafters to the plates. If roofs with different ridge heights intersect, the higher roof is framed first, and the valley is formed by nailing a plate over the higher roof surface. The jack rafters of the lower roof are then nailed to the plate. Or the valley rafter can be incorporated into the larger roof framing if attic communication is needed between the gables or hips (Fig. 12–101).

Hip Rafters   Like ridgeboards, hip rafters do not carry any loads and are therefore the same size as the common rafters (Fig. 12–102).

Gambrel-Roof Framing   This roof framing requires two sets of generally different length rafters. The first set of longer rafters receive a bird's-mouth cut and are nailed to the top plate of the exterior bearing walls. A knee wall or a full stud wall is used to support these steeply pitched rafters at the opposite end of the span. The second set of rafters with less pitch is framed into the top end of the longer rafters and to a ridge beam at the center line of the structure. Collar beams or ceiling joists are placed at the junction between the two sets of rafters (Fig. 12–103).

Dormer Framing   This framing is similar to gable-roof framing. Double headers and double trimmers must be framed around the opening in the roof. Studs are nailed to the top of the double trimmers to form the vertical

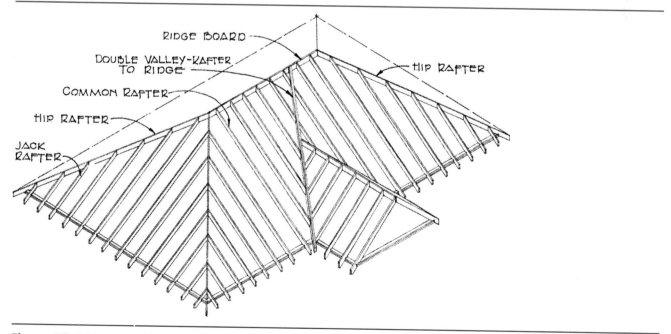

Figure 12–101 Hip-roof framing.

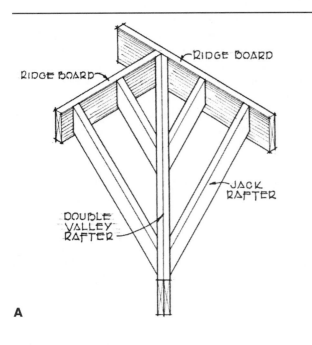

A

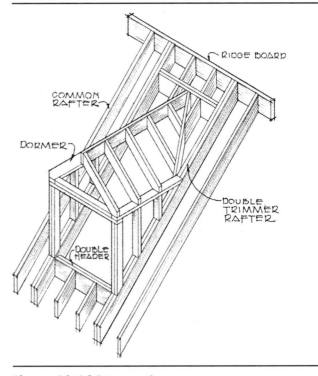

Figure 12-104 Dormer framing.

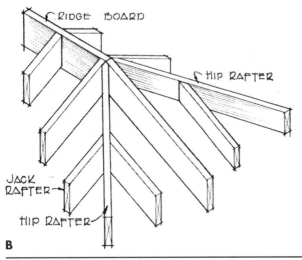

B

Figure 12-102 Roof framing. (A) Valley rafter. (B) Hip rafter.

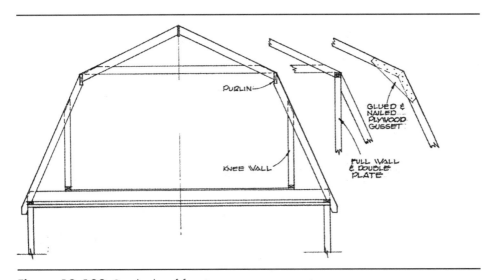

Figure 12-103 Gambrel-roof framing.

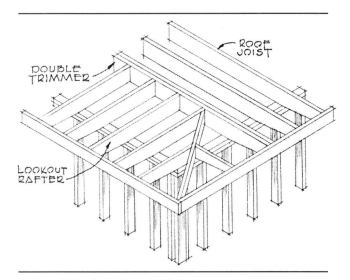

Figure 12–105 Flat-roof framing at corners.

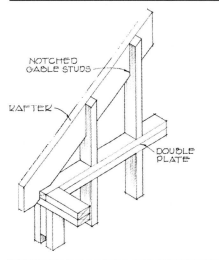

Figure 12–106 Gable studs.

sides of the dormer. A ridge beam and rafters frame the gable roof structure of the dormers (Fig. 12–104).

Flat-Roof Framing Flat-roof framing consists of roof joists bearing on the top plate of stud walls. The roof joists may cantilever over the bearing wall to create an eave or an overhang. When an overhang is desired around the perimeter of the structure, a double trimmer is placed to support cantilevered lookouts. These lookouts are perpendicular to the roof joists and form the overhang at the side of the structure (Fig. 12–105).

Collar Beams Collar beams are short lengths of 2 × 4s or 1 × 6s nailed across opposite gable rafters about ¹⁄₃ the distance down from the ridge (Fig. 12–100). They stiffen the roof against wind pressures. If the attic is used for occupancy, the collar beams act as ceiling joists as well as rafter ties and must be placed at ceiling height. When the roof pitch is low and the rafter span is long, collar beams are nailed to each pair of rafters; otherwise, to every third pair.

Gable Studs Gable studs must be framed with the same spacing as studs in the lower walls; they are diagonally cut and notched to fit the underside of the end rafters and butt to the double wall plate (Fig. 12–106). If a boxed overhang is required above the gable wall, a ladder-type framing can be cantilevered over the gable wall (Fig. 12–107). If wood siding is used on the gable wall, and if brick veneer is used below, 2 × 8 gable studs can be used to bring the gable siding out flush with the brick veneer.

Roof Venting Building codes generally require that the vent area be 1/150 of the horizontal roof area. The area may be 1/300 of the horizontal roof area if 50% of the

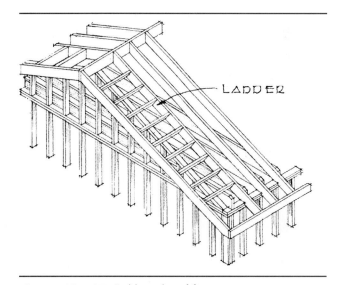

Figure 12–107 Gable-end roof framing.

required ventilating area is located at least 3′ above the eave or the cornice vents, with the balance provided by eave or cornice vents. The vent area may be 1/300 if a vapor barrier with a transmission rate not exceeding 1 perm (see Section 12.16.2) is installed on the warm side of the attic insulation. Venting should be equal between eaves and ridges to provide active flow. Gable-end louvers and other mechanical systems interfere with the natural venting process and should not be combined with traditional eave/ridge venting. For proper venting, an open ridge should be used (Figs. 12–108, 12–109, and 12–110).

In the past several years, shingle problems have been attributed to the lack of venting often in "compact" roofs, that is, roofs with exposed deck ceilings and no attics.

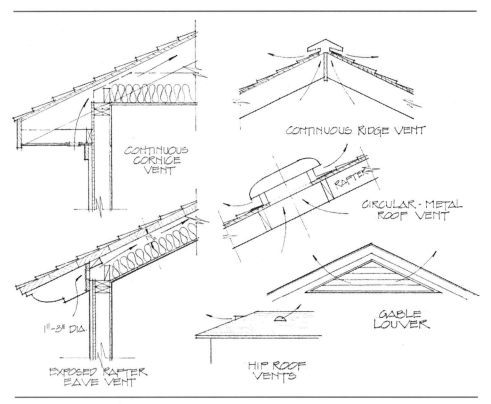

Figure 12–108 Methods of roof venting.

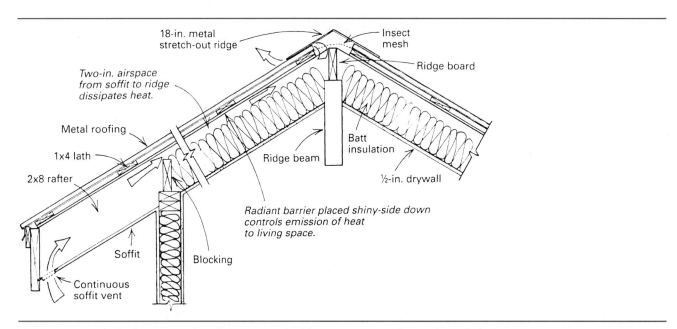

Figure 12–109 Cooling a metal roof. (*Fine Homebuilding* magazine, February/March 1995)

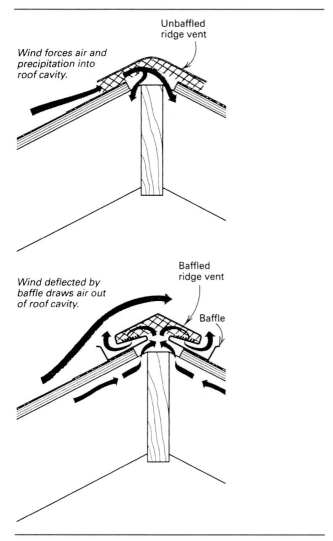

Wind forces air and precipitation into roof cavity.

Unbaffled ridge vent

Wind deflected by baffle draws air out of roof cavity.

Baffled ridge vent

Baffle

Figure 12–110 A baffled versus an unbaffled ridge vent. (*Fine Homebuilding* magazine, June/July 1996)

Compact roofs have the shingles applied directly onto insulation covered by plywood or onto foam insulation bonded to OSB. In some cases the shingles may be nailed directly to a roof deck that serves as an exposed cathedral ceiling. Some manufacturers have claimed that excessive heat loads on unvented roofs cause shingles to buckle and split, but recently the Owens-Corning Company has made available a 10-year warranty on shingles installed over unvented attics or cathedral ceilings.

Roof Decking   There are several methods of covering roof framing, depending on the materials used and on the type of roof covering. Nominal 1″ matched boards, usually 6″ wide and nailed perpendicularly to the rafters, can be used. They are started at the edge of the roof overhang and are laid up tight to the ridge. Use of shiplap or tongue-and-groove lumber will result in a tighter deck and more solid surface for the roof covering. Splices should be well staggered and made over the center of the rafters. Long pieces of lumber are desirable for roof decking.

Softwood plywood makes a satisfactory roof deck; it goes on fast and makes a rigid roof when properly nailed down. Plywood joints should fall on the rafters, and vertical joints should be midway on succeeding rows. The face grain of the plywood should be laid perpendicularly to the rafters (Fig. 12–111). Plywood that is 1/2″ thick can be used as decking when light roof coverings are used; 5/8″ or 3/4″ thickness is necessary for the heavier roof coverings. All plywood edges should be well covered with molding or trim at the edges of the roof. As soon as the roof deck is completed, it is covered with a good grade of roofing felt, preferably 2 layers of 15-lb felt. The felt is 2″ side lapped and is end lapped 4″ to 6″, depending on the roof covering. This covering is very important because it is rainproof and therefore protects the structure while the work is in progress.

Wall Sheathing   Wall sheathing is the outer subwall, which is nailed directly to the studs and the framework. Wall sheathing adds both strength and insulation to the building and forms a base for the exterior siding. The following types of materials are used for wall sheathing:

1. **Wood boards**, 1″ nominal thickness, usually 6″ or 8″ wide; wider lumber will shrink more and leave wider gaps. Wood sheathing can be applied either horizontally or at a 45° angle. The horizontal method is more economical but requires diagonal let-in bracing at the corners. Diagonal sheathing adds greatly to the rigidity of the wall and eliminates the need for corner bracing. It also provides an excellent tie between the framing and the sill plate. Wood sheathing can be either square edged, shiplap, or tongue-and-groove. End joints should be staggered and made over studs; two 8-penny nails are used at each stud bearing. Number 2 common lumber is usually specified.

2. **Fiberboard asphalt-coated sheathing** is available in 4′ widths and 8′, 10′, and 12′ lengths. It can be applied either horizontally or vertically. The latter method usually requires a 9′ long sheet to cover a wall that has 8′-0″ ceilings. Fiberboard sheathing 1/2″ thick is easy to handle by carpenters, and it goes on quickly compared to wood sheathing. Other types of fiberboard measure 2′ × 8′ and have shiplap or tongue-and-groove edges; they are applied horizontally. Fiberboard must be carefully nailed with galvanized roofing nails.

3. **Plywood** used for sheathing usually comes in 4′ × 8′ sheets and should be a minimum of 5/16″ thick over studs spaced 16″ o.c. and 3/8″ thick over studs spaced 24″ o.c. Six-penny nails should be used and should be spaced not more than 6″ apart for edge nailing and 12″ apart for intermediate nailing. Plywood is usually applied vertically to permit perimeter

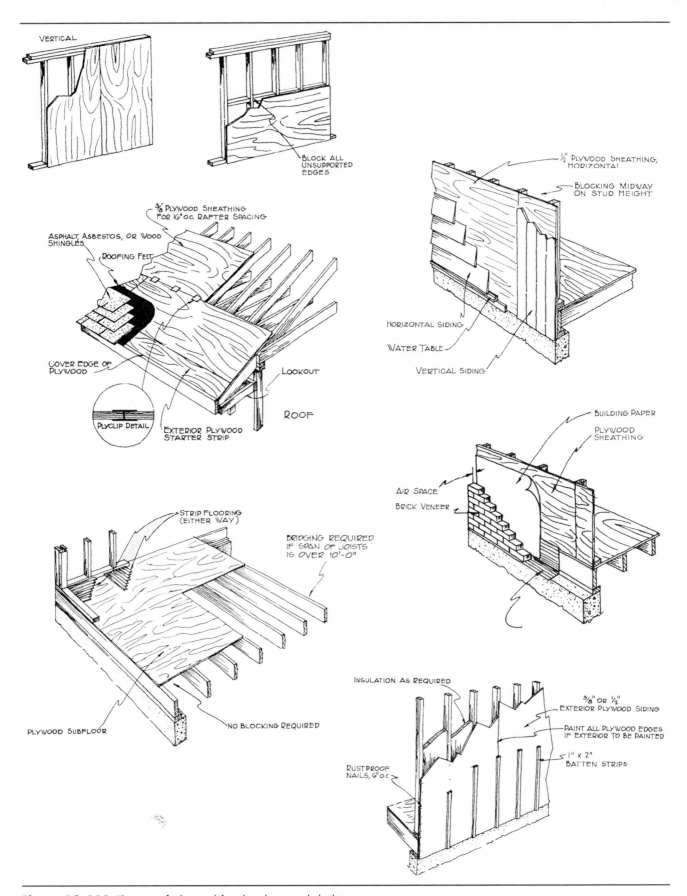

Figure 12-111 The use of plywood for sheathing and decking.

nailing without additional blocking. When the 8′ dimension is horizontal, blocking is desirable along the horizontal joints between the studs as a base for the finish-wall nailing. When the finish-wall material requires nailing between the studs (as with wood shingles), the plywood should be at least 3/8″ thick. Wood shingles must be nailed to stripping if 5/16″ plywood or other non-nail-holding sheathing is used. Plywood can be applied with power-driven staplers using galvanized wire staples. Asphalt building felt is placed over plywood sheathing in some locations.

4. **Gypsum sheathing,** composed of moisture-resistant gypsum filler faced on both sides with lightweight paper, can be used as a subwall. It is available in 2′ × 4′ and 2′ × 8′ sizes and 1/2″ thick. Some gypsum sheathing has V-joint edges for easier application and better edge ties; this type is applied horizontally and is nailed securely with galvanized roofing nails. Vertical joints are always staggered and made on the center of the studs. If finish siding is used that requires both nailing strips and a smooth base, the nailing strips are fastened directly to the studs, and the gypsum sheathing is nailed over the nailing strips.

5. **Styrofoam® and urethane sheathings,** usually 3/4″ or 1″ thick, some foil-covered on one or both sides for added reflective insulation and resistance to moisture features, are light but fragile. They are obtainable in sheet sizes similar to those of plywood, and they must be carefully fastened to framing with large-headed nails or staples. Either has excellent insulating but poor structural qualities.

6. **Particleboard,** available in 1/4″ and 7/16″ thicknesses and in 4′ × 8′ sheets, is used for sheathing. This material can be applied vertically or horizontally over studs. Particleboard is fastened in the same manner as plywood.

12.4.2 Balloon Frame Construction

The distinguishing feature of balloon construction (Fig. 12–112) is that the studs run from the sill to the top plate. These vertical lengths of wall framing minimize the amount of wood shrinkage in the total height of the building. Because the stud lengths are long, there is less chance that the wood frame will reduce in height when used with brick or stone veneer or stucco in 2-story buildings. Houses built after 1930 are more likely to be platform framed due to the advent of plywood and an increasing scarcity of framing lumber. If a house is more than 60 years old and has a wood frame, it probably has a balloon frame.

Wood shrinks considerably more across grain than on end grain; therefore, the absence of horizontal members, with the exception of the sill and the plate, provides end-

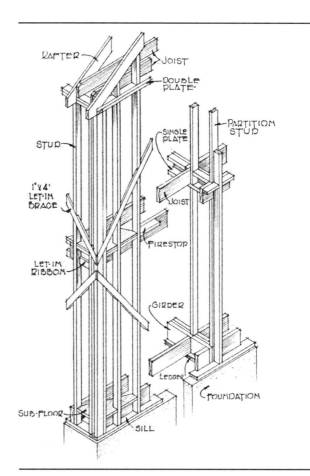

Figure 12-112 Balloon (eastern) frame construction.

grain fiber almost throughout the entire height of the building. Even though balloon framing is not used as often as a platform frame, it has features that are noteworthy, and it is sometimes used in combination with platform construction.

Sill   The sill plate, as in the platform frame, is well anchored to the foundation with 1/2″ diameter anchor bolts. If the top of the foundation wall needs trueing, a bed of grout or mortar is placed below the sill. The sill is usually a 2 × 6 or a 2 × 8; sometimes a double sill of well-spiked members is used. Both the studs and the first-floor joists rest on the sill plate, forming the solid sill. Usually, the sill plate is set the thickness of the sheathing inside the outer edge of the foundation wall (Fig. 12–51B).

Floor Joists   As in the platform frame, the size and the spacing of the floor joists in the balloon frame depend on their loads and span. Generally, the joists are spaced 16″ o.c. and are toenailed to the sill plate and the beam within the building. Joists can be butted to the beam and anchored, or they can be lapped over the beam. A row of bridging should be used if the joists span more than 10′-0″. There are no headers at the ends of the joists.

Studs   The studs are usually spaced 16″ o.c., as in other framing, and they run to the top plate. Each stud carries its own load from plate to sill without any lateral distribution by headers or beams. For this reason, studs must be carefully selected, especially in balloon frame, 2-story houses, where the studs must be considerably longer. The spacing of the studs must be compatible with the spacing of the floor joists and ceiling joists to allow the joists in all floors to be nailed directly to the studs. Openings for doors and windows must be surrounded with double studs, and headers or lintels must be carefully selected. Openings in first-story walls must have lintels or headers over them that are capable of carrying the entire weight of the studs above the openings.

One of the shortcomings of balloon framing is the ease with which a fire in the basement can burn through the sill and enter the cavities between the studs, causing flue action and rapid spread of the conflagration through the entire structure. To eliminate this hazard, codes require 2″ blocking between the joists and studs above the sill plate and often prescribe that the cavity at the sill be filled with masonry or a noncombustible material (Fig. 12–113).

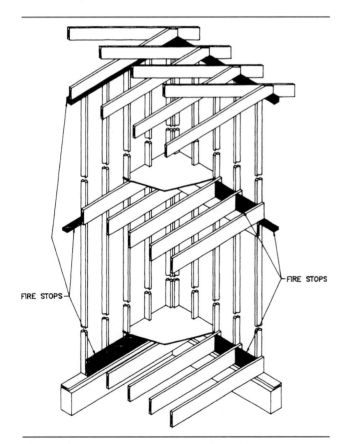

Figure 12–113 Firestopping in balloon framing.

Ribbon   Another typical feature of the balloon frame is the method of supporting second-floor joists. Ceiling or second-floor joists rest on a 1 × 6 board, let into the studs, called a *ribbon*. The joists are also spiked to the studs. The ribbon is necessary only on those walls taking the ends of the joists. Firestopping and fireblocking are also needed at the intersection of second-floor joists and the walls. In addition, horizontal blocking between the studs is required midway between room heights.

Posts   Corner studs and posts at intersections of partitions and walls are built up, similarly to the platform frame, so that nailing surfaces will be found on both inside and outside corners.

Braces   If horizontal board sheathing is used for the subwall, 1 × 4 let-in bracing must be used throughout the framing to resist raking stresses. The use of diagonal wood sheathing or plywood eliminates the need for the bracing. Diagonal 2 × 4 blocking at the corners does not have the strength that the let-in braces have, especially if the blocking is not tightly fitted into the stud spacings.

Subfloor   Boards, measuring 1 × 6 and laid diagonally, or plywood is generally used for the subfloor. The subfloor extends to the outer edge of the wall framing and is notched around the studs.

Top Plate and Roof Framing   The top plate consists of a double 2 × 4, lapped at the corners and the intersections of partitions and well-spiked together. If nailing surfaces are necessary for the application of gypsum board or other ceiling coverings, blocking must be inserted between the ceiling joists and spiked to the top plate. Rafters rest on the top plate and are nailed to both the plate and the ceiling joists, similar to platform framing. Metal framing anchors can be used to fasten rafters to the top plate if more rigidity is desired. The roof framing is done similarly to the method indicated in platform framing, and the treatment of the overhang varies with the type of architectural effect required. Some houses have very little overhang at the eaves; others may have as much as a 3′ overhang. Rafters can be left exposed at the eaves, or they can be boxed in with a horizontal soffit. The treatment of the detail at the junction of the wall and the roof plays an important part in the general character of the building (see Fig. 15–71).

12.4.3 Wood Roof Trusses

Lightweight, prefabricated wood trusses offer many advantages in roof framing of small- or average-size residences (Fig. 12–114). Wood trusses can be used with ei-

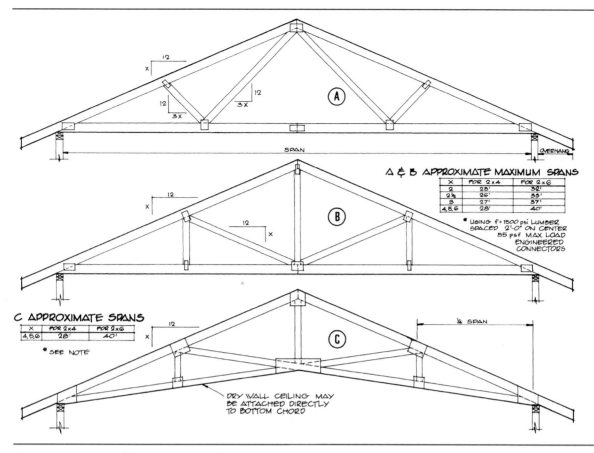

Figure 12-114 Table content:

A & B APPROXIMATE MAXIMUM SPANS

X	FOR 2x4	FOR 2x6
2	23'	32'
2½	26'	35'
3	27'	37'
4,5,6	28'	40'

* USING f = 1500 psi LUMBER SPACED 2'-0" ON CENTER 55 psf MAX LOAD ENGINEERED CONNECTORS

C APPROXIMATE SPANS

X	FOR 2x4	FOR 2x6
4,5,6	28'	40'

* SEE NOTE

DRY WALL CEILING MAY BE ATTACHED DIRECTLY TO BOTTOM CHORD

Figure 12-114 Prefabricated trusses for light construction.

ther platform or balloon framing; when spaced 24″ o.c., which is usually the case in small buildings, they become compatible with sheet sizes of plywood and other materials. Trusses can be lifted into place and quickly anchored on the job. Complete freedom can be exercised in the placement of partitions within the building. They can be made thinner, and when they are not load bearing, the foundation can be simplified, which constitutes a saving.

The basic shape of the truss is a rigid, structural triangle, and with the addition of strut members, the basic shape subdivides into smaller triangles, each reducing the span of the outer members or chords. This reduction of individual member spans allows narrower members, and less material is required to carry similar loads as compared with conventional framing.

The fasteners used to join the members of the truss can be nails, bolts, split rings, barb-pointed plates, or plywood gusset plates fastened with nails and glue. Many lumber dealers stock engineered trusses of standard lengths and roof pitches. Volume builders prefabricate their own trusses with simple jigs and production-line efficiency. Fabrication of trusses requires careful workmanship and attention to detail; the joints between members are critical, especially the heel joint, which is subjected to more lateral stresses. The heel joint should bear on the wall plate below (Fig. 12-115). If an overhang

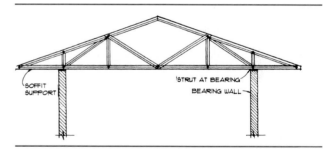

Figure 12-115 An alternate overhanging truss that provides for stable overhang.

is required, the top chords can be extended beyond the heel joint to support the overhang (Figs. 12-114 and 12-115).

Modified trusses are available for complete framing of hip-type roofs. Occasionally, conventional framing is combined with trussed-rafter framing; the trusses then replace the common rafters of the pitched roof, and conventional framing is used at the intersections of sloping roofs and around the hips of hip roofs. Trusses are as varied as the houses they go on and can be combined to create complex roof shapes (Fig. 12-116).

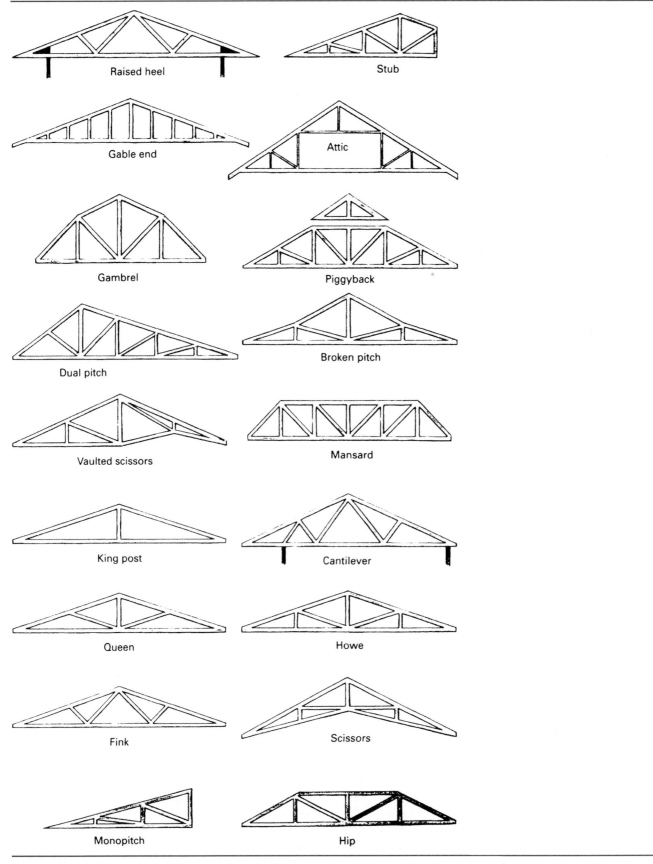

Figure 12-116 Roof choices. (*Fine Homebuilding* magazine, June/July 1994)

Residential roof trusses range from 15′ to 50′ long and from 5′ to 15′ high. Length and height are determined by roof pitch and roof span plus cantilever, if any. Special trusses can be made for virtually any situation. Attic trusses provide room-sized central openings, and vaulted scissor trusses provide a raised ceiling effect where wanted. Common trusses are fabricated with a variable top chord overhang, a variety of soffit returns, and details for box and closed cornicles. Unlike with conventional framing, trusses cannot be modified in the field without radically altering their strength. Truss manufacturers typically provide all engineering services required for fabrication, along with the placement diagrams for construction.

12.5

PLANK-AND-BEAM FRAME CONSTRUCTION

Plank-and-beam wood framing (Figs. 12–117 and 12–118), now an established method of residential construction, has been adapted from the older type of mill construction that utilized heavy framing members and thick wood floors and roofs. Plank-and-beam framing is similar to steel skeleton framing in concept: It develops structural stability by concentrating loads on a few large members, which in turn transfer their loads directly to the foundation by the use of posts or columns. In structures, these lines of load transfer from the beams should not be interrupted unless special provision has been made for their placement. Posts needed to support beams in light construction must be at least 4″× 4″. Where heavy, concentrated loads occur in places other than over main beams or columns, supplementary beams are necessary to carry the loads.

One structural advantage of plank-and-beam framing is the simplicity in framing around door and window openings. Because the loads are carried by posts uniformly spaced throughout the walls, large openings can be easily framed without lintels. Large window walls, which are characteristic of contemporary construction, can be formed by merely inserting fixed glass between the posts. However, a window wall should have several solid panels in appropriate places to stabilize it against lateral movement. All walls must be braced with diagonal bracing or suitable sheathing.

Plank-and-beam framing requires disciplined planning; savings will be realized by employing modular dimensions to the framing members and to the wall and

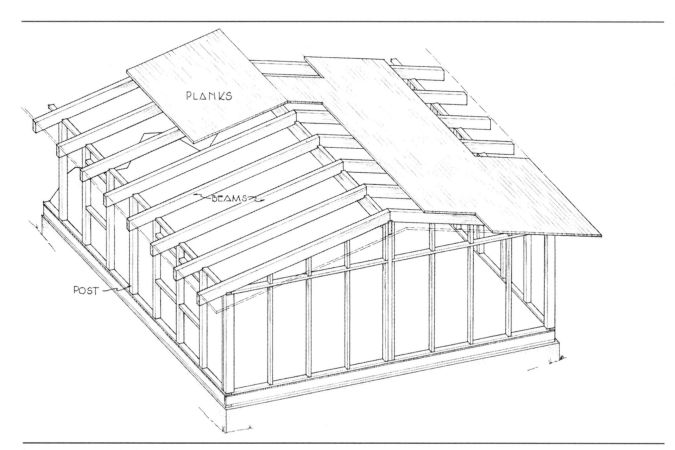

Figure 12–117 Plank-and-beam construction with transverse beams.

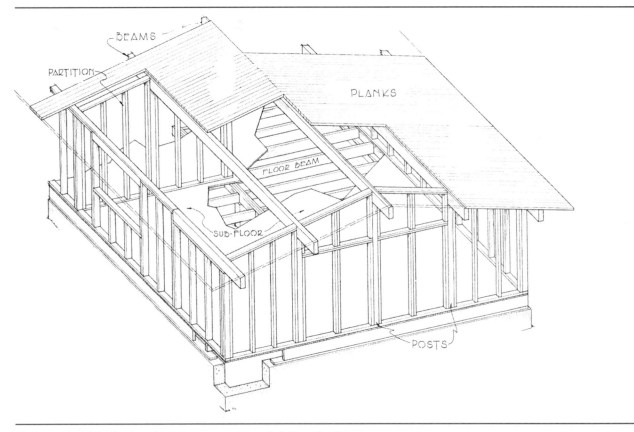

Figure 12-118 Plank-and-beam construction with longitudinal beams.

roof coverings. (See Section 5.4, "Dimensions on Drawings.") It lends itself to the use of plywood. Stock plywood sizes can be used with a minimum of cutting and waste. Massive, exposed structural members within a house are instrumental in giving it a feeling of stability and rhythmic beauty, attributes that testify to the architect's competence.

With rising relative costs of labor in home building, plank-and-beam framing, which requires the handling of fewer pieces than does conventional framing, offers structural economies to home builders. In a study made by the National Lumber Manufacturers Association, the plank-and-beam method saved 26% on labor and 15% on materials over conventional joist framing. Bridging is eliminated between joists, nails are fewer and larger, and inside room heights are increased with no increase in stud or building heights. Masonry piers can be used instead of continuous foundation walls for economy. (The piers must be located under the structural posts.) Eave details and roof overhangs are simpler; usually, the 2″ roof decking is cantilevered over the wall, and no material is needed for boxing in the cornice.

Conventional framing can be successfully combined with plank-and-beam framing within the same building. Sometimes a plank-and-beam roof is used in only one room of the house, often the living or family area (Fig. 12–119). If so, sufficient column bearing must be provided within the stud walls for the beams of the one roof. If plank-and-beam wall framing supports conven-

Figure 12-119 Plank-and-beam interiors.

tional roof framing, a continuous header should be used at the top of the walls carrying the roof loads to transfer the load from the closely spaced rafters to the posts (Fig. 12–94).

There are limitations to plank-and-beam framing, and attention to special details is necessary at these points of construction: Partitions must be located over beams, or else supplemental framing must be installed; partitions perpendicular to the beams must rest on a 4 × 4 or 4 × 6 soleplate to distribute the load across the beam of the floor. Additional beams are needed under bathtubs and other concentrated loads. Because of the absence of concealed air spaces in floors and roofs, wiring and plumbing are more difficult to install. Often built-up cavity beams are used through which wiring or piping can pass. Flexible wiring cable can often be laid in the V joint on the top of 2″ roof decking. Surface-mounted electrical raceways may have to be used for electrical outlets. Plumbing, especially in 2-story houses, may have to be hidden within false beams or furred spaces. Moldings and raceways may have to be employed, and top surfaces of beams may have to be grooved to conceal wiring or piping.

Plank roofs with no dead air space below often require additional insulation and vapor barriers. Insulation is usually applied above the decking and must be rigid enough to support workers on the roof. The thickness of the insulation must conform to local climatic conditions. Vapor barriers should be installed between the decking and the insulation. Prefinished insulation board, 2″ to 4″ thick, is available in large sheets to replace decking and insulation combinations. It is light and can be installed rapidly by workers; no decorating of the underside ceiling is necessary. Prefinished insulation tile or panels can also be applied to the underside of regular wood decking. They can be attached directly to the decking, or furring strips can be used between the underside of the decking and the insulation. When rigid insulation is used above the decking, wood cant strips of the same thickness are fastened around the edges of the roof for application of flashing or gravel stops (Fig. 12–120). Light-colored coverings for plank roofs also keep the house cooler in summer and warmer in winter.

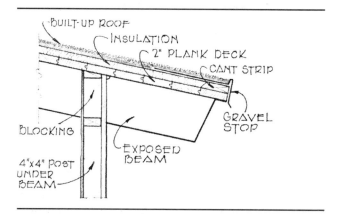

Figure 12–120 Typical plank-and-beam cornice detail.

Another consideration in plank-and-beam framing is the quality of the exposed structural lumber. Higher grades of framing lumber are required, because large knots, streaks, and resin blemishes are objectionable if painted and are even more unsightly if natural finishes are used. Workmanship in installing the structural pieces is especially important. Finishing nails must be used; metal anchors, if used, must be carefully fastened; and no hammer marks should be left on the exposed structural surfaces. Planking should be accurately nailed to the beams; joints should occur above the beams; and the tongue-and-groove planking should be nailed up tightly so that shrinkage between planks will not be objectionable. Kiln-dried lumber should be specified, especially for visible ceilings. Blocking, or filler wall, between beams on exterior walls must be carefully fitted and detailed, or else leaky joints and air infiltration will result when the blocking dries and shrinks. Particular care must be given to connections where beams abut each other and where beams join the posts. These connections must be strong and neatly done. Where gable roofs are used, provision must be made to absorb the horizontal thrust produced by the sloping roof beams (Fig. 12–121). Partitions and supported ridge beams help relieve this thrust. Metal straps fastened over the ridge beam or metal ridge fasteners are effective in resisting the lateral stresses.

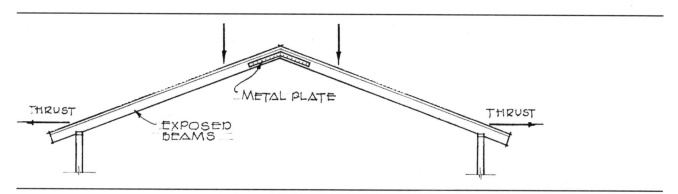

Figure 12–121 Horizontal thrust of a beam ceiling.

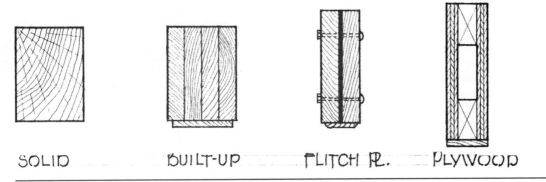

SOLID   BUILT-UP   FLITCH PL.   PLYWOOD

Figure 12–122 Types of beams used in plank-and-beam construction.

Plank-and-beam roofs can be supported with either transverse beams or longitudinal beams (Figs. 12–117 and 12–118).

12.5.1 Transverse Beams

Transverse beams are usually supported at the ridge by a heavy ridge beam and by the long, exterior walls of the building. The planks then span the beams perpendicularly to the slope of the roof, if it is a gable type. Beams may be solid, glue laminated, or built up (Fig. 12–122) of 2″ lumber, securely fastened together (fasteners are usually concealed). Strength and rigidity of the planks are improved if the planks are continuous over more than one span. Tests show that a plank that is continuous over two spans is nearly 2½ times as stiff as a plank extending over only one span, under the same conditions. Planks, usually 6″ or 8″ wide, can be tongue-and-grooved or splined (Fig. 12–123) and are designed to support moderate loads. If beams are spaced more than 6′ or 7′ apart, noticeable deflection of common species of nominal 2″ planking will result; 3″ T & G planking will span up to 12′.

12.5.2 Longitudinal Beams

Longitudinal beams running the entire length of the building can be used instead of transverse beams. In that case, the planking is applied parallel to the slope of the roof—down the slope. For either method, tables are used for the proper selection of beam size, span, and spacing, after the other architectural considerations have been resolved. Exterior beams at corners of gable ends are usually false or short beams, which are cantilevered out from the corner of the long, exterior walls (Fig. 12–118). Beam sizes can be reduced when interior partitions become supports for the beams.

When either method of plank-and-beam framing is employed, interior partitions are often more difficult to frame than when conventional framing is used. If ceilings are sloping, the top of the partition must meet the underside of the slope and must be made diagonal. Other partitions, running perpendicularly to the slope of the ceiling, will have horizontal cap plates but must be built to different heights. Interior partitions, using 2 × 4 studs, can be framed by several methods. First, they can be framed with the studs cut the same length as the exterior walls, and a double cap plate can be put over all of the exterior and interior walls, tying them securely together as in platform framing. Short studs are then attached above the plate to a single top plate that conforms to the slope of the ceiling and is attached to it. Or the partitions can be erected after the roof is completed. The various-length studs are cut to fit the slope of the ceiling, and each partition with a single 2 × 4 cap is fitted into its proper place.

Finish strip flooring should be nailed at right angles to the plank subfloor, similarly to conventional construction. Care must be taken to be sure that flooring or roofing nails do not penetrate through the planks when the planks serve as a ceiling. The inside appearance should be considered when the beams are spaced; moving a beam several inches one way or another near a partition will often be an improvement. Partitions, on the other hand, should be placed in relation to beams so that the beams seem to form a uniform pattern throughout the room. The appearance of exposed beams on the exterior of the building is also of importance. Plank-and-beam framing lends itself to open planning, a characteristic of contemporary houses.

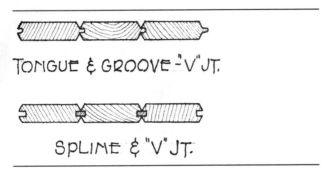

TONGUE & GROOVE - "V" JT.

SPLINE & "V" JT.

Figure 12–123 Wood decking used on plank-and-beam ceilings.

In most areas, the structural design of plank-and-beam framing will be controlled by the local building code to the extent of specifying live load requirements. A live load of 40 lb/sq ft is usually specified for floors. For roofs, some codes specify 20 lb/sq ft; others, 30 lb. Beam sizes, beam spacing, and the wood species necessary for various spans of both roofs and floors can be taken from the plank-and-beam tables found in Chapter 13.

12.5.3 Glue Laminated Structural Members

Glue laminated structural members offer many design possibilities in plank-and-beam construction that otherwise would be restricted by the structural limitations of sawed wood members (Fig. 12–124). Many stock beam shapes are available (Fig. 12–125), providing variations in conventional building silhouettes. Many church buildings successfully employ either laminated beams or arches for interesting ceilings and roof structures. Special waterproof glues are used for exterior exposures. Usually, comparative sectional sizes span one-third more than sawn timbers. Consult manufacturers' catalogs for specific shapes, loads, and spans.

In larger buildings, structural systems such as rigid frames, two-hinged barrel arches, and three-hinged arches, as well as bowstring arches and domes, use laminated members. Rigid frames are made of separate leg and arm members carefully joined at the haunch to maintain the desired roof slope. The two-hinged barrel arches must have provision for counteracting outward thrust at their bases with the use of foundation piers, tie rods, or concrete foundation buttresses. Each support must be engineered to resist the horizontal thrust of the arch.

Figure 12–124 A California redwood home with an interesting roof of planks over laminated beams. (Courtesy Dandelet)

Three-hinged arches such as Tudor and Gothic are popular in contemporary church architecture because of their beauty. Their spans vary typically from 30′ to 100′. Their base construction is similar to that of the two-hinged arches. Because of the distance between arches, purlins are commonly placed perpendicularly between arches to carry the roof decking. Arch design data vary with the roof slope and span distance. Metal connectors at critical points ensure neat, engineered joining of the laminated members (Fig. 12–126).

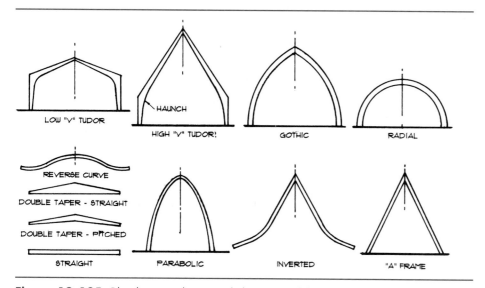

Figure 12–125 Glue laminated structural shapes available in wood.

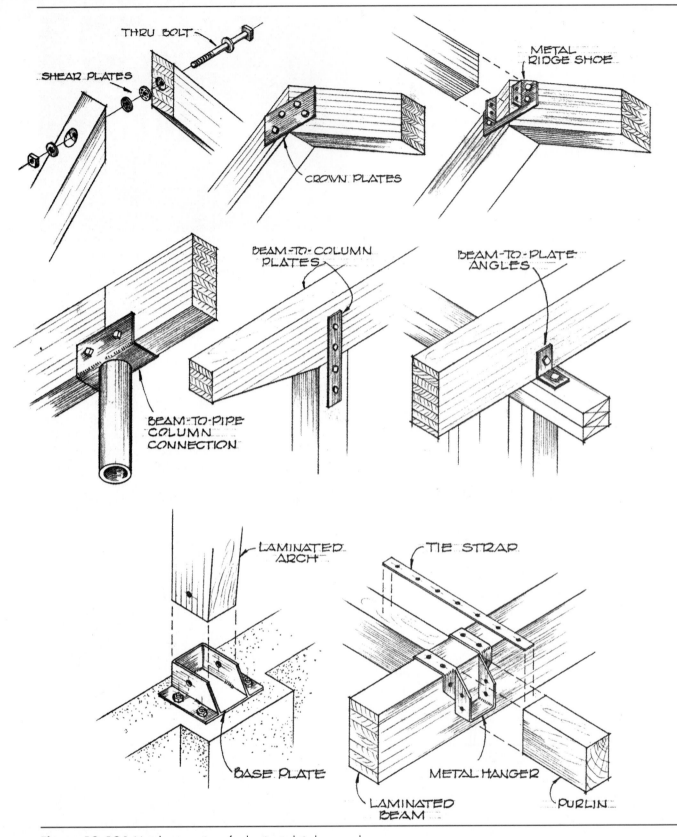

Figure 12–126 Metal connections for laminated timber members.

BRICK AND STONE VENEER CONSTRUCTION

In areas of the country where brick or building stone is economical, veneered frame dwellings are popular. Masonry veneer over wood framing is more economical than solid masonry; the combination wall (masonry over wood) utilizes the good qualities of both materials. The exterior masonry is durable and requires no maintenance; the wood is economical, more flexible, and a good insulator. Any of the previously mentioned types of wood framing can be used with veneer; the wood frame supports the entire weight of the floors and the roof—no loads are put on the masonry veneer. The single wythe of brick or stone is merely an exterior wall covering for the structure (Fig. 12–127).

Bricks or 4" or 5" maximum cut stone can be used for veneering. Hard-burned, moisture-resistant brick should be used. Sandstones and limestones are commonly used for stone veneers. Standard-size brick as well as oversized brick (jumbo, utility, or Norwegian) is available (Fig. 12–138). Bricks are manufactured in a wide range of colors and finishes. A better-quality brick has the same clay throughout the brick, whereas a less expensive brick may have a color or sand finish applied only to the face of the brick. Wood-mold brick is available, which resembles the handmade brick of colonial times. Adobe brick is popular in the Southwest.

It is common practice to start the veneer at the grade line. Provision must be made on the foundation for the width of the veneer and a 1" minimum air space between

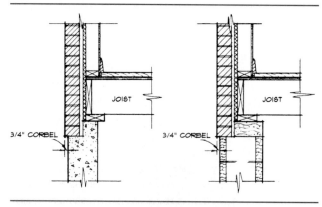

Figure 12–128 Brick veneer may be corbeled on foundation.

the veneer and the wall sheathing; at least 5" bearing width on the foundation wall is allowed (Fig. 12–128). Good construction usually starts the veneer lower than the wood sill on the foundation wall. Brick may be corbeled 3/4" beyond the edge of the foundation wall. Condensation often develops within the air space, and provisions must be made to keep the moisture from penetrating the wood members. Veneer is tied to the wood frame wall with rustproof, corrugated metal ties (Fig. 12–129). The ties are nailed through the sheathing to every other stud and are laid into the mortar joints of every sixth course (about every 16" to 20" of vertical masonry).

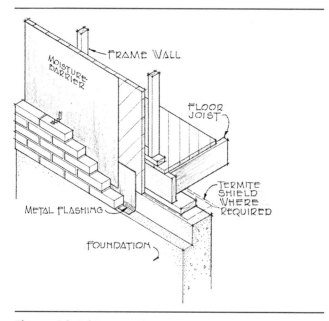

Figure 12–127 Wood frame construction with brick veneer.

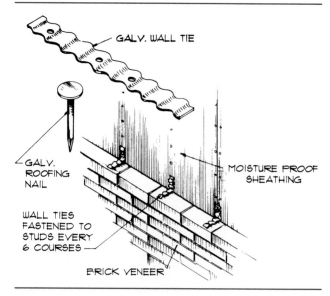

Figure 12–129 Brick wall ties.

Wall ties must be corrosion resistant and either sheet metal or wire. If sheet metal, they must be a minimum size of 0.030″, No. 28 galvanized sheet gauge by ¾″. If wire, they must have a minimum diameter of 0.148″. Wall ties must be spaced so they support no more than 2 sq ft of wall area but no more than 24″ o.c. horizontally. Areas of the country subject to special seismic considerations require additional reinforcement (Fig. 12–130).

The exterior wood sheathing must be covered with a moisture barrier, such as roofing felt. If asphalt-coated sheathing board is used, the asphalt coating acts as a moisture barrier. Styrofoam and urethane sheathing are water-resistant materials and do not require a moisture barrier.

At the bottom of the air space, a rustproof base flashing must be inserted under the felt and extended into the lower mortar joint of the veneer. This flashing prevents any moisture from penetrating the wood sill members after gathering at the bottom of the air space. Weep holes can be provided at the bottom course by leaving mortar out of vertical joints about 4′ apart (Fig. 12–131).

Brick or masonry sills under window and door openings must be flashed in the same manner. Water running off from windows and doors onto their sills must be kept from entering the air cavity. Masonry will absorb moisture unless it is waterproofed (Figs. 12–132 and 12–133).

Sloping sills of brick, set on edge, are usually used under all openings of brick veneer, and a similar sloping sill must cap brick veneer when it extends only partially up

WEEPHOLE	RECOMMENDED MAXIMUM SPACING (Center to Center)
Wick Material	16 in.
Open Head Joints	24 in.

Other:
  Inserts
  Tubes
  Oiled Rods (removable)
	24 in.

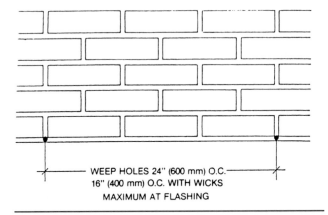

WEEP HOLES 24″ (600 mm) O.C.
16″ (400 mm) O.C. WITH WICKS
MAXIMUM AT FLASHING

Figure 12–131 Recommended spacing of weepholes.

an exterior wall. Combination brick and wood siding exteriors require a bead of caulking where the two materials join together. Stone veneer usually has a cut stone sill under all openings. It also extends out from the face of the veneer, similarly to brick, and has a sloping top surface, called a *wash*, and a drip underneath to prevent water from running back to the wall. All mortar joints in veneers should be well tooled to prevent moisture penetration.

Steel angle iron, usually $3\frac{1}{2}″ \times 3\frac{1}{2}″ \times \frac{5}{16}″$ or $6″ \times 3\frac{1}{2}″ \times \frac{3}{8}″$ depending on loads and spans, is used for lintels above openings in brick veneer (Fig. 12–134 and Table 12–10). Often brick *soldier courses* (Fig. 12–135) are put directly above openings. Some builders eliminate the use of steel lintels above veneer openings in 1-story houses by using a wide-trim frieze board and molding above all window and door openings. If a low, boxed-in cornice is used, no masonry is necessary above the openings. The wide frieze board continues around the entire house below the cornice, and blocking is used to fur out to the face of the veneer above the windows and doors.

Brick or stone is often used to veneer concrete masonry units (Fig. 12–136). In low walls, corrugated metal ties can be used to tie the two materials together; in larger walls, more positive bonds are attained with *header courses* called "bonding" of brick set into the masonry backing.

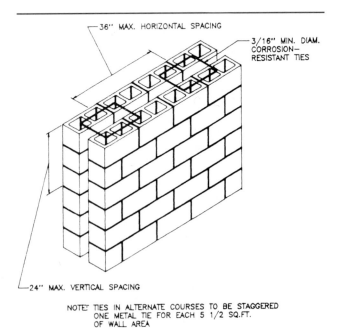

36″ MAX. HORIZONTAL SPACING

3/16″ MIN. DIAM. CORROSION-RESISTANT TIES

24″ MAX. VERTICAL SPACING

NOTE: TIES IN ALTERNATE COURSES TO BE STAGGERED ONE METAL TIE FOR EACH 5 1/2 SQ.FT. OF WALL AREA

Figure 12–130 Masonry bonding metal ties.

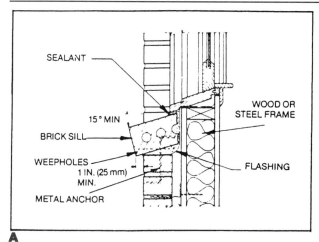

A

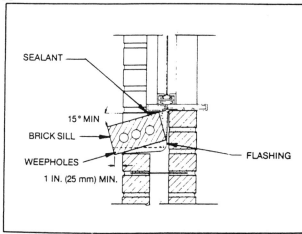

B

Figure 12–132 (A) Sill in frame/brick veneer construction. (B) Sill in cavity wall construction.

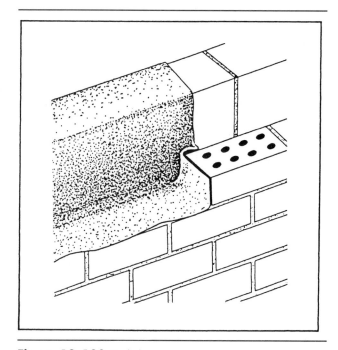

Figure 12–133 End dams.

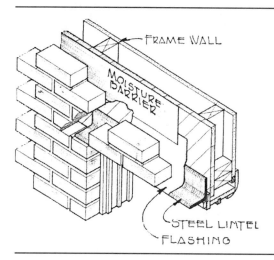

Figure 12–134 Steel lintel above brick veneer openings.

Table 12–10 Steel angle lintels for light frame, brick veneer construction.

Span	Size
0'–5'	$3^{1}/_{2}'' \times 3^{1}/_{2}'' \times {}^{5}/_{16}''$
5'–7'	$4'' \times 3^{1}/_{2}'' \times {}^{5}/_{16}''$
7'–9'	$5'' \times 3^{1}/_{2}'' \times {}^{3}/_{8}''$
9'–10'	$6'' \times 3^{1}/_{2}'' \times {}^{3}/_{8}''$

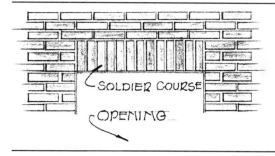

Figure 12–135 Brick soldier course above opening.

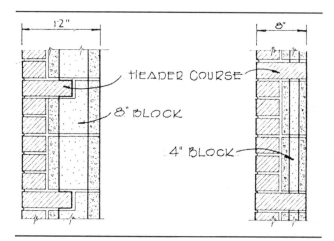

Figure 12–136 Concrete block and brick veneer wall details.

12.7

SOLID MASONRY AND CAVITY WALL CONSTRUCTION

Solid masonry, especially brick and stone, has served people well for centuries as a desirable construction material. Many structures still standing and still beautiful provide evidence of its pleasing as well as durable characteristics. Both brick and stone are available in many different colors, textures, and sizes to challenge the creativity and artistry of modern designers and builders. Although many solid masonry buildings have been built of these materials in the past, the present-day trend is to use these materials as exterior veneers only, with wood frame, metal studs, concrete block, or reinforced concrete as the structural, load-bearing material. Solid stone in residential work is limited to small walls; in contemporary residences it is often used to create contrasting walls where the texture and the color of the brick or stone are exposed both inside and outside. Many architects feel that the natural beauty of stone provides sufficient textural interest to make any further decoration of the walls unnecessary.

Local codes in some residential areas require solid masonry exterior walls in all structures. Usually 8″ thick brick walls are used in 1-story houses; local codes must be invoked if higher walls are involved. This type of construction, where solid masonry walls carry the weight of the floors and roof, is known as *bearing-wall construction*. The floors and the roof are constructed of wood framing members, similarly to conventional framing, except that the floor joists are set into the masonry walls and transfer their loads to them; the rafters rest on wood plates well anchored to the top of the walls. All of the wood structural members of the floors, partitions, and roof are surrounded by the shell-like exterior masonry wall. This shell of incombustible material has the advantage of restricting the spread of fire to surrounding buildings. Floor joists that are in the masonry wall must have a diagonal cut, called a *firecut*, on their ends to prevent the joists from rupturing the wall should a fire burn through the joists and cause them to drop to the bottom of the building (Fig. 12–137). The firecut allows full bearing at the bottom surface of the joist, yet it is self-releasing in case of fire.

Because of the porous and noninsulating nature of solid brick or masonry, some method must be used to prevent moisture penetration and to insulate the exterior walls. The conventional method is to waterproof the inside surface of the masonry with hot asphalt or tar and then fasten 1 × 2 furring strips to the inside masonry surface, to which the interior wall covering is applied. Rigid insulation can also be inserted between the furring strips. Another method of furring the masonry wall, to create an air space between the exterior and interior surfaces, is by using a self-furring metal lath over which ¾″ plaster is applied.

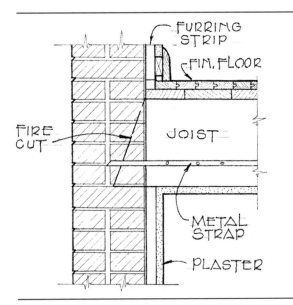

Figure 12–137 Wood joists in solid-brick wall detail.

In economical 1-story buildings with wall heights not exceeding 9′, 15′ at the gable peak, a single-wythe brick called *Norwegian brick* can be used for the masonry load-bearing walls (Fig. 12–138). Since Norwegian brick is larger than common brick, a wall can be laid up much faster. It is 5½″ wide, 2³⁄₁₆″ high, and 11½″ long and can be furred with 2″ × 2″ furring strips, fastened to metal clips that are set into the mortar joints throughout the wall. Interior finish materials can be fastened to the furring strips.

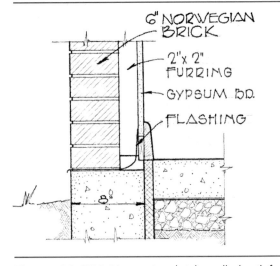

Figure 12–138 Norwegian brick wall detail for 1-story walls.

Many other hollow brick units are available up to 8″ thick that are suitable for single-unit masonry walls. Many small commercial buildings are built with this method. For residential construction, however, small brick units are preferable because their texture imparts more interest than that of the larger units.

If a brick surface is required for both the outside and the inside walls, a *brick cavity* wall can be used (Fig. 12–139). This method of construction was introduced from Europe in the late 1930s and has been widely accepted in this country. A 2″ air cavity is used between the wythes of common brick, making the wall usually 10″ thick. Rigid metal ties are placed in the mortar joints spaced 16″ o.c. horizontally and vertically to tie the two wythes together. No furring is necessary, since the air cavity prevents moisture penetration and insulates the wall. In colder climates, air stops are used at corners to minimize air circulation throughout the cavity. Added insulation can be given by the application of rigid insulation to the inside wall of the air cavity; insulation should not touch the outside wythe of brick. Flashing must be used at floor levels, sills, heads of openings, and other critical places. Weep holes must be used in the outer base course to dissipate moisture accumulating in the lower part of the air cavity. Such weep holes can be accomplished by omission of the mortar from header joints at intervals of 3′ to 4′.

Wood floor joists can rest on the inside wythe of brick; additional metal ties must be used in the course below the joists, and the joists must not project into the cavity. Girders and beams must be supported by solid, 8″ wide pilasters that bond the inner and outer wythes together. Roof and ceiling construction must be supported by a 2″ thick wood plate that rests on both wythes and is anchored with 1/2″ anchor bolts 6′-0″ o.c. The anchor bolts must extend down into the air cavity about 15″, and a 3″ × 6″ × 1/4″ plate is welded to the head of the bolt and is anchored in the mortar joints of the brick (Fig. 12–139). Mortar joints should be well tooled on both the outer and the inner finish surfaces.

12.7.1 Concrete Masonry Units (Concrete Block)

Concrete blocks have become the most economical building material for small buildings, even for above-ground construction. It is estimated that a C.M.U. wall costs about 10% less than conventional stud wall with insulation construction. It is a widely available material that combines structure, insulation, and exterior and interior finish surfaces. Many various shapes and sizes of blocks are manufactured; as previously mentioned, the most used sizes are nominal 8″ × 8″ × 16″ and 8″ × 12″ × 16″. (Refer to Fig. 12–33 for other sizes and shapes.) The units are made either of dense concrete mixes, which give them good structural qualities, or of lightweight aggregate mixes, which result in blocks that are weaker structurally, but that provide better insulation. Lightweight blocks of foamed concrete are available in some areas; they have the same insulating qualities as stud walls with wood sheathing, siding, and 2″ of insulation.

Improperly cured units often result in cracks in C.M.U. walls. The blocks should be air cured for at least 28 days, or they should be cured under high-pressure steam, which is the better method for reducing ultimate shrinkage. Steam-cured units will shrink from 1/4″ to 3/8″ in a 100′ wall; air-cured units will shrink about twice that amount. Cracking can often be eliminated by the introduction of wire mesh or lightweight C.M.U. steel reinforcement into every second or third horizontal mortar joint (Fig. 12–140). Extra reinforcement is required above and below all wall openings. In long C.M.U. walls, control joints must be used to relieve contraction and other stresses; they are continuous, vertical joints through the wall and are spaced at 20′-0″ intervals.

Concrete masonry unit walls are not successful in areas subjected to earthquakes, unless heavy reinforcement is used extensively at each floor and ceiling level. Refer to local seismic codes if blocks must be used.

In warm areas such as Florida, Arizona, and parts of Texas, concrete masonry units are a very popular above-ground construction material, even in fine houses. Units can be waterproofed on the exterior with waterproof,

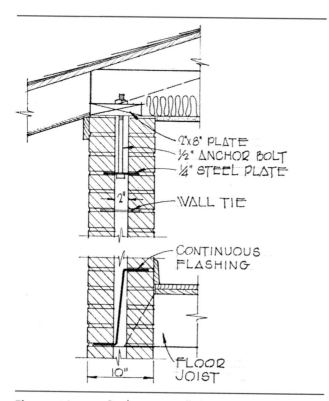

Figure 12–139 Brick cavity wall detail.

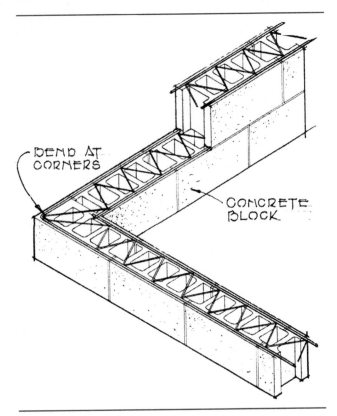

Figure 12-140 Horizontal reinforcement in C.M.U. courses.

12.7.2 Steel and Concrete Joists

Steel joists are often used in small commercial buildings for the structural support of roofs and light-occupancy floors. Several different types of open-web steel joists are manufactured; details and allowable spans can be obtained from manufacturers' catalogs.[2]

Total dead and live toads must be calculated, and preliminary considerations must be studied, before open-web steel joists are adopted. After they have been found adaptable in the preliminary planning, a thorough engineering calculation is made of all aspects of their use before the steel joist construction is finally selected for the structure.

H-series steel joists are available in standard depths of 8″, 10″, 12″, 14″, 16″, 18″, 20″, 22″, 24″, 26″, 28″, and 30″, each in different weights that can span up to 60′ for roofs. Maximum permissible spacing of the joists for floors is 24″ o.c.; for roofs it is 30″ o.c. In flat roof construction with 2″ thick tongue-and-groove decking, a 2″ × 4″ nailing strip is fastened to the top of the joists with metal clips or bolts to receive the decking. Some joists are made with wood nailing strips already attached to both the top and the bottom chord. If a ceiling is required below the roof joists, ³⁄₈″ ribbed metal lath can be clipped to the bottom chord, and plaster applied over it (Fig. 12–141).

cement-base paints, or they can be directly covered with stucco. The inside is usually furred and plastered. Florida hurricane codes require continuous, reinforced concrete bond beams above each story level of C.M.U. work, which must be tied to the foundation with vertical corner reinforcement (Fig. 12–35).

Different architectural effects can be achieved by the use of different-sized units in various combinations; however, the *half-lapped bond* is used in structural walls. A *stacked bond* is occasionally used for novelty effects, but horizontal reinforcement must then be used in the horizontal joints. As mentioned in the material on foundations, use unit modules for lengths, heights, and opening sizes, as well as for placement of the openings (Fig. 12–27). Hollow units must be capped with solid cap blocks where framing joists or beams rest on them Non-load-bearing partitions can be made with 4″ thick hollow units.

Ribbed or split-faced concrete masonry units in a wide variety of styles and colors are available from certain manufacturers. Although more expensive than standard units, the split-faced units eliminate the need for painting or furring the exterior or interior faces of the walls to apply other finishes.

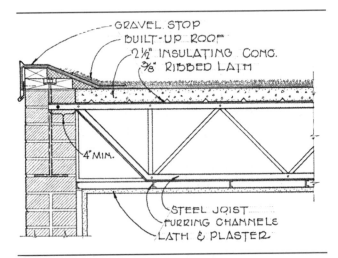

Figure 12-141 Bearing wall and steel joist detail.

² Specifications and details can also be obtained from the Steel Joist Institute, 1346 Connecticut Avenue NW, Washington, DC 20036.

This construction will give the roof a ¾-h fire rating according to the National Bureau of Standards Report TRBM-44.

Flat-type roof construction with steel joists also employs a 2″ concrete slab poured over paper-backed steel mesh or a 2″ gypsum slab as the deck. Usually, built-up roofing is applied directly over the slabs, unless rigid insulation is needed. Lightweight, insulating concrete makes a desirable roof. The steel joists rest on the load-bearing masonry walls or, in wider buildings requiring a steel girder through the center, they rest on the girder as well as on the walls. Metal bridging is attached to joists between spans. Suspended ceilings of acoustical tile can be supported with hangers and metal channels below the steel joists; the open-web character of the joists allows pipes, ducts, and wiring to be run through the dead air space above the ceiling.

In floor construction with steel joists, the joists are supported in the masonry wall similarly to wood framing, as well as on steel girders through the center if the building is wider than the span of one joist. A 2″ or 2½″ poured concrete slab is placed over one of several types of permanent forming material—paper-backed 6″ × 6″ No. 10 wire mesh, corrugated steel sheets, or ⅜″ expanded metal rib lath. The forming material is left in place. Floor slabs are reinforced with ¼″ (No. 2) bars, 12″ o.c. both ways, or with the 6″ × 6″ welded wire mesh; generally, no other reinforcing is used.

Precast, reinforced concrete slabs of similar thickness can also be used on floors. Asphalt tile or other floor finishes can be put over the concrete slabs.

12.7.3 Reinforced Concrete Joists

Occasionally, reinforced concrete joists are used in light construction (Fig. 12–142). In the case of either a concrete floor with a basement below or a fireproof flat roof, the use of concrete joists is appropriate. These roofs are manufactured in many areas, so a source can usually be found near any given building site. Depth and reinforcing size are custom designed according to the loads and the spans required. Concrete joists are generally spaced 24″ o.c. and bear on concrete block walls. The slab can be poured over wood forming placed between the joists, or the permanent forming materials, mentioned above, can be used.

Another method utilizes filler block laid tightly between the concrete joists and resting on the joists. A 2″ slab is poured over the filler block to form the floor or the roof. Precast concrete joists are designed for each specific job and are delivered to the site, ready to be installed.

12.8

STEEL FRAME CONSTRUCTION

The American Iron and Steel Institute estimates that steel will capture 25% of the building market by the year 2000. Just as today's standard for design and construction of wood frame houses has evolved over time, the best techniques for engineering a steel frame house are still being developed.

A steel frame house is similar to a stick frame house but with some significant differences. Steel framing members consist of two basic components that are C-shaped in section. Studs, joists, and rafters are made with members that have flanges folded inward about ¼″ at their open corners. Studs come with prepunched holes in their webs for electrical conduits. Joists and rafters can have solid webs, or they can be prepunched. The other basic components are tracks; they have solid webs and do not have the folded corners on the legs. Tracks are used both as sill plates and top plates and as part of posts or headers when combined with studs.

Steel studs, joists, rafters, and tracks come in various dimensions that differ in regular increments. Unlike wood dimensions, steel dimensions are a net figure; 6″ means 6″. Studs are measured from their outside dimensions, and tracks are measured from the inside of their legs. A 6″ stud fits snugly into the track. The assembly is held together by self-tapping screws driven through the legs of the track and the legs of the stud.

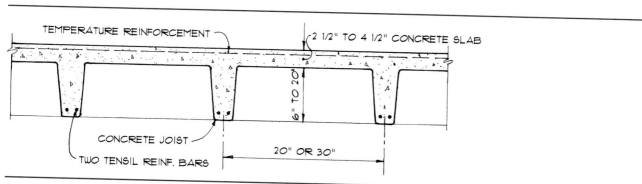

Figure 12–142 Reinforced concrete joist detail.

Steel components are surprisingly light for their strength. Studs are often shipped nested together in bundles of ten, and one person can pick up and carry a bundle of 10′ studs with relative ease. Another attractive aspect of steel framing is its strength. You can increase the strength and the load-bearing capabilities of the member simply by increasing the thickness or gauge of the material. You do not have to increase the dimensions of the member; 12 gauge is stronger and heavier than 20 gauge.

A steel house is put together in small pieces, just as the typical platform frame house of wood. The foundations are the same for wood or steel, but with steel it is important to trowel the top of the foundation smooth because the steel track is not as forgiving as a wood mudsill atop a layer of sill sealer. A flat, level foundation is essential.

Accurate layout of framing members is important in any system, but with steel, layout accuracy takes on greater importance. Rafters must be over either a post or a double stud. If the house is more than 1 story tall, posts and studs bear on a plate that is supported either by a joist or by a short piece of stud called a *web stiffener* placed in the joist track directly under the post or the stud (Figs. 12–143 through 12–146). Loads are transferred to

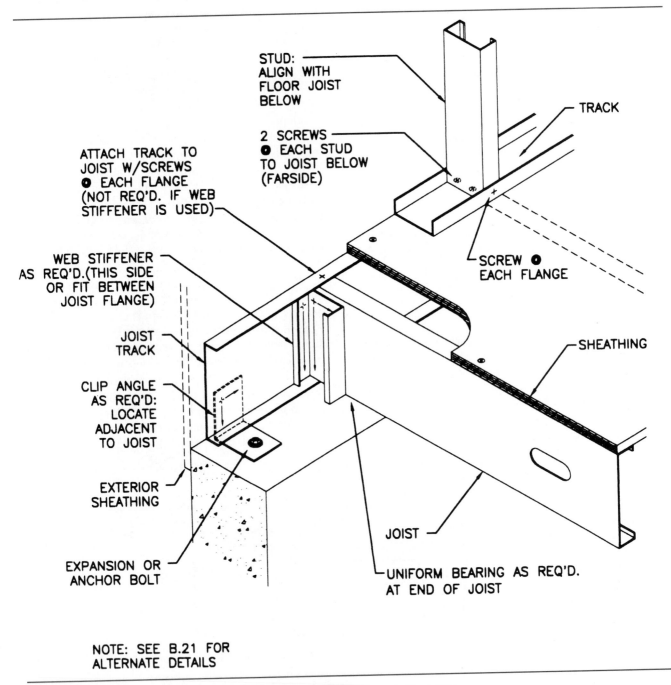

Figure 12-143 Floor joists bearing on the foundation.

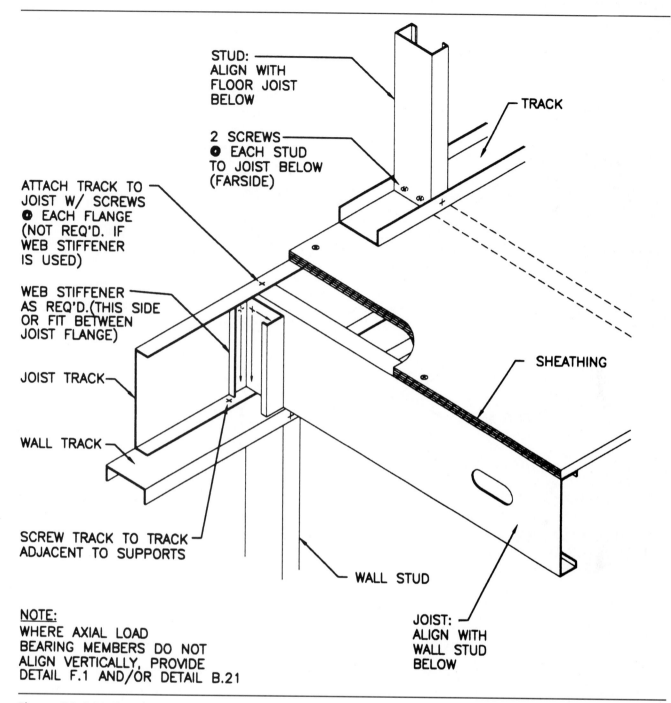

STUD:
ALIGN WITH
FLOOR JOIST
BELOW

TRACK

2 SCREWS
⊕ EACH STUD
TO JOIST BELOW
(FARSIDE)

ATTACH TRACK TO
JOIST W/ SCREWS
⊕ EACH FLANGE
(NOT REQ'D. IF
WEB STIFFENER
IS USED)

WEB STIFFENER
AS REQ'D.(THIS SIDE
OR FIT BETWEEN
JOIST FLANGE)

JOIST TRACK

WALL TRACK

SCREW TRACK TO TRACK
ADJACENT TO SUPPORTS

SHEATHING

WALL STUD

JOIST:
ALIGN WITH
WALL STUD
BELOW

NOTE:
WHERE AXIAL LOAD
BEARING MEMBERS DO NOT
ALIGN VERTICALLY, PROVIDE
DETAIL F.1 AND/OR DETAIL B.21

Figure 12–144 Floor framing at an exterior wall.

the foundation by corresponding posts and studs in the lower floors. When you lay out first-floor walls, you are laying out the roof framing. Screws for steel frame walls must be driven from both sides of the track into the legs of the stud. Screwing steel framing components to one another is not difficult, but it is time-consuming.

Complicated roofs, low-pitch roofs, and most hip roofs can be built with wood for less cost than steel because wood truss assemblies are standardized and factory built. However, almost all roofs can be framed with steel if you have the time and the budget.

Potentially severe energy penalties can accompany steel framing. Steel has very high conductive capabilities which provide a direct flow of heat out of a building in heating climates and into a building in cooling climates. Steel framing conducts 10 times more heat than

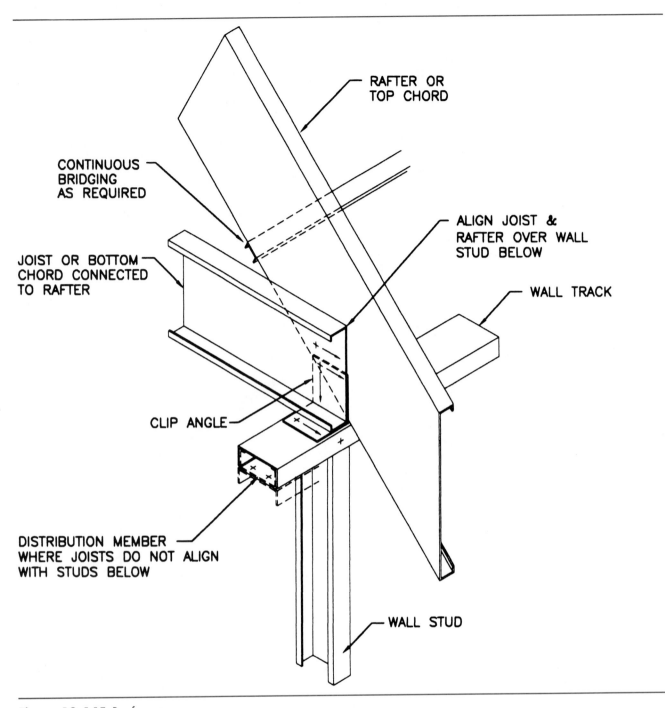

RAFTER OR TOP CHORD

CONTINUOUS BRIDGING AS REQUIRED

ALIGN JOIST & RAFTER OVER WALL STUD BELOW

JOIST OR BOTTOM CHORD CONNECTED TO RAFTER

WALL TRACK

CLIP ANGLE

DISTRIBUTION MEMBER WHERE JOISTS DO NOT ALIGN WITH STUDS BELOW

WALL STUD

Figure 12–145 Roof eave.

wood framing. When a building envelope is insulated only in the cavities between framing members, the steel framing reduces the insulation's performance by more than 50%, versus 10% to 15% for wood framing (Table 12–11).

In an attempt to understand the energy loss problems associated with steel framing, the American Iron and Steel Institute, in conjunction with the NAHB Research Center, ran tests on a variety of wall systems. The results

of these tests are published in the *Thermal Design Guide for Exterior Walls*, which includes tables of expected R values for several steel stud wall sections (Table 12–12 and Fig. 12–147). These tests looked at simple walls built with steel studs without openings, headers, corners, and partition leads, all of which increase heat loss. The tests also did not study the conductive heat loss from stud walls through their connections to concrete foundations and steel roof trusses.

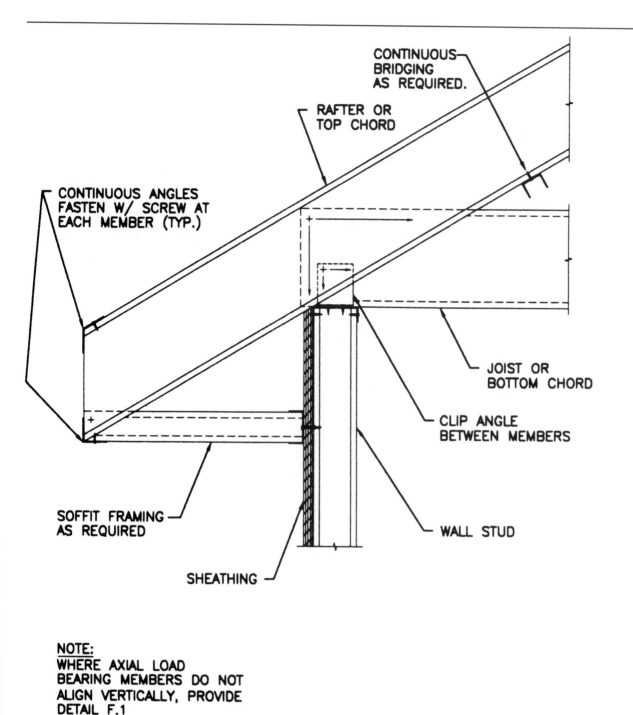

CONTINUOUS BRIDGING AS REQUIRED.

RAFTER OR TOP CHORD

CONTINUOUS ANGLES FASTEN W/ SCREW AT EACH MEMBER (TYP.)

JOIST OR BOTTOM CHORD

CLIP ANGLE BETWEEN MEMBERS

WALL STUD

SOFFIT FRAMING AS REQUIRED

SHEATHING

NOTE:
WHERE AXIAL LOAD BEARING MEMBERS DO NOT ALIGN VERTICALLY, PROVIDE DETAIL F.1

Figure 12-146 Roof eave and soffit.

While performing infrared thermography of a steel frame home, the Bonneville Power Administration (BPA) staff found interior wall surface temperatures of 45° over steel studs when the outside temperature was 40°. These temperatures occurred even though the exterior of the wall was sheathed with R-6 foam and the 6″ wall cavities were filled with insulation. This study discovered that heat trav-eled vertically down through the steel to the foundation. The engineers also found that where steel walls connected directly to steel roof trusses, exterior foam on the studs only redirected heat flow up through studs and through the uninsulated truss tails. Another finding of these studies was that R-6 foam located inside the studs worked more effectively than foam on the exterior.

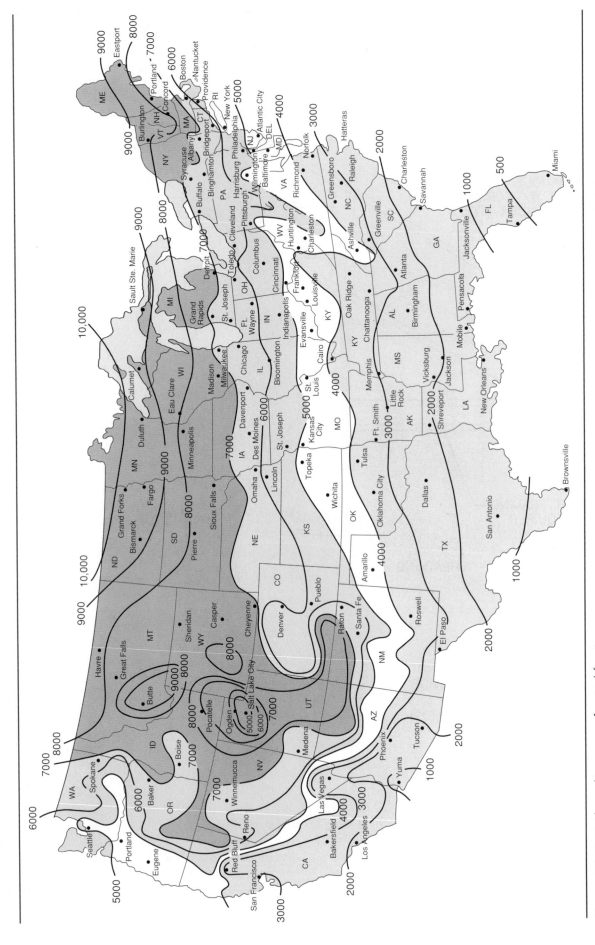

Figure 12-147 Thermal zone map for steel framing.

Table 12-11 Framing and wall *R* values.

Framing and Spacing	Nominal Cavity Insulation	Wood Frame[c]	Steel Frame[d]
2 × 4, 16″ o.c.[a]	R-11 R-13 R-15	R-9.0 R-10.1 R-11.2	R-5.5 R-6.0 R-6.4
2 × 4, 24″ o.c.[b]	R-11 R-13 R-15	R-9.4 R-10.7 R-11.9	R-6.6 R-7.2 R-7.8
2 × 6, 16″ o.c.[c]	R-19 R-21	R-15.1 R-16.2	R-7.1 R-7.4
2 × 6, 24″ o.c.[d]	R-19 R-21	R-16.0 R-17.2	R-8.6 R-9.0
2 × 8, 16″ o.c.[a]	R-25	R-20.1	R-7.8
2 × 8, 24″ o.c.[b]	R-25	R-21.2	R-9.6

SOURCE: *Environmental Building News,* Vol. 3, No. 4.
NOTES:
C-channel metal studs are 16 gauge or thinner.
Wall *R*-values are without sheathing or air films.
[a]Assumes that 11.9% of wall area is framing.
[b]Assumes that 8.9% of wall area is framing.
[c]Values for wood are calculated using the parallel path method.
[d]Values for steel are from ASHRAE Standard 90.1.

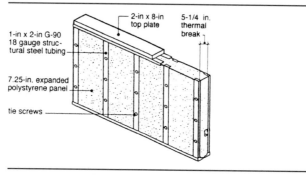

Figure 12-148 The Thermotech wall is one of a number of alternatives that use a thermal break to reduce energy loss. (*Architectural Record,* September 1995)

Table 12-12 Expected *R* values for steel stud wall sections.

(A) Suggested Insulation Levels for Steel Frame Construction (Refer to Thermal Zone Map of Fig. 12-147 for Steel Framing)					
Zone	Annual Degree-Day Demand	Cavity Insulation	Wall Construction	Exterior Insulated Sheathing	Effective Wall *R* Value
Blue	> 7000	R-19 R-11	2 × 6 2 × 4	R-10 R-13	20 21
Green	5000–7000	R-19 R-11	2 × 6 2 × 4	R-5 R-7	15 15
Yellow	4000–5000	R-19 R-11	2 × 6 2 × 4	R-2.5 R-5.0	12.5 13.5
Red	< 4000	R-19 R-11	2 × 6 2 × 4	0 0	10 8.5

(B) Equivalent Insulation Systems for Steel Frame Construction								
Wall *R* value	13		13.5		15		20	
Framing	Wood	Steel	Wood	Steel	Wood	Steel	Wood	Steel
ASHRAE ref.	RW55		RW45B		RW13A		RW58	
Cavity insulation R value	12	11	11	13	11	11	12	15
Exterior sheathing	½″ plywood + alum.siding	1″ XPS	⅝″ cedar + 4″ brick	1″ XPS	1″ polyiso. + stucco	1½″ XPS	1½″ polyiso. + alum. siding	2″ XPS

NOTES:
1. Steel studs are 3⅝″ × 1⅝″ @ 24″.
2. Wood studs are nominal 2″ × 4″ @ 16″.
3. Interior sheathing is ½″ gypsum wallboard for both.
4. Information on the wood system is from the *ASHRAE Handbook.*

Recommendations to address energy concerns include using foam sheathing (*R*-5 to *R*-10) over steel frame walls and filling wall cavities with *R*-11 batts that are a full 16″ or 24″ wide. Spacing studs 24″ o.c. rather than 16″ o.c. can reduce the amount of thermal loss by about 10%. The use of steel from the foundation wall up requires that exterior foam insulation extends below grade. The bottoms of cantilevers and steel floor joists must be covered when they are over unheated garages, attic spaces, and vented crawl spaces. If you use foam sheathing, you must install steel bracing in the frame; also, you should fill headers and corners with fiberglass before assembling them. Attaching siding can be a problem because it cannot be attached to the foam sheathing directly. You may use a layer of plywood or OSB underneath the foam and nail it through to the sheathing. Some steel framers reduce heat loss by using wood trusses and wood sills instead of steel.

Some companies are designing proprietary framing systems using steel in more efficient configurations. For example, Techbuilt Systems, Inc., of Cleveland, Ohio, has installed Thermotech 21® in over two hundred structures. Only four bolts penetrate the thermal break that separates 1″ × 2″ steel tubes at the exterior and interior of the wall (Fig. 12–148).

12.9

INSULATING CONCRETE FORM (ICF) SYSTEMS

Stay-in-place, insulating concrete forms (ICFs) made from expanded or extruded polystyrene foam have become a much-discussed alternative to wood framing. ICFs are hollow blocks, panels, or planks that are made of rigid foam and that are erected and filled with concrete to form the structure and insulation of exterior walls. ICFs are cost competitive with frame construction and deliver a product that is more energy efficient, comfortable, durable, and resistant to natural elements. Virtually any house design that can be built with wood or steel frame can be built with ICFs, although an ICF wall has greater thickness, greater weight, and a higher *R* value.

Table 12–13 lists the ICFs available in the United States as of the summer of 1995. They all work on the same principle—filling forms with concrete—but there are three major differences:

1. the size of the form unit and the ways they connect to one another,
2. the shape of the cavities into which the concrete goes, and
3. whether the formwork has surfaces to which materials can be fastened with a screw or nail.

This table classifies the systems according to size and connection method: (1) panel, (2) plank, and (3) block (Fig. 12–149). It also classifies the systems by the type of concrete wall that is created: flat, grid, and post-and-beam (Fig. 12–150).

Panel systems are the largest units, as big as 4′ × 8′. Plank systems consist of long (usually 8′), narrow (8″–12″) planks of foam held a constant distance apart by steel or plastic ties. Block systems include units ranging from standard concrete block size, 8″ × 16″, to a much larger 16″ high by 4′ long.

Each system has one of three distinct cavity shapes: flat, grid, or post-and-beam (Figs. 12–151 through 12–154). These systems produce different shapes of concrete beneath the foam. Flat cavities produce a concrete wall of constant thickness. Grid cavities are "wavy," both horizontally and vertically. Post-and-beam cavities are filled with concrete only every few feet horizontally and vertically. Many of the systems have a fastening surface, which is of material other than the foam, embedded in the units that can be nailed into, similarly to a stud or a furring strip (Fig. 12–155).

Detailed information on ICF systems and construction procedures is contained in the text *Insulating Concrete Forms: A Construction Manual* by Pieter A. VanderWert and W. Keith Munsell. This manual, published by McGraw-Hill, was sponsored by the Portland Cement Association and is available through the association or the publisher.

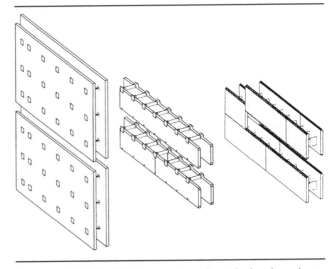

Figure 12–149 ICF formwork made with the three basic units: panel (left), plank (center), and block (right).

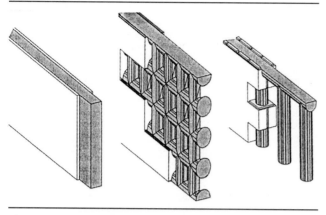

Figure 12–150 Cutaway diagrams of ICF walls with the three basic cavity shapes: flat, grid, and post-and-beam.

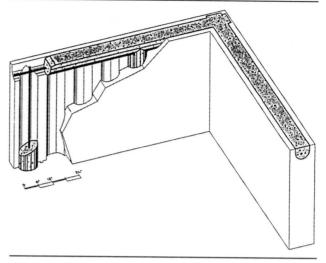

Figure 12–153 Cutaway diagram of a post-and-beam panel wall.

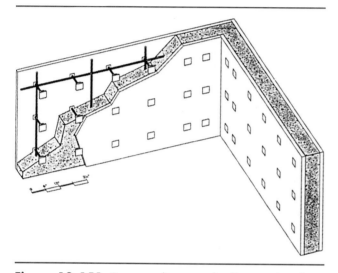

Figure 12–151 Cutaway diagram of a flat panel wall.

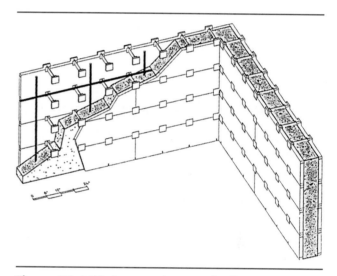

Figure 12–154 Cutaway diagram of a flat plank wall.

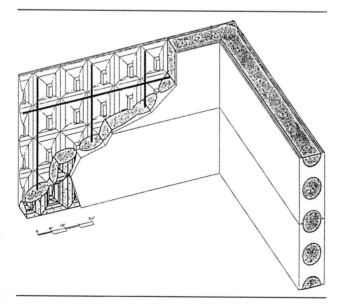

Figure 12–152 Cutaway diagram of a grid panel wall.

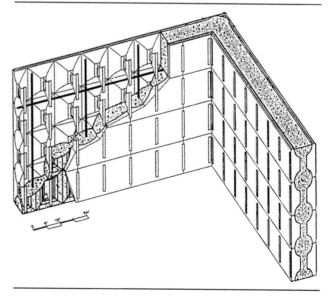

Figure 12–155 Cutaway diagram of a grid block wall with fastening surfaces.

Table 12–13 Available ICF systems.[a]

System	Dimensions[b] (Width × Height × Length)	Fastening Surface	Notes
1. Panel systems:			
Flat panel systems			
R-FORMS	8″ × 4′ × 8′	Ends of plastic ties	Assembled in the field; different lengths of ties available to form different panel widths.
Styroform	10″ × 2′ × 8′	Ends of plastic ties	Shipped flat and folded out in the field; can be purchased in larger/smaller heights and lengths.
Grid panel systems			
ENER-GRID	10″ × 1′3″ × 10′	None	Other dimensions also available; units made of foam/cement mixture.
RASTRA	10″ × 1′3″ × 10′	None	Other dimensions also available; units made of foam/cement mixture.
Post-and-beam panel systems			
Amhome	9³⁄₈ × 4′ × 8′	Wooden strips	Assembled by the contractor from foam sheet. Includes provisions to mount wooden furring strips into the foam as a fastening surface.
2. Plank systems			
Flat plank systems			
Diamond Snap-Form	1′ × 1′ × 8′	Ends of plastic ties	
Lite-Form	1′ × 8″ × 8′	Ends of plastic ties	
Polycrete	11″ × 1′ × 8′	Plastic strips	
QUAD-LOCK	8″ × 1′ × 4′	Ends of plastic ties	
3. Block systems			
Flat block systems			
AAB	11.5″ × 16³⁄₄″ × 4′	Ends of plastic ties	
Fold-Form	1′ × 1′ × 4′	Ends of plastic ties	Shipped flat and folded out in the field.
GREENBLOCK	10″ × 10″ × 3′4″	Ends of plastic ties	
SmartBlock Variable Width Form	10″ × 10″ × 3′4″	Ends of plastic ties	Ties inserted by the contractor; different-length ties available to form different block widths.
Grid block systems with fastening surfaces			
I.C.E. Block	9¼″ × 1′4″ × 4′	Ends of steel ties	
Polysteel	9¼″ × 1′4″ × 4′	Ends of steel ties	
REWARD	9¼″ × 1′4″ × 4′	Ends of plastic ties	
Therm-O-Wall	9¼″ × 1′4″ × 4′	Ends of plastic ties	
Grid block systems without fastening surfaces			
Reddi-Form	9¼″ × 1′ × 4′	Optional	Plastic fastening surface strips available.
SmartBlock Standard Form	10″ × 10″ × 3′4″	None	
Post-and-beam block systems			
ENERGYLOCK	8″ × 8″ × 2′8″	None	
Featherlite	8″ × 8″ × 1′4″	•None	
KEEVA	8″ × 1′ × 4′	None	

[a]All systems are listed by brand name.
[b]"Width" is the distance between the inside and outside surfaces of the foam of the unit. The thickness of the concrete inside will be less, and the thickness of the completed wall with finishes added will be greater.

AUTOCLAVED, AERATED CONCRETE BLOCK

A solid but lightweight concrete block that has an *R* value of about 30 is now being distributed in the United States by Hebel Southeast of Atlanta, Georgia.[3] The Hebel Group of Germany, the parent company, has been producing the system for sale outside the United States, for more than fifty years. (Fig. 12–156).

The block, called *precast, autoclaved, aerated concrete*, is 25″ long and 10″ tall, and it comes in three widths: 4″, 8″, and 10″. The blocks, which can be cut with a bandsaw, are made of sand, lime, cement, and water. The wider blocks can be used for exterior load-bearing walls, and the narrower 4″ version is designed for partition walls.

[3] The company distributing the material is Hebel Southeast, 3340 Peachtree Road NE, Suite 150, Atlanta, GA 30326 (1–404–812–7400).

Figure 12–156 Workman lifts a lightweight Hebel unit into place.

The blocks have a high *R* value, exceeding 30 for the 8″ width, because they contain a lot of air. According to Michael Sweeney, Hebel's marketing manager, the blocks are about 80% air by volume. The air is contained in what Hebel calls *noncapillary pockets*, or *bubbles*. The air comes from an expanding agent added to the raw materials while they are in liquid form. The mix is then poured into a mold and allowed to rise. The blocks are cut to dimension and are then steam cured under pressure in a huge oven called an *autoclave*.

The company says that the 8″ blocks have a 4-h fire rating and the 4″ blocks have a 2-h rating. The blocks are available in three strength categories, and they have an average compressive strength ranging from 363 psi to 1090 psi. They are commonly covered with stucco but can be left unfinished. The Southern Building Code Congress has evaluated the material and has found it suitable for use in building load-bearing exterior walls.

12.11

STRAW BALE CONSTRUCTION

Baled straw is an inexpensive and environmentally sound material for the construction of low-scale buildings. Straw bale construction dates back 400 years in Germany, but it was first used for housing in the United States only a century ago in the Sand Hills of southwestern Nebraska.

The density of straw bales makes them a highly energy efficient building material. Straw buildings are less expensive to cool in the summer and heat in the winter, and their walls tend to absorb excess sound. Bale sizes range from 18″ deep by 35″ to 40″ long with an *R* value of 43, to 23″ deep by 43″ to 47″ long with an *R* value of 55.

Straw bale construction can be of two types: (1) load-bearing, in which the straw bales carry roof, wind, and other loads (Figs. 12–157 and 12–158); and (2) non-load-bearing, in which a post-and-beam structural system supports building loads with the straw bales acting as infill (Fig. 12–159).

A concrete foundation wall is poured, and reinforcing steel rods are embedded vertically around the perimeter. The first row of straw bales is impaled on the reinforcing steel. Subsequent rows are stacked like large bricks in a running bond pattern. Interior and exterior walls are typically finished with cement or adobe plaster.

The roof of a load-bearing straw bale building is connected to a wood roof plate or a concrete bond beam that rests on the stacked straw bales. The roof plate is also tied to the building foundation with threaded rods driven through the bales. The rods are connected at the bottom to anchors embedded in the foundation, and they are bolted at the top to the roof plate or the bond beam.

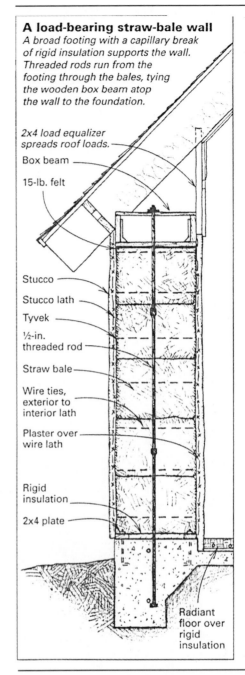

A load-bearing straw-bale wall
A broad footing with a capillary break of rigid insulation supports the wall. Threaded rods run from the footing through the bales, tying the wooden box beam atop the wall to the foundation.

2x4 load equalizer spreads roof loads.

Box beam

15-lb. felt

Stucco

Stucco lath

Tyvek

½-in. threaded rod

Straw bale

Wire ties, exterior to interior lath

Plaster over wire lath

Rigid insulation

2x4 plate

Radiant floor over rigid insulation

Figure 12–157 Example of load-bearing straw bale construction.

Foundations for post-and-beam straw bale systems are similar to those in load-bearing construction except that the structural frame and the roof are assembled before the straw bale walls are raised. The bales are attached to

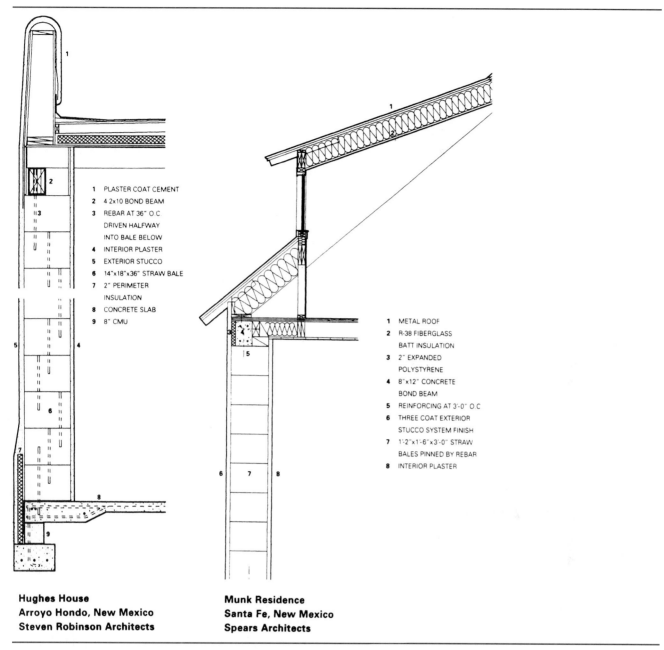

1. PLASTER COAT CEMENT
2. 4 2x10 BOND BEAM
3. REBAR AT 36" O.C
   DRIVEN HALFWAY
   INTO BALE BELOW
4. INTERIOR PLASTER
5. EXTERIOR STUCCO
6. 14"x18"x36" STRAW BALE
7. 2" PERIMETER
   INSULATION
8. CONCRETE SLAB
9. 8" CMU

1. METAL ROOF
2. R-38 FIBERGLASS
   BATT INSULATION
3. 2" EXPANDED
   POLYSTYRENE
4. 8"x12" CONCRETE
   BOND BEAM
5. REINFORCING AT 3'-0" O.C
6. THREE COAT EXTERIOR
   STUCCO SYSTEM FINISH
7. 1'-2"x1'-6"x3'-0" STRAW
   BALES PINNED BY REBAR
8. INTERIOR PLASTER

Hughes House
Arroyo Hondo, New Mexico
Steven Robinson Architects

Munk Residence
Santa Fe, New Mexico
Spears Architects

Figure 12–158 Example of load-bearing, straw bale construction.

vertical posts at every course with strips of expanded metal lath which are nailed to the post. Metal lath strips attach the bales to the horizontal roof plate or bond beam.

Building code officials generally prefer post-and-beam straw bale houses over load-bearing structures, although both types have been approved by Arizona and California building departments under the Uniform Building Code provisions for experimental structures.

The greatest danger to straw bale construction is moisture. If bales become wet and are not permitted to dry out, they will rot. It is important to start with dry bales and to keep them dry during construction. A vapor barrier is installed on the foundation below the first course of bales to keep water from wicking up into the straw. Vapor barriers are usually not applied to wall surfaces because they do not allow the straw to breathe.

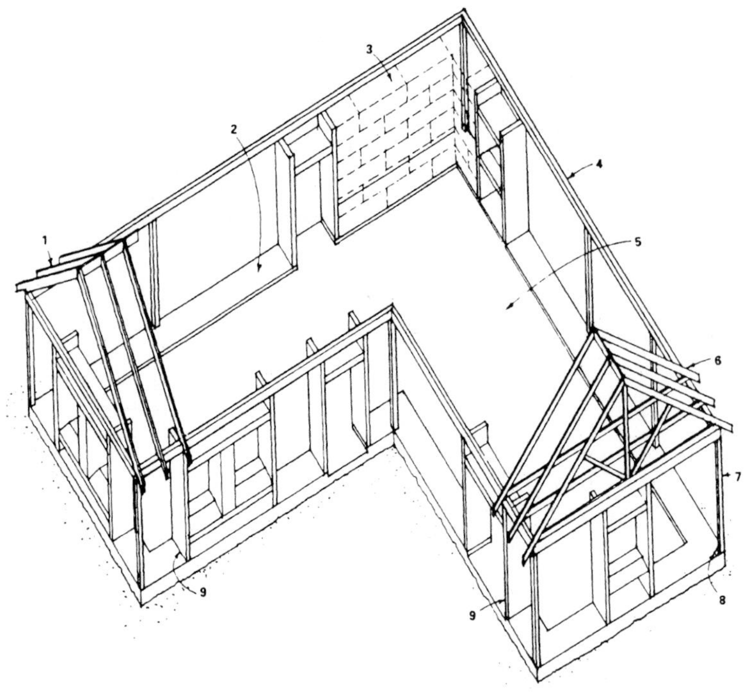

1 2x10 ROOF JOISTS AT 24" O.C.
  WITH NONBEARING RIDGE BOARD
  AND COLLAR TIES AS REQUIRED
2 4" CONCRETE LEDGE AT PERIMETER
  WITH WATERPROOFING AS REQUIRED
3 STRAW BALE INFILL BETWEEN DOOR
  AND WINDOW ASSEMBLIES
4 PERIMETER BEAM ABOVE TOP BALE
  COURSE, SIZE AS REQUIRED
5 POURED CONCRETE FOUNDATION
  AND FLOOR SLAB
6 PRE-ENGINEERED WOOD
  TRUSSES AT 24" O.C.
7 4x4 WOOD POST
8 1/8-INCH STEEL
  POST-BASE CONNECTOR
9 WINDOW AND DOOR ASSEMBLIES
  CONSTRUCTED OF 2x4s AND PLYWOOD
  OR PRESSED STRAWBOARD

Tillman House
Tucson, Arizona
Paul Weiner, Architect

Figure 12–159 Example of non-load-bearing, straw bale construction.

12.12

ROOF TYPES
AND ROOF COVERINGS

The roof is a very important part of any building, large or small, and it should therefore receive careful consideration. We might say that a house is no better than the roof that covers it. Like an umbrella, it protects the structure from the elements; in some climates it is subjected to intense heat, almost everywhere to rain and dampness, and in many areas to driving winds. Not only is the roof extremely functional, but its shape and nature play a vital role in the architectural appearance and styling of a house, which is often designated by its roof type. In commercial buildings, however, the roof is often flat and obscured by parapet walls, and it is therefore less important as far as the building's appearance is concerned.

In years past, the snow load of buildings in northern areas required steep roofs. However, with the improvement of roofing materials and more engineering design in buildings, the design of modern roofs has become less restrictive. Because architects and builders now have available a generous selection of roof coverings for any desired architectural treatment, we find successful roofs of all types and slopes in almost every area of the country, regardless of climatic conditions.

All roofs should be built to provide positive drainage. Flat roofs should have a minimum pitch or have tapered rigid insulation to provide a minimum slope to roof drains.

12.12.1 Roof Types

The Flat Roof   The flat roof is one of the simplest and most economical roofs as far as material is concerned (Fig. 12–160). The roof joists become the framing members of both the roof and the interior ceiling. Sizes and spans of joists must be adequate to support both the live load of the roof and the dead loads of the roofing and the finished ceiling. The spacing of wood roof joists is generally 16" o.c. Comparatively little air space is provided in a flat roof; therefore, batt-type or blanket insulation, with the vapor barrier near the interior side, is recommended between roof joists. All of the air cavities between the roof joists must be ventilated. Solid bridging should not be used in flat roofs; the blocking would prevent uniform air circulation.

Overhangs, for reason of weather protection or appearance, vary in width from 1' to 4'. The usual over-

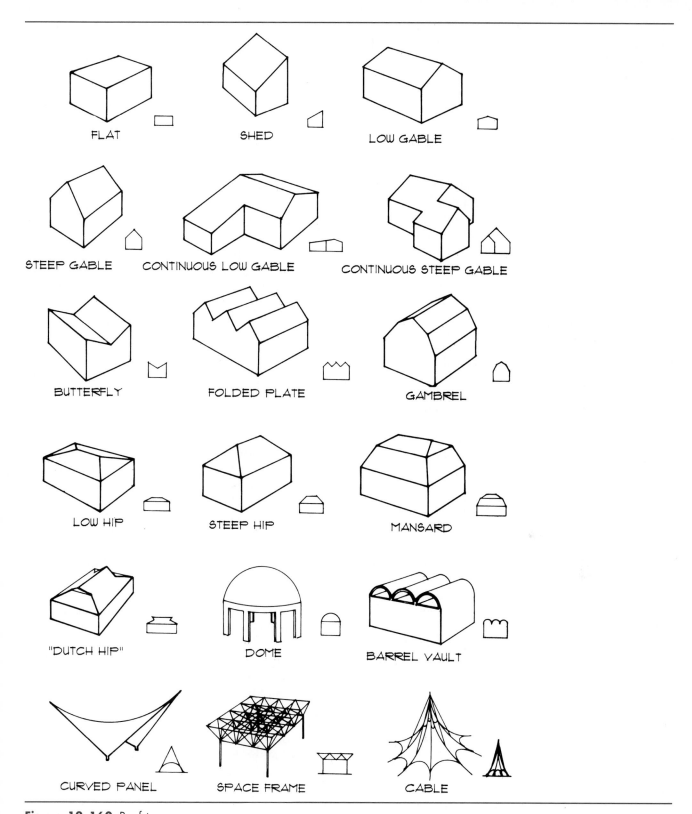

Figure 12–160 Roof types.

hang is 2′, so 4′-0″ wide plywood panels can be conveniently ripped and used for soffits. Overhangs of the flat roof are simply cantilevered over the exterior wall plates; the joists are toenailed to the plates, or metal framing anchors can be used. If a boxed-in overhang is used, a header joist is recommended around the periphery of the roof (Fig. 12–161). The header joist becomes the backing for the finish fascia board, provides blocking for the soffit material, and tends to straighten the roof line. Overhangs from exterior walls that are parallel to the joists must be framed with *lookout joists*, sometimes called *tail joists* (Figs. 12–105 and 12–161). Lookout joists are short and butt into a double-long joist. Generally, the distance from the double-long joist to the wall line is the same as to the overhang. If wide joists are necessary, because of the span, the joist width is often tapered from the exterior wall to the outside end of the joist. This tapering of the joists prevents the fascia board from appearing too wide and heavy. Occasionally, to avoid wide fascias, narrow outlookers are lap nailed to the roof joists. Blocking is required on the top wall plate between each roof joist to provide nailing strips for the soffit and crown molding.

Flat roofs have the advantage of ease and flexibility of framing over odd building shapes and offset exterior walls. Other types of roofs often restrict the general shape of the building. Many architects resort to a flat roof when living spaces within the building do not conform to tra-ditional shapes and relationships. Space planning can be almost unrestricted if a flat roof is used. However, the straight, horizontal roof line of a flat roof often appears hard and monotonous; variety and interest can be gained by combining other roof shapes with the flat roof, or by using several roof levels throughout the building. A watertight, built-up or single-ply membrane roof covering must be used on a flat roof, and metal gravel stops are put around the edges.

The Shed-Type Roof The shed-type roof is framed very similarly to a flat roof, yet it has a definite slope (Fig. 12–160). As a rule, the slopes are made low, unless the shed is used in combination with gable-type roofs. Ceilings can be applied directly to the rafters to provide interest and variety to interior living spaces. The span of a sloping rafter is the horizontal distance between its supports. Shed rafters should be well anchored to the plates of the exterior wall.

The Gable Roof Used for centuries, the gable roof is the traditional shape of most roofs. It consists of two inclined planes that meet at a ridge over the center of the house and slope down over the side walls. The inclined planes form a triangular shape at the ends of the house called a *gable* or *gable end* (Fig. 12–107). Framing of the gable roof is discussed in Section 12.4.1, "Platform Frame Construction." Many different slopes or pitches can be used with the gable roof, depending on the architectural treatment and roofing materials.

Roof pitches must be indicated on drawings, and there are several methods of expressing the slope (refer to Section 14.3).

Details of the framing and architectural treatments of the intersection of the roof and the walls (Fig. 12–162) are shown in *cornice details*. Steep gable roofs allow enough headroom for second-story living space, even though the floor area may not be as large as that of the first floor. Light and ventilation can be provided by dormers with windows, which are framed out from the main roof (Fig. 12–104). In traditional, colonial-type houses, dormers must be symmetrically spaced on the roof; they should have good proportions so that they will appear light and neat. Massive dormers often ruin an otherwise attractively designed colonial-type house.

Low-slope gable roofs are characteristic of the newer, rambling ranch-type house; they usually have wide overhangs. The space below the roof can be used, if at all, for minor storage.

The Hip Roof The hip roof is another conventional-type roof that is popular in many areas of the country (Fig. 12–160). The inclined roof planes slope to all outside walls, and the treatment of the cornice and roof overhang is identical around the entire house. It eliminates gable-end walls but requires more roofing material than

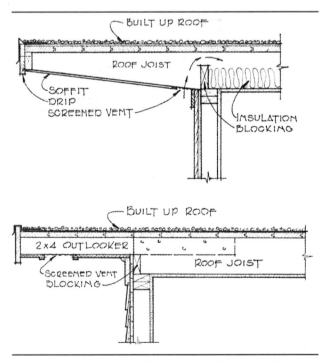

Figure 12–161 Flat roof with narrow fascia details.

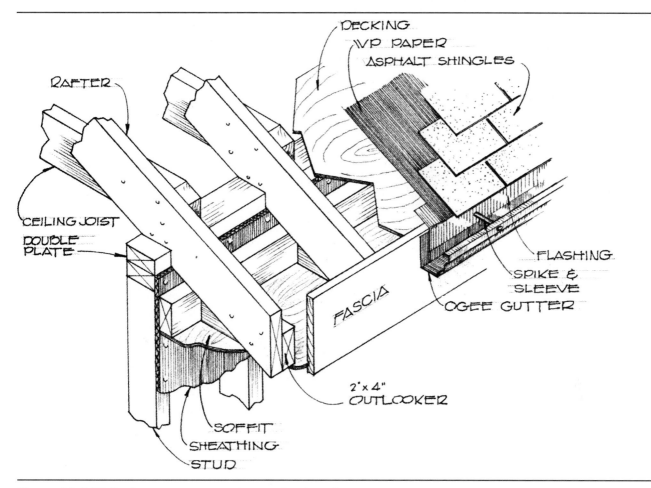

RAFTER

CEILING JOIST
DOUBLE PLATE

DECKING
WP PAPER
ASPHALT SHINGLES

FLASHING
SPIKE & SLEEVE
OGEE GUTTER

FASCIA

2"x 4" OUTLOOKER

SOFFIT
SHEATHING
STUD

Figure 12–162 Roof construction at overhang with a boxed-in eave.

a gable roof of comparative size. Usually, the plan is rectangular, in which case common rafters are used to frame the center portion of the roof. Hip rafters, which carry very little of the load, are used at the exterior corners of the sloping intersections. Valley rafters (Figs. 12–101 and 12–102), on the other hand, carry considerable loads and are used at the interior intersections of the roof planes. Short jack rafters join either the hip rafter and the wall plate or the valley rafter and the ridge.

More complex framing is required in hip roofs; hip, valley, and jack rafters require compound diagonal cuts at their intersections. For simple hip roofs, a roof layout is not necessary on the working drawings, but more complex roofs with ridges of different heights and other offsets would definitely require a roof layout drawing.

Trussed rafters are available in some areas for framing standard span and pitch hip roofs. Regular trusses are used in the center section along the ridge, and modified trusses are made to conform to the end slope of the hip.

Like other roofs, hip roofs must be properly ventilated; because the roof covering is continuous over the higher areas of the roof, small metal vents or louvers are often employed on the rear side near the ridge to allow

warm air to escape. Air inlets are provided by vents under the eaves.

Steep hip roofs often allow additional living space under the roof, similar to steep gable roofs. To be consistent, dormers on hip roofs usually have hip roofs as well.

The Dutch Hip Roof A variation of the hip roof, the Dutch hip roof is framed similarly to the gable roof throughout the center section to the ends of the ridge, where double rafters are used (Fig. 12–160). The end hips are attached to the double rafters, are started below a louver, and extend over the corners of the building. The louvers at each end of the ridge allow more positive ventilation under the roof.

Because a hip roof can become more complex to frame than other roofs, the following method is offered *to lay out a working drawing for a hip roof*:

Using Fig. 12–163 as an example, draw the outline of the building, including all ells and offsets. Draw the main roof formed by the rectangle *ABCD*; the 45° hip lines are drawn from the main corners and corners of the ells. The intersections of the 45° hip lines will locate the ridge lines.

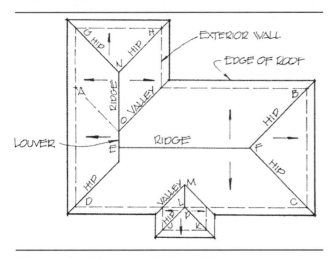

Figure 12-163 Layout of a hip roof.

Draw the 45° valley lines connecting all roof intersections. A Dutch hip is indicated at point *E*; the ridge end can be indicated at any location along the diagonal hip lines, depending on the size of the louver used on the end elevation of the building.

Uniform overhangs can be drawn around the exterior wall lines, and hip lines can be extended to the corners of the overhangs.

This layout can be used regardless of the pitch that has been selected—low or steep slope. Hip roofs are more pleasing if the same pitch has been used on all planes of the roof. Rafters are drawn using either 16″ or 24″ o.c. spacing.

The Gambrel Roof

The gambrel roof is typical of Dutch colonial houses, and it is also found on many farm buildings. Each slope has a break or change of pitch; the lower part is always steeper than the upper (Fig. 12–103). From a practical standpoint, this roof allows more headroom compared with a gable roof with the same ridge height. More labor is involved in framing a gambrel roof; as a result, it is seldom used in modern construction except on strictly colonial houses.

The Butterfly Roof

The butterfly roof is similar to an inverted gable roof (Fig. 12–160). Caution should be exercised in the selection of a butterfly roof; roofers indicate that the intersection of the two inclined planes produces severe strain on the roof covering. Drainage of the valley is difficult, yet this drainage can be accomplished with the use of tapered cant strips under the roofing, which act very much as a saddle that is used to spread water away from a chimney on a sloping roof. A bearing partition should be used below the valley for a stable support for the rafters.

The Barrel-Vault Roof

This roof employs precured plywood panels or regular plywood attached to curved ribs (Fig. 12–160). The panels are custom designed and are made by authorized plywood manufacturers. With relatively thin cross sections, the curved, stressed-skin panels permit spanning long distances because of their arching action. Tie rods are usually required to counteract the thrust action when thinner panels are used. Insulation can be incorporated into the panels, and the underside of the panels can be used as the finished ceiling. Supported beams are uniformly spaced below the intersections of the vaults.

The Folded-Plate Roof

This roof creates a roof line that is truly contemporary in profile (Fig. 12–160). Its form is the result of engineered functional design. Reinforced-concrete, folded-plate roofs have been in use for years. Corrugated metal, on a small scale, obtains its stiffness similarly because of its folded characteristic. However, it has been largely through the efforts of the American Plywood Association that the shape has been adapted to wood construction, mainly because of the high shear strength and rigidity of plywood.

The construction of a plywood folded-plate roof differs from conventional pitched roof construction in that the roof sheathing and framing are designed to act together as a large inverted V-shaped beam to span from end to end of the building. A multiple-bay roof is made of several such beams connected together side by side.

The inclined planes of the roof transfer their vertical loads to adjoining ridge and valley intersections, very much as crossbridging in conventional floor construction transfers a concentrated load to surrounding floor joists. Loads on the inclined planes are carried in shear by the plywood diaphragm action to the ends of the building and by horizontal thrust action to the sides of the building.

Vertical supports are necessary at the ends of the building under the valleys and along the sides of the building. Either a beam or a bearing wall can be used along the sides. Horizontal thrust that is transferred to the sides of the building must be counteracted by horizontal ties of steel or wood.

The analysis of a multiple-bay folded plate differs slightly in that at the interior valleys, additional support is gained by each plane supporting the other; for this reason, walls or vertical beams are not required, provided that the valley chords are adequately connected. Omission of supports below valleys, of course, increases the shear that must be resisted by the interior sloping planes. Horizontal chord stresses are greater then as well.

The following lists point out the advantages and disadvantages of the folded-plate roof as compared with other methods of roof construction:

Advantages

1. Trusses or other members that span from valley to valley are eliminated, thereby resulting in clear, uncluttered interiors.
2. Inasmuch as the plywood sheathing constitutes part of the structural unit, long spans are possible with small framing units.
3. A plywood ceiling can be used as a finish as well as for additional resistance to shear.
4. Assembly can be with nails only; no gluing is required, although it will improve rigidity.
5. Either conventional site construction or prefabrication can be employed. Components of the folded plate can be stressed-skin panels, or one or more full-length plates can be prefabricated.
6. Interesting form can be achieved economically for buildings of all sizes, from residences to large industrial structures.

Disadvantages

1. Drainage requires special attention on multiple-bay structures.
2. A well-designed folded-plate roof necessitates more roofing material. At least a 5:12 slope is recommended, since shear in the roof diaphragm diminishes with increase in slope.
3. Folded plates do not readily lend themselves to incorporation of a flat ceiling.

12.12.2 Roof Coverings

Many roofing materials are available for small buildings, and each has its advantages and limitations. Consequently, selection of the proper roofing material is of paramount importance. Thus, in selection of roofing materials, the following points should be considered:

1. slope of the roof, if any;
2. quality or permanence required of the roof;
3. inherent architectural features of the roofing in relation to the rest of the structure; and
4. cost of the roofing material.

In areas subject to wind-driven snow or to the buildup of roof ice, which forms ice dams, the underlayment application in the eave area is modified to prevent ice dams from forcing water under the roofing, thereby damaging ceilings, walls, and insulation (Fig. 12–164). Two layers of underlayment consisting of nonperforated Type 15 felt should be cemented together with asphalt cement from the edge of the roof up to a point that is at least 24″ inside the interior wall line of the building (Fig. 12–165). For wood shingle or wood shake roofs, this double underlayment should extend 36″ inside the interior wall line of the building.

The environment within the envelope of the building provides adequate warmth to prevent ice dams from forming above the heated space.

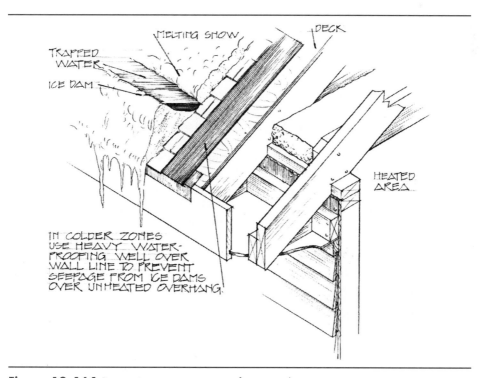

Figure 12–164 Preventing water seepage from ice dams.

Built-Up Roofing A flat roof or a roof of very little slope must, of necessity, be covered with a completely watertight material. The majority of flat roofs are now covered with built-up roofing (see Fig. 12–166), which, if properly installed, is durable. On a wood deck, the first heavy layer of felt is nailed to the decking with galvanized roofing nails. Each succeeding layer of felt is mopped with either hot asphalt or hot coal-tar pitch. The top layer is then well mopped and covered with a thin layer of pea gravel, fine slag, or marble chips. This mineral surface protects the roof to some degree from the elements, and the lighter colors reflect the sun's rays from the building. For every 100 sq ft of roof, the use of 400 lb of gravel or 300 lb of slag is recommended.

Over smooth concrete decks, fewer piles are required, and the first layer of felt is mopped directly to the slab. Asphalt pitch can be used on steeper slopes than can coal-tar pitch, but the limiting slope for a built-up roof is 2:12, if gravel is expected to stay on the roof. Even on a 2:12 slope, gravel will lose its grip during heavy rains. Metal gravel stops, preferably copper, are fastened to the wood deck surrounding the entire roof. Proper use of base and cap flashing is shown in Figs. 12–167 and 12–168.

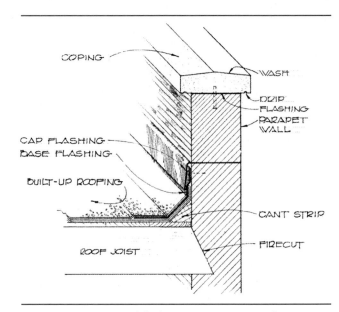

Figure 12-167 Roof flashing at a parapet wall.

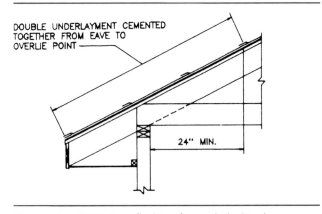

Figure 12-165 Eave flashing for asphalt shingles.

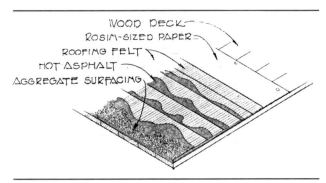

Figure 12-166 Built-up roof on wood deck (level to 2:12 pitch).

On flat roofs as well as pitched roofs, substantial and rigid roof decks play an important part in the success of the roof covering. Decks should be well nailed, clean, and dry before coverings are applied.

In roofing terminology, a *square* is an area of 100 sq ft; thus, a square of roofing material is the amount necessary to cover 100 sq ft. Requirements for built-up roof construction are summarized in Table D–6.

Membrane Roofing In recent years, major advances have been made in the membrane roofing industry. High-strength, reinforced, synthetic sheets suitable for single-ply roofing systems have been developed. Although many membrane systems are installed over concrete roof decks or rigid insulation on metal decks, they can also be applied over wood decks. Most systems use mechanical fasteners to anchor the sheets to the roof deck. Special flashings are installed at roof penetration. Seams are usually heat welded. Some systems require a stone ballast to stabilize the membrane during high winds. The majority of such systems are easily repaired, if damaged, and have a warranty.

Shingle Roofing When a building has a sloping roof, more consideration must be given to the appearance of the roofing material. Shingles or multiple-unit type roofing materials are designed for sloping roofs; the incline allows water to drain off quickly without readily challenging the watertightness of the roof. The small, individual units of such roofing are also not subjected to stresses or shock from severe temperature changes; each piece can expand and contract individually without producing a noticeable effect, as in a built-up roof.

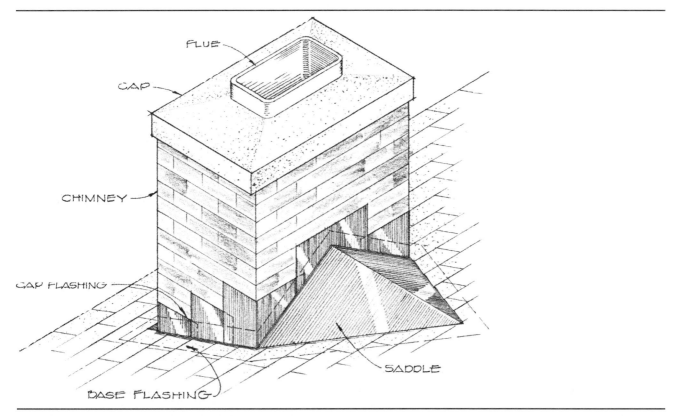

Figure 12-168 Chimney flashing on a sloping roof.

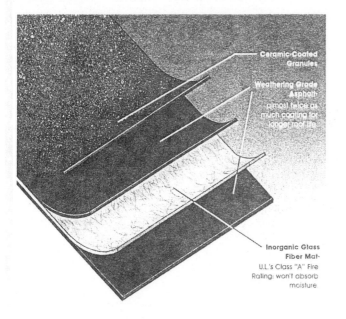

OWENS-CORNING'S FIBERGLAS® SHINGLE

Ceramic-Coated Granules

Weathering Grade Asphalt·
almost twice as much coating for longer roof life

Inorganic Glass Fiber Mat·
U.L.'s Class "A" Fire Rating: won't absorb moisture

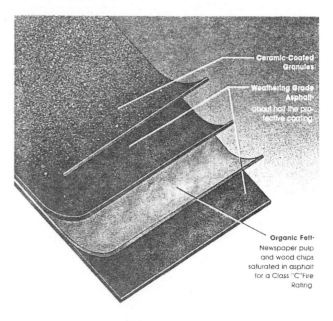

THE ORGANIC SHINGLE

Ceramic-Coated Granules

Weathering Grade Asphalt·
about half the protective coating

Organic Felt·
Newspaper pulp and wood chips saturated in asphalt for a Class "C" Fire Rating

Figure 12-169 Fiberglass and organic shingles. (Owens-Corning Product Information brochure)

The degree of resistance to water penetration of shingle materials varies directly with the slope of the roof and indirectly with the amount of exposure of each unit to the weather. Usually, steeper slopes will allow more exposure of the units than flatter slopes because the water drains off more quickly. Strong winds tend to drive moisture under roof units unless some provision is made for sealing the edges or holding them down. This tendency is especially noticeable on lower slopes. Exposure to the weather is specified in inches. Some roofing materials are exposed 4″ between laps; other materials, for example, can be allowed 10″ exposure. However, manufacturers' recommendations should be followed in application of all roofing materials.

Currently, almost all shingles are based on either organic or glass fiber felt (Fig. 12–169). Each has its advantages. Glass fiber shingles are more fire resistant than organic shingles. They may lower insurance costs and may be required by code in certain locales. They are also more mildew resistant, are less affected by moisture absorption, and weigh less than organic shingles.

On the other hand, glass fiber shingles have poorer wind resistance and lower tear strength than organic shingles, and they are therefore more likely to crack in cold weather. They also show underlayment irregularities more than organic shingles. Requirements for asphalt shingle construction are summarized in Table D–1.

Sloping roofs in light construction are generally covered with one of the following roofing materials:

1. Asphalt shingles
2. Fiberglass shingles
3. Wood shingles
4. Wood shakes
5. Slate shingles
6. Clay tile units
7. Cement tile units
8. Metal shingles or panels

Asphalt Shingles Asphalt-saturated felt, coated with various-colored mineral granules, is manufactured into strip shingles of different shapes and sizes. The most popular shape is the square-butt strip shingle, 12″ wide and 36″ long, with slotted butts to represent individual shingles at the exposure. Strips with hexagonal exposures are also made. The amount of 4″ or 5″ is the usual exposure to the weather, and weights per square vary according to the quality of the shingles. Quality shingles should weigh at least 235 lb/square.

The shingles are attached to the wood deck with 1¼″ galvanized roofing nails; four nails are needed for each strip. Each nail is driven about ½″ above each slot on the square-butt shingles. Two layers of shingles are then fastened with each nail, and the nailhead is just covered with the succeeding course. A double course is fastened at the eave. Occasionally, a starter strip of 53-lb roll roofing is used under the double layer of strip shingles at the eave.

Along the gable edge of the roof, a strip of wood bevel siding is often nailed to the deck with the thin edge in, to guide water from the edge as it drains along the incline. Metal edge flashing can also be used at gable edges. Roof decks should be first covered with 15-lb asphalt felt, lapped 2″ horizontally and 4″ vertically, before asphalt shingles are applied.

Asphalt strip shingles can be used on low-slope roofs if self-sealing shingles are used. This type has asphalt adhesive under the exposed edges, which seals the edges against driving rains and capillary action. Often, starter strips near the eaves are full mopped with adhesive, rather than nailed, to prevent this moisture penetration.

To cover ridges and hips, strip shingles are cut into smaller pieces and are nailed over the ridges in a lapped fashion called a *Boston lap* (Fig. 12–170). Intersections of roofs and walls must be flashed with durable metal flashing. Valleys are flashed with either 28-gauge (minimum), galvanized, corrosion-resistant sheet metal or woven asphalt shingles. Both require an underlayment of Type 15 felt. Metal shall extend at least 8″ each way from the center line of the valley, with a minimum section overlap of 4″. The underlayment below a woven shingle valley must extend 18″ each way from the center line of the valley. A woven valley performs well when installed by a skilled worker. However, with some of the premium-grade laminated shingles, the adjacent shingles do not lie flat, so woven valleys should not be used. In an open valley, the shingles must be cemented, and exposed corners must be clipped, to avoid lateral pick-up of water that runs down-slope (Figs. 12–171 and 12–172).

Although they burn readily when ignited, asphalt shingles are more resistant to combustion than wood shingles, and many local codes that prohibit wood shingle roofs will allow the use of asphalt shingle materials. Shingles are available in a variety of patterns (Fig. 12–173).

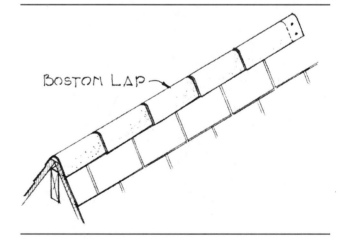

Figure 12–170 Covering a ridge with fiberglass shingles.

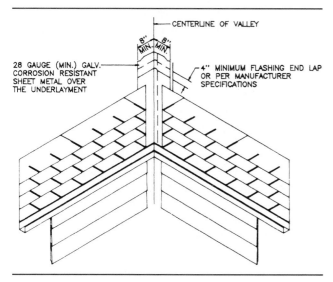

Figure 12-171 Flashing for an open roof valley.

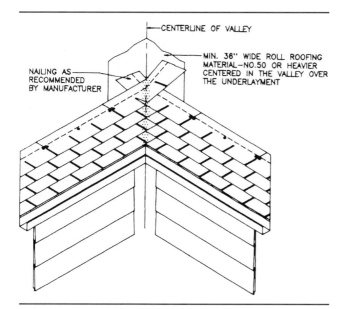

Figure 12-172 Flashing for a woven valley.

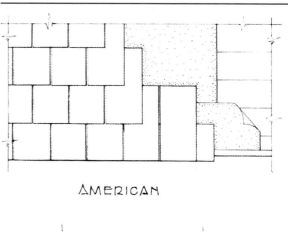

AMERICAN

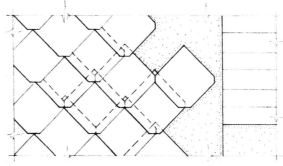

HEXAGONAL

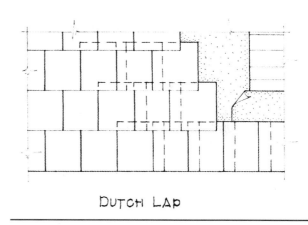

DUTCH LAP

Figure 12-173 Shingles are available in a variety of patterns.

Fiberglass Shingles   A variation of the asphalt shingle is known as the fiberglass roof shingle. It has better fire- and weathering-resistance qualities than the conventional asphalt shingle because of the inorganic fiberglass-mat base. With these additional properties, it is rapidly replacing the organic-based asphalt shingle in the marketplace. Those with self-sealing adhesive strips below each shingle can be used on low-slope (2:12 minimum) roofs if additional underlayment is provided. The self-sealing adhesive is activated by the heat of the sun to seal down the shingle tabs and help to prevent wind damage.

The worst problem with fiberglass shingles is thermal splitting. Fully adhered, self-sealing shingles with a low tear strength often split when cold shrinks them. Some splits are horizontal, between the line of shingle fasteners and the fully adhered bottom edge of the shingle, and some are vertical, usually directly over abutting shingles or slots. Organic-based shingles have about twice the tear strength of glass-fiber-based shingles and rarely, if ever, split.

Wood Shingles Wood shingles have been used for many years in residential construction. Because of their inherent combustibility, standard wood shingles or shakes are prohibited by many local fire codes. Recently, a chemical treatment process has been developed that makes it possible to obtain a wood shingle or shake with a Class B or Class C fire rating. In other than restricted areas, the standard wood shingle and shake is being used; they make handsome, durable roofs when properly installed. Wood weathers to a soft, mellow color after exposure to the elements and, if properly installed, can be expected to last 25 years.

Wood shingles are made chiefly of western red cedar, redwood, or cypress, all highly decay-resistant woods. Wood shingles are manufactured in 24″, 18″, and 16″ lengths and are graded into three categories. Table 12–14 describes each grade, and Figure 12–174 shows an example label. Each bundle of shingles should have a label specifying the grade in accordance with the rules of the Red Cedar Shingle and Handsplit Shake Bureau.

Shingles may be applied to roofs with solid or spaced sheathing. Shingles should be laid with a side lap of at least 1½″ in adjacent courses. Each shingle should be laid approximately ¼″ from an adjacent shingle and nailed with two corrosion-resistant nails only, placed not less than ¾″ from each edge and 1″ above the exposure line (Fig. 12–175). To prevent cupping after exposure, the better grades are cut from the logs so that the annular rings of the log are perpendicular to the flat surfaces of the shingles. Butt ends vary in thickness from ½″ to ¾″.

Wood Shakes Wood shakes are manufactured in three different types: (1) handsplit and resawn, (2) taper split, and (3) straight split. The shakes are cut into 15″, 18″, and 24″ lengths. Each bundle should have a label specifying the grade in accordance with the requirements of the Red Cedar Shingle and Handsplit Shake Bureau. Table 12–15 describes each grade, and Figure 12–176 shows an example label.

Wood shakes are split from the log to reveal a natural, irregular, wood-grain surface. Their butts are thicker,

ranging from ⅝″ to 1¼″, which produces strong horizontal course lines that give the roof a rustic quality. Thick, natural butt shingles are popular on ranch-type houses. In dry areas, wood shingles are usually put directly on plywood or tight wood decks. In damp areas, installation of nailing strips, spaced the same distance apart as the shingle exposure, provides ventilation directly below the shingles (Fig. 12–177). Air space below the shingles prolongs the life of the roof.

Shakes may be applied to roofs with solid or spaced sheathing. In areas with wind-driven snow, sheathing should be solid, and the shakes should be applied over an underlayment of not less than Type 15 felt applied single fashion. Shakes should be laid with a side lap of not less than 1½″ between joints in adjacent courses. Spacing between shakes should not be less than ⅜″ or more than ⅝″. Spacing of preservative-treated shakes can range from ¼″ to ⅜″. Shakes are fastened to sheathing with two nails positioned approximately 1″ from each edge and 2″ above the exposure line. The starter course at the eave is doubled. Shakes should be laid with not less than 18″ interlayment of Type 30 felt shingled between each course so that no felt is exposed to the weather below the shake butts and between the shakes.

More course-line texture can be obtained by doubling each course, usually with No. 2 grade shingles below. Careful nailing technique must be used to ensure that joints in preceding and successive courses are staggered or are offset at least 1½″ to allow good drainage and eliminate water penetration into the roof along the edges of the shingles. Shingles are loosely spaced to allow for swelling and shrinking, and zinc-coated or copper three-penny nails should be used; thicker shingles require larger nails.

Valley flashing for both wood shingles and wood shakes must be a minimum No. 28, galvanized, corrosion-resistant metal over an underlayment of at least Type 15 felt. Sections of flashing must overlap a minimum of 4″ and must extend 8″ from the center line each way for shingles and 11″ for shakes (Fig. 12–178). Maximum weather exposures are given in Table D–3, and requirements for wood shingle and shake construction are summarized in Table D–2. Hips and ridges can be covered by several methods, but the alternating Boston lap is the most popular as it retains the shingle texture to the edge of the ridge.

Many variations can be used in laying shingles. Sometimes every fifth or sixth course is doubled to vary the shadow line throughout the roof; sometimes the shingles are laid at random with no set course. Sometimes they are staggered so that alternate shingles project beyond adjoining ones; sometimes the course lines are made wavy to stimulate thatch roofs, with even the rakes and eaves turned over slightly to emphasize the effect of the thatch. Stained shingles are occasionally used to give definite color to the roof; staining also prolongs the life of the shingles. Regardless of the treatment and variations that are possible with wood shingles, the basic water-shedding principles of application must be followed.

Figure 12–174 Example label of wood shingle grade.

Table 12–14 Description of grades of wood shingles.

Grade	Length	Thickness (at Butt)	No. of Courses per Bundle	Bdls. or Cartons per Square	Description
No. 1 BLUE LABEL	16″ (Fivex) 18″ (Perfections) 24″ (Royals)	0.40″ 0.45″ 0.50″	20/20 18/18 13/14	4 bdls. 4 bdls. 4 bdls.	The premium grade of shingles for roofs and side walls. These top-grade shingles are 100% heartwood, 100% clear, and 100% edge grain.
No. 2 RED LABEL	16″ (Fivex) 18″ (Perfections) 24″ (Royals)	0.40″ 0.45″ 0.50″	20/20 18/18 13/14	4 bdls. 4 bdls. 4 bdls.	A good grade for many applications. Not less than 10″ clear on 16″ shingles, 11″ clear on 18″ shingles, and 16″ clear on 24″ shingles. Flat grain and limited sapwood are permitted in this grade.
No. 3 BLACK LABEL	16″ (Fivex) 18″ (Perfections) 24″ (Royals)	0.40″ 0.45″ 0.50″	20/20 18/18 13/14	4 bdls. 4 bdls. 4 bdls.	A utility grade for economy applications and secondary buildings. Not less than 6″ clear on 16″ and 18″ shingles, 10″ clear on 24″ shingles.
No. 4 UNDER-COURSING	16″ (Fivex) 18″ (Perfections)	0.40″ 0.45″	14/14 or 20/20 14/14 or 18/18	2 bdls. 4 bdls. 2 bdls. 4 bdls.	A utility grade for undercoursing on double-coursed side-wall applications or for interior accent walls.
No. 1 or No. 2 REBUTTED-REJOINTED	16″ (Fivex) 18″ (Pefections) 24″ (Royals)	0.40″ 0.45″ 0.50″	33/33 28/28 13/14	1 carton 1 carton 4 bdls.	Same specifications as above for No. 1 and No. 2 grades, but machine-trimmed for parallel edges with butts sawn at right angles. For side-wall application where tightly fitting joints are desired. Also available with smoothly sanded face.

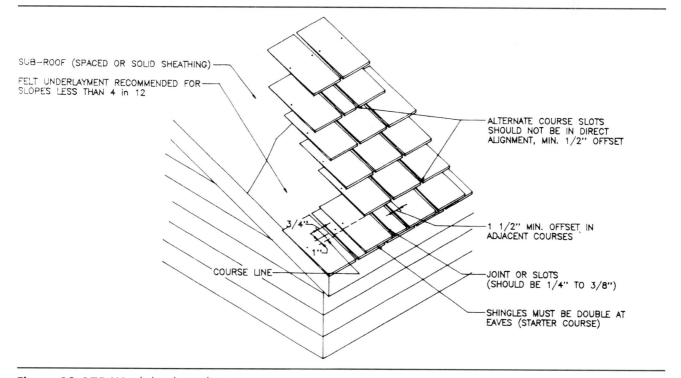

Figure 12–175 Wood shingle application.

Table 12–15 Description of grades of wood shakes.

Grade	Length and Thickness	18" Pack Courses per Bdl.	Bdls. per Square	Description
No. 1 HANDSPLIT AND RESAWN	15" Starter-Finish	9/9	5	These shakes have split faces and sawn backs. Cedar logs are first cut into desired lengths. Blanks or boards of proper thickness are split and are then run diagonally through a bandsaw to produce two tapered shakes from each blank.
	18" × 1/2" Mediums	9/9	5	
	18" × 3/4" Heavies	9/9	5	
	24" × 3/8"	9/9	5	
	24" × 1/2" Mediums	9/9	5	
	24" × 3/8" Heavies	9/9	5	
No. 1 TAPER SAWN	24" × 3/8"	9/9	5	These shakes are sawn both sides.
	18" × 3/8"	9/9	5	
No. 1 TAPER SPLIT	24" × 1/2"	9/9	5	Produced largely by hand, using a sharp-bladed steel froe and a wooden mallet. The natural, shingle-like taper is achieved by reversing the block, end-for-end, with each split.

		20" Pack		
No. 1 STRAIGHT SPLIT	18" × 3/8" True-Edge	14 Straight	4	Produced in the same manner as taper-split shakes, except that because they are split from the same end of the block, the shakes acquire the same thickness throughout.
	18" × 3/8"	19 Straight	5	
	24" × 3/8"	16 Straight	5	

Slate Roofing Natural slate taken from quarry beds and split into thin sheets has provided builders with one of our more aristocratic roofing materials. The marked cleavage of the slate rock imparts a natural, irregular, and pleasing surface to a roof, and the natural color variations of slate from different quarries produce subtle shades ranging from blacks and grays to blues and greens, and even browns and reds. Some slate changes in color after weathering. A well-laid slate roof is very durable and will outlast the life of the building. Slate is used only on the finest homes and more formal commercial buildings, yet its use has diminished in recent years.

Slate is split, cut into sheets, and punched for nails at the quarry. Common commercial sizes of the sheets are 12" × 16" and 14" × 20" and 3/16" or 1/4" thick. Random sheets vary in size and thickness. Each sheet is attached with at least two copper nails; pieces near ridges and hips are often secured with copper wire and roofing cement. Slate roofs are heavy, and therefore the roof-framing members must be more substantial than those for wood or asphalt shingles or other light roofing materials. Because of the weight of the slate, the sheets hug the roof well against driving wind. Damaged slate sheets are difficult to replace after the roof has been completed; ex-perienced slate roofers are required to install properly a slate roof.

Under most conditions, at least a 6:12 pitch is necessary for slate. Roof boards are nailed up tight, over which a heavy roofing felt is applied and secured with roofing nails. A cant strip is used at the eaves to give the starter course the same slant as that of successive courses. Usually, the starter course is laid along the eave with the long dimension parallel to it, and the doubled course is started flush with the edge. Depending on the slope of the roof, slate should be laid with a 3" or 4" head lap. The head lap is the amount that a lower edge of a course laps or covers the top edge of the second course underneath it. To determine the amount of slate exposure, deduct the lap from the total length of the sheet, and divide the difference by 2. For example, the exposure of a 16" long sheet with a 4" head lap is $1/2(16 - 4) = 6$.

Slate can be laid with different exposures on the same roof, however. As with wood shingles, all joints must be staggered to shed water effectively. After random sizes are sorted, it is preferable for the slater to attach the large, thick slates near the eaves and the medium slates near the center, and to leave the smaller, thin pieces for the ridge area.

Figure 12-176 Example labels of wood shake grades.

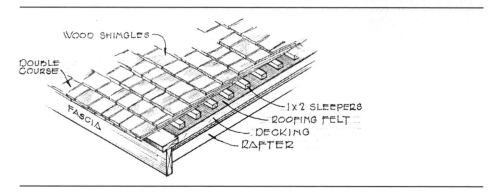

Figure 12-177 Installing wood shakes.

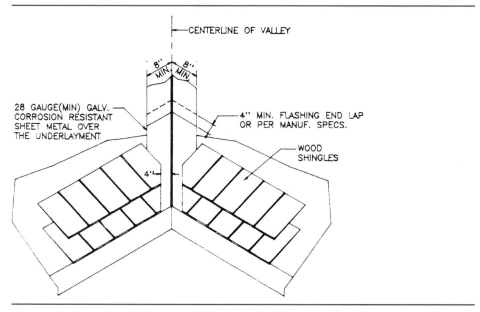

Figure 12-178 Wood shingle valley flashing.

This method produces rather interesting slate textures on a roof.

Hips and ridges are generally finished with the Boston-lap method. Both flashing and roofing cement are used to make the ridges and hips water resistant. Closed valleys, made with trapezoidal pieces of copper flashing placed under each course and lapped at least 3″, are the most popular on slate roofs.

Simulated slate shingles manufactured of fiber-reinforced cement resemble natural slate shingles in appearance. The simulated slate is lighter in weight and stronger than natural slate. The installation procedure for these shingles is similar to that for natural slate. Simulated slate carries a Class A fire rating and can be guaranteed up to 30 years.

Clay Tile Roofing   Many variations of clay roofing tile are manufactured. They are generally very durable, made from different types of clay, and burned in kilns, as are many other types of terra-cotta building products. Roofs of clay tile are characteristic of Mediterranean and Spanish architecture. The early Jesuits of Mexico first brought tile roof construction to this country when they built the old Spanish missions of California and the Southwest.

Traditionally, clay tiles were available in natural, burnt-orange colors. Today a wide selection of natural colors, ranging from light tan to dark brown, is available. Some manufacturers produce a glazed ceramic roofing tile that comes in a spectrum of colors from white to dark blue.

Although there are various patterns or shapes, all clay tile can be divided into two general categories: flat and roll type (Fig. 12–179). The flat tiles vary from simple pieces to pieces with interlocking edges and headers. Roll tiles are available in semicircular shapes, reverse-S shapes, and pan and cover types. Each type requires a tightly sheathed roof covered with heavy roofing felt. On low-pitched roofs and on all ridges and hips, the paper should be doubled; cant strips are needed at the eaves, as in slate roofs. Although some flat tile is fastened by hanging it on wooden cleat strips across the roof, the majority is secured with copper nails through prepunched holes. Semicircular roll tiles require wooden strips running from the eaves to the ridges between the pans.

Clay tile application requires careful planning inasmuch as tile pieces must fit together, and very little latitude is allowed in their placing. Allowance must be made for expansion and contraction of the roofing at intersections of walls and projections. Because clay tile is one of the heavier roofing materials, roof framing must be adequate to support it.

Open valleys are usually used with tile, and copper flashing is placed below (Fig. 12–180). Manufacturers furnish special shapes and pieces for covering hips and ridges, as well as pieces for rake edges and starter strips at the eaves (Figs. 12–181, 12–182, and 12–183). Often, starter pieces at the eaves, as well as ridge and eave pieces, are set in mastic or cement, both of which are available in colors to match the tile. Ridge rolls often require wood

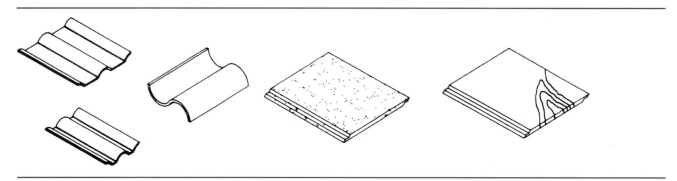

Figure 12–179 Examples of roll and flat tiles.

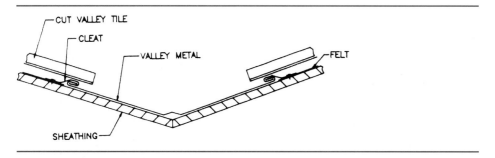

Figure 12–180 Tile roof flashing at a valley.

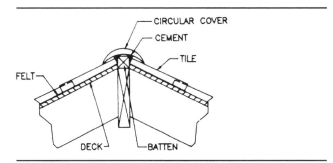

Figure 12–181 Tile roof ridge section.

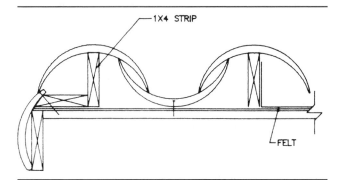

Figure 12–182 Tile roof gable rake section.

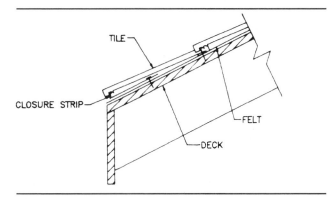

Figure 12–183 Tile roof eave section.

strips along the ridge for fastening the tile. Some tile shapes tend to look bulky on residential roofs; however, if carefully chosen to complement the style of the house, tile will help in creating a picturesque effect.

Cement Tile Roofs   This type of roof has become popular in Florida for residential construction. Precast concrete tiles with wide exposures are attached to the roof and are neatly grouted with mortar. Heavy roofing felt is used below, and white waterproofing, cement-base paint

is applied as a finish over gray cement tiles. Cement tile can be used on low pitches, and the roof becomes resistant to sun and heat.

Cement tiles are now available in a number of colors. Colored cement is used in the manufacture of the tiles, eliminating the need for painting the tiles after installation. Requirements for roofing tile construction are summarized in Tables D–4 and D–5.

Metal Roofing   Occasionally, metal roofing is the most appropriate material to cover residences or light construction buildings. The principal metals used in roofing are copper, terneplate, aluminum, painted steel, and galvanized iron. Because of the cost of material and labor, metal is generally restricted to use in finer residences and more expensive structures. On moderately priced residences, we often see metal employed for small dormers, projections over front entrances, and minor roofs, in combination with other roofing material on the major roof. Metal must be applied by experienced roofers, who are aware of the individual properties of the different metals.

Copper is the most durable as well as one of the more expensive metals for roofing. It is malleable, tenacious, and easy to work, and it does not require maintenance after installation. It weathers to a gray-green color and blends well with other building materials.

Terneplate is made of sheet iron or steel coated with an alloy of 25% tin and 75% lead. Metal roofers purchase the metal in small sheets that have been shop-coated with an oil-base paint. After application, the top surface must be painted with two coats of paint long in oil. If properly maintained, terneplate will last from 30 to 50 years. One advantage of terneplate is its light weight; also, it expands less than copper.

Both copper and terneplate are applied on tight board roof decks covered with rosin-sized paper. On roof pitches from 3″ to 12″, either the standing-seam or the batten-seam method is used (Fig. 12–184). After the small sheets have been joined into long strips, the standing seam is formed by turning up one edge 1½″ and the adjoining edge 1¼″. The edges are bent and are locked together without soldering. Cleats are nailed to the roof deck and are incorporated into the seam to hold the metal to the deck, but no nails are driven through the sheets.

Batten seams are formed by nailing wood strips, ranging in size from 2″ × 2″ to 4″ × 4″, to the deck; the strips run from the eaves to the ridge and are parallel to each other. Metal lengths are put between the battens and are bent up along the edges of the battens to form pans down the slope of the roof. Cleats of the same metal are nailed to the edges of the battens and are incorporated into lock joints, which are formed by the edges of metal caps that rest on the battens, as well as the edges of the metal pans. Batten strips are used on the ridges and the hips where necessary. The ends of wood batten strips at the edges of the roof must be covered with metal.

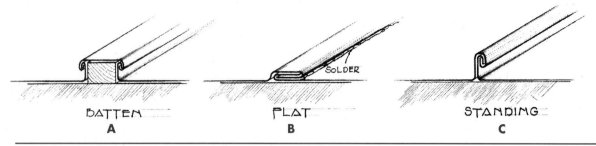

Figure 12–184 Metal roof seams.

Several types of aluminum and steel roofing are used on residences and small buildings. Both aluminum and steel sheets, flat or corrugated, with baked enamel and other patented finishes are popular metal roofing materials. Some are laid beginning at the ridge and continuing down to the eave; others are laid beginning at the eave and proceeding up to the ridge, similarly to the method of applying copper sheets. Snap-on battens usually seam the joints between sheets, which are commonly 12″, 16″, or 24″ wide. Special caps are designed to cover ridges and hips. Pressed aluminum or steel shingles and tiles are available that resemble wood shakes and roll tiles. Like metal sheets these roofing materials have baked enamel or special finishes. Available in strips, the roofing is nailed in place; concealed locking devices connect one strip to the other at each end. Metal sheets, shingles, and tiles can be installed on wood furring strips nailed perpendicularly to the roof framing members. These roofing materials can be installed over roofing felt on a plywood deck.

Corrugated aluminum sheets are employed mainly for temporary buildings. They are fastened directly to the deck with aluminum nails and neoprene washers; the fastening is done through the high portion of the lapped corrugations.

On economy roofs, corrugated galvanized iron sheets can be used. However, galvanized iron has a limited service life as well as an unpleasing appearance. This material should be reserved for temporary or less important structures.

12.13
EXTERIOR FINISH MATERIALS

The exterior finish material plays a very important part in the esthetics of a building, and if maintenance is a factor, the selection of the finish is especially important. Many finish materials are available; some of the more common materials for wood construction are wood siding, wood shingles, plywood, hardboard, aluminum, stucco, exterior insulation and finish systems (E.I.F.S.s), and masonry veneer (see Section 12.6 on masonry veneer).

Wood siding has been used successfully for years and is typical of American-built houses. Much of it requires periodic repainting; some of the durable wood species, such as redwood, western red cedar, and cypress, can be left natural with merely an application of several coats of water repellent. Kiln-dried lumber is preferred, and select grades with a minimum of knots and defects should be used. Figures 12–185 and 12–186 illustrate some of the more common types of wood siding available.

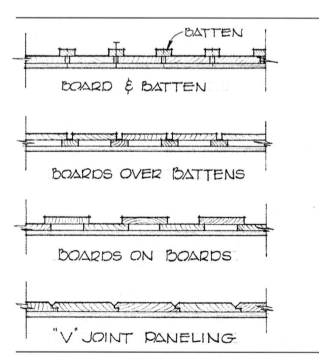

Figure 12–185 Various vertical wood siding as shown through horizontal sections.

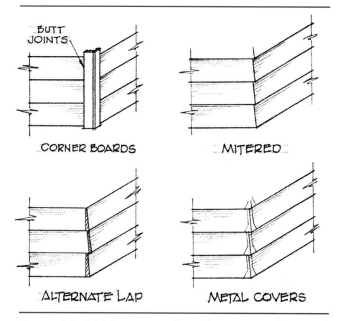

Figure 12–186 Treatments of outside corners on horizontal wood siding.

12.13.1 Bevel Siding

Bevel siding is widely used in present-day construction. It is cut into widths varying from 4″ to 12″, and butt thicknesses range from $7/16$″ to $3/4$″. The usual amount of lap is from 1″ to 2″. For example, a bevel siding 6″ wide is usually applied with $4\frac{1}{2}$″ to the weather, and a siding 10″ wide is usually applied with 8″ to the weather. Wider sizes have a tendency to cup and warp.

Much of the warping in siding can be prevented by the application of a prime coat of paint to all sides of the siding, including ends, before installation. Rustproof nails should be used, because rust stains bleed through the paint film. Figure 12–187 describes nailing recommendations for a number of types of siding. Notice in the figure that the nails do not pass through the underpiece of the lap. This method allows for expansion in the lap without the risk of splitting the thin portion of the underpiece, should the nails be put through both pieces of the lap.

Strong, horizontal shadow lines are characteristic of bevel siding. Various corner treatments are shown in Fig. 12–186. A starter strip must be nailed to the bottom of the sheathing so that the first course of siding will slope the same amount as succeeding courses. The bottom course is started about 1″ below the top of the foundation wall to weatherproof and hide the wood and masonry joint. Bevel siding exposures are worked out exactly by carpenters so that horizontal butt lines coincide with sill and head edges of windows and doors, and courses are similar throughout the wall height. Minor variations of exposure on a wall are permissible inasmuch

as they are not readily noticeable. Water tables should be used above all windows and doors; a rabbeted trim member is generally used at the top of the wall to cover the thin edge of the top course. Rabbeted bevel siding is also available in various widths and thicknesses.

On economy walls, drop siding, also known as *novelty siding*, is acceptable without sheathing. Building felt is tacked directly to the studs, and the drop siding is nailed over the felt. Corners can be treated as mentioned previously, or square-edged corner boards can be nailed directly over the siding.

12.13.2 Board-and-Batten Siding

Board-and-batten siding can be used to obtain a rustic texture on exterior frame walls (Fig. 12–185). Either surfaced or unsurfaced, square-edged boards can be used. The wide boards are nailed vertically with at least $1/8$″ space between them for expansion. Battens of the same type of lumber, usually only 2″ wide, are nailed over the vertical joints of the wide boards. The 2″ batten strips are also used for trim around doors and windows, as well as along the top of the wall. Joints between the wide boards are often caulked before the batten strips are applied. Horizontal blocking must be installed in the stud wall if insulation board sheathing has been used so that the boards and the battens can be nailed at a minimum of four levels of vertical height throughout a 1-story wall. If nominal 1″ wood or plywood sheathing has been used, the boards and the battens can be secured by nailing into the sheathing. Care should be exercised in nailing the batten strips; the nails should pass between the wide boards in the joint, or the nails can pass through the batten and one of the wide boards *only*. This method allows expansion of the joint without splitting either the boards or the batten. Only rustproof nails should be specified.

Another interesting effect can be gained by applying the wide boards over the narrow batten strips. This application becomes a reverse board-and-batten method, which produces narrow grooves and a slightly different overall effect. Redwood lumber is frequently applied in this manner.

12.13.3 V-Joint, Tongue-and-Groove Paneling

V-joint, tongue-and-groove paneling can also be used for vertical siding. Adequate wood sheathing or horizontal blocking must also be installed to provide proper nail anchoring; blind nailing is generally used along the tongue to hide the nails within the joint. Kiln-dried lumber with minimum shrinkage characteristics should be selected, and joints should be nailed up tight so that moisture will not penetrate between the boards. Either random-width or uniform-width paneling can be used. Bottom cuts at the base of the wall should be beveled to provide a drip. Because of cupping, paneling over 8″ wide should be avoided.

Nailing recommendations

Nailing patterns vary depending on the profile of the siding. But no matter what the pattern, one thing remains the same: Don't nail through overlapping pieces. To do so will eventually split the siding as it seasonally expands and contracts. Nails should penetrate at least 1½ in. into studs or blocking (1¼ in. for ring-shank or spiral-shank nails). Spacing should be no more than 24 in. o. c. Use box or siding nails for face nailing and casing nails for blind nailing.

Estimating coverage

To calculate the amount of material required to side a house, first figure the square footage of the walls minus any openings. Add 10% for trim and waste. Now multiply your answer by the appropriate board-ft. factor or linear-ft. factor to tally the amount.

Bevel

6 in. and narrower:  1 in. overlap. One nail per bearing, just above the 1 in. overlap.

8 in. and wider: 1 in. overlap. One nail per bearing, just above the 1 in. overlap.

Nominal width	Dressed width	Exposed face	Factor for linear feet	Factor for board feet
4	3½	2½	4.8	1.6
6	5½	4½	2.67	1.33
8	7¼	6¼	1.92	1.28
10	9¼	8¼	1.45	1.21

Shiplap (Dolly Varden)

6 in. and narrower:  One nail per bearing, 1 in. up from bottom edge.

8 in. and wider: One nail per bearing, 1 in. up from bottom edge.

Nominal width	Dressed width	Exposed face	Factor for linear feet	Factor for board feet
4	3½	3	4	1.33
6	5½	5	2.4	1.2
8	7¼	6¾	1.78	1.19
10	9¼	8¾	1.37	1.14
12	11¼	10¾	1.12	1.12

Channel rustic

6 in. and narrower:  One nail per bearing, 1 in. up from bottom edge.

8 in. and wider: Use two siding or box nails, 3 in. to 4 in. per bearing.

Nominal width	Dressed width	Exposed face	Factor for linear feet	Factor for board feet
4	3⅜	3⅛	3.84	1.28
6	5⅜	5⅛	2.34	1.17
8	7⅛	6⅞	1.75	1.16
10	9⅛	8⅞	1.35	1.13

Drop

6 in. and narrower: 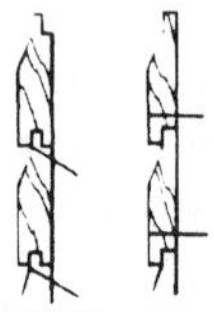 T&G pattern; Shiplap pattern. Blind nail T&G patterns; lace nail shiplap patterns, 1 in. up from bottom edge.

8 in. and wider: T&G pattern; Shiplap pattern. Two nails 3 in. to 4 in. apart to face nail, 1 in. up from bottom edge.

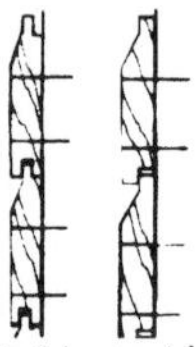

Nominal width	Dressed width	Exposed face	Factor for linear feet	Factor for board feet
4	3⅜	3⅛	3.84	1.28
6	5⅜	5⅛	2.34	1.17
8	7⅛	6⅞	1.75	1.16
10	9⅛	8⅞	1.35	1.13

Figure 12–187 Nailing recommendations for wood siding. (*Fine Homebuilding* magazine, September 1995)

In recent years, rough-sawn cedar paneling has been widely used in contemporary-style homes and small commercial buildings, in many applications with diagonal coursing. The paneling can be stained to gain various color treatments or left to weather with natural wood appearance. Both cedar and redwood are soft and require careful nailing with rustproof nails.

12.13.4 Wood Shingles

Wood shingles produce warm and charming exterior finishes that are popular in many areas of the country. The shingles are versatile in that many effects can be obtained by various applications. They can be painted, stained, or left to weather naturally. Unfinished shingles weather un-

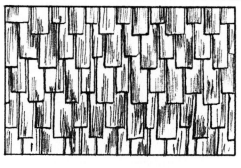

RANDOM COURSING

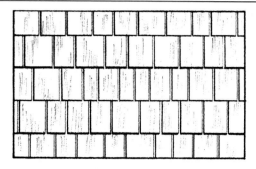

REGULAR COURSING

Figure 12–188 Wood shingle textures.

evenly because of projections, overhangs, and the like. Handsplit shakes can be used for rustic effects; double courses of regular shingles can be used for strong horizontal shadow lines; or regular shingles can be applied in staggered course lines for unusual effects (Fig. 12–188). More exposure widths can be allowed on wood shingle walls than is necessary when applying similar shingles to a roof. Table D–8 lists wood shingle and shake side-wall exposure requirements.

Western red cedar is the most popular wood used for commercial shingles, and the method of application to walls is similar to the methods mentioned under Section 12.12.2, "Roof Coverings." For siding, No. 1 grade is usually specified; No. 2 or 3 grade can be used for underlays of double courses. Wood shingles must have substantial wood backing for nailing; if insulation board sheathing is used on the wall, horizontal nailing strips, usually 1 × 3, are nailed over the insulation board. The strips are spaced the same distance apart as the exposure width of the shingles.

12.13.5 Exterior Plywood

Exterior plywood is now used extensively for siding. Plywood must be of the exterior type and not less than $3/8''$ thick. Panel siding must be installed according to Table D–9. Plywood can be applied vertically with 1 × 2

batten strips placed vertically either 16″ or 24″ o.c. to hide the joints of the plywood (Fig. 12–189). Plywood must be well painted or stained. Also, texture one-eleven plywood, $5/8''$ thick, with rabbeted joining edges, can be nailed vertically on stud walls to resemble vertical paneling. Plywood is also available in strips to be applied horizontally similar to shingles.

12.13.6 Sheet Hardboard

Sheet hardboard is a popular exterior siding material which, if properly applied and painted, will give years of satisfactory service. Hardboard is made from natural wood fibers that are subjected to heat and pressure to form sheets or strips of various sizes and thicknesses. It comes in various surface textures: smooth, grooved, and ribbed. Hardboard must have solid backing support and is durable if kept painted. Strips can be applied horizontally to resemble bevel siding, or larger panels can be applied vertically to resemble board-and-batten siding.

The horizontally applied panels are usually 8″ to 1′ wide by 16′ long by $3/8''$ to $7/16''$ thick, applied generally with 7″ or 11″ exposures. Starter strips at the base of the wall and corner treatments are handled similarly to wood bevel siding. Vertical joints should be made over studs, and nailing is done through the lap at each stud location.

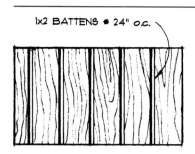

1x2 BATTENS ● 24″ o.c.

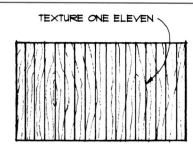

TEXTURE ONE ELEVEN

Figure 12–189 Plywood siding. Battens can be spaced either 24″ or 16″ o.c.

For vertical application, 4' wide panels are used. Lengths are obtainable up to 16', and recommended thicknesses are ¼" and ⁵⁄₁₆". Vertical wood battens spaced 16" o.c. can be used to hide vertical joints on smooth hardboard edges; a ⅛" space must be left between panels, and joints are caulked before batten strips are applied. Grooved-and-ribbed panels have shiplap edges for concealing the joints. The builder should follow manufacturers' directions in applying either type of hardboard exterior panels. Hardboard siding must be installed according to Table D–11.

12.13.7 Aluminum Siding

Aluminum siding is a low-maintenance material used for both residential and commercial construction; it is also popular for remodeling exterior walls. Two types of aluminum panels are most popular: those applied horizontally to resemble either narrow- or wide-bevel siding, and those applied vertically to resemble board-and-batten siding. Both styles are available with baked-on enamel finishes of various colors, and both can be painted if necessary, similarly to wood siding. Aluminum siding is light in weight, easy to apply, and resistant to weathering.

The horizontal panels are started at the base of the sheathing with a wood cant strip nailed below the bottom edge of the first course, and each course is nailed to the sheathing along its top edge. Courses are held to the adjacent course with a lock joint. Matching corner pieces, backer strips for end joints, and trim pieces are furnished by the manufacturers for each type of panel. Horizontal panels are made 12½' long, with either a bevel exposure 8" wide or two 4" bevel exposures. Some panels have foam insulation backing.

To apply vertical aluminum panels, first nail a starter corner strip vertically along one corner; then lock the first panel into the strip, and nail the panel along the edge opposite the lock joint. This procedure is continued across the wall. You can cut aluminum with a hacksaw or tinsnips when fitting it around openings and such; appropriate pieces are furnished by manufacturers to form water tables and trim members. Vertical panels are made in lengths to cover 1-story heights so that joints are eliminated.

12.13.8 Vinyl Siding

Vinyl siding is available in a wide range of colors and does not require painting. Manufactured from polyvinyl chloride sheets, the siding comes in 10' and 12' long panels. The panels are preformed to resemble either horizontal or vertical wood siding. The surface of the siding may be smooth or embossed to replicate wood graining. Both the vertical and the horizontal panels have perforated edge strips for nailing. Galvanized nails should be used to attach the siding to solid, nailable sheathing. A variety of vinyl trim pieces are available to complement the siding. These pieces are especially convenient for finishing corners, cornices, soffits, and fascias and around doors and windows (Fig. 12–190).

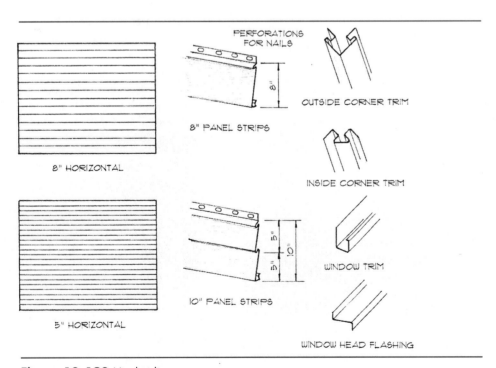

Figure 12–190 Vinyl siding.

12.13.9 Stucco

Stucco has been used for years as a satisfactory exterior siding material, but present-day use is limited largely to direct application over concrete block and other masonry materials. Some application for stucco over wood frame walls in residential construction is still found, however.

On masonry walls of concrete block, tile, or brick, two coats of portland cement stucco are applied like plaster to form a coat ¾″ thick. This particular stucco should be made of 1 part portland cement, 3 parts sand, and 10% by volume hydrated lime. It must have a minimum compressive strength of 1000 lb/sq in. Or a prepared portland cement stucco may be used and applied according to the manufacturer's directions. The masonry wall is first cleaned with a wire brush, and a scratch coat is applied. After the scratch coat is thoroughly dry, the finish coat is applied and is given whatever surface texture is desired. A finish may be smoothed with a metal trowel finish, sand-finished with a wood float, or stippled by application of various tooling or stroking effects. To render the finish waterproof, cement-base paints are usually applied after the finish coat is completely dry. Colored stucco can be obtained by mixing mineral pigments into the finish coat, or colored cement-base paints may be applied to the final stucco surface.

On wood frame walls, stucco is applied over ribbed metal lath that is backed up with heavy asphalt-treated paper (Fig. 12–191). Wood or metal furring strips 15″ o.c. can also be used under galvanized wire fabric, or galvanized furring nails can be used to keep the fabric at least ¼″ away from the paper and sheathing. Foundations must be adequate to prevent settling and subsequent cracking of the stucco. Stucco-covered, 2-story buildings should be constructed with balloon framing to minimize vertical shrinkage. Usually, a rustproof metal ground strip is placed at the base of the wood wall to close the opening between the stucco and the sheathing. Flashing must be used above all doors and windows, and the tops of stucco walls must be protected by overhangs and trim members. When applied over wood frame walls, a three-coat stucco job is standard (consisting of scratch coat, brown coat, and finish coat) to make the stucco ⅞″ thick.

Stucco is used to produce imitation, half-timber-type exterior walls used mainly in Normandy-style houses. Treated boards representing the timbers are fastened directly to the sheathing, and stucco is applied to the furring lath in the areas between the timbers. In this construction, attention must be given to the prevention of moisture penetration between the timbers and the stucco. Flashing must be used liberally over horizontal timber pieces as well as joints in the timber pieces, which are especially susceptible to the accumulation of moisture.

All of the aforementioned exterior siding materials must be applied over a suitable moisture-resistant building paper that has been fastened to the wall sheathing. This paper is important in preventing moisture penetration under and between siding pieces and in preventing air infiltration through the exterior walls from wind pressures.

12.13.10 Exterior Insulation and Finish Systems (E.I.F.S.s)

Exterior insulation finish systems were introduced into the United States in 1969. The system originated in post–World War II Europe as a finish system applied over masonry and concrete in building renovations. American manufacturers have adapted the European system for use on metal and wood frame walls (Fig. 12–191). E.I.F.S.s were initially used in commercial applications but have been used more extensively in the residential market in the last 10 years. There are now an estimated 250,000 houses clad in E.I.F.S.

E.I.F.S. can be applied to smooth masonry walls or to stud walls that have exterior-grade gypsum sheathing. The application is generally a four-step process. First, an adhesive is applied to one side of an insulating panel such as Styrofoam®, which is then attached to the masonry wall or gypsum sheathing. Next, an adhesive with an admixture of portland cement is troweled over the insulating panel. Third, a fiberglass mesh reinforcement is embedded into the wet adhesive, which is allowed to cure. Last, an acrylic polymer finish coat is troweled over the exterior wall surface. Color tints can be added to the finish coat. The tints are usually blended into the acrylic polymer at the factory.

The E.I.F.S. treatment adds insulating value to the wall system, will not crack as will stucco, and is a low-maintenance material. Sharp objects or strong impacts can damage the soft surface; however, a heavy-duty reinforcing mesh with substantial impact resistance can be used. Creative designs and trims are easily achieved by varying the thickness or cutting sculptural shapes into the Styrofoam.

Recently, North Carolina's New Hanover County Inspection Department has observed significant problems with E.I.F.S.-clad houses. An investigation of 72 randomly selected E.I.F.S.-clad houses less than three years old revealed that 70 of them suffered from water infiltration. A subsequent study of 209 E.I.F.S.-clad houses, conducted by the Wilmington section of AIA North Carolina, drew similar conclusions: 90% registered high moisture levels.

In most cases, water had managed to get into the walls and through joints between the cladding and the windows, doors, and other penetrations such as roof flashings. Improperly caulked windows, poor sealants, and inadequate flashing will ruin an E.I.F.S. wall by allowing moisture to penetrate. Once moisture enters, it is not able to escape. A traditional E.I.F.S. has no internal drainage provision if moisture gets past the system membrane.

Another recently discovered problem concerns termite infestation of the expanded polystyrene foam in E.I.F.S. The termite problem and the wood rot problem are both exacerbated by the same cause—moisture. In typical

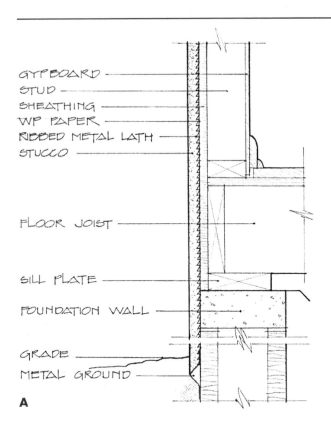

GYPBOARD
STUD
SHEATHING
WP PAPER
RIBBED METAL LATH
STUCCO

FLOOR JOIST

SILL PLATE

FOUNDATION WALL

GRADE
METAL GROUND

A

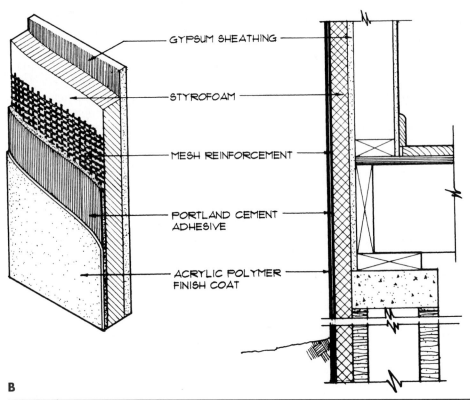

GYPSUM SHEATHING

STYROFOAM

MESH REINFORCEMENT

PORTLAND CEMENT
ADHESIVE

ACRYLIC POLYMER
FINISH COAT

B

Figure 12–191 Stucco and exterior insulation finish system.

E.I.F.S. installations, the exterior foam insulation extends from the above-grade walls down into the ground, providing a convenient path for termites to get almost anywhere in the above-grade structure. If water gets trapped in the E.I.F.S., termites do not have to return to the ground, because they have ready access to food, moisture, and warmth.

In response to these problems, the E.I.F.S. Industry Members Association (EIMA) has implemented a third-party builder certification program. The National Association of Home Builders Research Center has agreed to certify qualified participants based on EIMA-specified installation techniques. EIMA is also revising 10 basic construction details that were published in 1994. These details cover generic installation conditions at foundations, windows, parapet caps, dissimilar substrates, and expansion points. A supplementary set, published in 1996, addresses more complex circumstances such as a balcony wall intersection and penetrations.

12.14

INTERIOR FINISHES AND TRIM
12.14.1 Gypsum Wallboard

Gypsum wallboard (commonly called gypsum drywall) construction is now used as the interior finish in the majority of American homes under construction. The paper-covered gypsum sheets are made in $4' \times 8'$, $4' \times 10'$, and $4' \times 12'$ sizes. Standard thicknesses are $3/8''$, $1/2''$, and $5/8''$. Gypsum has good thermal and fire-resistance qualities. The sheets can be easily cut and nailed to framing; joints and nailheads can be smoothly finished with joint compound and perforated tape; and very little moisture is introduced into the structure from the material, which is the case when plaster is used (Figs. 12–192 and 12–193).

Building codes require inspection of gypsum wallboard installation before board joints and fasteners are taped and finished. Also, gypsum wallboard may not be installed until weather protection is provided.

Water-resistant gypsum board is used as a base for tile or wall panels for tub, shower, or water closet compartment walls. Water-resistant gypsum board cannot be used over a vapor barrier, in areas subject to continuous high humidity such as saunas, or on ceilings where frame spacing exceeds $12''$ o.c.

Wallboard can be fastened with nails, screws, or adhesive. Supports and nailings for wallboard are indicated in Table D–20 for single-ply applications, and in Table D–21 for two-ply applications.

Two-ply installation is usually used where extra thickness is required for fire resistance because it is much easier to lay up two pieces of $3/8''$ wallboard than one piece that is $3/4''$ thick. One advantage of two-ply installation is that you can get a smoother wall. You can secure it with adhesive rather than nailing it, thereby reducing the amount of taping required.

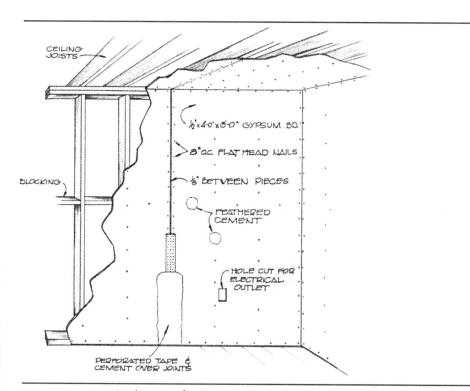

Figure 12–192 Application of gypsum board applied vertically for interior walls.

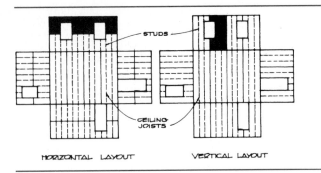

Figure 12–193 Two methods of gypsum board layout.

HORIZONTAL LAYOUT   VERTICAL LAYOUT

STUDS

CEILING JOISTS

12.14.2 Plaster

Two types of plaster are used in construction today: (1) portland cement plaster for both indoor and outdoor work and (2) gypsum plaster only on interior work. Both require three coats when applied over metal lath or wire fabric lath. You must apply at last two coats over any other material allowed by code, but you can never apply plaster directly to fiber insulation board.

12.14.3 Prefinished Plywood Sheets

Prefinished plywood sheets in various natural wood finishes are becoming popular in many room interiors. They are blind nailed directly to studs and never require painting or finishing. Other manufactured sheet materials as well as solid wood paneling are also available for various wall coverings. To straighten nailing surfaces when using sheet materials or ceiling acoustical tile, you can apply furring strips perpendicularly to studs or ceiling joists.

12.14.4 Trim

Trim is the general term given to molding, base, casing, and various other finish members that must be carefully fitted by finish carpenters to complete the appearance of a structure (Fig. 12–194). Many stock pieces and patterns are available at lumber dealers. In higher-class construction, custom-made moldings and trim pieces are often shown in detail on the working drawings and therefore must be specially milled. Care must be taken to select consistent patterns to be used throughout a building so that harmony in the trim results. Colonial interiors traditionally require ogee profiles, whereas contemporary styles usually appear best with simpler tapered or teardrop profiles. Casing and finishing nails are used to apply trim members; the nails are driven below the surface of the wood with a nail set so that the holes can be filled, and care must be exercised to prevent hammer marks from injuring the surface of the wood.

12.15

INSULATION AND MOISTURE CONSIDERATIONS IN FRAME STRUCTURES

Builders are fully aware that indoor comfort is of primary importance to the individual. Consequently, proper insulation becomes a major consideration in warm climates as well as in cold. Many types of insulation are available. The most suitable usually depends on the details of the construction; the amount depends on the extent of comfort desired. When care in selection and installation is taken, optimum benefit will be realized.

Heat-loss calculations and the design of heating and cooling systems are not discussed in this material; however, the general aspects of insulation, such problems as water vapor within buildings, and the ventilation of unheated air spaces are of concern to the drafter if details of construction are to be correctly conceived and drawn.

All construction materials can be classified more or less as either insulators or conductors, depending on their porosity or density. Still air is an excellent insulator if confined into small spaces such as the matrix of insulation materials or small cavities within walls. On the other hand, dense materials, such as glass or masonry, are relatively poor insulators. All materials, however, resist the flow of heat to a certain extent, the resistance of the given material being directly proportional to its thickness (Fig. 12–195).

Heat may be transmitted by three different methods: conduction, convection, and radiation.

- **Conduction** is the transfer of heat by direct molecular contact. Metals, for instance, conduct heat more readily than wood, yet all materials conduct heat if a temperature difference exists between their surfaces. Remember that *heat flows from warm to cold surfaces.*
- **Convection** is the transfer of heat by air or another agent in motion. Although air is a good insulator, when it circulates, it loses part of its value as an insulator. Air spaces about 3/4" wide within walls or ceilings are the most restrictive to circulation. In larger spaces, air acts as a conveyor belt, taking heat from warm surfaces and depositing it on cold surfaces.
- **Radiation** is the transfer of heat through space from a warm surface to a cold surface, very much as light travels through space. Effective resistance to radiation can be provided by shiny surfaces, such as aluminum foil; the more actual surfaces a heat ray has to penetrate, the more effective the reflective insulation becomes.

Actual heat transfer through walls and ceilings usually employs all three of the above-mentioned methods of transfer, to various degrees. Of course, some heat is also

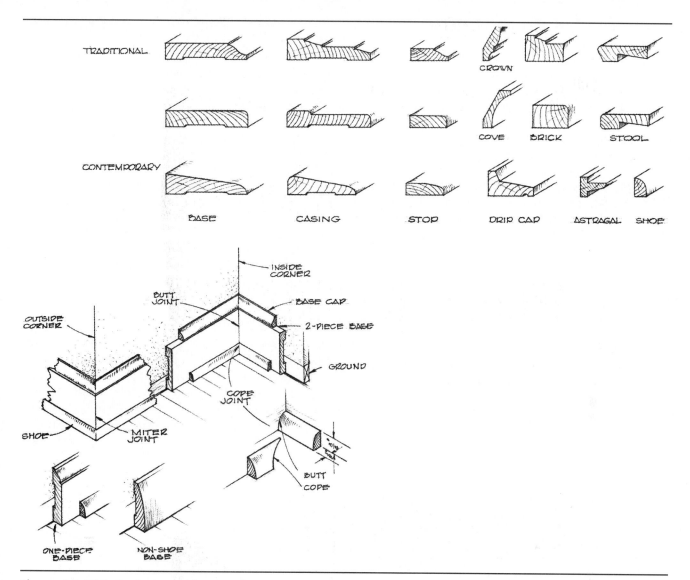

Figure 12–194 Stock trim and base applications.

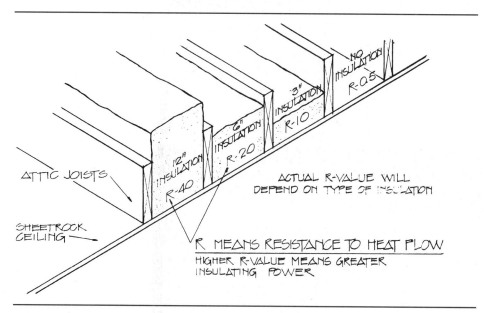

Figure 12–195 Insulation R values.

dissipated from buildings through openings around doors and windows.

To compare the suitability or insulating qualities of different building materials and insulations, a standard of reference must be used. By accurate experimentation, the thermal qualities of individual building materials have been determined. When these materials are incorporated into typical combinations found in walls, floors, roofs, and so on, including occasional air spaces, the rate of heat flow or the coefficient of transmission, known as the *U factor*, can be calculated. This factor can be defined as the number of Btu's (British thermal units—heat units) that will flow through 1 sq ft of the structure from one air region to another due to a temperature difference of 1°F in 1 h.

Easy reference can be made from tables giving the *U* factors with and without insulation; these tables are published by the American Society of Heating, Refrigeration, and Air Conditioning Engineers (A.S.H.R.A.E.). Tables for *U* factors of typical constructions and various insulation combinations can also be found in the *Forest Products Laboratory Report R1740*. Comparisons of these *U* factors make it possible to evaluate different combinations of materials and insulations on the basis of overall heat loss, potential fuel savings, influence on comfort, and cost of installation.

Other than the reflective insulators, commercial insulations are usually made of glass fibers, glass foam, mineral fibers, organic fibers, polystyrene, and polyurethane. The best materials should be fireproof and vermin-proof, as well as resistant to heat flow. The cellular materials utilize tiny, isolated air cells to reduce conduction and convection; the fibrous materials utilize tiny films of air surrounding each fiber. These efficient insulating materials are manufactured into blankets, batts, and sheets in sizes and shapes that fit conveniently into conventional structural spaces. Some have tabs or edges for stapling to wood frame members; others are sized so that they can be tightly wedged into conventional structural cavities. Proper selection depends on initial cost, effectiveness, and the adaptation of the insulation to the construction features. Figure 12–196 illustrates the proper placement of insulation within typical frame buildings.

12.15.1 Insulation

Heat transfer through the building envelope is probably most affected by the way that the insulation is used and installed. Insulated spaces must be completely filled with material. There should be no gaps or voids, and the material must be kept dry. Insulation materials available include batt-type, loose fill, boardstock, spray-type, and radiant barriers.

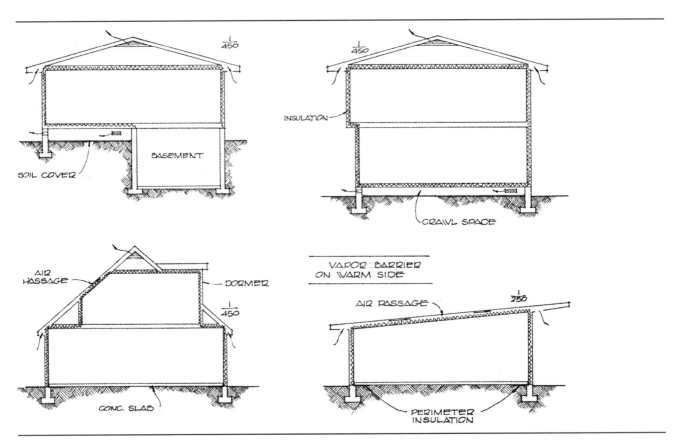

Figure 12–196 Placement of insulation in wood frame buildings. Ventilation of attic and crawl spaces is necessary.

Batt Type Batt-type insulation is made from glass or mineral fibers. These materials are acceptable for interior use such as on inside exterior walls and on foundation walls. Insulation values of the various products depend on the density of the material. Increased density generally increases the resistance of the material and reduces convective air movements within a building cavity. Performance of batt-type insulation products is directly related to installation practices. Gaps around wiring and plumbing must be prevented, and batt materials must fill cavities completely and evenly.

Loose Fill Glass fiber and mineral wool insulations are also available as loose fill. Loose fill insulation is blown into place using special equipment. It can be used in attics and inside above-grade exterior walls. It is important that it be installed at the correct density in order to provide acceptable performance. Cellulose fiber insulation is made from shredded, recycled newsprint which has been treated with chemicals to prevent the growth of molds, to keep rodents out, and to control flammability.

Boardstock Boardstock insulations can be rigid or semirigid and are available in five types:

1. **Expanded polystyrene** is made by expanding polystyrene beads in a mold. Large blocks are then cut into sheets of various thicknesses. Low-density polystyrene can be used as a sheathing material for above-grade exterior walls and can be used to insulate interior basement walls, flat roofs, and cathedral ceilings. High-density polystyrene can be used to insulate below-grade exterior walls because it will withstand high pressures. Both materials can be used to insulate interior walls above and below grade, but, due to their combustibility, they must be covered with a fire-protective covering such as 1/2″ drywall if used to insulate living spaces.

2. **Extruded polystyrene** is manufactured by extruding a hot mass of polystyrene through a slit. It expands upon exposure to atmospheric pressure and creates a closed-cell foam material. It is acceptable for use in all situations described for high-density expanded polystyrene. A higher-density extruded polystyrene can be used in built-up roofing applications. These materials must be covered with a fire-protective material if used in living spaces.

3. **Rigid glass fiber insulation** can be used in built-up roofing applications, as exterior wall sheathing, and as below-grade exterior wall insulation. The fibers in this product are aligned vertically so that water runs down the fibers. Product densities are three to five times higher than those of standard batt-type products.

4. **Polyurethane** and **polyisocyanurate** are insulations made in continuous slabs which are cut with hot wires. They can be used in all of the applications noted earlier for expanded polystyrene. They are also combustible and must be covered with fire-protective materials when used in living spaces.

5. **Phenolic insulation** is manufactured in much the same way as polyurethane and polyisocyanurate insulations. However, it is much less combustible and is suitable for use as wall sheathing and for use inside, both above and below grade.

Spray Type Spray-type insulations are recent innovations in the residential construction industry. Three different types are presently available:

1. **Spray cellulose insulation** is available in a variety of types. The material is applied using special applicators that mix the insulation with materials. They allow it to hold together and adhere to the surface to which it is applied. Wet spray materials offer thorough cavity coverage and reduce envelope air leakage characteristics.

2. **Two-component isocyanurate foam** is best suited for use in exterior stud wall cavities, in perimeter joist spaces, and in the shim spaces around windows and doors. Special applicators are used to mix the chemicals in the correct proportions.

3. **Polyurethane formulations** are available in a variety of spray applications. The material is mixed on site using special foaming equipment for large projects. Single-component polyurethane foam is available in cans with "gun-type dispenser" canisters for sealing shim spaces around windows and doors.

Radiant Barriers A radiant barrier is a sheet of reflective material that is installed between a heat-radiating surface and a heat-absorbing surface. The reflective material stops the radiant transfer of heat between the two surfaces. Radiant barriers reduce cooling loads in summer by reducing the radiation of heat from the attic through the ceiling. The *R* value of a radiant barrier depends on the direction of heat flow.

If the restriction of heat flow were the only consideration in the application of insulation within buildings, it would be a simple matter to provide enough insulation around heated spaces to theoretically reduce fuel bills sufficiently and thereby offset the initial cost of the insulation. However, air contains water vapor, and because water vapor acts as a gas to penetrate porous materials and always flows from areas of high temperature to areas of lower temperatures, many types of insulation soon lose their effectiveness if they become saturated with water vapor.

Air is saturated or has a relative humidity of 100% when it contains as much moisture as it will hold. Warm air has the ability to retain more moisture than cold air.

If saturated air is lowered in temperature, some of its moisture will be given off as condensation. This point in temperature change at which a specimen of air gives off moisture in the form of condensation is known as its *dew point*. When high temperature differences exist between insulated wall surfaces (between inside and outside air), the dew point often occurs in the insulation itself, where the resistance to heat flow is greater than in the structural members. The condensation that can gather in the insulation from the cooling of the vapor as it passes through a wall not only may reduce the value of the insulation but may also eventually cause permanent damage to the structural members. This problem generally arises in winter, during the heating season, especially when the humidity within the building is high.

12.15.2 Moisture Considerations

Water moves from place to place in one or all of four ways: (1) gravity, (2) capillary action, (3) air flow, and (4) diffusion.

Gravity Gravity causes water to seek its lowest point by traveling the path of least resistance. If improper drainage directs surface runoff toward a basement wall, water will enter the building through cracks in the wall. Measures to prevent such leaks include proper grading to direct water away from the wall and appropriate dampproofing and footing drains to direct the flow.

Capillary Action Capillary action causes water to move upward, downward, or sideways through thin tubes in materials. It affects foundations made of concrete or other porous materials. The pores in the concrete are like long, thin tubes. Water will rise up the foundation wall until it is released into an area with a low water-vapor pressure. This action usually occurs inside the basement or the crawl space.

If the water is not allowed to diffuse into the basement or crawl space, it will rise to the top of the foundation wall where it may rot the sill plate, the headers, and the floor joists if they are not treated lumber. Capillary action can be minimized by sealing the pores in the foundation materials with coatings or membranes. The use of coarse materials, rather than sand or pit-run gravel, below floor slabs can also help.

Dampproofing foundation walls does not prevent water from rising through the footing. A water-impermeable membrane or foundation coating should be installed over footings. A through-the-footing drain tile will help to reduce water pressure under the slab (Fig. 12–197).

Air Flow Moisture enters attics and wall cavities through air flow. Figure 12–198 shows potential air leakage sites where warm, indoor air can carry water vapor into the shell of the building. All of these sites must be sealed to prevent warm, moist air from exfiltrating into the building envelope, where condensation can occur.

Diffusion Diffusion is a process in which water vapor passes through a seemingly solid material. Water vapor will flow from an area of high vapor pressure to an area of low vapor pressure.

Some materials slow the diffusion of moisture better than others. Since vapor diffusion is never stopped, but

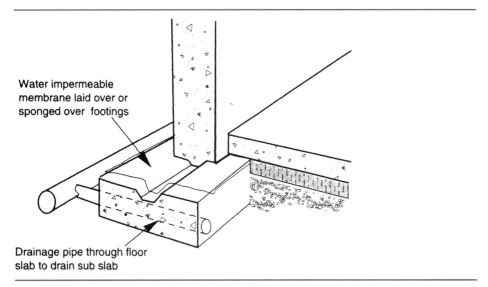

Water impermeable membrane laid over or sponged over footings

Drainage pipe through floor slab to drain sub slab

Figure 12–197 Footing detail to prevent capillary action.

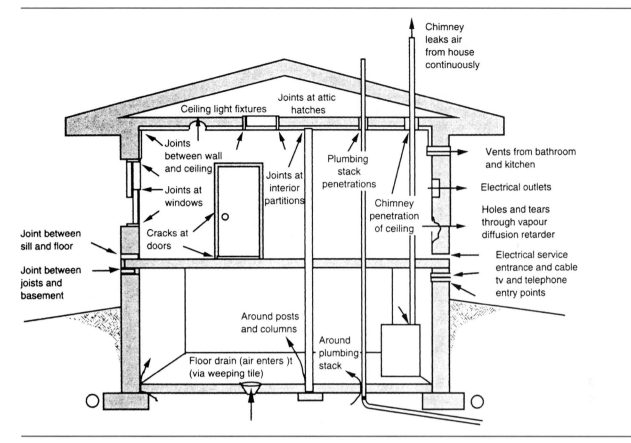

Figure 12-198 Air leakage sites.

merely retarded, the term *vapor diffusion retarder* (VDR) should be used instead of the term *vapor barrier*. A VDR must be placed on the warm side of a wall in order to ensure that water vapor does not reach a temperature at which it can condense.

Moisture levels are considerably higher inside a heated house than outside under normal winter conditions because warm air can hold more water vapor than cold air. Moisture is generated by normal daily activities such as cooking, bathing, washing and drying clothes, and breathing. Furniture, drywall, and the framing materials inside the house absorb moisture from humid air in the summer and release it to the interior during the winter. This process is termed *seasonal storage*. Moisture also enters a house from concrete in the basement and from crawl space foundations and floor slabs. An earthen floor in a crawl space or a basement of an older house can be a significant source of moisture.

Table 12–16 shows the amount of moisture that various activities contribute in a home. In a well-built home, indoor humidity levels are controlled by a combination of better building practice and the use of a properly designed and operated mechanical ventilation system.

CONSTRUCTION STRATEGIES

Houses are subjected to a range of weather conditions. Building assemblies must include elements to protect the structural members from exposure to excessive humidity levels and to ensure that wind does not adversely affect the performance of thermal insulation materials. The four most common approaches include the use of (1) air barriers, (2) vapor diffusion retarders, (3) weather barriers, and (4) moisture barriers/dampproofing.

12.16.1 Air Barriers

In cold climates, air barriers are designed to reduce the outward migration of moisture-filled air (Fig. 12–199). The air barrier is one of the most critical elements of all building envelope components because almost all water vapor is carried by air movement, with only a small amount diffusing through materials. An air barrier must be continuous at all corners, all partition walls, all floors, and all ceiling-wall junctions.

Table 12-16 Typical moisture sources per day in houses.

Occupant-Related Sources		Building-Related Sources	
Four occupants	5 liters (1.3 gal)	Seasonal building storage (i.e., framing, drywall, concrete)	8 liters (2.1 gal)
Clothes drying indoors	1.2 liters (0.32 gal)		
Floor washing 10 m³ (107 sq ft)	1 liter (0.26 gal)	Exposed, uncovered earth crawl space	40–50 liters (10.5–13.2 gal)
Cooking (three meals/day)	1 liter (0.26 gal)	Drying and burning of firewood	5 liters (1.3 gal)
Dishwashing (three meals/day)	½ liter (0.13 gal)	New construction—drying of framing and concrete over first 18 months	4–5 liters (1–1.3 gal)

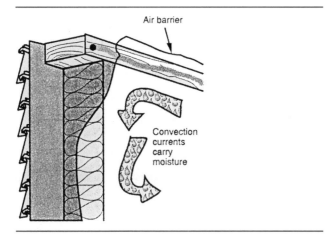

Figure 12–199 Air barrier.

An effective air barrier is impermeable to air flow, continuous over the entire envelope, and durable over the expected lifetime of the building. The smallest cracks, holes, or tears can greatly reduce the effectiveness of an air barrier. All joints and seams must be carefully sealed.

Several different building systems are commonly used to achieve a continuous air barrier. Polyethylene is used as the primary interior air barrier in combination with sealants and header wraps. A second approach, known as the *airtight drywall air barrier (ADA)*, uses drywall in combination with sealants, gaskets, framing members, and other rigid materials. Exterior air barrier systems are a third option.

Membrane Air Barrier Polyethylene is one of the most common air barrier systems. Its primary advantage is its dual function in providing the air barrier and the vapor diffuser retarder (VDR). Seams in the polyethylene should occur over solid backing and should be overlapped, caulked with an acoustical caulking, and sandwiched between rigid materials. Detailing at penetrations through the envelope, such as electrical boxes, pipes, and ducts, require special attention. In many instances, an exterior air barrier housewrap is used to provide continuity of the air barrier between floors.

The membrane air barrier cannot be accessed after construction, and certain steps should be taken to ensure its longevity (Fig. 12–200). These steps include the following:

Locate the membrane on the warm side of the wall.

Use UV-stabilized, 6-mil polyethylene.

Overlap seams by a mininum of 6″.

Locate seams over rigid backing such as framing members.

Caulk seems with a flexible, nondrying sealant.

Staple seams through the sealant into the solid backing.

Sandwich polyethylene seams between two solid materials such as the stud and the drywall.

Rigid Air Barrier The rigid air barrier approach was developed as an alternative to the use of polyethylene. It stresses the use of rigid materials such as drywall as components of the air barrier system. An effective air barrier can be achieved by sealing the drywall at all junctions.

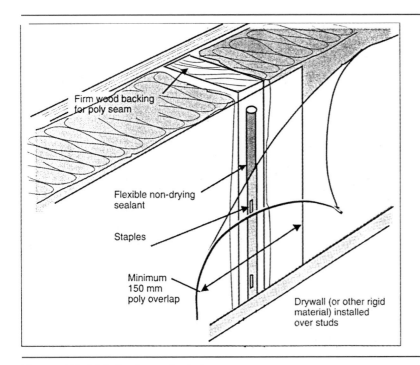

Figure 12–200 Sealing polyethylene.

The rigid air barrier offers several advantages. It is not easily damaged in construction. If damaged, it can easily be repaired. It will stand up to high wind pressures on the walls and ceilings. The vapor diffusion retarder can be a number of materials such as polyethylene, foil-backed drywall, or low-permeance interior paints.

Seams between all components of the air barrier must be adequately sealed. Drywall is usually sealed to the framing using adhesive-backed foam tape. Most commonly used is a $1/2'' \times 3/16''$, low-density, closed-cell PVC foam tape. This tape is stapled to wood framing to ensure that it will stay in place during drywalling. It is important that the drywall be screwed or nailed at an 8″ spacing over the tape to provide an airtight seal (Fig. 12–201). A rigid air barrier does not act as a vapor diffusion retarder. Most rigid air barriers are installed with a vapor barrier primer or paint used as a VDR.

Exterior Air Barrier An exterior air barrier can consist of certain rigid insulation materials or more often a housewrap such as spun-bonded polyolefin or spun-bonded polypropylene sheeting. These materials must have the ability to allow adequate vapor transmission while stopping air movement, or they must be located so they prevent condensation of moisture in the wall. Exterior air barriers have the advantage of eliminating the labor required in sealing interior air barrier penetrations on walls and ceilings.

Seams in sheets should be taped with a compatible material as specified by the manufacturer. The membrane air barrier can be sandwiched between rigid materials to provide protection against wind pressure. Typically, fiberboard is used for this purpose. The air barrier must be made continuous and must be properly sealed to both the sill plate and the roofing members (Fig. 12–202).

12.16.2 Vapor Diffusion Retarders

A vapor diffusion retarder (VDR) is a membrane, material, or coating that slows the diffusion of water vapor. VDRs are classified according to their permeability and are measured in *perms*. The lower the perm rating, the more effectively the material will retard diffusion. One perm represents a transfer of one grain (0.002285 oz) of water per square foot of material per hour under a pressure difference of 1″ of mercury (1.134 ft of water). Polyethylene, aluminum foil, and certain kinds of paint can be used as VDRs.

Polyethylene film is very resistant to the flow of water vapor. If the film is made from virgin polyethylene resins, it will be clear. If it is cloudy, it probably contains reused resins. Choose crosslaminated film for best performance.

Paint can also be used as a vapor diffusion retarder. Varying numbers of coats of different types of paints will provide the required resistance to the flow of water

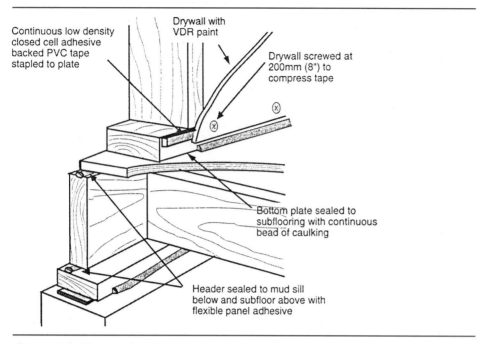

Figure 12–201 Rigid air barrier.

vapor. Using paint as the VDR can have advantages because it is applied at the end of the construction process and serves as the finish as well. Care is required to ensure adequate coverage to attain a suitable perm rating.

12.16.3 Weather Barriers

Weather barriers are materials placed on the exterior of the wall or ceiling assembly to protect the interior components of the wall from the effects of wind, rain, snow, and sun. They resist the harmful effects of rain and snow, keeping the insulation and structural members dry. They also improve the thermal performance of the system by keeping wind out. If wind penetrates the insulation cavity, it may reduce the effectiveness of fiber insulations such as glass fiber batts (Fig. 12–203). If rain or snow penetrates the weather barrier, the insulation may become wet, and structural members may rot.

Housewrap materials have gained considerable market share over the last several years, both as sheet-applied materials and as a product laminated to rigid fiberglass sheathing. Spun-bonded polyolefin and woven polypropylene shed liquid water and resist air flow, but they are permeable to the diffusion of water vapor. They are tough enough to withstand severe wind and the rough handling that they might receive during the construction process. They should be covered within 60 days to prevent potential deterioration caused by ultraviolet radia-

tion. Special sheathing tape is used to seal seams, holes, and openings around windows, doors, and service penetrations. Housewrap applications may not provide the desired airtightness unless considerable attention is paid to air sealing the house interior. For most purposes housewrap materials should be considered as a weather barrier, not an air barrier.

Asphalt-impregnated building papers have been used for decades. They must be lapped and secured to perform adequately. Both solid and perforated building papers are available.

A weather barrier should be resistant to the flow of air, be able to shed rain or snow, be able to withstand forces that may act on it, and last the lifetime of the building.

12.16.4 Moisture Barriers/Dampproofing

Soil vapor pressure can cause moisture to move from the ground through the foundation wall by diffusion, where it is released into the air in the basement or crawl space. To avoid this, dampproofing is required on the exterior of foundation walls located below grade. The material most commonly used for dampproofing is a spray- or mop-applied bituminous substance. It fills the pores in concrete, reducing the penetration of liquid water due to pressure against the wall and capillary action.

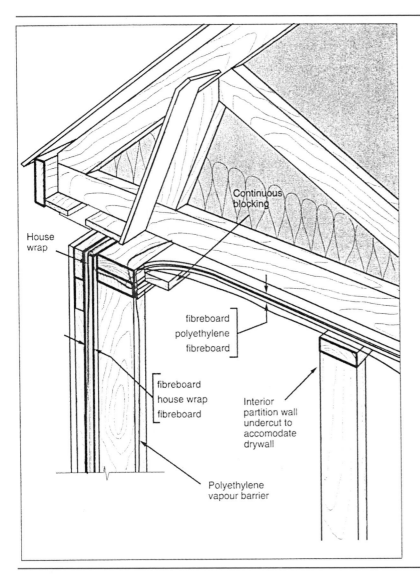

Figure 12–202 External air system elements.

Labels within figure:
- House wrap
- Continuous blocking
- fibreboard / polyethylene / fibreboard
- fibreboard / house wrap / fibreboard
- Interior partition wall undercut to accomodate drywall
- Polyethylene vapour barrier

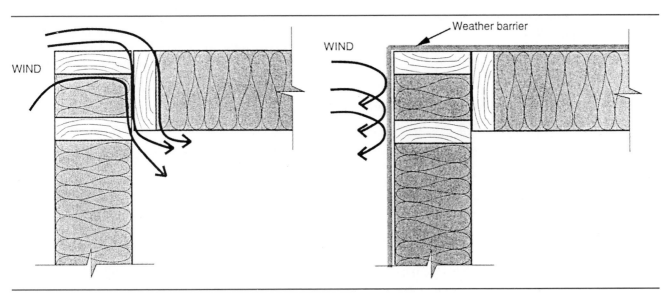

Figure 12–203 Weather barrier.

Labels within figure:
- WIND
- WIND
- Weather barrier

A moisture barrier is a material, membrane, or coating used to keep moisture which might come through the foundation wall from penetrating the interior foundation insulation or the wood members supporting the insulation and interior finishes. It is installed only on the below-grade portion of the foundation walls or under floor slabs. Asphalt emulsions applied to the interior surface of a wall provide a continuous membrane. Sheet materials, such as 2-mil polyethylene or building papers, can be fastened to the wall using adhesives or a cold asphalt mastic such as plastic roof cement. All seams must be overlapped and sealed. Newer, specially designed foundation membranes provide protection from moisture and also promote free-flow drainage.

Even good vapor barriers do not prevent some moisture from penetrating through minor breaks in the barrier around pipes, electric outlets, joints, and the like; some moisture will even permeate the barrier itself, not necessarily through minor breaks. Water vapor that has possibly found its way into walls can be released to the outside by *cold side venting*. This condition requires sheathing paper to be weather-resistant but not vapor-proof. Lightweight roofing felt is satisfactory—it can breathe. Siding materials must not be airtight, yet they must have the ability to shed water and resist driving rains. Many of the newer, water-base exterior paints possess this feature of allowing water vapor from the inside to escape through the paint film.

Another suggestion that will help minimize the vapor problem in homes is the installation of forced-air fans in rooms of excess humidity, such as bathrooms, kitchens, and utility areas. The use of exhaust fans directly to the outside relieves the home of much of the otherwise destructive concentrations of humidity.

EXERCISES

The following exercises should be done on 8½″ × 11″ drawing paper with a ½″ border. Some of the drawings require the use of instruments; others may be done freehand.

1. Using guidelines ¼″ high and ⁵⁄₁₆″ between each set, divide the sheet into three equally spaced columns. Using architectural style lettering, number 1 through 39 vertically (each vertical row will contain thirteen ¼″ high guidelines and thirteen numbers). Using Fig. 12–1, cover the written terms with opaque paper, read the numbers on the drawing of the house, and letter on the numbered lines the proper term for the items identified by numbers 1 through 39.

2. Using the scale ¼″ = 1′-0″, draw dashed lines to represent the outside wall of a 24′-0″ long by 12′-0″ wide building. Illustrate in plan view the placement of batter boards, string lines, and diagonal measurements that are required to locate the foundation wall and footings for this small structure.

3. Lay out a floor framing plan at a scale of ¼″ = 1′-0″ with 2 × 8 floor joists at 16″ o.c. for the 24′-0″ × 12′-0″ building. Show headers and two rows of bridging. Label materials and sizes on the drawing.

4. Using the scale ¼″ = 1′-0″, lay out a hip-roof framing plan with 2 × 6 rafters at 16″ o.c. for the 24′-0″ × 12′-0″ building using 2′-0″ projection beyond the exterior wall to form an overhang eave. Label materials and sizes on the drawing.

5. Draw the elevation view of a standard W-pattern truss with an 8:12 pitch and a clear span of 28′-0″ at a scale of ¼″ = 1′-0″. Top and bottom chords are 2 × 6s, web members are 2 × 4s, and gussets are ½″ plywood. Eaves extend 1′-6″ beyond the face of the exterior wall. Label materials and sizes on the drawing.

6. Draw freehand sketches of the front elevation and one side elevation of small buildings with these roof types: gable, hip, shed, mansard, folded-plate, and gambrel. Scale the sketches to fit neatly on the sheet.

7. Draw a freehand section sketch through a 1-story wood frame structure with crawl space below the floor framing and a pitched gable roof with an attic. Using arrows, graphically show where the crawl space and attic should be ventilated. Label vents, and show the ratio of vent area to surface areas.

8. Draw nine 2″ wide by 1½″ high rectangles equally spaced on the sheet; sketch within each rectangle an elevation view of the following types of exterior siding: horizontal bevel, board-and-batten, novelty, boards-on-boards, V-joint, boards-over-boards, wood shingles with random coursing, wood shingles with regular coursing, and stucco.

9. Sketch a front elevation view of a 1-story gable-roof residence with a brick chimney penetrating the roof surface. Draw one side elevation of the same house, and show a saddle on the roof at the chimney.

10. Sketch four elevations of the end portion of a building with a hip roof. Show a different roofing material on each of the four sketches. Include the following roofing materials: fiberglass shingles, standing seam copper, Spanish clay tile, and wood shakes.

1. Why must you consider local frost depths when drawing footings?

2. What is a good rule of thumb for sizing footings in light construction?

3. Describe a stepped footing, and indicate its application.

4. What condition may require accurate calculation of footing sizes?

5. List three materials that are commonly used for foundation walls.

6. What is a pier?

7. Indicate the logical application of grade beam foundation construction.

8. What are the advantages of a slab-on-ground-floor construction in comparison with crawl space construction?

9. What are the identifying characteristics of platform (western) frame construction?

10. Why are trussed rafters widely used in small-home construction?

11. In plank-and-beam framing, how is the load on roof beams transmitted to the foundation?

12. Why should the location of beams in plank-and-beam framing be shown on the working drawings?

13. What part of construction limits the spacing of beams in plank-and-beam framing?

14. What is a transverse beam?

15. List the good points of brick veneer construction.

16. Explain modular sizing of concrete block walls.

17. What are the important factors in the selection of a roofing material?

18. What is the purpose of a vapor barrier in a wood wall, and why would it be placed facing the interior side?

19. What are metal framing anchors, and where are they used in construction?

20. What is flashing? Identify two places where flashing is used.

PRODUCT RESEARCH USING COMPUTERS

As we described in Chapter 2, "Computer-Aided Drafting and Design," CD-ROM resources provide relatively fast access to a wealth of information about construction-related products. The figures on pages 384–387 show screens generated for two different building products from the *SweetSource* CD. With *SweetSource*, you can collect and organize the products that you select. Information on a specific product may contain text describing the technical features, CAD drawings, tables of product data, and photographs. You can copy or export these elements to other software products, including word processors, graphics applications, and CAD programs.

The first three figures contain information about Hebel's aerated concrete products. Figure 1 shows an introductory page which contains a digitally reproduced photograph of a residence built using the Hebel products and a written description of the product. Either window can be enlarged to nearly full screen. Figure 2 shows three more windows containing construction details, product specifica-

tions, and a product spreadsheet table. Each window can be enlarged and printed independently, as shown in Figure 3.

Figure 4 describes a rigid Styrofoam® exterior foundation drainage insulation manufactured by Dow Chemical Company. The initial screen contains three windows: a product photo, a product description and specifications, and a product properties table. The bottom of the figure shows enlarged portions of the product description window and the entire product properties table.

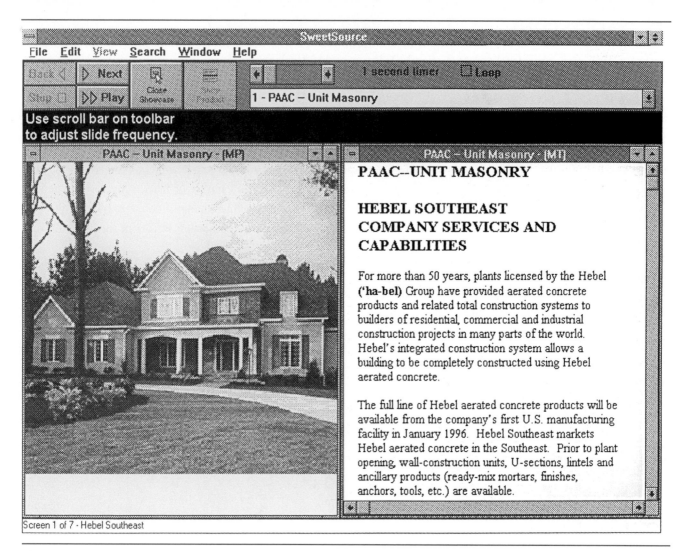

Figure 1 (SweetSource CD-ROM, McGraw-Hill)

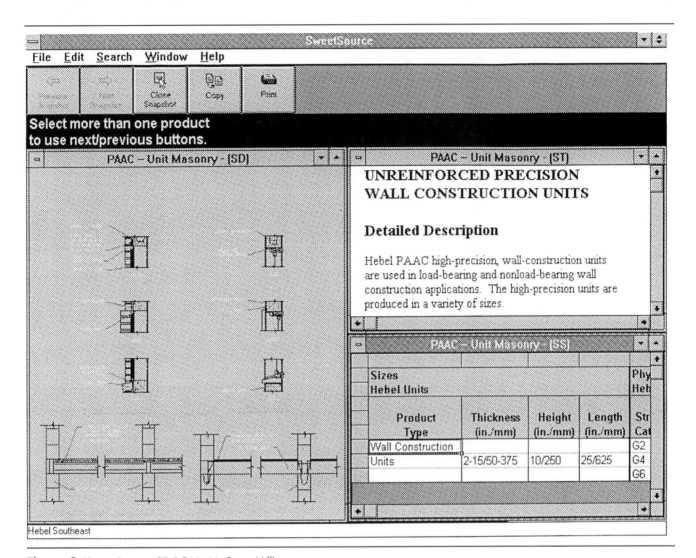

Figure 2 (SweetSource CD-ROM, McGraw-Hill)

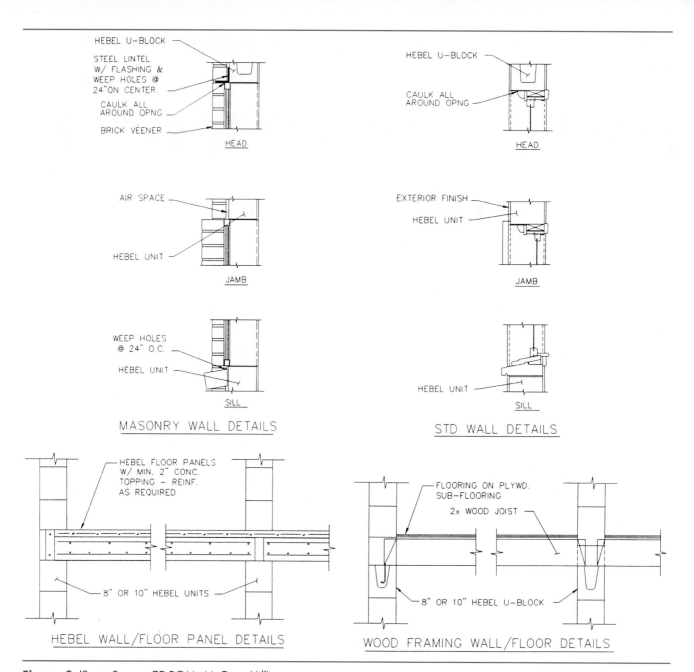

MASONRY WALL DETAILS

STD WALL DETAILS

HEBEL WALL/FLOOR PANEL DETAILS

WOOD FRAMING WALL/FLOOR DETAILS

Figure 3 (SweetSource CD-ROM, McGraw-Hill)

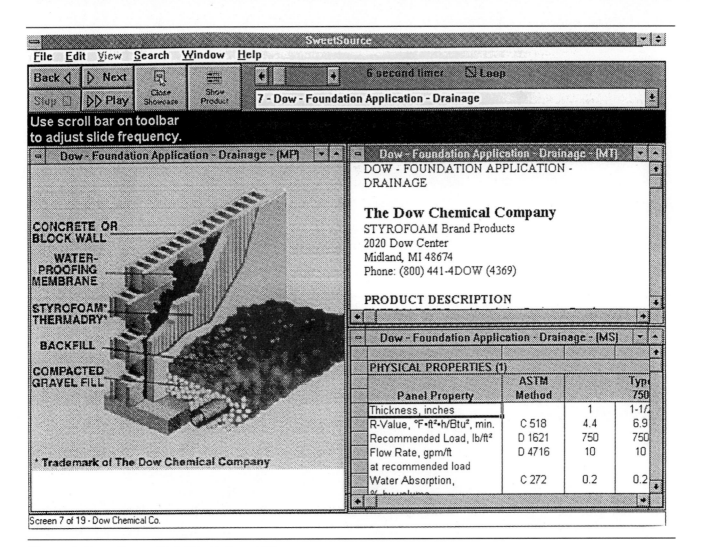

Figure 4 (SweetSource CD-ROM, McGraw-Hill)

"There is always a best way of doing
everything, if it be to boil an egg."

—RALPH WALDO EMERSON

STRUCTURAL MEMBER SELECTION 13

It is common practice to indicate the sizes of structural members on working drawings. This chapter is concerned with this information and the span and loading tables that must be used to obtain the sizes. The tables are readily available to drafters and designers. On residential drawings this information is placed on the details, floor plans, or other views that seem the most appropriate. In a set of working drawings for a commercial building, structural information is found on separate structural drawings included with the architectural set—usually completed by a structural engineer or other specialist in the area involved.

The tables are compiled and published by various lumber manufacturing associations (for example, the National Forest Products Association) as well as by federal government departments involved with housing and construction. Some are included as part of local code manuals, which often include a nationally recognized building code as a part and which are available at local building inspection offices. Drafters should consult these local code restrictions before making any decisions on structural members.

Notice that each species of framing lumber has its own loading table; a number of the tables are included in this chapter. The tables give minimum sizes, and sometimes good judgment may require that sizes taken from the tables be modified for structural soundness. Only experience in wood framing can provide this valuable background. Occasionally, it becomes feasible to use steel members in critical situations in wood framing; several steel tables are included here for that purpose. Many steel shapes and sizes are available to builders; consult the latest *Manual of Steel Construction* by the American Institute of Steel Construction for detailed information on steel shapes. Often a structural problem can be solved by the introduction of a steel member where wood may not be entirely adequate. For unusual framing problems, it is safest to employ the services of a competent structural engineer experienced in wood structures.

However, the drafter or designer should be familiar with the span tables and other structural information available for specifying and selecting the correct sizes of structural members needed in residential working drawings—mainly so that the builder will not be confronted with a hit-or-miss situation and so that the building will satisfactorily serve its intended purpose.

13.1

STRESSES IN MEMBERS

Let us begin by discussing lumber and its ability to support loads. Wood is a natural material as opposed to human-made, quality-controlled materials such as steel or aluminum; thus, it has inherent structural variations and limits. These variations are caused by species characteristics, natural defects, manufacture, moisture content, and so on. Some species, for example, oak and maple, have excellent resistance to the wear qualities needed on finish floors; other species resist weathering, insect and fungi attack, and decay, such as redwood, cedar, and cypress; yet they possess less desirable structural qualities. The widely cut Douglas fir and southern pine are especially suited for structural use. The majority of structural lumber manufactured today is made from these two species, although considerable amounts of lumber are made from several other minor species. Notice from the tables in this chapter that even those species that have less favorable load-bearing qualities are readily used if need be; yet their sizes must be increased or spans reduced to carry comparable loads.

Normal safety factors are reflected in some tables. Final selection of lumber, however, is usually based on economy and availability of the lumber locally. Higher grades of similar species of lumber can be expected to carry greater loads, as indicated by the tables. Although the innate character of the wood does not change with lower grades, the fewer defects in the better grades will no doubt result in more reliability (Table 13–1) and, as a result, longer spans are given on the tables.

For economy reasons, residential framing lumber is usually visually graded rather than machine graded as is needed for select structural lumber where critical strength and loading properties are needed.

The several structural species of lumber mentioned previously are usually stiffer than other types, especially under loads. The tendency for a member to bend when loaded is known as *deflection* (Fig. 13–1). In some situations within the structure, excessive deflection can cause serious damage, for example, in a plastered ceiling supported by undersized ceiling joists. The deflection under load could cause plaster cracking. Therefore, many tables show several spans, each limited by a different deflection factor. The standard for maximum allowable deflection where very little deflection can be tolerated is $^1/_{360}$ of the span; for less critical situations, $^1/_{240}$ of the span is used. Deflection on long beams can be significant. For example, a horizontal beam or joist limited to, say, an $^1/_{240}$ deflection and selected to span 20'-0" would possibly deflect 1" near its center and still support its designed load. The less restrictive deflection factor is usually specified for rafters or other members where the actual deflection is not as visible as in floor or ceiling framing. Deflection varies inversely with the depth of the beam; so, if suit-

Table 13–1 Basic grade classifications for yard lumber.

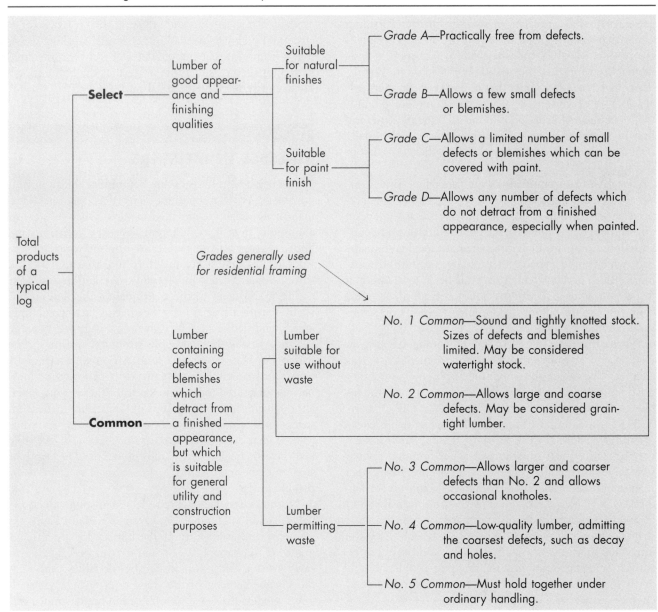

able, you should select sizes with maximum depth. Members loaded with continuous heavy loads are affected by long-term deflection, often referred to as *live-load creep*. Over a period of years the deformation may be substantial. We see evidence of this creep in older buildings where floors are under continual heavy load; the loaded wood beams often sag considerably. To overcome creep, stiffer and larger members, or steel, can be used.

Elasticity is another important characteristic in selecting framing lumber; the greater the modulus of elasticity (*E*), the stiffer the lumber. Notice in the tables that both southern pine and Douglas fir rate high in this quality.

The stress to be considered in column selection is *buckling* (Fig. 13–1). The fiber in wood columns is under *compression* (parallel to grain) when loaded; some species can resist this squeezing-type stress better than others. *Tension* is a pulling-type stress, not of major concern in most light frame members, except those in triangular-shaped trusses. Whereas columns resist mainly compressive forces parallel to the grain, flexural members, such as beams and joists, are subject to a combination of stresses as shown by Fig. 13–1. This bending (flexure) produces longitudinal stress. For simply supported members, this stress results in compression at the top fiber and tension at the bottom fiber. Vertical applied loads are also resisted by internal shear stresses, which produce a sliding or cutting action. Where joists or beams are supported, compression stresses perpendicular to the grain are developed; if the bearing length is too short, the bearing area is reduced, thus increasing the unit stresses to the extent that the wood fiber collapses.

MOISTURE CONTENT

Moisture content also affects the strength of lumber. It is the weight of the water in the wood, expressed as a percentage of the completely dry weight of the wood. Green lumber is not as strong as dried lumber, and we must remember that wood, especially dry wood, is a very good insulator from heat transfer. Lumber, however, tends to take on or give off moisture according to its environment. When it gives off moisture, it shrinks; when it takes it on, it swells. Lumber will shrink more in the dry, southwestern United States than it will in, for example, the coastal and southeastern states. The swelling and shrinking in various amounts is inevitable in wood structures, and the designer should be aware of it. Wood fiber, however, tends to shrink and swell less on end grain than across grain. Therefore, critical supports would be best placed with stresses against end grain (such as with studs and posts) wherever possible. Temperature changes, on the other hand, affect the structural properties of wood very little.

For economy and availability, residential framing lumber is universally specified at 19% moisture content; structural lumber used for more critical design loads is often specified at 15% moisture content. Check these criteria on

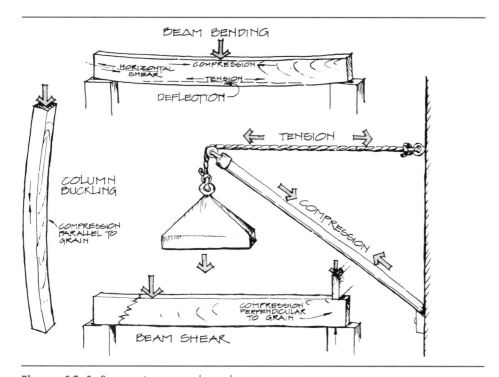

Figure 13–1 Stresses in structural members.

tables, because values on identical members will vary. Finish lumber is usually specified at 12%; lumber with moisture content above 19% is considered unseasoned.

LOCAL CODES

Since every building must conform to building codes and restrictions, drafters must be familiar with these documents when doing residential drawings. The purpose of the codes is to provide for safety and to protect the health and welfare of the occupants. Live loads on floors and roofs are especially important in wood framing (Fig. 13–2). Local conditions throughout the United States vary considerably, and therefore local restrictions must be used to reflect these conditions. However, most local codes also include one of the national codes—the Uniform Building Code, the Standard Building Code, and others—as part of the document. Some areas use the F.H.A. Minimum Property Standards as a supplement to their codes. Remember that codes indicate minimum values, and any departures should be on the restrictive side.

FIRST DETERMINE THE LOADS

Before using the tables in this chapter to size framing members, calculate the total load that will be imposed on each supporting member. The weights of materials can be found in Table 12–2. More complete building material weights can be obtained from Architectural Graphic Standards. Include all materials that will form the tributary load (see Fig. 12–6). Roof loads are usually transmitted directly to exterior walls, especially roofs that are supported with prefabricated trusses, and they therefore do not need to be considered in column and girder loading. The total applied load supported by a structural member includes both the *dead load* and the *live load*. Refer to Table 13–2 for typical minimum live loads in residential framing.

Figure 13–3 presents an example of residential dead and live loads. The total design load at the base of the foundation wall or the footing of a typical single-family, wood frame house does not generally exceed 2500 lb per square foot (psf) for 1-story structures or 3000 psf for 2-story structures. The total design loads on center girder support columns generally do not exceed 7000 lb for 1-story structures or 10,000 lb for 2-story structures. Usually the local building code establishes the design snow load due to the regional variations in seasonal snowfall.

13.4.1 Columns

Indicate the size and the type of basement columns on the *basement plan*. Because steel columns are readily available and their smaller size allows more usable floor space, they have become more popular than wood columns, especially in 1-story heights (Fig. 13–2). Welded ¼" steel plates are attached to each end for securing (see Fig. 12–50). The hollow center may be filled with concrete to increase their strength.

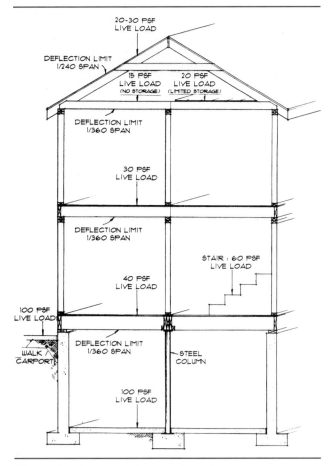

Figure 13–2 Live loads in residential construction.

Table 13–2 Minimum residential live loads.

Location	Live load (psf)
Dwelling rooms (other than sleeping)	40
Dwelling rooms (sleeping only)	30
Attics (served by any type of stairs)	30
Attics (limited storage)	20
Attics (no storage)	15
Stairs	60
Public stairs, corridors (2-family duplex)	60
Garage or carport (passenger cars)	100
Walks	100

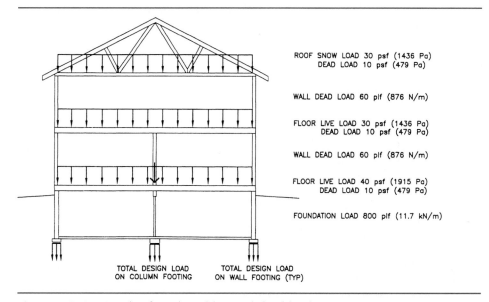

ROOF SNOW LOAD 30 psf (1436 Pa)
DEAD LOAD 10 psf (479 Pa)

WALL DEAD LOAD 60 plf (876 N/m)

FLOOR LIVE LOAD 30 psf (1436 Pa)
DEAD LOAD 10 psf (479 Pa)

WALL DEAD LOAD 60 plf (876 N/m)

FLOOR LIVE LOAD 40 psf (1915 Pa)
DEAD LOAD 10 psf (479 Pa)

FOUNDATION LOAD 800 plf (11.7 kN/m)

TOTAL DESIGN LOAD ON COLUMN FOOTING

TOTAL DESIGN LOAD ON WALL FOOTING (TYP)

Figure 13–3 Example of residential live and dead loads.

Try to locate columns at load points if possible. When the floor area to be framed is established, make several sketches on overlap paper above the plan to determine the best framing and support method possible. Columns used to support wood or steel girders are typically 3″ steel pipe or 4 × 4 or 6 × 6 timber. Table 13–3 shows the maximum column spacing for typical wood and steel columns that have a maximum unsupported length of 8′.

The American Iron and Steel Institute, in their publication RG-936, *Residential Steel Beam and Column Load/Span Tables* (1993), has developed load and span tables for column and beam sections commonly used in residential construction. The tables identify recommended spacing for columns of various sizes depending on the width of the house and the tributary area supported by the column. Roof loads are not included. It is assumed that the roof system spans between exterior bearing walls. Table 13–4 contains an example for supporting one floor only.

13.4.2 Beams

Indicate the size and the type of beams that support first-floor joists on the basement or foundation plan. Usually, a heavy broken line is a sufficient symbol. Again, if living space is important in the basement area, it may be advantageous to use a steel beam rather than wood to allow columns to be spaced farther apart. In crawl spaces use of steel beams is not critical; C.M.U. piers are uni-

versally used at ground level since they can be located at critical points under the framing and they can be somewhat closer together to minimize beam sizes.

Wood beams can be either solid or built up from 2″ framing lumber well spiked together. Notice that built-up beams span slightly shorter spans than similar-size solid members. This fact is reflected in the tables because more sides have been planed during finishing, which results in a smaller cross section. In small homes it is usually more practical to build up the wood beam using the same-size framing material as that used for joists, shown in Fig. 12–46D. If headroom or an uninterrupted ceiling is important below, or if a beam must be introduced into the joist framing and must be at the same level, ledger strips or hangers can be used (Fig. 12–46D) to support the joists against the beam rather than rest on the beam. Table 13–5 indicates allowable spans for built-up wood girders for 1-, 2-, and 3-story structures of varying widths.

Steel beams are usually W sections; allow for a wood plate to be fastened above the beam for securing the wood joists. The wood joists can also be framed to the side of the steel beam as shown in Fig. 12–46E. Steel column cap plates are welded or fastened to the beam with bolts (see Fig. 12–49). As mentioned previously, steel beams usually allow more headroom below.

Steel angle lintels, usually needed over brick veneer openings, are listed in Table 12–10. Table 13–6 shows allowable spans for steel girders for 1-, 2-, and 3-story structures of varying widths.

Table 13-3 Maximum spacing for typical wood and steel columns for support of center beams.

Width of Structure in Feet (Meters)	Column Size	Maximum Spacing					
		1 Story		2 Story		3 Story	
		ft-in.	m	ft-in.	m	ft-in.	m
24 (7.32)	4 × 4 wood	8-4	2.53	4-7	1.40		
	6 × 6 wood	20-2	6.16	11-2	3.41	7-10	2.38
	3″ diameter STD	25-2	7.68	13-7	4.15	9-4	2.83
	TS 3 × 3 × 0.1875	27-4	8.32	14-10	4.51	10-1	3.08
	3.5″ diameter STD	34-7	10.55	18-8	5.70	12-10	3.90
28 (8.53)	4 × 4 wood	7-1	2.16	4-0	1.22		
	6 × 6 wood	17-4	5.27	9-7	2.93	6-7	2.01
	3″ diameter STD	21-8	6.61	11-10	3.60	8-1	2.47
	TS 3 × 3 × 0.1875	23-6	7.16	12-10	3.90	8-10	2.68
	3.5″ diameter STD	29-10	9.08	16-2	4.94	11-1	3.38
32 (9.75)	4 × 4 wood	6-4	1.92				
	6 × 6 wood	15-1	4.60	8-5	2.56	5-10	1.77
	3″ diameter STD	19-1	5.82	10-5	3.17	7-1	2.16
	TS 3 × 3 × 0.1875	20-8	6.31	11-4	3.44	7-8	2.35
	3.5″ diameter STD	26-2	7.99	14-4	4.36	9-10	2.99
36 (10.97)	4 × 4 wood	5-7	1.71				
	6 × 6 wood	13-5	4.08	7-6	2.29	5-2	1.58
	3″ diameter STD	17-0	5.18	9-4	2.83	6-5	1.95
	TS 3 × 3 × 0.1875	18-5	5.61	10-1	3.08	6-11	2.10
	3.5″ diameter STD	23-5	7.13	12-10	3.90	8-10	2.68

SOURCE: For pipe and tube columns: *Residential Steel Beam and Column Load/Span Tables—Pipe and Tube Columns,* American Iron and Steel Institute, Washington, DC, 1993. Used by permission of AISI.
NOTES:
1. Live load for the first floor is 40 psf (1915 Pa); for the second and third floors, 30 psf (1436 Pa).
2. Dead load is 10 psf (479 Pa) per floor. A clear-span, trussed roof is assumed; no attic or roof loads are included.
3. The unbraced length of a column is 8 ft (2.44 m).
4. Wood columns are limited by column compression buckling. Values are not applicable to adjustable-height columns.

Table 13-4 Steel column spacing in residential construction.

Residential Steel Column Load/Spacing Tables - Pipe and Tube Columns
MAXIMUM COLUMN SPACING, when C = D or MAXIMUM TRIBUTARY LENGTH, (C+D)/2, when C ≠ D; ft.
ONE FLOOR ONLY (no roof or attic loads) - Unbraced Length of Column = 8 feet

DL (psf) + 10   LL (psf)* 40

COLUMN	Column Properties					TRIBUTARY WIDTH SUPPORTED BY THE CENTER BEAM - (A+B)/2								
SIZE	Weight/Ft.	Fy	A	Pe	Pa	8'-0	10'-0	12'-0	14'-0	16'-0	18'-0	20'-0	22'-0	24'-0
3″dia. STD.	7.58	36	2.23	16	34	35.0	30.0	25.2	21.7	19.1	17.0	15.3	14.0	12.8
TS 3x3x0.1875	6.87	46	2.02	17	35	35.0	32.5	27.3	23.5	20.7	18.4	16.6	15.2	13.9
3.5″dia. STD.	9.11	36	2.68	22	44	35.0	35.0	34.6	29.8	26.2	23.4	21.1	19.2	17.7
TS 3x3x0.2500	8.81	46	2.59	21	44	35.0	35.0	33.7	29.1	25.5	22.8	20.6	18.7	17.2

DL (psf) + 15   LL (psf)* 40

COLUMN	Column Properties					TRIBUTARY WIDTH SUPPORTED BY THE CENTER BEAM - (A+B)/2								
SIZE	Weight/Ft.	Fy	A	Pe	Pa	8'-0	10'-0	12'-0	14'-0	16'-0	18'-0	20'-0	22'-0	24'-0
3″dia. STD.	7.58	36	2.23	16	34	33.8	27.3	23.0	19.8	17.4	15.5	14.0	12.7	11.7
TS 3x3x0.1875	6.87	46	2.02	17	35	35.0	29.6	24.9	21.4	18.8	16.8	15.2	13.8	12.7
3.5″dia. STD.	9.11	36	2.68	22	44	35.0	35.0	31.6	27.2	23.9	21.3	19.2	17.5	16.1
TS 3x3x0.2500	8.81	46	2.59	21	44	35.0	35.0	30.8	26.5	23.3	20.8	18.7	17.1	15.7

DL (psf) + 20   LL (psf)* 40

COLUMN	Column Properties					TRIBUTARY WIDTH SUPPORTED BY THE CENTER BEAM - (A+B)/2								
SIZE	Weight/Ft.	Fy	A	Pe	Pa	8'-0	10'-0	12'-0	14'-0	16'-0	18'-0	20'-0	22'-0	24'-0
3″dia. STD.	7.58	36	2.23	16	34	31.1	25.2	21.1	18.2	16.0	14.2	12.8	11.7	10.7
TS 3x3x0.1875	6.87	46	2.02	17	35	33.8	27.3	22.9	19.7	17.3	15.4	13.9	12.7	11.6
3.5″dia. STD.	9.11	36	2.68	22	44	35.0	34.6	29.0	25.0	22.0	19.6	17.7	16.1	14.8
TS 3x3x0.2500	8.81	46	2.59	21	44	35.0	33.7	28.3	24.4	21.4	19.1	17.2	15.7	14.4

Column loads are based on a maximum eccentricity of 1″ between the resultant (total) load and the centerline of the column.
Supported beams may be single span or continuous with a maximum eccentricity of 1″ for the resultant load.
Fy = Minimum design yield stress per the AISC Specification, ksi.   K = 1.0
A = Gross cross-sectional area of column per the AISC Manual, in.
Pe = Maximum axial load with an eccentricity of 1″, per the AISC Manual, kips.
Pa = Allowable axial load values from the 1989 AISC - ASD Manual, Allowable Concentric Load Tables, kips.
* No live load reductions have been included.   + DL is in addition to beam weight.
(C+D)/2 has been limited to 35 feet to correspond with the beam tables.
Column bearing design must be per the AISC Specification. Guidance on base plate design can be found on pages 3-106 through 3-111 of the 1989 AISC ASD Manual.

SOURCE: Copyright 1993 by the American Iron and Steel Institute.

Table 13–5 Allowable spans for built-up wood center girders.

Width of Structure in Feet (Meters)	Girder Size	Maximum Clear Span					
		1 Story		2 Story		3 Story	
		ft-in.	m	ft-in.	m	ft-in.	m
24 (7.32)	3- 2 × 8	6-7	2.02	4-11	1.50	4-1	1.25
	4- 2 × 8	7-8	2.33	5-8	1.74	4-9	1.44
	3- 2 × 10	8-5	2.57	6-3	1.92	5-3	1.60
	4- 2 × 10	9-9	2.97	7-3	2.22	6-1	1.84
	3- 2 × 12	10-3	3.13	7-8	2.33	6-4	1.94
	4- 2 × 12	11-10	3.61	8-10	2.69	7-4	2.24
26 (7.92)	3- 2 × 8	6-4	1.94	4-9	1.44	3-11	1.20
	4- 2 × 8	7-4	2.24	5-6	1.67	4-7	1.39
	3- 2 × 10	8-1	2.47	6-1	1.84	5-0	1.53
	4- 2 × 10	9-4	2.86	7-0	2.13	5-10	1.77
	3- 2 × 12	9-10	3.01	7-4	2.24	6-1	1.87
	4- 2 × 12	11-5	3.47	8-6	2.59	7-1	2.15
28 (8.53)	3- 2 × 8	6-2	1.88	4-7	1.39	3-10	1.16
	4- 2 × 8	7-1	2.16	5-3	1.61	4-5	1.34
	3- 2 × 10	7-10	2.38	5-10	1.78	4-10	1.48
	4- 2 × 10	9-0	2.75	6-9	2.05	5-7	1.71
	3- 2 × 12	9-6	2.90	7-1	2.16	5-11	1.80
	4- 2 × 12	11-0	3.35	8-2	2.49	6-10	2.08
32 (9.75)	3- 2 × 8	5-9	1.75	4-3	1.30	3-7	1.08
	4- 2 × 8	6-7	2.02	4-11	1.50	4-1	1.25
	3- 2 × 10	7-4	2.23	5-5	1.66	4-6	1.38
	4- 2 × 10	8-5	2.57	6-3	1.92	5-3	1.60
	3- 2 × 12	8-11	2.71	6-8	2.02	5-6	1.68
	4- 2 × 12	10-3	3.13	7-8	2.33	6-4	1.94

NOTES:
1. Values are for a clear-span, trussed roof, a load-bearing center wall on the first floor in 2-story construction, and a load-bearing center wall on the first and second floors in 3-story construction.
2. Spans are based on a species and grade of lumber having an allowable bending stress $F_b = 1000$ psi (6895 kPa) for repetitive members.

Table 13–6 Allowable spans for typical steel center girders.

Width of Structure in Feet (Meters)	Beam Size	Maximum Center-to-Center Span					
		1 Story		2 Story		3 Story	
		ft-in.	m	ft-in.	m	ft-in.	m
24 (7.32)	W 6 × 9	11-4	3.44	8-10	2.68	7-4	2.23
	W 8 × 10	14-0	4.27	10-5	3.17	8-7	2.62
	W 10 × 12	16-10	5.12	12-4	3.75	10-2	3.11
	W 12 × 14	19-7	5.97	14-5	4.39	11-11	3.63
	W 14 × 22	26-0	7.92	20-0	6.07	16-6	5.03
28 (8.53)	W 6 × 9	10-10	3.29	8-2	2.50	6-10	2.07
	W 8 × 10	13-2	4.02	9-8	2.96	8-0	2.44
	W 10 × 12	15-7	4.75	11-5	3.47	9-6	2.90
	W 12 × 14	18-2	5.55	13-5	4.08	11-0	3.35
	W 14 × 22	24-8	7.53	18-7	5.67	15-4	4.66
32 (9.75)	W 6 × 9	10-4	3.14	7-8	2.35	6-4	1.92
	W 8 × 10	12-5	3.78	9-1	2.77	7-6	2.29
	W 10 × 12	14-7	4.45	10-8	3.26	8-11	2.71
	W 12 × 14	17-0	5.18	12-6	3.81	10-5	3.17
	W 14 × 22	23-7	7.19	17-5	5.30	13-6	4.11
36 (10.97)	W 6 × 9	9-10	2.99	7-2	2.19	6-0	1.83
	W 8 × 10	11-8	3.57	8-7	3.11	7-1	2.16
	W 10 × 12	13-10	4.21	10-1	3.08	8-5	2.56
	W 12 × 14	16-1	4.91	11-10	3.60	9-8	2.96
	W 14 × 22	22-4	6.80	16-6	5.03	12-0	3.66

SOURCE: *Residential Steel Beam and Column Load/Span Tables—Wide Flange Beams,* American Iron and Steel Institute, Washington, DC, 1993. Used by permission of AISI.
NOTES:
1. Live load for the first floor is 40 psf (1915 Pa); for the second and third floors, 30 psf (1436 Pa).
2. The spans assume clear-span, trussed roof construction (no attic or roof loads are included).
3. Dead load is 10 psf (479 Pa) per floor.

Table 13-7 Relative strengths of engineered wood products.

Engineered Wood Product	Fiber Stress in Bending, F_b		Tension Parallel to Grain F_t		Shear Parallel to Grain, F_v		Compression Perpendicular to Grain, $F_{o\perp}$		Compression Parallel to Grain $F_{o\parallel}$		Modulus of Elasticity, E	
	psi	kPa	psi	kPa	psi	kPa	psi	kPa	psi	kPa	psi	kPA
Parallel-strand lumber[a]	2950	20,340	2400	16,547	290	1999	600	4137	2900	19,995	2,000,000	13,789,520
Laminated veneer lumber[b]	2875	19,822	1850	12,755	285	1965	500	3447	2700	18,616	2,000,000	13,789,520
Glue laminated lumber[c]	1600-2400	11,032-16,547	650-1300	4482-8963	90-200	621-1379	300-600	2068-4137	900-1750	6205-12,066	1,100,000-1,800,000	7,584,236-12,410,568

SOURCE: *Alternatives to Lumber and Plywood in Home Construction*, NAHB Research Center. Prepared for the U.S. Department of Housing and Urban Development, Washington, DC, 1993.
[a]Parallam PSL, 2 × 10.
[b]MicroLam, 2 × 10.
[c]Consult National Design Specification by American Forest and Paper Association (AFPA) for design values and adjustment factors for specific grades of glue laminated lumber.

13.4.3 Engineered Wood Girders

Engineered wood products such as glue laminated lumber (glulam), laminated veneer lumber (LVL), and parallel-strand lumber (PSL) are also available. These products generally have more uniform structural properties and a lower moisture content. Table 13–7 provides a comparison of the relative strengths of some of these products. Table 13–8, developed by the American Institute of Timber Construction (AITC), compares equivalent glulam sections for solid beams, steel beams, and LVL beams. It is important to consult the manufacturer of any of these products for design analysis for any particular situation.

Table 13-8 Equivalent glulam sections.

Equivalent Glulam Sections for Solid Sawn Beams

Sawn Section Nominal Size	Roof Beams				Floor Beams			
	Select Structural		No. 1		Select Structural		No. 1	
	Douglas Fir	Southern Pine	Douglas Fir	Southern Pine	Douglas Fir	Southern Pine	Douglas Fir	Southern Pine
3 × 8	3¹/₈ × 6	3¹/₈ × 6⁷/₈	3¹/₈ × 6	3¹/₈ × 5¹/₂	3¹/₈ × 7¹/₂	3¹/₈ × 6⁷/₈	3¹/₈ × 7¹/₂	3¹/₈ × 6⁷/₈
3 × 10	3¹/₈ × 7¹/₂	3¹/₈ × 8¹/₄	3¹/₈ × 6	3¹/₈ × 6⁷/₈	3¹/₈ × 9	3¹/₈ × 9⁵/₈	3¹/₈ × 9	3¹/₈ × 9⁵/₈
3 × 12	3¹/₈ × 9	3¹/₈ × 9⁵/₈	3¹/₈ × 7¹/₂	3¹/₈ × 8¹/₄	3¹/₈ × 12	3¹/₈ × 11	3¹/₈ × 10¹/₂	3¹/₈ × 11
3 × 14	3¹/₈ × 9	3¹/₈ × 11	3¹/₈ × 7¹/₂	3¹/₈ × 8¹/₄	3¹/₈ × 13¹/₂	3¹/₈ × 13³/₄	3¹/₈ × 13¹/₂	3¹/₈ × 12³/₈
4 × 6	3¹/₈ × 6	3¹/₈ × 6⁷/₈	3¹/₈ × 6	3¹/₈ × 5¹/₂	3¹/₈ × 6	3¹/₈ × 6⁷/₈	3¹/₈ × 6	3¹/₈ × 6⁷/₈
4 × 8	3¹/₈ × 7¹/₂	3¹/₈ × 8¹/₄	3¹/₈ × 6	3¹/₈ × 6⁷/₈	3¹/₈ × 9	3¹/₈ × 8¹/₄	3¹/₈ × 7¹/₂	3¹/₈ × 8¹/₄
4 × 10	3¹/₈ × 9	3¹/₈ × 9⁵/₈	3¹/₈ × 7¹/₂	3¹/₈ × 8¹/₄	3¹/₈ × 10¹/₂	3¹/₈ × 11	3¹/₈ × 10¹/₂	3¹/₈ × 9⁵/₈
4 × 12	3¹/₈ × 10¹/₂	3¹/₈ × 12³/₈	3¹/₈ × 9	3¹/₈ × 9⁵/₈	3¹/₈ × 12	3¹/₈ × 12³/₈	3¹/₈ × 12	3¹/₈ × 12³/₈
4 × 14	3¹/₈ × 10¹/₂	3¹/₈ × 12³/₈	3¹/₈ × 9	3¹/₈ × 9⁵/₈	3¹/₈ × 15	3¹/₈ × 15¹/₈	3¹/₈ × 13¹/₂	3¹/₈ × 13³/₄
4 × 16	3¹/₈ × 12	3¹/₈ × 13³/₄	3¹/₈ × 10¹/₂	3¹/₈ × 9⁵/₈	3¹/₈ × 16¹/₂	3¹/₈ × 16¹/₂	3¹/₈ × 16¹/₂	3¹/₈ × 16¹/₂
6 × 8	5¹/₈ × 7¹/₂	5¹/₈ × 6⁷/₈	5¹/₈ × 6⁷/₈	5¹/₈ × 7¹/₂	5¹/₈ × 7¹/₂	5¹/₈ × 8¹/₄	5¹/₈ × 7¹/₂	5¹/₈ × 8¹/₄
6 × 10	5¹/₈ × 9	5¹/₈ × 8¹/₄	5¹/₈ × 9	5¹/₈ × 8¹/₄	5¹/₈ × 10¹/₂	5¹/₈ × 9⁵/₈	5¹/₈ × 10¹/₂	5¹/₈ × 9⁵/₈
6 × 12	5¹/₈ × 10¹/₂	5¹/₈ × 9⁵/₈	5¹/₈ × 9	5¹/₈ × 9⁵/₈	5¹/₈ × 12	5¹/₈ × 12³/₈	5¹/₈ × 12	5¹/₈ × 12³/₈
6 × 14	5¹/₈ × 12	5¹/₈ × 11	5¹/₈ × 10¹/₂	5¹/₈ × 11	5¹/₈ × 13¹/₂	5¹/₈ × 13³/₄	5¹/₈ × 13¹/₂	5¹/₈ × 13³/₄
6 × 16	5¹/₈ × 13¹/₂	5¹/₈ × 13³/₄	5¹/₈ × 12	5¹/₈ × 12³/₈	5¹/₈ × 16¹/₂	5¹/₈ × 16¹/₂	5¹/₈ × 16¹/₂	5¹/₈ × 16¹/₂
6 × 18	5¹/₈ × 15	5¹/₈ × 15¹/₈	5¹/₈ × 13¹/₂	5¹/₈ × 13³/₄	5¹/₈ × 18	5¹/₈ × 17¹/₈	5¹/₈ × 18	5¹/₈ × 17¹/₈
6 × 20	5¹/₈ × 18	5¹/₈ × 16¹/₂	5¹/₈ × 16¹/₂	5¹/₈ × 15¹/₈	5¹/₈ × 19¹/₂	5¹/₈ × 19¹/₄	5¹/₈ × 19¹/₂	5¹/₈ × 19¹/₄
8 × 10	6³/₄ × 9	6³/₄ × 8¹/₄	6³/₄ × 9	6³/₄ × 8¹/₄	6³/₄ × 10¹/₂	6³/₄ × 9⁵/₈	6³/₄ × 10¹/₂	6³/₄ × 9⁵/₈
8 × 12	6³/₄ × 10¹/₂	6³/₄ × 9⁵/₈	6³/₄ × 10¹/₂	6³/₄ × 9⁵/₈	6³/₄ × 12	6³/₄ × 12³/₈	6³/₄ × 12	6³/₄ × 12³/₈
8 × 14	6³/₄ × 12	6³/₄ × 12¹/₈	6³/₄ × 12	6³/₄ × 11	6³/₄ × 13¹/₂	6³/₄ × 13³/₄	6³/₄ × 13¹/₂	6³/₄ × 13¹/₄
8 × 16	6³/₄ × 13¹/₂	6³/₄ × 13³/₄	6³/₄ × 13¹/₂	6³/₄ × 12³/₈	6³/₄ × 16¹/₂	6³/₄ × 17⁷/₈	6³/₄ × 16¹/₂	6³/₄ × 17⁷/₈
8 × 18	6³/₄ × 16¹/₂	6³/₄ × 15¹/₈	6³/₄ × 15	6³/₄ × 13³/₄	6³/₄ × 18	6³/₄ × 17⁷/₈	6³/₄ × 18	6³/₄ × 17⁷/₈
8 × 20	6³/₄ × 18	6³/₄ × 17⁷/₈	6³/₄ × 16¹/₂	6³/₄ × 15¹/₄	6³/₄ × 19¹/₂	6³/₄ × 20⁵/₈	6³/₄ × 19¹/₂	6³/₄ × 20⁵/₈
8 × 22	6³/₄ × 19¹/₂	6³/₄ × 17⁷/₈	6³/₄ × 18	6³/₄ × 17⁷/₈	6³/₄ × 22¹/₂	6³/₄ × 22	6³/₄ × 22¹/₂	6³/₄ × 22

(continued)

Table 13–8 *(continued).*

Equivalent Glulam Sections for Steel Beams

Steel Section	Roof Beams Douglas Fir	Roof Beams Southern Pine	Floor Beams Douglas Fir	Floor Beams Southern Pine
W 6×9	$3\frac{1}{8}\times10\frac{1}{2}$ or $5\frac{1}{8}\times7\frac{1}{2}$	$3\frac{1}{8}\times9\frac{5}{8}$ or $5\frac{1}{8}\times8\frac{1}{4}$	$3\frac{1}{8}\times10\frac{1}{2}$ or $5\frac{1}{8}\times9$	$3\frac{1}{8}\times11$ or $5\frac{1}{8}\times9\frac{5}{8}$
W 8×10	$3\frac{1}{8}\times12$ $5\frac{1}{8}\times9$	$3\frac{1}{8}\times12\frac{3}{8}$ $5\frac{1}{8}\times9\frac{5}{8}$	$3\frac{1}{8}\times13\frac{1}{2}$ $5\frac{1}{8}\times10\frac{1}{2}$	$3\frac{1}{8}\times12\frac{3}{8}$ $5\frac{1}{8}\times11$
W 12×14	$3\frac{1}{8}\times16\frac{1}{2}$ $5\frac{1}{8}\times13\frac{1}{2}$	$3\frac{1}{8}\times16\frac{1}{2}$ $5\frac{1}{8}\times12\frac{3}{8}$	$3\frac{1}{8}\times18$ $5\frac{1}{8}\times15$	$3\frac{1}{8}\times17\frac{7}{8}$ $5\frac{1}{8}\times15\frac{1}{8}$
W 12×16	$3\frac{1}{8}\times18$ $5\frac{1}{8}\times13\frac{1}{2}$	$3\frac{1}{8}\times17\frac{7}{8}$ $5\frac{1}{8}\times13\frac{1}{2}$	$3\frac{1}{8}\times19\frac{1}{2}$ $5\frac{1}{8}\times16\frac{1}{2}$	$3\frac{1}{8}\times19\frac{1}{4}$ $5\frac{1}{8}\times16\frac{1}{2}$
W 12×19	$3\frac{1}{8}\times19\frac{1}{2}$ $5\frac{1}{8}\times15$	$3\frac{1}{8}\times19\frac{1}{4}$ $5\frac{1}{8}\times15\frac{1}{8}$	$3\frac{1}{8}\times21$ $5\frac{1}{8}\times18$	$3\frac{1}{8}\times20\frac{5}{8}$ $5\frac{1}{8}\times17\frac{7}{8}$
W 10×22	$3\frac{1}{8}\times21$ $5\frac{1}{8}\times16\frac{1}{2}$	$3\frac{1}{8}\times20\frac{5}{8}$ $5\frac{1}{8}\times16\frac{1}{2}$	$3\frac{1}{8}\times19\frac{1}{2}$ $5\frac{1}{8}\times16\frac{1}{2}$	$3\frac{1}{8}\times20\frac{5}{8}$ $5\frac{1}{8}\times16\frac{1}{2}$
W 12×22	$5\frac{1}{8}\times16\frac{1}{2}$ $6\frac{3}{4}\times15$	$5\frac{1}{8}\times16\frac{1}{2}$ $6\frac{3}{4}\times15\frac{1}{8}$	$5\frac{1}{8}\times18$ $6\frac{3}{4}\times16\frac{1}{2}$	$3\frac{1}{8}\times22$ $5\frac{1}{8}\times19\frac{1}{4}$
W 14×22	$5\frac{1}{8}\times18$ $6\frac{3}{4}\times16\frac{1}{2}$	$5\frac{1}{8}\times18$ $6\frac{3}{4}\times16\frac{1}{2}$	$5\frac{1}{8}\times21$ $6\frac{3}{4}\times18$	$5\frac{1}{8}\times20\frac{5}{8}$ $6\frac{3}{4}\times17\frac{7}{8}$
W 14×26	$5\frac{1}{8}\times21$ $6\frac{3}{4}\times18$	$5\frac{1}{8}\times19\frac{1}{4}$ $6\frac{3}{4}\times17\frac{7}{8}$	$5\frac{1}{8}\times21$ $6\frac{3}{4}\times19\frac{1}{2}$	$5\frac{1}{8}\times22$ $6\frac{3}{4}\times19\frac{1}{4}$
W 12×26	$5\frac{1}{8}\times19\frac{1}{2}$ $6\frac{3}{4}\times18$	$5\frac{1}{8}\times19\frac{1}{4}$ $6\frac{3}{4}\times16\frac{1}{2}$	$5\frac{1}{8}\times21$ $6\frac{3}{4}\times18$	$5\frac{1}{8}\times20\frac{5}{8}$ $6\frac{3}{4}\times19\frac{1}{4}$
W 16×26	$5\frac{1}{8}\times21$ $6\frac{3}{4}\times19\frac{1}{2}$	$5\frac{1}{8}\times20\frac{5}{8}$ $6\frac{3}{4}\times17\frac{7}{8}$	$5\frac{1}{8}\times22\frac{1}{2}$ $6\frac{3}{4}\times21$	$5\frac{1}{8}\times23\frac{3}{8}$ $6\frac{3}{4}\times20\frac{5}{8}$
W 12×30	$5\frac{1}{8}\times21$ $6\frac{3}{4}\times19\frac{1}{2}$	$5\frac{1}{8}\times20\frac{5}{8}$ $6\frac{3}{4}\times17\frac{7}{8}$	$5\frac{1}{8}\times21$ $6\frac{3}{4}\times19\frac{1}{2}$	$5\frac{1}{8}\times22$ $6\frac{3}{4}\times19\frac{1}{4}$
W 14×30	$5\frac{1}{8}\times22\frac{1}{2}$ $6\frac{3}{4}\times19\frac{1}{2}$	$5\frac{1}{8}\times22$ $6\frac{3}{4}\times19\frac{1}{4}$	$5\frac{1}{8}\times22\frac{1}{2}$ $6\frac{3}{4}\times21$	$5\frac{1}{8}\times23\frac{3}{8}$ $6\frac{3}{4}\times20\frac{5}{8}$
W 16×31	$5\frac{1}{8}\times24$ $6\frac{3}{4}\times21$	$5\frac{1}{8}\times23\frac{3}{8}$ $6\frac{3}{4}\times20\frac{5}{8}$	$5\frac{1}{8}\times25\frac{1}{2}$ $6\frac{3}{4}\times22\frac{1}{2}$	$5\frac{1}{8}\times24\frac{3}{4}$ $6\frac{3}{4}\times22$
W 14×34	$5\frac{1}{8}\times24$ $6\frac{3}{4}\times21$	$5\frac{1}{8}\times23\frac{3}{8}$ $6\frac{3}{4}\times20\frac{5}{8}$	$5\frac{1}{8}\times24$ $6\frac{3}{4}\times22\frac{1}{2}$	$5\frac{1}{8}\times22$ $6\frac{3}{4}\times22$
W 18×35	$5\frac{1}{8}\times27$ $6\frac{3}{4}\times24$	$5\frac{1}{8}\times26\frac{1}{8}$ $6\frac{3}{4}\times22$	$5\frac{1}{8}\times27$ $6\frac{3}{4}\times25\frac{1}{2}$	$5\frac{1}{8}\times27\frac{1}{2}$ $6\frac{3}{4}\times24\frac{3}{4}$
W 16×40	$5\frac{1}{8}\times28\frac{1}{2}$ $6\frac{3}{4}\times25\frac{1}{2}$	$5\frac{1}{8}\times27\frac{1}{2}$ $6\frac{3}{4}\times23\frac{3}{8}$	$5\frac{1}{8}\times27$ $6\frac{3}{4}\times25\frac{1}{2}$	$5\frac{1}{8}\times27\frac{1}{2}$ $6\frac{3}{4}\times24\frac{3}{4}$
W 18×40	$5\frac{1}{8}\times30$ $6\frac{3}{4}\times25\frac{1}{2}$	$5\frac{1}{8}\times27\frac{1}{2}$ $6\frac{3}{4}\times24\frac{3}{4}$	$5\frac{1}{8}\times28\frac{1}{2}$ $6\frac{3}{4}\times27$	$5\frac{1}{8}\times28\frac{7}{8}$ $6\frac{3}{4}\times26\frac{1}{8}$
W 21×44	$5\frac{1}{8}\times33$ $6\frac{3}{4}\times28$	$5\frac{1}{8}\times30\frac{1}{4}$ $6\frac{3}{4}\times27\frac{1}{2}$	$5\frac{1}{8}\times33$ $6\frac{3}{4}\times30$	$5\frac{1}{8}\times31\frac{5}{8}$ $6\frac{3}{4}\times28\frac{7}{8}$
W 18×50	$5\frac{1}{8}\times34\frac{1}{2}$ $6\frac{3}{4}\times30$	$5\frac{1}{8}\times31\frac{5}{8}$ $6\frac{3}{4}\times27\frac{1}{2}$	$5\frac{1}{8}\times31\frac{1}{2}$ $6\frac{3}{4}\times28\frac{1}{2}$	$5\frac{1}{8}\times31\frac{5}{8}$ $6\frac{3}{4}\times28\frac{7}{8}$
W 21×50	$5\frac{1}{8}\times34\frac{1}{2}$ $6\frac{3}{4}\times30$	$5\frac{1}{8}\times33$ $6\frac{3}{4}\times28\frac{7}{8}$	$5\frac{1}{8}\times34\frac{1}{2}$ $6\frac{3}{4}\times31\frac{1}{2}$	$6\frac{3}{4}\times31\frac{5}{8}$
W 18×55	$6\frac{3}{4}\times31\frac{1}{2}$	$6\frac{3}{4}\times30\frac{1}{4}$	$5\frac{1}{8}\times33$ $6\frac{3}{4}\times30$	$6\frac{3}{4}\times30\frac{1}{4}$
W 24×55	$6\frac{3}{4}\times34\frac{1}{2}$	$6\frac{3}{4}\times31\frac{5}{8}$	$6\frac{3}{4}\times34\frac{1}{2}$	$6\frac{3}{4}\times34\frac{3}{8}$
W 21×62	$6\frac{3}{4}\times36$	$6\frac{3}{4}\times34\frac{3}{8}$	$6\frac{3}{4}\times34\frac{1}{2}$	$6\frac{3}{4}\times34\frac{3}{8}$

Equivalent Glulam Sections for LVL Beams

LVL Section	Roof Beams Douglas Fir	Roof Beams Southern Pine	Floor Beams Douglas Fir	Floor Beams Southern Pine
2 pcs. $1\frac{3}{4}\times9\frac{1}{2}$	$3\frac{1}{8}\times12$ or $5\frac{1}{8}\times9$	$3\frac{1}{8}\times11$ or $5\frac{1}{8}\times9\frac{5}{8}$	$3\frac{1}{8}\times10\frac{1}{2}$ or $5\frac{1}{8}\times9$	$3\frac{1}{8}\times11$ or $5\frac{1}{8}\times9\frac{5}{8}$
2 pcs. $1\frac{3}{4}\times11\frac{7}{8}$	$3\frac{1}{8}\times13\frac{1}{2}$ $5\frac{1}{8}\times12$	$3\frac{1}{8}\times13\frac{3}{4}$ $5\frac{1}{8}\times11$	$3\frac{1}{8}\times13\frac{1}{2}$ $5\frac{1}{8}\times12$	$3\frac{1}{8}\times13\frac{3}{4}$ $5\frac{1}{8}\times11$
2 pcs. $1\frac{3}{4}\times14$	$3\frac{1}{8}\times16\frac{1}{2}$ $5\frac{1}{8}\times13\frac{1}{2}$	$3\frac{1}{8}\times16\frac{1}{2}$ $5\frac{1}{8}\times12\frac{3}{8}$	$3\frac{1}{8}\times16\frac{1}{2}$ $5\frac{1}{8}\times13\frac{1}{2}$	$3\frac{1}{8}\times15\frac{1}{8}$ $5\frac{1}{8}\times13\frac{3}{4}$
2 pcs. $1\frac{3}{4}\times16$	$3\frac{1}{8}\times18$ $5\frac{1}{8}\times15$	$3\frac{1}{8}\times17\frac{7}{8}$ $5\frac{1}{8}\times15\frac{1}{8}$	$3\frac{1}{8}\times18$ $5\frac{1}{8}\times15$	$3\frac{1}{8}\times17\frac{7}{8}$ $5\frac{1}{8}\times15\frac{1}{8}$
2 pcs. $1\frac{3}{4}\times18$	$3\frac{1}{8}\times21$ $5\frac{1}{8}\times16\frac{1}{2}$	$3\frac{1}{8}\times20\frac{5}{8}$ $5\frac{1}{8}\times16\frac{1}{2}$	$3\frac{1}{8}\times19\frac{1}{2}$ $5\frac{1}{8}\times16\frac{1}{2}$	$3\frac{1}{8}\times20\frac{5}{8}$ $5\frac{1}{8}\times16\frac{1}{2}$
3 pcs. $1\frac{3}{4}\times9\frac{1}{2}$	$3\frac{1}{8}\times13\frac{1}{2}$ $5\frac{1}{8}\times10\frac{1}{2}$	$3\frac{1}{8}\times13\frac{3}{4}$ $5\frac{1}{8}\times11$	$3\frac{1}{8}\times12$ $5\frac{1}{8}\times10\frac{1}{2}$	$3\frac{1}{8}\times12\frac{3}{8}$ $5\frac{1}{8}\times11$
3 pcs. $1\frac{3}{4}\times11\frac{7}{8}$	$3\frac{1}{8}\times18$ $5\frac{1}{8}\times13\frac{1}{2}$	$3\frac{1}{8}\times17\frac{7}{8}$ $5\frac{1}{8}\times13\frac{3}{4}$	$3\frac{1}{8}\times15$ $5\frac{1}{8}\times13\frac{1}{2}$	$3\frac{1}{8}\times15\frac{1}{8}$ $5\frac{1}{8}\times12\frac{3}{8}$
3 pcs. $1\frac{3}{4}\times14$	$3\frac{1}{8}\times21$ $5\frac{1}{8}\times16\frac{1}{2}$	$3\frac{1}{8}\times20\frac{5}{8}$ $5\frac{1}{8}\times15\frac{1}{8}$	$3\frac{1}{8}\times18$ $5\frac{1}{8}\times15$	$3\frac{1}{8}\times17\frac{7}{8}$ $5\frac{1}{8}\times15\frac{1}{8}$
3 pcs. $1\frac{3}{4}\times16$	$5\frac{1}{8}\times18$ $6\frac{3}{4}\times16\frac{1}{2}$	$5\frac{1}{8}\times17\frac{7}{8}$ $6\frac{3}{4}\times15\frac{1}{8}$	$3\frac{1}{8}\times21$ $5\frac{1}{8}\times18$	$5\frac{1}{8}\times17\frac{7}{8}$ $6\frac{3}{4}\times16\frac{1}{2}$
3 pcs. $1\frac{3}{4}\times18$	$5\frac{1}{8}\times21$ $6\frac{3}{4}\times18$	$5\frac{1}{8}\times20\frac{5}{8}$ $6\frac{3}{4}\times17\frac{7}{8}$	$5\frac{1}{8}\times19\frac{1}{2}$ $6\frac{3}{4}\times18$	$5\frac{1}{8}\times19\frac{1}{4}$ $6\frac{3}{4}\times17\frac{7}{8}$

Source: American Institute of Timber Construction.

13.4.4 Machine Stress-Rated (MSR) Lumber

Machine stress-rated (MSR) lumber is dimension lumber that has been evaluated by mechanical stress-rating equipment. Each piece of MSR lumber is nondestructively evaluated for bending stiffness (F_b) and is then sorted into modulus of elasticity (E) classes. One of the prime uses for MSR lumber is trusses. However, this product can also be used as floor and ceiling joists, as rafters, and for other structural purposes where assured strength is a primary concern. MSR lumber produced under an approved grading agency's certification and quality-control procedures is accepted by regulatory agencies and all major building codes.

Specifying MSR lumber is simple: Specify machine-rated, grade-stamped lumber, and list the strength value (F_b) and the corresponding modulus of elasticity (E) values. Also indicate nominal sizes and lengths required. Refer to Fig. 13–4 and Table 13–9 for examples and interpretations of a grade stamp and design values, respectively.

Machine-rated lumber

Lumber that bears the stamp "machine rated" or "MSR" (machine stress rated) has been tested by a machine that measures the wood's stiffness. The designation "E" stands for modulus of elasticity and is listed in millions of pounds per square inch (psi). "Fb" stands for extreme fiber stress in bending and is measured in psi. Machine-rated lumber is also inspected visually.

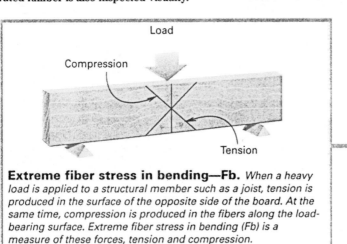

Extreme fiber stress in bending—Fb. *When a heavy load is applied to a structural member such as a joist, tension is produced in the surface of the opposite side of the board. At the same time, compression is produced in the fibers along the load-bearing surface. Extreme fiber stress in bending (Fb) is a measure of these forces, tension and compression.*

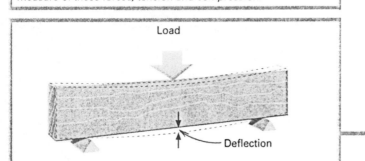

Modulus of elasticity—E. *Modulus of elasticity is a measure of stiffness. It is a ratio of how much a piece of lumber will deflect, or bend, in proportion to an applied load. E-values provide a conservative prediction of how much deflection might occur in a wall, floor or roof.*

How MSR lumber compares with visually graded lumber. *The table below shows four grades of visually graded lumber and their corresponding design values. The figures shown are base values that must be adjusted according to the width of the lumber. The figures represent the average that can be expected of a piece of lumber in a given grade. Each piece of MSR lumber is measured separately and marked with its particular strength and stiffness values.*

Species group	Grade	Extreme fiber stress in bending (Fb) (psi)	Modulus of elasticity (E) (psi)
Douglas fir-larch	Construction	1,000	1,500,000
	Standard	550	1,400,000
	Utility	275	1,300,000
	Stud	675	1,400,000

Figure 13–4 Interpretation of MSR lumber grade stamp. (Western Wood Products Association)

Table 13–9 MSR lumber design values for lumber 2″ and less in thickness and 2″ in width and wider.

Grade Designation[a]	Extreme Fiber Stress[b] in Bending, F_b	Modulus of Elasticity, E	Tension Parallel to Grain, F_t	Compression Parallel to Grain, $F_{c\parallel}$
	Single			
2850 F_b–2.3E	2850	2,300,000	2300	2150
2700 F_b–2.2E	2700	2,200,000	2150	2100
2550 F_b–2.1E	2550	2,100,000	2050	2025
2400 F_b–2.0E	2400	2,000,000	1925	1975
2250 F_b–1.9E	2250	1,900,000	1750	1925
2100 F_b–1.8E	2100	1,800,000	1575	1875
1950 F_b–1.7E	1950	1,700,000	1375	1800
1800 F_b–1.6E	1800	1,600,000	1175	1750
1650 F_b–1.5E	1650	1,500,000	1020	1700
1500 F_b–1.4E	1500	1,400,000	900	1650
1450 F_b–1.3E	1450	1,300,000	800	1625
1350 F_b–1.3E	1350	1,300,000	750	1600
1200 F_b–1.2E	1200	1,200,000	600	1400

SOURCE: Grades are described in Section 52.00 of *Western Lumber Grading Rules*.
NOTE: Design values are in psi. Design values for compression perpendicular to grain ($F_{c\perp}$) and horizontal shear (F_v) are the same as those assigned to visually graded lumber of the appropriate species, unless indicated on grade stamp.
[a]For any given value of F_b, the average modulus of elasticity (E) and the tension value (F_t) may vary depending upon the species, the timber source, and other variables. The E and F_t values included in the Fb–E grade designations in the table are those usually associated with each F_b level. Grade stamps may show higher or lower values if the machine rating indicates that the assignment is appropriate. If the F_t value for the MSR grade is different from that shown in the table for the same F_b level, the assigned F_t value shall be included on the grade stamp. When an E or F_t varies from the designated F_b level in the table, the tabulated $F_{c\parallel}$, $F_{c\perp}$, and F_v values associated with the designated F_b value are applicable.
[b]The tabulated F_b values are applicable only to lumber loaded on edge.

13.4.5 Floor Joists

Floor framing requires careful consideration in residential construction. Often the type of subfloor (plywood or 1″ lumber), the span required, and the species of lumber available play a part in the selection of a satisfactory solution. Codes require 40 psf in living areas, but it is common practice to maintain the same live load throughout the building. Indicate joist size, spacing, and direction with a note and a short line with arrows at each end. Plan the joist layout so that the joists span the shorter dimension, which should be reflected in the overlap sketch.

Normally, joists are spaced 16″ apart, yet situations may arise where 12″ or 24″ spacing may be more suitable. Plan the joist layout also so that a minimum of cutting is necessary; that is, plan joist lengths in 2′ increments if possible. Table 13–10 indicates allowable spans for floor joists, and selected floor joist span tables for southern pine and western wood species are given in Appendix E. Table 13–11, published by the Southern Pine Council (SPC), provides maximum span comparisons for joists made from a number of different lumber species.

As an example of using the tables, a living room floor with a 40-lb live load and a span of 15′-0″ requires 2 × 10 joists spaced 16″ o.c. if No. 2 southern pine is used. Notice that the joists are capable of spanning 16′-1″ if necessary. Should a span of 17′-0″ be needed in another area of the house, for instance, it may be best to space this set of joists 12″ o.c. rather than go to larger 2 × 12 joists spaced at the same 16″ o.c. spacing. This solution results in neater and less complex framing unless a higher grade of lumber is used. Other situations occur where some latitude in joist selection must be available to arrive at the most economical yet sound arrangement. It is usually most practical to stay with one grade and one cross-sectional size throughout the entire floor framing. Ceiling joists that support another floor above must be considered as floor joists as well.

Table 13-10 Allowable spans for floor joists.

Size	Joist Spacing in.	mm	1.0 psi ft-in.	6.9 kPa m	1.2 psi ft-in.	8.3 kPa m	1.4 psi ft-in.	9.6 kPa m	1.6 psi ft-in.	11.0 kPa m	1.8 psi ft-in.	12.4 kPa m	2.0 psi ft-in.	13.8 kPa m
					Modulus of Elasticity, E, in 1,000,000 psi or kPa									
							30 psf (1436 Pa) Live Load							
2 × 6	16	406	9-2	2.79	9-9	2.97	10-3	3.12	10-9	3.28	11-2	3.40	11-7	3.53
	24	610	8-0	2.44	8-6	2.59	8-11	2.72	9-4	2.84	9-9	2.97	10-1	3.07
2 × 8	16	406	12-1	3.68	12-10	3.91	13-6	4.11	14-2	4.32	14-8	4.47	15-3	4.65
	24	610	10-7	3.23	11-3	3.43	11-10	3.61	12-4	3.76	12-10	3.91	13-4	4.06
2 × 10	16	406	15-5	4.70	16-5	5.00	17-3	5.26	18-0	5.49	18-9	5.72	19-5	5.92
	24	610	13-6	4.11	14-4	4.37	15-1	4.60	15-9	4.80	16-5	5.00	17-0	5.18
2 × 12	16	406	18-9	5.72	19-11	6.07	21-0	6.40	21-11	6.68	22-16	6.96	23-7	7.19
	24	610	16-5	5.00	17-5	5.31	18-4	5.59	19-2	5.84	19-11	6.07	20-8	6.30
F_b psi	16	406	889	6129	1004	6922	1112	7667	1216	8384	1315	9067	1411	9729
(kPa)	24	610	1018	7019	1149	7922	1273	8777	1392	9598	1506	10,384	1615	11,135
							40 psf (1915 Pa) Live Load							
2 × 6	16	406	8-4	2.54	8-10	2.69	9-4	2.84	9-9	2.97	10-2	3.10	10-6	3.20
	24	610	7-3	2.21	7-9	2.36	8-2	2.49	8-6	2.59	8-10	2.69	9-2	2.79
2 × 8	16	406	11-0	3.35	11-8	3.56	12-3	3.73	12-10	3.91	13-4	4.06	13-10	4.22
	24	610	9-7	2.92	10-2	3.10	10-9	3.28	11-3	3.43	11-8	3.56	12-1	3.68
2 × 10	16	406	14-0	4.27	14-11	4.55	15-8	4.78	16-5	5.00	17-0	5.18	17-8	5.38
	24	610	12-3	3.73	13-0	3.96	13-8	4.17	14-4	4.37	14-11	4.55	15-5	4.70
2 × 12	16	406	17-0	5.18	18-1	5.51	19-1	5.82	19-11	6.07	20-9	6.32	21-6	6.55
	24	610	14-11	4.55	15-10	4.83	16-8	5.08	17-5	5.31	18-1	5.51	18-9	5.72
F_b psi	16	406	917	6322	1036	7143	1148	7915	1255	8653	1357	9356	1456	10,039
(kPa)	24	610	1050	7239	1186	8177	1314	9060	1436	9901	1554	10,714	1667	11,494

SOURCE: *Span Tables for Joists and Rafters*, American Forest and Paper Association, Washington, DC, 1993. Used by permission of AFPA.
NOTES:
1. The required bending design value F_b in Psi (kPa) is shown at the bottom of each section of the table and is applicable to all lumber sizes shown.
2. Allowable spans are based on a deflection of $^L/360$ at design load.

Table 13-11 Maximum span comparisons for joists.

Species and Grade	40 psf Live Load, 10 psf Dead Load, $^L/360$ 2 × 10 16" o.c.	2 × 10 24" o.c.	2 × 12 16" o.c.	2 × 12 24" o.c.	30 psf Live Load, 10 psf Dead Load, $^L/360$ 2 × 10 16" o.c.	2 × 10 24" o.c.	2 × 12 16" o.c.	2 × 12 24" o.c.
SP No. 1	16'-9"	14'-7"	20'-4"	17'-5"	18'-5"	16'-1"	22'-5"	19'-6"
DFL No. 1	16'-5"	13'-5"	19'-1"	15'-7"	18'-5"	15'-0"	21'-4"	17'-5"
SP No. 2	16'-1"	13'-2"	18'-10"	15'-4"	18'-0"	14'-8"	21'-1"	17'-2"
HF No. 1	16'-0"	13'-1"	18'-7"	15'-2"	17'-8"	14'-8"	20'-10"	17'-0"
SPF Nos. 1 and 2	15'-4"	12'-7"	17'-10"	14'-7"	17'-2"	14'-0"	19'-11"	16'-3"
DFL No. 2	15'-4"	12'-7"	17'-10"	14'-7"	17'-2"	14'-0"	19'-11"	16'-3"
HF No. 2	15'-2"	12'-5"	17'-7"	14'-5"	16'-10"	13'-10"	19'-8"	16'-1"
SP No. 3	12'-2"	9'-11"	14'-5"	11'-10"	13'-7"	11'-1"	16'-2"	13'-2"
DFL No. 3	11'-8"	9'-6"	13'-6"	11'-0"	13'-0"	10'-8"	15'-1"	12'-4"
HF No. 3	11'-8"	9'-6"	13'-6"	11'-0"	13'-0"	10'-8"	15'-1"	12'-4"
SPF No. 3	11'-8"	9'-6"	13'-6"	11'-0"	13'-0"	10'-8"	15'-1"	12'-4"

SOURCE: Southern Pine Council.
NOTES:
1. These spans were calculated using published design values and are for comparison purposes only. They include the repetitive member factor, $Cr = 1.15$, but do not include the composite action of adhesive and sheathing. Spans may be slightly different from other published spans due to rounding.
2. SP = southern pine; DFL = douglas fir–larch; HF = hem–fir; SPF = spruce–pine–fir.

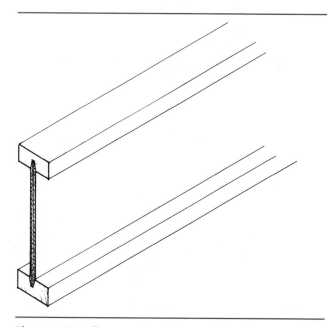

Figure 13-5 Truss joist.

In addition to sawn lumber floor joists, manufactured-type members are available. One is a wood I-shaped member, as shown in Fig. 13–5. It is made in depths of 9½″ to 24″ and in lengths to 40′. The flanges are made of microlaminated wood or solid wood, and the webs are made of plywood or oriented strand board. These members are lightweight, easy to handle, and dimensionally stable, and they have the capacity to support large loads. Such members as the residential TJI® joists are fabricated by Trus Joist MacMillian. See Fig. 13–6 for examples of plywood I-beam uses in residential floor construction.

Another member is the fabricated wood flat truss. It is built up of 2 × 4 material and is assembled with metal truss plate connectors. It is usually available from roof trussed rafter suppliers. Large load capacity on longer spans means that load-bearing partitions and interior beams, posts, and footings can frequently be eliminated (Fig. 13–7). Table 13–12 lists floor truss spans for fabricated wood flat trusses. Table 13–13 contains span comparisons for wood trusses, I joists, and dimension lumber.

13.4.6 Ceiling Joists

Indicate the ceiling joist size and direction with a symbol and a note on the floor plan that requires the ceiling above; always indicate framing information that is overhead or over the plane of the floor plan. Notice that ceiling joists may be subjected to various types of loads. Ceilings where storage above is necessary must be adequately framed to prevent objectional deflection. If ceiling joists are to tie with the rafters to counteract lateral thrust, consideration must be given to their direction; otherwise, joists are placed to result in the shortest span for economy. Table 13–14 indicates allowable spans for ceiling joists. Selected ceiling joist span tables for Southern Pine and Western Wood species are given in Appendix E.

13.4.7 Rafters

As we mentioned previously, roof loads are usually transmitted directly to exterior walls; occasionally, partial loads are transmitted to interior partitions. Rafter size can be affected by roof slope, with lower slopes generally requiring larger rafters. The span of a rafter is the horizontal distance from the wall plate to the ridgeboard (see Figs. 13–9 and 14–8). Collar beams, usually 1 × 4 or 1 × 6 ties, fastened midway between the ridge and the plate to opposing rafters, are meant to stiffen the roof. They are conventionally applied to every second or third set of rafters.

Design loads are figured from span to collar beam only, but in actual practice the full horizontal span of the rafter is used. The weight of roofing material is also important. Fiberglass or wood shingles are considered light roofing, whereas slate or mission tile is considered heavy roofing. Select rafter size from the table after roofing materials have been determined. Table 13–15 indicates allowable spans for roof rafters. Selected rafter span tables for Southern Pine and Western Wood species are given in Appendix E. Table 13–16, published by the Southern Pine Council (SPC), provides maximum rafter span comparisons for a number of different lumber species.

13.4.8 Headers

Structural headers are required to carry roof or floor loads over openings in load-bearing walls. Usually double 2 × lumber headers are installed above a wall opening with short "cripple" studs set between the header and the top plate. Table 13–17 and Figure 13–8 present allowable spans for conventional lumber headers with different roof and floor load conditions. Specific manufacturers will provide header span tables for other types of material such as parallel-strand lumber (PSL) and glue laminated lumber.

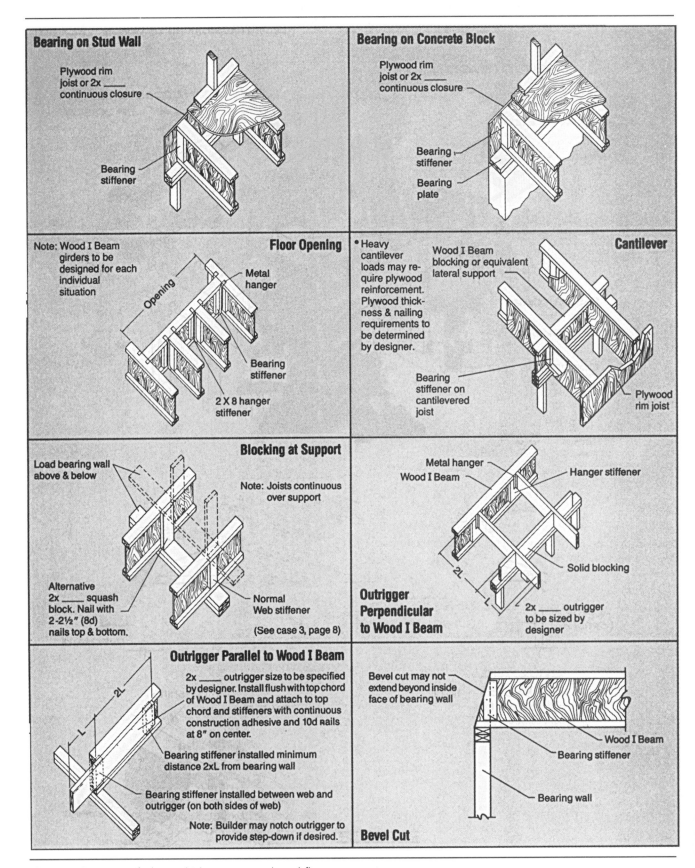

Bearing on Stud Wall

Plywood rim joist or 2x ____ continuous closure

Bearing stiffener

Bearing on Concrete Block

Plywood rim joist or 2x ____ continuous closure

Bearing stiffener

Bearing plate

Note: Wood I Beam girders to be designed for each individual situation

Floor Opening

Opening

Metal hanger

Bearing stiffener

2 X 8 hanger stiffener

• Heavy cantilever loads may require plywood reinforcement. Plywood thickness & nailing requirements to be determined by designer.

Wood I Beam blocking or equivalent lateral support

Cantilever

Bearing stiffener on cantilevered joist

Plywood rim joist

Blocking at Support

Load bearing wall above & below

Note: Joists continuous over support

Alternative 2x ____ squash block. Nail with 2-2½" (8d) nails top & bottom.

Normal Web stiffener

(See case 3, page 8)

Metal hanger

Wood I Beam

Hanger stiffener

Solid blocking

Outrigger Perpendicular to Wood I Beam

2x ____ outrigger to be sized by designer

Outrigger Parallel to Wood I Beam

2x ____ outrigger size to be specified by designer. Install flush with top chord of Wood I Beam and attach to top chord and stiffeners with continuous construction adhesive and 10d ʀails at 8" on center.

Bearing stiffener installed minimum distance 2xL from bearing wall

Bearing stiffener installed between web and outrigger (on both sides of web)

Note: Builder may notch outrigger to provide step-down if desired.

Bevel cut may not extend beyond inside face of bearing wall

Wood I Beam

Bearing stiffener

Bearing wall

Bevel Cut

Figure 13–6 Use of plywood I beams in residential floor construction.

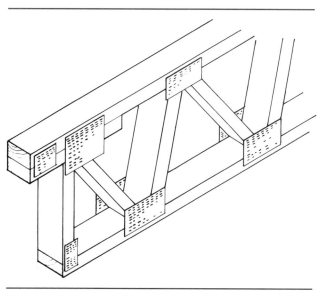

Figure 13–7 Factory-fabricated wood trusses.

Table 13–12 System 42 floor truss span tables.

TDO	Spacing	Allowable Total Loads (psf)														
		Clear Span (ft)														
		17'	18'	19'	20'	21'	22'	23'	24'	25'	26'	27'	28'	29'	30'	31'
12"	16.0"	97	87	78	71	64										
	19.2"	80	72	65	59	53										
	24.0"	64	58	52	47	42										
14"	16.0"	101	97	94	85	77	70	64								
	19.2"	96	87	78	71	64	58	53								
	24.0"	77	69	62	56	51	46	42								
16"	16.0"	120	115	110	99	89	82	75	69	63	58	54				
	19.2"	112	101	91	82	74	68	62	57	53	49	45				
	24.0"	90	81	73	66	59	54	49	45	42						
18"	16.0"	120	120	120	111	102	93	85	78	72	67	62	58	54	50	47
	19.2"	120	112	104	94	85	78	71	65	60	55	51	48	45		
	24.0"	102	92	83	75	67	61	56	52	48	44	41				
20"	16.0"	120	120	120	117	114	105	96	89	82	75	69	64	60	56	53
	19.2"	120	118	117	106	95	87	80	74	68	63	58	54	50	47	44
	24.0"	115	104	93	84	76	69	63	58	54	50	46	43	40		
22"	16.0"	120	120	120	120	120	113	106	98	91	84	77	72	67	63	59
	19.2"	120	120	120	112	105	96	88	82	76	70	64	60	56	52	49
	24.0"	120	112	104	94	84	77	70	65	60	55	51	48	45		

NOTES:
1. These overall spans are based on NDS 91. Allowable design load is limited by deflection to $^L/_{360}$ under live load. Subtract the total dead load from the loads shown to determine the allowable live load.
2. TDO is overall truss depth in inches. Spacing of trusses is center to center (in inches). Other depths are available.
3. Allowable total loads are in psf. Top chord dead load = 10 psf; bottom chord dead load = 5 psf. Center line chase = 24" max.
4. Lumber basic design values are as follows: F_b = 2000 psi, F_t = 1100 psi; F_c = 2000 psi; E = 1,800,000 psi; duration of load = 1.00.

Table 13-13 Span comparisons.

	Truss	I Joist	Dimensional Lumber Joists				
				10" Nominal Depth			
	SP No. 1 Dense	I Joist	SP No. 1	DF No. 1	SP No. 2	DF No. 2	SPF No. 1 and No. 2
16" o.c.	19'-3"	17'-7"	16'-9"	16'-5"	16'-1"	15'-5"	15'-5"
24" o.c.	16'-10"	15'-5"	14'-7"	13'-5"	13'-2"	12'-7"	12'-7"
				12" Nominal Depth			
	SP No. 1 Dense	I Joist	SP No. 1	DF No. 1	SP No. 2	DF No. 2	SPF No. 1 and No. 2
16" o.c	22'-2"	21'-0"	20'-4"	19'-1"	18'-10"	17'-10"	17'-10"
24" o.c.	19'-3"	17'-11"	17'-5"	15'-7"	15'-4"	14'-7"	14'-7"
				14" Nominal Depth			
	SP No. 1 Dense	I Joist					
16" o.c	24'-11"	23'-10"					
24" o.c.	21'-1"	17'-11"					

NOTES:
1. SP = southern pine; SPF = spruce–pine–fir; DF = Douglas fir.
2. Values are based on a 40-psf live load, a 10-psf dead load, and an $L/360$ deflection limitation.
3. This table is for comparison purposes only. All engineered products must be designed specifically for field conditions.
4. These spans do not include composite action of adhesive and sheathing.

Table 13-14 Allowable spans for ceiling joists.

	Joist		Modulus of Elasticity, E, in 1,000,000 psi or kPa							
	Spacing		1.0 psi	6.89 kPa	1.2 psi	8.27 kPa	1.40 psi	9.65 kPa	1.60 psi	11.03 kPa
Size	in.	mm	ft-in.	m	ft-in.	m	ft-in.	m	ft-in.	m
				10 psf (479 Pa) Live Load (No Attic Storage)						
2×4	16	406	9-8	2.95	10-3	3.12	10-9	3.28	11-3	3.43
	24	610	8-5	2.57	8-11	2.72	9-5	2.87	9-10	3.00
2×6	16	406	15-2	4.62	16-1	4.90	16-11	5.16	17-8	5.38
	24	610	13-3	4.04	14-1	4.29	14-9	4.50	15-6	4.72
2×8	16	406	19-11	6.07	21-2	6.45	22-4	6.81	23-4	7.11
	24	610	17-5	5.31	18-6	5.64	19-6	5.94	20-5	6.22
2×10	16	406	25-5	7.75						
	24	610	22-3	6.78	23-8	7.21	24-10	7.57	26-0	7.92
F_b psi	16	406	909	6267	1026	7074	1137	7839	1243	8570
(kPa)	24	610	1040	7171	1174	8094	1302	8977	1423	9811
				20 psf (958 Pa) Live Load (Limited Attic Storage)						
2×4	16	406	7-8	2.34	8-1	2.46	8-7	2.62	8-11	2.72
	24	610	6-8	2.03	7-1	2.16	7-6	2.29	7-10	2.39
2×6	16	406	12-0	3.66	12-9	3.89	13-5	4.09	14-1	4.29
	24	610	10-6	3.20	11-2	3.40	11-9	3.58	12-3	3.73
2×8	16	406	15-10	4.83	16-10	5.13	17-9	5.41	18-6	5.64
	24	610	13-10	4.22	14-8	4.47	15-6	4.72	16-2	4.93
2×10	16	406	20-2	6.15	21-6	6.55	22-7	6.88	23-8	7.21
	24	610	17-8	5.38	18-9	5.72	19-9	6.02	20-8	6.30
F_b psi	16	406	1145	7895	1293	8915	1433	9880	1566	10,797
(kPa)	24	610	1310	9032	1480	10,204	1640	11,307	1793	12,362

SOURCE: *Span Tables for Joists and Rafters*, American Forest and Paper Association, Washington, DC, 1993. Used by permission of AFPA.
NOTES:
1. Allowable spans are based on a deflection of $L/240$ at design load.
2. Dead load is 10 psf (479 Pa). The required bending design value, F_b, in psi or kPa is shown at the bottom of each section of the table.

Table 13–15 Allowable spans for roof rafters.

Size	Rafter Spacing in.	mm	Design Value in Bending 800 psi ft-in.	5516 kPa m	1000 psi ft-in.	6895 kPa m	1200 psi ft-in.	8274 kPa m	1400 psi ft-in.	9653 kPa m
20 psf (958 Pa) Live Load										
2 × 6	16	406	10-0	3.05	11-3	3.43	12-4	3.76	13-3	4.04
	24	610	8-2	2.49	9-2	2.79	10-0	3.05	10-10	3.30
2 × 8	16	406	13-3	4.04	14-10	4.52	16-3	4.95	17-6	5.33
	24	610	10-10	3.30	12-1	3.68	13-3	4.04	14-4	4.37
2 × 10	16	406	16-11	5.16	18-11	5.77	20-8	6.30	22-4	6.81
	24	610	13-9	4.19	15-5	4.70	16-11	5.16	18-3	5.56
E in 10⁶psi	16	406	0.58	4.00	0.82	5.65	1.07	7.38	1.35	9.31
(kPa)	24	610	0.48	3.31	0.67	4.62	0.88	6.07	1.10	7.58
30 psf (1435 Pa) Live Load										
2 × 6	16	406	8-8	2.64	9-9	2.97	10-8	3.25	11-6	3.51
	24	610	7-1	2.16	7-11	2.41	8-8	2.64	9-5	2.87
2 × 8	16	406	11-6	3.51	12-10	3.91	14-0	4.27	15-2	4.62
	24	610	9-4	2.84	10-6	320	11-6	3.51	12-5	3.78
2 × 10	16	406	14-8	4.47	16-4	4.98	17-11	5.46	19-4	5.89
	24	610	11-11	3.63	13-4	4.06	14-8	4.47	15-10	4.83
E in 10⁶psi	16	406	0.57	3.93	0.80	5.52	1.05	7.24	1.32	9.10
(kPa)	24	610	0.46	3.17	0.65	4.48	0.85	5.86	1.08	7.45
40 psf (1915 Pa) Live Load										
2 × 6	16	406	7-9	2.36	8-8	2.64	9-6	2.90	10-3	3.12
	24	610	6-4	1.93	7-1	7.08	7-9	2.36	8-5	2.57
2 × 8	16	406	10-3	3.12	11-6	3.51	12-7	3.84	13-7	4.14
	24	610	8-4	2.54	9-4	2.84	10-3	3.12	11-1	3.38
2 × 10	16	406	13-1	3.99	14-8	4.47	16-0	4.88	17-4	5.28
	24	610	10-8	3.25	11-11	3.63	13-1	3.99	14-2	4.32
E in 10⁶psi	16	406	0.54	3.72	0.76	5.24	1.00	6.89	1.26	8.69
(kPa)	24	610	0.44	3.03	0.62	4.27	0.81	5.58	1.03	7.10
50 psf (2394 Pa) Live Load										
2 × 6	16	406	7-1	2.16	7-11	2.41	8-8	2.64	9-5	2.87
	24	610	5-10	1.78	6-6	1.98	7-1	2.16	7-8	2.34
2 × 8	16	406	9-4	2.84	10-6	3.20	11-6	3.51	12-5	3.78
	24	610	7-8	2.34	8-7	2.62	9-4	2.84	10-1	3.07
2 × 10	16	406	11-11	3.63	13-4	4.06	14-8	4.47	15-10	4.83
	24	610	9-9	2.97	10-11	3.33	11-11	3.63	12-11	3.94
E in 10⁶psi	16	406	0.52	3.59	0.72	4.96	0.95	6.55	1.20	8.27
(kPa)	24	610	0.42	2.90	0.59	4.07	0.77	5.31	0.98	6.76

SOURCE: *Span Tables for Joists and Rafters*, American Forest and Paper Association, Washington, DC, 1993. Used by permission of AFPA.
NOTES:

1. The live load plus the dead load of 10 psf (479 Pa) determines the required bending design value.
2. The modulus of elasticity, E, in 1,000,000 psi (kPa) is shown at the bottom of each section of the table.
3. Allowable spans are based on a deflection of ¹/₂₄₀ at design load.

Table 13-16 Maximum span comparisons for rafters.

Species and Grade	30 psf Live Load, 15 psf Dead Load, L/180, C_d=1.15, 6:12 Slope						20 psf Live Load, 10 psf Dead Load, L/240, C_d=1.25, 3:12 Slope					
	2×6		2×8		2×10		2×6		2×8		2×10	
	16" o.c.	24" o.c.	16" o.c.	24" o.c.	16" o.c.	24" o.c.	16" o.c.	24" o.c.	16" o.c.	24" o.c.	16" o.c.	24" o.c.
SP No. 1	13'-6"	11'-1"	17'-0"	13'-11"	20'-3"	16'-6"	14'-4"	12'-6"	18'-11"	16'-6"	24'-1"	21'-1"
DFL No. 1	12'-0"	9'-10"	15'-3"	12'-5"	18'-7"	15'-2"	14'-4"	12'-6"	18'-11"	15'-10"	23'-9"	19'-5"
SP No. 2	11'-9"	9'-7"	15'-3"	12'-5"	18'-2"	14'-10"	14'-1"	12'-3"	18'-6"	15'-10"	23'-2"	18'-11"
HF No. 1	11'-9"	9'-7"	14'-10"	12'-1"	18'-1"	14'-9"	13'-9"	12'-0"	18'-1"	15'-6"	23'-1"	18'-11"
DFL No. 2	11'-3"	9'-2"	14'-3"	11'-8"	17'-5"	14'-3"	14'-1"	11'-9"	18'-2"	14'-10"	22'-3"	18'-2"
SPF Nos. 1 and 2	11'-3"	9'-2"	14'-3"	11'-8"	17'-5"	14'-3"	13'-5"	11'-9"	17'-9"	14'-10"	22'-3"	18'-2"
HF No. 2	11'-1"	9'-1"	14'-0"	11'-6"	17'-2"	14'-0"	13'-1"	11'-5"	17'-3"	14'-8"	21'-11"	17'-11"
SP No. 3	9'-1"	7'-5"	11'-7"	9'-6"	13'-9"	11'-3"	11'-8"	9'-6"	14'-10"	12'-2"	17'-7"	14'-4"
DFL No. 3	8'-6"	6'-11"	10'-9"	8'-10"	13'-2"	10'-9"	10'-10"	8'-10"	13'-9"	11'-3"	16'-9"	13'-8"
HF No. 3	8'-6"	6'-11"	10'-9"	8'-10"	13'-2"	10'-9"	10'-10"	8'-10"	13'-9"	11'-3"	16'-9"	13'-8"
SPF No. 3	8'-6"	6'-11"	10'-9"	8'-10"	13'-2"	10'-9"	10'-10"	8'-10"	13'-9"	11'-3"	16'-9"	13'-8"

NOTES:
1. These spans were calculated using published design values and are for comparison purposes only. They include the repetitive member factor, C_r = 1.15, but do not include the composite action of adhesive and sheathing. Spans may be slightly different from other published spans due to rounding.
2. SP = southern pine; DFL = Douglas fir–larch; HF = hem–fir; SPF = spruce–pine–fir.
3. C_d = load duration factor.

Table 13-17 Maximum header spans based on house width and load condition (see Fig. 13–8).

Load Condition	Header Size	House Width									
		24 ft ft-in.	7.32 m m	26 ft ft-in.	7.92 m m	28 ft ft-in.	8.53 m m	30 ft ft-in.	9.14 m m	32 ft ft-in.	9.75 m m
1. Roof only	2-2 × 3	2-1	0.63	2-0	0.61						
	2-2 × 4	2-11	0.89	2-10	0.85	2-8	0.82	2-7	0.79	2-6	0.77
	1-2 × 6	3-3	0.99	3-1	0.95	3-0	0.91	2-11	0.88	2-10	0.86
	1-2 × 8	4-3	1.30	4-1	1.25	4-0	1.21	3-10	1.16	3-8	1.13
	2-2 × 6	4-7	1.40	4-5	1.34	4-3	1.29	4-1	1.25	4-0	1.21
	1-2 × 10	5-5	1.66	5-3	1.60	5-1	1.54	4-11	1.49	4-9	1.44
	2-2 × 8	6-0	1.84	5-10	1.77	5-7	1.70	5-5	1.65	5-3	1.59
	1-2 × 12	6-8	2.02	6-4	1.94	6-2	1.87	5-11	1.81	5-9	1.75
	2-2 × 10	7-9	2.35	7-5	2.26	7-2	2.18	6-11	2.10	6-8	2.03
	2-2 × 12	9-4	2.86	9-0	2.75	8-8	2.65	8-5	2.56	8-1	2.47
2. Roof, second-story floor and wall	2-2 × 4	2-2	0.67	2-1	0.65	2-1	0.63				
	1-2 × 6	2-5	0.75	2-4	0.72	2-3	0.69	2-3	0.67	2-2	0.65
	1-2 × 8	3-3	0.98	3-1	0.95	3-0	0.92	2-11	0.89	2-10	0.86
	2-2 × 6	3-6	1.06	3-4	1.02	3-3	0.98	3-1	0.95	3-0	0.92
	1-2 × 10	4-1	1.26	4-0	1.21	3-10	1.17	3-9	1.13	3-7	1.10
	2-2 × 8	4-7	1.39	4-5	1.34	4-3	1.30	4-1	1.25	4-0	1.22
	1-2 × 12	5-0	1.53	4-10	1.47	4-8	1.42	4-6	1.38	4-5	1.33
	2-2 × 10	5-10	1.78	5-7	1.71	5-5	1.65	5-3	1.60	5-1	1.55
	2-2 × 12	7-1	2.16	6-10	2.08	6-7	2.01	6-5	1.95	6-2	1.89
3. Roof, second- and third-story floors and walls	1-2 × 6	2-1	0.64	2-0	0.62						
	1-2 × 8	2-9	0.84	2-8	0.82	2-7	0.79	2-6	0.76	2-5	0.74
	2-2 × 6	3-0	0.91	2-10	0.87	2-9	0.85	2-8	0.82	2-7	0.80
	1-2 × 10	3-6	1.08	3-5	1.04	3-4	1.01	3-2	0.97	3-1	0.95
	2-2 × 8	3-11	1.19	3-9	1.15	3-8	1.11	3-6	1.08	3-5	1.05
	1-2 × 12	4-4	1.31	4-2	1.26	4-0	1.22	3-11	1.19	3-9	1.15
	2-2 × 10	5-0	1.52	4-10	1.47	4-8	1.42	4-6	1.38	4-5	1.34
	2-2 × 12	6-1	1.85	5-10	1.79	5-8	1.73	5-6	1.68	5-4	1.63
4. Second-story floor only	2-2 × 3	2-8	0.80	2-6	0.77	2-5	0.74	2-4	0.72	2-3	0.70
	2-2 × 4	3-8	1.12	3-6	1.08	3-5	1.04	3-4	1.01	3-2	0.97
	1-2 × 6	4-1	1.25	3-11	1.20	3-9	1.16	3-8	1.12	3-7	1.08
	1-2 × 8	5-5	1.65	5-2	1.58	5-0	1.52	4-10	1.47	4-8	1.43
	2-2 × 6	5-10	1.77	5-7	1.70	5-4	1.64	5-2	1.58	5-0	1.53
	1-2 × 10	6-11	2.10	6-7	2.02	6-5	1.95	6-2	1.88	6-0	1.82
	2-2 × 8	7-8	2.33	7-4	2.24	7-1	2.16	6-10	2.08	6-7	2.02
	1-2 × 12	8-5	2.56	8-1	2.46	7-9	2.37	7-6	2.29	7-3	2.21
	2-2 × 10	9-9	2.97	9-4	2.86	9-0	2.75	8-9	2.66	8-5	2.57
	2-2 × 12	11-10	3.61	11-5	3.47	11-0	3.35	10-7	3.23	10-3	3.13

NOTES:
1. End splits may not exceed one times the header depth.
2. For design load conditions, see Fig. 13–8.
3. Minimum allowable bending stress F_b = 1000 psi (6895 kPa).

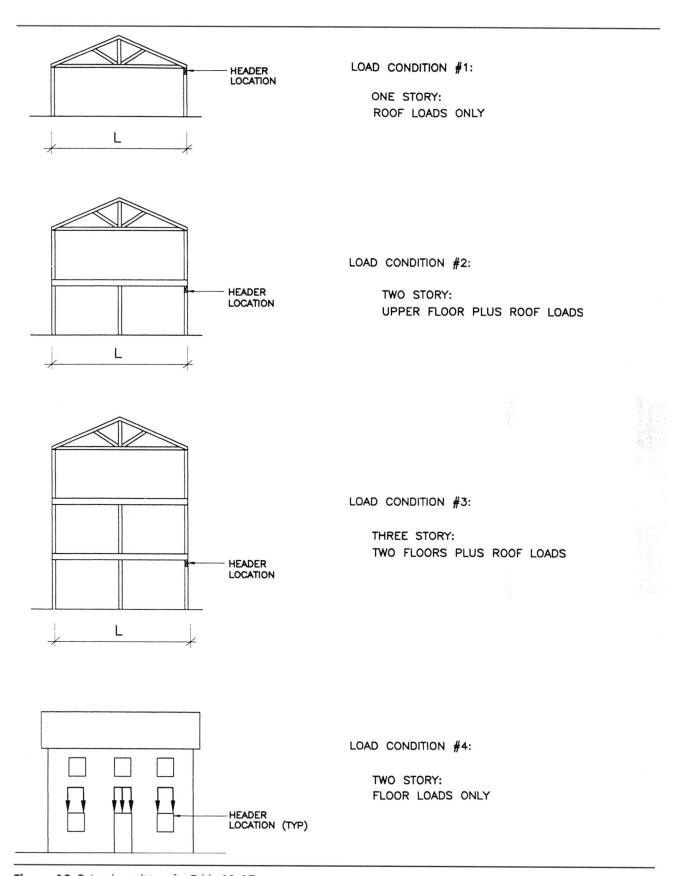

LOAD CONDITION #1:

ONE STORY:
ROOF LOADS ONLY

LOAD CONDITION #2:

TWO STORY:
UPPER FLOOR PLUS ROOF LOADS

LOAD CONDITION #3:

THREE STORY:
TWO FLOORS PLUS ROOF LOADS

LOAD CONDITION #4:

TWO STORY:
FLOOR LOADS ONLY

Figure 13-8 Load conditions for Table 13–17.

13.4.9 Roof Trusses

Wood trusses are easy to install, and they provide a means of rapidly enclosing a building shell. Trusses easily adapt to most roof designs and are normally designed for 24″ o.c. spacing. Truss fabricators normally have the capability of supplying the necessary engineering for each unique design condition. Maximum spans for selected Fink, Howe, scissors, and king post truss types at 24″ spacing are provided in Appendix E.

13.5

PLANK-AND-BEAM FRAMING[1]

Section 12.5 has additional information on plank-and-beam frame construction.

13.5.1 Design Data for Planks

Design data for plank floors and roofs are included in Table 13–18. Computations for bending are based on the live load indicated, plus 10 psf of dead load. Computations for deflection are based on the live load only. The table shows four general arrangements of planks as follows:

Type A—Extending over a single span
Type B—Continuous over two equal spans
Type C—Continuous over three equal spans
Type D—A combination of types A and B

On the basis of a section of planking 12″ wide, the following formulas were used in making the computations:

For type A:

$$M = \frac{wL^2}{8} \quad \text{and} \quad D = \frac{5wL^4(12)^3}{384EI}$$

For type B:

$$M = \frac{wL^2}{8} \quad \text{and} \quad D = \frac{wL^4(12)^3}{185EI}$$

For type C:

$$M = \frac{wL^2}{10} \quad \text{and} \quad D = \frac{4wL^4(12)^3}{581EI}$$

For type D:

$$M = \frac{wL^2}{8} \quad \text{and} \quad D = \frac{1}{2}\left(\frac{5wL^4(12)^3}{384EI} + \frac{wL^4(12)^3}{185EI}\right)$$

Notations In the preceding formulas and in the tables of Appendix E, the symbols have the following meanings:

w = load in lb/lin ft
L = span in ft
M = induced bending moment in lb-ft
f = fiber stress in bending in psi
E = modulus of elasticity in psi
I = moment of inertia in inches to the fourth power
D = deflection in in.

To use Table 13–18, first determine the plank arrangement (types A, B, C, or D), the span, the live load to be supported, and the deflection limitation. Then select from the table the corresponding required values for the fiber stress in bending (f) and the modulus of elasticity (E). The plank to be used should be of a grade and species that meets these minimum values. You can determine the maximum span for a specific grade and species of plank by reversing these steps.

For those who prefer to use random-length planks (instead of arrangements type A, B, C, or D), similar technical information is included in *Random Length Wood Decking*, a publication of the National Forest Products Association.

In addition to the nominal 2″ plank described above, decking is available in nominal 3″ thickness and in various laminated deck systems.

---

[1] From *Plank-and-Beam Framing for Residential Buildings* (Manual No. 4). By and with permission from the National Forest Products Association.

Table 13-18 Nominal 2″ plank: required values for fiber stress in bending (*f*) and modulus of elasticity (*E*) to support safely a live load of 20, 30, or 40 psf within a deflection limitation of $l/240$, $l/300$, or $l/360$.

Plank Span in Feet	Live Load (psf)	Deflection Limitation	Type A *f* (psi)	Type A *E* (psi)	Type B *f* (psi)	Type B *E* (psi)	Type C *f* (psi)	Type C *E* (psi)	Type D *f* (psi)	Type D *E* (psi)
6′	20	$l/240$	360	576,000	360	239,000	288	305,000	360	408,000
		$l/300$	360	720,000	360	299,000	288	381,000	360	509,000
		$l/360$	360	864,000	360	359,000	288	457,000	360	611,000
	30	$l/240$	480	864,000	480	359,000	384	457,000	480	611,000
		$l/300$	480	1,080,000	480	448,000	384	571,000	480	764,000
		$l/360$	480	1,296,000	480	538,000	384	685,000	480	917,000
	40	$l/240$	600	1,152,000	600	478,000	480	609,000	600	815,000
		$l/300$	600	1,440,000	600	598,000	480	762,000	600	1,019,000
		$l/360$	600	1,728,000	600	717,000	480	914,000	600	1,223,000
7′	20	$l/240$	490	915,000	490	380,000	392	484,000	490	647,000
		$l/300$	490	1,143,000	490	475,000	392	605,000	490	809,000
		$l/360$	490	1,372,000	490	570,000	392	726,000	490	971,000
	30	$l/240$	653	1,372,000	653	570,000	522	726,000	653	971,000
		$l/300$	653	1,715,000	653	712,000	522	907,000	653	1,213,000
		$l/360$	653	2,058,000	653	854,000	522	1,088,000	653	1,456,000
	40	$l/240$	817	1,829,000	817	759,000	653	968,000	817	1,294,000
		$l/300$	817	2,287,000	817	949,000	653	1,209,000	817	1,618,000
		$l/360$	817	2,744,000	817	1,139,000	653	1,451,000	817	1,941,000
8′	20	$l/240$	640	1,365,000	640	567,000	512	722,000	640	966,000
		$l/300$	640	1,707,000	640	708,000	512	903,000	640	1,208,000
		$l/360$	640	2,048,000	640	850,000	512	1,083,000	640	1,449,000
	30	$l/240$	853	2,048,000	853	850,000	682	1,083,000	853	1,449,000
		$l/300$	853	2,560,000	853	1,063,000	682	1,354,000	853	1,811,000
		$l/360$	853	3,072,000	853	1,275,000	682	1,625,000	853	2,174,000
	40	$l/240$	1067	2,731,000	1067	1,134,000	853	1,444,000	1067	1,932,000
		$l/300$	1067	3,413,000	1067	1,417,000	853	1,805,000	1067	2,415,000
		$l/360$	1067	4,096,000	1067	1,700,000	853	2,166,000	1067	2,898,000

13.5.2 Design Data for Beams

Design data for beams along with tables for computations are included in Appendix E. Table 13–19 provides section properties of standard dressed (S4S) sawn lumber based on the American Forest and Paper Association's National Design Specifications Supplement, 1991. Tables 13–20 and 13–21 list base design values for visually graded dimension lumber for southern pine and other selected species. the National Design Specification for Wood Construction Supplement provides various tables of adjustment factors that are to be used in conjunction with the design values in the previous tables. Consult the National Design Specifications for the proper use of the adjustment factors. Expanded tables of design values for a wide range of lumber species are given in Appendix E.

Table 13–19 Properties of sections.

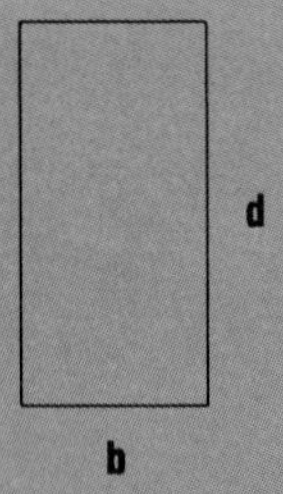

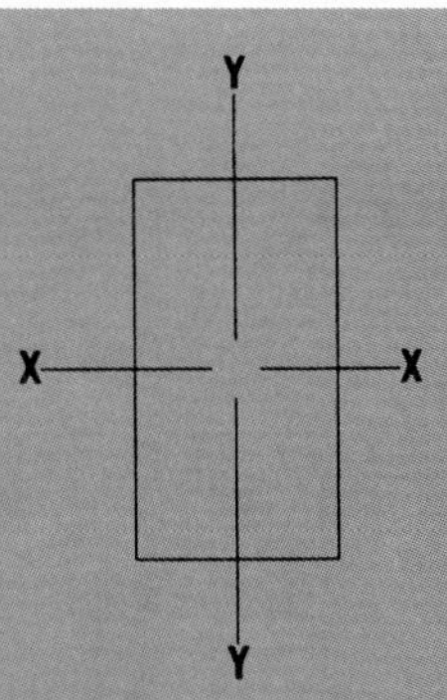

Nominal Size (in.) b	d	Actual Size (in.) b	d	Area (in.²)	Axis XX S (in.³)	I (in.⁴)	Axis YY S (in.³)	I (in.⁴)	Board Measure per Lineal Foot	Weight per Lineal Foot (lb)
2 × 2		1½ ×	1½	2.250	0.563	0.422	0.563	0.422	0.33	0.73
	3		2½	3.750	1.563	1.953	0.938	0.703	0.50	1.10
	4		3½	5.250	3.063	5.359	1.313	0.984	0.67	1.47
	5		4½	6.750	5.063	11.391	1.688	1.266	0.83	1.83
	6		5½	8.250	7.563	20.797	2.063	1.547	1.00	2.20
	8		7¼	10.875	13.141	47.635	2.719	2.039	1.33	2.93
	10		9¼	13.875	21.391	98.932	3.469	2.602	1.67	3.84
	12		11¼	16.875	31.641	177.979	4.219	3.164	2.00	4.60
	14		13¼	19.875	43.891	290.775	4.969	3.727	2.33	5.59
3 × 3		2½ ×	2½	6.250	2.604	3.255	2.604	3.255	0.75	1.80
	4		3½	8.750	5.104	8.932	3.646	4.557	1.00	2.30
	6		5½	13.750	12.604	34.661	5.729	7.161	1.50	3.45
	8		7¼	18.125	21.901	79.391	7.552	9.440	2.00	4.60
	10		9¼	23.125	35.651	164.886	9.635	12.044	2.50	6.00
	12		11¼	28.125	52.734	296.631	11.719	14.648	3.00	7.20
	14		13¼	33.125	73.151	484.626	13.802	17.253	3.50	8.40
4 × 4		3½ ×	3½	12.250	7.146	12.505	7.146	12.505	1.33	3.19
	6		5½	19.250	17.646	48.526	11.229	19.651	2.00	5.00
	8		7¼	25.375	30.661	111.148	14.802	25.904	2.67	6.68
	10		9¼	32.375	49.911	230.840	18.885	33.049	3.33	8.33
	12		11¼	39.375	73.828	415.283	22.969	40.195	4.00	10.00
	14		13¼	46.375	102.411	678.476	27.052	47.341	4.67	11.68
*6 × 6		5½ ×	5½	30.250	27.729	76.255	27.729	76.255	3.00	11.40
	8		7½	41.250	51.563	193.359	37.813	103.984	4.00	15.20
	10		9½	52.250	82.729	392.964	47.896	131.714	5.00	19.00
	12		11½	63.250	121.229	697.068	57.979	159.443	6.00	22.80
	14		13½	74.250	167.063	1127.672	68.063	187.172	7.00	26.60
*8 × 8		7½ ×	7½	56.250	70.313	263.672	70.313	263.672	5.33	20.25
	10		9½	71.250	112.813	535.859	89.063	333.984	6.67	25.35
	12		11½	86.250	165.313	950.547	107.813	404.297	8.00	30.40
	14		13½	101.250	227.813	1537.734	126.563	474.609	9.33	35.45
*10 × 10		9½ ×	9½	90.250	142.896	678.755	142.896	678.755	8.33	31.65
	12		11½	109.250	209.396	1204.026	172.979	821.651	10.00	38.00
	14		13½	128.250	288.563	1947.797	203.063	964.547	11.67	44.35
*12 × 12		11½ ×	11½	132.250	253.479	1457.505	253.479	1457.505	12.00	45.60
	14		13½	155.250	349.313	2357.859	297.563	1710.984	14.00	53.20
*14 × 14		13½ ×	13½	182.250	410.063	2767.922	410.063	2767.922	16.33	62.05

SOURCE: *National Design Specification Supplement*, 1991 Edition. American Forest and Paper Association, Washington, DC, 1991. Used by permission of AFPA.
*Properties are based on minimum dressed green size, which is ½" off nominal in both *b* and *d* dimensions.

Table 13-20 Base design values for visually graded dimension lumber, selected species.

Species	Commercial Grade	Size Classification	Bending, F_b	Tension Parallel to Grain, F_t	Shear Parallel to Grain, F_v	Compression Perpendicular to Grain, $F_{c\perp}$	Compression Parallel to Grain, $F_{c\parallel}$	Modulus of Elasticity, E
						Design Values in psi		
Douglas fir–larch	Select structural		1450	1000	95	625	1700	1,900,000
	No. 1 and better	2″–4″ thick	1150	775	95	625	1500	1,800,000
	No. 1		1000	675	95	625	1450	1,700,000
	No. 2	2″ and wider	875	575	95	625	1300	1,600,000
	No. 3		500	325	95	625	750	1,400,000
	Stud		675	450	95	625	825	1,400,000
	Construction	2″–4″ thick	1000	650	95	625	1600	1,500,000
	Standard		550	375	95	625	1350	1,400,000
	Utility	2″–4″ wide	275	175	95	625	875	1,300,000
Douglas fir–larch (north)	Select structural	2″–4″ thick	1300	800	95	625	1900	1,900,000
	No. 1/No. 2		825	500	95	625	1350	1,600,000
	No. 3	2″ and wider	475	300	95	625	775	1,400,000
	Stud		650	375	95	625	850	1,400,000
	Construction	2″–4″ thick	950	575	95	625	1750	1,500,000
	Standard		525	325	95	625	1400	1,400,000
	Utility	2″–4″ wide	250	150	95	625	925	1,300,000
Hem–fir	Select structural		1400	900	75	405	1500	1,600,000
	No. 1 and better	2″–4″ thick	1050	700	75	405	1350	1,500,000
	No. 1		950	600	75	405	1300	1,500,000
	No. 2	2″ and wider	850	500	75	405	1250	1,300,000
	No. 3		500	300	75	405	725	1,200,000
	Stud		675	400	75	405	800	1,200,000
	Construction	2″–4″ thick	975	575	75	405	1500	1,300,000
	Standard		550	325	75	405	1300	1,200,000
	Utility	2″–4″ wide	250	150	75	405	850	1,100,000
Hem–fir (north)	Select structural	2″–4″ thick	1300	775	75	370	1650	1,700,000
	No. 1/No. 2		1000	550	75	370	1450	1,600,000
	No. 3	2″ and wider	575	325	75	370	850	1,400,000
	Stud		775	425	75	370	925	1,400,000
	Construction	2″–4″ thick	1150	625	75	370	1750	1,500,000
	Standard		625	350	75	370	1500	1,400,000
	Utility	2″–4″ wide	300	175	75	370	975	1,300,000
Spruce–pine–fir	Select structural	2″–4″ thick	1250	675	70	425	1400	1,500,000
	No. 1/No. 2		875	425	70	425	1100	1,400,000
	No. 3	2″ and wider	500	250	70	425	625	1,200,000
	Stud		675	325	70	425	675	1,200,000
	Construction	2″–4″ thick	975	475	70	425	1350	1,300,000
	Standard		550	275	70	425	1100	1,200,000
	Utility	2″–4″ wide	250	125	70	425	725	1,100,000

Source: *National Design Specification Supplement*, 1991 Edition. American Forest and Paper Association, Washington, DC, 1991. Used by permission of AFPA.

NOTES:

1. Tabulated design values are for normal load duration and dry service conditions. See National Design Specification 2.3 for a comprehensive description of design value adjustment factors.

2. *Lumber dimensions:* Tabulated design values are applicable to lumber that will be used under dry conditions such as in most covered structures. For 2″ to 4″ thick lumber, the DRY dressed sizes shall be used (see NDS® Table 1A) regardless of the moisture content at the time of manufacture or use. In calculation of design values, the natural gain in strength and stiffness that occurs as lumber dries has been taken into consideration, as well as the reduction in size that occurs when unseasoned lumber shrinks. The gain in load-carrying capacity due to increased strength and stiffness resulting from drying more than offsets the design effect of size reductions due to shrinkage.

3. *Stress-rated boards:* Stress-rated boards of nominal 1″, 1¼″, and 1½″ thickness, 2″ and wider, of most species, are permitted the design values shown for select structural, No. 1 and better, No. 1, No. 2, No. 3, stud, construction, standard, utility, clear heart structural, and clear structural grades as shown in the 2″ to 4″ thick categories herein, when graded in accordance with the stress-rated board provisions in the applicable grading rules. Information on stress-rated board grades applicable to the various species is available from the respective grading rules agencies. Information on additional design values may also be available from the respective grading agencies.

Table 13–21 Base design values for visually graded southern pine dimension lumber.

| Species | Commercial Grade | Size Classification | Design Values in psi | | | | | |
			Bending, F_b	Tension Parallel to Grain, F_t	Shear Parallel to Grain, F_v	Compression Perpendicular to Grain, $F_{c\perp}$	Compression Parallel to Grain, $F_{c\parallel}$	Modulus of Elasticity, E
Southern pine	Dense select–	Structural	3050	1650	100	660	2250	1,900,000
	select structural		2850	1600	100	565	2100	1,800,000
	Non-dense select–	Structural	2650	1350	100	480	1950	1,700,000
	No. 1 dense		2000	1100	100	660	2000	1,800,000
	No. 1	2"–4" thick	1850	1050	100	565	1850	1,700,000
	No. 1 non-	Dense	1700	900	100	480	1700	1,600,000
	No. 2 dense	2"–4" wide	1700	875	90	660	1850	1,700,000
	No. 2		1500	825	90	565	1650	1,600,000
	No. 2 non-	Dense	1350	775	90	480	1600	1,400,000
	No. 3		850	475	90	565	975	1,400,000
	Stud		875	500	90	565	975	1,400,000
	Construction	2"–4" thick	1100	625	100	565	1800	1,500,000
	Standard		625	350	90	565	1500	1,300,000
	Utility	4" wide	300	175	90	565	975	1,300,000
	Dense select–	Structural	2700	1500	90	660	2150	1,900,000
	select structural		2550	1400	90	565	2000	1,800,000
	Non-dense select–	Structural	2350	1200	90	480	1850	1,700,000
	No. 1 dense		1750	950	90	660	1900	1,800,000
	No. 1	2"–4" thick	1650	900	90	565	1750	1,700,000
	No. 1 non-	Dense	1500	800	90	480	1600	1,600,000
	No. 2 dense	5"–6" wide	1450	775	90	660	1750	1,700,000
	No. 2		1250	725	90	565	1600	1,600,000
	No. 2 non-	Dense	1150	675	90	480	1500	1,400,000
	No. 3		750	425	90	565	925	1,400,000
	Stud		775	425	90	565	925	1,400,000
	Dense select	Structural	2450	1350	90	660	2050	1,900,000
	Select structural		2300	1300	90	565	1900	1,800,000
	Non-dense select–	Structural	2100	1100	90	480	1750	1,700,000
	No. 1 dense	2"–4" thick	1650	875	90	660	1800	1,800,000
	No. 1		1500	825	90	565	1650	1,700,000
	No. 1 non-dense	8" wide	1350	725	90	480	1550	1,600,000
	No. 2 dense		1400	675	90	660	1700	1,700,000
	No. 2		1200	650	90	565	1550	1,600,000
	No. 2 non-	Dense	1100	600	90	480	1450	1,400,000
	No. 3		700	400	90	565	875	1,400,000
	Dense select	Structural	2150	1200	90	660	2000	1,900,000
	Select structural		2050	1100	90	565	1850	1,800,000
	Non-dense select–	Structural	1850	950	90	480	1750	1,700,000
	No. 1 dense	2"–4" thick	1450	775	90	660	1750	1,800,000
	No. 1		1300	725	90	565	1600	1,700,000
	No. 1 non-dense	10" wide	1200	650	90	480	1500	1,600,000
	No. 2 dense		1200	625	90	660	1650	1,700,000
	No. 2		1050	575	90	565	1500	1,600,000
	No. 2 non-dense		950	550	90	480	1400	1,400,000
	No. 3		600	325	90	565	850	1,400,000
	Dense select	Structural	2050	1100	90	660	1950	1,900,000
	Select structural		1900	1050	90	565	1800	1,800,000
	Non-dense select–	Structural	1750	900	90	480	1700	1,700,000
	No. 1 dense	2"–4" thick	1350	725	90	660	1700	1,800,000
	No. 1		1250	675	90	565	1600	1,700,000
	No. 1 non-dense	12" wide	1150	600	90	480	1500	1,600,000
	No. 2 dense		1150	575	90	660	1600	1,700,000
	No. 2		975	550	90	565	1450	1,600,000
	No. 2 non-dense		900	525	90	480	1350	1,400,000
	No. 3		575	325	90	565	825	1,400,000

NOTES:
1. Tabulated design values are for normal load duration and dry service conditions. See NDS® 2.3 for a comprehensive description of design value adjustment factors.
2. *Lumber dimensions:* Tabulated design values are applicable to lumber that will be used under dry conditions such as in most covered structures. For 2" to 4" thick lumber, the DRY dressed sizes shall be used (see Table 1A in NDS®) regardless of the moisture content at the time of manufacture or use. In calculation of design values, the natural gain in strength and stiffness that occurs as lumber dries has been taken into consideration, as well as the reduction in size that occurs when unseasoned lumber shrinks. The gain in load-carrying capacity due to increased strength and stiffness resulting from drying more than offsets the design effect of size reductions due to shrinkage.
3. *Stress-rated boards:* Information for various grades of southern pine stress-rated boards of nominal 1", 1¼", and 1½" thickness, 2" and wider, is available from the Southern Pine Inspection Bureau (SPIB) in the "Standard Grading Rules for Southern Pine Lumber."
4. *Spruce pine:* To obtain recommended values for spruce pine graded to SPIB rules, multiply the appropriate design values for mixed southern pine by the corresponding conversion factor shown in Table 13–22, and round to the nearest 100,000 psi for E; to the next lower multiple of 5 psi for F_v and $F_{c\perp}$; to the next lower multiple of 50 psi for F_b, F_t, and F_c if 1000 psi or greater, 25 psi otherwise.
5. For mixed southern pine, and wet and dry service condition design values, see the manufacturer's tables or the National Design Specification.

Table 13–22 Conversion factors for determining design values for spruce pine.

	Bending, F_b	Tension Parallel to Grain, F_t	Shear Parallel to Grain, F_v	Compression Perpendicular to Grain, $F_{c\perp}$	Compression Parallel to Grain, F_c	Modulus of Elasticity, E
Conversion factor	0.784	0.784	0.965	0.682	0.766	0.807

SOURCE: Derived from the *National Design Specification Supplement*, 1991 Edition. American Forest and Paper Association, Washington, DC, 1991. Used by permission of AFPA.

EXERCISES

The following exercises should be done on 8½" × 11" drawing paper with a ½" border, using drawing instruments; drawings do not have to be drawn to a particular scale.

1. Draw an elevation view and a section of a wood beam. Graphically illustrate which fibers are in tension and which fibers are in compression. Show where the neutral axis is located to illustrate that the stress changes from tension to compression.

2. Draw the elevation view of a beam that is supported by two columns and cantilevers beyond the end of one of the columns. Graphically illustrate which fibers are in tension and which fibers are in compression.

3. Draw the elevation view of a W-pattern truss. Identify the members of the truss that are in tension and the ones that are in compression. Use arrows to indicate tension and compression.

4. Draw a section of each of the following structural members: a wide flange beam (W section), a glue laminated beam, a flitch beam, and an I-shaped floor joist. Poché and label the materials.

5. Draw an elevation view of a flat floor truss with metal truss plate connectors. Show a bearing wall at one end of the trusses and a steel beam at the other end.

REVIEW QUESTIONS

1. What lumber grades are generally used for the major amount of residential framing throughout the country?

2. What physical characteristics must structural lumber possess?

3. List the two major species of wood that are now used for framing houses.

4. What is the main use in home construction for oak lumber?

5. Explain the word *deflection* as used in structural terminology.

6. To what type of stresses will wood columns used to support a girder in a structure be subjected?

7. What is the maximum amount of moisture allowed in seasoned framing lumber?

8. What minimum live load do local building codes require on living room floors?

9. Why must you consider both the live load and the dead load when determining the size of a structural framing member?

10. If you are using No. 2 Douglas fir for framing a kitchen floor, what size joists will be necessary if they are spaced 16" o.c. and span 11'-6"?

11. Specify the ceiling joist size for the following situation: no attic storage, No. 2 southern yellow pine, 16" o.c. spacing, and 13'-0" span.

12. Define the term *safety factor*.

13. Differentiate between light roofing and heavy roofing when sizing rafters.

14. Why is it important in conventional roof framing to have ceiling joists parallel to the rafters?

15. In planning a plank-and-beam roof with wide spacing of the beams, why is the stiffness of the planking an important consideration?

16. Is the rafter length or the rafter span used in selecting a proper size for a rafter from Table 13–16?

17. What is the maximum allowable span for 2 × 10 Douglas fir floor joists spaced at 16″ o.c. if the live load is 30 lb/sq ft?

18. What will be the length of a rafter if the roof slope is 9 : 12 and the horizontal span is 16′-0″? (See Fig. 13–9.)

19. What advantage is gained by the use of a flitch beam?

20. What size of standard steel pipe column is required to support 50 kips if the unsupported length is 10′-0″?

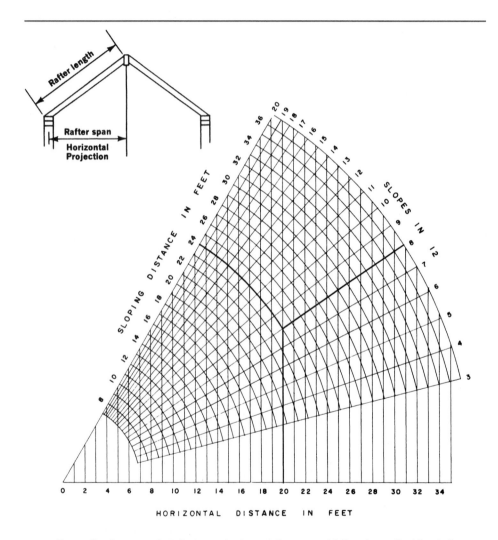

To use the diagram select the known horizontal distance and follow the vertical line to its intersection with the radial line of the specified slope, then proceed along the arc to read the sloping distance. In some cases it may be desirable to interpolate between the one foot separations. The diagram also may be used to find the horizontal distance corresponding to a given sloping distance or to find the slope when the horizontal and sloping distances are known.

Example: With a roof slope of 8 in 12 and a horizontal distance of 20 feet the sloping distance may be read as 24 feet.

Figure 13-9 Conversion diagram for rafters.

COMPUTER-BASED BUILDING STRUCTURAL DESIGN RESOURCES

Two examples of computer applications related to building structural design or components are included. The first example, Fig. 1, from Alpine Engineered Products, Inc., shows a sample output sheet for the design of a wood truss roof system. The second example, Fig. 2, from ParaVision Technologies, Inc., was downloaded from an Internet site.

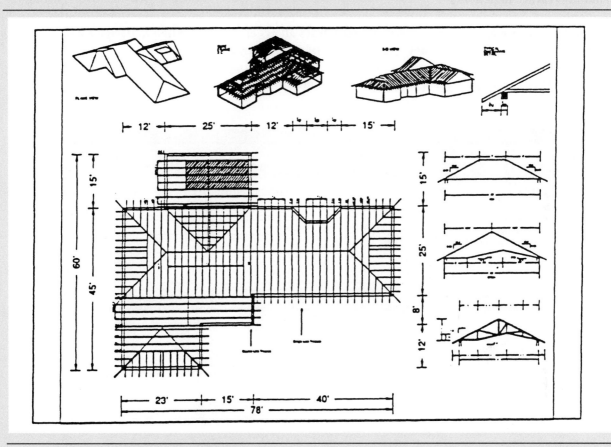

Figure 1

Truss Design and Manufacturing

Truss Design

The design and drafting of trusses can be a time-consuming and costly task. For manufacturers of trusses used in residential and commercial construction, the ability to quickly design, draft and estimate the cost of a project provides an important competitive advantage. With ParaGrafixtm the logic for design and drafting standards can be incorporated in the system so that fabrication information is available in minutes rather than hours.

ParaGrafix's ease of use and parametric capabilities allows easy optimization of design solutions. Better solutions means greater efficiency and a competitive advantage.

Here's a look at the application running under Windowstm 95:

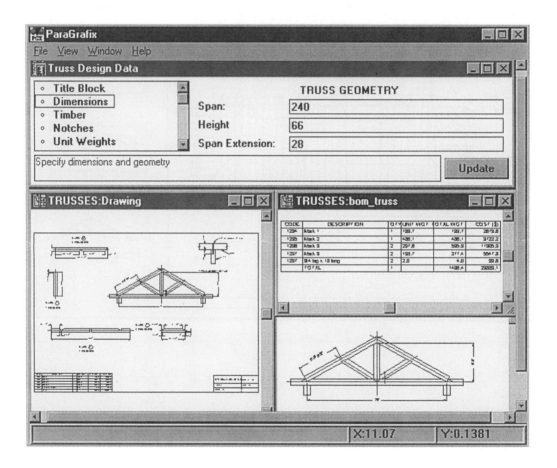

 Return to previous page.

For more information, phone (604)733-8600, fax (604)733-8622 or E-mail info@paravision.ca

Copyright © 1995 ParaVision Technologies Inc

Figure 2

"Genius is the ability to reduce the
complicated to the simple."
—C. W. CERAN

The word *details* covers the broad category of the isolated and enlarged drawings that, together with the plans and the elevations, form a complete set of working drawings. Each detail furnishes specific information at key points throughout the construction. Almost every set of drawings requires a different set of details, the choice of which is determined by the architect or the drafter. Only a few standard details may be necessary, however, on drawings of modest structures incorporating conventional construction. In fact, some sets of drawings contain very few details, allowing the builders considerable latitude in the choice of construction, whereas other sets contain details for almost every construction arrangement in the building.

This chapter is limited to the basic general details found in sets of drawings for average residential and light construction. Other minor details will be discussed in Chapter 18.

unimportant areas of long members. This method of condensing details to show only the significant points is common practice in architectural drawings since many members are necessarily large, and reducing the scale would only make smaller pieces difficult to show. However, if space permits, the true heights and lengths of important members of a detail are definitely more informative to workers. You should resort to break symbols only when limited space prevents drawing the entire detail.

The choice of scale may also be affected by the units in the detail. For example, if concrete block coursing is used, by selecting the 3/4" = 1'-0" scale, you can easily lay out the 8" coursing with the 1/2" = 1'-0" architect's scale. See Table 14–1 for a list of typical drawing scales.

14.1

SELECTING THE CORRECT SCALE FOR DETAILS

Before starting a detail drawing, take a minute or two to decide on the most appropriate scale. Suitable scale depends on the size of members in the detail, the amount of detail that must be shown, and the space available in the layout of the sheet; all of these factors must be considered. Plan the sheets so that details are placed for their most logical association with other drawings and so that all sheets of a set are utilized for economy as well as appearance. Scales for details range from 1/2" = 1'-0" to full size. Small pieces, such as moldings, are usually drawn full size to show clearly profiles and small detail. Larger members or assemblies can be drawn at smaller scales. Note that the preferable larger scales can be used in minimum spaces if *break symbols* are employed to remove

Table 14–1 Drawing scale table.

Drawing	Range	Description
Site plan	1:30–1:20	Site plan and survey should be drawn at the same scale.
Detail of site plans	1/8"–1/2" 1 1/2"–3"	Large areas Small areas
Floor plans, exterior building sections	1/16"–1/4"	Very large to small
Detail plans	1/4"–1/2" 1 1/2"–3"	Large areas Small areas
Wall sections	3/8"–3/4"	Small-scale drawing should be sketchy.
Interior elevations	1/4", 3/8", 3/4"	1/4"–3/8" are most common.
Details	1 1/2"–3"	Use the same scale for details in a family.
Schedules: door, window, and cabinet	1/4", 3/8", 3/4"	1/4"–3/8" are most common.

14.2

DRAWING TYPICAL WALL DETAILS

If the construction of a house is identical throughout all exterior walls, only one wall section may be necessary; if the construction varies, several sections should be drawn. A typical wall section conveniently incorporates the *footing detail*, *sill detail*, and *cornice detail*—all properly in line for relative reference (Figs. 14–1, 14–2, and 14–3).

To draw a wall section, start with the poured footing (usually twice as wide and the same height as the thickness of the foundation wall), and center the foundation wall directly above. If footing depths vary below grades, and if a basement is required, the foundation wall can be drawn with a break symbol, and only a convenient amount of the wall need be shown.

The construction at the top of the foundation, known as the *sill detail*, is important, and various arrangements of the framing members at this point determine the height of the finished floor from grade. If the foundation is concrete masonry units, show the courses, and insert an anchor bolt through the sill plate into the C.M.U. Wood members should be kept at least 8″ from grade. If a wood floor over a crawl space has been decided on, allow at least 18″ clearance between floor joists and the grade line. This condition usually places the finished floor at least 28″ or 30″ above grade (unless excavation is done within the foundation). Exterior steps to the floor surface will then be necessary. If a basement is planned, such as in the house of Fig. 18–3, the foundation must extend far enough above grade to meet local codes and to allow installation of basement windows. If slab-on-ground construction is required, the finish floor level will usually fall about 8″ above the grade level, requiring one step height for entry. Other variations of the sill detail may result in still other floor heights; usually, it is advantageous to keep the house as low as possible. Draw the grade line after the sill is completed. Be sure that the wood joists have at least 3″ bearing surface on the sill plate. In western frame construction, the box sill requires the soleplate and the studs to be placed over the subfloor.

In brick veneer construction, a better sill detail results if the brick starts at, or just below, grade and if the wood plate is raised above the bottom of the veneer and frame air space (Fig. 14–2B). This prevents moisture, which usually forms at the bottom of the air space during cold weather, from penetrating the wood framing. Otherwise, the wood members are identical to that used in frame construction. Brick rowlock sills (Figs. 14–4F and 14–5) must have flashing below.

In solid masonry construction, several methods can be used to support wood floor joists. A diagonal firecut can be used at the ends of the floor joists resting within the solid masonry walls (Fig. 14–4D), or the masonry wall can be widened to provide bearing for the floor joists.

The cornice detail shows all of the construction at the intersection of the walls and the roof, including the roof overhang. In conventional frame construction, both the notched bird's-mouth of the rafters and the ceiling joists bear on the double wall plate, and the joists are usually shown parallel to the rafters. A bird's-mouth cut provides horizontal bearing surface for the rafters, and because it supports the main weight of the roof, the surface should be a minimum of 3″ wide. Remember that members in details should be drawn *actual size rather than nominal size*.

Extend the rafters to form the overhang (often as much as 2′ on some homes). The overhang can be given various treatments, depending on the decisions made on the design sketches. If a wide, boxed-in cornice seems advisable, draw 2 × 4 horizontal outlookers from the studs to the rafter ends (Fig. 14–2A). Attach soffit material to the underside of the outlookers, and show a screened vent under the eave when applicable. At the ends of the rafters, draw either a vertical or inclined fascia board, making it wide enough to allow about a ¾″ drip below the surface of the soffit material. If a sloping soffit is desirable, the soffit material is attached directly to the underside of the rafter overhang; on other treatments, the rafters can be left exposed on the exterior. Cornice details showing exposed rafters, in either conventional or post-and-beam construction, must have blocking and exterior wall covering shown between the rafters. This detail must be carefully conceived to produce an airtight joint between the wall and the underside of the roof decking; exterior trim at this point also requires consideration.

Complete the decking, roofing material, gravel stops, gutter, and the like, and include all hatching, notes, and applicable dimensions. Avoid scattering descriptive notes on section views in a disorderly fashion. Plan the notes in a logical sequence, with noncrossing leaders touching the identifying part. Line up the notes on a vertical guide line insofar as possible, and make the spacing between notes uniform when they occur in groups. Keep the notes as close to their counterparts as practical to minimize misinterpretation. Outline the section with *bold lines*. See Fig. 14–6 for cutting plane indications. Roof slopes, as discussed below, should be indicated with a roof slope diagram.

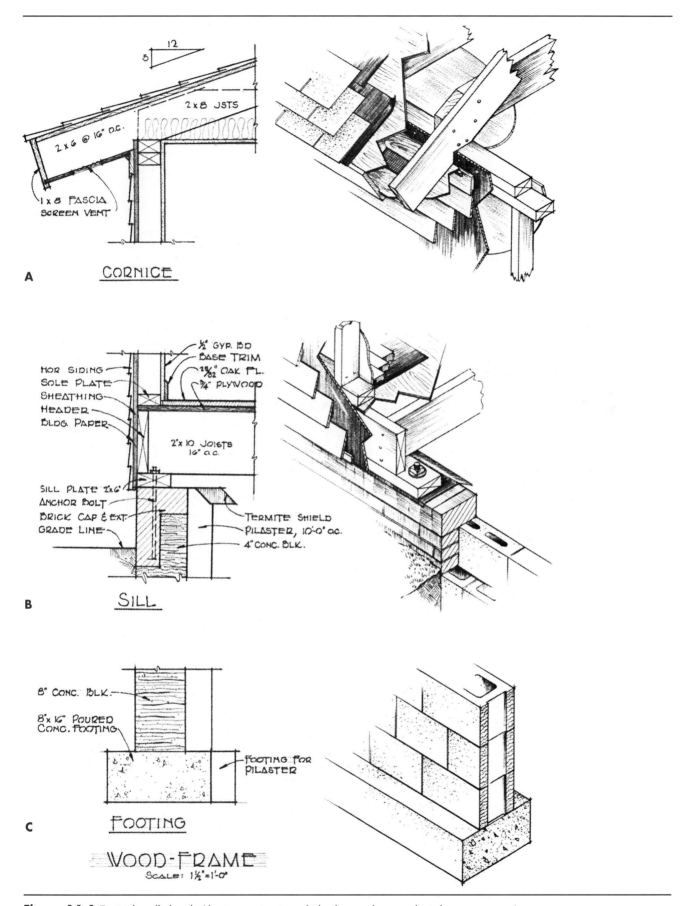

A CORNICE

12
5

2x6 @ 16" O.C.

2x8 JSTS

1x8 FASCIA
SCREEN VENT

B SILL

HOR SIDING
SOLE PLATE
SHEATHING
HEADER
BLDG. PAPER

½" GYP. BD
BASE TRIM
²⁵⁄₃₂" OAK FL.
¾" PLYWOOD

2"x 10 JOISTS
16" o.c.

SILL PLATE 2"x6"
ANCHOR BOLT
BRICK CAP & EXT.
GRADE LINE

TERMITE SHIELD
PILASTER, 10'-0" o.c.
4" CONC. BLK.

C FOOTING

8" CONC. BLK.

8"x 16" POURED
CONC. FOOTING

FOOTING FOR
PILASTER

WOOD-FRAME
SCALE: 1½"=1'-0"

Figure 14–1 Typical wall details (the isometric views help the reader visualize the construction).

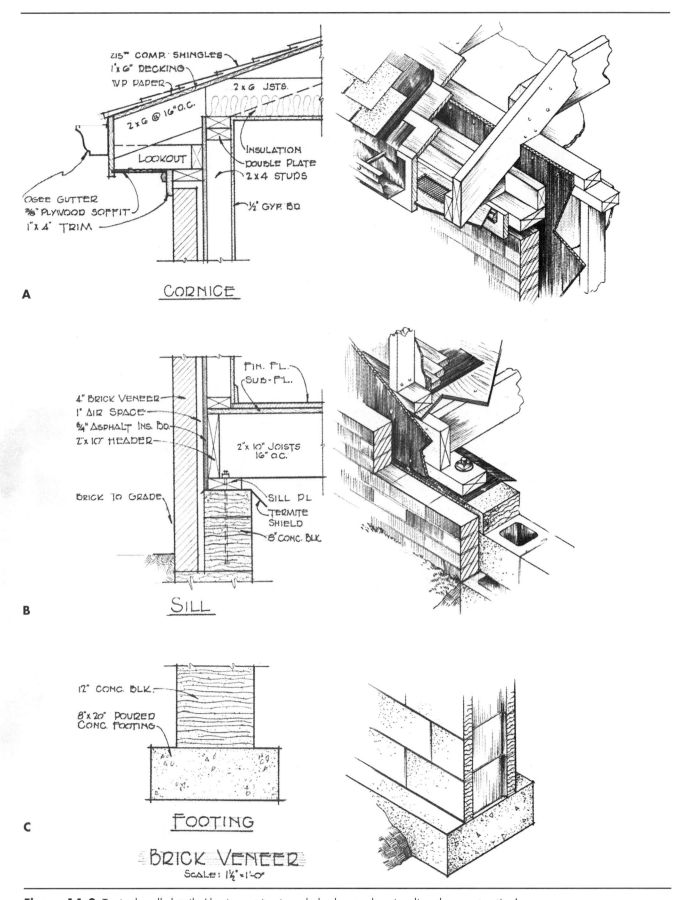

A CORNICE

- 215# COMP. SHINGLES
- 1"x 6" DECKING
- WP PAPER
- 2 x 6 JSTS.
- 2 x 6 @ 16" O.C.
- LOOKOUT
- INSULATION
- DOUBLE PLATE
- 2 x 4 STUDS
- ½" GYP. BD.
- OGEE GUTTER
- ⅜" PLYWOOD SOFFIT
- 1" x 4" TRIM

B SILL

- 4" BRICK VENEER
- 1" AIR SPACE
- ¾" ASPHALT INS. BD.
- 2"x 10" HEADER
- FIN. FL.
- SUB-FL.
- 2"x 10" JOISTS 16" O.C.
- BRICK TO GRADE
- SILL PL.
- TERMITE SHIELD
- 8" CONC. BLK.

C FOOTING

- 12" CONC. BLK.
- 8"x 20" POURED CONC. FOOTING

BRICK VENEER
SCALE: 1½"=1'-0"

Figure 14–2 Typical wall details (the isometric views help the reader visualize the construction).

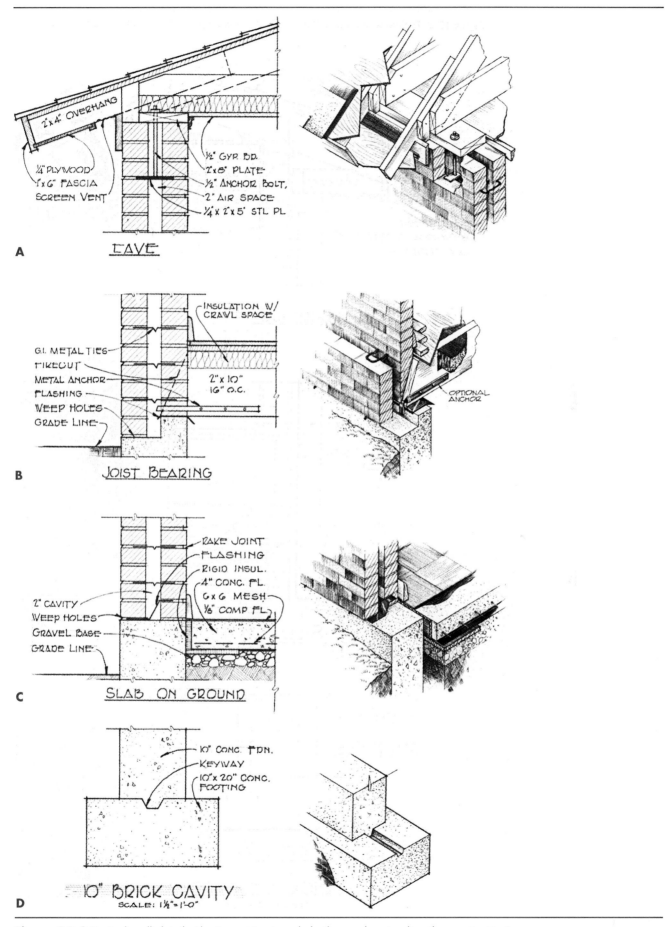

A EAVE

- 2"x4" OVERHANG
- ¼" PLYWOOD
- 1"x6" FASCIA
- SCREEN VENT
- ½" GYP. BD.
- 2"x8" PLATE
- ½" ANCHOR BOLT
- 2" AIR SPACE
- ¼" x 2" x 5' STL PL

B JOIST BEARING

- G.I. METAL TIES
- FIRECUT
- METAL ANCHOR
- FLASHING
- WEEP HOLES
- GRADE LINE
- INSULATION W/ CRAWL SPACE
- 2"x10" 16" O.C.

C SLAB ON GROUND

- 2" CAVITY
- WEEP HOLES
- GRAVEL BASE
- GRADE LINE
- RAKE JOINT
- FLASHING
- RIGID INSUL.
- 4" CONC. FL.
- 6x6 MESH
- ⅛" COMP FL.

D 10" BRICK CAVITY
SCALE: 1½"=1'-0"

- 10" CONC. FDN.
- KEYWAY
- 10"x 20" CONC. FOOTING

Figure 14–3 Typical wall details (the isometric views help the reader visualize the construction).

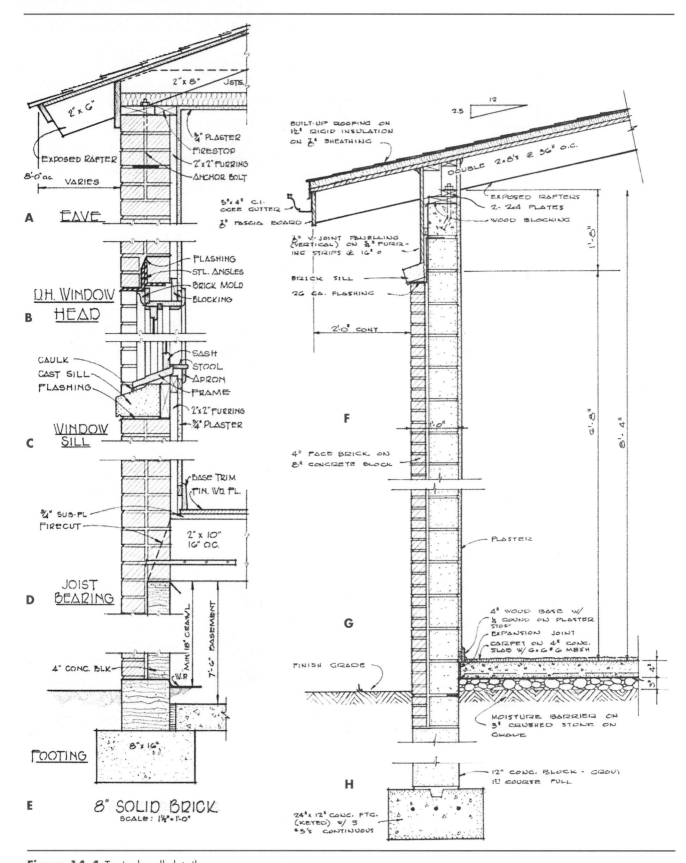

A EAVE

EXPOSED RAFTER
8'-0"o.c. VARIES

2" x 6"
2" x 8" JSTS.
¾" PLASTER
FIRESTOP
2"x2" FURRING
ANCHOR BOLT

B D.H. WINDOW HEAD

FLASHING
STL. ANGLES
BRICK MOLD
BLOCKING

C WINDOW SILL

CAULK
CAST SILL
FLASHING

SASH
STOOL
APRON
FRAME
2'x2" FURRING
¾" PLASTER

D JOIST BEARING

BASE TRIM
FIN. WD. FL.
¾" SUB-FL.
FIRECUT

2" x 10"
16" O.C.

E 8" SOLID BRICK
SCALE: 1½" = 1'-0"

FOOTING

4" CONC. BLK.
W.P.
7'-6" BASEMENT
Min 18" CRAWL
8" x 16"

BUILT-UP ROOFING ON
1½" RIGID INSULATION
ON ⅞" SHEATHING

2.5 12

DOUBLE 2x8's @ 36" O.C.
EXPOSED RAFTERS
2- 2x4 PLATES
WOOD BLOCKING

5"x 4" C.I. OGEE GUTTER
8" FASCIA BOARD

⅜" V-JOINT PANELLING
(VERTICAL) ON ¾" FURR-
ING STRIPS @ 16" O

BRICK SILL
26 GA. FLASHING

2'-0" CONT.

F

4" FACE BRICK ON
8" CONCRETE BLOCK

1'-0"
6'-8"
8'-4"

G

PLASTER

FINISH GRADE

4" WOOD BASE W/
¾ ROUND ON PLASTER
STOP
EXPANSION JOINT
CARPET ON 4" CONC.
SLAB W/ 6x6 #6 MESH

4"
3"

MOISTURE BARRIER ON
3" CRUSHED STONE ON
GRADE

H

12" CONC. BLOCK - GROUT
1ST COURSE FULL

24"x 12" CONC. FTG.
(KEYED) W/ 3
#5's CONTINUOUS

Figure 14–4 Typical wall details.

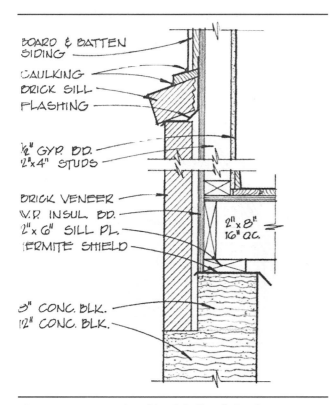

Figure 14–5 Wood and brick half-wall detail.

Labels on figure:
BOARD & BATTEN SIDING
CAULKING
BRICK SILL
FLASHING
½" GYP. BD.
2"x 4" STUDS
BRICK VENEER
V.P. INSUL. BD.
2"x 6" SILL PL.
TERMITE SHIELD
8" CONC. BLK.
12" CONC. BLK.
2"x 8" 16" O.C.

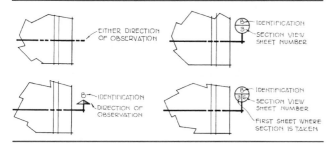

Figure 14–6 Methods of indicating cutting planes for sections.

Labels on figure:
EITHER DIRECTION OF OBSERVATION
IDENTIFICATION
SECTION VIEW SHEET NUMBER
IDENTIFICATION
DIRECTION OF OBSERVATION
IDENTIFICATION
SECTION VIEW SHEET NUMBER
FIRST SHEET WHERE SECTION IS TAKEN

INDICATING ROOF PITCHES ON DRAWINGS

The slope of a roof largely determines the kind of roofing; this in turn influences preliminary planning when pleasing roof lines have been established. Although any slope can be made workable, it is advisable to modify the roof lines of the sketch slightly if necessary so that the working drawings will have roof slopes that are not too difficult to measure. This modification affects the appearance very little and makes the construction of the roof somewhat simpler (see Fig. 14–7 for typical roof slopes).

Roof slopes or pitches can be shown by any of three methods on drawings:

1. slope-ratio diagram,
2. fractional pitch indication, and
3. angular dimension.

The *slope-ratio triangle* is drawn with its hypotenuse parallel to the roof profile. Its opposite sides, representing the rise-and-run slope ratio, are drawn vertically and horizontally. The horizontal leg of the triangle is usually measured 1″ long for convenience and is given a numerical value of 12. The rise value shown on the vertical leg of the triangle can then be easily measured off on the 1″ = 1′-0″ scale. This ratio of the rise of the roof to its run is stated, for example, as 3 : 12, 7 : 12, or 1½ : 12, making it consistent with conventional references associated with roofing materials and other construction applications. In Fig. 14–8, the 4 : 12 right-triangle symbol means, briefly, that for every 12″ of horizontal run, the roof rises 4″. Most architects use this method for slope indications.

The *fractional pitch* of a roof is derived from a standard formula:

$$\text{Pitch} = \frac{\text{rise}}{\text{span}}$$

The span is the total distance between top plate supports (twice the run). Fractional pitches are infrequently found on working drawings; yet carpenters have traditionally used fractional pitches for many of their rafter layouts.

Angular slopes are shown with an arc dimension line, revealing the angular dimension in degrees from the horizontal (Fig. 14–7). Use of an arc dimension line is limited mainly to minor construction features.

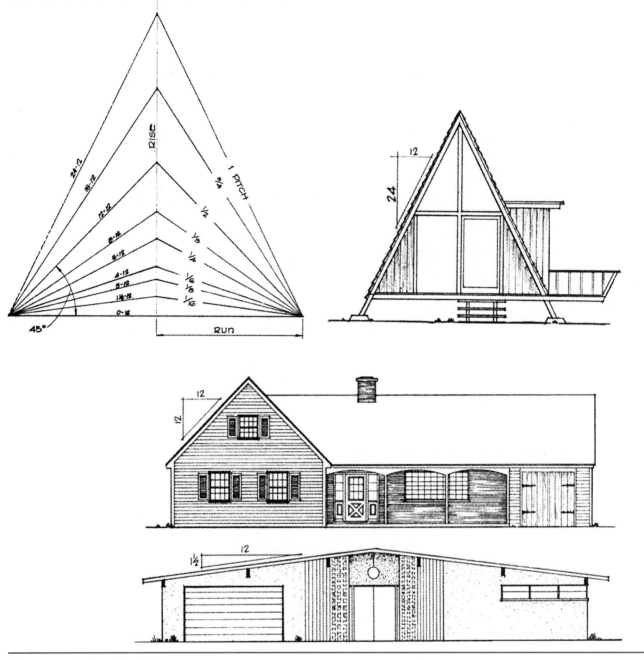

Figure 14–7 Typical roof slopes.

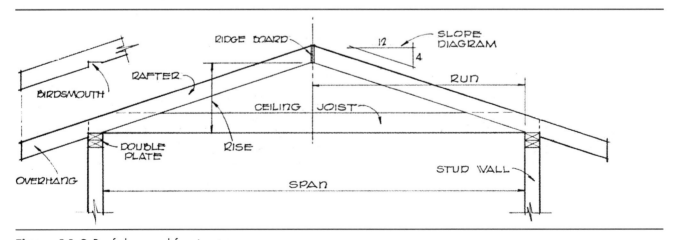

Figure 14–8 Roof slope and framing terms.

14.4

WINDOWS AND THEIR REPRESENTATION

Windows provide, first of all, the light and ventilation necessary in dwellings; they are the eyes through which the surrounding landscape can be enjoyed, bringing in the outdoors to make the interior more spacious and less confined. Windows must be carefully selected as to size, type, and placement if they are to provide the desired architectural character to the structure. Thus, both practical and esthetic considerations are important in their selection. As a rule, the total window area should be at least 10% of the floor space. Choosing windows can be a difficult task unless you pay close attention to three characteristics: (1) energy performance, (2) operation, and (3) service.

14.4.1 Energy Performance

Heat loss (the U value), solar heat gain, air infiltration, and condensation resistance are important areas to investigate. Whole-unit U values rather than middle-of-glass values should be studied. This information, found on the labels of the National Fenestration Rating Council (NFRC), provides a good comparison of the combined effectiveness of all of a window's components, including glazing, frame construction, and weatherstripping. Glazings that maximize solar gain in cold climates and minimize solar gain in warm climates should be chosen in order to reduce heating and cooling loads. Review the manufacturer's air infiltration ratings. Be sure that weatherstripping seals the entire perimeter and compresses or interlocks when the sash is closed and locked. Weatherstripping material should resist damage caused by ultraviolet radiation. Features that minimize potential condensation include thermal breaks in metal frames, warm-edge seals between multiple glazings, and glass coatings and air-space gases that improve performance.

The NFRC has developed a fenestration energy rating system based on whole product performance. This system accurately accounts for the energy-related effects of all of a product's component parts, and it prevents information about a single component from being compared in a misleading way to other whole product properties. At this time, NFRC labels on window units give ratings for the U value, the solar heat gain coefficient, and visible light transmittance. Soon labels will include air infiltration rates (FHR) and an annual fenestration cooling rating (FCR). Figure 14–9 is a sample NFRC label. Table 14–2 notes characteristics of typical windows and annual energy performance ratings.

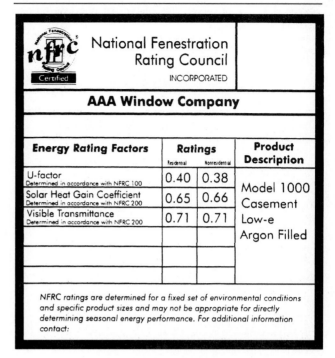

Sample NFRC Label

Figure 14–9 Sample NFRC label.

14.4.2 Operation

Windows should operate smoothly and should be easy to clean; they should also have good-quality screens and storm sashes. Crank mechanisms should operate smoothly: Latches should pull the sash closed without being forced. Weatherstripping should seal tightly but not bind or catch. Windows should also have a sash that you can tilt in order to clean the windows; easy-to-use sash removal mechanisms; a sliding sash that glides with little effort; and hardware, glass, and a sash that can be easily replaced if necessary.

14.4.3 Service

Purchase units from suppliers and manufacturers who support their products and who have representatives available to come to your site to address problems. Check out the warranties. Parts are usually covered for 1 year; glass and seals, from 1 to 20 years, with 5 years the most common. Factory-applied exterior finishes may be covered by a separate warranty. Be sure to follow the manufacturer's finishing instructions. Check on the company's replacement parts policy. Some manufacturers provide parts only as long as existing inventories last, whereas others try to provide lifetime service for any window that they have ever made.

Table 14–2 Characteristics of typical window types and annual energy performance ratings.

Window Description	Overall Window Characteristics					
	U-Factor (Btu/h-ft²-°F)	Solar Heat Gain Coefficient	Visible Transmittance	Air Leakage (cfm/ft²)	FHR	FCR
1. Single-glass, aluminum-frame, no thermal break	1.30	0.79	0.69	0.98	0	0
2. Single-glass, bronze; aluminum-frame; no thermal break	1.30	0.69	0.52	0.98	−2	8
3. Double-glass, aluminum-frame, thermal break	0.64	0.65	0.62	0.56	19	12
4. Double-glass, bronze; aluminum-frame; thermal break	0.64	0.55	0.47	0.56	17	20
5. Double-glass, wood- or vinyl-frame	0.49	0.58	0.57	0.56	24	18
6. Double-glass, bronze; wood- or vinyl-frame	0.49	0.48	0.43	0.56	22	25
7. Double-glass; low-E, argon; wood- or vinyl-frame	0.33	0.55	0.52	0.15	32	19
8. Double-glass; low-E, argon; wood- or vinyl-frame	0.30	0.44	0.56	0.15	32	27
9. Double-glass; selective low-E, argon; wood- or vinyl-frame	0.29	0.32	0.51	0.15	30	36
10. Double-glass; selective low-E, argon; wood- or vinyl-frame	0.31	0.26	0.31	0.15	27	40
11. Triple-glass; low-E (2), krypton; insulated-vinyl-frame	0.15	0.37	0.48	0.08	38	33
12. Triple-glass, wood- or vinyl-frame	0.34	0.52	0.53	0.08	32	22

Glassed areas can be either preassembled window units or *fixed glass*, which is directly affixed to the framing. Fixed glass allows large window areas to be installed with a minimum of expense, inasmuch as no operating sash, hardware, or screens are necessary. If ventilation is needed, occasional-opening sash or panels can be incorporated into window walls. Today's market offers a wide assortment of preassembled window units; most of them are made of wood, aluminum, or vinyl cladding, which eliminates the need for painting.

Wood windows make up about 50% of the residential market. Although all of the parts of a wood window are treated with a preservative prior to assembly, wood windows do require maintenance. Some manufacturers paint the exterior of their wood windows in the factory.

Many manufacturers make clad windows, which can have a relatively maintenance-free exterior with a wood interior. There are a number of different ways to clad a window. Some manufacturers attach the cladding by gluing it onto a wood frame. Others have designed their cladding so that it snaps onto a wood frame. Other companies make a vinyl or an aluminum frame to which a wood interior is attached.

All-vinyl windows are typically much less expensive than wood windows, and they never need painting. All-vinyl windows should not be confused with vinyl-clad windows. In the past, some all-vinyl windows experienced problems. Vinyl expands and contracts at a different rate than glass, and on some windows, thermal cycling caused the vinyl to distort and to pull away from the seals around the glazing. Faulty seals affect a window's ability to withstand air and water infiltration. Vinyl is a brittle material that is 80% salt. In order to make vinyl pliable, a plasticizer is added so that the vinyl can be molded. With time, as the plasticizer evaporates, the vinyl becomes brittle. New vinyl formulas, called uPVCs (unplasticized polyvinyl chlorides), are supposed to be more stable and more resistant to distortion and movement. An important feature to look for in vinyl windows is welded corners rather than mitered and screwed corners.

In recent years some people have steered away from aluminum windows because they are not energy efficient. Aluminum is a good conductor and thus a poor insulator. Aluminum windows are typically less expensive than wood windows or vinyl windows, and they should not require much maintenance. Aluminum can be painted, and manufacturers offer their products with a factory-applied colored coating. Another advantage of aluminum is that, because of its inherent strength, aluminum windows have a much larger glass area per window size. Aluminum-clad windows do not suffer from the energy disadvantages of all-aluminum windows because the heat-conducting properties of the aluminum are broken by the wood interior.

Fiberglass has advantages as a material for both window frames and sashes. Fiberglass expands and contracts at almost the same rate as window glass, so there should not be the kinds of problems associated with older all-vinyl windows. Other benefits include corrosion resistance and dimensional stability.

When choosing windows for a house, you will want to decide three things: (1) the type of windows—casement, double-hung, fixed, awning, or slider; (2) the kind of glass to use; and (3) the material used for the frames.

Insulating glass—two panes of hermetically sealed glass separated by a spacer—came into widespread use in the 1970s. Dead air space between two pieces of glass improves the energy efficiency of a window. The wider the air space between panes, the higher the insulating value of the window. Filling the space between the pieces of glass with gases that are less conductive than air, such as krypton or argon, improves a window's insulating ability.

Low-E glass is covered with a heat-reflective, low-emissivity coating. The direction in which low-E glass reflects heat is determined by the glass surface to which the coating is applied. In a cooling environment, low-E glass reflects the sun's heat away from the house. In a heating environment, low-E glass reflects heat back into the house instead of letting it pass through the window.

Heat mirror is a low-E film suspended in the space between the panes in a piece of insulating glass. Different types of heat mirror films perform different functions. Some are better at reflecting exterior heat away from a house; others are better at keeping heat inside the house. All heat mirror films are superb at keeping ultraviolet (UV) light out of a house. Window systems developed using the heat mirror system can cost up to twice as much as other systems using regular insulating glass.

14.4.4 Types of Windows

The two basic components of all windows are a sash and a frame. The sash is the part of the window that holds the glass, and the frame is the part that holds the sash. Windows are categorized by the method and direction in which the sash moves, or does not move, in the frame. Figure 14–10 shows the major types of window units currently used in residential construction.

Double-Hung Windows   These windows, also called *vertical-slide windows*, are the most familiar, having two sashes that operate vertically. A double-hung window allows a maximum of 50% ventilation and is typically American. The unit, including the two sashes, the frame, weatherstripping, and hardware, is set into the rough opening; then it is plumbed and is nailed to the double studs and header. Inside trim is applied after the interior

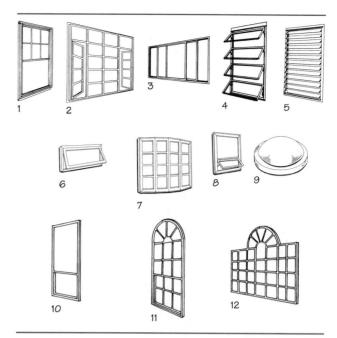

Figure 14–10 Major residential window types: (1) double-hung; (2) casement; (3) horizonal sliding; (4) awning; (5) jalousie; (6) basement; (7) bow or bay; (8) hopper and fixed; (9) skylight; (10) single hung; (11) arch top; (12) Palladian.

walls are complete. If units are combined into groups, a structural mullion is shown between the units. The slopping sill member rests on the lower member of the rough opening (Fig. 14–11).

Casement Windows   Casement windows have their sash hinged on the side. Most of them open out, but some that open in are available. Casement sash windows have the advantage of directing breezes into the room when open, and 100% ventilation is possible. Operating cranks open and close the sash.

Horizontal Sliding Windows   Horizontal sliding windows operate horizontally, like sliding doors. Like double-hung windows, horizontal sliding windows allow only 50% ventilation when open. Units are often combined with fixed picture-window sash.

Awning Windows   These windows open out horizontally, with the hinges of each sash at the top and the cranking hardware operating all of the sash in each unit simultaneously. Their major advantage is that they can be left open during rain without adverse effects. Dust collects on the opened windows readily, however.

Jalousie Windows   Jalousie window units have small, horizontal glass panels that operate horizontally, much as an awning sash window does. The individual glass panels do not have frames. The advantages are similar to those

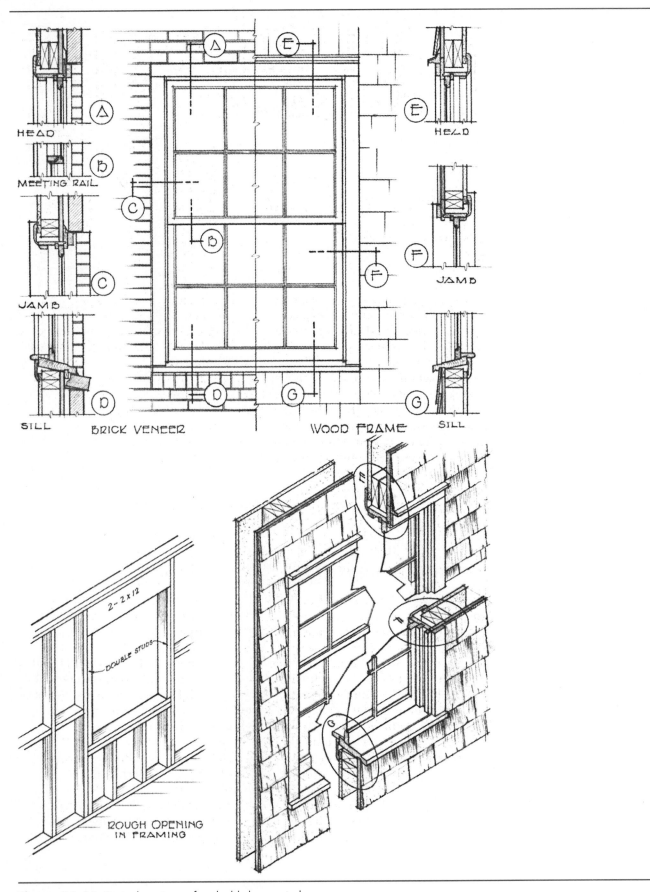

Figure 14–11 Typical sections of a double-hung window.

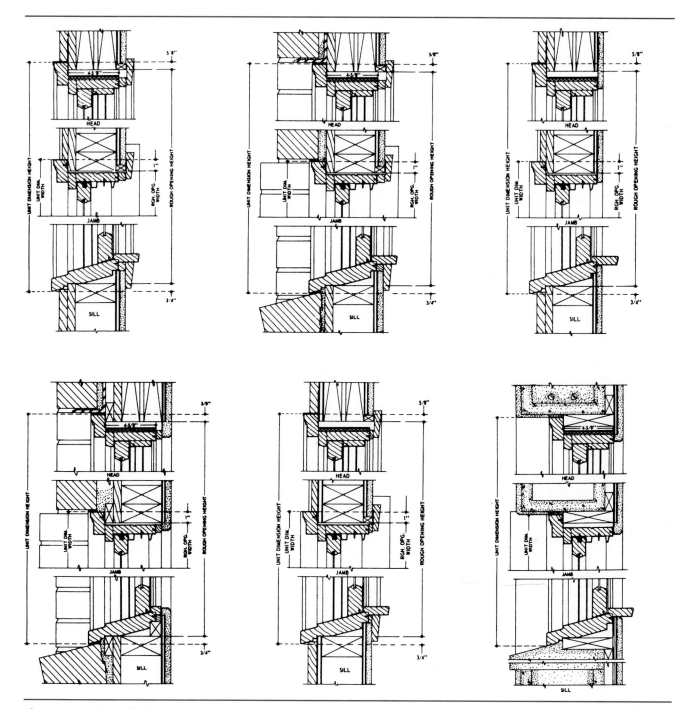

Figure 14–12 Double-hung window details in various walls.

of awning windows, except that jalousie windows are seldom as weathertight when closed. Therefore, they should be reserved mainly for glass-enclosed porches or for residences in warmer climates.

Hopper Windows Hopper windows have small sashes, opening either in or out, which are combined with larger fixed windows. The complete unit is set into the rough opening like other window units.

Palladian Windows Palladian windows are a comparitively recent adaptation from the sixteenth century Italian architect Andrea Pallidio. They are tall, semicircular center windows combined with straight-headed side windows; they have become popular in many American homes within the last 20 years. Often classical pilasters and heads are trimmed out in the interior to form very elegant windows, especially in rooms with high ceilings.

The Palladian window should be carefully proportioned to complement other windows and features. They must be properly placed without crowding other elements and spaces, and they should be used sparingly to avoid dominating the main architectural features. Usually small lights within the combined units are most pleasing.

These windows are generally used to provide added light to large rooms with high ceilings, as we mentioned, and often to illuminate stairways where a focal point might contribute style to the exterior. Manufacturers frequently refer to these arched windows as "cathedral windows." Stock windows properly sized can be combined using a larger central unit and semicircular head and two smaller side windows to form very satisfactory combinations.

Skylights Skylights are actually roof windows that can be placed over openings in roofs of little or no slope and sealed into the roofing material. Residential skylights are usually bubble-shaped, plastic units with flanges or sealing rims around the base. Small skylights are convenient for giving natural light to interior hallways or window-less bathrooms. However, in warm climates, the direct glare of the sun through a skylight can be troublesome unless some manner of light control is provided. Caution should be exercised in the use of skylights on roofs.

Clerestory Windows Clerestory windows are so named because of their placement—usually high in a wall above a lower roof level. They are often a series of small windows; their height affords privacy and permits them to cast dramatic light effects within the room.

Picture Windows So called because usually no muntins interfere with the framed view, picture windows are fixed-glass units (as a rule rather large), which often become the center unit of several regular windows.

Study the many manufacturers' catalogs for more detailed information of various windows, and observe the recommended method of attaching each window in different wall openings (Fig. 14–12).

14.4.5 Window Unit Measurements

There are four basic measurements associated with any one window unit: (1) unit size, (2) rough opening size, (3) sash size, and (4) glass size (Table 14–3 and Fig. 14–13). It is important to identify which one of the four measurements is being used when window dimensions are given. Window sizes are usually noted with the width first and the height second.

The *unit dimension* represents the overall outside dimensions of the window. This size can be taken from different places, depending on the type of window being described. A wood double-hung unit normally has preattached brick mold trim, so the unit size is the out-to-out size of the brick mold. Clad casement windows often come with plastic nailing flanges, and the unit size will be the same as the out-to-out dimensions of the jambs.

The *rough opening size* is the dimension required in the rough frame wall in order to receive the window. This size is most important for the builder. It determines the framing size for the exterior wall opening. It is usually 1/4″ larger per side than the outside-of-jamb-to-outside-of-jamb measurement. It provides extra space needed for leveling and plumbing the window in the rough framing of the wall.

The *sash size* is the overall measurement of the window sash. The window sash fits between the two jambs and is the same as the distance between the jambs for the width size. The height dimension will vary, depending on the type of window. A casement window sash extends from the head to the sill, whereas a double-hung window has two or more sashes between the head and the sill.

The *glass size* is the measurement of the glass area as viewed in the sash. It is the sash minus the stile size (for the width) or minus the rails (for the height).

Figure 14–13 and Table 14–3 illustrate the sizes and their relationship to each other for a 24″ × 24″, double-hung window unit.

In Fig. 14–11, note the method of framing window openings in wood frame construction. As mentioned previously, the units are supported by the wood framing, and the section views reveal the relationship of the unit to the framing: how they are attached and what trim is necessary for completion. These details should be shown on working drawings for proper installation of the windows.

Table 14–3 Opening sizes of windows in inches.

Glass width (glass size)	24
Sash stiles (2″ each)	4
Side jambs (3/4″ each)	28
Plumbing or fitting allowance (1/4″ per side)	1 1/2
	1/2
Rough opening width	30
Glass height (glass size)	48
Window rails	6
Head jamb	3/4
Window sill	2
Plumbing or fitting allowance (1/4″ top and bottom)	1/2
Rough opening height	57 1/4

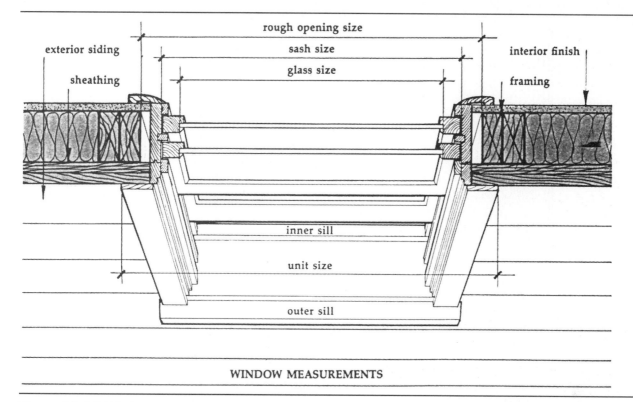

exterior siding
sheathing
rough opening size
sash size
glass size
interior finish
framing

inner sill
unit size
outer sill

WINDOW MEASUREMENTS

Figure 14–13 Window unit measurements.

To understand better the details of a double-hung window, observe the cutting planes shown on the elevation view of Fig. 14–11. The head, jamb, and sill sections to the left of the elevation clearly explain these important points in the construction. A meeting rail section is also shown merely to familiarize you with the relationship of the two sashes. Note that, for more information, the right side of the elevation appears to be in a brick veneer wall, and the same sections to the right show typical construction around the double-hung window in brick veneer construction. The isometric view in Fig. 14–11 describes the pictorial representations of the same head, jamb, and sill sections in wood frame. Study these details carefully. Details of windows vary with different manufacturers; therefore, after window selections have been made, details of the windows must be taken from the manufacturers' catalogs.

To draw a window detail, first establish the wood frame members, or the rough openings of masonry if the window is to go into solid masonry. Use large scales such as $3'' = 1'-0''$ or $1\frac{1}{2}'' = 1'-0''$ if small members are present in the section. Then draw the sheathing and inside wall covering. Next locate the window frame in relation to the rough opening surface, and complete the sash, trim, and exterior wall covering. Use the conventional break symbol so that only the necessary amount of detail needs to be drawn. Be sure that the structural members line up vertically and that the head section is drawn at the top, the jamb in the center, and the sill section at the bottom.

DOORS AND THEIR REPRESENTATION

Because of their economy and the wide assortment of available sizes, stock doors are universally used in residential and light construction. Two major types are manufactured: flush and panel (Fig. 14–14).

14.5.1 Flush Doors

Flush doors are built of solid wood stiles and rails with various wood veneering glued to both sides, producing smooth, easy-to-paint surfaces. This type is popular in contemporary architecture because of its clean-cut appearance and easy-to-maintain features. Various-shaped lights can be introduced to flush doors for lighting dark entrances, and decorative moldings can be applied directly to the veneered surfaces for style treatments. Flush doors can be subdivided into two types: *hollow-core* and *solid-core*. The hollow-core have grillage-filled cavities, making them lighter and ideal for interior use. The solid-core have their cavities filled with wood blocking or other dense materials, making them heavier and more appropriate for exterior entrances. The heavier construction reduces warping during extreme temperature differences between inside and outside. Many decorative wood grain veneers are available on flush doors to match natural wood interior wall coverings. See Fig. 14–15 for typical sizes.

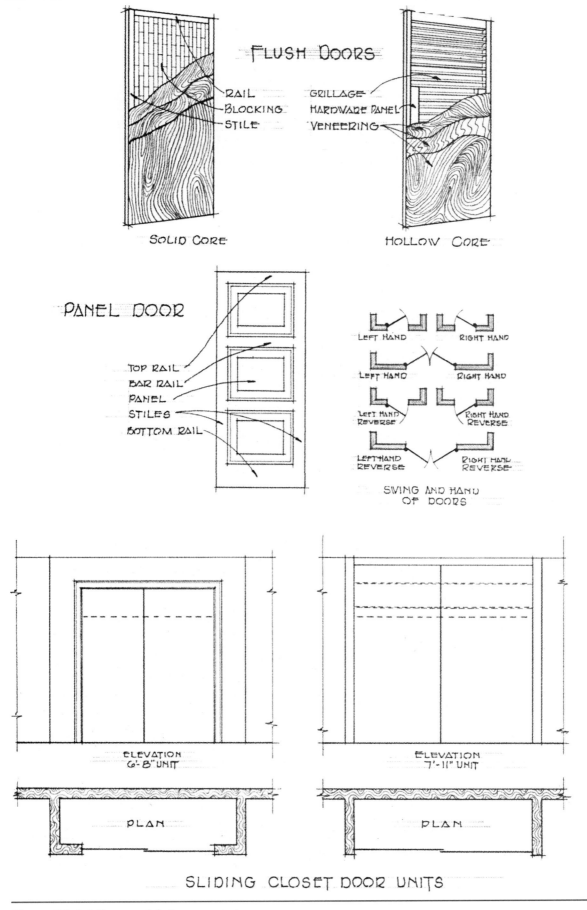

Figure 14–14 Doors and their representation.

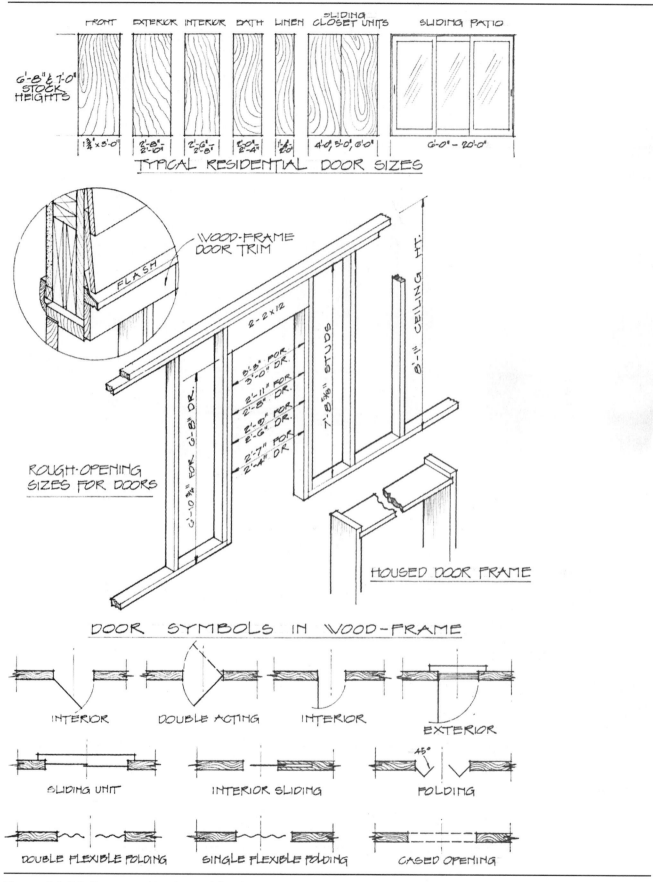

FRONT   EXTERIOR   INTERIOR   BATH   LINEN   SLIDING CLOSET UNITS   SLIDING PATIO

6'-8" & 7'-0" STOCK HEIGHTS

1³⁄₄" x 3'-0" | 2'-8", 2'-10" | 2'-6", 2'-8" | 2'-0", 2'-4" | 1'-6", 2'-0" | 4'-0", 5'-0", 6'-0" | 6'-0" – 20'-0"

TYPICAL RESIDENTIAL DOOR SIZES

WOOD-FRAME DOOR TRIM

FLASH

2 - 2x12

6'-8" FOR 3'-0" DR.
2'-11" FOR 2'-8" DR.
2'-0" FOR 2'-6" DR.
2'-7" FOR 2'-4" DR.

8'-1" CEILING HT.

7'-8³⁄₈" STUDS

6'-10³⁄₄" FOR 6'-8" DR.

ROUGH-OPENING SIZES FOR DOORS

HOUSED DOOR FRAME

DOOR SYMBOLS IN WOOD-FRAME

INTERIOR   DOUBLE ACTING   INTERIOR   EXTERIOR

SLIDING UNIT   INTERIOR SLIDING   FOLDING (45°)

DOUBLE FLEXIBLE FOLDING   SINGLE FLEXIBLE FOLDING   CASED OPENING

Figure 14–15 Door representation and framing.

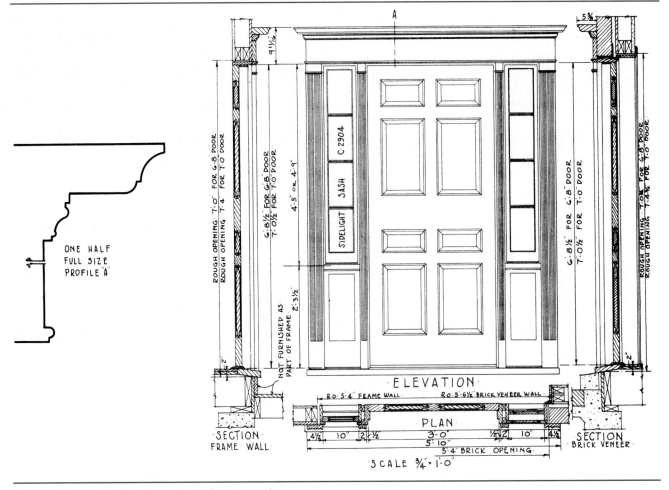

Figure 14–16 Details of a traditional entrance door.

14.5.2 Panel Doors

Manufactured in a wide variety of styles, panel doors are generally made of ponderosa pine or Douglas fir. Various arrangements of the panels and lights are surrounded with stiles and rails, usually of solid wood, to produce handsome and sturdy doors. Panel doors have been used for many years in American homes. Some panel doors are manufactured with glue and veneer stiles and rails. They are sometimes preferred for exterior installation. For colonial and traditional homes, panel doors are the most authentic in character. Elaborate panel door and frame units are available for classic entrances (Fig. 14–16). Variations of the panel door include the following styles:

- **French doors** have glazed lights separated with muntins throughout, rather than panels. When additional light is needed, or when a view is desired between rooms, these doors find application, especially in traditional homes. Often they are used in pairs with an astragal molding between.

- **Dutch doors,** also typical of some colonial styles, are cut in half horizontally, and each half operates on its own set of hinges. The bottom half can act as a gate while the top remains open.
- **Louver doors,** which have horizontal strips placed on the diagonal for vents instead of panels, are attractive in both traditional and contemporary settings. They are excellent for closet doors because of their venting characteristics, but they are time-consuming to paint.
- **Jalousie doors,** usually suitable for enclosed porches or exterior doors in warmer climates, have operating glass panel units inserted within the stiles and the rails.

14.5.3 Wood Stock Door Sizes

Both flush and panel doors are available in $1\frac{3}{8}''$, $1\frac{3}{4}''$, and $2\frac{1}{4}''$ stock thicknesses. Stock heights are 6'-8" and 7'-0"; narrow doors can be obtained in 6'-6" heights. Widths are available in 2" increments, from 1'-6" to 3'-0"; a few styles are made in 4'-0" widths. See Fig. 14–15 for the conventional door widths applicable to residential construction.

14.5.4 Folding- or Sliding-Door Units

With the increase in private outdoor living areas, large sliding glass doors have found widespread acceptance. Sliding units are manufactured of aluminum, steel, or wood (Fig. 14–17). Stock heights are usually 6'-10", and unit widths, ranging from 6'-0" to 20'-0", utilize various sliding- and fixed-glass panel arrangements. Like other component sizing indications, door unit widths are given first, such as 8'-0" × 6'-10" or 12'-0" × 6'-10". Exact sizes and specifications should be taken from manufacturers' literature. Because of serious accidents involving large glass doors, many local codes require either tempered glass or protection bars across the doors.

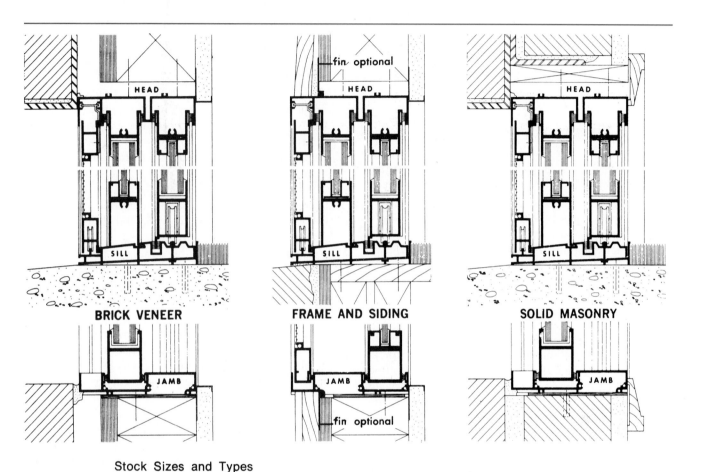

Stock Sizes and Types

Door Types Viewed from exterior		Door Sizes Width	Height	Stock Numbers	Glass Sizes Width only All heights 76¾"	Rough Openings All heights 6'–10½"
▶	XO	5'–11"	6'–10"	G2- 6	33"	5'–11½"
		7'–11"	6'–10"	G2- 8	45"	7'–11½"
		9'–11"	6'–10"	G2-10	57"	9'–11½"
◀	OX	5'–11"	6'–10"	G2- 6	33"	5'–11½"
		7'–11"	6'–10"	G2- 8	45"	7'–11½"
		9'–11"	6'–10"	G2-10	57"	9'–11½"
▶	OXO	8'–10¼"	6'–10"	G3- 9	33"	8'–10¾"
		11'–10¼"	6'–10"	G3-12	45"	11'–10¾"
		14'–10¼"	6'–10"	G3-15	57"	14'–10¾"
◀ ▶	OXXO	11'– 8⅞"	6'–10"	G4-12	33"	11'– 9⅜"
		15'– 8⅞"	6'–10"	G4-16	45"	15'– 9⅜"
		19'– 8⅞"	6'–10"	G4-20	57"	19'– 9⅜"

Figure 14–17 Details of an aluminum sliding glass door.

For interior use, many types of folding- and sliding-door units are appropriate for closet doors and room dividers. Stock sizes conform to the 6'-8", 7'-0", and 8'-0" heights, and custom-built units are available from some manufacturers for special situations. The 8'-0" high closet door units (Fig. 14–14) allow an additional amount of accessible storage space in a closet in comparison to lower heights, and these units are highly recommended. Their use reduces the amount of framing around the door as well.

Accordion doors, available in 6'-8" and 8'-0" stock heights and in widths ranging from 2'-4" to 8'-2", are made of flexible plastic material and require little space at the jambs when open. Distinctive wood folding doors are also made of various wood panels. Sliding closet units come in 3', 4', 5', and 6' widths.

14.5.5 Fireproof Doors

Flush wood doors filled with heat-resisting materials and treated with fire-resisting chemicals (1-hour fire rating) are available for light construction. Some codes require fireproof doors between the house and its attached garage. Many types of metal doors having hollow, grillage-filled, or wood cores are manufactured mainly for commercial and industrial application and are labeled according to the amount of time of fire resistance: C = ³/₄, B = 1½, A = 2 hours.

14.5.6 Doorframes

Exterior wood doorframes are usually made and assembled at the mill and then delivered to the site, ready for installation into the rough openings. In wood frame floor construction, the header joist and the subflooring below an entrance must be cut out slightly to receive the sloping sill member of a doorframe (Fig. 14–16). Frames must be heavy enough to carry the weight of the doors and to take the strain of door closures. Generally, exterior frames are 1¹/₁₆" thick, and the sill member is made of hardwood to resist wear. Heavy doors may require thicker jamb members. A planted threshold of either hardwood or metal is necessary to hide the joint between the sill and the finished floor below entrance doors.

Interior doorframes are purchased in sets and are usually assembled on the job. Usually 1" or 1⅛" thick lumber is satisfactory for interior use. Other dimensions depend on the size of the door and the thickness of the walls in which they are installed. To provide simple door installation, planted doorstops are preferred on interior frames.

Metal doorframes, called *door bucks*, are being used by some housing developers. They are prefitted and predrilled, ready for the installation of the door hardware. Both wood and metal frames are available with prehung doors.

14.5.7 Overhead Garage Doors

Because of their size and weight, garage doors must be operated with counterbalances or springs on overhead tracks. Stock sizes range from 8'-0" × 7'-0" to 9'-0" × 7'-0" for single garage widths; double-garage openings require a 16'-0" × 7'-0" door size. Their symbol on plans is merely a line through the opening, accompanied by a dimensional note and the manufacturer's number.

14.5.8 Drawing Door Details

Although accurate details of door sections are seldom shown on working drawings involving conventional construction, the drafter must choose the size, type, and style to best fit the situation. If doors are incorporated into special framing, or if they are placed in walls of different construction, section views of the jambs and the surrounding construction, as well as the trim, must be drawn. To show the proper installation methods in various types of walls, several door details are included in this material (also refer to Fig. 5–10).

In Fig. 14–16, note that the cutting planes on the door elevation indicate sections similar to window details, described previously. Profile A reveals the construction through the head jamb and sill, and the plan section reveals the construction through the side jambs. The sections are placed in a position for convenient projection of, as well as for analysis of the relationships among, the members. In drawing the sections, start by laying out the frame walls, then the wall coverings, and finally the door-jambs and surrounding trim. In solid masonry construction, the masonry outlines are drawn first; then the jambs and trim are added. Steel angles or reinforced concrete lintels must be used to support the weight of the masonry above the door opening and must be shown on the head section.

14.6

FIREPLACE DESIGN AND DETAILS

For centuries, fireplaces were the only method of heating homes, and even with our modern fuels and central heating systems, the fireplace is still as popular as ever because we have not found a substitute for the sheer fascination of an open fire, which provides warmth and a peaceful gathering place in the home.

Because the old-time fireplace mason, who knew by experience the necessary dimensional requirements of successful fireplaces, is slowly disappearing, complete working drawings of fireplaces must be included in plans to ensure their correct construction. If an honest house is to be built, the fireplace must be honest as well and must be designed to be workable.

Residential fireplaces of various forms and styles are being constructed (Fig. 14–18). Some are merely openings in massive interior masonry walls, with no mantel or decor other than the texture of the masonry. Some have provision for log storage near the opening; others may have outdoor barbecue grills incorporated into the chimneys. Many are masonry walls with metal hoods above to concentrate the draft and carry off hot gases and smoke. For general purposes, fireplaces can be divided into the following groups:

1. Single-face fireplaces (conventional)
   (a) Wall
   (b) Corner
   (c) Back-to-back
2. Multiple-face fireplaces (contemporary)
   (a) Two adjacent faces (corner)
   (b) Two opposite faces (through opening)
   (c) Three faces: two long, one short
   (d) Three faces: one long, two short
   (e) Hooded
   (f) Freestanding

The successful operation of a fireplace depends on its construction. Even special fireplaces, which incorporate various shapes and materials, can be made workable if the designer follows the basic principles of fireplace design instead of resorting to chance or mere luck. The critical points in single-face fireplace design and construction are discussed in the following sections (Fig. 14–19).

14.6.1 Size of Fireplace Opening

Not only should the fireplace be given a prominent position in a room, away from major travel routes, but the fireplace opening should also be in keeping with the room size. Small rooms should have small fireplace openings; large rooms, large ones. The opening should have pleasing proportions: A rectangle is more desirable than a square, and the width should exceed the height. A fireplace width of, for instance, 28″ to 36″ would be adequate for a room of 250-sq ft floor space. In Fig. 14–20, notice some of the more common opening sizes. A raised hearth, usually 16″ above the floor, brings the opening nearer to eye level and often makes the fireplace appear larger.

Keep wood trim at least 8″ from the opening, and use a steel angle-iron lintel to support the masonry above the opening.

Figure 14–20 shows that fireplace openings have relative proportions of height, width, and depth; however, you can make minor variations to meet the various masonry courses and other restrictions of layout. The size of the opening becomes the starting feature in the design of a fireplace.

14.6.2 Flue Size and Chimney Height

Each fireplace in a building should have an independent flue, unconnected to other vents or flues, and it must start on the center line of the fireplace opening. A chimney can take care of a number of fireplaces, and each flue must continue to the top with as little offset as necessary. Each flue must be large enough in cross-sectional area to create the proper draft through the fireplace opening. Unless the sectional flue area is at least $\frac{1}{10}$ of the fireplace opening area (a rule of thumb), a troublesome condition may result (see the table that accompanies Fig. 14–22 for exact sizes). Chimney heights under 14′ may require even larger flue sizes, as the height of the chimney, surrounding buildings, and prevailing winds affect flue velocities. The higher the chimney, the more draft velocity. For fire protection of the roofing materials, the height of the chimney should be as noted in Fig. 14–21. Flue sizes are important and therefore should be clearly indicated on the drawings; a flue that is too large is better than one that is too small. Vitrified clay flue linings are used to resist high chimney temperatures and to provide smooth flue interiors. Without the lining, cracks tend to develop in the chimney walls. Round linings operate more efficiently, but rectangular types can be more easily fitted into chimney spaces.

14.6.3 Shape of Combustion Chamber

The combustion chamber (Fig. 14–22) is lined with 4″ thick firebrick set in fireclay. It is shaped not only to reflect heat into the room, but also to lead hot gases into the throat with increased velocity. If the fireplace is too deep, less heat will be radiated into the room. The sections in Fig. 14–22 indicate the conventional shape of combustion chamber walls. The back and end walls should be at least 8″ thick. Slight variations are permissible in the layout of the chamber, but long experience has shown that the conventional shapes and proportions are the most satisfactory as far as operation is concerned.

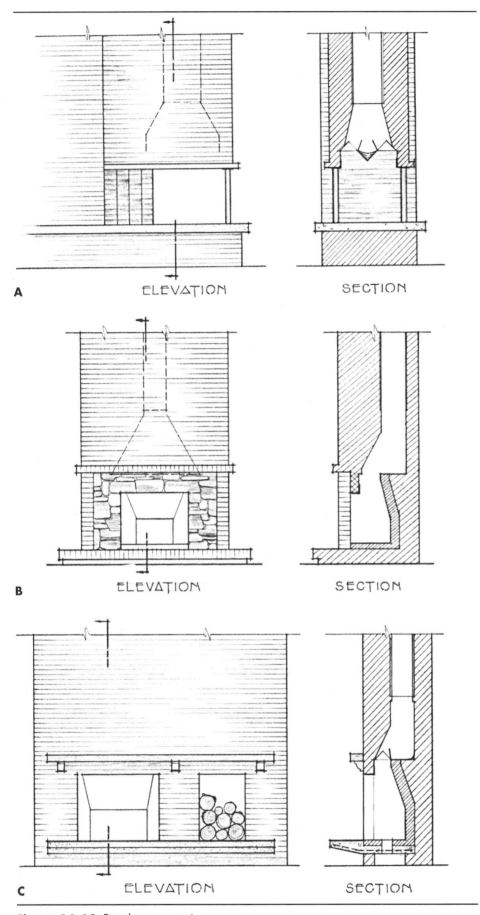

A

ELEVATION   SECTION

B

ELEVATION   SECTION

C

ELEVATION   SECTION

Figure 14–18 Fireplace suggestions.

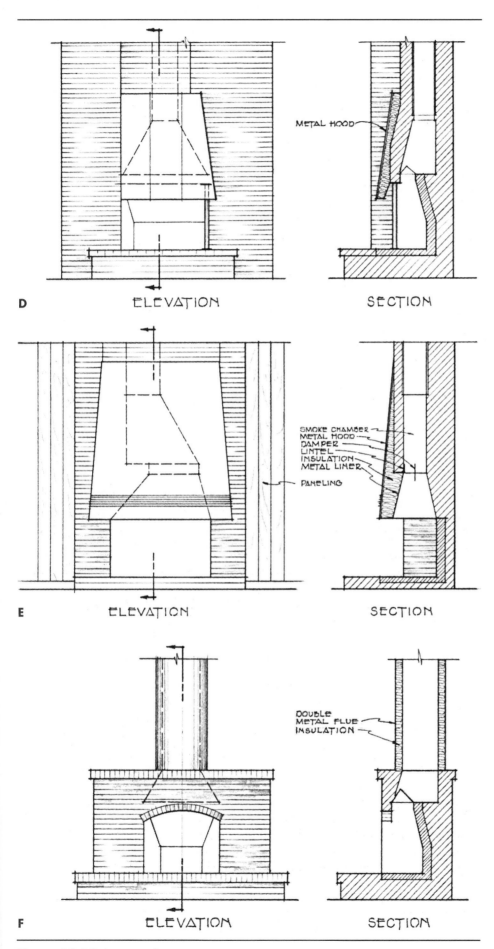

D ELEVATION   SECTION

METAL HOOD

E ELEVATION   SECTION

SMOKE CHAMBER
METAL HOOD
DAMPER
LINTEL
INSULATION
METAL LINER

PANELING

F ELEVATION   SECTION

DOUBLE
METAL FLUE
INSULATION

Figure 14–18 *(continued).*

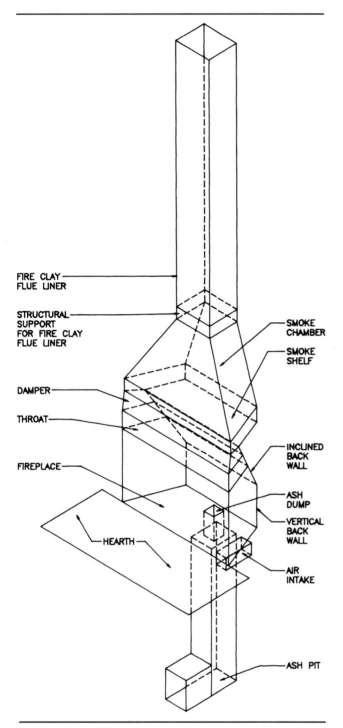

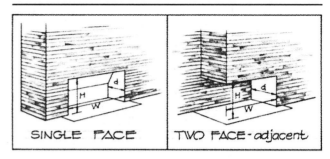

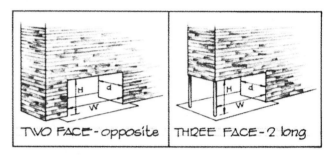

Figure 14-19 Masonry fireplace.

FIRE CLAY FLUE LINER

STRUCTURAL SUPPORT FOR FIRE CLAY FLUE LINER

DAMPER

THROAT

FIREPLACE

HEARTH

SMOKE CHAMBER

SMOKE SHELF

INCLINED BACK WALL

ASH DUMP

VERTICAL BACK WALL

AIR INTAKE

ASH PIT

Fireplace Dimensions

Fireplace Type	Opening Height h, in.	Hearth Size w by d, in.	Modular Flue Size, in.
Single Face	29	30 × 16	12 × 12
	29	36 × 16	12 × 12
	29	40 × 16	12 × 16
	32	48 × 18	16 × 16
Two Face—adjacent	26	32 × 16	12 × 16
	29	40 × 16	16 × 16
	29	48 × 20	16 × 16
Two Face—opposite	29	32 × 28	16 × 16
	29	36 × 28	16 × 20
	29	40 × 28	16 × 20
Three Face—2 long, 1 short	27	36 × 32	20 × 20
	27	36 × 36	20 × 20
	27	44 × 40	20 × 20

Figure 14-20 Typical fireplace openings and workable dimensions.

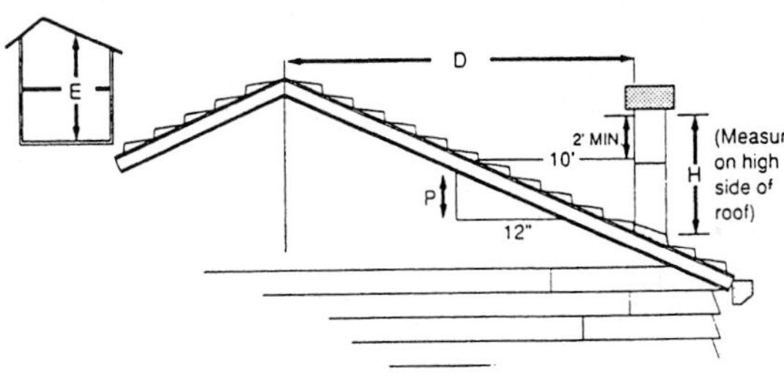

"P" Dim. Roof Pitch Inches	Minimum Required Chimney Height (H Dimension) (Do not include terminal cap in Dimension H) "D" Dimension (Feet) 1'	2'	3'	4'	5'	6'	7'	8'	9'	10' +
1"	3'	3'	3'	3'	3'	3'	3'	3'	3'	3'
2"	3'	3'	3'	3'	3'	3'	3'2"	3'4"	3'6"	3'8"
3"	3'	3'	3'	3'	3'3"	3'6"	3'9"	4'	4'3"	4'6"
4"	3'	3'	3'	3'4"	3'8"	4'	4'4"	4'8"	5'	5'4"
5"	3'	3'	3'3"	3'8"	4'1"	4'6"	4'11"	5'4"	5'9"	6'2"
6"	3'	3'	3'6"	4'	4'6"	5'	5'6"	6'	6'6"	7'
7"	3'	3'2"	3'9"	4'4"	4'11"	5'6"	6'1"	6'8"	7'3"	7'10"
8"	3'	3'4"	4'	4'8"	5'4"	6'	6'8"	7'4"	8'	8'8"
9"	3'	3'6"	4'3"	5'	5'9"	6'6"	7'3"	8'	8'9"	9'6"
10"	3'	3'8"	4'6"	5'4"	6'2"	7'	7'10"	8'8"	9'6"	10'4"
11"	3'	3'10"	4'9"	5'8"	6'7"	7'6"	8'5"	9'4"	10'3"	11'4"
12"	3'	4'	5'	6'	7'	8'	9'	10'	11'	12'
13"	3'1"	4'2"	5'3"	6'4"	7'5"	8'6"	9'7"	10'8"	11'9"	12'10"
14"	3'2"	4'4"	5'6"	6'8"	7'10"	9'	10'2"	11'4"	12'6"	13'8"
15"	3'3"	4'6"	5'9"	7'	8'3"	9'6"	10'9"	12'	13'3"	14'6"
16"	3'4"	4'8"	6'	7'4"	8'8"	10'	11'4"	12'8"	14'	15'4"
17"	3'5"	4'10"	6'3"	7'8"	9'1"	10'6"	11'11"	13'4"	14'9"	16'2"
18"	3'6"	5'	6'6"	8'	9'6"	11'	12'6"	14'	15'6"	17'
19"	3'7"	5'2"	6'9"	8'4"	9'11"	11'6"	13'1"	14'8"	16'3"	17'10"
20"	3'8"	5'4"	7'	8'8"	10'4	12'	13'8"	15'4"	17'	18'8"
21"	3'9"	5'6"	7'3"	3'9	10'9"	12'6"	14'3"	16'	17'9"	19'6"
22"	3'10"	5'8"	7'6"	9'4"	11'2"	13'	14'10"	16'8"	18'6"	20'4"
23"	3'11"	5'10"	7'9"	9'8"	11'7"	13'6"	15'5"	17'4"	19'3"	21'2"

Figure 14–21 Relationship between roof pitch and required chimney height.

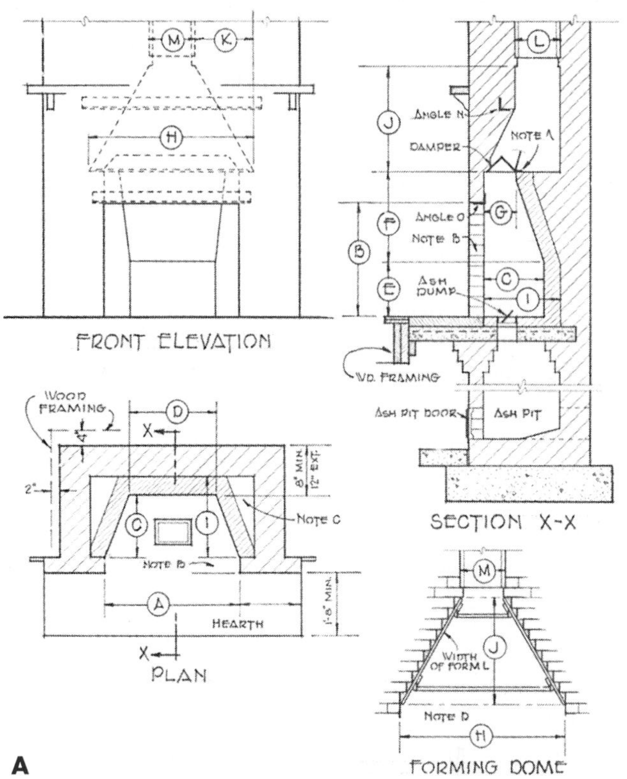

A

Fireplace Dimensions, Inches [1]

Rough Brick Work and Flue Sizes																		
Finished Fireplace Opening							New sizes[3]							Old sizes			Steel angles[2]	
A	B	C	D	E	F	G	H	I	J	K	L	M	R[4]	K	L	M	N	O
24	24	16	11	14	18	8¾	32	20	19	10	8×12		8	11¾	8½ ×	8½	A-36	A-36
26	24	16	13	14	18	8¾	34	20	21	11	8×12		8	12¾	8½ ×	8½	A-36	A-36
28	24	16	15	14	18	8¾	36	20	21	12	8×12		10	11½	8½	×13	A-36	A-36
30	29	16	17	14	23	8¾	38	20	24	13	12×12		10	12½	8½	×13	A-42	A-36
32	29	16	19	14	23	8¾	40	20	24	14	12×12		10	13½	8½	×13	A-42	A-42
36	29	16	23	14	23	8¾	44	20	27	16	12×12		12	15½	13	×13	A-48	A-42
40	29	16	27	14	23	8¾	48	20	29	16	12×16		12	17½	13	×13	A-48	A-48
42	32	16	29	14	26	8¾	50	20	32	17	16×16		12	18½	13	×13	B-54	A-48
48	32	18	33	14	26	8¾	56	22	37	20	16×16		15	21½	13	×13	B-60	B-54
54	37	20	37	16	29	13	68	24	45	26	16×16		15	25	13	×18	B-72	B-60
60	37	22	42	16	29	13	72	27	45	26	16×20		15	27	13	×18	B-72	B-66
60	40	22	42	16	31	13	72	27	45	26	16×20		18	27	18	×18	B-72	B-66
72	40	22	54	16	31	13	84	27	56	32	20×20		18	33	18	×18	C-84	C-84
84	40	24	64	20	28	13	96	29	61	36	20×24		20	36	20	×20	C-96	C-96
96	40	24	76	20	28	13	108	29	75	42	20×24		22	42	24	×24	C-108	C-108

[1] See Fig. 14–22 A.

[2] Angle sizes: A. 3" x 3" x 3/16"; B. 3½" x 3" x ¼"; C. 5" x 3½" x 5/16".

[3] New flue sizes: Conform to modular dimensional system. Sizes shown are nominal. Actual size is ½" less each dimension.

[4] Round flues.

Note A. The back flange of the damper must be protected from intense heat by being fully supported by the masonry. At the same time, the damper should not be built in solidly at the ends, but given freedom to expand with heat.

Note B. The thickness of the fireplace front will vary with the material used: brick, marble, stone, tile, etc.

Note C. The hollow, triangular spaces behind the splayed sides of the inner brickwork should be filled to afford solid backing. If desired to locate a flue in either space, the outside dimensions of the rough brickwork should be increased.

Note D. A good way to build a smoke chamber is to erect a wooden form consisting of two sloping boards at the sides, held apart by spreaders at the top and bottom. Spreaders are nailed upward into cleats. The form boards should have the same width as the flue lining.

B

Figure 14–22 Successful fireplace construction. (Courtesy Donley Bros. Co.)

14.6.4 Design of the Throat

The throat offsets the draft above the chamber, and if a metal damper is used, the door of the damper acts as a valve for checking downdrafts while the fireplace is in operation. The inclusion of a damper is a convenient method of preventing heat loss and outside drafts when the fireplace is not in use. Notice that the door of the damper hinges so that it opens toward the back of the fireplace. The opening of either the damper or the throat should be larger than the area of the flue lining. Also, the damper should be the same length as the width of the fireplace and should be placed at least 6″ (preferably 8″) above the top of the fireplace opening. Both steel and cast-iron dampers are available; check manufacturers' catalogs for types and sizes.

14.6.5 Shape of the Smoke Shelf and Smoke Dome

The location of the damper establishes the height of the smoke shelf, which is directly under the flue and which stops downdrafts with its horizontal surface. The smoke dome is the area just above the shelf. Notice that the back wall is built vertically; the only offset on the back is a corbeled brick course that supports the flue lining, where a corbeled brick course is also shown for flue support. Usually, the sloping walls are represented with lines drawn 60° from the horizontal.

14.6.6 Support for the Hearth

The hearth is best supported with a reinforced concrete slab resting on the walls of the fireplace foundation and extending in front at least 18″. (Previously, a brick trimmer arch was used to support the forward hearth against the chimney foundation in conventional construction.) Number 3 reinforcing rods are placed near the upper part of the slab, and brick, stone, or other masonry material is applied directly on the slab for a textured hearth covering. If an ash dump is feasible and is accessible to a basement ash pit or an outdoor cleanout door, the reinforced slab may simply be formed with ribbed metal lath having a hole cut for the ash dump and any upcoming flues; the concrete is poured over the lath.

In homes with slab floors, the fireplace and the hearth are built up from an isolated footing to the floor level. For easy cleaning, the fireplace floor can be made 1″ higher than the outer hearth.

14.6.7 Wood Framing Precautions Around the Fireplace and Chimney

Precautions must be taken when fireplaces or their chimneys pass through wood frame floors, roofs, and the like. Fire codes prohibit direct contact between framing members and chimney surfaces (Fig. 14–22). Usually a 2″ air space is provided between the masonry and wood; this space is filled with incombustible insulating material. Trimmers and headers around wood openings are doubled to give support to the cut joists or rafters, and the subflooring or decking is applied closer to the masonry. The hearth slab, which is independent of the wood framing, usually has its outer edge resting on a ledger strip fastened to the headers or the floor joists.

14.6.8 Steel Fireplace Boxes

Prefabricated, one-piece, factory built fireboxes are available (see Figs. 14–23, 14–24, and 14–25 and Sweet's Architectural Catalog File) that combine the combustion chamber, the smoke dome, and even a steel lintel for the masonry opening. The firebox is merely set on a masonry hearth, and the brick or stone enclosure is built up around it, thus eliminating any necessity for fireplace design other than the correct flue size. Because they are hollow, the fireboxes can utilize warm- and cold-air ducts and vents to improve heating efficiency. After the desired openings have been determined, correct catalog numbers can be established.

14.6.9 Special Fireplaces

Special fireplaces include variations of the multifaced and hooded types, often referred to as "contemporary." Because of their unusually large openings and their complexity of shapes, special fireplaces occasionally cause smoking in the room and have poor draw through their throats. Generally, if these fireplaces operate poorly, the flue is too small, the damper throat is too narrow, the chimney height is too low, or possibly nearby buildings or trees prevent proper draft through the flues. Also, fireplaces with the sides of the combustion chamber exposed to allow room drafts to pass directly through, such as freestanding or open-back fireplaces, will tend to allow smoke into the room unless room drafts are prevented or glass sides are installed. Special metal smoke dome and damper boxes are available to improve multiface fireplace operation (see Sweet's File). Hooded fireplaces require heat-resisting insulation below the metal hoods.

Freestanding, complete fireplace units are sold that require neither masonry enclosures nor a masonry chimney. The units, complete with chimney and hearth, can be installed in existing structures.

In drawing special fireplace details, you can take design data from the fireplace dimensions table and the accompanying drawings (Fig. 14–22); further information can be found in Ramsey and Sleeper's Architectural Graphic Standards. However, the best source for information about special fireplaces, as well as the equipment needed for regular fireplaces, is from manufacturers' catalogs and literature.

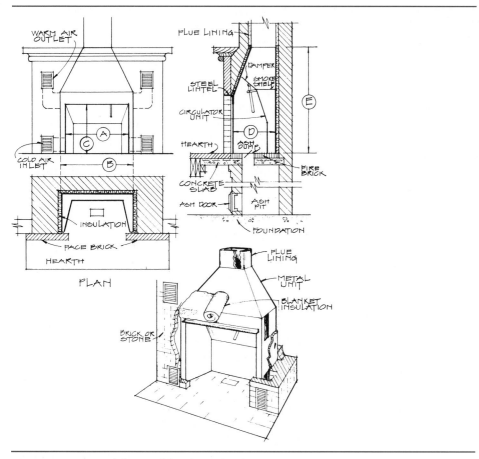

Figure 14-23 Prefabricated fireplace detail.

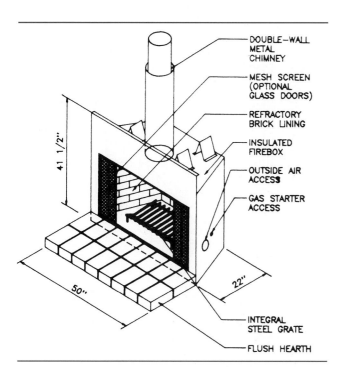

Figure 14-24 Traditional factory-built fireplace.

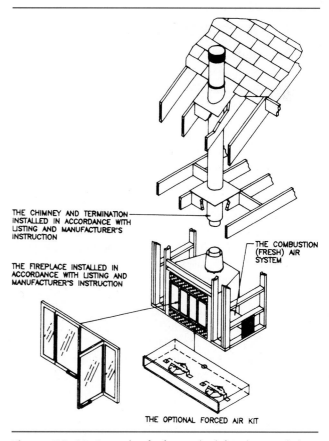

Figure 14-25 Example of a factory-built fireplace and chimney.

14.7

STAIR DESIGN AND LAYOUT

Let us now consider the problems encountered in stair construction. A flight of stairs is part of the system of hallways communicating between the occupied floor levels of a building. In multistory buildings, the stairs, both interior and exterior, must be given careful consideration in the total design. Main stairways in homes, as well as in public buildings, have long been objects of special ornament by architects. Many beautiful staircases are evident in scores of buildings throughout the country. The present trend, however, is toward simplicity and comfort in residential stairs, although dramatic stairways are occasionally found in contemporary public buildings.

14.7.1 Stair Terminology

Stairs have their own terminology in construction work; the following terms are commonly encountered:

- **Balusters** The thin vertical supports for the handrail of open stairs.
- **Bullnose** The first step on an open stair; it has been extended out, forming a semicircle and often receiving the newel post.
- **Carriage** The rough structural support (usually $2'' \times 12''$) for treads and for risers of wood stairs, sometimes called *string* or *stringer*.
- **Closed stringer** The visible member of a stairs that abuts the risers and treads and that is not cut to show the profile of stairs.
- **Handrail** The round or decorative member of a railing which is grasped with the hand during ascent or descent.
- **Headroom** The narrowest distance between the surface of a tread and any ceiling or header above.
- **Housed stringer** The stringer that has been grooved to receive the risers and the treads.
- **Landing** The floor between flights of stairs or at the termination of stairs.
- **Newel** The main post of the railing at the bottom of a stair or at changes in direction of the railing.
- **Nosing** The round projection of the tread beyond the face of the riser.
- **Open stringer** The stringer that has been cut to fit the profile of the stairs; the riser cut is mitered, and the tread cut is square.
- **Platform** The intermediate landing between various parts of the stair flight.
- **Railing** The handrail and the baluster forming the protection on open stairs.
- **Rise** The total floor-to-floor vertical height of a stairs.
- **Riser** The vertical face of the step.

- **Run** The total horizontal length of a stairs, including the platform.
- **Stairwell** The enclosed chamber into which the stairs are built.
- **Step** The combination of one riser and one tread.
- **Stringer** The inclined member supporting the risers and treads; sometimes a visible trim member next to the profile of the stairs.
- **Tread** The horizontal surface member of each step, usually hardwood.
- **Winder** The radiating or wedge-shaped treads at turns of stairs.

14.7.2 Types of Stair Construction

Previously, stair building was a highly specialized craft, especially in home construction; many stairs required considerable hand labor. Today, stairways are often ordered directly from a mill or shops, fabricated according to details furnished by the designer. Usually, first-quality construction accepts only shop-made stair units (Fig. 14–26) which have expert crafting and which are delivered to the site and quickly installed by carpenters. Figure 14–27 shows a more economical stair that is built on the job. Usually, 2×12 carriages are cut to the profile of the stairs, and the treads are attached. Three carriages are preferred construction if the stair width is 3′-0″ or over. Double trimmers and headers must surround the stairwell framing, and the carriages must have sufficient bearing at both ends.

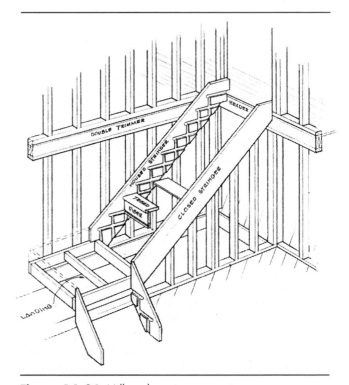

Figure 14–26 Mill-made stair construction.

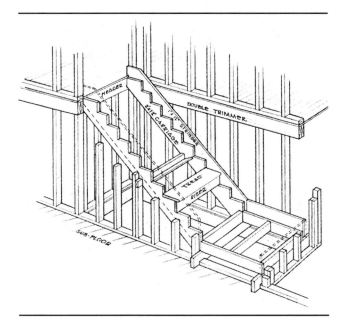

Figure 14–27 On-the-job stair construction.

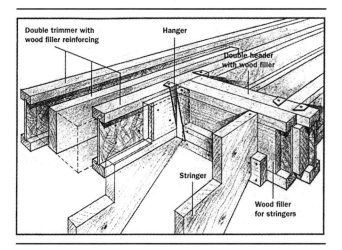

Figure 14–28 Stair framing with wood I-beam joists.

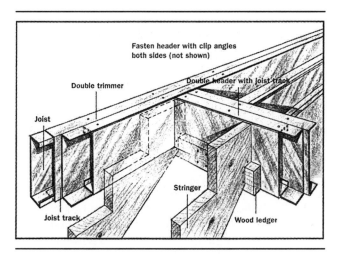

Figure 14–29 Stair framing with light-gauge steel.

Other structural methods of stair support have been devised by architects to produce both novel and sturdy stair flights.

Figures 14–28 and 14–29 illustrate familiar details for framing around stair openings but with wood I-beam joists and light-gauge steel. In these cases, special attention must be paid to doubling up trimmers and headers. Wood filler pieces are used between webs to reinforce doubled wood I-beam joists near headers (Fig. 14–28), and steel joists are doubled by "tubing" them into pieces of joist track (Fig. 14–29). Be sure to use the recommended fasteners and fastening schedules.

14.7.3 Riser and Tread Proportions

Because considerable effort is expended in ascending and descending stairs, regardless of type, comfort and safety should be the first considerations in their design. In designing a comfortable stair, you must establish a definite relationship between the height of the risers and the width of the treads; all stairs should be designed to conform to established proportions. If the combination of riser and tread is too great, the steps become tiring, and a strain develops on the leg muscles and the heart. If the combination is too short, the foot has a tendency to kick the riser at each step in an attempt to shorten the stride, also producing fatigue. Experience has shown that risers 7″ high with an 11″ tread result in the most satisfactory combination for principal residential stairs.

14.7.4 Stair Formulas

The following three formulas have been devised for checking riser and tread proportions, exclusive of molding. Each will be satisfactory:

Two risers + 1 tread = between 24″ and 25″

Riser × tread = between 72″ and 77″

Riser + tread = between 17″ and 18″

As an example of formula application, suppose that an 11″ tread is found suitable on a preliminary stair layout. After examining each of the stair formulas, you will see that a riser height of 7″ satisfies any formula. If a 12″ tread is selected, the riser will have to be 6″; and so on. Minor variations would still make the stairs workable, but treads are seldom made less than 11″ or more than 12″ wide.

14.7.5 Angle of Stairs

Stair flights should be neither too steep nor too flat in incline. An angle from 30° to 33° from the horizontal is the most comfortable. Long flights are also tiring; usually a platform about midway in a long flight helps relieve fatigue. There will always be one fewer tread than there are risers in all stairs.

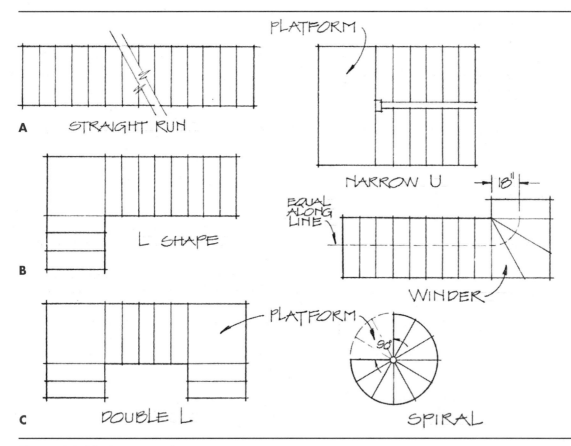

Figure 14-30 Various stair layouts.

14.7.6 Plan Layout Variations

Various stair plan layouts are shown in Fig. 14–30. If space on the plan prohibits the use of a straight run, other layouts are shown to increase flexibility in fitting stairs in restricted spaces and to lend stairs various character treatments. In limited space, a winder stair (Fig. 14–30) can be used if necessary, yet it is more hazardous than a platform arrangement. The winder-stair profile should be considered at a line 1′-4″ from the winder corner. Remember that a landing counts as a tread and should conform to the width of the rest of the stairs.

14.7.7 Headroom

Headroom should be between 6′-8″ and 7′-4″ over major stairs; 6′-6″ is usually sufficient over minor flights to the basement or attic. Individual conditions may necessitate slight variations of headroom. Disappearing stairs to attic storage spaces are usually prefabricated units, which can be shown on the plan with a dashed rectangle and a note.

14.7.8 Railings

Whether a stair is open or is enclosed between walls, a railing must be installed to be within easy reach in case of stumbling or loss of balance. It is general practice to use a continuous handrail from floor to floor. It can be plain or ornamental—in keeping with the design features

of the building—but it should be smooth and sturdy. The most comfortable handrail height is 32″ above the tread surface measured from the edge of the tread to the top of the handrail. Landings in residences should have 36″ high railings, landings in commercial building, 42″ high.

14.7.9 Exterior Stairs

Exterior stairs are usually designed with smaller riser heights and therefore with wider treads than interior stairs. A popular proportion is a 6″ riser and a 12″ tread for maximum safety on long flights. Landings should be provided every 16 risers on continuous stairs. If slopes are gradual, ramps (up to 12°) are often preferable to steps for outside use. Usually, masonry with reinforcing and well-anchored footings is the most practical exterior material.

Requirements of all classes of stairs are given in the publication *NFPA 101, Life Safety Code,* published by the National Fire Protection Association. The design of all stairs should meet their specifications.

14.7.10 Drawing Stair Details (Figs. 14–31 and 14–32)

Step 1: *To draw a stair detail,* first you must know the finished-floor-to-finished-floor height, although carriages rest on rough framing. You can arithmetically determine the number of risers necessary to ascend the height by dividing the heights by 7″ (typical risers). If the

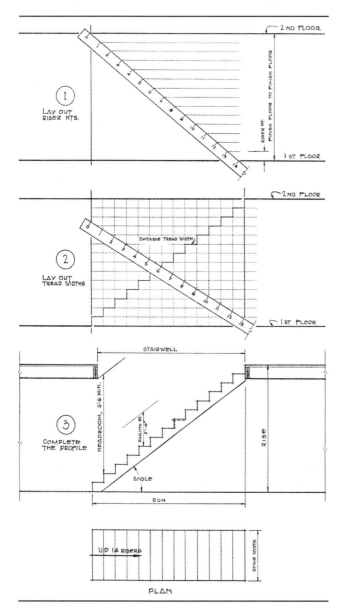

Figure 14-31 Drawing a stair detail.

Total number of risers	16
Total number of treads	15
Riser height	6.75"
Tread width	11.25"
Total rise	108"
Total run	168.75"

Step 2: Because odd fractions are difficult to measure on a drawing, the graphic method of stair profile layout is recommended. Scale the floor-to-floor height, and draw the floor lines between which the stair is to be drawn. Divide the between-floor space into 16 divisions graphically. You can use divisions on the scale by adopting 16 convenient divisions, adjusting them diagonally across the space, and marking them to form the tread surfaces (see the left side of Fig. 14–31). Draw the 15 vertical riser surfaces similarly, or measure them if feasible. The top riser meets the second-floor surface. The result is the correct stair profile from floor to floor. Strengthen the profile and erase the construction lines.

Step 3: Locate the ceiling or soffit surface above the stairs, keeping in mind that the headroom for important stairs should be 6'-8" to 7'-4" high, if possible; slightly less is permissible if the stairwell above must be made smaller. This headroom line locates the header framing and the sloping ceiling above. In basement stairs, less headroom is satisfactory. To conserve stairwell space, usually two or more flights of stairs are superimposed directly above one another in multistory buildings. Openings for the stairwell can now be dimensioned for the framing, and final adjustments on second-floor walls near the stairs can be made.

Step 4: If the carriage outline is to be drawn, the thickness of the treads (usually 1¹/₁₆") is shown below the profile surface, and the thickness of the risers (usually ³/₄") is shown within the vertical lines. The original profile will remain the surface of the risers and the treads. Usually, the carriage is cut from 2 × 10 or 2 × 12 lumber, and at least 3" of solid material must remain below the depth of the cuts for stability. Nosings commonly extend 1¹/₈" beyond the surface of the risers. Basement stairs seldom require covered risers.

Step 5: For safety, railings are a necessary part of the stairs and should be drawn next. If necessary, details of railings, ornamental newel posts, balusters, and trim can be taken from manufacturers' catalogs. Draw the railings 2'-6" above the surface of the treads. Major stairs should be made from 3'-0" to 3'-6" wide in residences and 3'-8" wide or wider in commercial buildings.

Complete the stair plan layout by projecting the risers from the elevation detail. Show the riser surfaces, and indicate the number of risers; include either an UP or a DOWN note and an arrow (Fig. 14–32). About half of the complete stair symbol is shown on each floor plan, and a conventional diagonal break line is drawn near the symbol center. A concrete stair detail is shown in Fig. 14–33.

floor-to-floor dimension is conveniently divisible by either typical riser height, calculation is simplified. However, seldom will it come out even, unless ceiling heights have been purposely figured. In Fig. 14–32, the total height or rise needed for the stairs is 108" (8'-1¹/₈" stud and plate height, plus 9¹/₄" joist height, plus 1⁵/₈" floor thickness). When we divide 108 by 7, the quotient is 15.42. Since all risers must be the same height, 16 risers can be adopted. Using 16 risers with 11¹/₄" tread width seems to satisfy the stair formulas:

$$\frac{108''}{16} = 6.75'' \text{ riser}$$

Formula 1: $2(6.75) + 11.25 = 24.75$
Formula 2: $6.75 \times 11.25 = 75.93$
Formula 3: $6.75 + 11.25 = 18$

Therefore, the stair data for Fig 14–33 would be as follows:

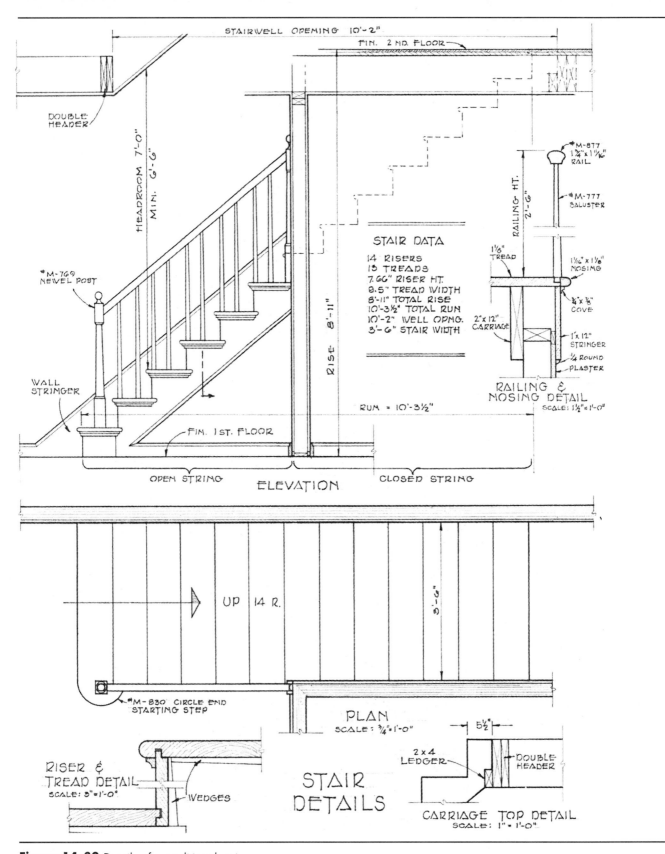

STAIRWELL OPENING 10'-2"

FIN. 2 ND. FLOOR

DOUBLE HEADER

HEADROOM 7'-0"
MIN. 6'-6"

#M-877
1¾" x 1¹¹⁄₁₆"
RAIL

RAILING HT.
2'-6"

#M-769
NEWEL POST

#M-777
BALUSTER

1⅛"
TREAD

1¹⁄₁₆" x 1⅛"
NOSING

STAIR DATA
14 RISERS
13 TREADS
7.66" RISER HT.
8.5" TREAD WIDTH
8'-11" TOTAL RISE
10'-3½" TOTAL RUN
10'-2" WELL OPNG.
3'-6" STAIR WIDTH

RISE 8'-11"

2" x 12"
CARRIAGE

¾" x ⅜"
COVE

1" x 12"
STRINGER

¼ ROUND
PLASTER

RAILING &
NOSING DETAIL
SCALE: 1½"= 1'-0"

WALL
STRINGER

FIN. 1ST. FLOOR

RUN = 10'-3½"

OPEN STRING   CLOSED STRING
ELEVATION

UP 14 R.

3'-6"

#M-830 CIRCLE END
STARTING STEP

PLAN
SCALE: ¾"= 1'-0"

5½"

2 x 4
LEDGER

DOUBLE
HEADER

RISER &
TREAD DETAIL
SCALE: 3"= 1'-0"

WEDGES

STAIR
DETAILS

CARRIAGE TOP DETAIL
SCALE: 1"= 1'-0"

Figure 14–32 Details of a traditional stairs.

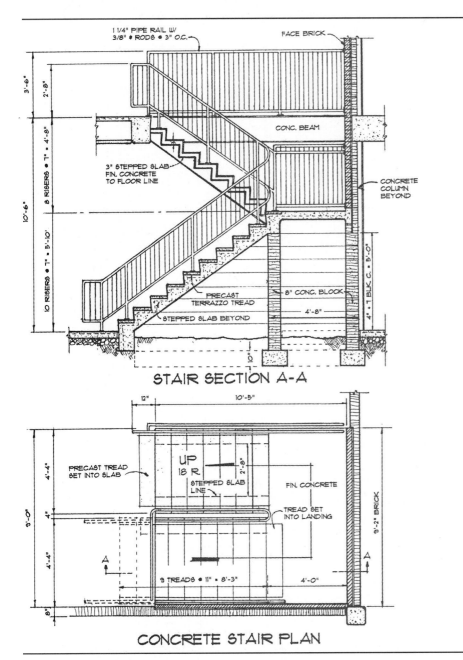

STAIR SECTION A-A

CONCRETE STAIR PLAN

Figure 14-33 Details of a commercial (concrete) stairs.

14.8

NEW CONSTRUCTION SYSTEMS AND MATERIALS

In the period after World War II, America built 73.8 million homes that set world standards for comfort, quality, and design. The approach of the twenty-first century brings new issues and opportunities, especially in the areas of innovative structural systems that can be used as alternatives to wood framing and that provide increased energy efficiency.

The new structural systems reflect innovations in the material sciences in the areas of engineered wood products, concrete, and steel. All of the systems have implications for energy efficiency. Other technologies with a relationship to energy efficiency include innovative heating and air-conditioning equipment and photovoltaics.

It is important for students to be aware of the changes in building materials and systems that are appearing in the construction industry and to be able to detail the construction situations associated with these new materials and systems. One of the best, current examples of the use of innovative construction materials and systems is the 21st Century Townhouses built by the National Association of Home Builders. The products and systems described in the accompanying boxed material, "The 21st Century Townhouses: The NAHB Research Home Program" (see pages 452–478), show significant promise for U.S. home building and are likely to enter the mainstream of housing construction during the balance of this decade or shortly thereafter.

The 21st Century Townhouses were built by the Research Center in the NAHB Research Home Park as part of the Research Center's research home program (Fig. 1). The purposes of the program are as follows:

- to test, demonstrate, and gain experience with innovative home products, systems, and technologies, for the purpose of assisting U.S. home building to maintain its position of world leadership;
- to disseminate information on the features of innovative systems and methods, and to assist promising new products, systems, and technologies to move into the mainstream of home construction; and
- to ensure that American home buyers get the highest quality and the greatest possible value when they purchase homes.

During construction as well as after completion, each research house is opened for a period of one to two years, for inspection, for tours, and for conducting research. The house is then placed on the market and sold. The location of the Research Home Park in a standard, attractive development, and the requirement that the research houses be sold on the commercial market, both assure that the new and innovative features are united in a home design that can meet the test of broad consumer acceptance.

ARCHITECT, PROJECT DIRECTOR, DEVELOPMENT MANAGER, AND PROJECT SUPERVISOR

Architect for the townhouses was John T. Stovall, President of John Stovall and Associates, Gaithersburg, Maryland. Daniel J. Ball, AIA, of Daniel Ball &

Figure 1 NAHB 21st Century Townhouses.

Associates, served as Project Director. Miles Haber of Monument Construction, Inc., served as Development Manager. J. Albert van Overeem of the NAHB Research Center staff served as Project Supervisor.

The 21st Century Townhouses were built with products that feature two themes:

1. innovative structural systems in home building and
2. approaches to achieving advanced residential energy efficiency.

LOT 7   LOT 8   LOT 9   LOT 10

FRONT ELEVATION
1/8" = 1'- 0"

INNOVATIVE STRUCTURAL SYSTEMS

Interest in alternative systems for home construction grew rapidly in the early 1990s, stimulated by sharp fluctuations in the price of dimensional lumber. Interest was also created by the performance characteristics of many innovative technologies that had not yet entered the mainstream of the marketplace. Many of these technologies show the promise of significant structural merits and excellent energy performance.

ADVANCED ENERGY EFFICIENCY

Increased energy efficiency of homes can make an important contribution to national energy conservation. The average U.S. home is a well-built structure that lasts for at least 75 years. Savings in energy consumption can add up to substantial amounts over the life span of even a single home. Energy-saving features that are not built into a home during construction can be difficult or impossible to introduce later.

In the 21st Century Townhouses, contributions to energy conservation are made by the structural systems, heating and air-conditioning equipment, wastewater heat reclamation, and the use of photovoltaics to generate electricity. The performance of the structural systems will be compared to the minimum requirements of the Model Energy Code.

STRUCTURAL WALL AND FLOOR SYSTEMS

Each townhouse features a different structural system. The four systems are (1) structural insulated panels; (2) I.C.E. Block™ concrete forming system; (3) steel; and (4) Hebel precast, autoclaved, aerated concrete units.

House 1: Structural Insulated Panels

Structural insulated panels (SIPs), used for the exterior structural system of House 1, consist of a form-core center that contains no environmentally harmful CFCs or HCFCs and is clad on both sides with oriented strand board (Fig. 2). The panels were precut, including some openings, before delivery to the site. SIPs were used for all walls above grade and for roof panels. Panels as large as 8′ × 24′ were incorporated into the exterior walls and roof structure (Figs. 3 and 4).

Benefits of structural insulated panels include the following:

Figure 2 SIP roof panel being lifted into place by a crane.

- SIPs contain relatively low levels of embedded energy. Manufacture involves moderate technological requirements and utilizes moderate amounts of energy.
- The joining systems that are used with SIPs reduce the use of dimensional lumber. The panels contain almost no conventional studs.
- SIP-clad houses present broad, seamless areas to the external environment, offering potential reduced air infiltration and energy savings.
- Construction with precut, structural insulated panels involves low waste generation at the site.

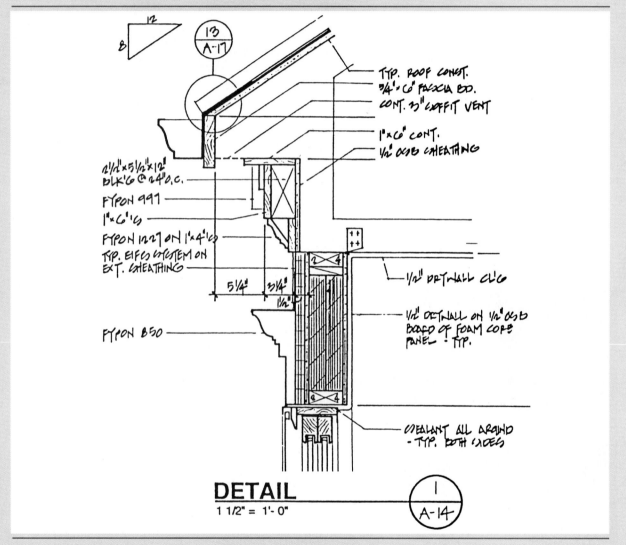

Figure 3 Head detail.

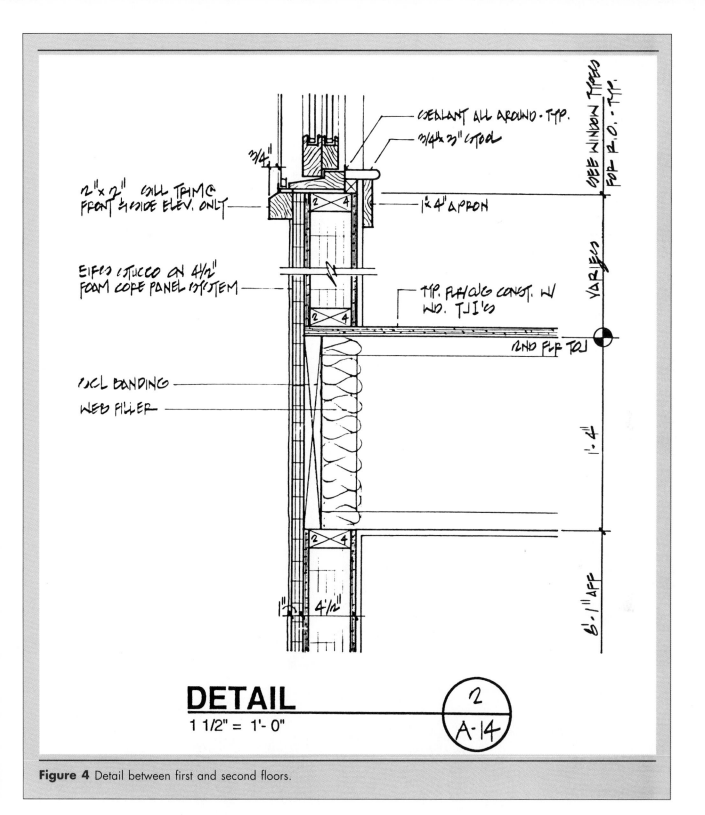

DETAIL
1 1/2" = 1'- 0"

2
A·14

Figure 4 Detail between first and second floors.

Concrete forming systems employing polystyrene forms originated 30 years ago in Europe and were introduced in the United States in modified and improved form in the 1990s. The concrete forming system used for the structural system of House 2, provided by I.C.E. Block™ Building Systems, Inc., utilizes forms that are made of expanded polystyrene (EPS) and that are stacked, reinforced with metal rebars, and filled with concrete. I.C.E. is an acronym for Insulate Concrete Efficiently. The system was used to create above- and below-grade exterior walls (Fig. 5).

The tongue-and-groove EPS forms used in the I.C.E. Block™ system are 48″ × 16″ and are either 9¼″ or 11″ thick. The concrete core thicknesses are 6″ or 8″, respectively.

When the concrete is poured, it produces a post-and-beam, grid-pattern concrete wall with 6″ or 8″ vertical columns, 12″ o.c., and horizontal concrete beams, 16″ o.c., with the horizontal links 2″ thick (Figs. 6, 7, and 8).

The blocks contain light-gauge steel sections, 1½″ wide and 13″ high, vertically embedded 12″ o.c. These sections provide attachment surfaces for interior and exterior finishing material. The sections are embedded ½″ inside the interior and exterior faces of the blocks, providing a thermal break. Marks on the inner and outer surfaces indicate the location of the sections, for attachment of drywall or siding.

In building with I.C.E. Block™, steel reinforcement bars are set, and the corners and window and door frames are braced. When stacking of the blocks reaches each floor level, concrete is poured. When the concrete sets, the bracing is removed. The forms remain in place and serve as exterior and interior insulation.

Concrete forming systems can be used both below and above grade. Standard designs can withstand sustained winds of hurricane force. The walls do not support combustion, soundproofing levels are high, and the manufacturer's product literature states that insulation value is also high. The system can be constructed at temperatures down to 0°F without special freeze protection in the concrete formulation.

Figure 5 I.C.E. Block™ foundation system.

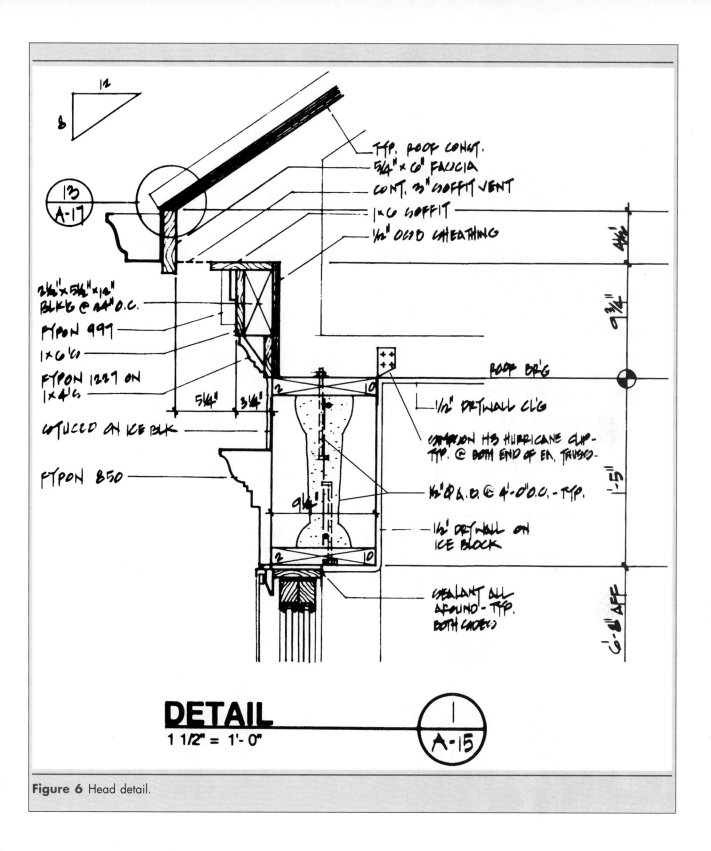

13
A-17

TYP. ROOF CONST.
5/4" x 6" FASCIA
CONT. 3" SOFFIT VENT
1x6 SOFFIT
1/2" OSB SHEATHING

2½"x5½"x12"
BLK'G @ 24" O.C.
FYPON 997
1x6's
FYPON 1227 ON
1x4's

5/4"   3/4"

STUCCO ON ICE BLK

FYPON 850

9¼"

ROOF BR'G

1/2" DRYWALL CL'G

SIMPSON H3 HURRICANE CLIP -
TYP. @ BOTH END OF EA. TRUSS-

½"Ø A.B. @ 4'-0" O.C. - TYP.

1/2' DRYWALL ON
ICE BLOCK

SEALANT ALL
AROUND - TYP.
BOTH SIDES

4½'

9¾"

1'-5"

6'-0" AFF

12
8

DETAIL
1 1/2" = 1'- 0"

1
A-15

Figure 6 Head detail.

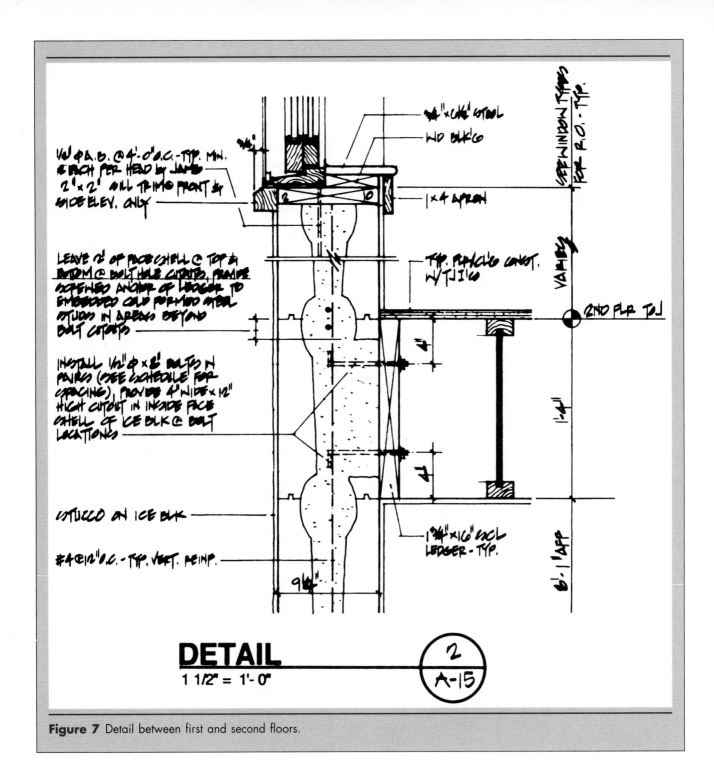

DETAIL
1 1/2" = 1'- 0"

2 / A-15

Figure 7 Detail between first and second floors.

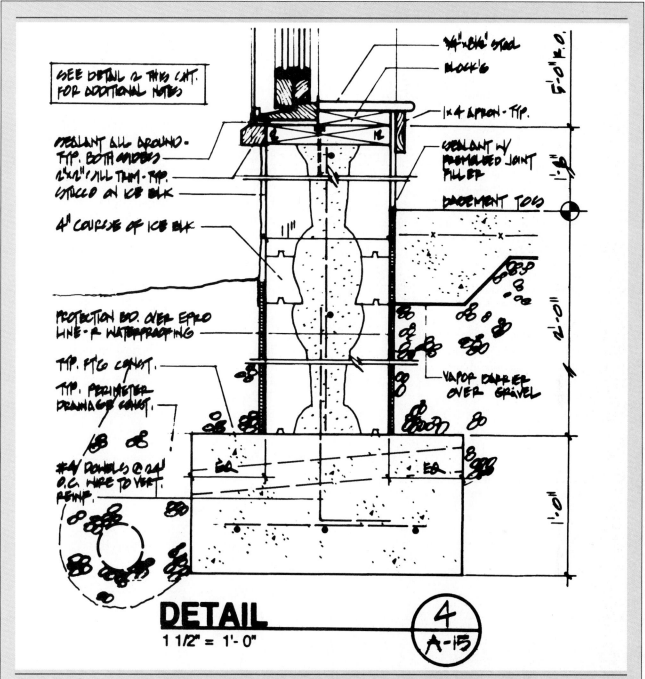

DETAIL
1 1/2" = 1'-0"

4 / A-15

Figure 8 Foundation and footing detail at basement.

House 3: Steel

Cold-formed steel provided by the American Iron and Steel Institute was used for the framing, the floor joists, and the roof trusses of House 3. Steel was also used for the interior partition walls and the roofs of all four townhouses. Steel used in the house's structural system was galvanized to provide corrosion protection (Fig. 9).

The framing consisted of 18-gauge (0.0043″) 2 × 6 studs, 24″ o.c. The same gauge was used for top and bottom wall tracks (the equivalent of plates in wood construction). Twenty-five-gauge (0.0018″) 2 × 4 studs were used in the interior non-load-bearing walls.

Studs were fastened to the tracks by No. 8 × ½″ drill-point, flat-head screws, to provide a smooth surface for drywall. A hex-drive system with No. 10 × ¾″ drill-point screws was used for structural connections such as clip angles, header connections, and truss connections (Figs. 10, 11, and 12).

Advantages of steel include the following:

- **Fire performance** Steel is noncombustible.
- **Durability** Steel is invulnerable to rot or termites and does not shrink, warp, or swell. Galvanized steel resists corrosion.
- **Quality control** Steel is manufactured to exacting specifications, providing uniform dimensions and highly consistent quality. It is free of twisting, warping, or similar defects.
- **Light weight and strength** The light weight of steel components makes them easy to handle and assemble. Steel has a high strength-to-weight ratio and utilizes framing screws that resist uplifting loads. It may perform better in earthquakes and hurricanes, although currently few field data are available.
- **Supply and pricing stability** Steel has a relatively stable price history.
- **Environmental considerations** A substantial proportion of steel is recycled, with old cars being a major source. Construction with steel produces small amounts of scrap, easily gathered and recycled.

An important consideration in the use of steel is its thermal conductivity. Cold-formed steel studs are much more conductive of heat and cold than wood studs. This problem is usually countered by the use of insulating sheathing which provides a thermal break. In House 3, ⅝″ Durock™ cement board was placed over the exterior of the steel frame. A layer of 1″ thick sheets of expanded polystyrene (EPS) was adhesively applied to the Durock™ cement board. These EPS sheets are part of an exterior insulation and finishing system.

Figure 9 Cold-formed steel frame construction.

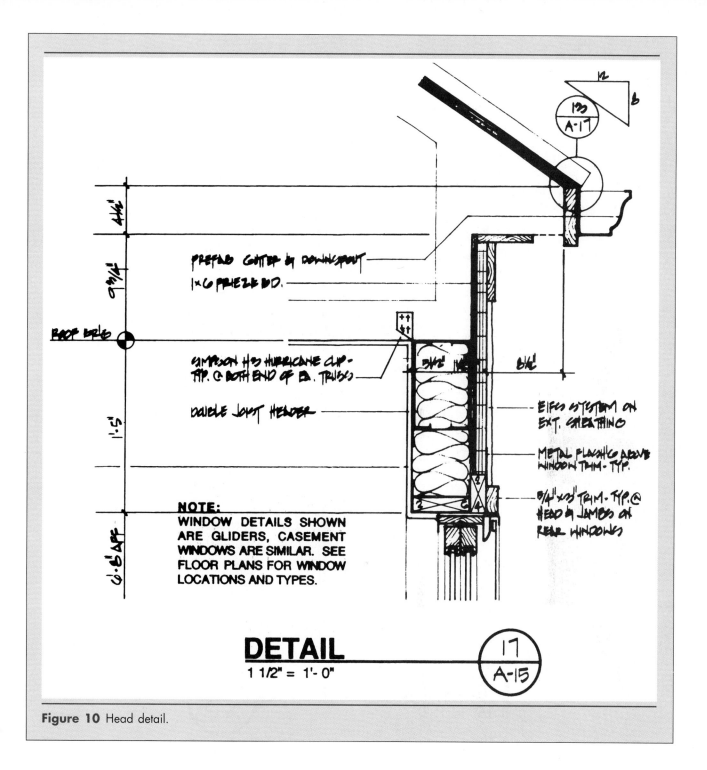

PREFAB GUTTER & DOWNSPOUT

1×6 FRIEZE BD.

SIMPSON H3 HURRICANE CLIP - TYP. @ BOTH END OF EA. TRUSS

DOUBLE JOIST HEADER

EIFS SYSTEM ON EXT. SHEATHING

METAL FLASHING ABOVE WINDOW TRIM - TYP.

3/4"×2½" TRIM - TYP.@ HEAD & JAMBS ON REAR WINDOWS

NOTE:
WINDOW DETAILS SHOWN ARE GLIDERS, CASEMENT WINDOWS ARE SIMILAR. SEE FLOOR PLANS FOR WINDOW LOCATIONS AND TYPES.

DETAIL
1 1/2" = 1'- 0"

Figure 10 Head detail.

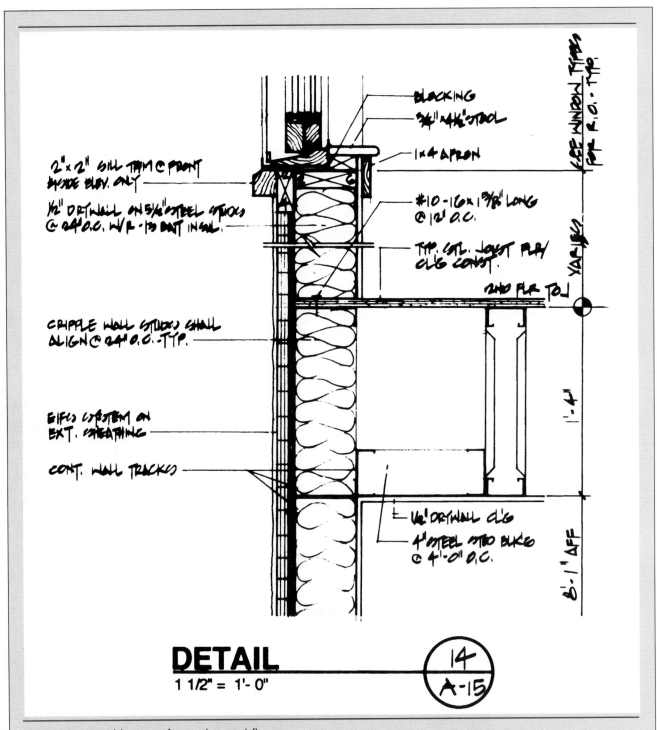

2" × 2" SILL TRIM @ FRONT
INSIDE ELEV. ONLY

½" DRYWALL ON 5/8" STEEL STUDS
@ 24" O.C. W/R-13 BATT INSUL.

CRIPPLE WALL STUDS SHALL
ALIGN @ 24" O.C. - TYP.

EIFS SYSTEM ON
EXT. SHEATHING

CONT. WALL TRACKS

BLOCKING

¾" × 4½" STOOL

1 × 4 APRON

#10-16 × 1⅜" LONG
@ 12" O.C.

TYP. STL. JOIST FLR/
CLG. CONST.

2ND FLR T.O.J.

SEE WINDOW TYPES
FOR R.O. - TYP.

VARIES

1'-4"

8'-1" AFF

½" DRYWALL CLG.
4" STEEL STUD BLKG.
@ 4'-0" O.C.

DETAIL
1 1/2" = 1'- 0"

14
A-15

Figure 11 Detail between first and second floors.

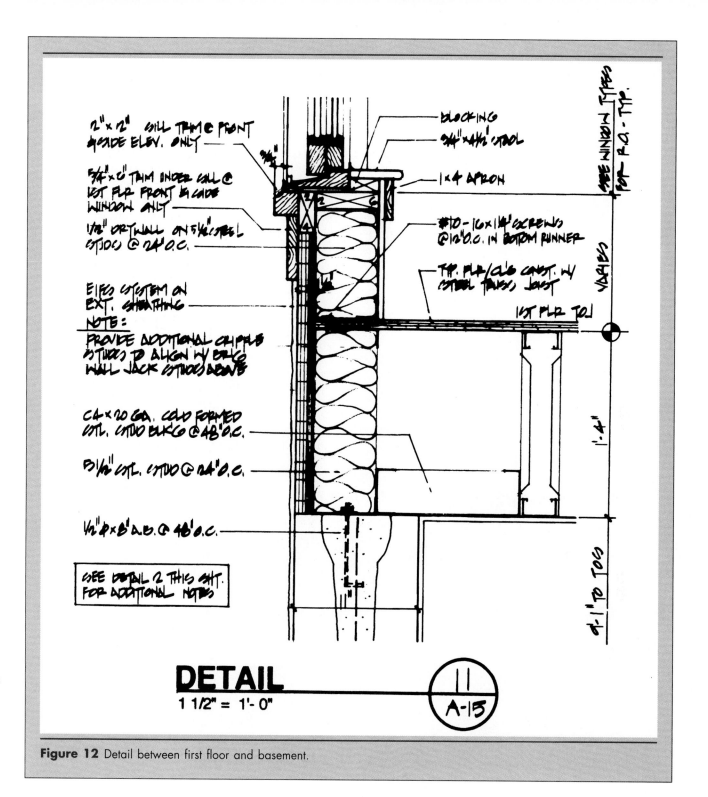

DETAIL

1 1/2" = 1'- 0"

11 / A-15

Figure 12 Detail between first floor and basement.

House 4: Precast, Autoclaved, Aerated Concrete Units

Precast, autoclaved, aerated concrete (PAAC) units provided by Hebel USA were used for the structure of House 4. Hebel is a German firm that has been producing and selling the product for about 50 years. Hebel USA was established in 1988, and the first U.S. manufacturing facility was recently established in Adel, Georgia.

Standard Hebel units measure 8" × 24" on their interior and exterior surfaces; they are 8" thick. The units are made with sand, cement, lime, water, and an ex-panding agent that, when combined, form a uniform cellular structure. Hebel units are about $\frac{1}{2}$ the weight of concrete blocks of the same size, and they provide built-in insulation (Fig. 13).

Hebel base course wall units are laid on a Type M mortar bed. A thin-bed mortar, similar to tile-setting mortar, is used for subsequent horizontal and vertical joints. The units can be trimmed accurately with normal woodworking tools at the site. They can be drilled, nailed, or screwed for attachment of finishes. Raceways or channels for wiring and plumbing can be made by routing the interior face of the wall.

Figure 13 Hebel USA precast, autoclaved, aerated concrete units being lifted into place.

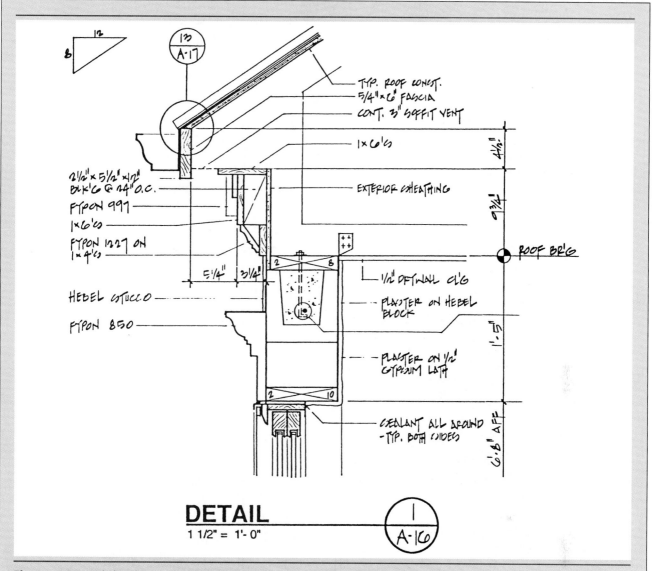

DETAIL

1 1/2" = 1'- 0"

⊙ 1 / A-16

Figure 14 Head detail.

Hebel U-units can be field cut or factory supplied. They are laid at each floor level to create a bond beam that is filled with standard concrete and reinforcing (Figs. 14 and 15).

Hebel units contain no combustible materials and exceed building code requirements for fire resistance.

They are also pest resistant and sound resistant, possess dimensional stability, and require little maintenance over the life of the house. In House 4, a plaster interior finish and a cement stucco exterior finish system were applied directly to the units.

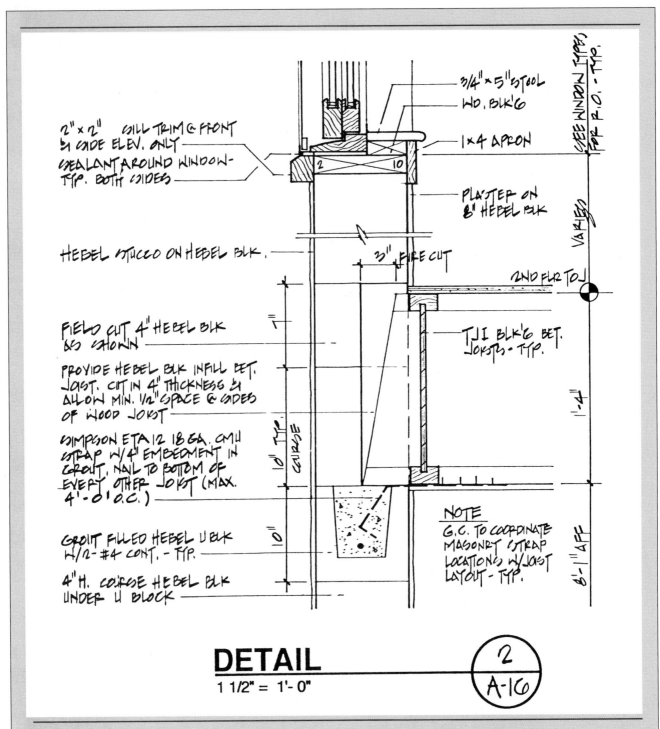

2" x 2" SILL TRIM @ FRONT
SIDE ELEV. ONLY

SEALANT AROUND WINDOW - TYP. BOTH SIDES

3/4" x 5" STOOL
WD. BLK'G

1 x 4 APRON

PLASTER ON 8" HEBEL BLK

HEBEL STUCCO ON HEBEL BLK.

3" FIRE CUT

FIELD CUT 4" HEBEL BLK AS SHOWN

PROVIDE HEBEL BLK INFILL BET. JOIST. CUT IN 4" THICKNESS & ALLOW MIN. 1/2" SPACE @ SIDES OF WOOD JOIST

TJI BLK'G BET. JOISTS - TYP.

SIMPSON ETA12 18 GA. CMU STRAP W/4" EMBEDMENT IN GROUT, NAIL TO BOTTOM OF EVERY OTHER JOIST (MAX. 4'-0' O.C.)

NOTE
G.C. TO COORDINATE MASONRY STRAP LOCATIONS W/JOIST LAYOUT - TYP.

GROUT FILLED HEBEL U BLK W/2-#4 CONT. - TYP.

4" H. COURSE HEBEL BLK UNDER U BLOCK

SEE WINDOW TYP'S FOR R.O. - TYP.

VARIES

2ND FLR T.O.J

1'-4"

8'-1" AFF

1"

8" TYP COURSE

10"

DETAIL
1 1/2" = 1'- 0"

2
A-16

Figure 15 Detail between first and second floors.

FOUNDATIONS

Houses 1, 2, and 3: I.C.E. Block™ System

The foundations of Houses 1, 2, and 3 utilize the I.C.E. Block™ system, which was also used for the exterior walls of House 2 (Figs. 8, 16, and 17).

House 4: Precast, Preinsulated System

House 4 utilizes a precast, preinsulated foundation system. The system, supplied by Superior Walls, Inc., utilizes a 1¾″ outer skin of high-strength, reinforced concrete, with a continuous sheathing layer of 1″ foam insulation on its inner surface and with steel-reinforced concrete studs to which wood nailing strips are affixed (Fig. 18).

Panels are made at the factory to fit the perimeter of the house. The panels are formed with door and window openings. Normally, 16′ is the maximum length. There are three standard wall heights: 4′, 8′-2″, and 10′.

Concrete footings are not required for installation of the Superior Walls system. The panels are placed directly over a gravel bed (Fig. 19). As they are installed, a sealant is applied between the panels to provide watertightness. The foam insulation panels have an R value of 5. Additional insulation can be added between the studs to provide a total of R-24.

The panels are made with 5000-psi concrete. According to the American Concrete Institute, a 4000-psi mix design is adequate to provide watertightness. It is therefore not necessary to parge or tar the exterior of the wall after installation.

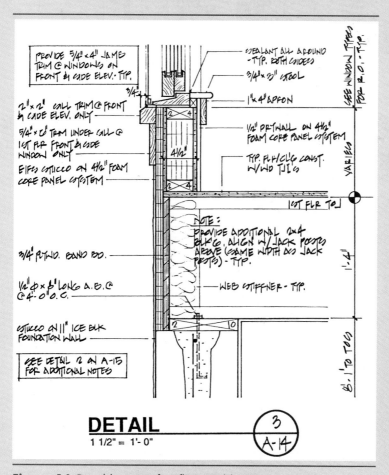

Figure 16 Detail between first floor and basement.

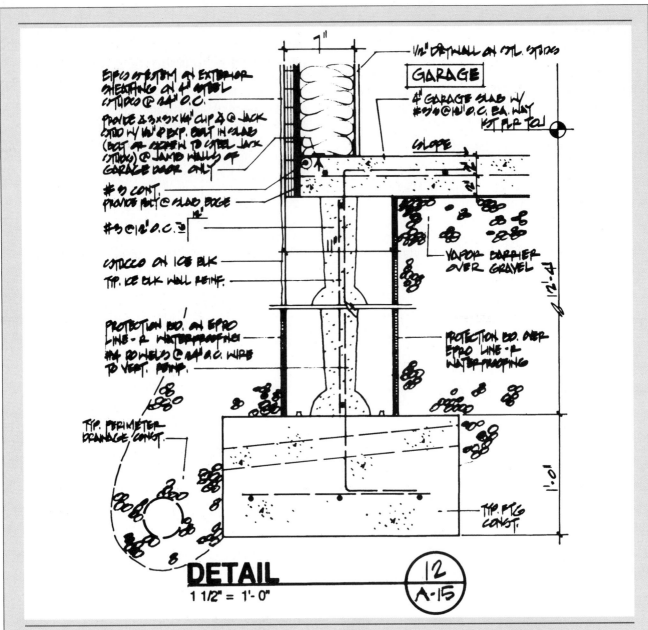

DETAIL
1 1/2" = 1'- 0"

12 / A-15

Figure 17 Foundation and footing at slab on grade.

Figure 18 Superior Walls, Inc., precast, preinsulated foundation system.

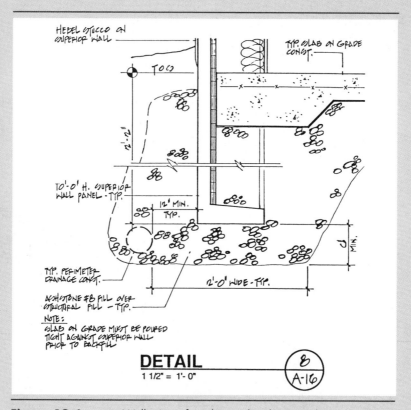

Figure 19 Superior Walls, Inc., foundation detail (no insulation required).

ROOFS

The townhouses have standing-seam steel roofs. The roofs are of various colors, which are applied as weather-resistant coatings during manufacture. The steel roof of House 1 is supported by structural insulated panels; of Houses 2 and 4, by wood trusses; and of House 3, by steel trusses.

The steel used to make the roofs is partially recycled. Steel roofs typically have a useful life of over fifty years, during which they require little or no maintenance.

EXTERIOR FINISHING SYSTEMS

Exterior Insulation and Finish System (E.I.F.S.)

Exterior Insulation and Finish Systems (E.I.F.S.s) are methods for covering and insulating the exterior of homes; these systems combine high insulating value with attractive external appearance. United States Gypsum Company (USG) provided two types of E.I.F.S.s for Houses 1, 2, and 3.

A USG Durock™ E.I.F. water-managed system was installed on House 3 (Fig. 12). The steps were as follows:

1. A water-resistive barrier was installed over the steel frame. For fire-rated party walls using gypsum wallboard, a vapor-permeable, water-resistive barrier is installed over the gypsum.
2. Durock™ cement board panels of ⅝″ thickness were installed.
3. A layer of 1″ thick sheets of expanded polystyrene (EPS) was adhesively applied to the Durock™ cement board.
4. The exterior surface of the EPS was rasped, and a ³⁄₃₂″ thick base coat of portland cement containing dry latex polymers was applied.
5. A glass fiber mesh was embedded in the base coat. The mesh was of 4.5-, 12.5-, or 24-oz/yd weight, depending on the location where it was installed.
6. A finish coat of USG Exterior Textured Finish was applied. Various tints and textures are available with the finish. For the townhouses, a coarse texture was utilized.

A standard E.I.F.S. application was provided over the structural insulated panels of House 1 (Fig. 20).

The walls of House 2, which was built with I.C.E. Block™ polystyrene forms, possess integral interior and exterior insulation (Fig. 21). The exterior surface of the forms was rasped, and a base coat, 12.5-oz/yd^2 glass fiber mesh, and finish were applied. The 12.5-oz. mesh was specified by USG for use with I.C.E. Block™.

Exterior Finish System for Hebel Units

House 4, built with Hebel PAAC units, received a ⅜″ base coat of a cement stucco which was spray applied and smoothed (Fig. 22). This application was followed by a trowel-applied finish coat of the same material. The material chosen for the finish possesses the same coefficient of expansion as that of the Hebel units.

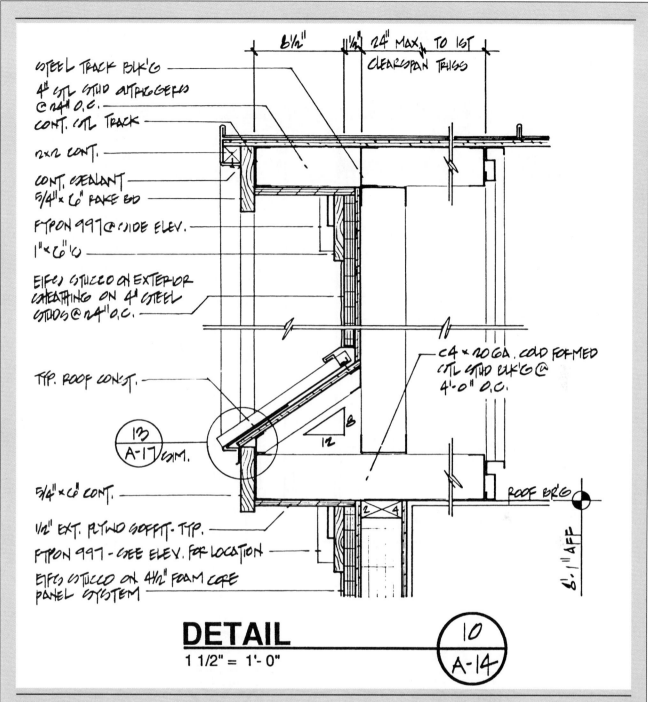

STEEL TRACK BLK'G

4" STL STUD OUTRIGGERS @ 24" O.C.

CONT. STL TRACK

2x2 CONT.

CONT. SEALANT
5/4" x 6" RAKE BD

FYPON 997 @ SIDE ELEV.

1" x 6" G

EIF(S) STUCCO ON EXTERIOR SHEATHING ON 4" STEEL STUDS @ 24" O.C.

TYP. ROOF CONST.

5/4" x 6" CONT.

1/2" EXT. PLYWD SOFFIT - TYP.

FYPON 997 - SEE ELEV. FOR LOCATION

EIF(S) STUCCO ON 4 1/2" FOAM CORE PANEL SYSTEM

6 1/2"   1 1/2" 24" MAX. TO 1ST
                  CLEARSPAN TRUSS

C4 x 20 GA. COLD FORMED STL STUD BLK'G @ 4'-0" O.C.

8
12

ROOF BR'G

8'-1" AFF

13
A-17 SIM.

DETAIL
1 1/2" = 1'- 0"

10
A-14

Figure 20 E.I.F.S. on exterior sheathing.

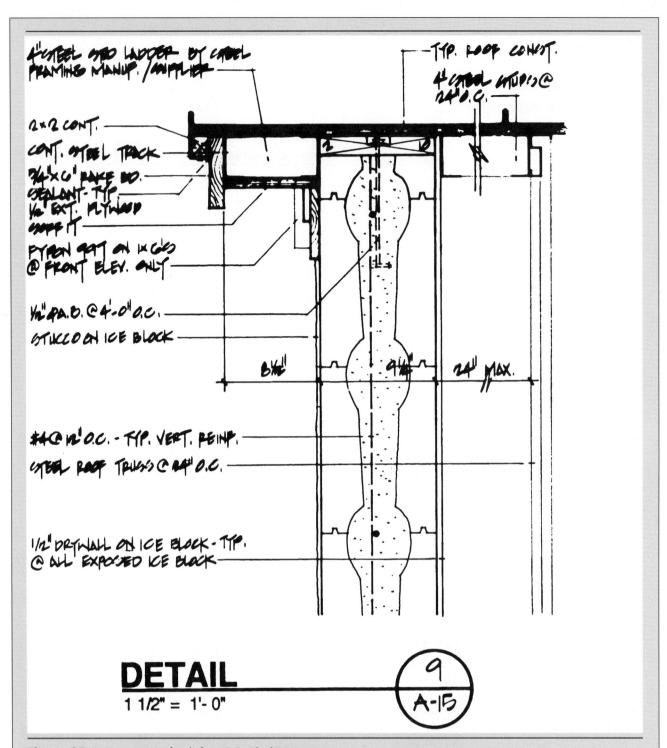

4" STEEL STUD LADDER BY STEEL FRAMING MANUF. / SUPPLIER

TYP. ROOF CONST.

4" STEEL STUDS @ 24" O.C.

2×2 CONT.

CONT. STEEL TRACK

3/4" × 6" RAKE BD.
SEALANT - TYP
1/2" EXT. PLYWOOD
SOFFIT

PTARN 947 ON 1×6'S
@ FRONT ELEV. ONLY

1/2" PA.B. @ 4'-0" O.C.

STUCCO ON ICE BLOCK

8 1/2"

9 1/4"

24" MAX.

#4 @ 12" O.C. - TYP. VERT. REINF.

STEEL ROOF TRUSS @ 24" O.C.

1 1/2" DRYWALL ON ICE BLOCK - TYP.
@ ALL EXPOSED ICE BLOCK

DETAIL
1 1/2" = 1'- 0"

9
A-15

Figure 21 Exterior stucco finish for I.C.E. Block™ system.

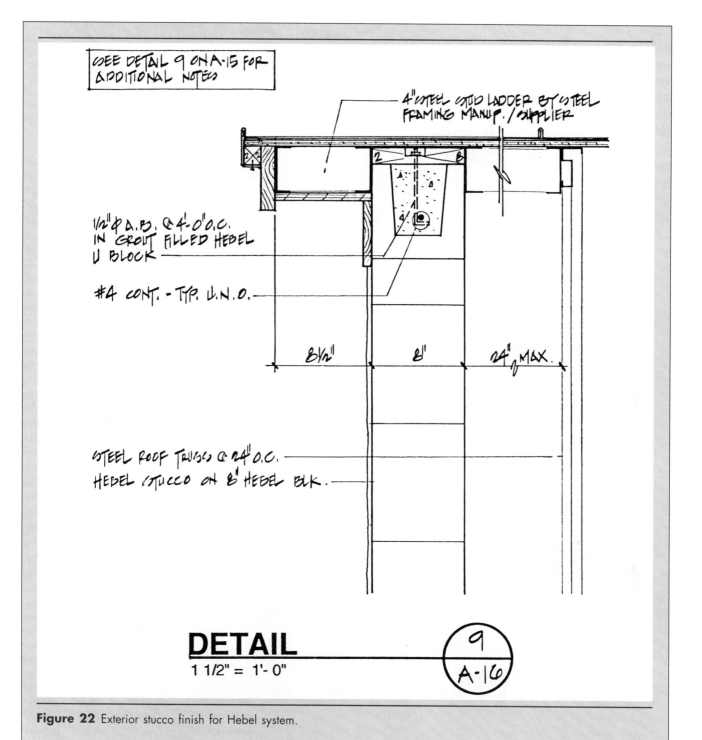

SEE DETAIL 9 ON A-15 FOR ADDITIONAL NOTES

4" STEEL STUD LADDER BY STEEL FRAMING MANUF. / SUPPLIER

1/2" Ø A.B. @ 4'-0" O.C. IN GROUT FILLED HEBEL U BLOCK

#4 CONT. - TYP. U.N.O.

8 1/2"

8"

24" MAX.

STEEL ROOF TRUSS @ 24" O.C.

HEBEL STUCCO ON 8" HEBEL BLK.

DETAIL
1 1/2" = 1'- 0"

9
A-16

Figure 22 Exterior stucco finish for Hebel system.

WINDOWS AND DOORS

Energy-efficient glider and casement windows installed in the townhouses have high-performance, argon-filled glazing panels. When the windows are closed, cam action pulls the sashes together to inhibit air infiltration. Finger-jointed wood is used for the frames, to make use of short pieces of wood that would normally be discarded. The windows have a U-value rating of 0.33.

Energy-efficient entry doors were installed in all of the units. The fiberglass doors are insulated and can be finished with stain to look like traditional six-panel wood doors.

SPRAY-IN FOAM INSULATION

House 3, which was framed with steel, was insulated with the Icynene Insealation System™ (Fig. 23). The system was also used in Houses 1 and 4 to insulate the exposed floor above the garage, and in all of the townhouses to insulate joist ends and to seal attics, cantilevered fireplaces, cantilevered floors, and gable ends.

In this system, the foam insulating material is sprayed as a liquid at a temperature of about 150°F, and it expands at a ratio of 100:1. The material cures dry within seconds, adheres to the surface onto which it is sprayed, and does not undergo loss of pliancy or subsequent shrinkage. After spraying, excess insulation is easily removed with a standard handsaw.

Water in the formulation reacts with other components to produce carbon dioxide, which serves as the blowing agent. The product contains no ozone-destroying CFCs or HCFCs, and no formaldehyde. After 30 days, it produces no detectable emissions.

The material fills cavities completely, providing a nearly perfect fit and drastically reducing air leakage, convection, and airborne moisture. The material has an R value of 3.6 per inch of thickness. It virtually eliminates the need for detailed special air seals, vapor barriers, or air barrier treatment such as building wraps, taping, and caulking.

PHOTOVOLTAICS

Photovoltaics (PV) is the generation of electricity from sunlight. The technology involves the installation of both PV modules where they are accessible to the sun and equipment for storing the electricity that is generated by the shining of the sun on the modules. Residential applications of photovoltaics represent an important frontier of energy efficiency and home building technology.

Figure 23 Icynen Insealation System™ being sprayed into steel stud wall cavities.

Figure 24 Flexible photovoltaic modules installed on the roof of House 4.

An innovative approach to residential photovoltaics, involving flexible modules that have the appearance of standard roofing products and that can serve both as roofing and as solar collectors, will be tested and demonstrated in House 4. The steel roofing was removed from more than 400 sq ft of the south-facing roof of this house, and the PV modules were mounted directly on the roof sheathing. Installation methods and products common to the installation of standard metal roofing were used to mount the modules (Fig. 24).

The power output from the modules, which is expected to exceed 2 kilowatts in the full sun, will be stored in batteries. The system is expected to contribute between 50% and 75% of the home's 120-volt requirements, exclusive of space conditioning, cooking, and water heating. The electricity can be used during on-peak periods, to reduce both energy demands of the home and energy cost to the homeowner. It can also be used as a source of backup power if there is a power outage.

The system operation is programmable and will involve an interactive relationship among the utility grid, the PV output, and the battery storage, which will enable the utility to use the system for demand-side management. The system will be able to feed excess electricity generated by the modules into the utility grid. Functioning of the system will be automatic, with minimum user involvement.

HOME AUTOMATION

SMART HOUSE, LP, provided home automation systems for the townhouses. SMART HOUSE coordinated contributions from SMART HOUSE; AMP, Inc.; and Molex, Inc., to provide a different system package for each house. Services provided by the systems include advanced energy management.

PROBLEMS AND SPECIAL CONSIDERATIONS ENCOUNTERED IN CONSTRUCTION

Working with Steel Framing

Interior steel framing was used in all of the houses. In order to avoid electrolysis, no copper piping and no electrical wiring can touch the steel. Piping and wiring must pass through openings in the steel studs that are insulated with plastic grommets. In addition, care must be taken regarding the condition of electrical tools and cords that are used in the vicinity of the steel frame. Electrical contact could result in shock to anyone who was touching the frame at any point. These requirements for working with steel generally translate to increased labor time.

Evaluation Criteria

As the townhouses were constructed, evaluation of the materials and systems was carried out in three areas:
1. **Technical performance** This included comparisons of features and characteristics of the innovative materials with those of standard materials currently in use.
2. **Buildability** This comprised evaluation of the capability of the average subcontractor work force to work with and install the innovative products and systems. What tools were needed? What training was required? How long did it take workers to master new methods and techniques?
3. **Infrastructure** This included easy availability of technical information from local building supply outlets, on-shelf availability of such items as fasteners, and availability of information in building codes.

Delays in Securing Building Permits

Substantial delays were encountered in securing permits for construction. Certain supporting data and fitness-for-use justifications that the county required did not exist for the innovative products that were to be used, or they were not in a form that was acceptable to the county officials, particularly with regard to fire performance.

The Prince George's County Fire Department conducted an extensive review of the project's design. In response to the department's requirements, the Research Center put available data into a form acceptable to the county, or it retained experts to help develop technical information and to obtain approvals. A cooperative effort between manufacturers and Research Center staff was involved. The compliance documents that were generated through this effort will be useful in securing approvals for the products in other jurisdictions.

Lack of Informational Infrastructure

For all of the basic structural products used to build the townhouses, far less informational infrastructure exists than is available for conventional construction. Technical information, recommendations, and advice for wood and masonry construction have evolved through nearly universal use over a long period of time, and all are available to builders at places that range from their local lumber supply outlet to the national building codes. A major challenge facing the manufacturers and suppliers of the innovative products is to develop an adequate informational infrastructure to support widespread use and adoption of the materials.

Integration of Products and Systems

Integrating nonconventional and dissimilar materials posed significant challenges. For example, special attention and engineering were required to develop anchorage details for attaching the steel framing of House 3 to the I.C.E. Block™ forms of House 2.

House-by-House Summaries

Other problems and special considerations involved in working with the innovative materials used to build the townhouses are described in the following house-by-house summaries:

- **House 1.** No structural insulated panel (SIP) system that was available for the project possessed a 1-h fire rating on both sides of the panels. This rating was required by county officials for use in the common wall between Houses 1 and 2. The Research Center retained a leading fire engineer to study the problem and devise a method for meeting the county's requirements. The method recommended by the engineer, which met with the county's approval, was installation of two layers of ⅝" drywall on each side of the wall.

- **House 2.** The initial design for House 2 involved the use of more reinforcing steel in the walls above the basement than was indicated in manufacturer's recommendations. The Research Center respecified the system to the manufacturer's recommendations, making it fully safe at less cost.
- **House 3.** The steel studs were light and were easy to handle and put into place. They were also easy to alter in the field. They had been made too long, and shortening them posed no difficulty. However, there were no commercially available hangers, and these had to be made at the site. In addition, it was necessary to train the workers in the use of the screw gun, and the training requirement for this skill had been underestimated. The screws themselves could not be purchased at the local home building supply outlet.

  The steel roof trusses that were installed on House 3 were light and easy to handle, but they were not as rigid as wood trusses. To set and brace the trusses required procedures different from those that are customarily used for wood trusses. It was not possible to stand either on the trusses or on the top of the interior steel frame. Auxiliary lifting equipment for the workers was therefore required.
- **House 4.** The learning curve of the workers was exhibited in their building of the structural system of House 4. It took experienced masons 2½ weeks, including minor weather delays, to lay the first-floor walls. It took 2½ days, including training of a new crew, to lay the second-floor walls.

  Carpenters built one Hebel wall section. Contrary to what might be expected, carpenters worked more quickly and more accurately than masons in erecting the Hebel units. A carpenter, rather than a mason, constructed the gable wall on the house. Working with Hebel units resembles carpentry in many respects. Carpenters are accustomed to precise measurements and cuts. The Hebel units can be measured to exact dimensions, cut with a saw, and fitted together snugly. Masons are accustomed to 3/8" to 1/2" mortar joints, which reduces the degree of accuracy required for fitting. Due to the dimensional accuracy of the Hebel PAAC units, the mortar joint is only 2 to 3 mm (1/16" to 1/12").

  The manufacturer notes that, whereas carpenters work well with Hebel PAAC units, construction has also been carried out successfully with masons or with crews made up of a lead mason or carpenter with three or four laborers.

RESEARCH ACTIVITIES

Research activities and programs that will be carried out in the townhouses include the following.

Energy and Thermal Performance

Analysis of the energy performance of the townhouses was initiated during construction and will continue after completion. Preliminary analysis included a review of the compliance of each of the townhouses with the Model Energy Code (MEC) minimum standards, using the *MECcheck* software package. Data on each house—such as structural dimensions, types of materials and their R values, fenestration, and HVAC equipment—will be integrated into the analysis.

An empirical analysis of the energy performance of each unit will be performed. This analysis will include compilation of data on outdoor air temperature, exterior wall surface temperature, indoor wall surface temperature, and HVAC gas and electric energy usage. A data acquisitions system has been installed that includes thermocouples in the outside south and north faces of the units to record surface temperatures. The thermocouples will be used to monitor diurnal temperature swings. The full set of data will be used to create a predictive model of the anticipated energy consumption of each house.

Focus Groups and Surveys

Focus groups in the townhouses will be conducted for the project's private-sector sponsors, the American Iron and Steel Institute, Hebel USA, the National Ready Mixed Concrete Association/Portland Cement Association, and United States Gypsum Company. These focus groups will provide a qualitative assessment of features and technologies used in the townhouses.

Two panel sessions will be held. One will include builders and members of trades directly associated with home building. The other will include real estate salespersons and potential home buyers. Participants will be briefed on the features of the houses, and their reactions and opinions will be elicited, recorded, and analyzed.

Two separate survey instruments will also be developed: one for home builders and associated tradespersons and one for real estate professionals and potential home buyers. All persons visiting and touring the houses will be asked to fill out a survey form. Results will be tabulated, and the results will be used to supplement the findings of the focus groups.

Study of Energy-Efficient Duct System

Under a contract with the Gas Research Institute, the Research Center is studying ways to improve duct energy efficiency in new residential construction. As part of the program, Research Center engineers will test the thermal efficiency of duct systems that are installed in the townhouses.

All of the units will be evaluated for distribution efficiency, with special studies being conducted in House 2. Redundant duct systems servicing the second floor were installed in this house. These systems consist of ceiling registers with ducting in the floor. The duct leakage and sealing techniques and the thermal distribution efficiency of ducts inside and outside the conditioned space will be evaluated. When the tests are completed, the information will be made available to builders, to enable them to make a cost-versus-benefits analysis of various alternatives for the benefit of their customers.

Photovoltaic System

Baltimore Gas & Electric will oversee the operation of the photovoltaic system and will monitor the output of the PV array as well as the operation of the battery storage. BG&E will also oversee the monitoring of the home loads and the interaction of the system with the utility grid.

The installation features of the photovoltaic roofing material will be evaluated, and input on its esthetic qualities will be secured from builders, code officials, and consumers.

ADDITIONAL PRODUCTS AND MATERIALS

Key Participants and Contributors

The following products and materials were provided by key participants and contributors:

- **Armstrong World Industries** Resilient flooring.
- **Benjamin Moore and Company** Interior coatings that are low in volatile organic compounds (VOCs). In these coatings, VOCs, which have an adverse environmental impact, have been virtually eliminated from the formulation.
- **Hearth Products Association** State-of-the-art technologies for fireplaces and free-standing stoves, including pellet stoves and gas-fired hearth products. Pellet stoves burn small, concentrated pellets, which are made from recycled waste products such as sawdust or cardboard. Gas-fired hearth products provide an esthetic simulation of a wood fire without the inconvenience, the environmental impact, and the sometimes high cost of burning wood.
- **International Paper Company, Masonite Division** Interior doors and garage doors, with facings simulating the appearance of dimension lumber, that are made from sawmill residue and wood scrap.

- **National Ready Mixed Concrete Association (NRMCA)** Driveways for Houses 2 and 3 that are certified in the Association's "Blue Ribbon" quality program. The driveways are designed to provide high durability. They are covered under a five-year limited warranty against major cracking and scaling.
- **Therma-Stor Products** Ventilation and dehumidification equipment, designated to prevent mold, mildew, dust mite infestations, and airborne allergens; dispel gaseous contaminates; and provide maximum indoor air quality through fresh air ventilation, humidity control, and air filtration.
- **Trus Joist MacMillan** The floor system for House 1, consisting of structural engineered lumber products that are designed to make efficient use of wood fibers.
- **Water Film Energy, Inc.** The drain-water, heat-recovery system, in which the heat from hot drain water is used to warm cold water circulating in a pipe wrapped around the hot-water drain pipe. According to the manufacturer, the system can typically recycle about half of the heat carried by a hot shower stream.
- **Wilsonart International** Solid sheet and molded sinks made of a structural material that can be carved and routed like wood, is renewable, and is produced by manufacturing processes designed to require minimal energy and produce minimal waste.
- **York International Environmental Systems** A natural gas heating and cooling system. Using a non-CFC refrigerant cycle, the system removes heat from the home in the summer and delivers heat in the winter. The system runs automatically at variable speeds in accordance with need, providing uniform temperature and humidity levels.

Other Products

Products donated for the project include the following:

- **Geothermal ground-source heat pump** This heat pump takes advantage of the stability of earth temperatures, as opposed to outdoor air temperatures, to operate at lower energy cost. Because the earth is a stable source of heat, the warmth delivered by ground-source heat pumps is not reduced when outdoor temperatures drop, as is the case with air-source heat pumps.
- **Pallet flooring** This flooring is made from old wooden pallets. The boards from the pallets are cut and shaped as tongue-and-groove flooring. After the flooring is laid, it is coated with a thin layer of filler, which fills the numerous nail holes in the boards. The floors are then sanded and finished.

EXERCISES

1. Draw a typical wall detail of a brick veneer, 1-story residence, using the scale 1″ = 1′-0″. Use a crawl space with concrete block foundation wall. Floor joists are 2″ × 10″; ceiling joists, 2″ × 8″; and roof slope, 4:12.

2. Draw a typical wall detail of a concrete block wall with a slab-on-ground floor, using the scale ¾″ = 1′-0″. Show a flat roof with 2″ × 8″ roof joists, and box in the 2′ overhang (see roof overhang details in Section 12.12 in Chapter 12).

3. Draw a typical wall detail of a frame wall for a house with a basement of 8″ poured concrete walls, using the scale ½″ = 1′-0″. Use a plank-and-beam roof with 3″ × 8″ beams. The roof slope is 3:12; board-and-batten siding is used.

4. Draw a section detail of an aluminum awning window, using the scale 1″ = 1′-0″. Select an actual window from *Sweet's File*.

5. Draw a section and an elevation of a wood hopper window, using the scale 1½″ = 1′-0″. Use information from a manufacturer's catalog.

6. Draw the head, jamb, and sill section of a door detail in a solid-brick (8″) wall, using the scale ¾″ = 1′-0″. Use 1¼″ wood jambs, suitable trim, and a 1¾″ flush door.

7. Draw the elevation, plan, and vertical section details for a contemporary fireplace, using the scale ¾″ = 1′-0″. Use stone masonry with a flagstone hearth. The size of the fireplace opening is 42″ × 28″.

8. Design and draw the details for a single flight of colonial stairs, using the scale ¾″ = 1′-0″. The finish-floor-to-finish-floor height is 9′-8″. Select a suitable scale, trim, and railing.

9. Draw a wall detail of a 2-story, contemporary, wood frame house with slab on ground, aluminum and glass sliding doors, 2 × 10 second-floor joists, 4′-6″ high aluminum casement windows on the second floor, 18″ deep flat roof trusses, membrane roofing with stone ballast, and a 2′-0″ high parapet. The exterior walls and parapet have 1½″ thick E.I.F.S.; the scale is ¾″ = 1′-0″.

10. Draw a section through a covered patio, slab on ground with 2″ stone flooring, 8″ square by 10′ high wood posts, a 6 × 12 wood beam between the posts, 4 × 8 wood rafters, 3″ laminated T&G decking, a Spanish tile roof with 5:12 slope, a 36″ high railing with 2 × 4 top and bottom rails, and 2 × 2 vertical balusters between the posts. Use a scale of 1″ = 1′-0″.

REVIEW QUESTIONS

1. What factors are used to determine the choice of scale when drawing details?

2. What information does a typical wall detail usually show?

3. Why are structural members in detail drawings drawn actual size rather than nominal size?

4. In gable-roof terminology, what is the difference between *span* and *run?*

5. Why must you consult manufacturer's catalogs when determining rough opening sizes for windows?

6. Differentiate between a window sash and a window frame.

7. Why are doorjambs usually the only sections needed when you are drawing door details?

8. What dimensions are first selected in fireplace design, and what should be the basis for their selection?

9. For a workable fireplace, what should be the proportion of the cross-sectional flue area to the fireplace opening area?

10. What two dimensions are the most critical in satisfactory stair design?

11. Describe how an architect's scale can be used to lay out the riser height and the tread width of stairs in elevation or section without measuring each riser and tread from floor to floor.

12. Write the three formulas for checking riser and tread proportions.

13. What is a fire door, and what do the labels A, B, and C designate?

14. What are anchor bolts, and where are they used in wood frame construction?

15. What is used to support brick masonry above a door or window opening?

16. What type of cut is usually placed on a rafter where it rests on a bearing wall?

17. Residential doors are usually 6′-8″ high. What is the common height of commercial doors?

18. What is the minimum headroom above stairs?

19. Where is a brick rowlock used in window detailing?

20. What is a clerestory window?

W. D. FARMER

*"The beautiful is as useful as the useful.
More so perhaps."*
—*VICTOR HUGO*

Because of the universal interest and the various aspects involved in home planning, the house is the most appropriate starting problem for students. It is appropriate not only because nearly everyone has been subjected to either the good or the bad qualities of residential planning at one time or another and has formed various opinions, but also because the principles of good residential planning can be adapted to many other types of structures.

The successful design of buildings requires a deep insight about abstract design, the many construction methods, the cost factors, and even human psychology. All are important, and few individuals are generously gifted with such varied qualifications. However, drafters who can grasp the basic planning principles as they work on the drawing board have a definite advantage, whether they contribute creatively or not. Minor decisions must continually be made during the development of working drawings. Prudent decisions reflect drafters' backgrounds and will certainly contribute to their advancement. Planning also provides an insight into the real meaning of architecture and challenges the ultimate thinking ability of the student. Good planning should start with correctly sized and oriented living spaces.

15.1

USE A PLANNING CHECKLIST

One inevitable aspect of residential planning is the unlimited variety of tastes and needs of different individuals. Almost every planning job brings together dissimilar combinations of these aspects. Realistic planning often becomes a matter of compromise—choosing between features that will genuinely contribute to the livability of the plan and those that the client feels he or she would like to have. Occasionally, choices are difficult to resolve. Within the limits of the budget, the custom-designed home should be a representation of the tastes and needs

of the occupants, and it should please them. The role of the designer often becomes that of an arbitrator with a practical-minded attitude, yet also with a flair for individuality, if he or she is to be successful. The designer must be able to reconcile the special requests of individuals with their actual needs and come up with satisfying solutions. As a rule, this role is not easy. The first step, then, must be a definite understanding of the tastes and needs of the client. To be certain that no important details have been overlooked or forgotten, a procedure must be followed that helps to organize these requirements before planning is started. One of the simplest methods for this purpose is a checklist similar to the following:

Planning Checklist
- ❏ **1.** Site:
    - ❏ Size of lot
    - ❏ Shape of lot
    - ❏ Type of contours (level, slope to rear, slope to front, slope to side)
    - ❏ Utilities available (water, electricity, gas, sewer)
    - ❏ Drainage (good, poor)
    - ❏ Types of trees, if any
    - ❏ Direction of most desirable view
    - ❏ North direction
    - ❏ Convenience to schools, shopping, churches, and so on
    - ❏ Satisfactory soil and topsoil
    - ❏ Wetlands
- ❏ **2.** Occupants:
    - ❏ Number and age of adults
    - ❏ Number and age of boys
    - ❏ Number and age of girls
    - ❏ Profession of owner
    - ❏ Provision for guests
    - ❏ Provision for servants, if any
    - ❏ Others (in-laws, and so on)
    - ❏ Pets

3. Individual requirements:
   - ❏ Formal entertaining
   - ❏ Separate formal dining area
   - ❏ Informal living areas
   - ❏ Outdoor living and eating areas
   - ❏ Supervised outdoor play area
   - ❏ Nursery
   - ❏ Recreation areas (billiards, swimming, tennis)
   - ❏ Hobby areas (music, sewing, woodworking, gardening, and so on)
   - ❏ Study or reading areas
   - ❏ Laundry area
   - ❏ Screened porch or deck
   - ❏ Entertainment center (TV, stereo, video)
   - ❏ Personal computer

4. General design:
   - ❏ One-story
   - ❏ Two-story
     - ❏ (Elaborate stairway)
   - ❏ Split-level
   - ❏ Crawl space
   - ❏ Concrete slab
   - ❏ Basement
     - ❏ (Exterior entrance)
   - ❏ Traditional exterior
   - ❏ Contemporary exterior
   - ❏ Type of roof
   - ❏ Exterior materials (brick, siding, stone, board-and-batten, plywood, stucco, E.I.F.S.)
   - ❏ Finish floor materials
   - ❏ Garage (number of cars)
   - ❏ Carport (number of cars)
   - ❏ Provision for eventual expansion
   - ❏ Open planning

5. Budget restrictions:
   - ❏ Size of house in sq ft (1000 and under, 1500, 2000, 2500 and over)
   - ❏ Financing of the house
   - ❏ Number of bathrooms (tub or shower, wall and floor material)
   - ❏ Fireplaces
   - ❏ Type of entrance
   - ❏ Quality of interior trim

6. Mechanical equipment required:
   - ❏ Central heating (warm air, hot water, radiant, heat pump, solar, and so on)
   - ❏ Air conditioning
   - ❏ Washer-dryer
   - ❏ Dishwasher
   - ❏ Range
   - ❏ Oven
   - ❏ Microwave
   - ❏ Refrigerator
   - ❏ Ironer
   - ❏ Food freezer
   - ❏ Garbage disposal
   - ❏ Exhaust fans
   - ❏ Water softener
   - ❏ Size of hot water heater
   - ❏ Solar panels
   - ❏ Sauna
   - ❏ Hot tub

7. Storage areas required:
   - ❏ Entrance closet
   - ❏ Bedroom closets
   - ❏ Linen closets
   - ❏ Toy storage
   - ❏ Kitchen equipment storage
   - ❏ Cleaning equipment storage
   - ❏ Tool storage
   - ❏ China storage
   - ❏ Gardening equipment storage
   - ❏ Hobby equipment storage
   - ❏ Others

15.2

BUILDING CODES

Building codes are the minimum requirements that should be utilized in the construction of a residence or other building. Their purpose is to protect the public's life, health, and welfare in the built environment. Before beginning a design, you should be familiar with requirements in the building codes. For example, the building code will place limitations on the area and the height of the structure permitted to be built based on the type of construction and the occupancy. In some instances, in order to maximize the limitations of a certain type of construction or occupancy, the code will require the installation of an automatic sprinkler system for fire protection. These and various other requirements will affect the design of a structure as well as impact economical considerations.

Presently in the United States, there are three model building codes: the National Building Code (BOCA), the Standard Building Code (SBCCI), and the Uniform Building Code (ICBO). These model codes tend to be in use regionally, and a particular edition is usually adopted by city, county, or state governments by ordinance for enforcement.

Model codes are not the only codes available for adoption. Certain government bodies have chosen to write their own building codes. For example, Miami-Dade County in Florida, New York City, and the states of North Carolina and Wisconsin all have written their own codes. Always consult with the building department that has jurisdiction over the location where the proposed structure is to be constructed to determine what code and which edition are under enforcement.

After researching building codes, you will find that

some of the construction methods in this book are not only good practices but also code requirements. For example, the use of natural light and ventilation in bedrooms, placing exhaust fans in bathrooms, using pressure-treated wood plates on slab-on-grade construction, installing joint reinforcement and wall ties in masonry construction, and ventilation of crawl spaces are all building code requirements. In addition, information concerning the minimum width permitted for an exit door, the fire rating of walls and doors, and the maximum safe loads that may be imposed on a floor or roof can all be found in the building codes.

REVIEW THE RESTRICTIONS AND LEGAL ASPECTS OF THE LOT

First, make a thorough investigation of the property on which the house is to be built. If possible, check for any restrictions created by easements; acquaint yourself with the zoning regulations of the property. All legal restrictions must be carefully followed. *Easements* are the legal means by which a party other than the owner has been

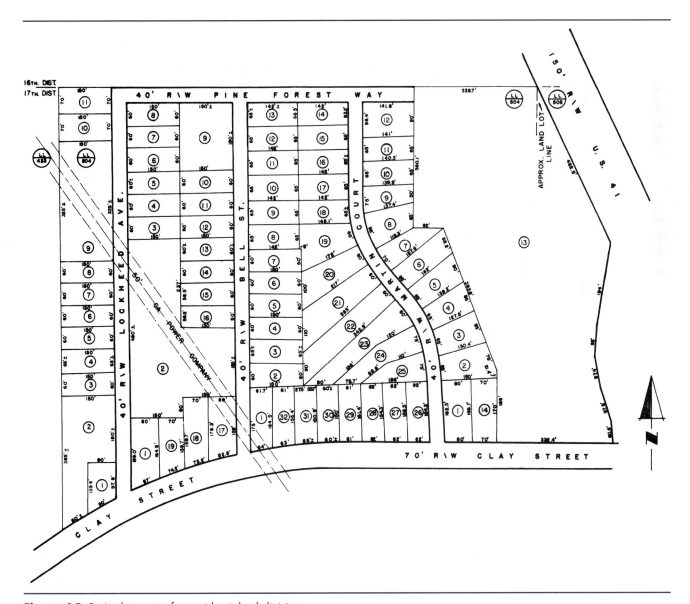

Figure 15-1 A plat map of a residential subdivision.

given access or certain rights to the property for various reasons. Utility companies have easements on property in many areas. For example, if electric power lines run across a piece of property, the power company no doubt has an easement on the property for the explicit purpose of running and maintaining the lines. The right is irrevocable and must be honored by the present and future owners. Occasionally, easements are granted to individuals for trespass privileges on property, and these rights must be honored as well.

Zoning regulations or ordinances, usually binding to residential property, restrict the minimum size of a dwelling that can be put on a lot in a given area. These restrictions protect property owners from encroachment of undesirable structures, which might devalue surrounding property values. From the standpoint of future resale of a house, it is wise to neither overbuild nor underbuild in a restricted area; it is advisable to maintain a consistent quality in keeping with neighboring homes. Overbuilding simply means the construction of a much larger or more expensive house than those in the immediate area, whereas underbuilding is the reverse—building a house much smaller or one in a lower price bracket than surrounding homes. Ordinances also usually restrict the placement of a house on a lot (the setback from the street as well as the side lot lines).

Of course, it is taken for granted that all conditions of lot selection—such as availability of utilities; convenience to shopping centers, schools, churches, and transportation; and suitable zoning regulations—have been considered by the prospective home builder. The selection of the lot and an investigation of the clear-title legality of the deed (usually done by title guarantee companies) are the responsibility of the owner.

If possible, make a personal inspection of the lot. No written description is as valuable as first-hand observation of the property.

A *plat*, which is a graphic description of a subdivision of a tract of land, is on file in county or city land registration offices. The plat and a written description constitute the legal documents of the property. Photocopies of registered plats may be obtained by the landowner for a small fee. The usual registered-plat drawing contains the following information (Fig. 15–1):

The names of the subdivision, district or section, county or city, and state

A north-pointing arrow

The scale of the plat

The names and widths of all streets

The bearing of street lines and lot lines

Lot dimensions and numbers

A description of any easements, setback lines, utility mains, and protective covenants

References to adjoining property

Certification as to the correctness of the plat by a registered land surveyor

A plat should not be confused with a *site plan*, which will be described later with working drawings. The site plan is drawn by the drafter to show the exact placement of the structure on the property, but it does not become part of the property's legal description.

Linear measurements describing the boundaries of land are usually given in feet and decimals of a foot carried to two places, such as 87.41′ or 264.92′. Bearing angles are indicated in degrees and minutes from either north or south, whichever results in the smaller angle, such as N31°17′E, S24°, or 45′W. Land surface heights shown on plats or survey maps can be distances in feet above sea level, or they can be distances above or below local permanent datum points, called *bench marks*. Either method is satisfactory for indicating various land surface heights.

Prior to construction planning, owners often have survey maps made of their property by a registered land surveyor. A survey map (Fig. 15–2) containing *contour lines* as well as other pertinent physical descriptions of the property will help the designer orient the structure properly. Contour lines indicate the elevation variations of the property. Each contour is plotted to represent similar elevations along the entire line. If you were to actually walk along a contour line, you would continually walk on the same level or elevation. The shoreline of a lake is a good example of a natural contour line. As the water's edge continues around the lake, it varies in shape and curvature depending on the nature of the terrain along the shore. If a small stream empties into the lake, the water's edge turns and follows the bed of the stream for a distance; then it crosses the stream and returns toward the mouth and continues around the edge of the lake. Each contour on the ground closes on itself, such as the complete shoreline of the lake, and eventually returns to itself. This fact is not always apparent on most contour maps, as the closures usually fall off the map.

Adjacent contour lines normally have similar characteristics because of natural surface formations (Fig. 15–3). Lines close together indicate a steep rise in elevation, whereas lines farther apart indicate a more level area. Lines will not cross each other except when indicating unusual conditions such as caves or overhanging cliffs. Nor will they touch each other except when a vertical wall or cliff is represented. On riverbeds, the contour lines point upstream. Figure 15–4 shows typical contour indications.

The vertical distance between contour levels is known as the *contour interval*. These contour intervals are usually 1′, 2′, 5′, or 10′, depending on the scale of the map. Heights above sea level or heights above or below local datum points are usually shown along every fifth line, which is often drawn darker to facilitate identification. Because of the uneven nature of land surfaces, contour lines are generally drawn freehand on topographic maps;

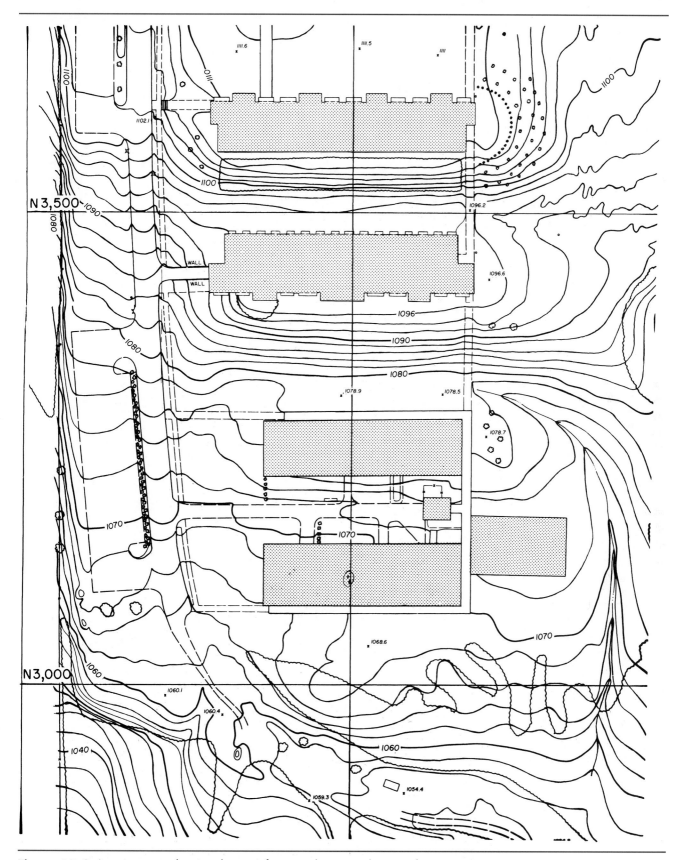

Figure 15–2 A survey map showing the use of contour lines to indicate surface variations.

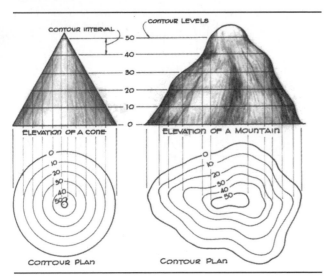

Figure 15-3 Contour lines showing the shape of a cone and a hill.

Figure 15–5 shows the contour lines of a typical building lot. To construct the contour profile of line A–A, project the intersections of line A–A and the contour lines that pass through it to the contour levels laid out horizontally below. If the lot is to be leveled for a more desirable construction area, finish contours would appear similar to those shown in the lower drawing. You could find the finish profile of line B–B by projecting the new contour intersections as shown. Fills and cuts are accurately defined by superimposing original and finish contour profiles. Notice that on contour maps showing required surface changes, the finish contour lines are usually made solid, and the original lines where the changes occur are shown as broken lines.

on plane surfaces they would appear more mechanical and therefore are occasionally drawn with instruments.

Slopes of land surfaces or roadbeds are generally expressed in percent of grade. A slope of 5%, for instance, rises vertically 5′ over a horizontal distance of 100′:

$$\frac{\text{Rise}}{\text{Run}} = \text{slope (expressed in percent)}$$

$$\frac{5}{100} = 0.05 = 5\%$$

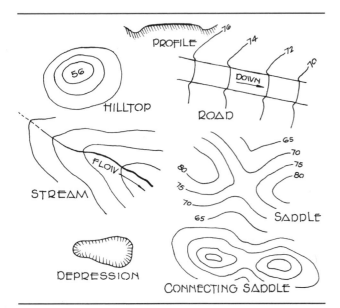

Figure 15–4 Contour representations of various land surfaces.

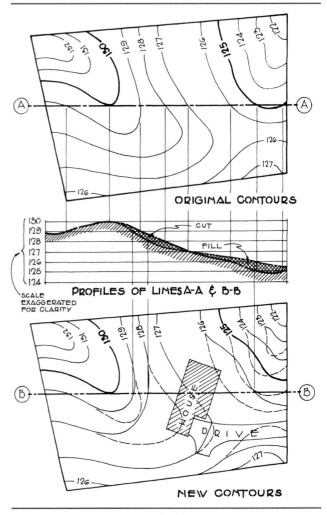

Figure 15–5 The method of plotting a contour profile and of indicating new contours on plot plans.

PREPARE A SITE ANALYSIS

An important predesign step is to conduct a comprehensive *site analysis*. The analysis focuses on existing as well as future conditions that may affect the site and the proposed residence. Extensive research is required to address all of the issues that may affect the site. Some information is easily obtained from the plat, the land registration office, the zoning office, the land surveyor, the utility companies, the city or county engineer's office, and weather centers. Other information can be assembled only from site visits, observations, and surveys at and around the vicinity.

As a first step, organize the site considerations into broad-scope headings. Second, list specific issues under the broad-scope headings. Sketches can be prepared to illustrate graphically the analysis data, such as those described in the following list:

- **Climate** Temperature highs and lows, average rainfall, average snowfall, relative humidity, prevailing winds, hurricanes, tornados, earthquakes, and the sun's orientation in summer and winter are all important climatic considerations (Fig. 15–6).
- **Circulation** Vehicular and pedestrian traffic patterns and frequency of movement are of prime concern. Automobile, truck, train, taxi, streetcar, bicycle, bus, aircraft, subway, and overhead rail

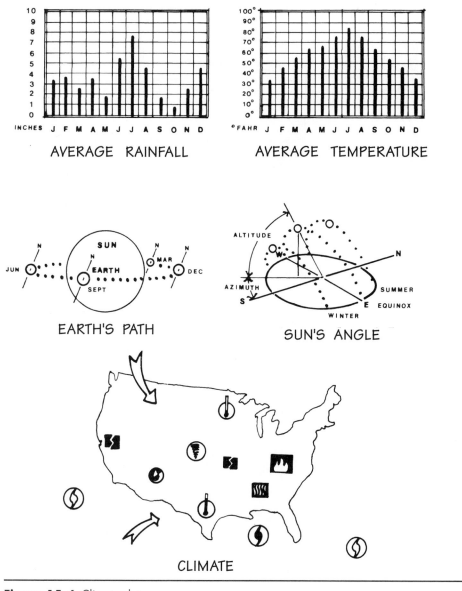

Figure 15–6 Climate data.

circulation patterns require study. Roads, sidewalks, and paths should be located on a site analysis drawing (Fig. 15–7).

■ **Neighborhood** The architectural, cultural, psychological, sociological, historical, ethnic, age, density, stability, recreational, safety, and economic factors of the neighborhood must be analyzed (Fig. 15–8).

■ **Natural elements** Trees, vegetation, topography, soil, rocks, lakes, ponds, rivers, streams, slopes, ridges, valleys, and drainage are some of the natural features present on many sites. These features become important site advantages or disadvantages (Fig. 15–9).

■ **Man-made additions** People often change the character of the land by adding fences, walls, pavements, streets, sidewalks, hard and soft landscaping, utilities, buildings, and outdoor furniture to the environment. Most such additions become major site considerations (Fig. 15–10).

■ **Sensory features** Views, sounds, and odors can repel or appease the senses. An effective site analysis should study views to and from the site in order to determine whether they are an asset or a liability. The same holds true for noises and odors within the neighborhood (Fig. 15–11).

■ **Zoning** Land utilization plans show a designated zoning or land use for all parcels of property in a community. Lot size, building density, setbacks, and type and size of building are determined by zoning (Fig. 15–12).

Composite site analysis diagrams can also be developed. These composite analyses combine a series of issues on a single graphic display (Fig. 15–13).

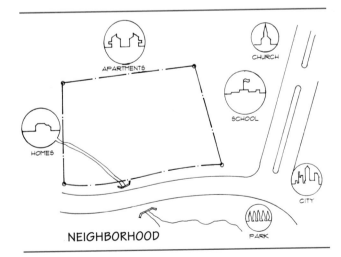

Figure 15–7 Circulation analysis.

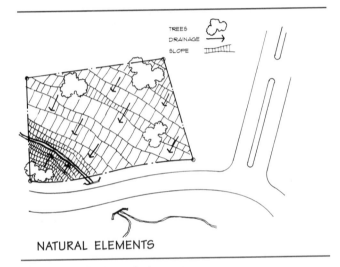

Figure 15–9 Natural elements.

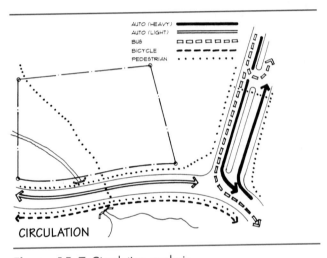

Figure 15–8 Neighborhood analysis.

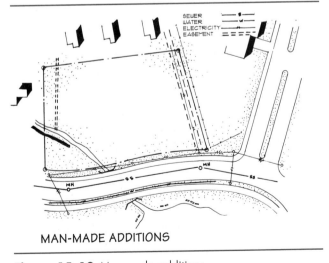

Figure 15–10 Man-made additions.

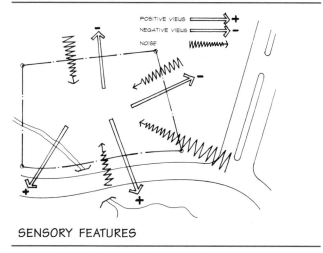

SENSORY FEATURES

Figure 15–11 Sensory features analysis.

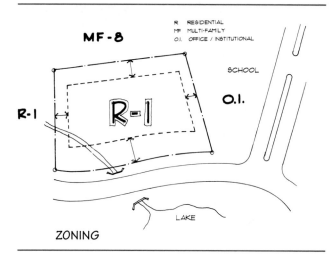

ZONING

Figure 15–12 Zoning analysis.

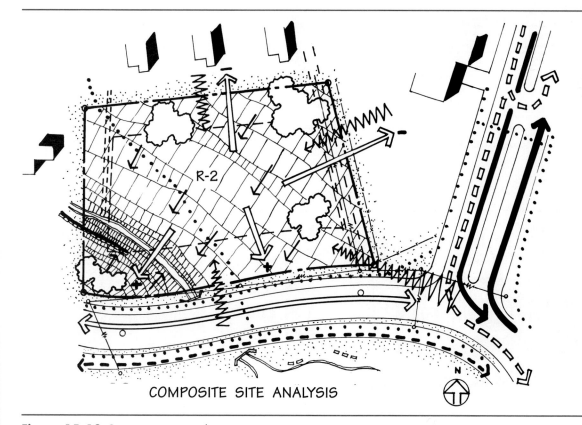

COMPOSITE SITE ANALYSIS

Figure 15–13 Composite site analysis.

PLAN THE HOUSE TO FIT THE SITE

After reviewing the legal restrictions and the site analysis of the lot, analyze the natural advantages of the area so that the house will become an integral and compatible part of its environment. Study the following points for their individual adaptability.

15.5.1 Plan a 1-Story House If You Have Enough Space to Spread Out

Houses on one level are easiest to build and maintain; there are no stairways that use up valuable floor space and constitute possible hazards for children and older people. Rambling 1-story houses are often referred to as *ranch type* (Fig. 15–14). If the contours of the lot will accommodate offset floor levels, plan a *split-level* house

(Fig. 15–15). If the space is limited, plan a *2-story house* (Fig. 15–16); build up instead of out. Less roof and less foundation will make the 2-story house more economical than a 1-story of comparative size. In areas where expensive land must be divided into very narrow lots, the best solution is *cluster houses* (Fig. 15–17). Their party walls afford more privacy than do facing windows 10′ or so apart on adjacent conventional houses. However, row houses can be considered only in tract or subdivision planning.

15.5.2 Make the House Part of the Land

By utilizing the entire lot for livability (Fig. 15–18), you can make even a modest-size home seem spacious if the inside is made to communicate with the exterior. Level areas adjacent to the house are the most inexpensive living space that your planning can provide. If possible, open the back to private outdoor areas by the use of glass and a view—easily created with landscaping if necessary. On

Figure 15–14 One-story house, ranch type.

Figure 15–15 Split-level house.

Figure 15–16 Two-story house.

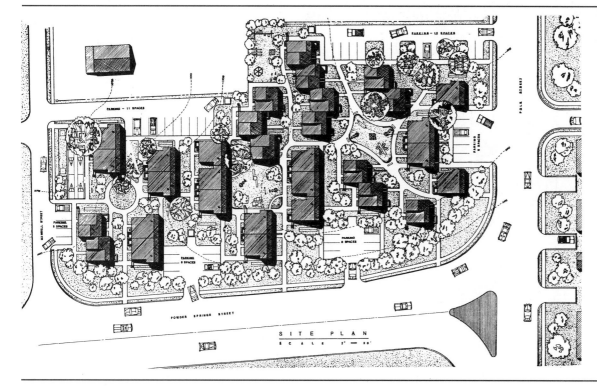

Figure 15–17 Layout of cluster houses.

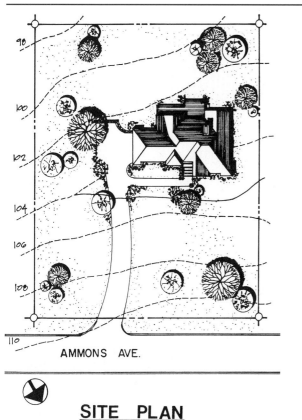

SITE PLAN

Figure 15–18 Landscaping sketch showing a good site utilization.

noisy streets, "turn your back" to the public by using few, if any, windows on the front.

15.5.3 Avoid Excessive Excavation

Usually, selection of the proper house type will minimize expensive excavation work. You will find that the less you disturb the natural contours of the land, the kinder nature will be to the future inhabitants of the home.

15.5.4 Orient the House to the Sun

Living areas are more pleasant with southern exposures. However, not all lots can easily accommodate this favorable exposure and still provide the most desirable view. On lots with the street side facing south, a walled patio near the front of the house may be the only method of obtaining a southern exposure and privacy as well (Fig. 15–19). To gain southern exposure on other lots, it may be necessary to turn the plan or place living areas on various sides of the building. With correctly designed overhangs and sunshades, window walls can harness the sun in summer and provide warmth and comfort within the house during the winter (Fig. 15–20). Sun charts, showing exact angles of the sun throughout various times of the day and year, should be consulted for accurate solar orientation of large glass walls, especially in warm cli-

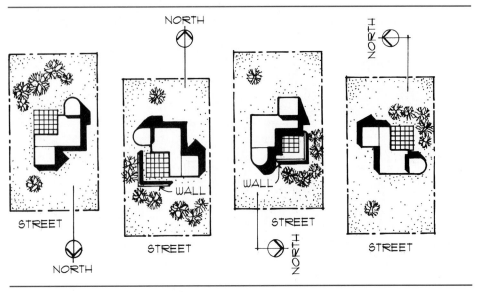

Figure 15–19 Orienting the house to the sun.

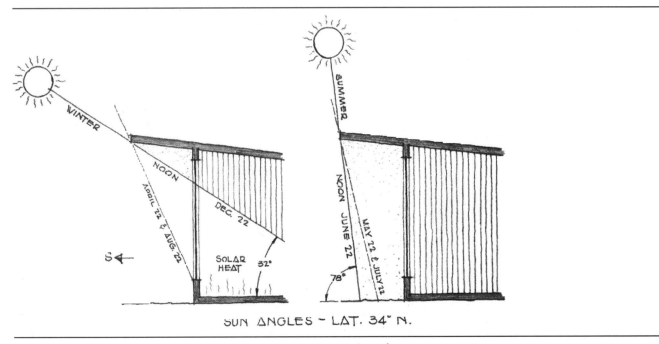

SUN ANGLES - LAT. 34° N.

Figure 15–20 Harnessing the sun using overhangs for year-round comfort.

mates. Plan the roof overhangs accordingly. Sunlight in moderation can be pleasant, but in excess it can be severe and damaging.

15.5.5 Make Maximum Use of the Existing Trees

Trees act as noise barriers and as shade producers during months of intense summer sun (Fig. 15–21). Shade on a house reduces inside temperatures considerably (as much as 10°) during the summer and reduces the maintenance of painted surfaces. Deciduous trees are especially effective; by dropping their leaves in winter, they allow desirable winter sunshine to penetrate a house when it is needed, and in summer they block the sun when it becomes destructive. For maximum utility, trees should be on the south side of the house. Evergreen trees, on the other hand, are more effective as windbreaks; they

1. Avoid the "lined up" look; vary the position and placement of the house on each lot.
2. Use curved streets; they prevent speeding.
3. Cut lots into a variety of shapes and sizes.
4. If possible use curved driveways; they are more pleasing than straight drives.
5. Control the view from windows by using plantings or walls to gain privacy.
6. Place living areas on the south side if possible; avoid a western exposure.
7. Combine a variety of housing styles, some 1-story, others 2-story, to prevent the uniform-roofline look.
8. Use deciduous trees on the south side for summer shade; use conifers for windbreaks.
9. Remember that a rolling terrain improves interest and ground line variety.
10. Plant in clusters to retain a natural look.
11. If possible, provide a turnaround drive so that cars will not have to back out into a busy street.
12. Place split-level houses on a sloping lot, a flat-roof house on a flat site.
13. Preserve natural contours to prevent harsh drainage problems.
14. To prevent obstruction of the view, avoid high plants near the intersection of the driveway and the street.
15. Provide greenery backgrounds for flowering shrubs.
16. Do not hide the front entrance; have plantings lead the eye to the door.

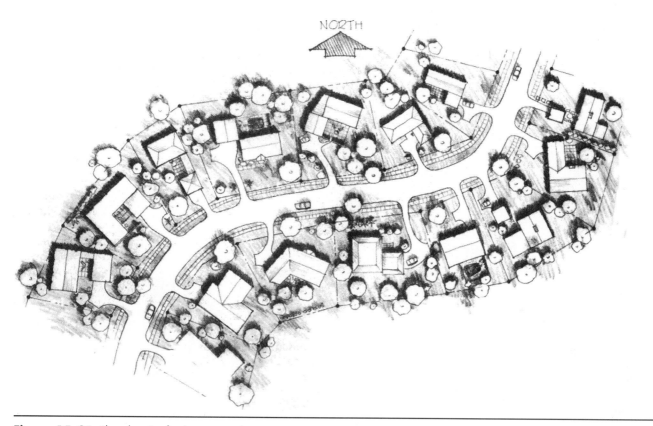

Figure 15–21 Plan the site for interest and privacy.

often make it possible to prolong the use of outdoor living areas well into the colder months of the year.

15.5.6 Consider Prevailing Winds

Through proper planning, breezes can be funneled through a house for additional comfort in warmer cli-

mates, and bitter winds can be diverted in colder climates (Fig. 15–22).

15.5.7 Keep Bedrooms and Quiet Zones Isolated

Bedrooms and other quiet areas should be isolated from noise-producing conditions. Cars, trains, play areas, and such, must be considered in house orientation.

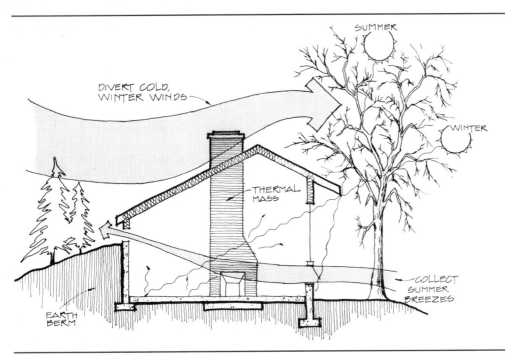

Figure 15–22 Using shape and orientation to provide natural ventilation in warmer climates and to divert bitter winds in colder climates.

15.5.8 Provide a Sense of Privacy

The design should provide a sense of privacy from possible observers on public streets or from surrounding buildings or property. The house should be a sanctuary where the occupants can relax and feel protected.

UTILIZE SOLAR PLANNING TO REDUCE ENERGY COSTS

During the mid-1970s, because of the oil shortage, the costs for heating and cooling homes in this country almost doubled. This fact placed a new emphasis on energy conservation and changed planning strategy in housing as well as in commercial structures throughout the nation.

It has been found that through careful planning in home construction, major savings in energy costs can be realized by capturing and controlling the heat from the sun. Many successful buildings have already been built that have verified the value of solar planning. Even adobe houses, built by the native Americans in the southwestern United States many years before the oil shortage, utilized heat from the sun to provide additional comfort within their dwellings (Fig. 15–23). Night-time temperatures in the area drop uncomfortably low, and daytime

Figure 15–23 Early adobe houses in the southwestern United States.

temperatures from the sun reach high levels; the massive adobe walls of their houses absorb the intense heat during the day and give it off from within during the night to maintain a more stable and comfortable temperature. Many of our contemporary homes are now keeping energy costs more manageable by the incorporation of solar concepts during their construction.

Although various solar collectors and distribution equipment are now available to builders to incorporate *active* solar features in solar homes, the most cost-effective savings can be realized by planning *passive* solar features into the construction. Passive features have few moving parts and no operating or maintenance costs, and the advantage is gained largely by the original design and the materials selected by the builder.

Planning for passive solar homes involves site orientation (Fig. 15–24), careful interior layout, sufficient glass for solar energy collection, proper storage of the energy until it is needed, and a means of energy distribution throughout the home. The use of superinsulating material and a tight shell to minimize heat loss is important.

To benefit from solar energy, plan the shape of the house so that large window areas are placed with southern exposure. Use deciduous trees on the south side, as previously mentioned, so that their leaves will provide shade during summer and will allow the sun to flood the glass during winter (Fig. 15–25A and B). Make use of evergreens for windbreaks from cold winter winds. Plan shade protection for the glass areas with accurately designed overhangs and shade louvers where necessary. Plan well-ventilated and light-colored roofs (Fig. 15–25C). Provide insulating glass in all windows (Fig. 15–25D and G). Plan some operable windows to take advantage of cooler breezes and temperatures, and specify adequate space within walls and roof to accommodate sufficient insulation (Fig. 15–25E). Place thermal-mass masonry features within the home where they can absorb the sunlight and store solar energy. For example, a massive masonry chimney within the house creates a flywheel effect by slowly taking on and giving off solar heat. Other features that can be used to store daytime solar energy are masonry floors of tile, brick, or concrete (Fig. 15–25D).

An interesting feature that can be useful in contemporary solar planning is a sunspace or a greenhouse (Fig. 15–25F) as part of the plan layout or attached to the house. These glassed-in areas provide ideal spaces for interior plants and greenery while collecting heat for human comfort. Although they are light and airy as part of the house, several problems must be overcome in their use. Depending on the extent of planting within, high levels of water vapor and window condensation may result. Also, considerable heat loss through the glass at night and during winter months can be expected unless provision is made for it. Even excessive heat gain during the summer months must be controlled with ventilation and sunshades. A sunspace provides a satisfactory way to heat a thermal-mass wall if the air circulation to the house proper can be controlled.

Another type of construction that might be considered for energy efficiency is an *earth-sheltered* home (Fig. 15–25H). This alternative is not inexpensive, yet it is an easy home to heat and cool. If parts of the design are placed within the earth, energy costs can be greatly reduced, the principle being that the earth itself maintains a more stable and moderate temperature than the air above. Therefore, less heat is needed in winter and less air conditioning in summer to maintain comfort temperatures for living. Site selection and orientation are ex-

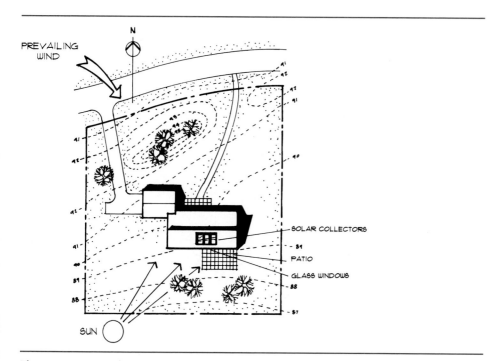

Figure 15–24 Solar planning involves correct house orientation.

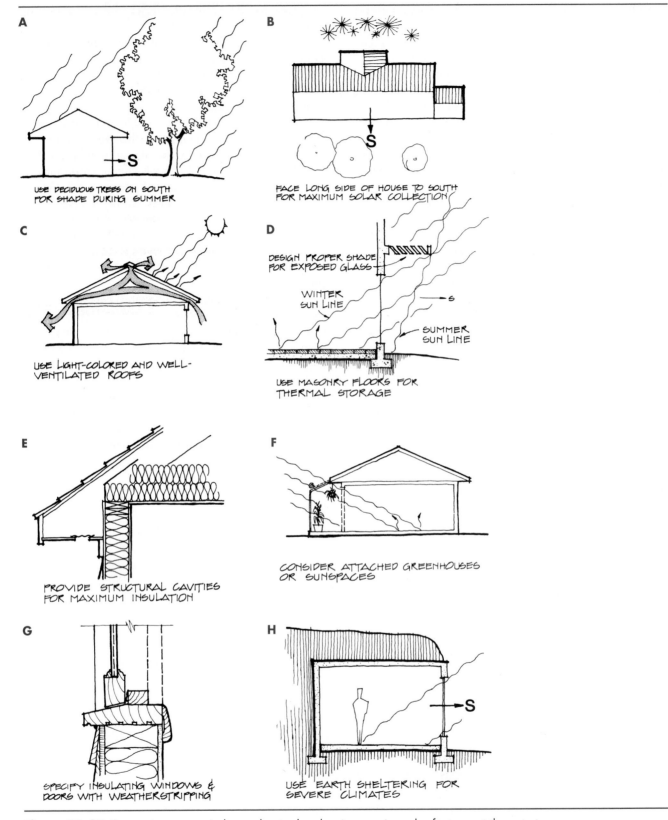

A

USE DECIDUOUS TREES ON SOUTH
FOR SHADE DURING SUMMER

B

FACE LONG SIDE OF HOUSE TO SOUTH
FOR MAXIMUM SOLAR COLLECTION

C

USE LIGHT-COLORED AND WELL-
VENTILATED ROOFS

D

DESIGN PROPER SHADE
FOR EXPOSED GLASS

WINTER
SUN LINE

S

SUMMER
SUN LINE

USE MASONRY FLOORS FOR
THERMAL STORAGE

E

PROVIDE STRUCTURAL CAVITIES
FOR MAXIMUM INSULATION

F

CONSIDER ATTACHED GREENHOUSES
OR SUNSPACES

G

SPECIFY INSULATING WINDOWS &
DOORS WITH WEATHERSTRIPPING

H

USE EARTH SHELTERING FOR
SEVERE CLIMATES

Figure 15–25 Conserving energy in homes begins by planning passive solar features at the outset.

tremely important. Several subtypes of earth-insulated houses might be considered. One is the atrium type, built around courtyards that give light and ventilation to the interior. Typically, it is usually entirely underground. This type fits a south-sloping lot well in cold climates, and a north-sloping lot in warm climates.

An earth-bermed type is one in which the windows and doors are placed in various locations in the walls with openings in the berms to allow for light and ventilation. Even the roof can be covered with earth in several of the types. This construction is dramatically effective in extremely cold climates where harsh winds prevail. Groundwater conditions and waterproof construction below grade (usually concrete) are premier considerations. Insulation must be placed below floor slabs and to the exterior of the walls and the roof. Other numerous problems must also be solved, such as draining exterior below-grade areas to outfalls, providing natural lighting to all interior living spaces, and selecting durable building materials.

If suitable, use a masonry Trombe wall, which can be glazed and ventilated to bring inexpensive heat into habited spaces. (Fig. 15–26). Water is another convenient medium to store heat within. To supplement other storage features, use water tubes or tanks placed near south-exposed windows to soak up the energy and store it for later use (Fig. 15–27).

Another economical heat-storage feature, if rocks are available near the site, is a bin filled with rocks placed underground below the house (Fig. 15–28). Solar-heated air circulated through the bin will give off its heat to the rocks; then air from within the house, when circulated through the bin, will bring the stored heat to where it is needed. Forced-air equipment (an *active* solar feature) must be installed with this arrangement.

With the continuing rise of energy costs, it is only logical that more solar planning should be included in all future home construction. Many houses will utilize many of the manufactured components, such as efficient solar collectors, air-circulating equipment, and other energy-saving features that are now available to home builders. These features include the active solar components, but wise planners will surely make the best use of the passive features mentioned here during the initial construction of future homes. Many of the solar features can also be applied to the remodeling of older homes.

15.7

CHOOSE THE MOST APPROPRIATE BASIC STRUCTURE

There are many basic structures that a designer must consider and analyze before deciding on the most suitable. Each of the following has both advantages and disadvantages, but after considering the needs carefully, the de-

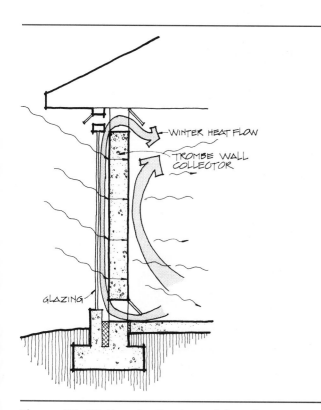

Figure 15–26 Use of a Trombe wall for solar energy collection.

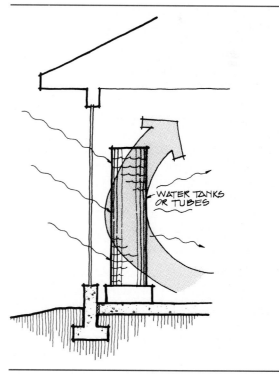

Figure 15–27 Water-filled tanks or tubes near south-exposed glass will store solar heat.

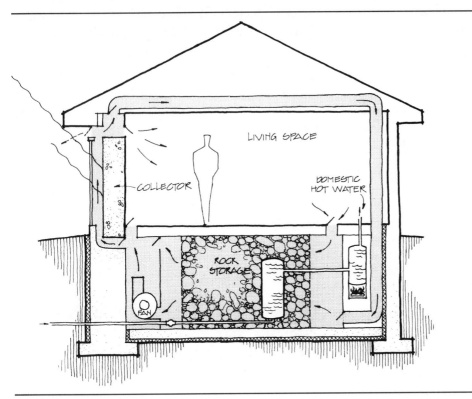

Figure 15–28 A rock bin underground can be used to store solar heat when air is circulated through it.

signer will know the one type that will prove to be the most appropriate.

15.7.1 One-Story with a Pitched Roof and Flat Ceilings

This popular house type (Fig. 15–29) is built with conventional rafter and ceiling joist framing methods, or the roof framing can be quickly erected with prefabricated roof trusses. It lends itself easily to prefabrication, which is gaining in acceptance throughout the United States. It also has the advantage of having its living areas on one level. This type of house can be built with or without a basement, with a crawl space or on a concrete slab; or it can be used in conjunction with multilevel designs to produce many of our traditional home styles.

15.7.2 One-Story with a Pitched Roof and Sloping Ceilings

This type (Fig. 15–30) has evolved from the plank-and-beam type framing, and its popularity is due mainly to the massive exposed beams and sloping ceiling in the interiors, as well as the interesting architectural effects produced by the extensions of the beams on the cornice exteriors. The type looks good with large overhangs, which produce expansive, unbroken roof lines. Many exciting examples of this structural type can be found in contem-

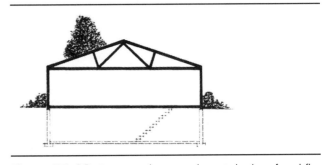

Figure 15–29 One-story house with a pitched roof and flat ceiling.

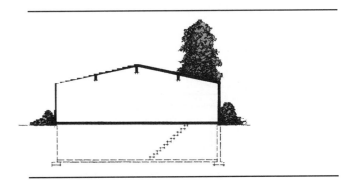

Figure 15–30 One-story house with a pitched roof and sloping ceilings.

porary designs. If a sloping ceiling with air space above for insulation is required, use the *scissors truss* as shown in Fig. 15–31. Gypsum board or other ceiling material can be fastened to the lower chords of the trusses.

15.7.3 One-Story with a Flat Roof and Flat Ceilings

With built-up roofing materials comparing favorably with conventional roofing, this type (Fig. 15–32) has become definitely contemporary in nature and is a very economical type of construction. (The same structural members are used for both the ceiling and the roof.) It requires skillful design of its elevation views to overcome a boxy appearance, and its strong horizontal roof lines often appear monotonous unless broken with offsets or various roof levels. The flat roof can be combined effectively with the shed-type roof (see the dashed lines in Fig. 15–32), or it can be used with gable roofs as carports, garages, or other attached wings for variety of roof lines.

15.7.4 One-and-One-Half-Story with Two Living Levels

This structural type (Fig. 15–33) has less living space on the top level. It is associated with the New England Cape Cod house. The basic simplicity of the true Cape Cod design should be retained, in proportion and in detail, if an approach to historical authenticity is desired. Although the shape is traditional, many contemporary-type houses have resulted from the use of the steep roof and various adaptations of the story-and-one-half principle.

15.7.5 Two-Story with Two Living Levels: Pitched Roof and Flat Ceilings

Numerous 2-story houses are rich in traditional charm (Fig. 15–34). Their heritage can be captured only by the correct proportion and detail usually associated with many of the New England houses. Either gable or hip

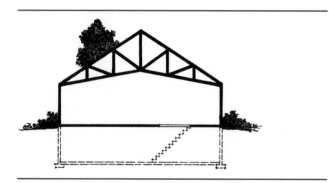

Figure 15–31 One-story house with a pitched roof and pitched ceilings, which allow the use of more insulation within the scissors truss area.

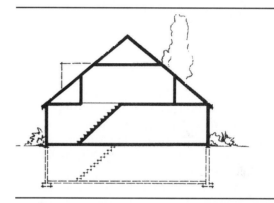

Figure 15–33 One-and-one-half-story house with a steep roof that allows only partial living space on the second level.

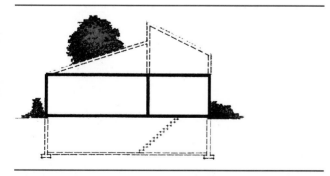

Figure 15–32 One-story house with a flat or shed-type roof and ceilings.

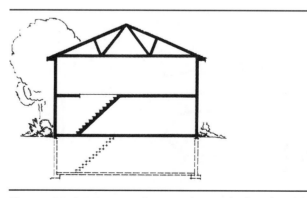

Figure 15–34 Two-story house with a pitched roof and flat ceilings.

roofs of various slopes can be used with the 2-story design; often lower-roofed additions to the main box seem to enhance the overall silhouette. Upper-story projections (see Fig. 18–15) not only increase the second-story living space but also tend to reduce the strongly vertical appearance.

15.7.6 Split-Level with Three Living Levels: Pitched Roof and Flat Ceilings

The majority of comparatively new split-level houses (Fig. 15–35) have three living levels and an optional basement level. All levels are connected with stair segments that combine in pairs to make one full-story height. Multilevels provide excellent zoning for different activities, and their exteriors can be given interesting variety. On sloping lots the split-level can be oriented to the site and still have full-story height exposure to outdoor living areas. Stair arrangements can become complicated, requiring considerable study to develop a successful design.

15.7.7 Split-Level with Three Living Levels: Continuous-Sloping Roofs and Various Ceilings

This second type of split-level (Fig. 15–36) is identified by low-pitched roofs (usually 2:12 or 3:12 slopes) that continue over several levels to create sloping ceilings in various upper and middle rooms. The continuous ceilings produce several ceiling heights, thereby adding interest to the interior living spaces. Roof slopes and floor levels must be carefully controlled to result in satisfactory wall heights and proper proportions. Both types of split-levels are similar in livability and popularity.

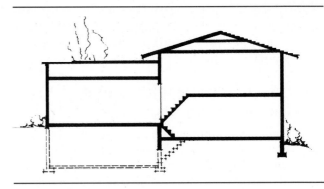

Figure 15–35 Split-level house with three living levels: pitched roof and flat ceilings.

USE A MATRIX TO ANALYZE SPACE RELATIONSHIPS

A *matrix* is a convenient aid in developing and understanding the relationships among the various spaces in a residence. First, as the drafter, you must prepare a list of all of the rooms, spaces, or areas that are essential to the design program. Then you must lay out a two-dimensional grid to form the matrix. The grid may be drafted in a format where the room names are listed in both a horizontal column and a vertical column. This system results in a square grid where all lines drawn in the matrix are perpendicular to each other (Fig. 15–37). Another option is to develop the matrix by listing the room names in a vertical column and drafting the grid with 45° lines (Fig. 15–38). The latter system eliminates the duplication of coding each space twice when completing the matrix.

A code must be established to differentiate between the relationships of adjacency of spaces listed on the matrix. A common scale used in defining the code, for ex-

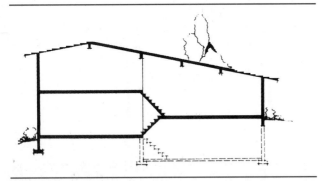

Figure 15–36 Split-level house with three living levels: pitched roof and various ceilings.

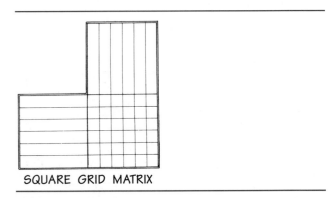

SQUARE GRID MATRIX

Figure 15–37 Square grid matrix.

DIAGONAL GRID MATRIX

Figure 15–38 Diagonal grid matrix.

ample, could be in a range through *essential, desirable, neutral,* and *unimportant* to *detrimental*. A graphic symbol is assigned to each term in the code (Fig. 15–39).

To complete the matrix, the user reads the names of the rooms and places a graphic symbol, which represents the relationship of adjacency between the two spaces, on the grid. For instance, in a residence, it would be essential for a foyer to be adjacent to a coat closet. The symbol for *essential* would be coded into the matrix where the grid lines for foyer and coat closet cross (Fig. 15–40 A and B). When completed, the matrix reveals the relationship of adjacency between all spaces in the program. Patterns will surface that indicate a common bond among several spaces, which can be clustered in close proximity.

15.8.1 Use Zone Diagrams to Establish Major Elements of Design

The clusters that become obvious from the matrix can be transformed to a zone diagram (Fig. 15–40C). Several zone diagrams could be developed for most residences. Living space used for family activities and entertaining guests could be zone 1. Eating areas where meals are prepared and enjoyed by family members and guests could be zone 2. Sleeping areas used primarily by the family

CODE

ESSENTIAL ●

DESIRABLE ◐

NEUTRAL ⊖

UNIMPORTANT ⦶

DETRIMENTAL ○

Figure 15–39 A common scale should be used to code the matrix.

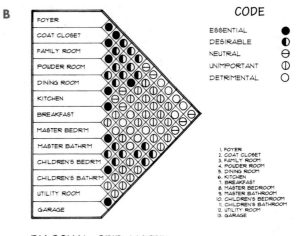

SQUARE GRID MATRIX

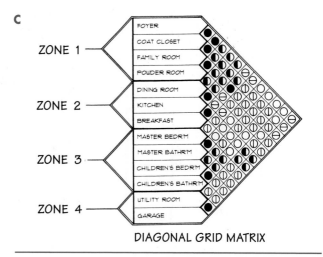

DIAGONAL GRID MATRIX

Figure 15–40 Matrix completed for a residence.

could be zone 3. To illustrate component spaces in zone 1, cluster the foyer, living room, and family room together. An example of zone 2 would be a grouping of the kitchen, dining room, and breakfast room. Zone 3, representing sleeping spaces, might include bedrooms, baths, and closets. Arrows are drawn to represent circulation among zones (Fig. 15–41).

Zone diagrams are useful tools in developing the mass and form of a residence. A cluster of space could easily represent a wing or a floor level (Fig. 15–42).

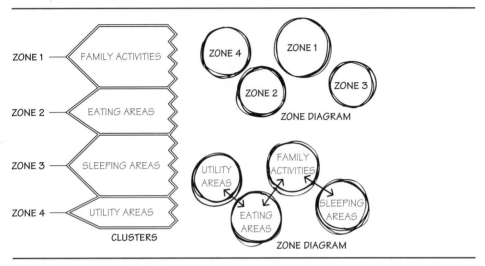

Figure 15–41 Zone diagrams.

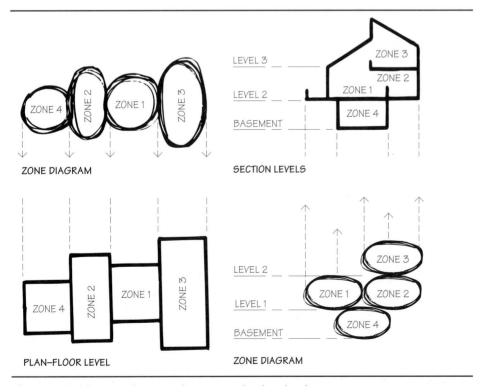

Figure 15–42 Using the zone diagram to develop the design.

15.8.2 Use Bubble Diagrams to Establish Adjacency Requirements

Although zone diagrams are useful, the main purpose of the matrix is to establish essential relationships between and among rooms. Once the degree of adjacency has been determined, the drafter can begin sketching *bubble diagrams*. Freehand circles or bubbles are drawn to represent each of the rooms listed in the matrix. The bubbles are placed in proximity according to their degree of adjacency indicated on the matrix. Arrows are used to depict circulation between and among spaces (Fig. 15–43).

Drafters frequently study several complete bubble diagrams before deciding which one best meets the criteria of the project. Once the appropriate diagram is determined, the drafter begins schematic site and floor plan studies. A good bubble diagram serves as the foundation for the development of a successful preliminary floor plan (Fig. 15–44).

15.9

PLAN THE INTERIOR SPACES TO REFLECT THE LIFE STYLE OF THE OCCUPANTS

Understanding the life style of the occupants of a home is important for effective interior space planning. Some occupants prefer casual living, whereas others desire a more formal life style. Analyze the needs of the client before preparing preliminary sketches.

Regardless of the client's life style, certain fundamentals are essential to all residential planning. Successful interiors have a definite relationship to successful exteriors.

The character of a traditional-style home should be reflected in both its exterior and its interior design. The same statement can be made for contemporary design. Frank Lloyd Wright (1867–1959) was a modern architect who

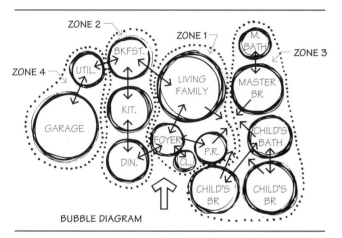

Figure 15–43 Bubble diagram sketches.

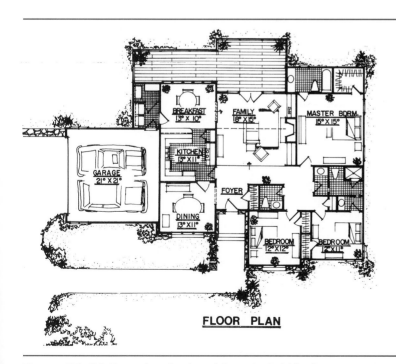

Figure 15–44 Floor plan developed from the bubble diagram in Figure 15–43.

masterfully blended exterior and interior forms and materials (Fig. 15–45). Thomas Jefferson (1743–1826), the colonial designer, tastefully integrated classical elements of design on the interior and the exterior (Fig. 15–46).

Monticello

Monticello

Figure 15–46 Thomas Jefferson's blending of classical elements between interior and exterior design. (Monticello, Thomas Jefferson Memorial Foundation)

Fallingwater

Fallingwater

Figure 15–45 Frank Lloyd Wright blended elements between interior and exterior design. (Western Pennsylvania Conservancy)

15.9.1 Allow the Correct Amount of Space for Each Major Activity

From a general standpoint, you can think of the basic activities in a home as being *eating*, *sleeping*, and *living*. Give each activity its proper allotments of space, according to the living habits of the inhabitants and the size of the plan. Use precious space generously where it counts most—in living and entertainment areas. Restrict floor space in individual-activity areas, such as kitchens and baths, where careful planning can make them adequately efficient. Living rooms set the interior atmosphere of the home and should not convey the impression of confine-

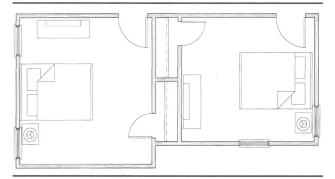

Figure 15–47 Using closets to prevent sound transmission between rooms.

Table 15-1 Typical room sizes (inside dimensions).

Living	14' × 20'
Dining	12' × 14'
Kitchen	8' × 16'
Bath	5' × 9'
Master bedroom	12' × 15'
Bedroom	11' × 13'
Entrance hall	7' × 8'
Powder room	4' × 5'
Utility	8' × 12'
Hall width	3'-6" to 4'
Stairs width	3'
Garage (single)	12' × 20'
Garage (double)	20' × 20'
Workshop	12' × 14'

ment. Isolate quiet zones—bedrooms, studies, and the like—from noisy family and recreation rooms. If possible, use closets for sound-barrier walls; a closet full of clothes is the simplest method of preventing sound transmission between rooms (Fig. 15–47).

15.9.2 Make Areas Serve Dual, Nonconflicting Purposes

This principle enables maximum space utilization and a flexible plan. Consider whether any of the following typical combinations might gain additional living space in the plan:

1. Living–dining
2. Living–family
3. Kitchen–laundry
4. Bath–laundry
5. Garage–laundry
6. Hall–laundry
7. Garage–workshop
8. Large bedroom with room divider
9. Carport adjacent to terrace

Plan the size of each room to fit the furniture to be used in it, especially the dining room and the bedrooms. Minimize partitions, if possible, by eliminating offsets and by keeping partition walls in line with each other. Unnecessary corners are expensive, and they make furniture arrangement difficult.

Refer to Table 15–1 to help you size the various home areas.

15.9.3 Keep Major Traffic Routes Short

Consider day-to-day activities so that the traffic patterns are minimized and hallways do not deprive the plan of important living space. Time and space, as well as maintenance of floors, can be saved by efficient traffic circulation. Traffic through rooms (especially through the cen-

ter) is a characteristic of many older houses, which were designed around formal exteriors; little concern was given to the routes between different activity areas. If the main entrance of a plan is placed near the center (Fig. 15–48), major traffic patterns can fan out in each direction and thus increase usable space within rooms and reduce hall lengths. The route from the carport or garage to the kitchen should definitely be short. If traffic must pass through a room, arrange the doors so that the traffic affects only a corner or an end. Often the correct placement of doors will reduce the length of routes. Avoid major traffic through an activity area, such as a kitchen work center. The activities of some families require many rooms, even if they are small; other families prefer fewer and larger rooms—commonly known as *open planning*. To sum up, to achieve good traffic flow in your plan, provide convenient relationships between and among rooms,

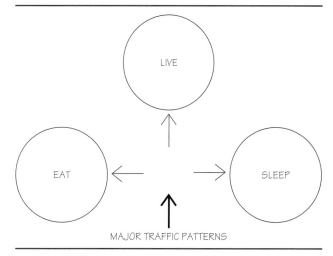

Figure 15–48 Defining major traffic patterns.

minimize traffic through rooms, and provide access to outdoor living areas (Fig. 15–49).

Good design begins with an identifiable entry. The front door to a residence should be easily accessible and inviting to both occupants and guests. Make the foyer or entrance hall a transitional space from the exterior to the interior. With minimum use of floor space, make the foyer large enough to remove outer garments and yet provide a sense of orientation for visitors. Place the coat closet near the front entrance and adjacent to the foyer. The foyer should set the architectural character for the home and serve as a space from which other areas of the home are accessible (Fig. 15–50).

All homes need both private and common spaces. Bedrooms and bathrooms, for example, are private spaces. Living rooms, dining rooms, and kitchens are common spaces. A good plan respects the needs of the occupants by recognizing the need for common space and private space. Clearly defined circulation among the areas of a home is most important.

15.9.4 Common Spaces Should Be Accessible from the Foyer

Since all occupants of the home and many visitors will use common spaces, these spaces should be accessible from the foyer. It is desirable to be able to move directly from the foyer into a living room or a family room. Larger homes may have both a living room and a family room. Living rooms are generally more formal in style and are frequently used for entertaining guests. The living room is usually located in the front of the house with windows facing the street. The size of the living room can be determined by the furnishings. A sofa, several lounge chairs, end tables for lamps, a coffee table, and a bookcase are

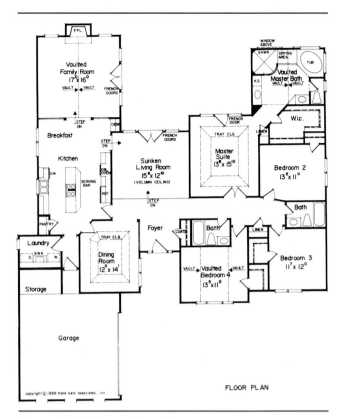

Figure 15–50 An efficient foyer.

usual furnishings that must be accommodated into the layout. The shape of the room must be considered as well so that conversation groups can be arranged with as much flexibility as possible (Fig. 15–51).

A family room is a common space where members of a family gather for entertainment, reading, and conversing. Most family rooms are places for relaxation.

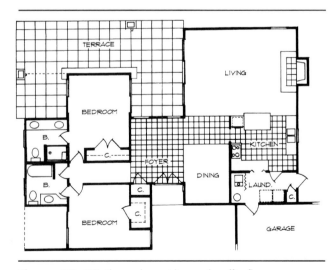

Figure 15–49 Floor plan with good traffic flow.

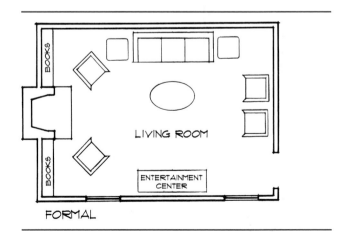

Figure 15–51 Placing furnishings in the living room.

The furnishings are usually casual, consisting of a sofa, lounge chairs, a coffee table, end tables, lamps, bookcases, entertainment centers, and game tables (Fig. 15–52). Placing the family room at the rear of the house with pleasant views and circulation to the exterior is desirable. The kitchen and dining space should be accessible to the family room (Fig. 15–53).

15.9.5 Bedrooms Are Private Spaces

Bedrooms are private areas generally occupied by one or two people. North, east, or south exposures are mainly desirable for bedrooms. The occupant of a bedroom may have a preference for orientation. Bedrooms on the north are usually cooler. The morning sun will brighten a bedroom with windows toward the east. A south exposure will allow the sun to warm the room during winter months. Due to high temperatures and direct solar radiation from the afternoon sun in the summer, it is not desirable to place bedrooms on the west.

Natural light and ventilation are required by building codes in all bedrooms. Consult local codes to verify specific requirements. Furnishings can vary from bedroom to bedroom; therefore, space requirements vary. Be sure that you are familiar with the actual sizes of twin beds, bunk beds, full-size beds, queen-size beds, and king-size beds before sizing the bedrooms. A chest, dresser, night tables, chairs, and desk are common bedroom furnishings (Fig. 15–54). Closets and bathrooms should be easily accessible to bedrooms. Provide suitable space near the closets that will serve as a convenient dressing area yet still offer access to the bed and other major furnishings (Fig. 15–55).

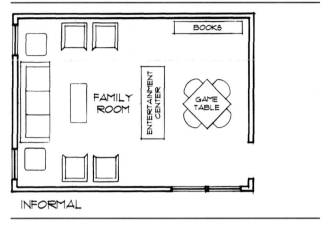

Figure 15–52 An efficient family room.

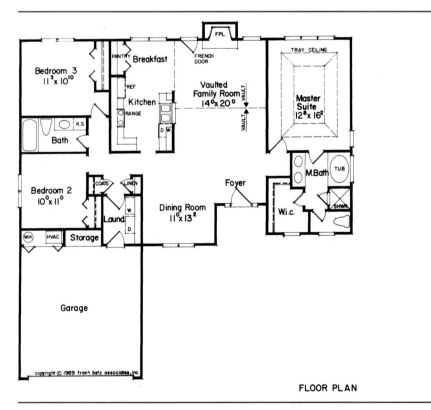

Figure 15–53 Relationship among family room, kitchen, and dining area.

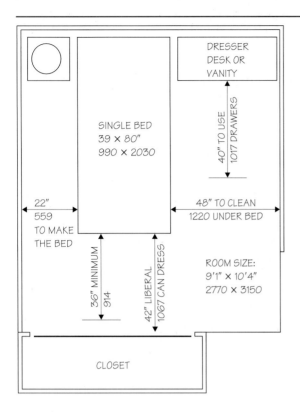

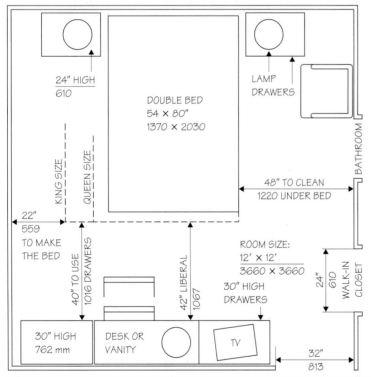

Figure 15–54 Placing furniture in bedrooms.

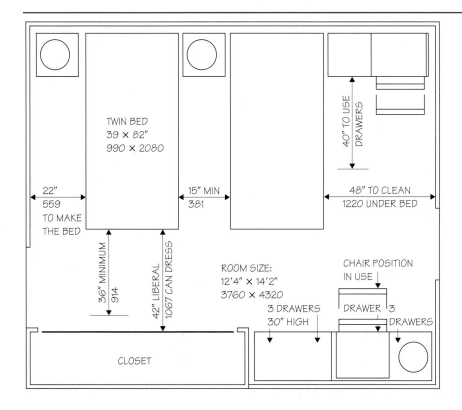

TWIN BED
39 × 82"
990 × 2080

40" TO USE DRAWERS

22"
559
TO MAKE
THE BED

15" MIN
381

48" TO CLEAN
1220 UNDER BED

36" MINIMUM
914

42" LIBERAL
1067 CAN DRESS

ROOM SIZE:
12'4" × 14'2"
3760 × 4320

CHAIR POSITION
IN USE

3 DRAWERS
30" HIGH

DRAWER 3
DRAWERS

CLOSET

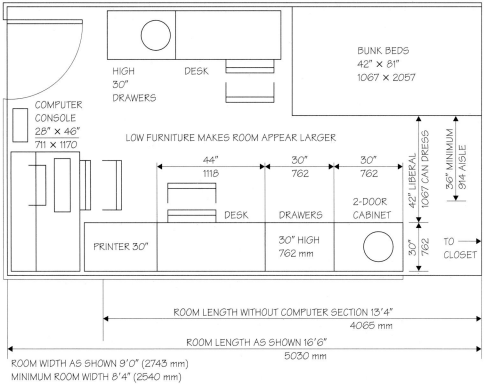

HIGH
30"
DRAWERS

DESK

BUNK BEDS
42" × 81"
1067 × 2057

COMPUTER
CONSOLE
28" × 46"
711 × 1170

LOW FURNITURE MAKES ROOM APPEAR LARGER

44"
1118

30"
762

30"
762

42" LIBERAL
1067 CAN DRESS

36" MINIMUM
914 AISLE

DESK

DRAWERS

2-DOOR
CABINET

30"
762

PRINTER 30"

30" HIGH
762 mm

TO
CLOSET

ROOM LENGTH WITHOUT COMPUTER SECTION 13'4"
4065 mm

ROOM LENGTH AS SHOWN 16'6"
5030 mm

ROOM WIDTH AS SHOWN 9'0" (2743 mm)
MINIMUM ROOM WIDTH 8'4" (2540 mm)

Figure 15–54 (continued)

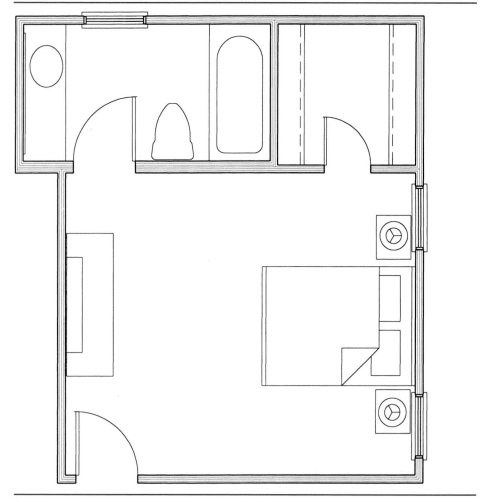

Figure 15–55 Provide space for dressing in the bedroom.

15.9.6 Bathrooms Are Essential for Health Care

Bathrooms are essential for the health care of the occupants of a home. These spaces are private areas where the individual can practice personal hygiene. The number of bathrooms should be determined by the size of the residence and the number of its occupants. All homes should have one full-size bathroom with essential plumbing fixtures. Many residences have more than one bathroom, and some have a half-bath or powder room. Two-story homes should have a bathroom on each floor.

Bathrooms in American homes are famous for their conveniences (Fig. 15–56). The half-bath usually has a lavatory and a water closet. The full bath, consisting of a lavatory, a water closet, and a tub, can be economically arranged in an 8'-0" × 5'-0" space (Fig. 15–57). Place the standard 5' tub across the end so that the other fixtures will fit along one plumbing wall. A 4" vent stack is nec-

essary in the wall near the water closet, which is usually placed between the other fixtures. To accommodate the vent stack, an 8" wet-wall is generally indicated. To shorten the plumbing runs, put baths back to back, and plan, if possible, a compact arrangement for the baths, sink, washer-dryer, and water heater (Fig. 15–57). However, do not sacrifice a good plan for the extra cost of separating the baths or other plumbing.

Locate the doors in bathrooms to give maximum privacy and isolation. Often the bathroom door is left open, and direct views from living areas are undesirable. Minor rearrangements of bathroom positions in relation to halls usually eliminate this situation. Doors should swing into the bathroom (never into halls) without hitting bathroom fixtures. Use narrow doors; usually 2'-0" to 2'-6" widths are sufficient. By isolating the water closet or tub into compartments, you can make the bathroom useful to more than one person simultaneously. With children in a home, twin lavatories will reduce peak demands on the bathroom.

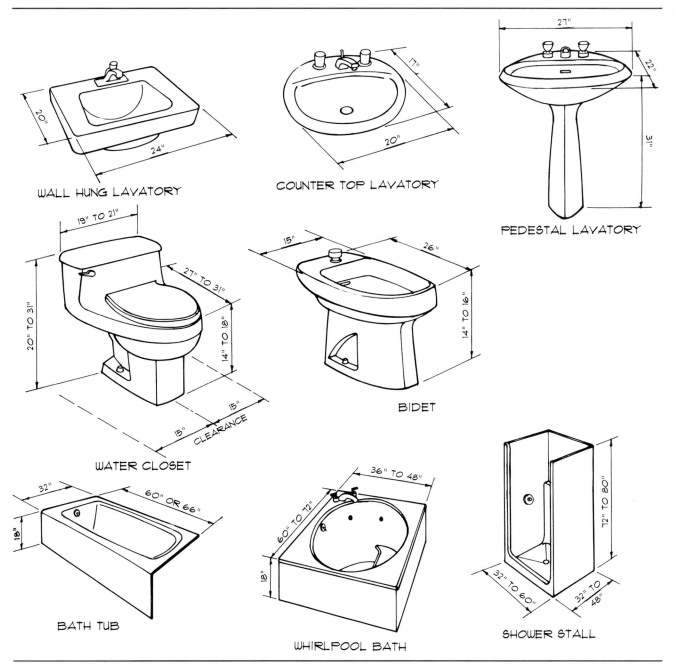

WALL HUNG LAVATORY

COUNTER TOP LAVATORY

PEDESTAL LAVATORY

WATER CLOSET

BIDET

BATH TUB

WHIRLPOOL BATH

SHOWER STALL

Figure 15–56 Standard bath fixture sizes.

Figure 15–57 Compact plumbing arrangement.

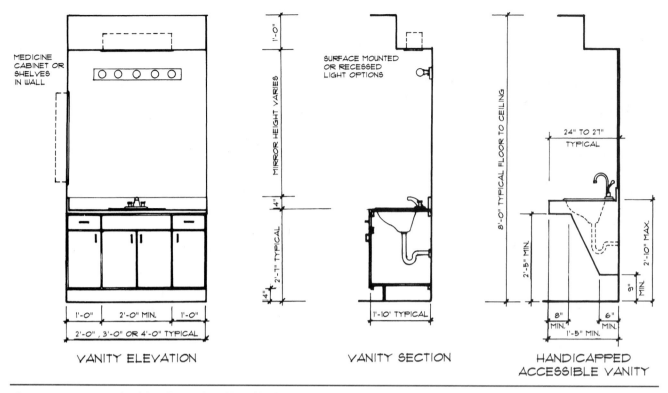

Figure 15-58 Standard heights and widths of bathroom vanities.

Built-in lavatory cabinets (Fig. 15–58) will result in more storage space and will allow countertop areas for grooming use. Large mirrors attached to the wall directly above the cabinets will further facilitate grooming and will also give the room a more spacious appearance. Do not center medicine cabinets directly over the lavatory; instead, use a mirror or a window above.

Many master bedrooms have accessible private baths and adjacent dressing facilities (with generous closets), resulting in complete master suites (Fig. 15–59). Shower stalls are preferred by some people and require less floor space than tubs. A shower should be a minimum size of 2'-6″ × 2'-6″. Tub and shower combinations are very popular. Versatility can be accommodated by the use of the tub or the shower in the same plumbing fixture. Luxury bathrooms may have both a tub and a separate shower stall. The use of a whirlpool bath or an exotic-shaped tub is common in many custom homes. The bidet, a plumbing fixture used in Europe for years, is now being manufactured in America and can be found in some modern bathrooms.

Ventilation is another important aspect of bathroom design. Forced-air exhaust fans are used in windowless bathrooms. Even bathrooms with windows should have an exhaust fan for more positive ventilation to the outside. Floors and wainscoting in bathrooms should be fin-

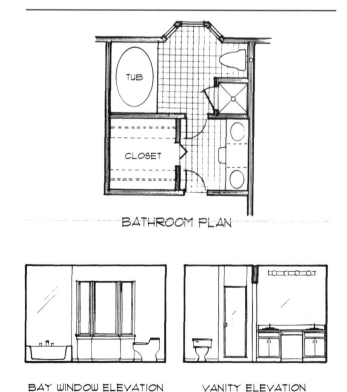

Figure 15-59 A combination dressing and bath area.

ished with durable, easily maintained materials that produce sanitary and pleasant interiors. Marble, tile, and vinyl are common flooring materials. Lighting should be soft and adequate, especially in front of mirrors.

15.9.7 Plan for Adequate Storage Space

Adequate storage space must be provided not only for clothes storage but also for the many items required in a modern home. Design storage space to fit the items, place it where it is needed, and make it easily accessible.

Many types of modern closet-door units are available for built-in storage spaces. Sliding doors eliminate swinging space needed within a room, and full, 8' high closet doors provide maximum storage space above clothes rods. Walk-in closets are satisfactory if clothes rods can be installed along two long walls; otherwise, they are inefficient (Fig. 15–60). Avoid a narrow closet with a door in the narrow end; the storage space has poor accessibility. Lining part of a carport or a garage with 1/4" pegboard provides inexpensive storage for garden tools. Built-in cabinets installed above the hoods of stored cars make efficient use of otherwise wasted space.

Diagonal areas below stairways can be devoted to storage. Often the attic can be utilized for dead storage by introducing a *scuttle*, or disappearing stair unit, in a convenient hall ceiling. If a basement is planned, a considerable amount of its space can be reserved for storage.

However, according to experts, in most cases basement storage space is more expensive than above-ground storage space. The average home must provide storage for the following items (excluding kitchen equipment): clothes, linens, cleaning equipment, toys, hobby and sports equipment, sewing needs, guests' coats, tools, and lawn and garden equipment. Well-designed storage units make for a better organized home and thereby help to simplify housekeeping.

15.9.8 Plan a Functional Kitchen

People who cook spend much time in the kitchen, planning meals, storing food, cooking and serving, and cleaning after meals. First of all, the kitchen should be easy to work in, conveniently located, and pleasant in appearance to make these tasks no more difficult than necessary. Unlike furniture, kitchen cabinets and appliances are permanently installed and cannot be rearranged by merely pushing them about to rectify planning errors. Kitchens can be placed at the front of the house, to be close to the garage and front door; or they can be placed in the back, to be accessible to outdoor eating and play areas.

Kitchens serve various purposes besides food preparation; many are designed to furnish room for snacking and informal meals. Others incorporate informal living and play areas for children. Some are even combined with entertainment centers. Many individuals, on the other hand, prefer a completely independent kitchen that can

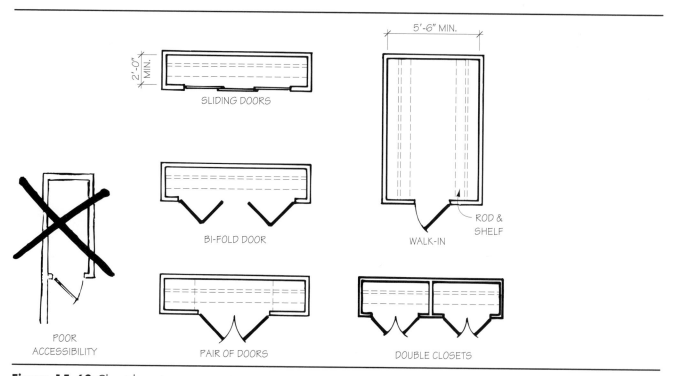

Figure 15–60 Closet layouts.

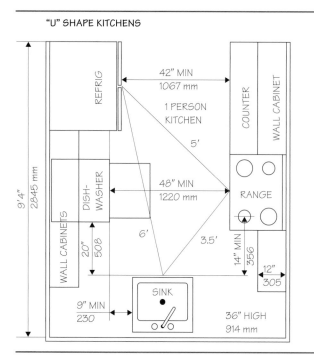

Figure 15–61 The efficiency of a kitchen depends on the size of the work triangle.

be closed off from other areas when cooking and eating are completed.

The layout of the actual work center of the kitchen deserves careful attention. The basic activity of this area is controlled largely by the placement of the sink, the cooking unit, and the refrigerator. If we were to connect these three major appliances with a triangle on the plan, it would represent the bulk of the travel within the kitchen; it comprises what is known as the *work triangle* (Fig. 15–61). Keep the perimeter of this triangle from 12′ to 20′ in length if you want an efficient kitchen; if it becomes longer, a rearrangement of the basic appliances should be made. Before progressing too far with the planning of the cabinets and other features, check manufacturers' literature for the actual size of the appliances to be included in your kitchen (to be sure that they will fit). According to the arrangement of the cabinets and appliances, kitchens can be made in these basic shapes: U shape, L shape, two-wall or corridor type, and the modest one-wall type (Fig. 15–62). Tests have shown the U shape to be the most efficient.

Occasionally, a peninsula or an island cabinet is incorporated into the L shape or the U shape to isolate the work center or to utilize excessive space within large kitchen ar-

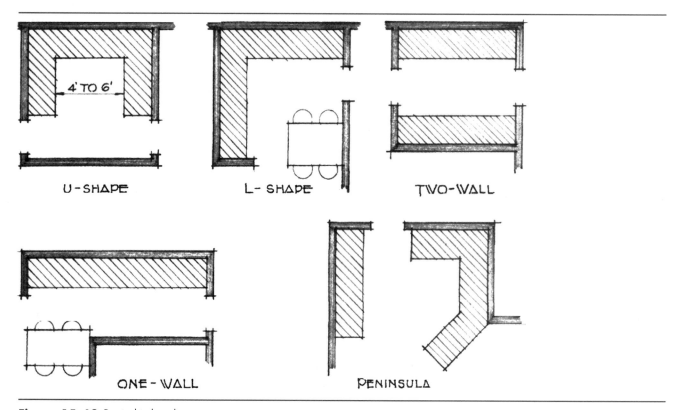

Figure 15–62 Basic kitchen layouts.

CORRIDOR KITCHENS

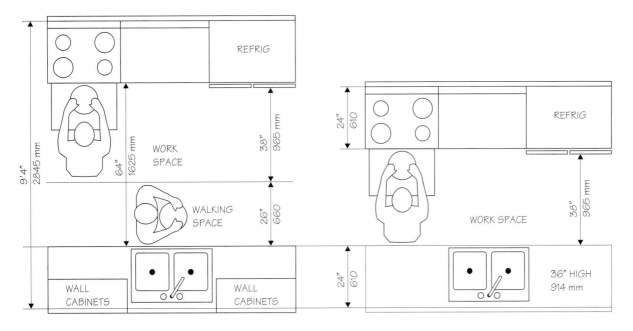

"L" SHAPE KITCHENS

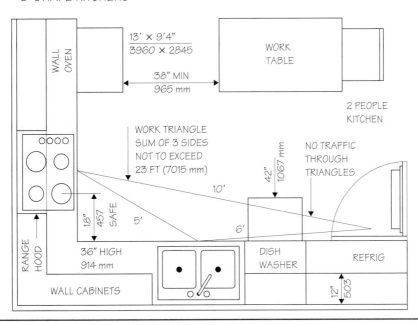

Figure 15–62 (continued)

eas. The ideal arrangement is to have the major appliances efficiently placed with continuous cabinets between and sufficient countertop space on each side of the appliances for satisfactory work surfaces (Fig. 15–63). Whether or not the kitchen is used for eating, sufficient countertop surface must be provided for many activities: menu planning, actual food preparation (mixing, chopping, and so on), cooking, serving, and cleaning up. These countertops should be surfaced with heatproof, easy-to-clean material. Through-traffic should bypass the kitchen work center to avoid interference with the preparation of meals.

Provide maximum kitchen storage with both base cabinets and wall cabinets where possible (Fig. 15–64). Run wall cabinets to the top of the standard 8′ high wall, and use the top 14″ for dead storage. Allow at least 40 sq ft of shelf space for general kitchen storage, with an additional 6 sq ft for each person living in the home. In U-shaped and two-wall types of kitchens, allow between 4′ and 6′ of floor space between faces of opposite cabinets. Various modifications of the basic arrangements can be made to create interest and to utilize the kitchen space effectively without destroying the function of the work center.

Water vapor, cooking odors, and heat should be controlled in a modern kitchen with the use of an exhaust fan. Such fans are especially important in open-plan kitchens. Range hoods usually incorporate quiet-running

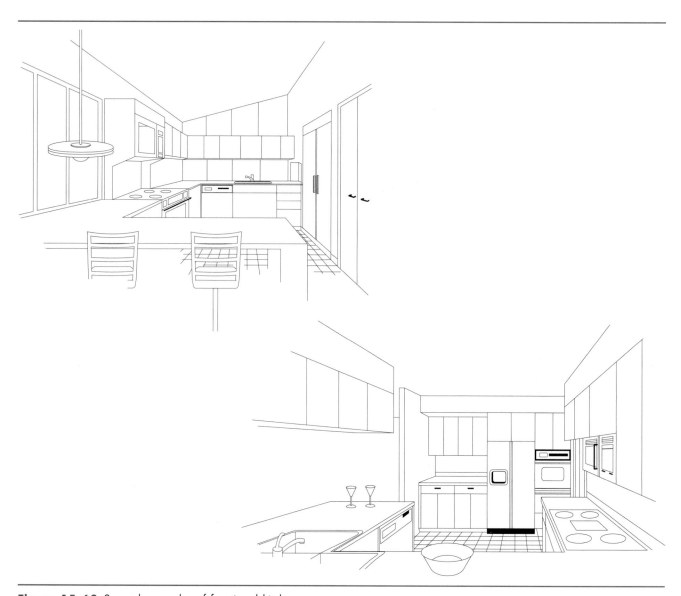

Figure 15–63 Several examples of functional kitchens.

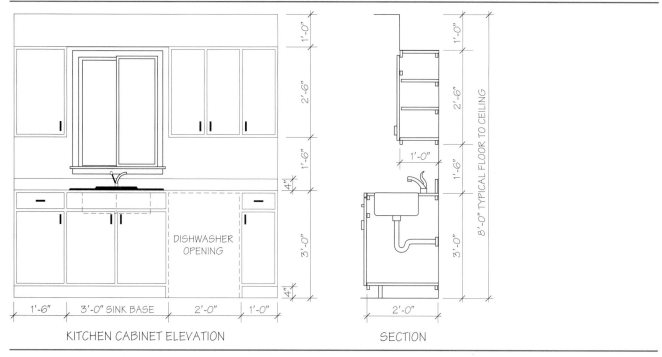

KITCHEN CABINET ELEVATION

SECTION

Figure 15–64 Standard kitchen cabinet heights and widths.

fans for the rapid disposal of the undesirable side effects of food preparation.

15.10

MAKE THE ELEVATIONS EXPRESS THE SPIRIT OF THE INTERIOR

The exterior of a house should evolve from the requirements of the inside. If a formal interior is planned, the exterior should reflect it; if the interior is informal, the exterior should create an impression of freedom and casual living. Traditional exteriors are usually based on the use of symmetry: identical features on both sides of a vertical center line, resulting in positive balance (Fig. 15–65A). Informality is gained by combining areas and features of dissimilar size and placement, unrestricted by identical balance, to produce a different type of visual balance (Fig. 15–65B). Either type should create a sense of tranquility. Both can be in good taste if they provide interest without too much busy "makeup" and if they reflect simplicity without being austere.

Unity, harmony, scale, balance, emphasis, and focal point are all essential elements of a pleasing design. Unity and harmony can be achieved by careful placement of repetitive elements such as windows. Scale relates one ob-

A

B

Figure 15–65 Formal and informal elevations.

ject to another relative to size and must be a major design consideration. All architecture should be sensitive to the human scale. Doors, steps, ceiling heights, furniture, and counters are elements that are designed around the scale of the human figure. Scale is also important in the proportions of the exterior and interior elements of the house. For example, a column's diameter in relation to its height is an element of scale (Fig. 15–66). Awkward proportions should be avoided in design. Emphasis can be achieved by contrasting colors, shades, and tones and by contrasting geometric shapes.

All pleasing houses should have a focal point on the exterior and the interior. A portico, entrance door, or steps can serve as a focal point on the façade. A fireplace, an entertainment center, stairs, or cabinets can create a focal point in interior space. It is important that the unity, harmony, scale, balance, emphasis, and focal point be considered in the design of a home.

15.10.1 Consider the Basic Structural Masses

Give basic structural masses good proportions and pleasing visual shapes. Unless its general form and silhouette are good, no amount of texture or ornament can improve a structure. Avoid extreme deviations from accepted building forms, unless you are creative enough to start a new trend in architecture. Traditional shapes of the past have lived on because their general proportions have been pleasing. Avoid awkward proportions.

15.10.2 Keep Important Roof Lines Simple

Not only do overly complex roofs look awkward, but they can also be expensive to build. A roof pitch must be proportional and related to the style and character of the building. Compare the height of exterior walls with the height of the roof. If the structure is L shaped or T shaped, make the wings the same width, if possible, so that the roof ridge line will be unified. If one wing must be narrower, be sure that the narrow span roof is received against the larger span (Fig. 15–67).

Consistency of roof pitch and type is important. When a roof pitch is selected, it is advisable to design all roof slopes on the building with the same pitch. A similar theory is true of roof styles. If a gable roof style is chosen,

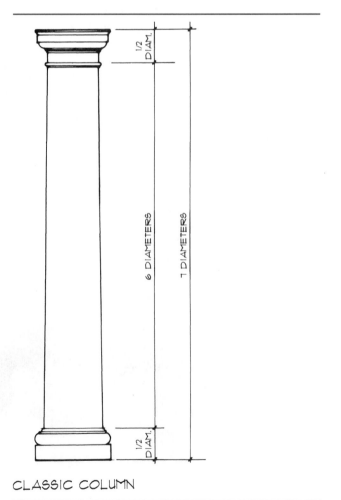

CLASSIC COLUMN

Figure 15–66 Elements of scale.

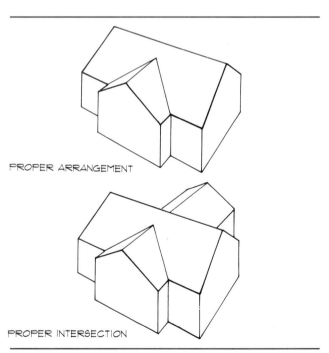

PROPER ARRANGEMENT

PROPER INTERSECTION

Figure 15–67 Arranging smaller gable roofs against larger ones.

stay with gables. Avoid artificial gables (Fig. 15–68). Hips, gables, sheds, mansards, and flat roofs are common styles. Complex roofs are most often the result of complicated floor plans; therefore, roof planning must be considered when the plan is developed. Simplicity is in many instances the proper approach.

15.10.3 Select Windows According to the Room Requirements

Room use, ventilation, and site orientation are the most important factors in determining the size, type, and placement of windows. In living areas where the view is important, windows can be large—to the floor if necessary. Incidentally, do not make the mistake of putting a large window in a wall where there is no view or where the view is undesirable. Few pieces of furniture are designed to be placed in front of large floor-to-ceiling windows. In dining areas and bedrooms where furniture must be placed against the wall, it may be desirable to design windowsills 3'-0" to 4'-0" above the floor. Windows planned above kitchen cabinets should be at least 3'-8" above the floor. In baths, where privacy is important, windowsills should be at least 4'-0" from the floor.

Window admit light and induce ventilation into the space. Light for illumination of the interior space is important. Too much light can create excessive glare or admit too much solar radiation. In the Northern Hemisphere, the best location for windows is on the south wall. Windows placed on the south allow sunlight to enter a room during winter months. A properly designed horizontal eave or overhang above the same window shades the glass from summer sun. Glass placed on the north provides excellent diffused light for reading but is an energy drain during the winter. Minimize glass on the east and west, or provide sun-control devices such as blinds to regulate unwanted sunlight. Many windows can be opened to provide natural ventilation. In warm climates, large windows that allow plenty of air circulation are desirable. In cold climates, small windows may be sufficient to ventilate a space adequately. A study of the prevailing summer and winter breezes is desirable. Windows can be oriented to maximize the effect of natural ventilation.

15.10.4 Coordinate the Window Sizes and Types

Almost every house requires windows of different sizes, yet good fenestration (placement and sizing of window openings) will unify the exterior if you select window sizes of similar proportions. Use the graphic method of drawing a diagonal through one of the selected window shapes to find the correct proportions of the other windows (Fig. 15–69).

Today's wide range of window sizes and styles permits almost unlimited freedom in correct window coordination. Double-hung windows with small lights are typical of traditional exteriors; large window walls are typical of contemporary exteriors. Line up the heads of windows and doors on the elevations. When making preliminary sketches, consider the appearance of the wall areas around window openings, as well as the windows themselves. If these shapes are not pleasing, or if the wall areas fail to produce the desired character, consider moving the windows one way or another without restricting their effec-

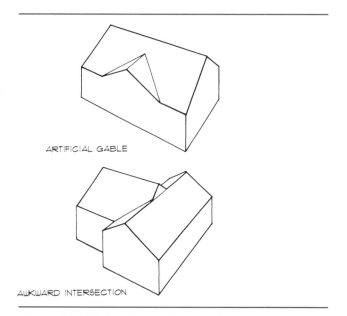

Figure 15–68 Avoiding artificial gables.

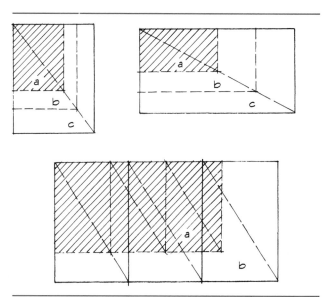

Figure 15–69 Graphic method of arriving at similar window proportions.

tiveness from the interior. Avoid windows so close together or so near to exterior corners that they create extremely narrow interjacent wall areas (Fig. 15–70A and B). These areas appear unstable, especially if the wall is masonry. Bring the windows together by joining them with a narrow mullion, or place a unifying panel between the windows, or spread the windows farther apart (Fig. 15–70C and D).

15.10.5 Select an Appropriate Front Door Style

Select an appropriate style for the front door (as well as for the other exterior doors), in keeping with the style of the house, and give this door the prominence it deserves. The treatment of the front entrance in traditional styles of the past has been highly ornamental. In present versions of these styles, various adaptations of the *panel door*, with or without lights, are appropriate. Otherwise, the *flush door*, painted or with a natural wood finish, is the best selection for most contemporary entrances.

If the front hall needs natural light, sash doors should be used, or side lights can be installed along the door frame (see Fig. 18–13). As a safety measure, it is advisable to place some type of small window in or near the front entrance so that the homeowner can identify callers before admitting them. Many types of *sliding glass doors*, framed in wood or metal, are available for easy access to outdoor living areas. To prevent serious accidents, sliding glass doors should be installed with tempered glass, or protection bars across the doors should be used. Refer to manufacturers' catalogs for details and availability of the various doors before you indicate the selection in your planning.

15.10.6 Keep Details in Scale

Many of the details found on large traditional homes will not be successful on a small home. A suitable cornice detail is especially important (Fig. 15–71).

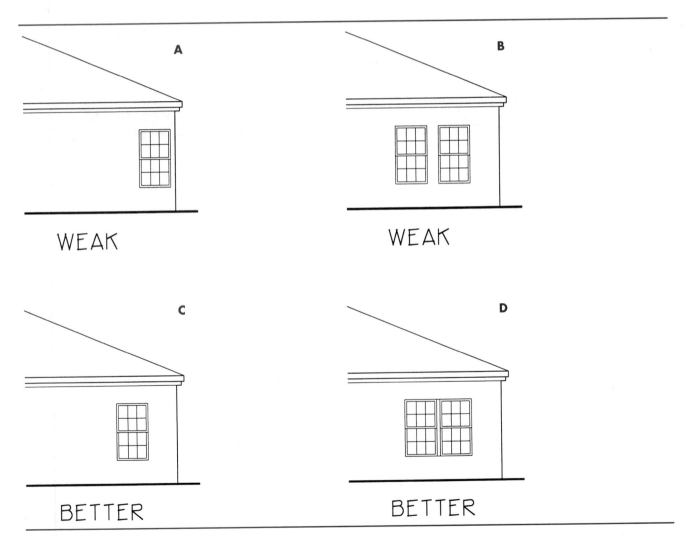

Figure 15–70 Placement of windows.

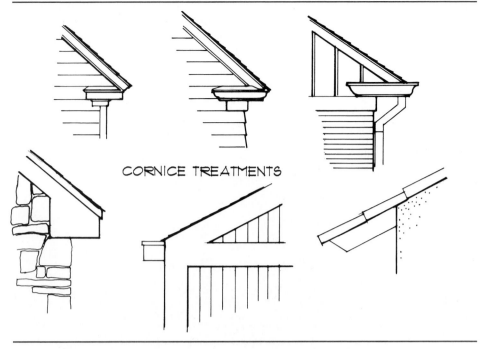

Figure 15–71 Cornice treatments should be consistent with the character of the elevations.

15.10.7 Relate the Four Elevations

Relate the four elevations by carrying out a repeating, interesting theme. Let some of the structural members be exposed, if appropriate, to contribute rhythmic enrichment. One exterior material should dominate in both texture and color; use few additional materials. Especially avoid the pretentious use of too many contrasting materials on the front and only plain materials on the sides and rear. Design all four sides. Above all, select materials for the exterior that require minimum maintenance throughout the years.

15.10.8 Accent Horizontal Lines

If you want a house to appear lower and longer, accent its horizontal lines. A lofty appearance is usually not pleasing. Wide overhangs of the roof and wall extensions from the house produce protected areas and help integrate the house to the land (Fig. 15–72).

15.10.9 Make the Chimney Massive

If you want this important design element to contribute character to the structure, make the chimney massive.

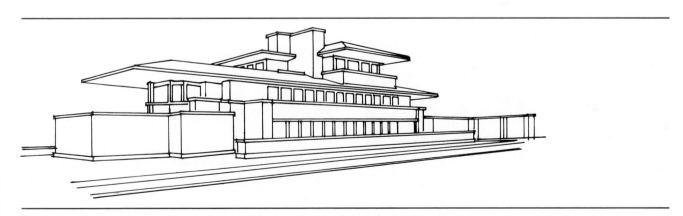

Figure 15–72 Accent horizontal lines if the house is to blend with the land. (Robie House by Frank Lloyd Wright)

15.10.10 Remember That Much of the Value of a House Is Created by the Design

If the design is simple and in good taste, the house will continue to be desirable for many years.

15.10.11 Study Prize-Winning Residential Designs

Leading architectural journals and home magazines publish prize-winning residential designs. Study these designs, and see whether you can determine why prominent architects and home specialists have chosen them over other designs. Study what makes a house, whether large or small, exceptional and appealing. This understanding should provide you with one of the important requirements of successful planning—good taste.

15.11
COMBINE EXTERIOR DESIGN ELEMENTS THAT ARE IN KEEPING WITH ACCEPTED HOME STYLES

Study the home photographs shown in Figs. 15–73 through 15–84 for their identification with a definite style. Like our culture, home styles have their roots in many other countries. However, because of local climate, materials, topography, and the needs of individuals, each style is strictly American in development.

Colonial styles have survived because they are honest houses, usually unpretentious and simple, yet having pleasing architectural features. Many were passed on to descendants after having given each generation a lifetime of enjoyment. Houses erected in early colonial days in New England were built by carpenters and masons skilled in traditions of England, the Netherlands, and Germany. The styles in Louisiana were influenced by the French; and in Florida, California, and Texas, by the Spaniards; and some areas give evidence of Oriental influences.

Figure 15–74 New England colonial styling. (Hedrich-Blessing Photos)

Figure 15–75 Cape Cod styling. (Hedrich-Blessing Photos)

Figure 15–76 Regency styling. (Hedrich-Blessing Photos)

Figure 15–73 Colonial saltbox styling. (Hedrich-Blessing Photos)

Figure 15–77 Dutch colonial styling. (Hedrich-Blessing Photos)

Figure 15–80 French provincial styling.

Figure 15–78 Southern colonial styling. (Hedrich-Blessing Photos)

Figure 15–81 Garrison styling. (Hedrich-Blessing Photos)

Figure 15–79 Ranch house styling. (Hedrich-Blessing Photos)

Figure 15–82 English half-timber styling.

Figure 15–83 Contemporary styling. (Frank Venning Photo)

Figure 15–84 Williamsburg styling.

Many of the most admirable American homes were built 150 to 200 years ago; some are still in use. The value of these early houses is evidenced by the degree to which they were copied as the population spread.

Today, very few houses exist that are pure examples of any one architectural style. All vary in proportion, window fenestration, or decorative treatment. These variations, however, give each house individuality and tend to show the stamp of a particular architect. However, many retain enough features to allow us to make positive identification of their basic style. This identification is important—a well-designed home should give evidence of a definite styling, and you should be able to recognize those characteristics that identify it with a particular style. Today's home designer would be wise to keep an eye on tradition while focusing on the new materials and innovations available in contemporary building.

EXERCISES

The following planning exercises should first be done freehand on tracing paper or ⅛″ coordinate paper. After you have gathered site data and various residential ideas from books, magazines, newspapers, and other sources, develop the problems as indicated in Chapter 8, under "Sketching Plans and Elevations." The plan sketch cannot be entirely developed before trials of elevation views, and possibly floor levels, have been established. Develop the plan and the elevations together; do not be discouraged if you need to do numerous sketches or try completely different approaches to the problem before attaining a workable layout.

1. Using the plat plan and the topographic survey shown in Fig. 15–85, prepare a site analysis showing a separate drawing for each of the seven site considerations described in Section 15.4. Lay out the plat, double size, on 8½″ × 11″ paper using ink or felt-tip pens. Assume that the property is in your hometown.

2. Using the plat plan and the topographic survey shown in Fig. 15–86, prepare a comprehensive site analysis. Using 11″ × 17″ paper, draw the plan four times larger than the illustration with ink or felt-tip pens. Two drawings will be required to illustrate the comprehensive site analysis. One drawing should depict slope, drainage, climate, prevailing winds, and solar orientation. The other drawing should show vegetation, natural features, circulation, views, noise, zoning, and setbacks. Assume that the property is located in the town where you are studying architectural drawing.

3. Prepare zone diagrams for a residence with the following space requirements:

   (a) Living room
   (b) Dining room
   (c) Kitchen
   (d) Family room
   (e) Powder room
   (f) Three bedrooms
   (g) Two baths
   (h) Two-car garage

4. Using the program and the zone diagram from Exercise 3, prepare bubble diagrams for the same residence.

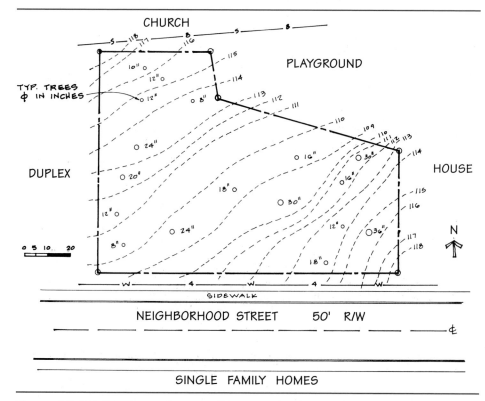

Figure 15-85 Plat plan and topographic survey for Exercise 1.

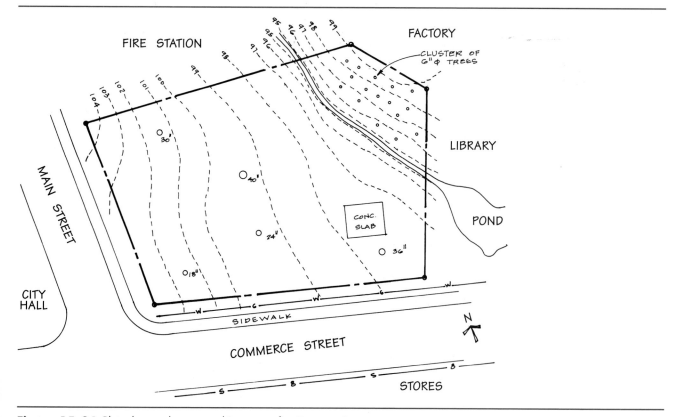

Figure 15-86 Plat plan and topographic survey for Exercise 2.

5. Prepare a freehand schematic plan of a residence using the spaces listed below for the site that you analyzed in Exercise 2. Use zone and bubble diagrams.

   (a) Foyer
   (b) Living room
   (c) Dining room
   (d) Breakfast room
   (e) Kitchen
   (f) Fitness room
   (g) Powder room
   (h) Two bedrooms
   (i) Two bathrooms
   (j) Library
   (k) Workshop
   (l) Two-car garage

6. Plan a *small beach house* for a young couple with a small daughter. The beach house should have minimum accommodations yet a festive spirit for vacation and weekend living. Provide for the following:

   (a) Lot size: 65′ ocean front, 200′ deep
   (b) 700-sq-ft floor space, contemporary design
   (c) Large living, dining, kitchen combination area with small kitchen utility unit behind folding doors
   (d) Set the house on preserved wooden piers to prevent damage from high water.
   (e) One easily accessible bath with shower
   (f) Two sleeping rooms
   (g) Fireplace, redwood exterior

7. Plan a bachelor's *retreat house*. The occupant is a young professor who requires space for books and study and space for occasional entertaining of large groups. Provide for the following:

   (a) Lot size: 100′ × 185′, level and wooded
   (b) 1500-sq-ft floor space, plank-and-beam construction
   (c) Master bedroom and guest bedroom
   (d) Large living–dining room, modest kitchen
   (e) One full bath
   (f) Carport for one car, outdoor terrace area

8. Plan a *home for a young family* with one child. They are interested in a 1-story, colonial-type house with the following specifications:

   (a) Lot size: 100′ × 200′; the lot slopes downward 10′ from the left to the right side.
   (b) 1800-sq-ft maximum floor space
   (c) Kitchen near family room with guest accommodations
   (d) Two bedrooms, 1½ baths
   (e) Full basement to be partially used as garage for one car

   (f) Provision for outdoor play area with supervision possible from kitchen
   (g) Brick-veneer construction with gable roof

9. Plan a *medium-priced home* for a family with two children: a boy and a girl. The clients indicate a preference for a ranch house on one level, with the following requirements:

   (a) Lot size: 150′ × 225′, with wooded view in rear
   (b) 2000-sq-ft maximum floor space, exclusive of double carport
   (c) Two full baths, one to be private for master bedroom
   (d) Three bedrooms, separate dining room, family room, living room, kitchen, and utility area
   (e) Outdoor living area, rear-entrance mud room
   (f) Adequate storage space

10. Plan a *traditional 2-story home* for a family of five (two boys and one girl), with the following specifications:

    (a) Lot size: 100′ × 175′
    (b) 2200-sq-ft maximum floor space plus double garage
    (c) Kitchen, dining room, entrance hall, powder room, and large living room on first floor (concrete slab on grade)
    (d) Three bedrooms and double bath on second floor
    (e) Outdoor terrace off living room with brick fireplace
    (f) Stone and/or wood exterior
    (g) Traditional-styled front entrance

11. Plan a *split-level* for a family of four (two teenage daughters). Provide the following:

    (a) Lot size: 125′ × 185′, rolling and wooded
    (b) 2000-sq-ft floor space plus double carport
    (c) Separate dining and living rooms, three bedrooms, two full baths
    (d) Utility and laundry area
    (e) Large recreation room with fireplace and access to outdoor living area
    (f) Brick-veneer and board-and-batten exterior

12. Plan a modest *duplex for elderly people*. Two retired couples are interested in homes that will be easy to maintain and convenient to live in. They specify the following:

    (a) Lot size: 90′ × 200′, with provision for gardening
    (b) 1000 sq ft in each unit
    (c) Frame construction, masonry party wall between units
    (d) Adequate storage, guest space, carport for one car for each unit

REVIEW QUESTIONS

1. Why is it important to use a planning checklist?
2. What is meant by the term *easement?*
3. Do contour lines crossing a small stream point upstream or downstream?
4. Why is a southern exposure generally the most satisfactory direction in solar-orienting a window wall?
5. Which part of a home should have the most isolation and quiet?
6. What geometric shape is the most economical for a basic plan?
7. Evaluate the *traffic flow* in your own home from the standpoint of convenience.
8. What is the maximum length of a satisfactory kitchen *work triangle?*
9. Define the word *fenestration.*
10. Why are simple, uncomplicated exteriors usually the most pleasing?

"This modest charm of not too much, part seen, imagined part."
—WILLIAM WORDSWORTH

BUILDING MODELS 16

There are some who may contend that model building does not belong in a book on drawing—that the construction of a model house, for instance, like the building of model airplanes, is mainly for those who enjoy tinkering with miniature things. However, model building is one of the more pleasant aspects of architecture, and it has become an important creative expression for professionals as well as students. Many architects rely on three-dimensional models for final evaluations of their designs, and they sometimes reserve work space in their offices exclusively for this interesting phase of their activity. When the scope of a project justifies it, some offices engage the services of professional model builders to construct impressive and realistic models. If promotional methods must be relied on to raise funds for a specific building project, executive boards or organization planning groups will find a model most advantageous. Also, lending institutions will usually approve loans more readily when shown an impressive model of the projected structure.

Landscape architects find that models are the most expedient method for site planning and landscaping. City planners, plant layout experts, and safety engineers all make invaluable use of models.

To the layperson, a model is usually the most honest means of visualizing all aspects of a building, including proportion and relationship of features, appearance of tentative materials, and other three-dimensional qualities. Often the designer catches a feature on the model that may have been overlooked on the working drawings. The model, then, is visual proof that the spatial concepts of a design are workable and pleasing.

Generally, the drafter is called on to construct a model after preliminary plans are drawn or occasionally after the final working drawings have been completed. As mentioned above, many architects rely on professional model builders to construct elaborate models of the proposed project. Such models are expertly detailed and are frequently used as advertising tools by the client.

For the student, models are unquestionable learning instruments. In teaching architectural drawing, the authors have found model building to be one of the most convincing methods of helping students to determine, for example, the correct choice of details to be included in their final drawings or to discover why the choice of a roof slope was not complementary to the building. No other means, except perhaps working on an actual construction job, offers the student such an effective way of relating all of the views of a drawing into a total building concept. Besides, the model has the advantage of small scale, which can be related to the drawing because of its similarity in size, and it allows the project to be viewed in its entirety.

16.1

TYPES OF MODELS

Depending on their purpose, several types of models can be constructed.

16.1.1 Architectural Models

Architectural models (Figs. 16–1 and 16–2) are intended primarily to show only the exterior appearance of a design. This type of model is usually constructed of heavy illustration board, thin balsa-wood panels, or, if necessary, hardboard, $1/8''$ thick, with surface textures applied directly to the board and scaled $1/4'' = 1'-0''$ ($1/8'' = 1'-0''$ for larger buildings). It is constructed directly from the drawings, and the elevations, roofs, walks, drives, landings, and so on, are cut out with a sharp model builder's knife and are attached to the base (Fig. 16–9). Contours are usually built up with layers of economical chip board; their edges are left to appear very much like contour lines on a site plan (Fig. 16–1). All exterior surfaces are neatly painted to represent the given materials; contours are made to look like grass or ground cover (Fig. 16–3A).

Figure 16-1 Architectural model showing a building complex and built-up contour levels.

Figure 16-2 Student architectural model constructed of balsa wood and basswood on a plywood base.

A

B

Figure 16-3 (A) Model of a contemporary residence. (B) Photograph of the residence after construction.

Trees and shrubs of the same scale are carefully glued in place. In addition, many commercially printed papers are available that simulate various material textures. On smaller-scale city planning models, neatly painted, solid wooden blocks may be used to show building masses.

16.1.2 Structural Models

For school or other projects requiring a visual analysis of structural systems, larger-scale models (usually $\frac{1}{2}'' = 1'-0''$ or $\frac{3}{4}'' = 1'-0''$) are built to correspond exactly to the construction of the actual building (Fig. 16–4). Framing members of balsa, fir, or soft pine are cut to scale and assembled to represent the building's actual framing system. When assembling, use quick-drying model builder's cement instead of nails or other fasteners. Considerable amounts of the exterior surfacings, especially rear elevations, are omitted to allow a full view of the structure as well as of the interior room arrangements. The terminations of the wall and the roof are made to show the layers of materials within the actual construction. Usually, front elevations are

A

B

Figure 16–4 Model of a contemporary residence showing structural framing.

completed in detail to reveal the important esthetic qualities of the design.

Models of residences also often include major interior cabinet work, plumbing fixtures, and even appliances or furniture. Leaving some areas open by omitting part of the roofing allows complete observation of these features. This type of model is the most rewarding for the student, since it requires solutions to structural problems that are not always evident on the drawings and that nevertheless must be considered when the final drawings are done.

Figure 16–5 A model showing realistic building materials.

The qualities necessary for a successful model builder are patience, care in attention to detail, and the ability to track down simple, everyday items that will faithfully represent building materials at the small scale (Fig. 16–5). Some items, such as sponges for shrubbery and small plastic figures and autos, are inexpensive; others, such as sandpaper, pins, bird gravel, and pieces of burlap or felt material, are common household items.

16.2

MODEL-BUILDING TOOLS

The following list indicates the basic equipment needed for model building:

1. Model builder's knife (Stanley, Exacto)
2. Small framing square
3. Architect's scale
4. Metal straightedge
5. Tweezers
6. Small hammer
7. Handsaw
8. Several small paintbrushes
9. Shears
10. Sandpaper and masking tape
11. Sturdy table
12. Hot-glue gun

Although a minimum number of tools is generally required, supplementing these tools with other small hand tools facilitates working with diverse materials. If they are available, more elaborate tools can save time and produce more professional work. For example, in constructing structural models, you may find a miter box useful. Power shop tools can be very helpful, especially a power table saw for cutting structural lumber to scale, as we will discuss later; a power sander is also very useful.

16.3

CONSTRUCTING ARCHITECTURAL MODELS

Working drawings, either preliminary or final, are necessary to construct a model, regardless of type. You must take dimensions from the drawings or prints of the drawings, and you must continually refer to them throughout the model construction.

Start with a substantial base, either ½" or ¾" plywood, upon which the model is to be attached. Use ¼" thick plywood or hardboard, if the model will be rather small, and size the base to allow for sufficient landscaping without making it unwieldy. Cut the walls from heavy illustration board or balsa-wood sheets of a thickness proportionate to the scaled wall thickness. Work from the elevation views of your drawing, and cut window and door openings to scale; allow for corner joints, either lapped or mitered. Avoid exposing the end grain of wood any more than necessary. If raised contours are to be applied to the base, allow for the height of the grade line above the base when cutting the wall pieces. It is best to glue the walls directly to the wood base, using triangles to keep them plumb. Occasionally, gluing small wooden blocks to the inside may be necessary to ensure a sturdy attachment to the base. If a detachable roof is planned, cut out partitions of a similar material, and glue them in place after exterior walls are finished and after the floor level has been constructed. The floor panel may have to be supported with blocking below. All work to be done on the interior should be completed before the roof is constructed.

Before gluing the exterior wall panels in place, apply the basic material texture to them. This is easier when the panels can be laid flat on the table. The following list describes ways to achieve exterior textures:

- **Natural wood**   Natural wood finishes can be represented with balsa, placed with the wood grain in the proper direction, stained with a suitable color, and wiped with a cloth. Avoid getting glue on a wood surface that is to be stained; the stain will not penetrate evenly. Batten strips, for example, should be glued in place *after* the board surfaces have been stained. Uniform grooves to represent paneling are easily scored in the soft balsa with a stylus or an empty ball-point pen.
- **Brickwork**   To simulate brickwork, paint the walls with a subtle variation of the basic brick color, usually earth colors, and occasional mortar joints drawn with white linework. Use white ink in a ruling pen.
- **Stonework**   Stonework can be represented in a similar manner. Paint subtle variations of the basic stone color—grays or dark ochres—with light-color mortar joints applied according to the desired stonework patterns; or you can use actual small stones (Fig. 16–6). Be sure to keep the textures in the same scale as the model.
- **Stucco and E.I.F.S. (exterior insulation finish system)**   To show these media, mix sand and paint, and apply this mixture as a final coat.

For the most part, use the subdued, opaque paints for model finishing. Tempera, acrylic, or casein colors are handy for color mixing, and latex-base wall paints are suitable for covering large areas. Avoid enamels and glossy paints, which are unsatisfactory because of light reflections. Because of its quick-drying qualities, lacquer is frequently used by professional model builders when spray painting is applicable.

After the walls have been fastened in place, the next step is the roof. Select a rigid material for roof panels to prevent warping. Work from a roof plan if necessary to ensure neat-fitting pieces; on complex roofs, such as hip roofs with offsets, it is simplest to project exact shapes carefully from roof drawings and lay them out on the material with drawing instruments before cutting. Use a sharp knife for clean, straight edges. Sloping roofs usually need several wood bracing members below, particularly on detachable roofs, if they are to remain rigid. Chimneys piercing the roof may be cut from wood blocks and merely glued to the roof.

Figure 16–6 Student architectural model made with molding plaster and actual small stones.

To represent a built-up type of roofing, apply coarse sandpaper to the roof with contact cement applied to both surfaces. Asphalt-type shingles may be represented with strips of construction paper or fine sandpaper, cut to appear as strip shingles and painted the desired color if necessary.

Next, work on the construction of grade levels or site contours. If the model requires no contouring and is to be shown on a perfectly flat site, the base can be painted dark green and covered with flock (tinted fibers that adhere to the wet paint, giving the surface a velvety appearance); or it can be covered with model builder's simulated grass (sawdust dyed green). Use a thick, oil-base paint to hold the grass material securely. If suitable, apply a stippled surface of texture paint, and finish it with green color.

Raised contours, if needed, are cut from heavy chip board (described below) and are glued over each other to the correct heights. This stepped surface looks best if painted a dull green; walks, drives, and the like, are cut from separate pieces, are painted, and then are glued over the contoured surface. The easiest way to make steep lot surfaces is to build them over wire screen or hardware cloth that has been blocked up to the correct height from below and covered with a layer of molding plaster or papier-mâché. After it has cured, paint and apply grass material as mentioned above (Fig. 16–7).

Exterior details, such as window and door trim and cornice moldings, are applied after the contour levels are fixed. Cut the trim members from thin balsa sheets using a metal straightedge and a sharp knife. Small dowels, toothpicks, and model builder's pieces may also be appropriate. Paint the trim (usually white) *before* gluing it in place to eliminate unsightly smearing and the running of paint on adjacent colors; masking tape may be necessary in some situations. Cut doors from thicker balsa, and add their trim details before gluing them in position; they can be left partially open rather than closed. Thin plastic may be fastened to the interior of window openings to represent glass, although windows are nearly as effective without it if the window trim has been carefully applied. On large glassed areas, muntin strips are usually glued directly to the plastic.

Complete the landscaping as the last step. Show fences, walls, terraces, swimming pools, and other outdoor features that will enhance the model. Water surfaces look realistic if represented with ripple window glass embedded in the surfacing material. Paint the area under the glass a light blue. For foundation planting, tear pieces of sponge (natural or synthetic) into various shrubbery shapes, dye them several shades of green, and glue them in natural-appearing groups near the model corners and entrances. Green lichen is also available at model supply stores for this purpose and is similarly secured with glue. To simulate trees, take actual twigs or small branches having the proper scale and formation, and glue pieces of the sponge or lichen throughout the small branches to

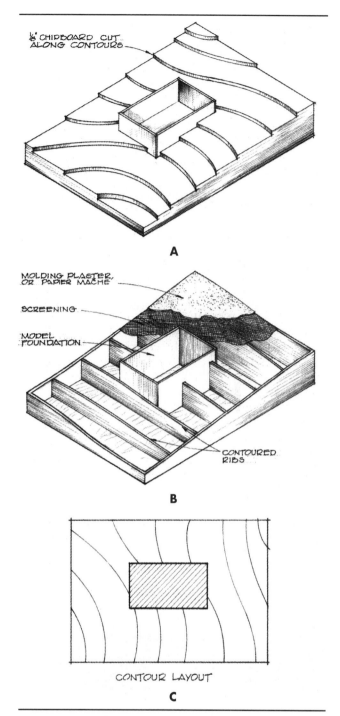

Figure 16–7 Building up contours on model bases.

shape the trees' foliage. Then drill a hole in the base and insert the tree trunk with glue.

Dress the edge of the base panel with suitable stock molding or trim; miter the molding neatly at the corners to give it a finished appearance. Incidental objects such as autos, people, furniture, and bath fixtures, if necessary to elaborate the model further, may be found generally at hobby shops or craft stores. Be sure that they are the right scale. These items can also be carved from white or colored soap. However, avoid superfluous items, which only clutter the model and detract from the architecture instead of supplementing it.

Many architectural offices use *chip board* or *foam core board* for the majority of their models. Chip board is a gray, economical paper board, about 1/16″ to 3/32″ thick, available at most drafting supply stores. All cuts are neatly made with a sharp knife, and pieces are glued together and held in place with straight pins, drafting tape, or rubber bands until the glue has set. The board has good stability; takes tempera, acrylic, or latex paint well; and does not seem to be greatly affected by humidity changes. Clean-cut models can be quickly made with this versatile model material (Fig. 16–8). Foam core board is white or gray and 1/16″ to 1/14″ thick.

In architectural planning, whether in an architectural office or for student projects, mass models are sometimes used to study space relationships, proportion, and general appearance of contemplated buildings. Only the general forms are needed, and they can be cut from balsa blocks or styrofoam and painted with a flat white latex to produce very effective planning models. Cut either of the materials with a sharpened putty knife or a slightly serrated bread knife after the scaled volumes have been laid out with pencil. Use a sturdy plywood base, and to ensure good adhesion, glue the blocks directly to the base *before* paint is applied. If thin sheets are needed in the model, use chip board or illustration board that has been carefully cut with a model knife.

For quick, less-definite mass models, use modeling clay pressed into the forms needed; even soap bars cut into accurate shapes may be satisfactory.

Another material that should be investigated by the model builder is basswood. The wood possesses a fine, uniform grain and can be cut, glued, finished, and painted similarly to other model materials. Textured basswood sheets representing many of the traditional architectural surfaces are available from many model supply stores and hobby centers.

Polystyrene sheets, available from similar sources, are also used in model building. This clear, colorless, plastic material can be scribed, tooled, sawed, scored, and textured, and it bonds quickly with the model cements mentioned in the next section. Other preformed plastic features such as windows, doors, people, and autos are available in several scales from dealers, but generally those that are custom-made give the model a more professional appearance.

CONSTRUCTING STRUCTURAL MODELS

To build a model as a true miniature of a tentative building requires a working knowledge of the various framing systems (Chapter 12). Because of the many small pieces involved in a structural model, scales must necessarily be larger (1/2″ = 1′-0″ or 3/4″ = 1′-0″) than architectural models, which are meant to show mainly exterior appearances. Small buildings such as residences, then, become the most practical projects when using these larger scales.

Structural models provide varied learning situations for students and lend themselves to classroom projects where groups can work together to solve the problems involved. The situations and problems encountered par-

Figure 16–8 Student architectural model made with chip board.

allel actual construction jobs and offer interesting and informative experience to drawing students. A foreperson, similar to an actual job superintendent, should be designated in the model crew to plan the individual tasks of the construction and keep the model moving along smoothly. Considerable time and work are involved. If a student's original design is used, the model definitely creates added enthusiasm.

Framing lumber for structural models is best cut from ponderosa or sugar pine stock on a power table saw. Equip the saw with a *hollow-ground blade* to produce neat cuts and smooth surfaces requiring no sanding. When cutting the lumber, you will find it most convenient to scale it down to its *nominal* size rather than its actual size. For example, 2′ lumber would be scaled exactly 2″ thick. Use convenient lengths (about 3′) of the stock, and cut the narrow dimension first. Although balsa wood is available for structural pieces, pine gives the model more stability.

Use model cement (for example, Testor's, du Pont Duco) for most of your gluing work. When small pieces can be quickly assembled, use the *extra-fast-drying* type of cement to save time in setting. For stronger wood joints, use white wood glue (Elmer's) with pressure, if possible; on thin paper, use contact cement or a glue stick to prevent wrinkling.

16.4.1 Base for the Model

First, build a sturdy base for the model, making it large enough to accommodate the floor plan, and leave sufficient space around it to show exterior features and landscaping. Remember that if the base is built too large, the model will become difficult to display and move about. Usually, ¼″ plywood with 1″ × 2″ framework (set vertically) is satisfactory if securely nailed (Fig. 16–9). Use diagonal bracing to ensure rigidity throughout the plywood.

16.4.2 Foundation Walls

Lay out the outline of the foundation on the base, and proceed with the foundation walls. If a basement or site contours are planned, consideration must be given to the floor-level heights, and the base surface as a reference level should be indicated on the typical wall section detail of the working drawings. Heights then can be quickly scaled from this reference point. If the house is to have a concrete slab floor, represent the slab with ¼″ plywood, and, according to the drawings, fasten it directly to the base or block it up to the required height. Slabs may also be cast of molding plaster or spackling compound.

Basements must have full-height foundation walls of the proper thickness, with raised lot contours to be finished as a later step. Construct crawl space foundations according to the foundation plan, and show walls and piers correctly placed. Build the foundation of wood, scaled to the correct thickness, and be sure that it is built

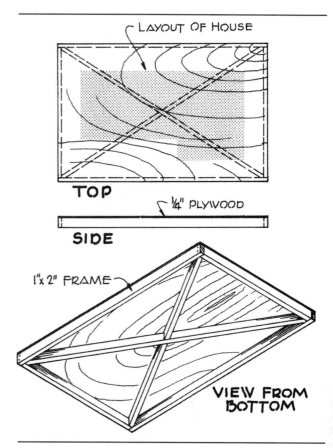

Figure 16–9 Details for constructing a satisfactory model base.

the same height above grade as indicated on the drawings. The foundation is a major step of the model construction and is important in getting it started correctly.

16.4.3 Wood Framing

Next, construct the wood framing (if applicable) according to the typical wall detail; follow good framing practice. Start with the sill plate, and glue all members together in the same manner and sequence that carpenters would use on the full-size construction. If a wood floor is indicated, insert all wood built-up girders or steel beams to support the floor framing. Steel beams can be made from three strips of thin wood, glued together and painted orange to resemble structural steel. Space the floor joists, insert them in the direction shown on the drawings, and cover them with the subflooring. To expose portions of the floor framing, cut curved, free-form holes in the subflooring near the center of several rooms. Then you can expose the flooring layers (subfloors, underlayment, and finish floor) by cutting similar holes just a little larger for each succeeding layer. Similar openings in ceilings and roofs are then planned directly above these floor openings so that the structure will be exposed completely through the model.

After the walls and partitions are laid out on the sub-floor according to the plan, construct each wall section with the use of a jig on the table top. In gluing the studs and plates together to form the wall sections, remember to place the studs 16″ o.c. (if conventional framing). Insert the proper posts at corners (corner posts) and intersections of walls (T posts); provide double studs and lintels around window and door openings. Cut the studs about 7″ to 9″ long for 8′ ceilings. Build the sections with the sill and only *one* top plate (the other 2 × 4 plate should be attached after the wall sections are set in place), and allow enough plate length for corner joints (see "Framing Details" in Chapter 12). Be sure that there will be nailing surfaces on both inside and outside wall corners. Glue the wall sections to the subfloor, using your triangles or tri-square to keep them plumb (Fig. 16–10A). Now apply the second top plate over the walls (Fig. 16–10B), overlapping the other top plate at corners and at intersections to produce a rigid wall framework. Do not forget diagonal braces at outside corners.

Complete all work to be done within room interiors before applying ceiling joists. Show suggestions of major

A

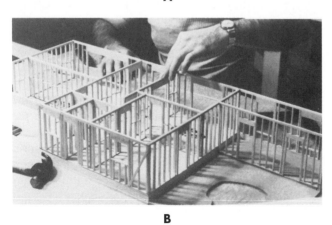

B

Figure 16–10 (A) Checking the plumb of wall framing. (B) Applying the double plate on structural models.

floor finishes, such as oak strip, parquet, tile, carpeting, ceramic, and flagstone. Usually, only portions of the various finishes need to be shown. Areas of the interior walls may be indicated with illustration board or wood paneling; however, avoid covering too much framing. Letter the room titles on small strips of white board, and glue them to the floor; place them so that they read from the front of the building.

While the interior is accessible, insert all permanent fixtures and built-in cabinet work. Bathroom fixtures may be carved from soap, as we mentioned earlier, and glued in place. Scaled fixtures of various types are available at hobby shops. Make cabinets from wood blocks. Cabinet doors may be made from thin balsa and glued to the blocks. Paint or stain the cabinets before installing them.

Next, attach ceiling joists to the wall framing according to the drawing (Fig. 16–11A). On gable roofs, joists are usually parallel to the rafter direction and are fastened to them above the plate. Cut common rafters from a pattern drawn out on paper, if necessary. Hip and valley rafters may require projection layouts as well. Place the sloping rafters in pairs opposite each other, and glue them to the plate adjacent to the ceiling joists to form a strong framing arrangement. If feasible, temporarily fasten the ridge board in place before attaching the rafters. Keep the edges of the overhangs straight. If truss rafters are employed, set up a triangular jig on the table, and build the identical trusses before setting them in place on the model.

In plank-and-beam framing, sand and stain the ceiling beams *before* gluing them in place. Use cripples above top plates to bring partitions up to sloping plank-and-beam ceilings; or you can build diagonal-top partitions with full-height studs after the ceiling slopes have been established. Use double studs or 4 × 4's within the walls to support the ceiling beams. Treat the rafter or beam overhang as shown on the drawings, using thin balsa for soffit material if a boxed cornice is called for. After the framing is complete, reinforce major corners and roof joints with common pins, shortened if necessary, to prevent observers from accidentally damaging the structure.

16.4.4 Exterior Walls

Apply exterior wall materials next; include subwall, building paper, and insulation, if applicable. Avoid covering walls around the entire building; usually the back walls can be left exposed so that the interior and the structure are easily visible (Fig. 16–11B). Complete the front in detail, on the other hand, to provide a basis for analyzing the design qualities (Fig. 16–11C).

Make brick veneer of balsa sheets that have been sized to fit the framing, cut for door and window openings, and mitered at outside corners. Use balsa strips glued in place to show brick sills extending out from the brick surface under windows. Before fastening the veneer in place, complete the brick texture by scoring the balsa with horizontal grooves to represent the brick coursing; do this with a blunt-tipped instrument such as a stylus.

A

B

C

Figure 16-11 (A) Construction of the wood framing in a residential model. (B) Completed view of the model from the rear, showing interior space and framing. (C) Completed view of the model from the front.

Provision must be made for the erection of the brick panels on the foundation. To paint the brick, first coat the entire surface with white color (tempera, acrylic, or latex works fine), making sure that the grooves are filled. Then apply the basic brick color with a drybrush technique, covering mainly the brick surfaces and leaving the mortar grooves white. Use subtle color variations to give the brick a realistic appearance. If you want to represent older, traditional brick, tone down the basic brick color by applying over it a wider variety of colors, primarily earthy, reddish-brown shades. This painting is best done in the drybrush technique also. Keep all of the panels of similar value, and check their development by viewing them at a little distance, rather than up close.

Stone textures are done in a similar manner; or you can achieve relief by cutting individual stones from thin balsa ($1/32''$ thick) and gluing them to the panels. Paint the panels white before applying the stones. Be sure that the stone pattern is pleasing and the pieces are glued in place with their wood grain in the same direction, usually horizontal. The stones are then painted in slight variations of gray or earth colors.

It may be worth mentioning that very attractive masonry walls and textures can be produced with molding plaster or spackling compound, available at builder's supply stores. First, build a wooden mold or form for each masonry wall or unit required. Then mix sufficient plaster and water in a container, carefully pour the mixture into one form at a time, and level it off. Add the plaster to the water when mixing; to prevent air bubbles, avoid vigorous mixing. You must work rather hurriedly, as the plaster sets quickly and cannot be left or worked too long. Avoid spilling the plaster, since it is difficult to remove from clothes and floor. After the plaster is sufficiently hard, remove it from the form without breaking. Repair any imperfections, and carve stone, brick, or other masonry textures on the surface with a suitable tool. The plaster carves easily if not entirely cured. When it is dry, coat it with white shellac to seal the surface, and paint the masonry units the desired colors, leaving the mortar joints white. Chimneys can be cast as a unit by this method, as well as replicas of many precast panels now in use in construction.

For even more textural stonework realism, small pea gravel or other aggregate can be placed in the bottom of forms to produce the face of the masonry panels, and a thick mixture of the molding plaster can then be poured over it. When these aggregate panels or units are cured and cleaned, they look very authentic and are extremely effective. Take precautions to construct the forms in the correct size and to scale. Join the panels at intersections with a thin mixture of the plaster as an adhesive, or use white wood glue.

Make the various types of wood siding from balsa wood. Whether painted or stained surfaces are required, balsa can be easily cut and finished to resemble many wood exterior finishes. Glue the siding to the exterior walls after the subwall materials have been applied. Be-fore attaching siding to the walls, carefully cut the openings for doors and windows, groove the wood if necessary, and then paint or stain it. As mentioned previously, be sure that the wood grain runs in the same direction as it would on the actual building and that stain is applied before glue is used. To simulate horizontal clapboard siding, apply thin strips of balsa in a lapped fashion, allowing the correct amount of exposure to the weather. Paint it after all of the siding is in place. Make wood shingle siding from small balsa pieces, which are cut and glued in place individually.

Build landings, steps, planters, and the like, from wood blocks, and veneer them with thin balsa if necessary for uniform masonry textures. Paint them off-white to represent concrete. Chimneys rising up through wood framing must be independent of the framing and may be made from cast-molding plaster or constructed of soft wood and veneered with thin balsa to obtain horizontal wood grain completely around the unit. Show flue extensions and a sloping wash at the top. If needed, fireplace openings can be carved out with a chisel or a knife, and the cavity painted a dull black. When the chimney is completed, fasten it to the base before framing the roof structure around it.

16.4.5 Roofing

Because of the comparatively small nature of models, their roofs are a predominant feature and therefore necessitate a great deal of care in finishing. Select a stable material such as balsa for the roof decking, rather than paper board, which has a tendency to warp when the roofing materials are glued on. At this point you should plan which parts of the roof surfaces are to be left exposed for viewing the framing members and the interior structure of the model. For this purpose, leave off most of the sloping surfaces in the rear. On gable or hip roofs, cover a considerable amount of the front slopes so that the front view gives the impression of completion. Roofing over important roof edges and ridges is necessary. On flat roofs, leave a large portion of the central area without covering, since it will not detract from the appearance of the elevations. Use black construction paper to represent roofing felt, usually cut to expose the multiple layers of a built-up roof. Fasten coarse sandpaper as the final aggregate surface to this type of roof, or use sand or bird gravel glued to the felt. Various abrasive sheets can be fashioned into strip fiberglass shingles and glued to the roof, placing them so that they overlap. For realistic wood shingles, use actual cedar shingles and cut the miniatures from the thin ends, or purchase them at a hobby shop. Copper foil (gasket stock) is attractive and easy to shape for metal roofs and flashing applications.

Many similar materials can be found to simulate other types of roofing; some will require painting. Half of the enjoyment of building models is the challenge of finding simple, inexpensive materials that will represent building textures at the correct scale.

16.4.6 Details

Minor details should be one of the last operations in the construction of the model, although it often becomes tempting to add a few during the construction to see how they will appear. By *details* we mean the small features such as doors, windows, cornice trim, and shutters that give the model its finished touch. When trim pieces are to be different in color from adjacent surfaces, paint the trim before attaching it. Avoid using aluminum paint on models; it generally appears tacky. If an aluminum window is specified, paint the pieces white. Positioning small pieces may require the use of a pair of tweezers. Cut window members, surrounds, sills, muntins, and the like, from thin balsa, making sure that the type and the scale of the window are accurately reproduced. Some of the muntin bars of a double-hung window, for instance, must be made very thin.

For exterior doors, cut each from balsa after the trim around their openings has been affixed, and give them their correct detail according to the door schedule. To represent various treatments, you may score a balsa door with a stylus to show panels, or glue stile and rail strips on the door, or glue small panels on the door. Flush doors are smooth except possibly for suggestions of the latch hardware. Take time to detail the front entrance carefully. Form wrought-iron work from soft wire, solder the joints, and spray it a dull black. Other minor details may be added if they contribute additional character to the model.

16.4.7 Landscaping

Landscaping on structural models is done in a similar manner as was suggested for smaller architectural models; the important thing is maintaining scale (Fig. 16–1).

If necessary, build up the lot contours with wire screen and blocking, covering the surface grade with molding plaster, spackling compound, or papier-mâché. Papier-mâché may be made from shredded newspapers soaked in water until they become a fibrous mass. Drain and mix the fibers with wallpaper paste and sawdust to make a sticky, plastic material for covering the screen. Large areas of papier-mâché require considerable time in drying and may split when dry unless shrinkage joints are provided.

Paint the grass surfaces green, and cover them with model builder's grass material (green sawdust), or use green flocking over the paint. When painting drives, walks, and the like, use masking tape to maintain neat edges between the adjacent colors. Drives and walks may be edged with small balsa strips to make them definitive and give them a clean-cut appearance.

For the trees and shrubs, use green sponges or lichen, properly shaped and sized (Fig. 16–2). For trees, cement the forms to the twigs or branches; to attach the trees, follow the same procedure as for architectural models. Fasten the shrubbery in interesting groups near corners of the model and in other areas where the architectural lines need to be softened. Also, use plantings to emphasize the front door, or place them as natural screens for privacy.

If edge grain shows around the periphery of the base, select a suitable molding or stock casing for covering the edge. Nail the trim about $1/8''$ above the surface of the grade line to prevent the grass material from dropping off. Miter the corners of the trim, and either paint or stain it with a suitable finish.

REVIEW QUESTIONS

1. What part does the model of a building play in architectural service?

2. What characteristics of an architectural proposal does a model most effectively show?

3. Why is balsa wood a favorite material for many model builders?

4. How are site contours usually represented on architectural models?

5. Why is the scale so important in the selection of model building materials?

6. Why must the model builder use particular care as to the direction in which wood grain is placed on exterior textures?

7. Why is soft pine recommended for structural lumber rather than balsa wood?

8. What is the disadvantage of using glossy enamels on model surfaces?

9. Name various materials convenient for representing roofing textures.

10. Why can landscaping problems be more easily solved on a model than on a landscaping plan?

"Nothing is achieved before it is
thoroughly attempted."
—SIR PHILIP SIDNEY

WRITING SPECIFICATIONS 17

In light of what has been said about graphic description, it will seem obvious that working drawings, however well drawn, cannot be entirely adequate in completely revealing all aspects of a construction project. Many things cannot be shown graphically. For example, how would you show on a drawing the quality of workmanship required on a kitchen cabinet, except by an extensive hand-lettered note?

The information that is necessary for the construction of any structure is usually developed by means of two basic documents: drawings and specifications. These two documents represent distinct methods of transmitting information. They should complement one another and should not overlap or duplicate one another. *Drawings* are a graphical portrayal, and *specifications* are a written description of the legal and technical requirements forming the contract documents. Working drawings describe the construction quantitatively, that is *how much*, whereas specifications describe the project qualitatively, or *how good/well*. Together, the working drawings and specifications communicate a complete picture of the construction.

This chapter examines the detailed instructions, often called *specs*, together with other written documents, and it shows their relationship to working drawings.

17.1

RELATIONSHIP OF DRAWINGS TO THE SPECIFICATIONS AND OTHER WRITTEN DOCUMENTS

We have seen the importance of architectural drawings in conveying the architect's intent. However, in the world of construction, several other written documents are just as important as the drawings in establishing the contract between the owner and the contractor. We must remember that the purpose of these documents is to provide communication that will allow the project (1) to be constructed by the general contractor and the subcontractors in a fashion and to the quality intended by the architect and perceived by the owner, as indicated in the contract documents, and (2) to meet building codes and other applicable laws.

The term *contract documents* is accepted as meaning those documents that taken together make up the contract between the owner and the contractor for building construction. These documents are as follows:

1. Working drawings
2. Specifications
3. Contract conditions (general and supplementary)
4. Addenda (changes made before contract signing)
5. Change orders (changes made after contract signing)
6. Agreement (or the contract between the owner and the contractor)

In projects for which contracts are awarded through the bidding process, additional documents called *bidding requirements* must be added to the contract documents. These include the following:

1. Invitation to bid
2. Instructions to bidders
3. Bid form
4. Information available to bidders

Notice that bidding requirements are not contract documents, and, since they are not, they do not actually make up a part of the documents upon which a contract or an agreement is signed between the owner and the contractor.

In the past, there was a tendency to develop *precedence* of one document over another in case of dispute. Under that scenario, specifications usually had precedence over drawings, assuming that the written word was more correct than graphics. This assumption proved not to be true, and it is recommended that precedence not be incorporated into documents. Discrepancies should be referred to the architect for clarification.

TYPES OF PROJECTS

There are a number of types of project delivery, depending on the conditions surrounding the project and, to a lesser extent, on the project size. The most common form is the *bid* project, in which the bidding documents are distributed by the architect to several general contractors, called *bidders*. These contractors, along with their subcontractors, estimate the construction cost and bid for a project, with the contract award going to the bidder submitting the lowest-cost bid. This type of contract is used for most government and public projects where bidding is required by law. Many owners prefer this method, as it assures them of the lowest initial construction cost.

Also in widespread use is the *negotiated* type of contract. A negotiated contract eliminates bidding and has the owner select a contractor with whom he or she negotiates the cost and time for construction. When using this type of contract, the contractor uses every effort to control costs and stay within the owner's budget and construction time frame as established before construction begins. Many owners protect themselves from cost overruns in negotiated projects by establishing a *guaranteed maximum cost* ceiling. This ceiling assures the owner that the contract amount cannot exceed the given amount unless the nature of the work is changed. A private owner might select a negotiated contract to get a contractor on line who has experience with a particular type of construction or who has expertise in delivery of a project within a short period of time. Naturally, the architect and the contractor work more closely together as a team during pricing and construction.

A variation of the negotiated type contract is the *fast-track* type of delivery process. The procedure is much the same as in a negotiated-type contract, except that the contractor is involved much sooner in the process, advising the owner and the architect of systems, products, costs, and methods during the development of the contract documents. Construction may actually start prior to completion of the working drawings and specifications to shorten the construction time. This factor may be important when an owner has a commitment to a tenant for an early move-in date or when construction financing costs make it advantageous to the owner to gamble in order to reduce this "soft cost." As with negotiated contracts, fast-track contracts usually have guaranteed maximum cost caps. The architect, the contractor, and the owner must work very closely in fast-track construction, as decisions must be made quickly and can require input from all. Since construction may already be underway as documents are developed, it is incumbent on all that the proper decisions be made in a timely manner.

The *design–build* type of project delivery stipulates that one entity, the contractor, has a contract with the owner to perform design as well as construction. This process is opposite to the normal process in which the architect has a separate design contract with the owner. In this case, the contractor has control of the design process and integrates it within his or her estimation, schedule, and construction processes to take total responsibility for design and construction.

THE PROJECT MANUAL CONCEPT

The term *specifications*, used to describe the book often accompanying the working drawings, is actually a misnomer, as the book called *Specifications* in most cases includes documents, such as bidding requirements and contract conditions, that are not specifications. The American Institute of Architects (A.I.A.) developed the idea of the *Project Manual* in 1964, as the volume that would contain the specifications and any other data conveniently bound into one volume. Although slow to gain acceptance, the term *Project Manual* has now come into wide usage. This definition is even more important with the increased inclusion of such items as door and finish schedules, traditionally designated as drawing items, in the Project Manual.

WHO WRITES SPECIFICATIONS

Regardless of which form a firm uses to define its specification-writing operation, the functions are essentially the same. The specifier has primary responsibility for producing the Project Manual, including performing the product research and the evaluation necessary to specify the right material to perform the right purpose. The specifier will prepare this text from a master specification that he or she maintains. The specifier must write specification sections from scratch when no master section exists. He or she sees product manufacturers' representatives who visit the office and discuss their products and the specifier's needs as they relate to the projects in the office. The specifier is generally the source for product and industry information for others in the office, and in many cases, he or she handles the field problems. The specifier is responsible for seeing that information gained is incorporated back into the system. The specifications may be written by one of several people.

In small architectural offices, *firm principals*, who have great experience with and appreciation for the liabilities of poorly written specifications, may do the specifying. In others, *project architects* and *construction administration personnel* who have had experience with materials and construction frequently compile the specifications.

In larger architectural offices, designated *specification writers* or *specifiers* often prepare specifications and related documents. They are registered architects or architectural personnel who have extensive experience both in material evaluation and selection and in-field applications and problems. In addition, specifiers usually have acquired an ability to write using correct English grammar and agreement and to use accepted words, spellings, and abbreviations consistently within established office policy.

Many multioffice firms maintain a *central specification department* in the home office, where specifications for projects the world over are prepared from the office's master specification system. Other firms prefer to keep specification personnel in various offices where they can be more conversant with local conditions, materials, and installations. At this time, there seems to be no clear-cut advantage to either system. Increasingly, specifiers are being assisted by office librarians, whose job it is to keep up with the building product catalogs and industry standards that are delivered to the office by mail and by manufacturers' representatives. It is virtually impossible for one person with several duties to be successful in maintaining a reference library used by an entire office.

In many firms, both large and small, the task of preparing project specifications from the firm's master system is the responsibility of the project architect. The master system is maintained by an in-house specification coordinator, who in a larger office performs the duties of the specifier but who leaves the preparation of the actual specification to the person who knows the most about the project—the project architect. This approach has grown greatly with the proliferation of commercially available master specification text and the personal computer.

A *specification consultant* specializes in the preparation of specifications and related documents for a number of architectural firms, while maintaining his or her independence as an outside consultant. The consultant's service makes expertly prepared specifications and related documents available to small- and medium-size firms so that these firms do not incur the cost of maintaining a full-time specification department. The specifier's expertise in material evaluation and selection, as well as knowledge of contract forms, makes him or her a valuable asset to firms that do not have or want in-house experience. Many specification consultants are members of the organization Specification Consultants in Independent Practice (S.C.I.P.), which deals with the particular problems and concerns of the consultant.

The correct preparation of construction documents is necessary in order to run economical and trouble-free projects. The construction industry today prefers that the person preparing the project specifications have demonstrated a minimum level of competency in the area of specification writing. The primary vehicles for demonstrating this competency are the certification exams sponsored by the Construction Specifications Institute (C.S.I.). The entry-level exam is the CDT (Certified Document Technician), and a person must pass this test before taking the advanced-level exam, the CCS (Certified Construction Specifier). C.S.I. also sponsors two additional certification programs: CCPR (Certified Construction Product Representative) and CCCA (Certified Construction Contract Administrator).

17.5

SPECIFICATION MATERIAL SOURCES

To write a completely new set of specifications for each job would be unnecessarily time-consuming. Instead, specifiers rely on various sources for reference material and standard specifications from which they can compile a set for each new job. They rely heavily on specifications that have repeatedly proved satisfactory in the past. Some specs may have to be modified to fit the conditions of a given job; many can be used word for word. Master specifications text such as A.I.A.'s *Masterspec*®, A.I.A.'s *SPECSystem*™, and C.S.I.'s *Spec-Text*® are often used to write new sections. These master systems contain guide specifications for many materials, allowing the specifier the luxury of editing out unnecessary text rather than generating new information each time. Master specifications also incorporate correct specification language and format for ease of specification preparation.

Most new specification sections are generated from master systems. Listed below are the major sources from which specification material is available:

1. City and national codes and ordinances
2. Manufacturers' catalogs (*Sweet's Catalog File*, *Man-U-Spec*, *Spec-data*)
3. Manufacturers' industry associations (Architectural Woodworking Institute, American Plywood Association, Door and Hardware Institute, Tile Council of America)
4. National standards organizations (American National Standards Institute, National Institute of Building Sciences)
5. Testing societies (American Society for Testing and Materials, Underwriters Laboratories)
6. Master specifications (*Masterspec*®, *SPECSystem*™, *Spec-Text*®)
7. Individual files of previously written specifications
8. Books on specifications
9. Federal specifications (*Specs-In-Tact*, G.S.A., N.A.S.A., N.A.F.V.A.C.)
10. Magazines and publications (*Construction Specifier*, *Architecture*, *Architectural Record*)

Not only must the specifier be familiar with construction methods, new materials, and building techniques, but he or she must also be able to accumulate reference material and organize it for easy access when it is needed.

SPECIFICATION ARRANGEMENT

The Construction Specifications Institute (C.S.I.) has for a number of years engaged in a program of standardizing specification format. After questioning its members, the institute concluded that a universal need exists for more uniformity in format arrangement and that a consistent national format would prove beneficial not only to the writers of specifications, but also to the contractors and material suppliers as well. After evaluating the various comments, the institute has compiled a format that provides the advantages of standardization yet has sufficient flexibility to allow writers throughout the country enough latitude in expression to accommodate local codes and trade practice variations. Only the arrangement in the format is restrictive, so that a set of specifications, like the alphabet or a numbering system, will be universally useful because of its consistency. The benefits of the format will be realized mainly from its widespread usage. Although nearly every set of specifications varies in content because of variations in circumstance, it has been found that arrangement of content can be identical.

The C.S.I. format consists of the following:

- A universal numbering system applicable to all possible items of work
- Breakdowns for each technical item and group of related items
- A three-part designation for each technical item
- A suggested page layout

C.S.I. breaks down construction into a group of *sections*, where each section is an item of work. Breakdown does not presume that the contractor will subcontract with regard to the section since the contractor has control over and charge of the construction, although the breakdown is so reasonable that in many cases this is possible. Each section can be *broadscope*, *mediumscope*, or *narrowscope* in nature, depending on how complex the project is and what the specifier chooses to include. An example is a section for acoustical ceilings. As a broadscope section, the section entitled "Acoustical Ceilings" might include several types of lay-in ceiling panels, a concealed-grid acoustical ceiling system, and metal suspension systems for all. In a mediumscope section, the specifier might specify only the lay-in ceiling panels and the related metal suspension system in a section entitled "Lay-in Acoustical Ceiling Systems." Another mediumscope section would specify the concealed-grid system. Finally, a narrowscope section might be limited to only the metal suspension systems, entitled "Ceiling Suspension Systems," assuming that related narrowscope sections would be written for the acoustical panels themselves.

The C.S.I. format groups sections of like construction into *divisions*. This format is often referred to as the *16-division format* since there are 16 static divisions. Since these divisions are always the same, any person familiar with this format can easily find his or her way through the Project Manual. In the table of contents, divisions that are not applicable to the project in question are simply designated as "Not Used."

The specification user is further aided by the breakdown of each section into three *parts*. The three-part format locates data within each section with regard to its basic relationships:

- **Part 1: General** Those items relating in general to an item, such as shop drawing requirements and delivery procedures.
- **Part 2: Products** Relates to the material itself, including mixing and fabrication.
- **Part 3: Execution** Relates to the installation of the material.

Breakdown of the section into these parts provides a skeleton for the specifier, thus making it simpler to prepare the specification in a consistent manner. It also makes it easy for the user to find exactly what he or she is looking for without laboriously reading the entire section. See Fig. 17–1 for suggestions of article titles within the parts.

C.S.I. further simplifies the use of the section by using the *page format*. This suggested layout provides article, paragraph, and subparagraph numbering; spacing, tabs and margins; and other similar data necessary to create the most readable text.

The five-digit numbering system unifies the 16-division, three-part format. Under this system, broadscope and many mediumscope section titles are given static numbers. The first two digits represent the division, 1 through 16. The other three digits represent a somewhat arbitrary sequence of sections, although by being organized into standard order, the sequencing promotes standardization. Creation of numbers for narrowscope sections rests with the specifier. The master list of titles and numbers is published by C.S.I. in conjunction with Construction Specifications Canada (C.S.C.) in *Masterformat*. See Fig. 17–2 for a list of divisions and broadscope section titles with appropriate five-digit numbers. *Masterformat* is available individually or as a part of the C.S.I. *Manual of Practice*, which contains a great deal more information regarding specification preparation and use than can be included in this text. It is recommended that the C.S.I. *Manual of Practice* be used by those having a need for greater detail.

A sample specification section incorporating the C.S.I. division, section, and page format is shown in Fig. 17–3.

In addition to the format described above, specifiers realize the need for other, shorter formats to accommodate smaller commercial projects and residential projects.

ARTICLE TITLES BY PART

Part 1 GENERAL

1.01 SUMMARY
 A. Section includes.
 B. Products furnished but not installed under this section.
 C. Products installed but not furnished under this section.
 D. Related sections.
 E. Allowances.
 F. Unit prices.
 G. Alternates.
1.02 REFERENCES
1.03 DEFINITIONS
1.04 SYSTEM DESCRIPTION
 A. Design requirements.
 B. Performance requirements.
1.05 SUBMITTALS
 A. Product data, shop drawings, and samples.
 B. Quality control submittals.
 C. Design data, test reports, certificates, manufacturer's instructions, field reports.
 D. Contract closeout submittals.
 E. Project record documents, operation and maintenance data, warranty.
1.06 QUALITY ASSURANCE
 A. Qualifications.
 B. Regulatory requirements.
 C. Certifications.
 D. Field samples.
 E. Mock-ups.
 F. Preinstallation conference.
1.07 DELIVERY, STORAGE, AND HANDLING
 A. Packing and shipping.
 B. Acceptance at site.
 C. Storage and protection.
1.08 PROJECT/SITE CONDITIONS
 A. Environmental requirements.
 B. Existing conditions.
 C. Field measurements.
1.09 SCHEDULING AND SEQUENCING
1.10 WARRANTY
 A. Special warranty.
1.11 MAINTENANCE
 A. Maintenance service.
 B. Extra materials.

PART 2 PRODUCTS

2.01 MANUFACTURERS
2.02 MATERIALS
2.03 MANUFACTURED UNITS
2.04 EQUIPMENT
2.05 COMPONENT
2.06 ACCESSORIES
2.07 MIXES
2.08 FABRICATION
 A. Shop assembly.
 B. Shop/factory finishing.
 C. Tolerances
2.09 SOURCE QUALITY CONTROL
 A. Tests, inspections.
 B. Verification of performance.

PART 3 EXECUTION

3.01 EXAMINATION
 A. Verification of conditions.
3.02 PREPARATION
 A. Protection.
 B. Surface protection.
3.03 ERECTION/INSTALLATION/APPLICATION
 A. Special techniques.
 B. Interface with other products.
 C. Tolerances.
3.04 FIELD QUALITY CONTROL
 A. Manufacturer's field service.
3.05 ADJUSTING/CLEANING
3.06 DEMONSTRATING
3.07 PROTECTION
3.08 SCHEDULES

Figure 17–1 Three-part format.

Outline specifications are often useful as communication tools for commercial projects and may be all that is necessary to construct simpler projects. A sample outline specification section is shown in Fig. 17–4.

DIVISION 1: GENERAL REQUIREMENTS

In examining the breakdown of the 16 divisions, note that all divisions except division 1 relate to the construction itself. Division 1, General Requirements, applies to all other divisions and acts as a bridge between the technical sections and other bidding and contract documents. Division 1 sections relate to administrative and procedural requirements and temporary facilities. Actual contract requirements might be specified by the agreement or contract conditions, but the rules for implementing would be specified in division 1.

GENERAL POINTS FOR PREPARING SPECIFICATIONS

Keep the following points in mind when writing specifications:

1. Use simple, direct language and accepted terminology, rather than abstract legal terminology.
2. Use brief sentences requiring only simple punctuation.
3. Specify standard items and alternates where possible, in the interest of economy, without sacrificing quality.
4. Avoid repetition; use cross-references if they apply and seem logical.
5. Avoid specifications that are impossible for the contractor to carry out; be fair in designating responsibility.
6. Avoid including specifications that are not to be part of the construction.
7. Clarify all terms that may be subject to more than one interpretation.
8. Be consistent in the use of terms, abbreviations, and format and in the arrangement of material.
9. Specify numbers, names, and descriptions of materials from the *latest editions* of manufacturers' catalogs.
10. Capitalize the following: major parties to the contract, such as Contractor, Owner, Designer, Architect; the contract documents, such as Specifications, Working Drawings, Contract, Supplementary Conditions; specific rooms within the building, such as Kitchen, Living Room, Office; grade of materials, such as No. 1 Douglas Fir, Clear Heart Redwood, FAS White Oak; and, of course, all proper names.
11. Differentiate between "shall" and "will"—"The Contractor shall. . ."; "The Owner or Architect will. . . ."

THE C.S.I. FIVE-DIGIT BROADSCOPE FORMAT

BIDDING REQUIREMENTS, CONTRACT FORMS, AND CONDITIONS OF THE CONTRACT

00010	PREBID INFORMATION
00100	INSTRUCTIONS TO BIDDERS
00200	INFORMATION AVAILABLE TO BIDDERS
00300	BID FORMS
00400	SUPPLEMENTS TO BID FORMS
00500	AGREEMENT FORMS
00600	BONDS AND CERTIFICATES
00700	GENERAL CONDITIONS
00800	SUPPLEMENTARY CONDITIONS
00900	ADDENDA

Note: The items listed above are not specification sections and are referred to as "Documents" rather than "Sections" in the Master List of Section Titles, Numbers, and Broadscope Section Explanations.

SPECIFICATIONS

DIVISION 1—GENERAL REQUIREMENTS

01010	SUMMARY OF WORK
01020	ALLOWANCES
01025	MEASUREMENT AND PAYMENT
01030	ALTERNATES/ALTERNATIVES
01035	MODIFICATION PROCEDURES
01040	COORDINATION
01050	FIELD ENGINEERING
01060	REGULATORY REQUIREMENTS
01070	IDENTIFICATION SYSTEMS
01090	REFERENCES
01100	SPECIAL PROJECT PROCEDURES
01200	PROJECT MEETINGS
01300	SUBMITTALS
01400	QUALITY CONTROL
01500	CONSTRUCTION FACILITIES AND TEMPORARY CONTROLS
01600	MATERIAL AND EQUIPMENT
01650	FACILITY STARTUP/COMMISSIONING
01700	CONTRACT CLOSEOUT
01800	MAINTENANCE

DIVISION 2—SITEWORK

02010	SUBSURFACE INVESTIGATION
02050	DEMOLITION
02100	SITE PREPARATION
02140	DEWATERING
02150	SHORING AND UNDERPINNING
02160	EXCAVATION SUPPORT SYSTEMS
02170	COFFERDAMS
02200	EARTHWORK
02300	TUNNELING
02350	PILES AND CAISSONS
02450	RAILROAD WORK
02480	MARINE WORK
02500	PAVING AND SURFACING
02600	UTILITY PIPING MATERIALS
02660	WATER DISTRIBUTION
02680	FUEL AND STEAM DISTRIBUTION
02700	SEWERAGE AND DRAINAGE
02760	RESTORATION OF UNDERGROUND PIPE
02770	PONDS AND RESERVOIRS
02780	POWER AND COMMUNICATIONS
02800	SITE IMPROVEMENTS
02900	LANDSCAPING

DIVISION 3—CONCRETE

03100	CONCRETE FORMWORK
03200	CONCRETE REINFORCEMENT
03250	CONCRETE ACCESSORIES
03300	CAST-IN-PLACE CONCRETE
03370	CONCRETE CURING
03400	PRECAST CONCRETE
03500	CEMENTITIOUS DECKS AND TOPPINS
03600	GROUT
03700	CONCRETE RESTORATION AND CLEANING
03800	MASS CONCRETE

DIVISION 4—MASONRY

04100	MORTAR AND MASONRY GROUT
04150	MASONRY ACCESSORIES
04200	UNIT MASONRY
04400	STONE
04500	MASONRY RESTORATION AND CLEANING
04550	REFRACTORIES

04600	CORROSION RESISTANT MASONRY
04700	SIMULATED MASONRY

DIVISION 5—METALS

05010	METAL MATERIALS
05030	METAL COATINGS
05050	METAL FASTENING
05100	STRUCTURAL METAL FRAMING
05200	METAL JOISTS
05300	METAL DECKING
05400	COLD FORMED METAL FRAMING
05500	METAL FABRICATIONS
05580	SHEET METAL FABRICATIONS
05700	ORNAMENT METAL
05800	EXPANSION CONTROL
05900	HYDRAULIC STRUCTURES

DIVISION 6—WOOD AND PLASTICS

06050	FASTENERS AND ADHESIVES
06100	ROUGH CARPENTRY
06130	HEAVY TIMBER CONSTRUCTION
06150	WOOD AND METAL SYSTEMS
06170	PREFABRICATED STRUCTURAL WOOD
06200	FINISH CARPENTRY
06300	WOOD TREATMENT
06400	ARCHITECTURAL WOODWORK
06500	STRUCTURAL PLASTICS
06600	PLASTIC FABRICATIONS
06650	SOLID POLYMER FABRICATIONS

DIVISION 7—THERMAL AND MOISTURE PROTECTION

07100	WATERPROOFING
07150	DAMPPROOFING
07180	WATER REPELLENTS
07190	VAPOR RETARDERS
07195	AIR BARRIERS
07200	INSULATION
07240	EXTERIOR INSULATION AND FINISH SYSTEMS
07250	FIREPROOFING
07270	FIRESTOPPING
07300	SHINGLES AND ROOFING TILES
07400	MANUFACTURED ROOFING AND SIDING
07480	EXTERIOR WALL ASSEMBLIES
07500	MEMBRANE ROOFING
07570	TRAFFIC COATINGS
07600	FLASHING AND SHEET METAL
07700	ROOF SPECIALTIES AND ACCESSORIES
07800	SKYLIGHTS
07900	JOINT SEALERS

DIVISION 8—DOORS AND WINDOWS

08100	METAL DOORS AND FRAMES
08200	WOOD AND PLASTIC DOORS
08250	DOOR OPENING ASSEMBLIES
08300	SPECIAL DOORS
08400	ENTRANCES AND STOREFRONTS
08500	METAL WINDOWS
08600	WOOD AND PLASTIC WINDOWS
08650	SPECIAL WINDOWS
08700	HARDWARE
08800	GLAZING
08900	GLAZED CURTAIN WALLS

DIVISION 9—FINISHES

09100	METAL SUPPORT SYSTEMS
09200	LATH AND PLASTER
09250	GYPSUM BOARD
09300	TILE
09400	TERRAZZO
09450	STONE FACING
09500	ACOUSTICAL TREATMENT
09540	SPECIAL WALL SURFACES
09545	SPECIAL CEILING SURFACES
09550	WOOD FLOORING
09600	STONE FLOORING
09630	UNIT MASONRY FLOORING
09650	RESILIENT FLOORING
09680	CARPETING
09700	SPECIAL FLOORING
09780	FLOOR TREATMENT
09800	SPECIAL COATINGS
09900	PAINTING
09950	WALL COVERINGS

DIVISION 10—SPECIALTIES

10100	VISUAL DISPLAY BOARDS
10150	COMPARTMENTS AND CUBICLES
10200	LOUVERS AND VENTS
10240	GRILLES AND SCREENS
10250	SERVICE WALL SYSTEMS
10260	WALL AND CORNER GUARDS
10270	ACCESS FLOORING
10290	PEST CONTROL
10300	FIREPLACES AND STOVES
10340	MANUFACTURED EXTERIOR SPECIALITIES
10350	FLAGPOLES
10400	IDENTIFYING DEVICES
10450	PEDESTRIAN CONTROL DEVICES
10500	LOCKERS
10520	FIRE PROTECTION SPECIALTIES
10530	PROTECTIVE COVERS
10550	POSTAL SPECIALTIES
10600	PARTITIONS
10650	OPERABLE PARTITIONS
10670	STORAGE SHELVING
10700	EXTERIOR PROTECTION DEVICES FOR OPENINGS
10750	TELEPHONE SPECIALTIES
10800	TOILET AND BATH ACCESSORIES
10880	SCALES
10900	WARDROBE AND CLOSET SPECIALTIES

DIVISION 11—EQUIPMENT

11010	MAINTENANCE EQUIPMENT
11020	SECURITY AND VAULT EQUIPMENT
11030	TELLER AND SERVICE EQUIPMENT
11040	ECCLESIASTICAL EQUIPMENT
11050	LIBRARY EQUIPMENT
11060	THEATER AND STAGE EQUIPMENT
11070	INSTRUMENTAL EQUIPMENT
11080	REGISTRATION EQUIPMENT
11090	CHECKROOM EQUIPMENT
11100	MERCANTILE EQUIPMENT
11110	COMMERCIAL LAUNDRY AND DRY CLEANING EQUIPMENT
11120	VENDING EQUIPMENT
11130	AUDIO-VISUAL EQUIPMENT
11140	VEHICLE SERVICE EQUIPMENT
11150	PARKING CONTROL EQUIPMENT
11160	LOADING DOCK EQUIPMENT
11170	SOLID WASTE HANDLING EQUIPMENT
11190	DETENTION EQUIPMENT
11200	WATER SUPPLY AND TREATMENT EQUIPMENT
11280	HYDRAULIC GATES AND VALVES
11300	FLUID WASTE TREATMENT AND DISPOSAL EQUIPMENT
11400	FOOD SERVICE EQUIPMENT
11450	RESIDENTIAL EQUIPMENT
11460	UNIT KITCHENS
11470	DARKROOM EQUIPMENT
11480	ATHLETIC, RECREATIONAL, AND THERAPEUTIC EQUIPMENT
11500	INDUSTRIAL AND PROCESS EQUIPMENT
11600	LABORATORY EQUIPMENT
11650	PLANETARIUM EQUIPMENT
11660	OBSERVATORY EQUIPMENT
11680	OFFICE EQUIPMENT
11700	MEDICAL EQUIPMENT
11780	MORTUARY EQUIPMENT
11850	NAVIGATION EQUIPMENT
11870	AGRICULTURAL EQUIPMENT

DIVISION 12—FURNISHINGS

12050	FABRICS
12100	ARTWORK
12300	MANUFACTURED CASEWORKS
12500	WINDOW TREATMENT
12600	FURNITURE AND ACESSORIES
12670	RUGS AND MATS
12700	MULTIPLE SEATING
12800	INTERIOR PLANTS AND PLANTERS

DIVISION 13—SPECIAL CONSTRUCTION

13010	AIR SUPPORTED STRUCTURES
13020	INTEGRATED ASSEMBLIES
13030	SPECIAL PURPOSE ROOMS
13080	SOUND, VIBRATION, AND SEISMIC CONTROL

Figure 17-2 The C.S.I. five-digit broadscope format.

13090	RADIATION PROTECTION	13900	FIRE SUPPRESSION AND SUPERVISORY SYSTEMS	15500	HEATING, VENTILATING, AND AIR CONDITIONING
13100	NUCLEAR REACTORS	13950	SPECIAL SECURITY CONSTRUCTION	15550	HEAT GENERATION
13120	PRE-ENGINEERED STRUCTURES			15650	REFRIGERATION
13150	AQUATIC FACILITIES			15750	HEAT TRANSFER
13175	ICE RINKS		DIVISION 14—CONVEYING SYSTEMS	15850	AIR HANDLING
13180	SITE CONSTRUCTED INCINERATORS	14100	DUMBWAITERS	15880	AIR DISTRIBUTION
13185	KENNELS AND ANIMAL SHELTERS	14200	ELEVATORS	15950	CONTROLS
13200	LIQUID AND GAS STORAGE TANKS	14300	ESCALATORS AND MOVING WALKS	15990	TESTING, ADJUSTING, AND BALANCING
13220	FILTER UNDERDRAINS AND MEDIA	14400	LIFTS		
13230	DIGESTER COVERS AND APPURTENANCES	14500	MATERIAL HANDLING SYSTEMS		DIVISION 16—ELECTRICAL
13240	OXYGENATION SYSTEMS	14600	HOIST AND CRANES	16050	BASIC ELECTRICAL MATERIALS AND METHODS
13260	SLUDGE CONDITIONING SYSTEMS	14700	TURNTABLES	16200	POWER GENERATION-BUILT-UP SYSTEMS
13300	UTILITY CONTROL SYSTEMS	14800	SCAFFOLDING	16300	MEDIUM VOLTAGE DISTRIBUTION
13400	INDUSTRIAL AND PROCESS CONTROL SYSTEMS	14900	TRANSPORTATION SYSTEMS	16400	SERVICE AND DISTRIBUTION
13500	RECORDING INSTRUMENTATION			16500	LIGHTING
13550	TRANSPORTATION CONTROL INSTRUMENTATION		DIVISION 15—MECHANICAL	16600	SPECIAL SYSTEMS
13600	SOLAR ENERGY SYSTEMS	15050	BASIC MECHANICAL MATERIALS AND METHODS	16700	COMMUNICATIONS
13700	WIND ENERGY SYSTEMS	15250	MECHANICAL INSULATION	16850	ELECTRICAL RESISTANCE HEATING
13750	COGENERATION SYSTEMS	15300	FIRE PROTECTION	16900	CONTROLS
13800	BUILDING AUTOMATION SYSTEMS	15400	PLUMBING	16950	TESTING

Figure 17–2 (continued)

12. Avoid the term *a workmanlike job* or similarly vague phrases; rather, describe the quality of workmanship or the exact requirements expected.

13. Use accepted standards when specifying quality of materials or workmanship required, such as "Lightweight concrete masonry units: ASTM C-90-85; Grade N, Type 1."

14. Number all pages within a section consecutively, and include a table of contents for each section.

15. Keep in mind that bidders and subcontractors of different trades will have to use the specifications to look up information in their respective areas; remember that the information dealing with their work should be stated in the logical section and not hidden throughout the various sections.

17.9

METHODS OF SPECIFYING

One of the first things a specifier must do when writing a master specification section or an individual project section from scratch is to decide what method he or she will use to communicate with the contractor. Four basic types of specifications can be prepared, and most items can be specified by one or more methods.

The first and easiest method of specifying is the *proprietary specification*, which basically names the manufac-

turers or products acceptable to the architect. A proprietary specification may be written as a *closed* proprietary specification, in which only one product is acceptable and the contractor is so advised by the nature of the specification. The specification may also be written as an *open* proprietary section, in which multiple manufacturers or products are named or alternatives solicited. Naturally, an open proprietary specification results in more competition among vendors and may result in a lower installed price. However, an open proprietary specification may not be applicable if one specific product is desired.

An example of a closed proprietary specification is one for brick for an addition to an existing building, where the brick must match existing brick. Only one product is desired. By comparison, an open proprietary specification might be applicable for building sealants when multiple products with the same characteristics can be specified. However, the specifier should be sure that the product specified is equivalent to avoid an "apples and oranges" specification.

A word of caution about proprietary specifications is due with regard to use of the phrase "or equal." Frequently used in the past to convey the concept that the architect would consider products other than those specified, which he or she considered "equal" in quality, the term is unclear and leads to disagreement as to what is equal and in whose eyes. Thankfully, the use of the "or equal" clause has almost disappeared from specification writing and is most often used by persons unwilling to do the research and product evaluation necessary to specify the desired products.

SECTION 07310

SHINGLE ROOFING

PART 1 GENERAL

1.01 SUBMITTALS
- A. Product data: Indicate material specifications and installation instructions.
- B. Samples: Submit full-size samples of shingles in selected color for architect's approval.

1.02 DELIVERY, STORAGE, AND HANDLING
- A. Stack bundles of shingles not more than 3'-0" high. Store roll goods on end.

1.03 QUALITY ASSURANCE
- A. Application standards: Standards of the following as referenced herein:
- 1. American Society for Testing and Materials (ASTM).
- 2. Underwriters Laboratories, Inc. (UL)

PART 2 PRODUCTS

2.01 SHINGLES
- A. Acceptable product: Celotex Corp. Presidential Shake.
- B. Characteristics:
  1. Type: Heavy-weight asphalt-saturated fiberglass laminate shingles meeting ASTM D3018-82 Type I and UL Class A rating for fire and wind resistance.
  2. Shingles shall be curved, double-laminated tabs, self-sealing, random tab style.
  3. Minimum weight: 365 lb./sq.
  4. Color: As selected by Architect from manufacturer's standard colors.

2.02 ACCESSORY PRODUCTS
- A. Roofing felts: Meeting ASTM D226-87, Type I, #15 felt, asphalt-saturated organic felt, 3'-0" width, unperforated.
- B. Nails: Hot-dipped galvanized, 10 ga. screw threaded shank, 3/8" head, length as required to penetrate roof deck 1" minimum.
- C. Underlayment, starter and ridge shingles: Types furnished by shingle manufacturer for use with specified shingles. Match color of selected shingle.

PART 3 EXECUTION

3.01 APPLICATION
- A. Install felt layer over roof sheathing, lapping each course over lower course 2" minimum horizontally and 4" minimum side laps at end joints. Lap 6" from both sides at ridge.
- B. Secure felt layer to deck with sufficient fasteners to hold in place until shingles are secured.
- C. Apply shingles in straight, even courses, parallel to eave and ridge line, with four nails per shingle; nail heads concealed. Drive nails tight without cutting into shingles.
- D. Apply a starter course using starter shingles in accordance with manufacturer's product data.
- E. Install shingles with exposure recommended by manufacturer, in regular pattern, in accordance with shingle manufacturer's product data.
- F. Construct ridges using two layers of ridge shingles.
- G. Furnish two extra squares of roofing shingles for Owner's maintenance use. Store where directed.

END OF SECTION

Smith Residence 07310-1
Atlanta, GA
October 1, 1992

Figure 17–3 Typical specification section.

The second method of specifying is the *descriptive specification*, which describes the product in detail without providing the product name. Descriptive specifications can be used when it is not desirable to specify a particular product. Some government agencies require that specifications be written in descriptive form to allow the greatest competition among product manufacturers. Descriptive specifications are more difficult to write than proprietary ones since the specifier must indicate all product characteristics for which he or she has concern in the specification.

DIVISION 7-THERMAL AND MOISTURE PROTECTION

Section 07310—Shingle Roofing

1. General requirements: Submit product data and samples for shingles. Provide 30-year warranty.
2. Materials: Fiberglass shingles shall be Certainteed Hallmark Shangle or similar by Celotex, Elk, or Tamco. Shingles shall be minimum 300 lb./sq weight in color selected by Architect from manufacturer's standard colors. Provide roofing felt and roofing nails.
3. Execution: Install shingle roofing in accorance with manufacturer's product data, in straight, even courses, parallel to eaves. Install one layer of 15# felt, shingle fashion over sheathing prior to shingle application. Attach each shingle with three roofing nails.

Smith Residence 07310-1
Atlanta, GA
January 15, 1991

Figure 17–4 Typical outline specification.

A third type of specification in prevalent use is the *reference standard*. This standard simply references an accepted industry standard as the basis for the specification and is often used to specify generic materials such as portland cement and clear glass. In using a reference standard, the specifier should have a copy of that standard; should know what is required by the standard, including choices that may be contained therein; and should enforce those requirements for all suppliers.

A fourth type of specification is the *performance specification*. The performance specification gives the greatest leeway to the contractor because it allows him or her to furnish any product that meets the performance criteria specified. It is the most difficult for the specifier to prepare since he or she must anticipate all products and systems that could be used and must specify only those characteristics he or she desires—specifically those related to test results. A performance specification must provide sufficient data to ensure that product characteristics can be demonstrated. Performance specifications have their greatest use in specifying complex systems, such as a curtain wall, which can be easily custom designed for a certain application and engineered to perform to required levels, say, for wind loading and air infiltration. Testing for compliance is often required in performance specifications.

Most product specifications actually end up using a combination of methods to convey the architect's intent. A specification for brick masonry would use a proprietary specification to name the product or products selected by the architect; a descriptive specification to specify size and color; and a reference standard to specify the ASTM standard, grade, and type required.

17.10

ALLOWANCES, ALTERNATES, AND UNIT PRICES

No discussion of specification types is complete without mentioning three methods that the architect may need to use to delay making a product selection during bidding or early construction: (1) allowances, (2) unit prices, and (3) alternates.

An *allowance* is defined as follows:

1. an amount of money (a *cash allowance*) or a quantity of material (a *quantity allowance*) for which the quantity can be determined from the actual drawings but for which a decision as to what material to use has not been reached; or
2. a *lump sum* for a system that will be required but that has not been designed at the time bids are solicited.

An example of a cash allowance is one for a building's carpet, where an allowance of *X* dollars per square yard is specified. The quantity required can be determined from the working drawings. Or there might be a lump-sum cash allowance established for landscaping not yet designed. A quantity allowance could be used for tenant work, where historically a certain number of doors or linear footage of wall might be anticipated, but the space has not yet been designed.

Unlike allowances, *unit prices* are prices solicited from the contractor per unit of material when the system is defined but quantity is not. Requests for unit prices are often included in the bid form for such items as rock removal, where the soils report indicates the presence of rock to be removed, but the quantity is unknown until the rock is exposed.

Alternates are often included in the architect's bidding documents in order to obtain a price for an alternative system or material from the contractor. They stem from the fact that the architect does not do detailed estimates and depends on the contractor's estimating ability to determine the affordability of some items. The architect is in effect asking, "How much more (or how much less) does *X* cost than *Y*?" The inclusion of alternatives in bids demonstrates that the architect may be willing to compromise if the budget is exceeded. Alternates may also be suggested by the contractor, especially in negotiated contracts, as a means of cost reduction. Such suggestions should usually be *value engineering* items, meaning that systems or products of equivalent or better value can be substituted for less or the same cost. Value engineering is an important concept best practiced by someone experienced in that discipline.

17.11

PRODUCT EVALUATION

Perhaps the most important part of specification writing is proper product evaluation, or determining the applicability of a particular system or product. Product evaluation takes into consideration items such as esthetics, acoustics, fire safety, personal safety, stain resistance, interface, and replaceability. Other considerations include the desires of the owner, the manufacturer's reputation, initial and maintenance cost, and long-term maintenance and service requirements.

17.12

MASTER SPECIFICATIONS

As already discussed, master specification systems are one source of information for specifiers. Specification writing is not about creating original text; it is about using standardized text, combined with product and project knowledge, to communicate clearly, completely, concisely, and correctly with others who use the information created by the specifier. The use of master specification systems is not new. What is relatively new is the ability, through the personal computer, to utilize and manipulate master systems to a greater extent than ever before.

Many architectural firms maintain master systems of their own. These systems represent the attitudes of the firm principals, define the level of quality normally desired by that firm, and contain master specifications for systems and products normally specified by that firm. For example, a firm that does low-rise hospital work would need a highly refined specification for hydraulic, hospital elevators, but it probably would not spend the time to develop a master section for gearless elevators.

Many other firms, particularly smaller firms and firms doing a wide variety of work, normally rely on one of the several commercial master specification systems available. The A.I.A.'s *Masterspec*® and SPECSystem™ and C.S.I.'s *Spec-Text*® are some of the most successful. All are available in hard-copy form as well as on disks. These systems contain master sections for a myriad of products and are available for those who perform only interior design or for those who practice mechanical and electrical engineering. Master systems usually come complete with drawing coordination checklists and explanation sheets for those uninitiated in a particular type of construction. The biggest disadvantage of these systems is that by their nature their scope is at least national. Some even try to anticipate construction in other parts of North America and other continents. Their biggest advantage is the uniformity that they bring to construction. In an effort to eliminate the disadvantage, some firms buy these master systems and custom tailor the sections to their own practice.

TODAY AND TOMORROW

Specification production and reproduction have come a long way in just a few years, due to advances in technology. In the area of production, the norm today is for the specifier to edit his or her master system directly from the computer screen. The commercially available master systems are available in disk form using a number of word processors. The specifier simply loads the master system onto his or her personal computer, and he or she has instant access to the master system, complete with drawing checklist and explanation sheets. The master text itself will contain notes that explain certain articles and paragraphs. Once the specifier has edited the section, he or she can print out a draft copy complete with an audit trail. The audit trail tells what has been deleted or what decisions must be made. After reviewing sections for the entire project, making changes and spelling corrections, and perhaps inserting the current date on every page, the specifier can send the entire Project Manual to print at eight pages per minute on a laser jet printer. The printing can be done in a relatively short time period, eliminating the need to print until very late in document development. Thus, changes can be made until the last minute. The entire Project Manual is printed, complete with the name of the project on every page, and the table of contents is self-generated. The quality of the copy is far better than that previously available by typing.

Now the Project Manual is ready for copying and binding. Most copying is done today using high-speed copiers. The cost for reproduction in this manner has been reduced to the point that only the needed number of copies must be made, as opposed to previous methods in which economy was achieved only through making many copies at the same time. Now, other copies can be economically made later, if needed.

Most offices use an $8\frac{1}{2}'' \times 11''$ page size, with printing on one or two sides as preferred. The C.S.I. page format accommodates either one- or two-sided printing. Af-

ter copying, the Project Manual copies are bound, usually with a heavy stock cover bearing the firm's name or logo. The manner of binding is an individual preference, with several options that a printer can explain. The basic idea in binding text is to use a system that will hold up as the Project Manual is used in the field.

The state of the art in specification preparation technique is the CD-ROM. *SPEC-SEARCH*, developed for C.S.I. by the Data Matic Systems Co., is designed to provide computer-based search and retrieval capabilities for SPEC-DATA® and MANU-SPEC® documents. The program is capable of displaying up to four product specifications simultaneously, enabling users to analyze up to four products at the same time. The program also allows users to copy specifications into a word processor and print or fax them. *SPEC-SEARCH* will work under Windows 3.1 and Windows 95, although at this time it is not an officially compatible product.

SweetSource is an interactive product selection tool available on CD-ROM. Developed by Sweet's Group/McGraw-Hill, Inc., it provides access to over 5500 manufacturers by C.S.I. classification. It is a complete reference to *Sweet's General Building & Renovation Catalog* file.

The National Institute of Building Sciences (N.I.B.S.), a government agency, has a CD-ROM that contains the *Masterspec* and *Specs-In-Tact* master specification systems; many Navy, NASA, and Army Corps of Engineers design and maintenance manuals; and ASTM standards and standards from many other industry associations. This process is continuing, as other manufacturers are providing master specification text as well as drawing details on disks to the design community.

As for future developments in the area of specification writing, as in the area of changes in the way we now design and draw, we cannot imagine the changes that will take place in just a few years. With CADD already in place in many architectural offices, the next step is to integrate specification production so that the first draft of the Project Manual is produced along with the drawings. Toward this end, we will see more sophisticated development of master specification systems.

REVIEW QUESTIONS

1. What qualifications are required of a construction specification writer?

2. Drawings depict the quantity of a certain material. What do specifications depict?

3. List five sources for specification reference material.

4. Why are construction material manufacturers' asso-

ciations interested in furnishing material to specification writers?

5. Who requires a copy of the specifications? How are these copies usually made?

6. How is the information in the general conditions related to the other technical sections?

7. How do dimensions given in the specifications differ from the dimensions required on the working drawings?

8. Why must requirements in specifications be made fair and just to all parties of the contract?

9. List the three recommended parts of a typical technical section.

10. What are the major advantages in using the five-digit numbering system in spec writing?

11. Name several architectural or construction books that follow the five-digit numbering format.

12. What organization originated the 16-division *uniform* construction specification system?

13. Describe the differences among broadscope, mediumscope, and narrowscope.

14. List and describe the *three-part section* in specification writing.

"Doing easily what others find difficult is talent."

—HENRI-FRÉDÉRIC AMIEL

This chapter is concerned mainly with typical working drawings of small, custom-designed homes and the steps necessary for the satisfactory completion of these drawings by the student. To show traditional methods of depicting technical instructions for building houses, several sets of house plans have been included in this chapter. Each set is typical of current architectural practice and should be studied carefully for its manner of showing construction information. Occasionally, modifications will be necessary to satisfy local conditions and code requirements.

The *2-story solar home* shown in Fig. 18–1 is a conservative, small home with three bedrooms for the average family; it should be economical to heat and cool even with rising home energy costs. Eighty percent of the total heating needs will be provided by passive features. The home was designed by the Solar and Design Section of the Tennessee Valley Authority (TVA) for use throughout the Tennessee Valley area. If it is used in other climates of the United States, some modifications must be made so that it conforms to local solar conditions. Larger blueprints and specifications are available from TVA, Knoxville, TN 37900, for a nominal fee.

Notice that the house is to be oriented on a south-facing sloping lot with the entry on the north. The two-level plan has 1008 sq ft on each level. The double garage

Figure 18-1 Perspective rendering of solar home. (Tennessee Valley Authority)

and storage feature are optional. Conventional building materials and construction skills that are universally available are used throughout. To maintain more heat in the living areas, the major bedroom area is in the lower level. Various passive solar features are used:

A sun porch to collect solar energy

A 12″ concrete-filled block Trombe wall to help store solar energy

Water-filled drums also for energy storage

Shaded overhangs to prevent solar penetration of glass areas during the summer

Earth berming of the lower back wall to temper year-round temperatures

Maximum use of insulation throughout ceiling and walls

The use of insulating glass in all windows

Everything about the design is ideal for contemporary living.

The *2-story home* shown in Figs. 18–9 through 18–14 is traditional in nature, resembling many homes built in New England during colonial times and still popular in many areas (Fig. 18–15). Built over a crawl space with a cantilevered second story, this house has about 3000 sq ft of floor space, including the double carport, which makes it suitable for a larger family. The sleeping area is organized on the second floor for isolation from the downstairs living and eating areas. Because of the overhang, brick veneer is used on the garage and the first-floor level, and bevel siding is used for the walls above.

Instructor assignments—limiting the size of a home, reflecting the tastes of a tentative client, and indicating the major types of construction and materials—are very much like the actual situations in an architectural office. Usually, the architect first consults with the client about his or her wishes. Then design sketches are prepared. Further consultation may be necessary to resolve information about the project. In fact, more sketches may have to be done before all ideas are clearly in mind and before working drawings can be started. Student problems should be developed in the same manner—first by sketches, preferably on ⅛″ coordinate paper, and then on tracing paper with instruments. In the architectural office, the sketches of a project, together with any written information, are given to an experienced drafter or project chief for their development. Some consultation may be necessary, yet one person usually assumes responsibility for the completion of a small project. Student drafting assignments provide similar learning situations.

Figures 18–2 through 18–8 show conventional drafting practice and should serve merely as guides for developing original ideas into complete working drawings. If only manipulative drafting exercise is advisable, the plates can be duplicated directly (they are shown approximately half-size). While working on the drawings, you are urged to refer continually to *Sweet's Catalogs*, Ramsey and Sleeper's *Architectural Graphic Standards*, and other reference sources for additional help.

START WITH THE FLOOR PLAN

In general, follow these steps in developing the first-floor plan.

Step 1: *Lay out the major exterior walls with light lines.* Use the scale ¼″ = 1′-0″, and take the dimensions from the sketch. A 3H or 4H pencil works best for preliminary construction lines. Extend the lines beyond their termination points during construction until definite measurements are established. The following wall thicknesses are satisfactory on small plans, unless special conditions exist:

Wood frame exterior walls	6″
Wood frame partitions	5″
Brick veneer	10″ (4″ brick, 6″ wood)
Concrete block (C.M.U.)	4″, 8″, 12″ (nominal)
Solid brick	8″, 12″ (plus furring)
Brick cavity	10″ (2″ air space)

Draw the light lines completely through window and door openings at this preliminary stage. Then lay out all partitions, including, if necessary, an 8″ wide wet-wall for the bathroom's 4″ vent stack enclosure. Block in the areas for stairs, fireplaces, door stoops, and so on. Give clothes closets a minimum of 2′ interior width.

Step 2: *Locate all windows and doors with center lines.* Refer to the elevation sketches, if necessary, for window placements, and check manufacturers' catalogs for available sizes (see Figs. 18–4 and 18–11 for various window symbols). Select window types for their practicality and for their harmony with the spirit of the exterior. If elevations are not definite and window changes seem inevitable, draw the window symbols with only light lines, and strengthen them later after the elevation fenestration is completed. After considering the proper door widths that become necessary in the different rooms (Fig. 14–15), lay out the door symbols. Scale the symbol the same width as the door.

Study the traffic flow through doors before indicating their swing. Since interior doors are frequently left open, be sure to provide sufficient clear space for doors in an open position. Exterior residential doors swing in; interior doors should not swing into hallways. For convenient accessibility, provide either sliding- or folding-door units in closets; louver doors are especially appropriate, since they allow ventilation for the clothes. Symbols are shown for doors (see Fig. 14–15). Erase the construction lines within the door symbols before completing the symbols. Indicate doorsills or waterproof thresholds below exterior doors. Cased openings or archways, which drop below ceiling heights, must be shown with broken lines.

A SOLAR HOME

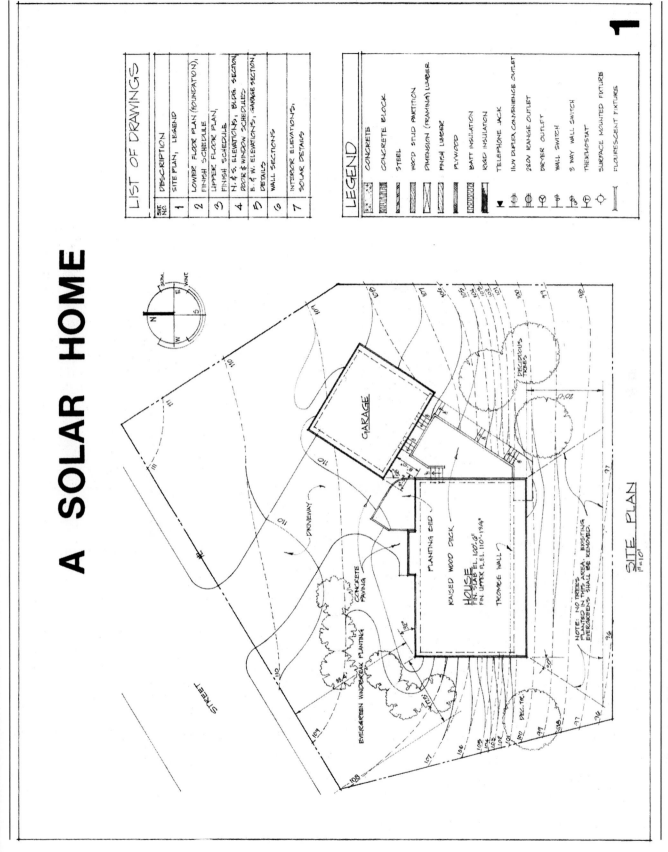

LEGEND

CONCRETE
CONCRETE BLOCK
STEEL
WOOD STUD PARTITION
DIMENSION (FRAMING) LUMBER
FINISH LUMBER
PLYWOOD
BATT INSULATION
RIGID INSULATION
TELEPHONE JACK
110V DUPLEX CONVENIENCE OUTLET
240V RANGE OUTLET
DRYER OUTLET
WALL SWITCH
3 WAY WALL SWITCH
THERMOSTAT
SURFACE MOUNTED FIXTURE
FLOURESCENT FIXTURES

SITE PLAN
1" = 10'

Figure 18-2 Site plan of the solar home shown in Fig. 18-1.

557

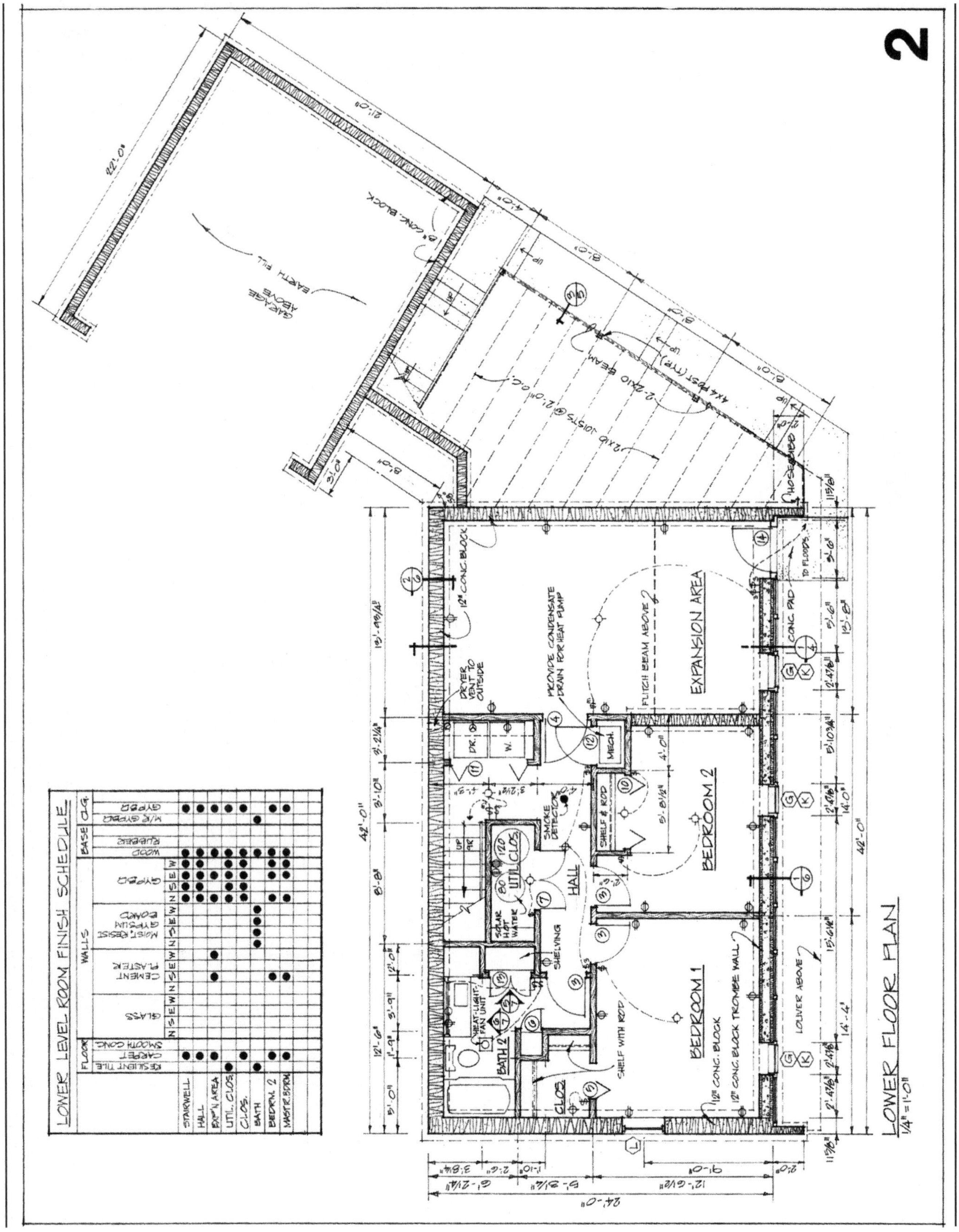

Figure 18-3 Lower-floor plan of the solar home.

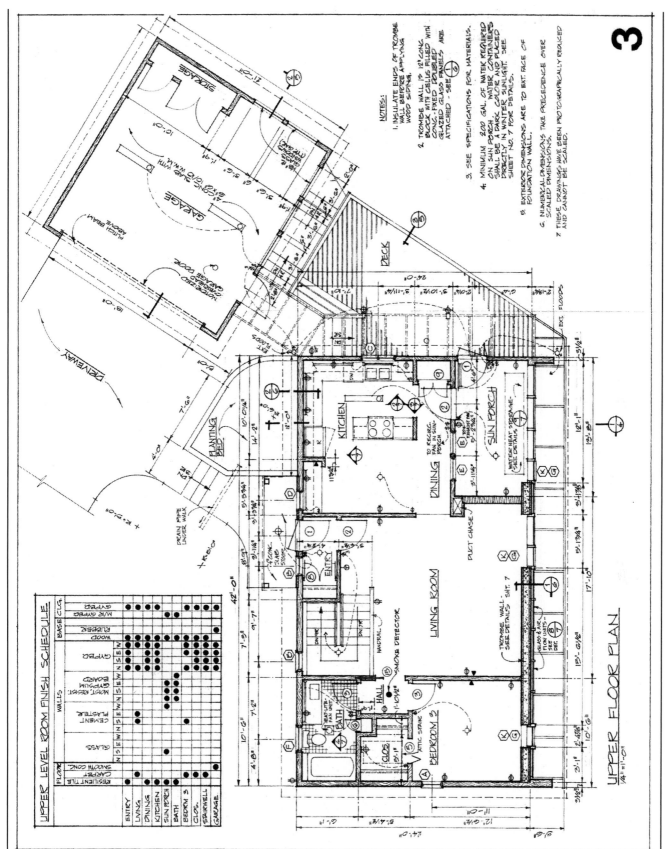

Figure 18-4 Upper-floor plan of the solar home.

559

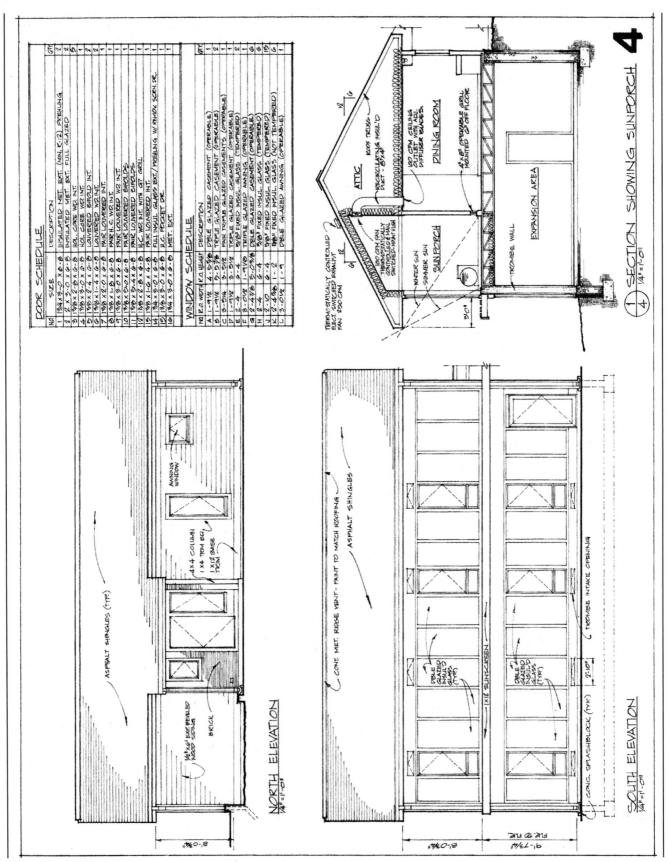

Figure 18-5 Elevations, section, and schedules of the solar home.

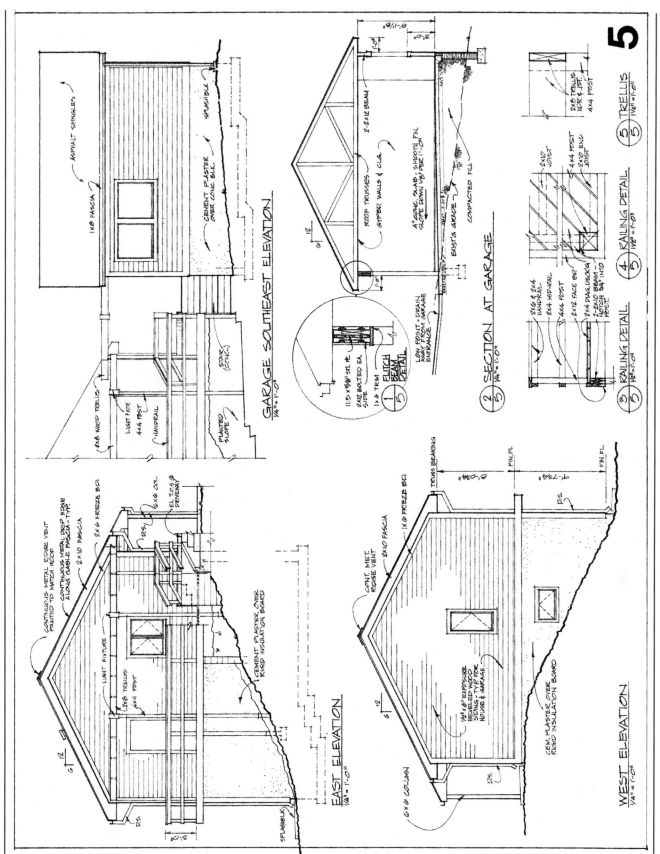

Figure 18-6 Elevations and details of the solar home.

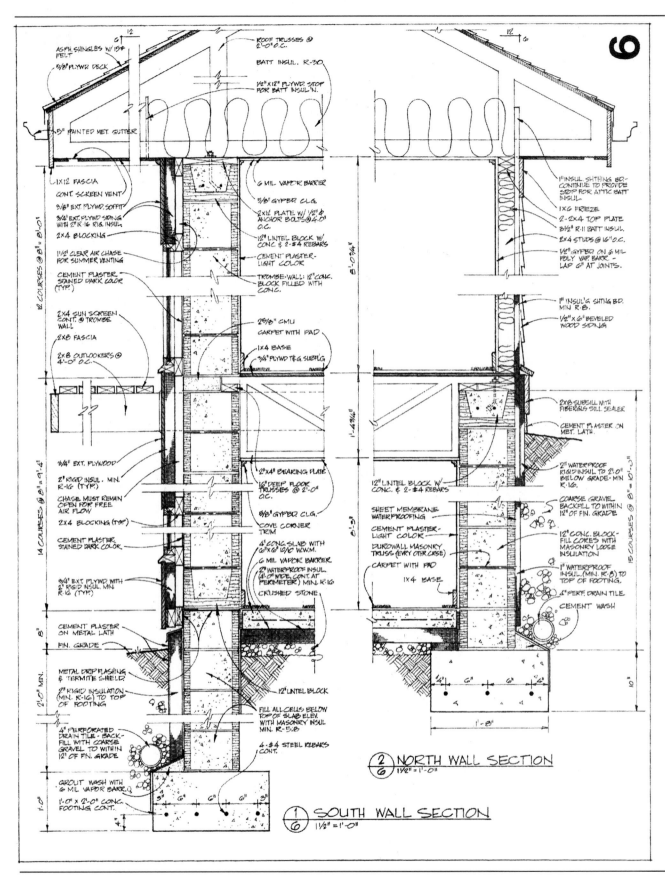

Figure 18-7 Wall sections of the solar home.

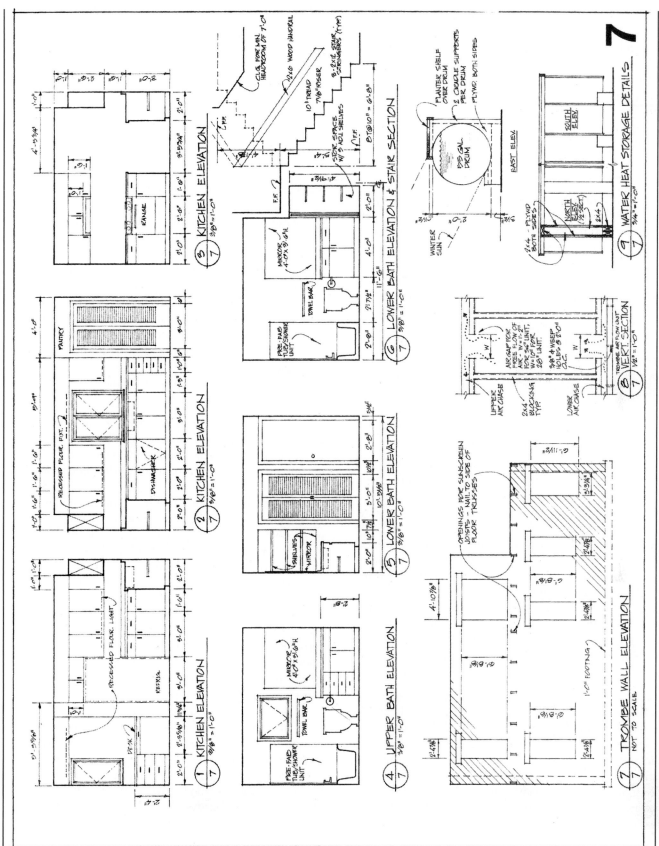

Figure 18-8 Kitchen, bath, and Trombe wall elevations of the solar home.

Step 3: *Draw the layout of the stairs.* If exact ceiling heights have not been established, stair indications on the plan may have to be delayed until stair details (see Fig. 14–32) have been completed. However, after stair information is complete, the risers are indicated with lines. An arrow indicating either up or down and a note showing the number of risers complete the stair symbol. A diagonal break line near the center of the stair run terminates the amount of stair symbol necessary on each floor. Usually an 11″ tread, requiring about a 7″ riser, results in a comfortable residential stair.

Step 4: *If a fireplace is required, draw the outline of the fireplace lightly.* The cutting plane of a fireplace representation passes through the lower part of the opening (see Fig. 14–22). Here again, details may require development on another sheet before the indication on the plan can be accurately drawn. Usually, a masonry hearth, 18″ or 20″ wide, is necessary in front of the opening. Size the fireplace opening and the masonry around it according to the proportions of the room in which it will serve. If a metal fireplace lining is used, the size and the number should be indicated with a note. Refer to Section 14.6 for fireplace designs, details, and methods of representation.

Step 5: *Lay out the kitchen cabinets and the bath fixtures.* Locate the sink, dishwasher, range, microwave, and refrigerator; show the kitchen base cabinets 24″ wide and the wall cabinets 12″ wide. Use broken lines for the wall cabinets as they are above the cutting plane of the plan. Carefully locate and draw the symbols for the bathroom fixtures. Sizes and clearances can be taken from Fig. 15–56 or from manufacturers' catalogs. Sizes of bath fixtures should be accurately scaled to indicate correct placement and workability of the arrangement. Bathroom templates save considerable time in this step. Indicate a medicine cabinet if no storage space is provided in built-in lavatory cabinets. Show all other built-in cabinet work throughout the house in its correct place. All built-in features, which become a permanent part of the construction, must be shown on the plan.

Step 6: *Draw all electrical outlets and switch symbols* (see Fig. 19–42). On ¼″ = 1′-0″ floor plans, use a circle ³/₁₆″ in diameter for the majority of the electrical symbols; a circle template will be handy. Locate ceiling outlets in the centers of room areas for general lighting. Long, rectangular rooms may require several ceiling outlets in order to distribute light evenly throughout the room. Living areas are best illuminated with table and floor lamps that can be operated from duplex outlets and convenient wall switches. Show a ceiling outlet above the tentative placement of a dining room table. Local lighting should be provided for work centers such as kitchen counters and sewing centers. Place light switches near doors (op-posite hinges) where entrance is made into the room. If a room has several entrances, use two 3-way switches (Fig. 19–42) to provide convenient light operation. If necessary, use two 3-way switches and one 4-way switch to operate the lights from three different locations. Place the switches so that the lighting can be turned on before a person enters the room. Show 3-way switches at the top and bottom of stairways. In addition, outside outlets are needed for entrances, yard floodlights, and terrace requirements.

When connecting switch symbols with their outlets, use an irregular curve as a ruling edge, and maintain a consistent broken-line technique. Place duplex wall outlets about 8′-0″ apart in living areas; then lamps can be positioned at will throughout the room. Halls should have outlets for vacuum cleaners. It is important to provide a sufficient number of outlets for the many electric appliances now being used in a modern home: refrigerator, food freezer, range, heating units, washer, dryer, motors, and diverse equipment for food preparation, grooming, and so on. Some appliances may require 220 volts, which is easily indicated with a special-purpose symbol. Waterproof outlets are specified for outdoor use. (Refer to Chapter 19 for further information on electrical drawings.)

Step 7: *Complete and verify all dimensions.* Several methods of dimensioning the framing can be used on wood frame construction. For example, in one method, applied in construction of a traditional house, the dimensions are taken from the outside surface of the exterior stud wall to the center lines of the partitions. This method is conventional, and it provides workers with the necessary dimensions for laying out the soleplates over the subfloors in western frame construction. Even on brick veneer construction, only the wood framing is dimensioned (refer to Chapter 4 for more details). As another method, the outside dimensions start at the surface of the sheathing, which coincides with the surface of the foundation wall on the detail. Continuous dimensions are carried throughout the house from partition surfaces, and the thickness of each partition is dimensioned. This method gives a better indication of inside room dimensions and is preferred by some architects.

In either method, dimensions are shown to the center lines of windows and doors. Generally, the obvious placement of interior doors in narrow areas, such as hallways, seldom needs dimensions for location. Notice that dimensions are placed on each drawing where they are legible, uncrowded, and related to their features. Make guidelines for all dimension figures no larger than ⅛″ high. Check cumulative dimensions to be sure that they equal the overall dimensions. Dimension solid masonry from wall surfaces, and indicate the thickness of each wall or partition so that masons will have definite surfaces to work from. They need to know the dimensions of door and window masonry openings as well. Avoid duplication of dimensions.

Remember that workers cannot construct frame buildings with the same precision found in, for instance, the machine trades. Finished buildings often show slight variations from dimensions indicated on working drawings. This condition, however, does not give license to careless and inaccurate dimensions on working drawings, but it is a situation that drafters should be aware of when dimensioning the drawings.

Step 8: *Letter the room titles, notes, and scale of the plan.* Place the room titles near the center of the room, unless dimensions or ceiling electrical symbols interfere and would result in a crowded appearance. Unless a finish schedule is to be used, the finish flooring should be indicated with a small note below the room title. All notes that clarify features on the drawing should be completed; however, avoid excessive notes that can better be included in the specifications. Door sizes must be given either in door schedules or on the door symbols. Scale indications are lettered very small.

Use a note to show the size, direction, and spacing of ceiling or floor joists above (Fig. 18–10). In conventional framing, ceiling joists must rest on bearing walls. To determine the size and spacing of the joists, based on their longest span, refer to the table on maximum spans for wood joists and rafters (Appendix E). Ceiling joists on 1-story houses should preferably run parallel with the rafters to act as lateral ties. On 2-story houses, the ceiling joists become the floor joists of the second floor, and their sizes must be computed as joists capable of supporting 40-lb/sq-ft live load. Number 2 common, southern yellow pine or standard- and utility-grade Douglas fir are widely used in house framing.

Step 9: *Complete all hatching of materials according to the symbols shown in Fig. 4–9.* If lettering or dimensions fall in the hatching areas, leave enough of the symbol out to make the lettering readable. Prepare a border and a title block similar to that shown in Figs. 18–9 through 18–14 or similar to those shown in Chapter 3.

Step 10: *Check the plan carefully for discrepancies or omissions.* For this step, the following list will prove helpful.

First-Floor Plan Information Checklist

- ❑ **1.** All necessary dimensions:
  - ❑ Outside walls
  - ❑ Interior partitions
  - ❑ Window and door center lines
  - ❑ Edges and thicknesses of solid masonry
  - ❑ Sizes of terraces, walks, and driveways
  - ❑ Special construction
  - ❑ Arrowheads on all dimension lines
- ❑ **2.** Window symbols, door symbols with swing
- ❑ **3.** Window and door identification
- ❑ **4.** Center lines in windows
- ❑ **5.** Type of passageways through partitions
- ❑ **6.** Stair symbol and note UP or DOWN, and number of risers
- ❑ **7.** Steps necessary at exterior doors, different floor levels
- ❑ **8.** Vent stack in 8″ wet-wall
- ❑ **9.** Window and door schedules
- ❑ **10.** Thresholds under exterior doors or between rooms of different flooring materials
- ❑ **11.** Symbols for all stationary kitchen, bath, and laundry fixtures
- ❑ **12.** Built-in millwork
- ❑ **13.** Shelves and rods in clothes closets
- ❑ **14.** Fireplace symbol including hearth
- ❑ **15.** Broken line indicating roof outline, on 1-story plans or on second-floor plans
- ❑ **16.** All lighting symbols:
  - ❑ Ceiling outlets
  - ❑ Wall fixture outlets
  - ❑ Switches with broken lines to outlets
  - ❑ Outside outlets and floodlights
  - ❑ Mechanical equipment outlets
  - ❑ Sufficient wall outlets
  - ❑ Special lighting
- ❑ **17.** Medicine cabinets or closets in bathroom
- ❑ **18.** Ventilating fan in interior baths
- ❑ **19.** Note indicating size, spacing, and direction of ceiling framing
- ❑ **20.** Dashed lines, showing scuttle to attic, if any
- ❑ **21.** Any special beams or structural members in overhead construction
- ❑ **22.** Any metal or wood columns
- ❑ **23.** Symbol for hot-water heater, central heating unit, or fixed mechanical equipment
- ❑ **24.** Hose bibbs
- ❑ **25.** Note indicating the drainage slope of concrete slabs (usually $1/8$ in./ft)
- ❑ **26.** Cutting-plane lines showing section detail locations
- ❑ **27.** Titles for all rooms, hallways, and areas
- ❑ **28.** Note indicating floor finish in each room (if no finish schedule is used)
- ❑ **29.** Correct symbol hatching for materials
- ❑ **30.** Floor plan title and scale

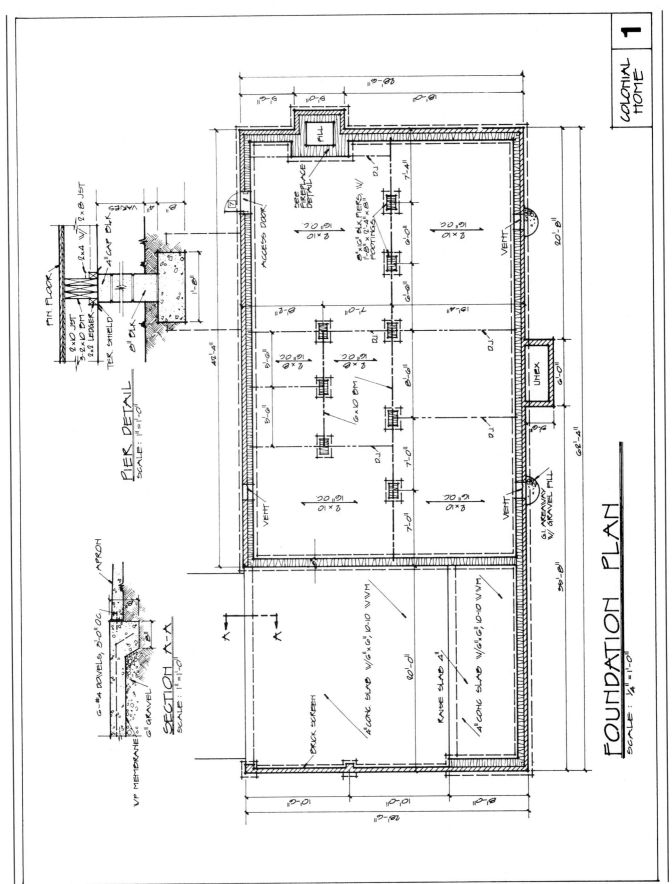

Figure 18-9 Foundation plan of a traditional 2-story home.

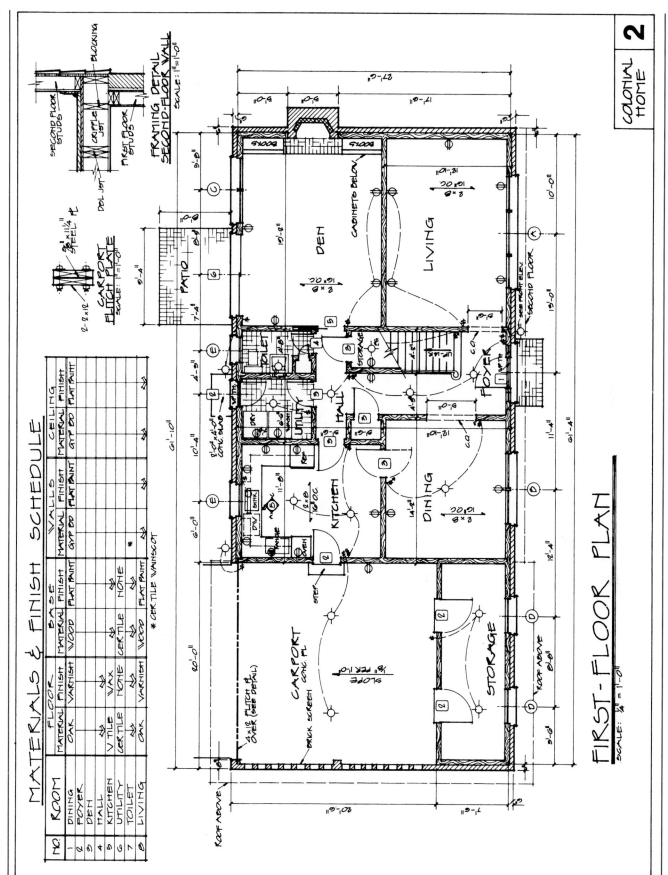

Figure 18–10 First-floor plan for the traditional home.

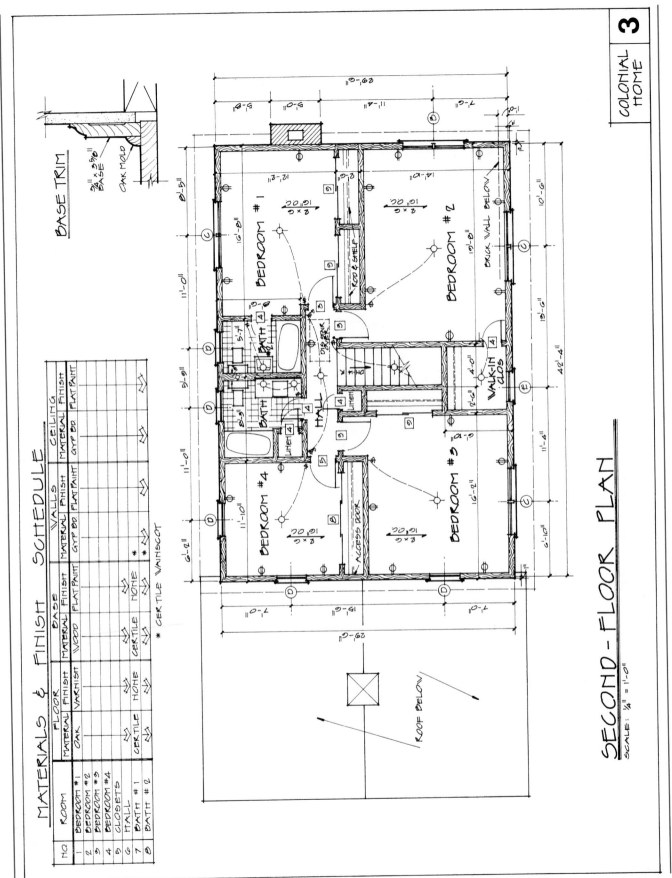

Figure 18-11 Second-floor plan for the traditional home.

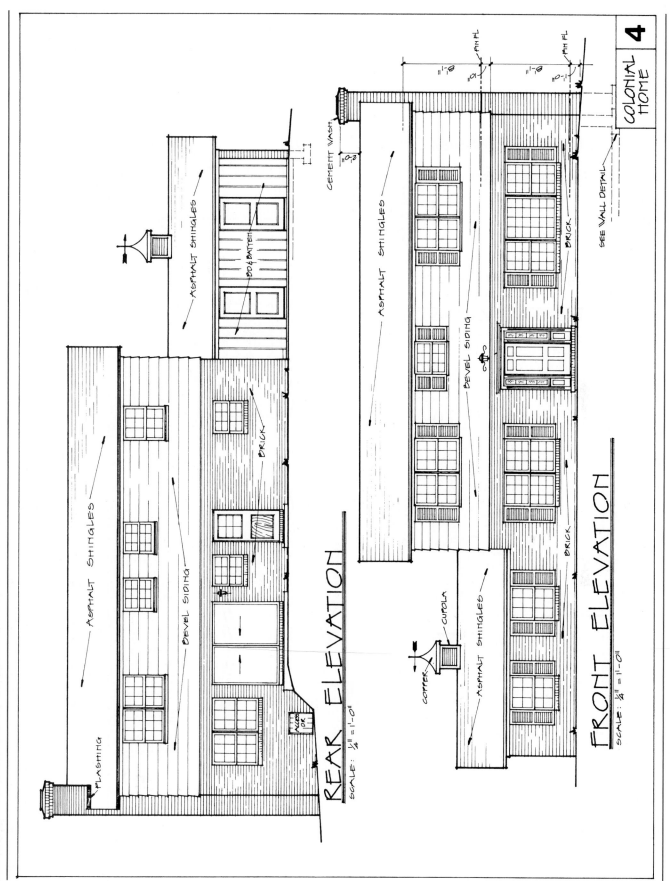

Figure 18-12 Elevation views of the traditional home.

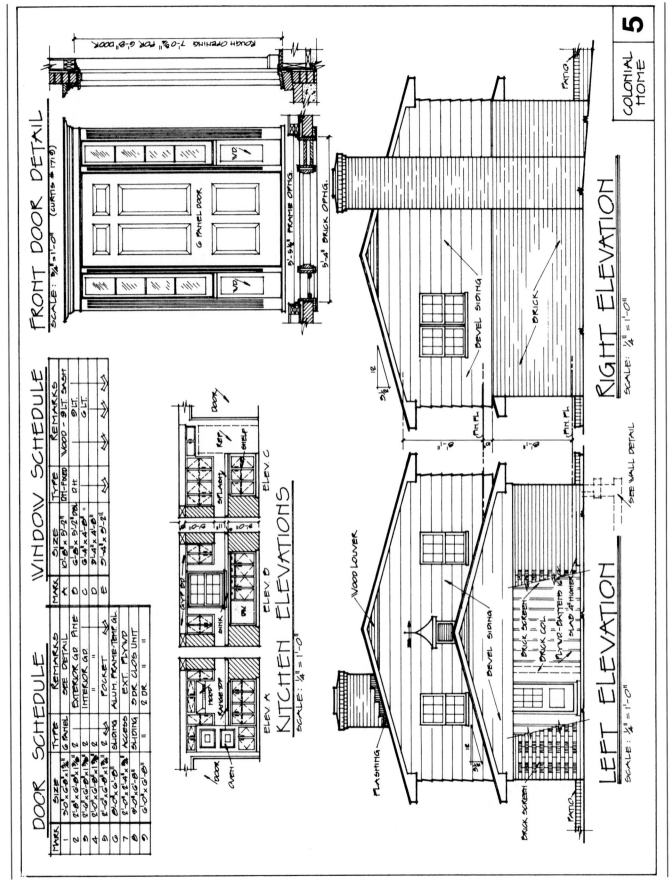

Figure 18–13 Elevation views, schedules, and details for the traditional home.

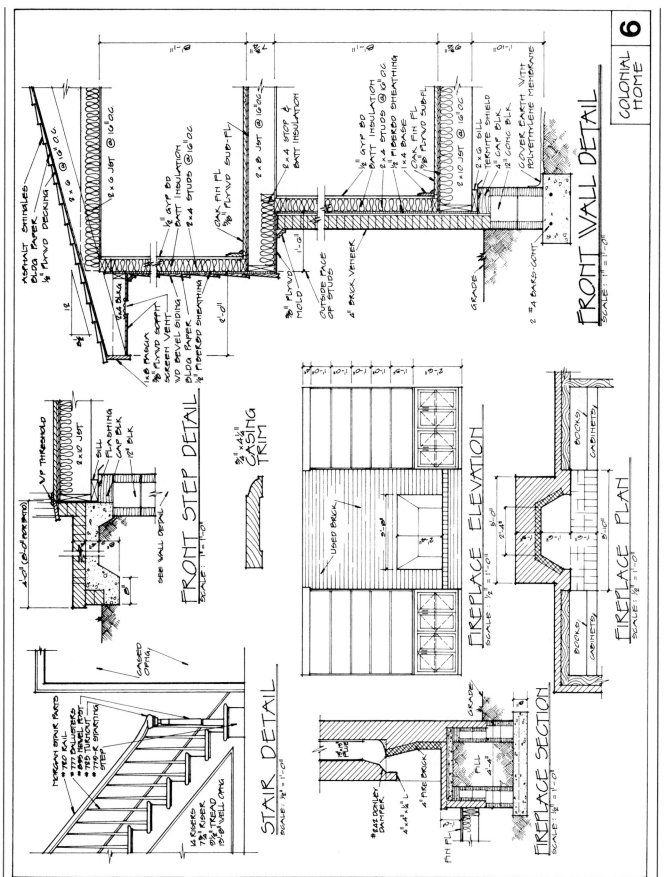

Figure 18-14 Wall section and details for the traditional home.

DEVELOP THE
SECOND-FLOOR PLAN OVER
THE FIRST-FLOOR PLAN

If a 2-story house, similar to those in this chapter, is to be drawn, the second-floor plan is developed directly over the first-floor drawing. Sometimes architects make a print of the first floor and use it instead of the tracing.

Start the second floor by first drawing the outside walls the same size as the walls below, unless a second-floor cantilever arrangement (Fig. 18–14) or a 1½-story house is being drawn. The knee walls under sloping roofs of typical 1½-story houses must be located on end elevations to determine the proper headroom in the living spaces. Knee walls should be a minimum of 4'-6" in height. Since light and ventilation are necessary in rooms with sloping walls, dormers must be planned so that the dormer windows will be on vertical planes. You can use the position of the knee wall to locate the position of the vertical dormer face, or you can place smaller dormers higher on the roof. Additional sketches may be necessary in developing neat-appearing dormers.

Partition layouts in second floors are somewhat restricted by the bearing walls below. Whenever possible, locate second-floor partitions directly over first-floor walls to avoid the need for additional support. When locating the walls, keep in mind that chimneys, stairs, and plumbing must be relative to the first floor and that the vent stacks from first-floor plumbing must come up through second-floor partitions. Stairways should begin and end in hallways. Draw the top half of the stair representation from the first-floor half, and include the proper note. Locate the center lines of second-floor windows; they are usually located directly above first-floor windows or are centered over first-floor wall spaces. Complete the window symbols and identification. Generally, window and door identifications are incorporated with first-floor schedules.

All rooms must communicate with halls. Locate and draw the door symbols with the correct swing. Usually, doors 2'-6" or 2'-8" wide are used in bedrooms, and bathroom doors can be 2'-0" to 2'-6" wide. Wider doors are required for wheelchairs. Partial, 1-story roofs, which are lower than the second-story roofs, should be shown to describe their relationships clearly. Also show scuttles or ceiling exhaust fans with broken lines. Locate and draw all bathroom fixtures and built-in cabinet work to scale. Show the lighting symbols, and dimension partitions, walls, and windows similarly to the first-floor plan. Draw in the hatching symbols for materials, where necessary, and complete all appropriate construction notes.

Because of the similarity between first- and second-floor plan requirements, use the first-floor plan checklist to verify all of the necessary second-floor plan information upon completion.

USE THE
FIRST-FLOOR PLAN TO
DEVELOP THE FOUNDATION
(BASEMENT) PLAN

In modern residential construction, basements are not as popular as they once were. Living space above-ground is usually more comfortable and may be more economical. Foundation plans of houses with crawl spaces appear similar to those with basements except that usually masonry piers are employed instead of metal columns for the floor framing supports (Fig. 18–9).

Develop the foundation plan, like the second-floor plan, on tracing paper taped down directly over the first-floor plan. As with other steps in the completion of working drawings, the foundation plan requires information about the total concept of the structure before it can be drawn. You must refer to a *typical sill detail* or a typical wall section in order to relate the framing of the first floor and its position on the foundation wall (Fig. 18–14). This information must be carefully observed if a satisfactory foundation is to be developed for the building. In a typical wall section of a 1-story house, the exterior sheathing surface is flush with the outside surface of the foundation walls, and the dimensions on the floor plan are given for the sheathing surface. Often the outside foundation lines are simply drawn exactly over the floor plan exterior lines. Relationships with other types of sill detail may not be as simple. For example, brick veneer construction must have a foundation surface about 5" to 6" wide beyond the wood framing upon which the brick must bear. Other sill details may require still different relationships between the wood framing and the foundation walls. For this reason, the typical wall construction detail must be worked out (at least on preliminary sketches) before the foundation plan can be started. Study various sill details, and observe the relative positions of the wood framing and the foundation wall. Western frame sills are commonly called *box sills*.

The general shape of the foundation is drawn first, and when economical 8" C.M.U. has been selected for the foundation, the thickness of the wall is scaled 8" wide. If the front entrance requires a concrete walk, which is exposed to the weather, this area cannot be excavated; a note should be added to indicate that the area is not excavated. Represent footing widths with broken lines. A satisfactory footing for 8" block walls is usually 16" wide. Although the footings for a carport and a terrace slab are poured integrally, the bearing surface of slabs should also be shown with a broken line.

One important problem involved in foundation plan development is the method of supporting the first floor if wood framing is employed. It becomes obvious that floor joists cannot span the entire width of the plan, even if they are placed parallel to the short dimension. To utilize best the lengths of available joists and to provide direct support under important points of load-bearing walls above, a 4 × 10 wood beam or girder is indicated along the stairwell through the entire length of the basement. This beam is supported by the foundation and by equally spaced metal columns. Each column has an independent footing, also shown by broken lines. Columns are also necessary on the opposite side of the stairwell to prevent overloading the header joists at the ends of the stairwell.

As mentioned previously, concrete block piers would be used instead of the metal columns if a similar wood frame floor were constructed above a crawl space with no basement. Ordinarily, the piers would be 8″ × 16″, hatched like the walls, and shown with independent footings and locating dimensions.

Another important consideration is the support for load-bearing walls above, other than those above the main girder. Outside walls are directly supported by the foundation walls, but interior walls must depend on floor joists for support. If a partition is perpendicular to the floor joists, no extra support is necessary, since its weight is distributed over a number of joists. However, if the partition is parallel to the joists, often no joist falls directly below, and therefore extra joists must be introduced to carry the load. The layout of the first-floor plan, then, is a definite aid in placing these additional members in the first-floor framing. Extra joists are also necessary below concentrations of bathroom fixtures and other special dead loads.

Place all notes and information pertaining to the first-floor framing on the foundation plans. Notes indicating details of the second-floor framing are included on the first-floor plan. In other words, the framing notes on any plan should pertain to the framing construction *directly overhead*. Use the conventional note for showing the size, spacing, and direction of first-floor joists (see Appendix E for joist sizes).

Windows should be installed in basements to allow light and ventilation. Screened vents and an access door are necessary in crawl spaces. Here, again, it may be necessary to refer to elevation sketches or typical sections for more information about the placement of basement windows. Until the elevations are completed, it is usually difficult to determine what the level of the grade line is and whether areaways around certain basement windows are necessary. If they are needed, include a section detail on the foundation plan sheet.

Dimension all masonry (to faces), partitions, columns or piers (to centers), slabs, and special framing. Complete all notes indicating the floor material, windows, doors, and framing sizes. Title the plan and show its scale.

Foundation (Basement) Plan Information Checklist

- ❑ **1.** All necessary dimensions:
  - ❑ Masonry walls, surfaces, and thicknesses
  - ❑ Partitions
  - ❑ Columns or pilasters
  - ❑ Girders or beams
  - ❑ Double joists
  - ❑ Outside slabs
  - ❑ Arrowheads on all dimension lines
- ❑ **2.** Window symbols, door symbols with swing
- ❑ **3.** Stair symbol and riser notes
- ❑ **4.** Footing indications under walls, columns, exterior stoops, and piers
- ❑ **5.** Special floor framing
- ❑ **6.** Areaways, if necessary
- ❑ **7.** Lighting symbols
  - ❑ Ceiling outlets
  - ❑ Switches with broken lines to outlets
  - ❑ Sufficient wall outlets
  - ❑ Mechanical equipment outlets
- ❑ **8.** Symbol hatching of materials
- ❑ **9.** Note indicating unexcavated areas
- ❑ **10.** Note indicating floor surfaces
- ❑ **11.** Exterior slabs
- ❑ **12.** Plumbing fixtures and heating units
- ❑ **13.** Foundation wall vents and access door
- ❑ **14.** Fireplace foundation
- ❑ **15.** Basement door
- ❑ **16.** Foundation plan title and scale

18.4

PROJECT ELEVATIONS FROM THE FLOOR PLAN

Elevations (Figs. 18–15 and 18–16) are the drawings of a house, not only revealing the height dimensions and exterior materials but also providing carpenters and workers a total view of the building. This total representation is more important than generally realized. Specific information about various parts of a building is shown in details that are drawn large enough to reveal small members; but unless these details can be related to the total concept of the building, they cannot be understood. The elevations help tie together these bits of information. Also, if minor construction details and dimensions are not shown on working drawings, workers must continually refer to elevation views for the specified finished appearance. Builders have slightly different methods of arranging members and trim during construction without violating given details. A graphic representation of the correct finished appearance, then, places considerable importance on the elevation views. Drawing these views is not difficult if the proper procedure is followed.

Although we have not discussed wall sections in this chapter (see Figs. 14–1 through 14–4), which are necessary in determining definite elevation view heights, you should at this point develop a typical wall section, if the drafter has

Figure 18–15 Pen-and-ink elevation rendering of a traditional home.

not already done so for the foundation plan (see Section 14.2). Consideration should also be given to the general roof layout and the amount of roof overhang, especially if offset exterior walls are encountered, which may require a roof plan to be drawn on another sheet of paper.

Draw the elevations by taking their widths from the plan view and their heights from the developed wall sections. If the house has a gable roof, start with an end elevation so that the profile of the roof can be established and transferred directly to the other elevations. Tape the plan to the board directly above the intended space for the elevation so that projection can be made with instruments. Or you can use a tick strip (a narrow strip of paper with the necessary markings) to transfer all outside walls, window and door openings, center lines, and even roof overhangs to the elevation view. Use the same scale as the plan (1/4" = 1'-0"), allowing direct transfer. Elevations of larger buildings are usually drawn to the 1/8" = 1'-0" scale; however, if possible, draw the elevations at the same scale as that of the floor plan.

Inasmuch as an elevation must have a horizontal line from which to spring, the tentative *grade line* is the logical beginning. Draw a light horizontal line (even if the finished grade must be made to slope later), and allow enough space on the paper for the foundation below and the height of the building above. From the sill detail or the wall section, lay out the *finish floor line* lightly—it will become an important reference line later. Various types of construction establish definite dimensions from the grade to the finished floor surface. Ceiling heights are indicated by distances to the top of the double wall plate, which usually are 8'-1" above the floor line. Draw two plate heights if two stories are required. Remember that both the ceiling framing and the roof rafters bear on a double plate on top of the wall.

Take roof pitches and cornice construction from details or sketches, and lay out the profile of the roof after the center line of the ridge is established. On sloping roofs, be sure that the roof line allows for the width of the rafters at the bird's-mouth on the top-plate bearing surface (see Figs. 14–1 and 14–2A). If trussed rafters are used, the lower chord of the trusses rests directly on the wall plates. Extend the roof lines beyond the walls to show the proper amount of overhang, draw the fascia board, and complete the exterior treatment of the cornice area. After critical examination, you may need to make slight changes in the roof profile to obtain the desired pleasing appearance.

To establish the heights of windows and doors on the elevation, measure the height of exterior doors (usually 6'-8" or 7'-0" stock doors) from the *finish floor surface*, and extend this height as the head for all windows and doors on the first-floor level. Correct window sizes, taken from manufacturers' literature, are then established from the head height. Features of windows and doors, including trim and brick sills if necessary, are carefully drawn on elevation views. If a number of smaller windows are needed, it usually becomes simpler to draw the window accurately on a small piece of paper first and then slip this drawing under each window position and trace it. Second-floor windows are located according to inside requirements and outside

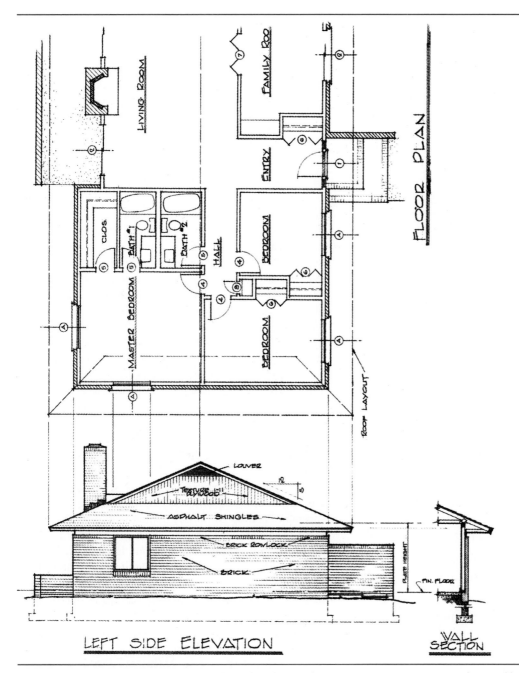

Figure 18-16 Project elevation views from the plan so that correct measurements can be quickly made.

appearance; generally, upper windows line up with lower windows, and higher windows are reduced in size. Place window identifications (letters or numbers within small circles) directly on the window symbol.

Project chimneys from the plan also, and extend their heights at least 2′ above the ridge line (see Fig. 14–21). Slight changes in position are possible through attic spaces if chimneys must appear symmetrical after piercing the roof.

Draw the footing and the foundation lines below the grade lines with broken lines. Basement heights are usually 7′ to 8′, and footings should extend down below local frost depths. The grade line can be changed, if necessary, to the desired finished level or slope. Strengthen all important elevation lines, especially the outline of the view and the eave line, and show the symbol for the various exterior materials. Portions of the hatching should be left out near the centers of large areas to prevent a mechanical appearance.

Dimensions on elevations need be only those that cannot be shown on details and plans. Dimension the finish floor; the ceiling heights, chimney heights, extended wall heights, window heights (other than standard door heights); and other features requiring a vertical dimension. To make the elevations informative without appearing too cluttered, simple notes should indicate exterior materials such as flashings, gutters, downspouts, and steps. Indicate the roof pitch with a graphic diagram (see Section 14.3).

Usually, four elevations must be drawn for a complete description of the building; project the other elevations from the correct wall of the floor plan and the elevation just completed. It is advantageous to construct subsequent elevations in line, either vertically or horizontally, so that the layout of major features can be easily projected, although features appear reversed in opposite elevations.

Occasionally, it becomes necessary, even for experienced drafters, to make minor changes in the elevations and plans as work progresses. As a student, you can surely expect changes to occur in early exercises until you attain more facility in handling the many interrelated conditions that exist in developing a complete set of working drawings. However, perseverance is rewarding—subsequent drawings will become much easier.

Elevation Views Information Checklist

- ❑ **1.** All necessary dimensions:
  - ❑ Floor-to-ceiling heights
  - ❑ Window heights
  - ❑ Chimney heights
  - ❑ Footing depths
  - ❑ Roof overhang
  - ❑ Special-feature heights or depths
- ❑ **2.** Grade line, floor lines, ceiling lines
- ❑ **3.** Correct window symbols and identification
- ❑ **4.** Correct door symbols and identification
- ❑ **5.** Footing and foundation lines (broken)
- ❑ **6.** Roof slope indications
- ❑ **7.** Exterior materials symbols
- ❑ **8.** Exterior materials notes
- ❑ **9.** Louvers: attic, crawl space, roof
- ❑ **10.** Notes indicating special features
- ❑ **11.** Section cutting-plane lines and identifications
- ❑ **12.** Exterior steps, stoops, roofed open areas, railings
- ❑ **13.** Columns, shutters
- ❑ **14.** Dormers
- ❑ **15.** Flashing, gutters, downspouts
- ❑ **16.** Elevation title and scale

18.5

SELECT SCALES SUITABLE TO THE DETAILS

To utilize remaining spaces on working drawings, details must be drawn to accommodate the layout. Usually, a detail is easier to read if it is drawn to a large scale. When there are many details, one or more additional sheets may be required. Follow the procedure indicated in Section 14.2 when drawing the *wall sections* as well as other ma-

jor details. For clarity, a typical wall section in a 2-story house has been drawn in its entirety in Fig. 18–14.

18.6

DRAWING TRANSVERSE AND LONGITUDINAL SECTIONS

When buildings have various floor levels or unusual interior construction, such as split levels, a full section throughout the entire house is often drawn. Such a section shows features in their relative positions, rather than as isolated details on the drawing, and this section becomes very informative to workers on the job. A full section taken through the narrow width of a house is known as a *transverse* section; through the long dimension, it is known as a *longitudinal* section. Either can be used, depending on which plane reveals the necessary information. The cutting plane can be straight across, or it can be offset to gather in features on several planes conveniently, thereby increasing the value of the drawing. Broken lines are also effective in showing, if necessary, important structural members beyond visible surfaces. For convenience, full sections are usually drawn to the same scale as the elevations.

To draw a transverse or longitudinal section, tape the tracing paper directly over the proper elevation view, and trace the outline of the house, the floor levels, the heights of windows and doors, the foundation, the grade line, and so on. You can save considerable time by working directly over the elevation. Interior partitions, stairs, and the like, are taken from the plans. Keep in mind the direction and spacing of ceiling and floor framing members; the placement of girders, headers, columns; and so on. Limit the notes on the section to important features within the building. Place the cutting-plane indication on the floor plan where offsets, if any, can be seen. Floor and ceiling heights, other vertical dimensions, footing depths, and roof pitch are easily dimensioned on the full section.

18.7

DRAWING THE KITCHEN DETAILS (INTERIOR ELEVATIONS)

Kitchens generally require built-in cabinets for work counters, fixture enclosures, and storage. Builders usually subcontract this millwork and have it installed prebuilt instead of building it on the job.

Kitchen details, showing the size, shape, and general cabinet design, are helpful to the subcontractors, who must not only estimate the cost but also fabricate and install the units as well. The heights and the widths of kitchen cabinets have been standardized according to body measurements (Fig. 15–64). However, special arrangements and conditions often exist in custom-designed kitchens, requiring accurately drawn details on the working drawings.

To draw the kitchen details, select a scale that will make the finished elevations neither too large nor too small; usually, 1/2″ = 1′-0″ is the best scale on residential drawings. Lay out each wall elevation upon which the cabinets are to be installed (keep them in a horizontal row if space allows), and lightly draw the horizontal heights of the cabinet features: 4″ toe space, 36″ base cabinet height, thickness of the countertop, heights of the wall cabinets, and so on. Notice that wall cabinets often go only to a point 14″ below the standard 8′ ceiling because of difficulty in reaching any higher. This area above wall cabinets can be furred out flush with the face of the cabinets, or cabinets can be extended to the ceiling to be used for additional storage. Sometimes a shelf is left above wall cabinets, but it becomes a dust collector. These features must be shown on the details.

When the kitchen is U shaped or L shaped, the cabinets must return toward the viewer at one or both corners. At this point a *profile* section of the cabinet work is drawn, and either the cut profile is hatched entirely, or it can be shown as a section with the top, doors, shelf construction, and the like, carefully described.

After drawing the returning sections, correctly lay out the remaining cabinet elevation surfaces to show the doors and drawer faces (they look best if symmetrical and if larger drawers are shown at the bottom of the cabinet), with hardware indications simply drawn. Use broken lines to indicate the door swing and the shelves within the cabinets. Show any windows or doors with their trim, which can be seen on the kitchen elevations. If necessary, draw an enlarged plan of the kitchen; otherwise, the arrangement of the cabinets shown on the floor plans is satisfactory. Include a symbol on the plan indicating the proper view of each kitchen detail, such as A, B, and C.

Indicate all installed kitchen equipment—sink, range, range hood, oven units, dishwasher, and so on—with a note. Strengthen important lines, and add appropriate dimensions. Do not forget a splash above the countertops, and indicate the countertop covering material.

DRAWING THE BATHROOM DETAILS (INTERIOR ELEVATIONS)

Here, again, details must be drawn of bathroom interiors if wall tile or other specially applied surfacing is to be used for the wall covering so that subcontractors can estimate and satisfactorily complete the work. Many bathrooms also have built-in cabinet work (Fig. 15–58).

To draw the bathroom details, use a scale similar to that of the kitchen details (usually 1/2″ = 1′-0″), and, if possible, lay out in a horizontal row the heights and the widths of the walls that are to receive special covering or cabinet work. Show the height of the wall wainscoting (6′ high around shower areas, and 4′ high or higher on other walls), and show all mirrors, cabinets, bathroom fixtures, tile fixtures, wall offsets, windows, and doors, all at their proper scale and heights. Relate each bathroom elevation to the floor plan layout with a direction indication. Use light horizontal and vertical lines to represent the elevation symbol for wall tile; the most common ceramic tile size is scaled 4″ square. Complete the necessary notes and dimensions, title each view, and show the scale.

DRAWING FIXED-GLASS DETAILS

With the increased use of glass in many homes, fixed glass has become a popular method of providing generous window openings without the use of prefabricated window units. Some prefabricated units, however, incorporating both fixed glass and opening sashes, can be installed into rough openings that have been sized according to the units. Fixed glass installed in framing on the job is not restricted by manufacturers' sizes and can be made satisfactory if well-conceived details are followed. Appearance may require finish lumber for the structural framing around the glass, or finish lumber must surround rough framing when the fixed glass is trimmed out.

In drawing fixed-glass details, you must draw sections through the sill, jamb, head, and adjoining mullions showing the following conditions.

First, the rough opening must be made substantial enough to eliminate any structural changes in the framing caused by shrinkage or bending of members from overhead wall loads. Proper lintels must be provided above fixed glass to prevent stresses, unless plank-and-beam framing relieves the window opening of this necessity.

Second, the surrounds within the rough window opening, which receive the glass, must have rabbets at least 1/2″ deep to allow a loosely sized piece of glass to be set in a bed of putty and be held firmly, yet allow slight flexibility. Window stops (usually attached on the interior) must be sized as thick as the rabbets so that their surface coincides with the exterior trim surface in contact with the glass. The trim member forming the exterior sill should be inclined slightly to allow water to drain properly from the sill surface. This principle of water drainage applies to all exposed materials forming ledges on walls. Drips should also be provided on head and sill exteriors to prevent water from running back under the members into the construction. Outside and inside wall edges must be kept in line so that inside casing trim will cover the joint between the surrounds and the interior wall surfacing and so that the exterior will work out with exterior surfacing materials. In brick veneer construction, the glass is placed within the stud wall, yet the trim must be wide enough to cover the air space between the stud wall and the brick. A masonry sill must be shown in masonry wall openings with fixed glass. Inside stools below the fixed glass must extend beyond the inside wall surface and must have proper apron trim. Each detail should be labeled, and dimensions should be applied to the different trim members. If prefabricated operating sashes for ventilation are to be incorporated with fixed windows, the glass surface of both should line up on the same vertical plane.

18.10

THE SITE PLAN ORIENTS THE HOUSE TO THE LOT

The site plan, showing important natural features, orients the building accurately on the property. Information about lot size, shape, elevation, and contours; roads; setback requirements; and the like; must be taken from a survey map or a legal description of the property and drawn to a convenient scale. Use the engineering scale to lay out and dimension all land measurements in feet and decimal parts of a foot. The dimensions on construction features, however, should be drawn in feet and inches. Usually, a scale of 1″ = 20′ or 1″ = 10′ is satisfactory, depending on the size of the lot; sometimes the scale 1/16″ = 1′-0″ is used, but the drawing should be large enough to show features and dimensions clearly.

After the outline of the lot is blocked in, determine the most favorable position of the house, and draw the outline of the foundation and all exterior slabs, terraces, walks, drives, and the like. (Occasionally, a roof plan is drawn instead of the foundation.) If contour lines are available from the survey map, show the original lines broken and the finish contour lines as solid lines. On the site plan of the solar home (Fig. 18–2), the contour lines are shown with 1′ intervals, and upon investigation you will notice that the lot slopes toward the left-rear corner. Excavation from the basement has been used to build up the terrace slab to its necessary level, and the finish contour lines indicate this surface change. Hatch the house area to give it prominence, and, if necessary, use other hatching variation to show relative contrast on other important indications. Often trees or other natural features are shown on the site plan to denote their removal or retention during excavation and grading. See Fig. 18–17 for drive layouts.

Use the following checklist, from FHA site plan information requirements, as a guide for dimensioning and completing the plot plan.

Site Plan Information Checklist

❏ **1.** Scale of the drawing and title
❏ **2.** Dimensions of the site and north point
❏ **3.** Lot and block number of the site
❏ **4.** Dimensions of front, rear, and side yards (setbacks from property line to the house)
❏ **5.** Location and dimensions of garage, carport, or other accessory buildings
❏ **6.** Location of walks, driveways, and approaches, with dimensions and materials
❏ **7.** Location of steps, terraces, porches, fences, and retaining walls
❏ **8.** Location and dimensions of easements and established setback requirements, if any
❏ **9.** Elevation level of first floor, floor of garage, carport, or other buildings
❏ **10.** Finish grade level at each principal corner of the structure

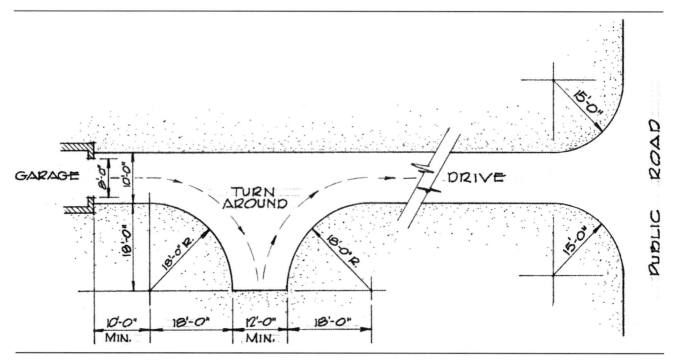

Figure 18–17 Layout of private drives and turning radii.

11. Finish grade at both sides of abrupt changes of grade, such as walls and slopes

12. Other elevation levels that may be necessary to show grading and drainage

13. Location of a reference bench mark and its elevation, from which the contractor can relate all grades

14. Existing trees, if any, to be removed or retained and their identification

15. Location and identification of utility lines or pipes that will service the structure

16. Septic tank and leaching field, if required

17. Obstructions such as utility poles, catch basins, and hydrants

EXERCISES
WORKING DRAWINGS OF SMALL HOMES

The following exercises should be done on 22″ × 34″ drawing paper with graphite lead or Mylar® with plastic lead. Lay out a ¹/₂″ border at the top, bottom, and right side, and a 1¹/₂″ border on the left side. Select a title block of your choice. All drawings are to be drawn to scale using instruments.

1. Draw the first-floor plan of the residence in Fig. 18–10; use the scale ¹/₄″ = 1′-0″.

2. Draw the foundation plan of the residence in Fig. 18–9; use the scale ¹/₄″ = 1′-0″.

3. Draw the front elevation from Fig. 18–12 and the left side elevation from Fig. 18–13; use the scale ¹/₄″ = 1′-0″. Draw the finish schedule (make room ti-

tle spaces ³/₈″ high) and the kitchen elevation at a scale of ¹/₄″ = 1′-0″.

4. Draw the front wall detail from Fig. 18–14 using a scale of 1″ = 1′-0″; draw the fireplace plan, section, and elevation at a scale of ¹/₂″ = 1′-0″.

5. Using the small beach house from Exercise 6 of Chapter 15, prepare working drawings of the following at a scale of ¹/₄″ = 1′-0″: floor plan, foundation plan, exterior elevations, cross section, and bath elevation. At a scale of 1″ = 1′-0″, draw a typical wall section. Show a room finish schedule. Combine views on a drawing to create good composition.

REVIEW QUESTIONS

1. When starting a floor plan, why must you begin with general shapes and light lines?

2. What is meant by the term *wet-wall?*

3. On wood frame construction drawings, why is it convenient for workers to have dimensions shown from the outside of exterior wall studs to the center lines of partitions?

4. Why are lengthy notes about construction commonly put in the specifications rather than on the working drawings?

5. Why is a floor plan more readable when symbol hatching has been used throughout the walls?

6. Does a joist note, giving size, spacing, and direction and shown on a floor plan, pertain to the framing above or below?

7. Why are piers within a crawl space generally masonry? How is this indicated?

8. What does the term *unexcavated* mean when shown on part of a foundation plan?

9. Of what value is the finish floor line indication on an elevation view?

10. What dimensions must necessarily be put on elevation views?

11. What is the minimum width for a residential driveway for one-way traffic?

12. At what scale are bathroom elevations usually drawn?

13. How can you eliminate columns and load-bearing walls from a basement and at the same time maintain the structural integrity of the floor above?

14. What is the standard height from the floor to the top of doors and windows in residences?

15. Why is the floor plan developed before the foundation plan in working drawings?

16. Place the following drawings in the order they would appear in a set of residential working drawings: exterior elevations, foundation plan, bathroom and kitchen details, site plan, typical wall sections, and floor plan.

17. Why are doors numbered on a working drawing floor plan?

18. How are section cuts referenced on floor plans?

19. What working drawing contains the most information about the construction of the residence, from the foundation to the roof?

20. How far should wood siding be placed above the exterior grade at the perimeter of the building?

"A comfortable house is a great source of
happiness. It ranks immediately after
health and good conscience."
—SYDNEY SMITH

Much of the comfort and livability of our present-day structures is provided by the wide variety of mechanical and electrical equipment now available to us. The various methods of conditioning the air within a building, the host of electrical appliances and lighting devices, the accommodations for plumbing facilities, the arrangements for use of public utilities—all must be given careful attention by the designer of both homes and commercial buildings. No structure is considered complete without adequate provision for this equipment and the human comfort that it affords (Fig. 19–1).

For large jobs, architects commonly employ the services of registered engineers to design layouts and furnish drawings dealing with mechanical and electrical systems. Engineering firms, who employ engineers specializing in this type of work, are equipped to handle the complexity of problems involved. Separate drawings, such as a Heating, Ventilation, and Air-Conditioning Plan, an Electrical Plan, and a Plumbing Plan, are prepared and are then combined with the architectural set of working drawings. However, much of the technical information needed on larger installations usually is not required for the majority of residential work; subcontractors can design and install conventional residential equipment with a minimum of information on working drawings. Commonly, this information can be included on the regular floor plan. Of course, when regular floor plans become overcrowded with notes, symbols, and information, it frequently is advisable to draw separate plans for each of these areas of information, especially for the heating system.

Although subcontractors usually lay out and design conventional residential heating-cooling, plumbing, and electrical systems, the student of architectural drawing, in order to plan for accommodating this equipment, should understand the general characteristics of the currently used systems and the problems involved in making them successful. Their provision in the structure contributes to the workability of the total concept and thereby justifies their careful study.

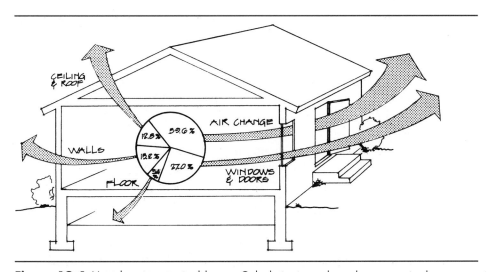

Figure 19–1 Heat loss in a typical home. Calculations are based on one air change per hour and on double-pane windows and doors.

PLANNING THE HEATING AND COOLING SYSTEMS

In homes, the selection of the most appropriate heating system is closely related to other aspects of the building, such as the necessity for a utility room, the feasibility of a basement, and the selection of floor construction materials. This selection probably requires more consideration during early planning than other equipment installations. The type and the cost of fuel available in the area are important factors, and the client's preference as to the method of heating the house also requires attention. In addition, local climate, degree of comfort desired, initial cost of equipment, efficiency, and amount of insulation should be given serious thought. In many areas, year-round space conditioning is now almost a requirement, and equipment capable of providing interior comfort in all seasons is a sound investment.

Local and national codes relating to heating and cooling installation must be carefully observed.

ARCHITECTURAL CONSIDERATIONS FOR HEATING AND COOLING

Comfort and efficiency of heating and cooling equipment begin with incorporating the following construction principles into the building during the design stage:

1. Protect the west walls of the house from direct afternoon sun by placing the carport or garage on the west side or by using trees, plantings, or fences for protection.
2. Use large glass areas on the south side, with sufficient overhangs or sunshades to protect the glass from direct sun during the summer months.
3. Protect the house, if possible, from cold winter winds. This is a matter of orientation and the intelligent use of windbreaks.
4. Use a light-colored roof to reduce heat absorption during summer for more cooling efficiency.
5. Provide a minimum R-13 insulation in 2×4 walls and a minimum R-21 in 2×6 walls. Provide a minimum R-30 attic insulation for flat ceilings and a minimum R-26 insulation for cathedral ceilings.
6. Provide for sufficient ventilation in attics, kitchens, baths, and laundries to eliminate excessive interior moisture.
7. Specify minimum double-glazed (R-1.8) window units. If using metal frames, be sure to specify only those with thermal breaks. If budgets allow, specify premium window units with gas-filled, low-emissivity glazing.
8. Locate the heating unit near the chimney, or provide as short and direct a vent pipe as possible.
9. For the most heating efficiency, plan a compact shape of the house perimeter.
10. In 2-story houses, provide coinciding first- and second-floor partitions for heat risers to the second story; chases may even be necessary.
11. Use insulated exterior doors, R-5 or better, with tight-fitting weatherstripping and thresholds.
12. Follow state energy codes regarding architectural choices for insulation, glass, and fresh air.

CURRENT RESIDENTIAL HEATING SYSTEMS

Employing various fuels and methods of installation, the following basic systems are now generally used for residential central heating:

1. *Forced-warm-air systems* heat the air in a furnace, force it through ducts with the use of a fan to all the heated areas of the house, return the cooled air, and filter it.
2. *Hot-water systems,* either one- or two-pipe, heat water in a central boiler and circulate it through piping to the different rooms where various types of convectors, radiators, baseboards, or radiant panels supply the heat to the air.
3. *Electric resistance systems,* using electricity as an energy source, provide a comfortable, radiant-type heat with the use of resistance wiring embedded in either floors or ceilings or mounted in metal baseboards.
4. *Heat pumps* use no primary fuel other than electricity to run the reverse cycle compressor; they operate as a combination heating and cooling system. However, supplementary electric resistance heating elements are necessary during extreme temperature drops. The application for heating and cooling purposes is recent; heat pumps have been found adequate primarily in mild climates. A heat-pump unit is basically a refrigeration system using electrical energy to pump the natural heat from air, ground, or water outside the house to produce a useful temperature level within the house. The procedure is reversed during summer cooling.

In warm climates, small wall or floor units are sometimes more satisfactory than central systems. Also, unit heaters can be combined with central systems when isolated areas are difficult to heat with a main system.

The following definitions are commonly used in the HVAC field.

Definitions

- **Btu: British thermal unit**   One Btu is the amount of heat it takes to raise the temperature of 1 lb of water 1°F at a specified temperature. The energy content of various fuels are listed below.

Heating oil	139,000 Btu/gal
Propane	91,300 Btu/gal
Kerosene	135,000 Btu/gal
Natural gas	1,027 Btu/cu ft

  Natural gas is metered in units of 100 cubic feet (ccf) or therms (1 therm = 100,000 Btu's).

- **AFUE annual fuel-utilization efficiency**   The measurement of furnace and boiler efficiency averaged over a heating season.
- **COP: coefficient of performance**   The ratio of the amount of energy delivered to the amount of energy consumed.
- **SEER: seasonal energy-efficiency rating**   The ratio of the seasonal cooling performance of heat pumps and air conditioners in Btu's, divided by the energy consumption in watt-hours, for an average U.S. climate.
- **EER: energy-efficiency rating**   Similar to the SEER and used for air-source heat pumps and air conditioners, except that cooling performance is not averaged over a cooling system.
- **HSPF: heating-season performance factor**   A ratio of the estimated seasonal heating output in Btu's, divided by the seasonal power consumption in watts; used for comparing air-source heat-pump performance.

Oversizing is one of the most serious problems with heating system design. Simplified rules of thumb developed when R-11 walls and single-glazed windows were common can result in heating systems that are two or three times larger than is needed for a given structure. An oversized furnace or boiler cycles on and off frequently and rarely operates long enough to reach optimal efficiency. A heating system for a house should be sized no more than 25% larger than the calculated heating load.

19.3.1 Forced-Warm-Air Systems

These systems utilize a motor-driven fan to circulate heated, filtered air from the central heating unit, called an *exchanger*, through a system of ducts to each part of the house. The air has been heated by hot water, gas, or oil. Often provision is made for the introduction of a small amount of fresh outside air to the system. Forced warm air has become the most popular heating method in the majority of current new homes throughout the country, mainly because of its economical initial cost and its compatibility and ease in combining with summer cooling systems. Heat distribution throughout the house is instantaneous, almost as soon as the furnace is turned on. Filters, introduced into the return ductwork, provide clean air throughout the operation.

This heating system has an additional advantage: Location of the furnace is not restricted to basements or utility rooms. Furnaces of various types and sizes can be conveniently installed in crawl spaces, attics, small closets, or adjoining garages. In addition, the design features of many modern furnaces make them acceptable for installation in living spaces, such as recreation and play rooms. Disadvantages include noise from the air handler (fan) and the ducts; increased air leakage in the house resulting from pressure imbalances; leakage through poorly sealed ducts; drafts generated by air circulation; and difficulty in zoning areas of the house separately.

The furnace in a forced-warm-air system can be either the conventional *upflow* or the reverse *counterflow* (*horizontal-flow*) *unit* (Fig. 19–2), depending on its placement and the necessary duct system. Counterflow furnaces must be used when ducts are embedded in concrete slab floors or when above-the-floor furnaces must supply the warm air through ducts located in crawl spaces.

Induced-draft gas furnaces use a fan to force the flue gases either up a chimney or out through a side-venting pipe. Improved heat exchangers extract more heat out of combustion gases; as a result, the flue gases are less buoyant, and the fan is needed to exhaust them (hence the term *induced draft*). Because the flue gases are relatively cool, plastic pipe is sometimes used for side venting. AFUEs for these furnaces are typically in the range of 78% to 85%.

Condensing gas furnaces are the highest-efficiency gas furnaces. The heat exchangers are so efficient that the gases cool down enough for water vapor to condense into liquid. When the water vapor condenses, it releases its latent heat, thereby boosting energy performance. AFUEs for condensing furnaces typically range from 90% to 97%. Flue gases are usually vented out through plastic piping, while condensate is piped to a floor drain.

Sealed combustion gas furnaces force outside air into the combustion chamber and force exhaust gases outside. There is no need for indoor air. Sealed combustion equipment is considered the safest, with a low risk of backdrafting. Sealed combustion products are available in both condensing and noncondensing models.

Various types of ductwork can be used with forced-warm-air furnaces. Circular sheet-metal or circular fiberglass ducts are usually the most economical. Fiberglass ducts are often used in attic installations, and PVC-coated steel ducts in concrete slab construction. Many systems utilize a perimeter arrangement, in both slabs and crawl

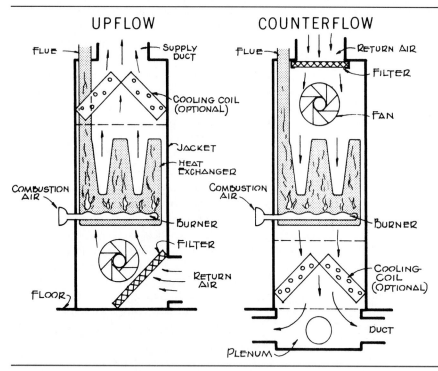

Figure 19–2 Section diagrams of typical gas-fired furnaces for forced-warm-air heating systems. Similar furnaces are made for other fuels, and various furnace shapes are available. Cooling coils are shown for the combination cooling-heating operation. The counterflow furnace is usually used in slab or crawl space construction.

spaces, as well as in some basement installations. In this arrangement, a continuous duct is placed near the periphery of the building, and a number of feeder ducts radiating out from the central plenum supply the perimeter duct with warm air (Figs. 19–3, 19–4, and 19–5). Room registers receive the warm air from the perimeter duct only. Another method utilizes a main trunk duct through the central part of the building, and smaller branch ducts carry the warm air to the different rooms. Rectangular sheet-metal ducts are usually used for this system, and a more compact and neater arrangement for basement installation results. Another system has a large central duct, called an *extended plenum*, with small circular ducts to each register (Fig. 19–6).

In all of these systems, return ducts, usually shorter and larger in cross section, bring the cooled air back to the furnace. The trend in residential installations is to give more consideration to return air duct sizing. Many systems now provide for the introduction of fresh air through adjustable outside registers capable of compensating for air not returned from kitchens, baths, and vent fans. This system maintains a continual supply of fresh air in the home and, because of infiltration, equalizes return air pressures.

We see, then, that various systems of warm-air distribution are available to the home designer and that each

system has slightly different characteristics and applications (Figs. 19–6 through 19–10).

The modern warm-air system, comprising the furnace, ductwork, fan, and registers, usually has automatic controls to make it almost self-sufficient in operation. The controls necessary to operate a typical warm-air system are

1. room thermostat,
2. fan thermostat, and
3. high-limit control.

Warm-air furnaces can be equipped with humidifiers that add water vapor to dry air in spaces, in which case a *humidistat*, or humidity detector, is used to control the correct humidity level.

The *room thermostat* automatically turns the burner on when inside temperatures drop below a pre-established setting, and it turns the burner off when the desired temperature has been attained. The thermostat should be located 5′ from the floor on an inside wall with free air circulation that will provide uniform temperature levels throughout the house. The *fan thermostat* is located within the furnace unit and turns the fan on when the heated air in the plenum reaches a comfortable temperature for circulation. When the room thermostat turns the burner off, and the air in the plenum drops to a tem-

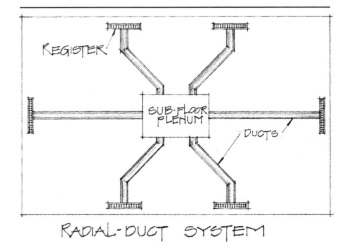

RADIAL-DUCT SYSTEM

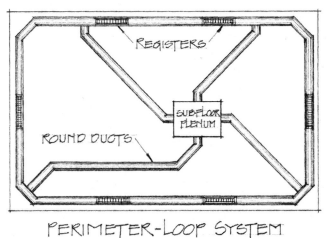

PERIMETER-LOOP SYSTEM

Figure 19–3 Warm-air duct systems used in slabs and crawl spaces.

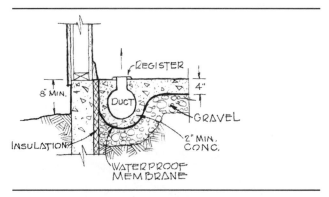

Figure 19–4 Sketch detail of slab construction and perimeter heat duct.

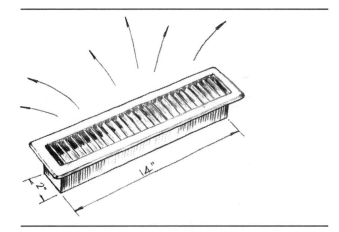

Figure 19–5 Typical floor register.

perature level too low for circulation, the fan is turned off. The *high-limit control*, also located in the furnace unit, is a safety device for the purpose of turning the furnace completely off should a malfunction occur, causing the temperature within the plenum to become dangerously high.

19.3.2 Hot-Water Systems (Hydronic Systems)

Heat distribution with the use of water has been in practice for many years and is still considered a very satisfactory method. Hydronic heating offers quiet operation, minimal drafts, and less heat loss from the distribution system when compared with forced-air heating. There is no air leakage from pressure imbalances in the house, and

it is much easier to zone different parts of the house. Negative aspects include high installation costs, possible interference with furniture placement, and an inability to have the distribution system serve other climate-control functions.

The central boiler (Fig. 19–11), utilizing gas, oil, or electricity as a fuel, heats water from a cold-water supply to the required temperature (usually 180°F) and pumps it throughout the rooms by means of various piping systems. A simple *one-pipe system*, having room converters or radiators connected in series, allows the water to be carried in the main pipe, diverted to the convectors, and then returned to the boiler for reheating. Special fittings allow restricted amounts of water into each convector so that the temperature of the water in the later convectors varies only slightly from the temperature of the water in the convectors closest to the boiler

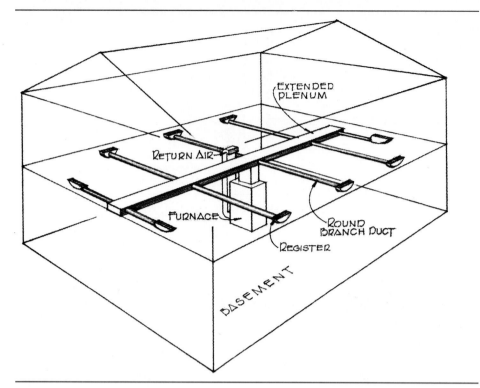

Figure 19–6 Warm-air distribution system with an extended plenum and round ducts to the room registers. The furnace is located in the basement.

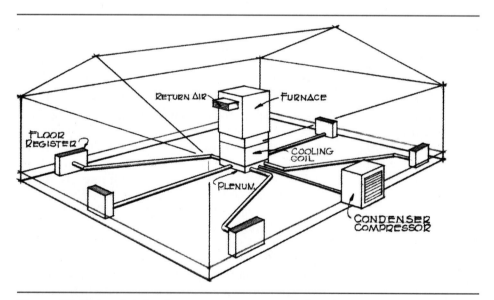

Figure 19–7 A radial distribution system using round ducts for a slab or a crawl space. If the system is used only for heating, the cooling coils and the compressor-condenser unit are omitted. The counterflow furnace is necessary in this installation.

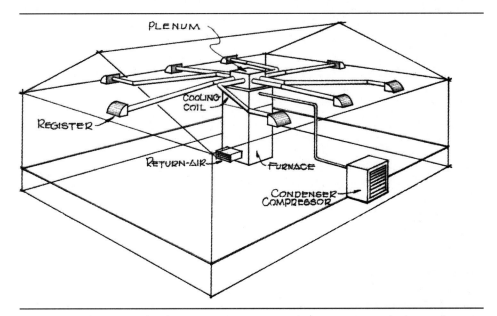

Figure 19–8 An attic distribution system using round ducts is practical when ceiling registers are used.

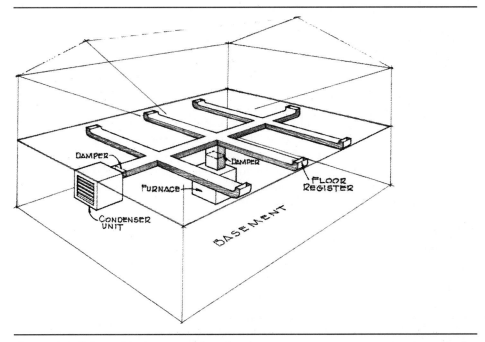

Figure 19–9 Forced-air heating-cooling distribution system using rectangular ducts. The furnace and the cooling unit are placed in the basement.

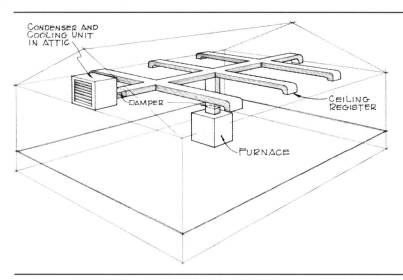

Figure 19–10 Forced-air heating-cooling distribution system using rectangular ducts to ceiling registers. The furnace is placed on the first floor, and the cooling unit is placed in the attic. Forced-warm-air furnaces can also be placed in the attic, and cooling units are available for placement outside.

(Fig. 19–12). A *two-pipe system* works more efficiently; of course, the initial installation cost is higher. One pipe supplies the rooms with hot water; the other, called the *return*, carries the cooled water back to the boiler (Fig. 19–13). The return can be either *direct* or *reverse*, referring to the order in which the cooled water returns to each convector in relation to the order in which the warm water enters each convector. The reverse-return method is usually preferred.

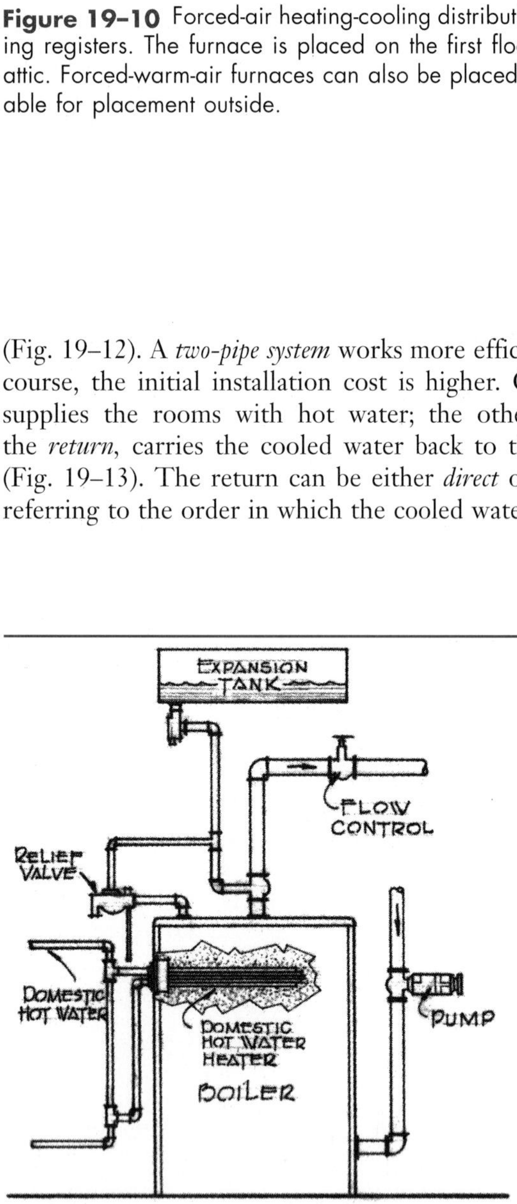

Another system, referred to as a *radiant system*, distributes the hot water through pipes embedded either in concrete floor slabs or in plastered ceilings (which, when warm, act as large radiators).

Various types of units can be used to dissipate the heat from the water to the air within each room. The conventional radiator has been used for years, but it is a dust collector and uses floor space unless recessed into walls. More popular convectors are enclosed heaters with extended surface coils or finned tubes that give off the heat readily when air circulates through the units. Various sizes are available for wall-hung, recessed, or completely concealed installation.

A popular adaptation of the convector is the baseboard unit. The finned piping is arranged along the base of the outside walls and is covered with a metal enclosure to resemble a baseboard. Heat is given off near the floor where it is needed, and various lengths of convectors can be combined with suitable trim to produce unobstructive heaters with pleasing appearances (Fig. 19–14). Forced-air convectors must be used with residential heating-cooling hydronic systems, and they are adaptable to many commercial applications.

Hot-water boilers or heaters must be located in the basement or in first-floor utility rooms of basementless homes. Gas boilers are available with venting configurations that achieve various efficiencies. Practical efficien-

Figure 19–11 Diagram of a boiler used with hot-water heating systems.

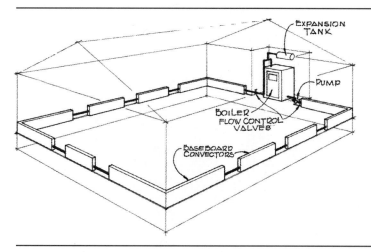

Figure 19–12 One-pipe hot-water heating system with baseboard convectors in the rooms.

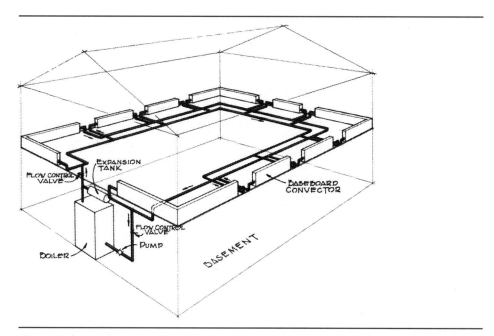

Figure 19–13 Two-pipe, reverse-return hot-water heating system with baseboard convectors. The boiler is installed in the basement.

ciencies are somewhat lower with boilers than with furnaces because the temperature of the return water from the hydronic heating loop is generally higher than the condensing point of the water vapor in the flue gases. As a result, it is difficult to squeeze out the extra efficiency represented by the latent heat of water vapor in the flue gases. Few gas boilers have efficiencies over 90%; most range from 80% to 85%.

Oil combustion is significantly different from gas combustion. Because the fuel is a liquid, it must be separated mechanically into tiny droplets and mixed with air for complete combustion. This process is accomplished with an injector-head burner that sprays air and atomized oil into the combustion chamber. Most oil burners today have flame-retention heads that increase turbulence in the combustion chamber, thereby improving combustion. As with gas-fired equipment, oil furnaces and boilers built since 1992 have been required to have AFUEs of at least 78%. Most oil furnaces and boilers today have AFUEs in the range of 78% to 82%.

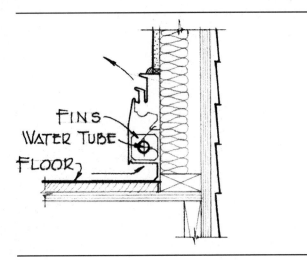

Figure 19–14 Sectional detail of a hot-water baseboard convector. Fins conduct the heat from the water tube to the air, creating air circulation through the baseboard.

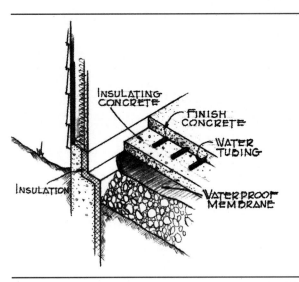

Figure 19–15 Floor-panel hot-water heating embedded in concrete slab construction.

If oil is used as a fuel, consideration must be given to storage space, which should be not too distant from the furnace. Each boiler is equipped with an expansion tank, either closed or open, to compensate for variations of the water volume at different temperature levels and to relieve air pockets from the system.

Another variation of a hot-water heating system is one that uses water from a high-efficiency water heater to supply heat to air handlers for forced-air heating. In many residences, hot water is used for bathing, washing dishes, and washing clothing only a couple of hours each day. Most of the time the domestic water heater is idle. By taking hot water from the water heater and passing it through a water to an air heat exchanger, this system forces warm air into the home through the duct system.

Radiant floor heating (Fig. 19–15) uses specialized plastic or rubber tubing which is embedded in a concrete floor slab and through which water, heated by a boiler, is circulated. Radiant floor systems provide a high level of comfort. Because of their comfortable average temperature and the fact that there is little stratification, homeowners are typically comfortable at air temperatures several degrees cooler than otherwise. Usually, a serpentine coil system is employed (Fig. 19–16).

Heat in floor slabs eliminates the usual coldness of concrete, yet because of its mass, concrete does not respond as quickly to heating as do ceiling panels. Higher surface temperatures (100°F to 120°F) are allowable on radiant ceilings than on slab surfaces (80°F to 85°F). Radiant surfaces heated to a temperature of 85°F or 86°F will keep the occupants comfortable, even with air temperatures slightly less than 60°. Balancing valves, usually located in closets or in other inconspicuous areas, allow complete control of heat within each room. The piping

is usually taken off main supply lines to each room and fed back to the boiler by return mains, similarly to typical two-pipe hot-water systems. Standard thermostatic controls cause the pump and the boiler burner to maintain comfortable room temperatures. Because of the heat

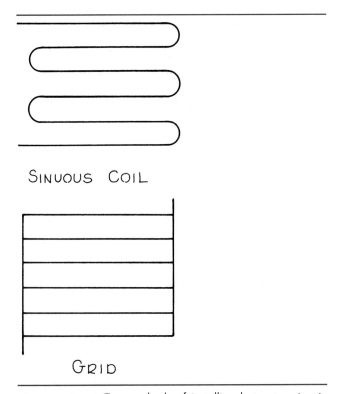

Figure 19–16 Two methods of installing hot-water pipe in radiant panel systems.

lag of slab radiant systems, it becomes necessary in many areas to include an exterior temperature change anticipator (an outdoor thermostat), which controls the heat of the water in the piping in the system, balancing it with the actual need within the house.

The advantages of radiant panel heating are as follows:

1. The heat is clean, creating little dust and reducing the need for frequent repainting.
2. There are no visible registers or heating units within the rooms.
3. There are no drafts or noise from blower equipment.

19.3.3 Electric Resistance Systems

Basically, electric heating systems convert electrical energy into heat with the use of resistance wiring installed in various ways throughout the structure. Both radiant- and convection-type heat are produced. In areas where electrical energy is economical, this type of fuel produces a very desirable method of heating residences. The heat is clean and easy to distribute and control, and installation of the system is economical. From a standpoint of safety, little damage can be done by fire since elements are designed to operate below 150°F. The heat can be placed exactly where it is needed as, for example, bathroom heaters or other auxiliary heaters.

Electrical resistance elements in baseboards, similar in appearance to hot-water baseboard units, can be installed along exterior walls to provide instantaneous and comfortable heat.

Radiant plaster ceilings are produced by resistance wiring attached to gypsum lath. The wiring is manufactured in specific lengths to provide exact wattage ratings for a room according to its heat loss. Wattage sizes are based on the conversion ratio of 1 watt = 3.41 Btu's. The wiring is insulated with a protective covering and should not be cut or spliced. After the circuit has been carefully tested, a ½" brown and finish plaster coat is applied. Connector wiring is regular nonheating type. Control thermostats are installed in each room. Care must be taken to keep the embedded wiring at least 6" from walls and at least 8" from any ceiling outlet boxes. It is recommended that 6" of mineral insulation be applied over all radiant ceilings. Resistance wiring can also be installed between two layers of gypsum wall board in drywall construction (Fig. 19–17). Spacing of the wiring on the first layer must allow for the nailing of the second layer, however, and a 3" wide space below each ceiling joist is left void of the resistance wiring for this purpose. Otherwise, the installation is identical to that of nonradiant plastered ceilings.

Resistance wiring can also be embedded in concrete slabs. Usually, 3" of insulating concrete (vermiculite aggregate) is used as a base, the wring is carefully stapled in place, and 1½" or 2" of surfacing concrete is applied

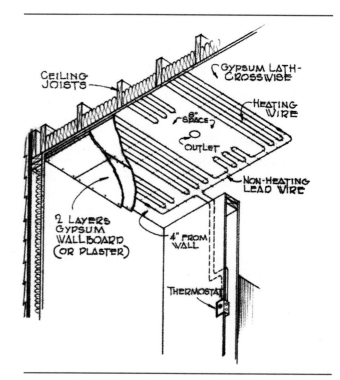

Figure 19–17 Installation of radiant heating electrical wire in drywall ceilings.

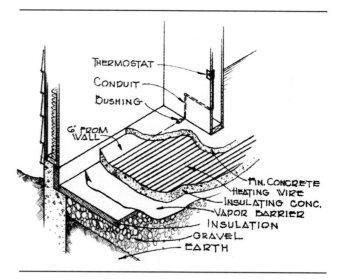

Figure 19–18 Installation of radiant heating electrical wire in concrete floor slabs.

over the wiring (Fig. 19–18). The perimeter of the slab should be well insulated, and it is advisable to place more wiring near the periphery to compensate for heat loss to the exterior. This method of central heating is not recommended except in mild climates where floor surface temperatures will not have to be made uncomfortably warm.

One architectural advantage of any of the electrical resistance–type heating systems is that no floor space needs to be reserved for a central heating unit, including chimney or flues. Another point, which is often overlooked, is that electrical heating installations are subject to minimum wear and therefore require less maintenance.

19.3.4 Heat Pumps

As mentioned earlier, the heat pump uses no combustion of fuel, such as gas, oil, or coal, as a primary heat source; rather, it operates on the principle that freely available outside air (or an economical water supply) contains useful heat, even during colder weather, and that this heat can be controlled and pumped into the house to maintain desirable temperature levels. Like electrical resistance heat, it is clean and free of soot particles. In summer the system operates in reverse, taking the heat from within the house and pumping it to the outside, working on a principle similar to that of the household refrigerator (Fig. 19–19). Actually, the system is more efficient for cooling than for heating, but since it performs a dual function, the heat pump is commercially feasible for year-round inside comfort in some climates.

Heat pumps circulate a fluid called a *refrigerant* through a cycle in which it alternately evaporates and condenses. In the process, the refrigerant absorbs and releases heat. The process is controlled by a *compressor*, which changes the pressure of the refrigerant, and by heat exchangers, which allow the refrigerant to absorb heat from or release heat to the surrounding air. In the heating mode, the system's efficiency drops considerably as the outside air drops in temperature. At a point just above freezing, it becomes necessary to supplement the heat supply, usually with resistance-type electric heating elements, in order to maintain a required heat supply. The additional expense for such supplementary electrical equipment may make operation of the pump impractical in many sections of the country unless electrical rates are low. Experience has shown that the most practical adoption of the heat pump for home heating has been in areas where outside temperatures seldom drop below 30°F.

Several heat-source and heat-delivery media have been found successful: water-source, ground-source, and air-source. With a water-source system, distribution of conditioned water within the house is handled with conventional piping systems. With an air-source system, the air distribution within the house resembles the forced-warm-air heating methods, and consideration must be given to duct sizing and register placement for cooling as well.

In appearance, either the heat pump is a single metal-covered unit, placed outside the house, or it is contained in two cabinets, called a remote unit, with one installed inside, usually in the basement, utility room, or attic near an outside wall, and the other outside. Some manufacturers make a unit that can be hung on an outside wall, but because of noise, the condensing unit is more satisfactory if it is not attached to the building in any way.

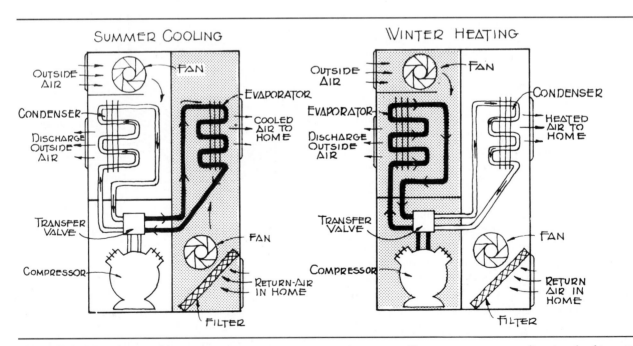

Figure 19–19 Diagram of heat-pump operation in summer and in winter. The unit acts as a two-directional refrigeration system, taking the heat from the inside air and discharging it outside in summer, and taking the heat from outside air in winter and discharging it inside the house.

In operation, the pump draws in outside air (or water), and the refrigerant in the evaporator absorbs the air's natural heat. A compressor pumps the refrigerant to a higher temperature and pressure level, and the condenser gives off the heat to the inside air, creating the heating cycle. The heat pump is considered efficient in operation up to the point where supplemental resistance heat must be introduced. This supplementary heat must be sufficient to supply at least 75% of the total calculated heat loss requirements (100% is recommended in case of compressor failure). Cooling is done with the fixed-refrigerant circuit by interchanging the air over the evaporator and the condenser. Automatic-reversing refrigerant valves make the changeover possible.

Because frost on the evaporator coils reduces efficiency, automatic hot-gas defrosting cycles or electric defrosters are necessary in air-source systems. Another required control is the inside air temperature–regulating thermostat, similar to that of other heating and cooling systems, which controls the inside temperature, the fans, and the heating and cooling changeover. Also, an outdoor temperature change anticipator is advisable in order to maintain uniform inside comfort during sudden weather changes. An emergency switch should be installed, however, so that outdoor thermostats can be bypassed and the auxiliary heat can be controlled with the indoor thermostat only.

Ground-source heat pumps and water-source heat pumps extract heat from the ground or from groundwater. Temperatures 8′ to 10′ underground are fairly uniform year-round. During the heating season, the temperature underground is considerably higher than that of the outdoor air; during the cooling season, underground temperatures are considerably cooler than the outside air.

In a ground-source heat pump, water or an antifreeze fluid is typically circulated through tubing that is buried underground. Heat is transferred from the water to the heat pump's refrigerant through a heat exchanger. A water-source heat pump typically circulates groundwater through a heat exchanger to transfer heat to the refrigerant. The result is much better performance than can be obtained when outside air is used as the heat source. With groundwater heat pumps, a coefficient of performance (COP) of 3 to 4 and summertime energy-efficiency ratings (EERs) as high as 20 can be achieved.

To measure the performance of an air-source heat pump, calculate its heating-season performance factor (HSPF). If outdoor air temperatures in winter commonly drop below 30°F, the seasonal-heating performance will be compromised. Thus, air-source heat pumps are most practical for moderate to warm climates.

In planning a heat-pump installation, you must provide a 220-volt service outlet. No provision, of course, is necessary for a chimney or a smoke vent, but it would be wise to plan for some other type of emergency heat in case of power failure.

York International Corporation manufactures a gas-fired heat-pump system. This system was developed by a consortium of natural gas suppliers, the Gas Research Institute, Battelle Memorial Institute, and several component suppliers. These pumps cost more than typical electrically driven air-source systems. Gas-fired heat pumps combine heating, cooling, and water heating. Gas engines replace the motors found on electric heat pumps, and a recuperating device recovers engine heat. An optional integrated boiler can provide supplemental heat for both space and hot water.

19.4

AIR-CONDITIONING METHODS

The process of cooling, dehumidifying, and filtering air to provide a desired level of comfort for occupants of the space is commonly called *air conditioning*. Summer cooling can be done most simply in residences with the use of individual room units placed in appropriate window openings. Or, in new construction, openings the size of window units can be provided in outside walls with electric outlets (usually 220 volts) nearby. This method of air conditioning may be the most economical in regions where summer temperatures are moderate and only one or two rooms in the home may require air conditioning. Basic planning for this type of cooling is not important and is usually not considered on working drawings, other than possibly a note on the floor plan where a unit is to be installed in an opening.

In areas where the cooling season is comparatively long, a central air-conditioning system, usually combined with the heating system equipment, is considered the most satisfactory for year-round inside climate control. As previously pointed out, these central systems, like heating systems, require consideration during initial planning of the structure; usually, the type of heating system selected will have a bearing on the most appropriate cooling system. If forced-warm-air heating is selected, cooling equipment utilizing the same circulating fan, ductwork, and registers will minimize the installation cost of the combination arrangement.

A mechanical refrigeration system, as mentioned in Section 19.3.4, "Heat Pumps," requires two basic units connected by closed-circuit refrigerant piping. One unit, called the *evaporator*, allows the refrigerant to vaporize under low pressure and absorb heat from the surrounding air or water. This unit must be placed within the house and is commonly combined within the central furnace jacket of combination systems. The other unit, called the *condenser*, dissipates the heat absorbed by the evaporator as well as the heat of compression. Higher pressures in the condenser change the vaporized refrigerant into a liquid, requiring the heat taken up in the evaporator to be given off as it circulates through the condenser. The condenser, then, must be placed outside the building where it can give off its heat; or, if installed inside, it must be adjacent to an outside wall opening.

Single-package cooling units are manufactured to fit conveniently on flat roofs, near louvers in attics, in crawl spaces, or in basement openings. Yet dual units are usually incorporated with heating systems requiring the condenser to be located on the exterior of the building. Many types of cooling units are available for combining with warm-air heating systems.

In dry regions, such as the southwestern part of the United States, where outside air contains comparatively little moisture, air-conditioning units utilizing *evaporative cooling* rather than mechanical systems are being used successfully for summer cooling. Dry air can be economically cooled by forcing it through a spray of recirculated water. This system is based on the fact that evaporation is a cooling process. Both individual room units and central systems working on this principle are manufactured.

If the home is to be heated with hot water, an entirely separate cooling unit and ductwork system (a *split* system) may be economically installed in an attic or a crawl space to cool the house with air and yet allow the hot-water system to heat the house independently with water. Also, an indirect cooling unit, called a *water chiller*, may be combined with a hot-water system to circulate chilled water through the same piping system. During the winter, heated water is piped through the system, and during the summer, chilled water is pumped for cooling. If a water-cooled type of condenser is used with the chiller, space must also be planned for a cooling tower near the rear of the building or on a level roof.

Usually the two-pipe, reverse-return piping system is the most popular with water heating-cooling combinations. One thermostatic control may be used to regulate both the boiler and the water chiller. Circulating chilled water through radiant floor slabs or ceiling panels is not considered practical for cooling, mainly because of condensation problems on cold surfaces.

19.5
RESIDENTIAL VENTILATING FANS

Circulation of fresh air in the home is not the major problem that it is in public buildings. Usually, a home is occupied by comparatively few people, doors are continually being opened and closed, and infiltration around windows furnishes natural ventilation. In public buildings, however, where many people congregate for extended periods, forced-air ventilating systems, usually designed by the same engineer who designs the heating-cooling systems, must be installed to provide a continual supply of fresh air.

Ventilation problems within the home consist mainly of eliminating excess moisture, odors, fumes, formalde-

hyde, and radon gas from specific areas such as kitchens, bathrooms, laundry rooms, basements, and attics. These areas should be serviced by locally placed exhaust fans, which are available in many models and sizes for almost any situation. Kitchen fans can be installed in ceilings or outside walls, incorporated in range hoods, or included within cabinet work. If access to the exterior is difficult, forced-air fans with electronic filters are installed to clean the air without requiring an outside outlet. Bathroom fans for ceiling installation commonly combine the exhaust fan with a light fixture and a supplementary resistance-type heater. Any source of excess moisture should have direct exhaust vents to the outside.

Central ventilating fans are effective as well as economical for cooling purposes when air-conditioning equipment is not installed. In many regions, pulling cool evening air through the house helps to make the interior more comfortable by alleviating the effects of the sun's radiation (the heat retained by the roof and the walls) on the structure. These quiet-running fans can be located in central hallway ceilings or other appropriate locations to draw outside air through windows of the lower floors and discharge it through the attic. Fan sizes, based on air movement capacities, should be selected to provide at least one air change (equal to the volume of the house in cubic feet) every 3 minutes. Some types of fans are made to operate in a horizontal position, and others in a vertical position.

The locations of attic fans are usually shown on floor plans by the use of broken lines, drawn to the approximate size; dimensions and type may be inserted with a note or may be included in the set of specifications. Manufacturers' catalogs must be consulted for detailed information about fans, which, like other types of equipment installed in a building, should be identified with model numbers and relevant size descriptions.

19.6
LOAD CALCULATION AND SIZING OF HEATING EQUIPMENT

To maintain a satisfactory air temperature within the home, the heating plant must be large enough to replace the heat loss of the house during average exterior temperatures; yet it should not be oversized if it is to operate efficiently and provide balanced heating. First, the *design temperature difference* must be established. Outside design temperature varies with the climate and can be determined for a given locality from the tables compiled by the American Society of Heating, Refrigeration, and Air Conditioning Engineers (A.S.H.R.A.E.) in their *Heating, Ventilating, Air-Conditioning Guide*; other books on heating design also provide this information. Usually 72°F is taken as a comfortable inside temperature in the winter. If the recommended outside temperature is, say, 5°F for a given area, then the design temperature difference for the heating system is 67°F. If an outside design temperature is 20°F, then the design temperature difference is 52°F. A correctly designed heating system would be able to furnish enough heat to the interior to maintain the 72°F inside temperature. Only after the rate at which the building loses heat during average extreme temperatures is calculated can the heating plant size be determined.

The next step is to find the entire heat loss of the building by individually computing the heat loss of each room to be heated. The number of square feet of cold partitions, cold ceilings, exposed walls, cold floors, windows, and exterior doors is calculated for each room. Then each surface area is multiplied by its *U factor*. Several different factors dealing with heat loss calculation are found in this information and various A.S.H.R.A.E. and ACCA tables available on home heating and cooling. The following definitions of the major ones encountered may clarify their meanings:

- **U factor** Coefficient of heat transmission through a combination of materials commonly found in walls, floors, and ceilings. It is the number of Btu's flowing from air surface to air surface through 1 sq ft of structure for each degree (F) difference in temperature between the two surfaces.
- **k Factor** Unit conductivity. The rate of heat flow in Btu/h through 1 sq ft of a homogeneous material 1″ thick per degree (F) difference in temperature between its two surfaces.
- **R factor** Thermal resistance of materials. It is obtained from the reciprocal of *U* factors and is expressed in degrees (F) per Btu/h per square foot. For example, a ceiling with a *U* factor of 0.20 would have an *R* factor of 1/0.20 = 5.0. The larger the *R* value, the more resistance to heat flow and therefore the better the heat barrier material.
- **HTM factor** Heat transfer multiplier as used in the ACCA *J Manual* method of heat loss calculation. It is similar to a *U* factor except that HTM factors are shown in their tables already multiplied by specific design temperatures and are used with the calculation forms shown in Appendix F. Window and door HTM factors also are modified to include infiltration heat losses, which simplifies the total calculation.

The larger *U* factors in the tables indicate comparatively poorer insulators. Heat transfer characteristics of the different materials and types of construction, as found in A.S.H.R.A.E. tables, have been established under controlled laboratory conditions.

The rate at which a building loses heat depends directly on the heat loss characteristics of its structural components and on the temperature gradient (the difference between inside and outside air temperatures). Remember that the areas of doors and windows in outside walls must be subtracted from the gross wall areas, and, usually, different *U* factors must be used for their calculations. As the tables show, windows and exterior doors have rather high transmission factors.

In heated basements, the masonry walls above grade have a *U* factor different from that of the walls below grade, since the ground absorbs very little heat compared to outside air. Partitions between heated rooms are not considered in heat loss calculations; one balances the other when uniform inside temperatures are maintained.

To sum up, heat loss by *transmission* will vary in proportion to the area in sq ft (A) of the ceiling and the outside wall areas, to the coefficients of their transmission values (U), and to the difference between the required inside temperature (t_i) and the outside design temperature (t_o). Thus, the following formula applies:

$$Q = AU(t_i - t_o)$$

where Q = Btu/h and where, after the heat loss of each different exposed surface of a room is determined, the result will be the heat loss of the room if it were airtight and doors were never opened. However, cracks around doors and windows allow heat to escape by what is called *infiltration*, which must be considered in an accurate heat loss calculation.

The amount of heat loss by infiltration varies directly with the wind pressure against the structure and the temperature difference between inside and outside air. In cold, windy climates, heat loss is considerable, especially if no weatherstripping is used around loose-fitting windows and doors. Of course, tight construction reduces infiltration. One method of determining heat loss by infiltration is by measuring the length of the cracks around all windows and doors in feet and then multiplying the totals by the appropriate infiltration factors, which also can be found in the previously mentioned heating manuals. This method is preferred for accurate calculations, especially in commercial building design.

An approximate *air change* method is satisfactory for residential heating systems. With this method, the number of air changes within a room per hour has been found by experience to vary from 1/2 to 2 changes. Under ordinary conditions, then, one air change per hour can be assumed. Therefore, a roomful of air must be reheated every hour to allow for infiltration loss, and the volume of the room in cubic feet is conveniently used. The density of air is taken to be 0.075 lb/cu ft. The specific heat of air is 0.24 Btu (the heat needed to raise the temperature of 1 lb of air 1°F). Therefore, $0.075 \times 0.24 = 0.018$ gives the Btu's of heat necessary to raise 1 cu ft of air 1°F. The infiltration heat loss can then be stated by the following formula:

$$Q = V(0.018)(t_i - t_o)$$

where V = volume of the air in the room, t_i = inside design temperature, and t_o = outside design temperature.

Totals of infiltration heat losses are combined with the transmission heat losses through walls, ceilings, floors, and so on.

In calculating the heat loss of concrete slabs on grade, usually only a 10′ wide area around the perimeter of the slab is considered (see the discussion on slab insulation in Section 12.3.2, "Foundation Types"). The heat loss to the ground within the interior of the slab is negligible. Transmission factors are available for both heated and unheated slabs.

After heat loss totals by both transmission and infiltration are determined for each room, the grand total provides the heat loss for the structure. Remember that, since U factors are given for a 1°F difference in inside and outside air, room heat loss calculations must include multiplication by the design temperature difference. The grand total of Btu's per hour of heat loss is the basis for sizing the heating unit after several factors have been taken into account. First, it usually is considered good practice to add 20% capacity to the unit to compensate for morning pickup when nighttime heating is allowed to drop as low as 55°F. Second, if plenum systems are placed in unheated crawl spaces or attics, the design temperature difference should be 30° greater than when ducts pass only through heated portions of the building. Third, heat loss in ductwork must be considered in sizing individual room ducts as well as in sizing the central unit. Heat loss of ducts passing through heated areas is not considered.

At present, the majority of all residential heating-cooling contractors throughout this country and Canada use the ACCA *J Manual* method of load calculation. Their tables and calculation forms have been devised from A.S.H.R.A.E. *U* factors and engineering information. Various shortcuts, such as infiltration factors included with door and window factors and specific design temperatures included with the factors, simplify either heating or cooling load calculation for residential buildings. The designer must make a thorough study of the step-by-step procedures given in the *J Manual* before attempting the calculations. Those students wishing to pursue this area of work are advised to take one of the short courses offered by ACCA several times each year in various metropolitan and university centers throughout the nation. Write to Air Conditioning Contractors of America, 1712 New Hampshire Avenue NW, Washington, DC 20009.

Heating calculations and system design for commercial buildings are required by law to be done by registered engineers engaged in this specialized field.

Hot-water boiler sizes must be taken from design tables distributed by the Institute of Boiler and Radiator Manufacturers (I.B.R.), after total heat loss calculations have been made. Sizes are usually indicated in Btu/h output and are determined after the following requirements have been considered:

1. maximum heat necessary for all connected radiator or convector units;
2. heat necessary for domestic hot water, if combined in the boiler;
3. heat loss in piping (piping factor); and
4. heat necessary for warming up the radiators or convectors and piping system after starting (picking factor)

Items 3 and 4 are often overlooked, inasmuch as the average boiler possesses enough reserve capacity to overcome the extra load, and in modern residences, piping and equipment are usually installed within insulated areas.

Boiler manufacturers' catalogs give boiler dimensions and vent pipe and flue sizes, according to rated outputs.

Room convectors or radiator sizes and capacities are based on individual room heat losses as well as on their distance from central boilers.

In hot-water radiant systems, the lineal feet and size of pipe that must be embedded in floors or ceilings are based on emission factors and individual room heat losses. You must consult tables for the emission rates in Btu/h per foot of piping in order to determine the amount and size of piping necessary to satisfy each room's heat loss. Water design temperatures adopted for the system must also be considered.

19.7

SIZING COOLING EQUIPMENT

In heating, the major problem is the addition of heat and moisture to the interior. In cooling during the summer, on the other hand, the problem becomes just the reverse: the *removal of heat and moisture*.

Many factors, both internal and external, affect the cooling load and become part of the cooling calculations. Internally, appliances, for example, exhaust fans, which remove heat and moisture from kitchen and baths, aid in reducing the cooling load. However, despite such reductions, the heat gain in the residential kitchen is rather large, so it is common practice to compensate for it by adding 1200 Btu's to the kitchen heat gain calculations. Another source of internal heat is that given off by the occupants. For this, 300 Btu's are usually allowed for each person living in the home. Larger factors up to 1400 Btu/h per person are used for calculations for commercial installations.

The major source of residential heat gain, however, is due to external factors: outside air temperature and the sun shining on glass, roof, and walls. Sun shining on fixed-glass units and windows contributes considerably to cooling loads and should be avoided if possible during early planning, as previously discussed. Also, the radiant heat of the sun, although not considered in heating loads, must be dealt with in air conditioning. Remember that heat has a natural tendency to travel from warm to cold temperatures.

All of these heat sources contribute to the cooling load, and the size of the cooling equipment is based on its capacity (Btu/h) to remove this heat.

To arrive at the correct cooling load for air systems, you must take design temperatures from tables similar to those used for heating design. Usually, 78°F (at 50% humidity) is used for the desirable inside temperature during the summer. The outside design temperature, including factors for daily temperature range, is taken from regional cooling tables. The design temperature difference between inside and outside air will then reveal the temperature reduction necessary for inside comfort.

Finally, the total heat gain within the structure must be calculated. Total exposed areas of walls, warm floors, warm ceilings, windows, and doors are multiplied by the heat gain factors. To find heat gain by air leakage, apply the correct factors from the tables. Multiply sizes of windows receiving direct sun by their factors. Add to the total heat gain through the structural components (external heat) the internal heat produced by people and appliances. Because moisture removal is part of comfort cooling, factors are used also for determining the heat gain from dehumidification (latent heat), which changes the moisture from a vapor to a liquid. Heat gain to ductwork in noncooled areas and other special heat sources are found with the use of factors from tables. All of these sources require attention in the determination of a heat gain total for each room. To determine the total for sizing the central cooling unit, add all of the calculated room Btu/h together.

Load calculations, design data, and code requirements for air systems can be taken from *Manual J* and *Manual 9* published by ACCA. A sample worksheet for job design and calculation is given in Appendix F.

Cooling units have been rated in tons of refrigeration capacity; 1 ton of refrigeration is the amount of cooling necessary to melt 1 ton of 32°F ice at 32°F in 24 hours. The latent heat of ice is 144 Btu/lb; then 1 ton (2000 lb) of ice absorbs 144×2000 or 288,000 Btu/24 h in changing ice to water. The heat gain in buildings is calculated in Btu/h; therefore, $288,000 \div 24 = 12,000$ Btu/h, which is the capacity of 1 ton of refrigeration. *To find the tonnage required, simply divide total heat gains by 12,000.* A ton of refrigeration does not necessarily equal 1 hp, although some units are rated in horsepower. Most manufacturers now rate their equipment in Btu/h, including both input and output.

Remember that if a water chiller is added to an existing hot-water system, new convectors must be installed in the rooms. Combination convectors, having fan, filter, condensate drains, and heating-cooling coils, as well as individual room thermostats, provide a more positive distribution of cooled air. Compared to residential cooling with air, hydronic systems have not become as universally popular.

Duct sizes for air distribution of combination heating-cooling systems are usually based on the total cubic feet of air per minute (CFM) necessary for the cooling. Total CFM circulating through a system is based on unit manufacturers' recommendations, usually from 300 CFM to 420 CFM per ton. (Heating ductwork need not be as large.) Tables indicate duct sizes for both main trunk and individual room ducts, based on total and room Btu heat gains. Minor factors, such as register distance from the central unit, duct offsets, and risers, are also considered. Ducts passing through unheated and uncooled spaces or outside walls must be insulated. Filters are installed in the return duct side of the system.

LOCATION OF OUTLETS IN ROOMS

Placement of registers and heat outlets in rooms requires careful consideration if the ultimate in comfort is to be attained from the system, especially in areas of extreme outside temperatures. Generally, it becomes advisable to place outlets in or near outside walls where the greatest amount of heat loss or heat gain occurs. To counteract cold downdrafts below windows, often uncomfortable in winter, low wall or floor outlets that blow the air up along the wall and windows are effective. However, when they are covered with floor-length window drapes, their maximum efficiency is decreased.

Registers can also be placed in partitions near the outside walls to blow the air along the exterior wall and thereby reduce downdrafts. Concrete floor slabs, utilizing the perimeter loop system, should have their registers in the slab. Sheet-metal boots for bringing the ductwork into wall registers placed near the floor may also be used. Generally, in northern regions where heating seasons are comparatively long, registers and heat outlets are located in the floor or low side wall; in southern areas where cooling seasons are longer, air is dispersed better with high side wall or ceiling outlets. Aside from these considerations, outlets should be placed so that air distribution is not noticeable to the occupants, no drafts are created, and the conditioned air will not have a tendency to stratify. Registers should be unobstructive and should fit into the decorative scheme of the interior; they should not restrict the placement of furniture. Since many types are available, it is possible to choose registers that fill the requirements of any situation.

Convectors for hot-water systems are mainly the baseboard type or wall units placed low near the floor.

Proper placement of return air grilles contributes to the dispersion of the conditioned air. In small homes, if centrally located returns are used, air will generally find its way through doorways, halls, and stairwells. Doors normally left closed are usually specified to be undercut 3/4" for an economical method of air return; or a grille may be installed in the door. Returns, however, must be large enough to prevent noise in the ductwork, and their sizes should be based on the volume of air delivered to the supply outlets. Properly sized and placed returns aid considerably in maintaining comfort throughout the building.

After studying the problems involved in providing comfort within a building during all seasons of the year, you should find it evident that each installation requires an individual application during planning. Selection of the most satisfactory equipment must necessarily follow a thorough analysis of all of the conditions involved. Heat loss factors of the various building materials and their combinations deserve attention. Initial costs of insulation should be compared in terms of fuel cost savings over the life span of the building. Of course, early costs of equipment and insulation generally will vary directly with the degree of comfort required.

19.9

SOLAR EQUIPMENT

Because of the extensive solar equipment now available to homebuilders, properly designed and integrated solar systems for homes can provide a large percentage of the energy needed for heating and cooling and domestic hot-water systems (Figs 19–20 and 19–21). When combining both passive and active features, solar systems can realistically and economically realize about 70% of space heating needs and about 90% of domestic hot-water needs. Mechanical solar cooling, on the other hand, while technically feasible at this time, requires additional research and development to achieve the same levels of efficiency and practical economy. This section will discuss mainly solar heating and solar domestic hot-water system equipment that is useful to students of residential construction.

Manufacturers of equipment are continually improving the efficiency of their products and introducing new components, including more efficient heat pumps, new and more efficient furnaces, automatic dampers, and controls that will make solar and backup equipment energy efficient without sacrificing comfort. New components should be evaluated on the basis of energy performance (EER), installed cost, reliability, maintenance costs, and warranties provided.

Passive solar planning, discussed in Chapter 15, should be an important aspect of systems that include any active solar equipment.

Several properties apply to all solar heating-cooling and domestic hot-water systems, whether they are simple or relatively complex. Each system includes these components: a collector, a storage medium, and a distribution system. It may also include a transport system, an auxiliary energy system, and controls. These components can vary widely in design and function, depending on needs and climate.

Solar radiation reaches the earth's surface by two methods: (1) by direct (parallel) rays and (2) by diffuse (nonparallel) sky radiation, which is reflected from clouds and atmospheric dust. The solar energy that reaches the surface of buildings includes not only direct and diffused rays but also radiation reflected from adjacent ground or building material surfaces. The amount of total radiation from these various sources varies widely according to climate—from hot, dry climates where clear skies enable a large amount of direct radiation to strike a building; to temperate and humid climates where up to 40% of the total radiation may be dispersed; to northern climates where snow reflection from the low winter sun may result in a greater amount of incident radiation than in

Figure 19–20 Solar collectors can be placed on south-facing roofs and made to contribute to the design of the exterior. (An RFT, Inc., mobile home design)

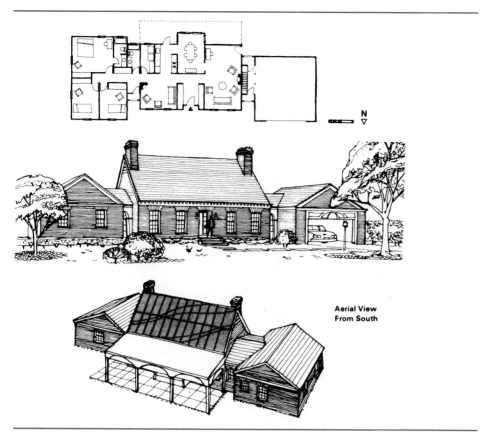

Figure 19–21 Use of solar panels on a traditional style house with a back-to-south orientation.

warmer but cloudier climates. As a result of the wide variation of solar availability in different locations, the design of as well as the need for solar systems varies widely. Selecting the most effective solar system and equipment begins with an analysis of local climatic conditions to determine the solar energy available. Before designing a solar house, take the time to review the subject further in several specialized texts and government publications.

19.9.1 Solar Collectors

As their name implies, *collectors* convert incident solar radiation into useful thermal energy. Transparent cover plates of glass or plastic are used to reduce convective and radiative cooling of the absorber. That is, these plates tend to trap more of the heat that would normally be reemitted from the absorbing surface. An efficient absorbing surface, usually dark, will have a high solar absorbance and a low emittance quality so that the maximum amount of energy will be conducted to the gas or liquid transporting medium.

The ideal slope of a collector is perpendicular to the sun's rays acting on it. However, this ideal situation is ordinarily not feasible, since the sun's relationship changes throughout the days and seasons. A stationary mounting is the most economical, but some mountings are flexible so that the most ideal angle can be variable. The majority of collectors are mounted directly to the roof for the best appearance, which requires some consideration in new construction on the selection of the most appropriate roof slope. Collectors may also be mounted on the ground near the building or mounted to appear as awnings above windows. Sun-angle charts are available for determining the most desirable collector tilt in local areas. As a rule of thumb, tilt should be the local latitude plus 15° (plus or minus 5° tolerance).

Experimental models of focusing collectors are made to focus the solar radiation to a concentrating point or area. Some are even made to follow the sun throughout the day. These models are comparatively more efficient, requiring an automatic tracking mechanism, than the fixed flat-plate type, yet their cost usually prohibits their use in residential systems.

Flat-Plate Solar Collectors   This relatively simple type of collector (Fig. 19–22) has now found the widest application. Possibly the reason for its popularity is its

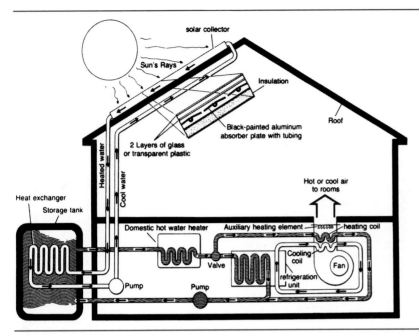

Figure 19–22 Solar-heated and -cooled home. Solar-heated water heats the storage tank water, which (1) heats domestic water, (2) heats water flowing through the solar heating coil, and (3) drives the refrigeration unit that chills water for the air-conditioning cooling coil.

economy of fabrication, installation, and maintenance. Of equal importance from an architectural standpoint is the fact that flat-plate collectors can be made to look good on gable roofs and can be made an integral part of the total design, rather than an unrelated appendage of the building. However, they are usually placed on the rear of the house if possible, regardless of type.

Flat-plate collectors utilize both direct and diffuse solar radiation. If the collectors are correctly designed and oriented, temperatures up to 250°F (121°C) can be attained, well above the moderate temperatures needed for space heating and domestic water heating. The collectors generally consist of an absorbing plate, usually metal, which can be flat, grooved, or corrugated and painted black (or chemically treated) to increase absorption. To minimize loss of heat, the backside is well insulated (with a type of insulation that is unaffected by intense heat), and the front is covered with a transparent cover sheet to trap heat and minimize convective cooling, as we mentioned earlier (Fig. 19–23).

The captured solar heat is removed from the absorber by a working medium, usually air or treated water, which becomes heated as it passes through or near the absorbing plate. The water must be in an antifreeze solution to prevent freezing. The heated medium must then be transported to the interior of the house or to storage, depending upon energy demand.

In selecting solar collectors, consider their thermal efficiency, the total area and orientation required, the durability of the materials, and, of course, the initial cost. There are innumerable variations, but the three basic flat-plate types are (1) the open-water collector, (2) the air-cooled collector, and (3) the liquid-cooled collector.

1. *Open-water collector* In this type (Fig. 19–24), water enters at the top, high side and trickles down over the corrugated absorbing plate. The heated water collects in a gutter on the lower side and is transported to interior storage. Some models have two corrugated plates spaced apart to allow the water to flow between; this configuration tends to reduce evaporation of the water and increase efficiency. Condensation forming on the inside of the cover plate reduces solar absorption, and therefore open-water collectors should be carefully evaluated before being used in cold climates.

    Factory-produced collectors have steadily improved in efficiency, yet on-site collectors are less expensive and have proved to be adequate if carefully constructed.

2. *Air-cooled collector* This type (Fig. 19–25) receives direct air through ducts from the building. The air

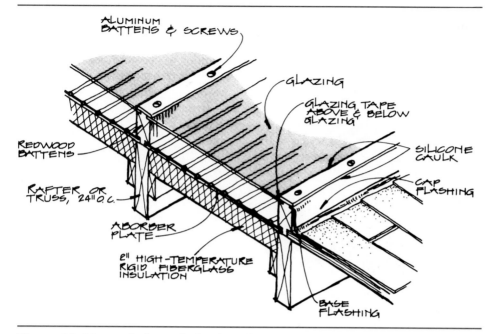

Figure 19–23 Detail of a site-built solar collector.

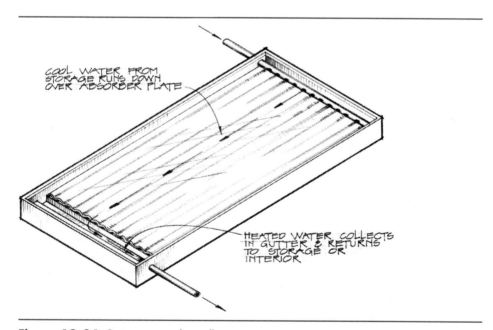

Figure 19–24 Open-water solar collector.

is heated within the covered collector and is forced back into the building or heat storage compartment (Fig. 19–26). Low maintenance and freedom from freezing problems are the major advantages. Disadvantages include the need for both relatively large ducts and electrical power for transport of the air, and the inefficiency of transfer of heat from the air to domestic hot water. Most experts believe that air-cooled collectors will increase in use in the future.

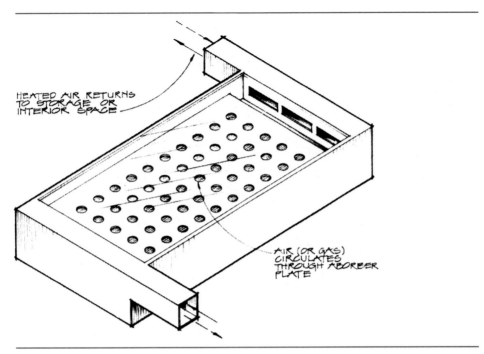

Figure 19–25 Air-cooled solar collector.

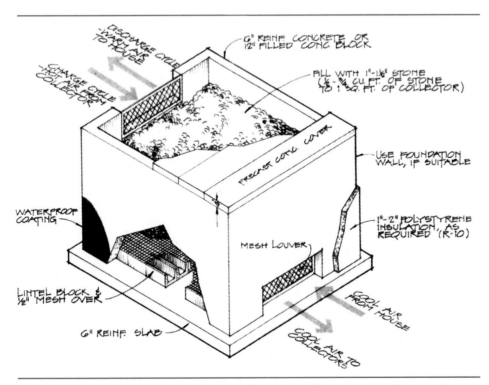

Figure 19–26 Suggested stone storage bin for an air collector system. Compared to water storage, rock storage requires 2½ times as much volume to store the same amount of heat over the same rise in temperature.

3. *Liquid-cooled collector* Water or an antifreeze solution is usually the liquid maintained within the piping of this type of collector (Fig. 19–27). The liquid is heated as it passes through the absorber piping in the collector and is then pumped back to a storage tank, transferring its heat to the storage medium. No condensate forms within the collector to interfere with solar absorption. Major problems have been the prevention of freezing, corrosion, and leaking; otherwise, these collectors are efficient. Oil that has been treated with corrosion inhibitors is also used as a transporting liquid.

19.9.2 Backup Heating and Cooling Units

Although solar systems can be relied on to provide the major amount of energy for heating a home and the domestic hot water needed, a backup system must be installed to provide the remainder of the necessary energy and to provide heat in extremely cold weather conditions. In some areas, codes require that buildings have conventional heating systems that are sized to provide 100% of the total load. Usually, it is advisable to select a system that is low in initial cost rather than one that is less expensive to operate (such as an electrical resistance baseboard), assuming that it will be used infrequently. All components should be carefully sized to meet only specific loads, rather than randomly selected. Refer to the earlier sections in this chapter for more information on systems that may be used for backup.

19.9.3 Controls

Controls on solar equipment are important, not only because they provide comfort and safety, but mainly because an active solar system must have controls in order to be workable. The best advice about controls as well as about the system itself is to keep them as simple as possible. Of course, solar components are subjected to extreme weather and temperature conditions, and therefore they should be of the highest quality—made specifically for solar installations rather than for run-of-the-mill plumbing. On the other hand, avoid hard-to-find or exotic controls that may cause time delays before replacements can be found should a breakdown occur. Controls make the system work, and only the most reliable controls should be accepted to avoid disappointment. Remember that collectors continue to collect heat even when controls fail to dispose of the energy, creating a troublesome situation.

Much solar equipment is now furnished by manufacturers in kit form to ensure that the controls are compatible with the other components of the system. For example, some air-handling kits are preengineered packages that include sensors, blowers, valves, and even ductwork to control the air completely throughout the system. These kits with fewer on-site fittings have proved to be the most trouble free. Other manufacturers recommend

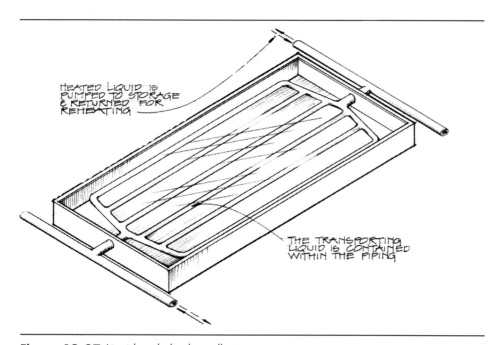

Figure 19–27 Liquid-cooled solar collector.

the exact type of controls that must be used with their components to have them function satisfactorily. Heretofore, the most trouble in active solar systems has been caused by the electrical controls and electrically operated valves. Manufacturers are now using solid-state technology and more innovation in their designing, which should result in more reliability.

Controls include thermostats, collector sensors, storage sensors, controllers, on-off switches, mixing valves, backflow stopping valves, circulators, and others.

19.9.4 Domestic Hot-Water Units

Various manufacturers now furnish domestic hot-water kits that are made to be installed during new construction or that may be installed in existing houses as retrofit applications. Even units that are made up of components furnished by different manufacturers can be satisfactory if installed by competent mechanics; many of the components, such as collectors or piping, can be constructed on-site. Many units are combined as part of the space heating system. The basic types of solar water heaters are closed-loop, thermosiphoning, recirculating, drain-back, and batch heaters. The majority of units now in use are closed-loop, mechanically circulated types that include roof-mounted, flat-plate collectors and covers, insulated piping, and individual controls, with various adaptations to satisfy local codes and weather conditions.

The *closed-loop* unit is controlled by a differential thermostat, one sensor at the collector, and one sensor on the side of the water heater (Fig. 19–28). When the early morning sun has warmed the collector sensor to a temperature of 15°F warmer than the tank sensor, the controller starts the circulator. On cloudy days or in the evening when the sun goes down, the collector quickly cools. When it reaches a temperature that is only 5°F warmer than the tank sensor, the controller turns off the circulator. Subsequent hot-water needs are automatically provided by the backup heater until the sun warms the collectors again. Many systems have high-limit controls that automatically stop the operation when the domestic water reaches 180°F to protect the tanks from overheating.

The other types of water heaters mentioned operate slightly differently, as their names imply, relating mainly to their method of freeze protection. Various conditions, such as availability, hot-water demand, architectural limitations, and type of backup unit, may restrict the selection.

Solar equipment may be combined with either gas-fired or electrical resistance–type water heaters for backup. Select units that are capable of furnishing 100% of domestic hot-water needs. In Fig. 19–28, the transport liquid in the closed loop could be water, propylene glycol, glycerin, Freon, or silicone oil. A nonfreezing liquid must be used in colder climates. Notice in Fig. 19–29 that the closed loop also includes a line through the wood-burning fireplace grate, which can be used to re-

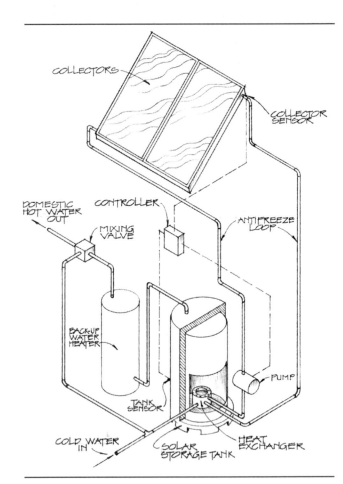

Figure 19–28 Closed-loop solar water heater.

claim some of its heat for the central heating system. Otherwise, the figure shows a typical space heating system with a water storage tank in the basement combined with a domestic hot-water unit. The stored solar heat is supplied to a conventional forced-air system trunk duct with a water-to-air heat exchanger.

19.10

DRAWING THE HEATING AND/OR COOLING PLAN

The Heating and Air-Conditioning Plan is basically a floor plan showing walls and structural features of the building with the layout of the central unit, the distribution system, and the room outlets in their correct position (Figs. 19–30 and 19–31). A plan of each floor involved is usually sufficient to show the important information, supplemented by several small details, if necessary. Information about simple heating or heating-cooling installations could be incorporated into the regular floor plan, but with more complex systems, it is usual practice to put the information on a separate plan for easier identification. Use the heating and

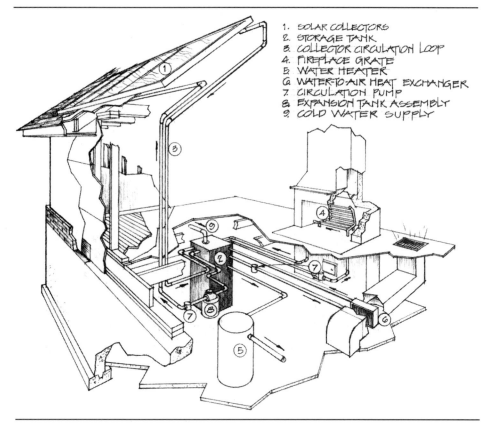

1. SOLAR COLLECTORS
2. STORAGE TANK
3. COLLECTOR CIRCULATION LOOP
4. FIREPLACE GRATE
5. WATER HEATER
6. WATER-TO-AIR HEAT EXCHANGER
7. CIRCULATION PUMP
8. EXPANSION TANK ASSEMBLY
9. COLD WATER SUPPLY

Figure 19–29 A solar system layout that includes a fireplace collector and a water-to-air exchanger.

piping symbols (Fig. 19–32) adopted by the American Standards Association, Inc. (A.S.A.), which have been standardized and are universally recognized. Large ductwork is scaled to show its horizontal dimension, and notes are added to indicate their cross-sectional sizes. In residential ductwork, it is standard procedure to use only 8″ vertical-duct depths with increased widths for increased air capacities. For example, a smaller duct might be 10″ × 8″, and a larger one may be 16″ × 8″. Notice that the horizontal dimension is given first in the specification. Piping is shown by single lines.

After heat loss calculations and design information are determined, start the plan with a freehand sketch of the entire layout. Add all pertinent sizes and information to the sketch. When ready to do the instrument drawing, fasten a clean sheet of tracing paper directly over the regular floor plan, and follow these general steps:

1. Trace the structural features (start with the basement plan, if applicable) using rather light linework. The linework of the heating or cooling equipment and the symbols, drawn later, should dominate the structural features.

2. Scale the floor size of the central unit, and locate it where planned; show the vent pipe to the chimney or vent. The unit output rating in Btu/h and manufacturers' catalog numbers may be added later.

3. Using their correct symbols, locate all room outlets, and indicate their sizes and capacities. Locate return outlets, and indicate their capacities.

4. If ceiling outlets are indicated, coordinate them with ceiling electrical outlets. Care must be taken near structural girders and plumbing piping.

5. If rectangular ductwork is to be drawn, lay out duct sizes to satisfy room Btu/h requirements by starting with the outlet farthest from the plenum. In general, one register will handle about an 8000-Btu heat loss in heating; for cooling, one register will handle only a 4000-Btu heat gain. Indicate each change of ductwork size with a note. Ductwork should be incorporated into the structure as simply and directly as possible. Show standard ductwork fittings with identification numbers.

6. Show circular ductwork with a single line, with a note giving its size.

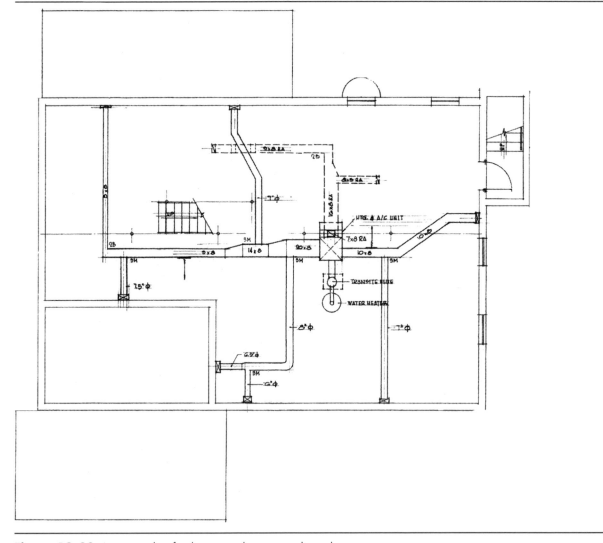

Figure 19–30 An example of a basement heating-cooling plan.

7. Locate risers for second-story outlets, if necessary. Riser duct sizes are made to pass through stud partitions. Be sure that second-floor partitions fall where risers are indicated.

8. In hot-water systems, use a single line for feeder pipes and a broken line for returns. Draw fittings, if necessary, with simple symbols.

9. Complete all notes, schedules, and structural details necessary to accommodate the system.

19.11

PLUMBING SYSTEMS AND DRAWINGS

Plumbing drawings are concerned mainly with the piping systems that supply hot and cold water, the pipes that carry waste materials to the sewer or disposal system, and the plumbing fixtures. With the exception of the fixtures, the systems are largely concealed within the structural cavities of the building.

For complex structures, the plumbing system design, like other mechanical or electrical design work, is usually done by registered engineers. Because these systems are often complicated, involving the action and control of water and air within pipes, engineers must be well qualified in the fields of hydraulics and pneumatics.

On the other hand, although small home systems involve similar basic principles, they seldom require the services of plumbing engineers. Usually, the responsibility for the design of residential systems falls on licensed subcontractors who are familiar with local and national plumbing codes. Much of residential plumbing is standardized, requiring little more than the location of

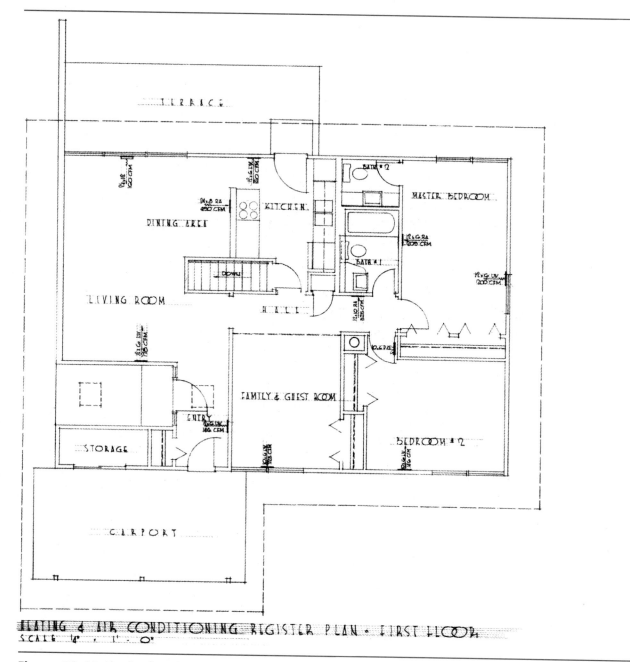

Figure 19–31 The first-floor heating-cooling plan showing the layout of room registers for the heating-cooling plan in Fig. 19–30.

plumbing fixtures on the working drawings. Simple structures do not always warrant separate plumbing plans, whereas complex commercial buildings may require a number of individual drawings devoted entirely to plumbing information.

Strict compliance with local plumbing codes is mandatory, regardless of the type of building. Plumbing codes, which may vary slightly from area to area, serve to ensure a sanitation standard to protect the health of the community. Students must understand the principles underlying good plumbing practices (Fig. 19–33) and must be acquainted with local code restrictions. The most up-to-date requirements can be found in the *National Plumbing Code* published by the American Society of Mechanical Engineers. Many of these recommendations have become part of local plumbing codes throughout the country. Some of the more common terms found in codes are illustrated in Fig. 19–34.

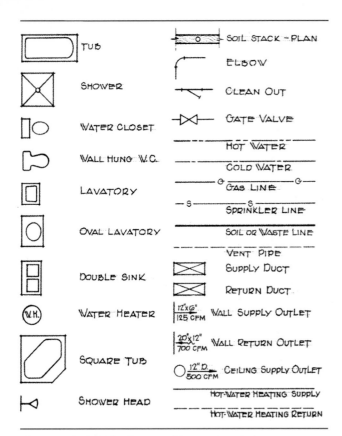

Figure 19-32 Plumbing and heating symbols.

Economy in plumbing costs is the direct result of careful planning in the arranging of the placement of rooms and fixtures requiring water outlets. Simplification of arrangement and the clustering of plumbing fixtures, usually into one wet-wall, should be the designer's objective in low-cost construction. In multistory construction, definite savings ensue if fixtures are located above each other to benefit from the use of common vents and soil stacks. However, in planning better-class homes, the designer should not sacrifice the workability of a good floor plan merely because extended plumbing lines constitute an added expense.

19.12

BASIC REQUIREMENTS OF THE WATER SUPPLY SYSTEM

Water supply piping within the home should be as direct as possible and sized large enough to provide an adequate supply at each outlet (Table 19–1). The entrance main pipe must be placed low enough below ground to prevent freezing or mechanical damage. Frequently, a copper pipe is used; a ¾" or 1" pipe is satisfactory. A shut-off valve near the entry of the line into the house is often a requirement, as is a pressure-reducing valve, a backflow preventer, and a pressure-relief valve.

One important consideration in water supply is the choice of piping material. Two types are commonly used today in residences: copper and plastic. Galvanized steel with its standard tapered pipe thread was used for many years. Although structurally rigid, steel pipe will eventually become corroded from oxygen in the water, sufficiently reducing the capacity of the pipe so that eventually it will require replacement. Plastic pipe is now gaining in popularity. A flexible butyl pipe with compression fittings has been approved by many building codes for water supply. Plastic pipe (PVC) is used for the water supply from the street main to the house. Wrought-iron pipe, although more expensive than steel, has less tendency to corrode. Copper pipe, often called *tubing*, eliminates the problem of corrosion altogether and is very durable. The diameter of the pipe can be smaller; simple, no-lead antimony–tin soldier joints are easy to make; and minimum water friction is created. Above ground, in exposed locations, hard-tempered copper produces the most acceptable installation; below ground, the flexible soft-copper tubing is simple to place. Because of its desirable attributes, architects usually specify copper for the better type of construction.

Hot- and cold-water lines are commonly run parallel to each other throughout the structure. However, care is taken to keep the pipes at least 6" apart, so that the water temperature in either pipe is not affected by the other. Insulating of hot-water lines is recommended, especially if the lines are long. Even cold-water lines in basements may require covering to prevent sweating. When water lines approach each fixture, hot water must be on the left, cold water on the right.

Another important point is the provision for "air cushions" in the water lines. These short, sealed-off extensions of the water pipes, attached just above the outlet in the lines, eliminate hammering noises when faucets are quickly turned off. Other types of shock-relief and expansion-chambers are available for incorporating into the lines. Much of the noise in plumbing walls can be isolated by having closets or storage rooms adjacent to baths; if this is not feasible, sound-deadening insulation or offset studding may be necessary in the wet-walls. Pipes should be sized for flow rates of 8' per second or less to avoid noise from rushing water in pipes.

19.13

BASIC REQUIREMENTS OF THE DISCHARGE SYSTEM

Whereas water is supplied into plumbing lines under pressure, drainage lines work on an entirely different principle—gravity. The designer must fully understand their operation to ensure that they function properly. Compared to the water lines, the piping for waste discharge must be larger, and branches that carry the wastes horizontally into vertical stacks must be properly sloped

The drainage system combines the use of TRAPS, VENTS, WASTE LINES, and SOIL STACKS

VENTS protect the water seals in traps and permit them to operate effectively. By admitting air to the system, vents permit atomspheric pressure on both sides of the trap seal to be maintained, and permit air to enter at the same time as gases escape the drainage system.

TRAPS permit waste and waste water to enter the drainage system and prevent any sewer gases from entering the house. The water seal utilizes a portion of the waste water to act as the barrier.

VENT THROUGH ROOF

STACK VENT

LOOP VENT

HOT WATER

COLD WATER

DRAIN

SOIL STACK

BATH DRAIN

CLOSET FLANGE

TO BUILDING DRAIN

WASTE LINES AND SOIL STACKS connect the plumbing fixtures to the traps and vents and eventually to the main disposal system. Waste piping is smaller in diameter than the main soil pipe, and the waste carried differs in content. Materials carried away by waste piping include such items as grease, hair, lint, food scraps, etc. Because of this function, cleanouts should be located so that the entire system can be opened up if necessary.

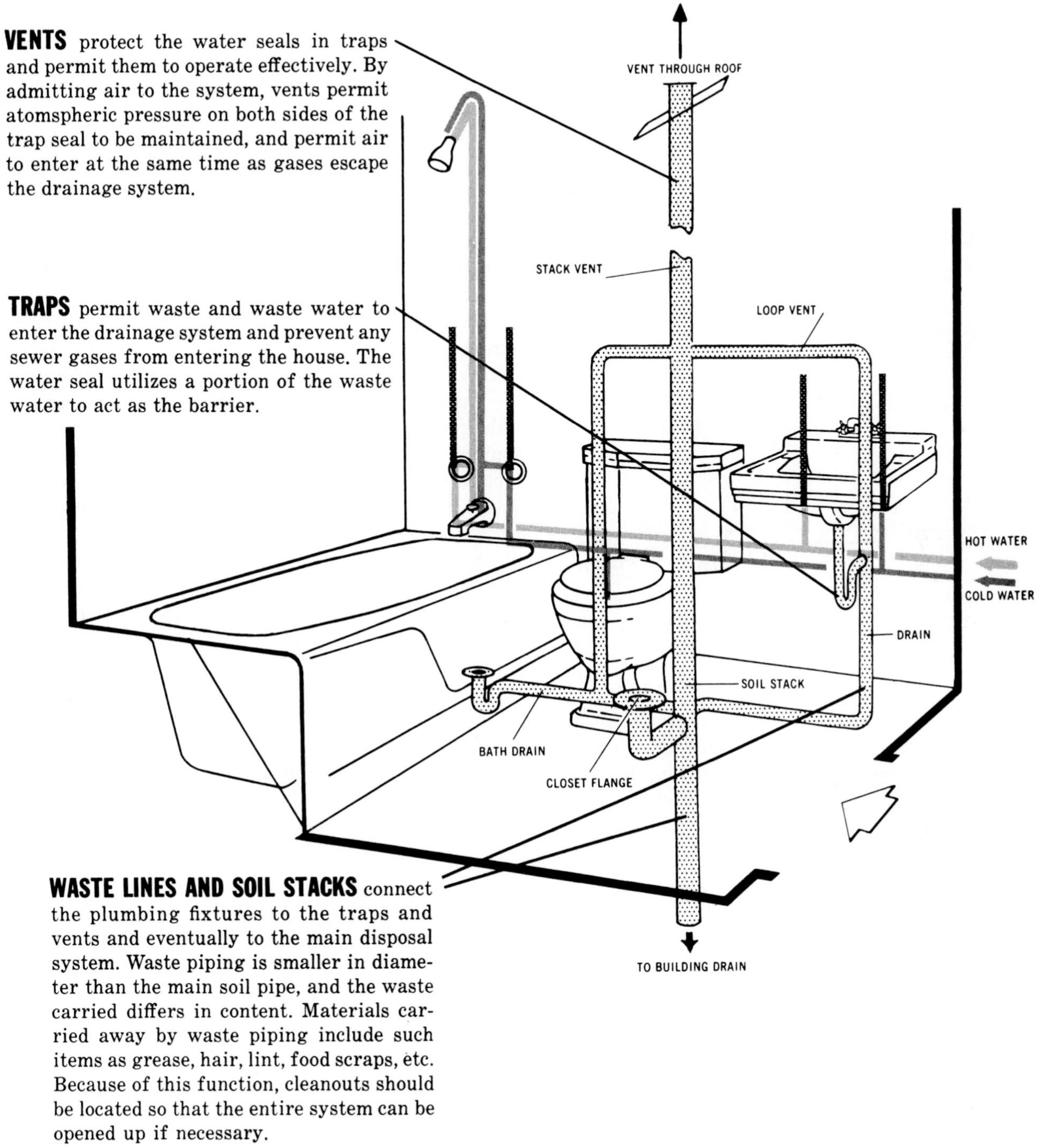

Figure 19-33 The basic parts of a bath plumbing system without reference to code or engineering limitations.

($1/8''$ to $1/2''$ per ft) to ensure complete and free drainage. Sewage lines produce offensive and harmful gases that will permeate the entire structure unless *traps* forming offsets in the pipe to retain a water seal are installed near each fixture. Water closets have a built-in trap. Vent stacks through the roof discharge sewer gases to the outside air. These vent stacks and their branch vents have other important functions. They allow a continuous atmospheric pressure to exist within the sewage system, which prevents the siphonage of water from fixture traps

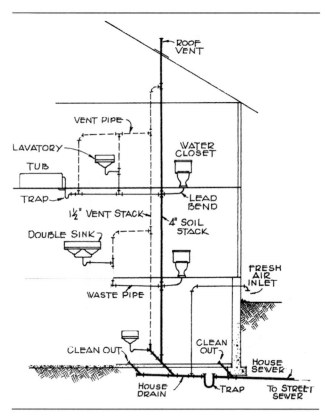

Figure 19-34 A drainage diagram for a multistory residence.

and reduces the decay action of bacteria on the sewage while it is in the piping. Sewer gases may possibly emerge through trap seals unless the vents allow the gases to pass readily to the open air. Vent stacks should be installed at least 6″ and not exceed code distances from the traps, and

they should extend at least 12″ above the roof. Although minor variations in practice exist, local codes carefully regulate methods of venting fixture traps and drainage lines. Some new types of air-relief vents used inside the dwelling are approved for use by local code officials.

Soil and vent pipe sizes vary according to the number of fixtures that they serve. In residential systems, the main soil stack and house drain are usually 4″ cast-iron soil pipe; waste branches and minor vent stacks generally are 1½″ or 2″ in diameter. The minimum size permitted for a soil stack pipe serving one water closet is 3″; the pipe for stacks handling two water closets must be at least 4″. Vent stacks that run parallel to soil stacks can be half as large as the soil stacks.

The majority of sewage line pipe used today is PVC plastic pipe, type DWV (drain, waster, and vent). Cast-iron soil pipe is still used in some situations, but plumbers can install plastic pipe more easily and more quickly. Other types of drainage pipe are also used: copper, galvanized steel and wrought iron, brass, or lead. The sewer pipe running below ground from the house to the public sewer may be cast iron, vitrified tile, concrete, or PVC plastic pipe; but codes usually require at least 5′ of the piping used within the building to extend beyond the foundation wall (house sewer). Most pipe exposed near fixtures is chrome-plated brass or flexible plastic. Pipe connections at fixtures allow slight flexibility in case of shrinkage or movement of wood framing without putting strain on the connected piping. Plumbing fixtures must be solidly supported by the structure. All drainage pipe should be as straight and direct as possible, fittings made to minimize friction, and cleanouts inserted at critical points of the system to allow unclogging if this becomes necessary.

Table 19-1 Minimum pipe sizes of fixtures (FHA).

Fixture	Branch Hot Water	Branch Cold Water	Soil or Waste Connections	Vent Connections
Water closet		½″	3″ × 4″	2″
Lavatory	½″	½″	1¼″	1¼″
Bathtub	½″	½″	1½″	1¼″
Sink	½″	½″	1½″	1¼″
Laundry tray	½″	½″	1½″	1¼″
Sink and tray combination	½″	½″	1½″	1¼″
Shower	½″	½″	2″	1¼″

DOMESTIC WATER HEATERS

To furnish a continual supply of hot water throughout the year, modern homes must be equipped with some type of automatic hot-water heater. The heater and storage tank combination unit is the most satisfactory, although heaters that supply hot water instantaneously as it is needed are available. Sizes of heaters and their tanks are based on the number of occupants. Figures show that the average residential occupant uses about 40 gallons of hot water per day. This estimate includes consumption for laundry and bath facilities, based on water heated to 140°F. Although tanks holding the minimum of 20 gallons are available, a 30-gallon tank is more satisfactory for the average family, and one holding 50 gallons is necessary for a larger family using automatic laundry equipment. Heaters with slow recovery rates require proportionately larger storage tanks. If a unit has a slow recovery rate, such as an electric water heater, then an 80-gallon tank unit will be necessary to furnish hot water at about the same rate as would a 30-gallon gas- or oil-fired unit.

In commercial buildings having many hot-water outlets, the consumption per fixture is usually not as great as in residences. However, the demands of the building and the consumption rate during peak hours must be known before the size and the type of hot-water heater can be indicated. Because hot-water lines tend to corrode more readily than cold-water lines, piping from hot-water heaters should be thick-wall copper. Heaters should be located near their principal outlets, and heaters having adjustable temperature settings are appropriate for such installations.

Indicate hot-water heaters on working drawings with a circle symbol and a W.H. note. Tank size can also be noted, if necessary. Also note the minimum efficiency desired.

PRIVATE DISPOSAL SYSTEMS

In most urban areas, buildings are serviced by public sewage lines that run along street thoroughfares. The house sewer connection is made by merely tapping the public sewer line. But in outlying areas where no public sewage facilities exist, provision must be made for satisfactory disposal of the sewage on the owner's property. The designs of individual disposal systems are closely regulated by local codes, and a number of conditions pertaining to the property must be analyzed before a permit is granted. Conditions affecting the disposal system design are number of occupants of the house, permeability of the soil, groundwater level in the area, size of the property, and relative location of wells or streams that may be affected by the sewage disposal.

The simplest system (often prohibited by local codes) is a cesspool. This is an underground cylindrical pit with porous walls built of stone or masonry to allow seepage of the discharge into the surrounding ground. The sewage is deposited directly into the pit. If the surrounding soil is permeable enough to carry off the liquid, and if the underground water table is low enough to prevent contamination, the system may be adequate for retaining a small amount of solid matter. However, experience has shown that cesspools become troublesome when the apertures of the masonry become clogged and the surrounding ground becomes saturated. Secondary cesspools may then be necessary.

Probably the best solution for individual sewage on property having a sufficient amount of space is a *septic tank and tile drain field* (Fig. 19–35). The sewage empties from the house sewer into a watertight concrete, fiberglass, or steel tank placed below ground. Cast-iron tees or wood baffles at the entrance and the discharge end of the tank prevent disruption of the surface scum within the tank and direct the solids to the bottom. Bacteria act on the sewage to break down the solids into a compact

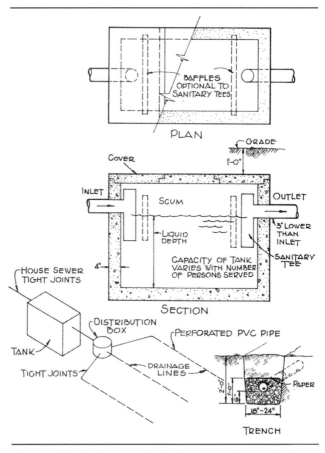

Figure 19–35 Septic tank details and a disposal field layout used on level ground. Perforated fiber or plastic pipe is commonly used for field lines.

sludge, which needs removal only after several years of use. Only after the sewage is about 70% purified will the effluent liquid leave the tank and flow through the seepage drains, thus eliminating the possibility of contaminating surrounding water tables.

Seepage lines consist of 4″ diameter, perforated plastic pipe (PVC) placed in trenches filled with coarse gravel to allow positive drainage into the surrounding soil. Clay agricultural tile with open joints can also be used. The lines are placed about 16″ below ground level and are sloped from 4″ to 6″ per 100′ to keep the drainage near the surface and yet maintain a gradual outflow. Much of the moisture near the surface evaporates into the air. Different arrangements of the tile lines are possible in restricted space and to accommodate various contours; if steep slopes are encountered, line drops to different levels are made with tight pipe. Lengths of drainage tile lines and sizes of gravel trenches are determined by local codes according to the size of the dwelling and the number of occupants, as well as by the absorption rate of the soil. Usually, 100′ to 300′ of drainage line is required for the average residence. Tests should be made to determine the rate of water absorption in the disposal area before reliable disposal fields can be designed. Sewage disposal systems should be installed by competent and reliable contractors. Because detergents interfere with the bacterial action in septic tanks, wastes from automatic laundry equipment should be emptied into isolated drainage lines. Also, storm drains should not empty into the septic tank.

Some local codes allow a *seepage pit*, similar in construction to a cesspool, to take the discharge from a septic tank and allow it to release the effluent liquid into the surrounding soil instead of drainage lines. Low underground water tables permit the use of a seepage-pit system.

19.16

DRAWING THE PLUMBING PLAN

On drawings of simple, noncomplex buildings, show the plumbing fixtures to scale, and locate any other water outlets on the regular floor plan. Usually, this plumbing information is all that is necessary. Include all exterior hose bibbs and automatic washer outlets from which piping lines can be planned. Use plumbing templates, if available, or simply scale the symbols with instruments. List fixture sizes and manufacturers' catalog numbers in the specifications, or list them in a plumbing fixture schedule. Public sewer lines and septic tank systems should be located on site plans.

If the plumbing information is more extensive, draw a separate floor plan showing the walls and the main structural features, traced from the regular plan. In general, follow these steps in completing the plumbing plan:

1. Draw all plumbing fixture symbols to scale.
2. Locate the cold-water supply line from the street, and lay out all cold-water lines to each outlet. Indicate valve, fitting, and pipe sizes.
3. Locate the water-heating equipment, and lay out the hot-water lines to all fixtures requiring hot water. Use a hot-water pipe symbol, and label pipe sizes and fittings.
4. Locate the house drain according to public sewer connections or location of the septic tank.
5. With heavy lines, lay out the soil lines after consulting local plumbing codes for pipe sizes and number of fixtures on each line. Show cleanout symbols and vents according to code. Make soil lines as short and direct as possible. Indicate size and description of pipe and fittings.
6. Locate the gas line entrance, and show the pipeline to the outlets. Give the pipe size.
7. Draw a vertical section through the stack and vent lines on drawings of multistory buildings. To show complex piping arrangements, use a single-line, isometric layout without standard fitting and fixture symbols.
8. Darken all plumbing information to make it more prominent than the structural outlines.
9. Add appropriate notes and titles.

19.17

HOME ELECTRICAL SYSTEMS

To represent electrical systems on the drawing board, the student should have a rudimentary knowledge of the principles of electrical distribution so that he or she may provide adequate wiring and lighting, not only to meet present demands, but also to anticipate future needs.

In the planning of an electrical wiring system, safety must be the primary consideration. Electricity passing through a conductor creates heat. Overtaxed wiring is not only inefficient but also dangerous. If the conductor is too small to accommodate the amount of current passing through it, the overload results in a breakdown and the melting of the protective insulation of the wiring, thus creating a hazard. Considerable destruction may be caused by fire due to outmoded electrical wiring. Minimum requirements for safety standards found in many local codes have been established by the *National Electrical Code* (N.E.C.). Compliance with all applicable codes is the first requisite of the electrical system. A copy of the local electrical code should be studied for requirements of local construction.

Like the design of mechanical equipment systems, the design of electrical systems in large buildings requires the services of engineering firms specializing in such work, whereas residential systems of a less complex nature are usually designed and installed by licensed electrical contractors who are familiar with the limitations and requirements of local and national codes. Proper planning for home electrical needs, however, must be done on paper when the total concept is conceived and the major planning is undertaken. A list of commonly used terms and their definitions follows.

Electrical Terminology

- *Ampere (A)*   The unit of current used to measure the amount of electricity passing through a conductor per unit of time. The flow of current in amperes can be compared to the flow of water in gallons per second.
- *Circuit*   Two or more conductors through which electricity flows from a source to one or more outlets and then returns.
- *Circuit breaker*   A device that breaks the flow of electricity in a circuit automatically and that performs the same safety function as does a fuse.
- *Conductors*   The common term for wires that carry electricity.
- *Fuse*   A safety device that contains a conductor and that melts when the circuit is overloaded, thus interrupting the flow of electricity.
- *Ground*   A connection between the electrical system and the earth—it minimizes damage from lightning and injury from shocks.
- *Horsepower (hp)*   A unit of work; 1 hp = 746 watts. (The term is diminishing in use.)
- *Outlet*   A point in a circuit that allows electricity to serve lights, appliances, and other equipment.
- *Receptacle*   A contact device to receive the plugs of electrical cords or portable lighting equipment and appliances.
- *Service entrance conductors*   The conductors that connect the service equipment to the service drop (overhead) or the service lateral (underground).
- *Service equipment*   The associated equipment, including switches or overcurrent devices, that makes up the main control of the power supply.
- *Short circuit*   A condition resulting from the unintentional connection between two ungrounded (hot) conductors or between an ungrounded conductor and a grounded conductor (neutral).
- *Volt (V)*   The unit used to measure the electrical potential difference.
- *Voltage drop*   The loss of voltage caused by the resistance in conductors or equipment too small to produce overloading.
- *Watt (W)*   The practical unit of electrical power equal to amperes times volts. Most appliances are rated in watts to indicate their rate of electrical consumption. Both voltage and amperage are considered:

  1 A with pressure of 1 V = 1 W
  1 W consumed for 1 h = 1 watt-hour (Wh)
  1000 Wh = 1 kilowatt-hour (kWh)

19.17.1 The Service Entrance

Electricity is furnished to a home by a three-wire service drop, installed by the local power company. This three-wire service supplies $^{120}/_{240}$-volt, single-phase, 60-Hz alternating current to the house, now considered standard procedure in most localities. Several methods of firmly anchoring service drops are shown in Figs. 19–36 and 19–37. If the service is to be attached to a low roof, a masthead (or weatherhead) installation prevents the service conductors from being attached too near the ground (Fig. 19–37). The service drop conductors should be at least 10' above the final grade. A rigid support for the service conductors must be furnished by the homeowner. Location of the point of attachment should be governed largely by the location of the service equipment. Minimum distances between the entrance and the service equipment result in wiring economy, less voltage drop, and more electrical efficiency. Generally, the entrance is made near the kitchen or utility area to provide short runs for the important energy outlets.

In some areas, service conductors are brought to the house below ground. This method conceals the conductors and fittings that would otherwise be unsightly. Electric meters installed on the exterior of the house are more convenient for both the power company and the homeowner. A protective device must be installed at the meter if the service conductors are extended more than 2' into the building.

The sizes of the service entrance conductors and the conduit are determined by the capacity of the service equipment (see the next section). A minimum 60-ampere service requires No. 6 THW conductors and 1" conduit; 100-ampere service requires No.3 THW conductors and 1¼" conduit; 150- or 200-ampere service requires No. 1/0 or No. 3/0 (type THW insulation) and 2" conduit.

19.17.2 Service Equipment

The service equipment (distribution panel) is the heart of the electrical system. Encased within a metal box, the equipment contains main disconnect switches, fuses or circuit breakers, and terminals for attaching the distribution circuit conductors. Panels can be either flush or surface mounted. Fuses in the panel are the protective devices in case of overloading or accidental short circuits, and they are sized according to each circuit capacity (replacements must be no larger than the original fuses). In place of fuses, similar protection may be provided by more expensive circuit breakers that automatically break the circuit when overloaded. No replacement is necessary when the breaker is cut off; a tripped circuit breaker is merely switched back to its closed position. Show the location of the service panel on the floor plan.

Every electrical system must be grounded to moist earth with the use of a conductor from the ground terminal in the service panel to a metal ground rod or a metallic cold-water pipe.

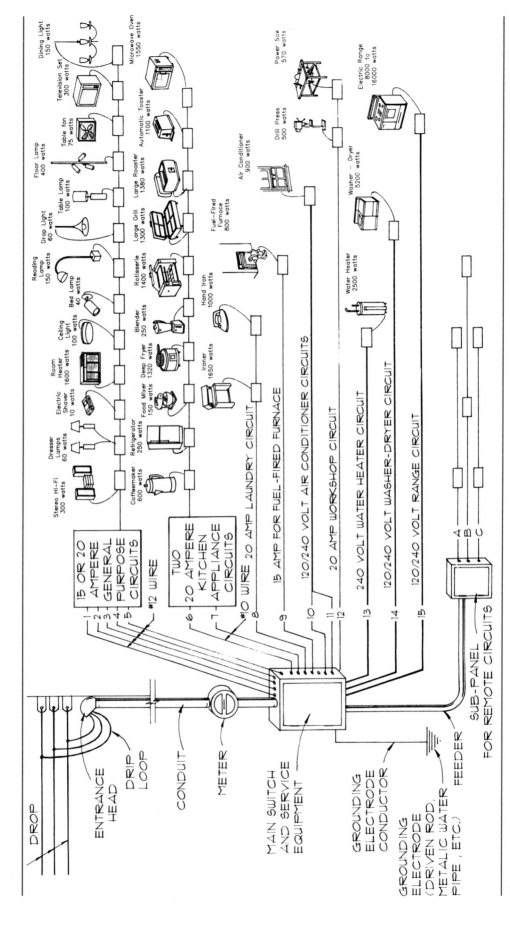

Figure 19-36 Wiring diagram of a residential electrical system.

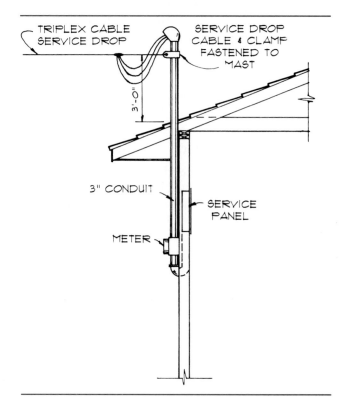

Figure 19-37 Entrance mast for low roofs.

Table 19-2 Wattage table.

Home Appliances	Watts
Automatic dryer (regular)	4500
Automatic dryer (high-speed)	8700
Automatic washer	700
Bathroom heater	1500–2500
Dishwasher and waste disposal	1500
Exhaust fan	300
Fuel-fired heating system	800
Hand iron	1000
Home freezer	350
Ironer	1650
Microwave oven	1500
Personal computer	250
Range	8000–13,500
Refrigerator	300
Roaster-broiler	1500
Room air conditioner (1 ton)	1500
Stereo	450
Television	300

The size of the service panel is determined by the estimated load requirements of the building, combined with the anticipated future needs. Lighting and appliance loads can be taken from Table 19–2 and converted into amperes. For example, if a total of 30,000 watts were found to be needed in the home, the minimum service would then be 125 amperes (30,000 ÷ 240 = 125). That is,

$$\frac{\text{watts}}{\text{volts}} = \text{amperes} \quad \text{or} \quad \frac{W}{V} = A$$

A 150-ampere service would be the logical selection to take care of future needs. A 20% additional capacity is commonly added for these contingencies.

The minimum rating for residential service entrance conductors is 60 amperes, provided that the total calculated load is less than 10 kVA and the total number of branch circuits is five or less. If either of these two conditions is not met, the minimum rating of the ungrounded service entrance conductors must not be less than 100 amperes. Few, if any, new electric services will be rated less than 100 amperes.

A *100-ampere service* is recommended as the minimum for all new homes up to 3000 sq ft in size. This service will be adequate for lighting and appliances requiring up to about 24,000 watts total load (see Table 19–2). If an electric range, a water heater, an electric dryer, and central air conditioning, along with lighting and the usual small appliances, are included, a *150-ampere service* will be needed. If electric resistance heating or other unusual loads are to be included with the above loads, a *200-ampere service* should be selected. Large homes of 5000 sq ft and over should have at least the 200-ampere service. Larger services are also available for commercial installations. Panels should be sized in accordance with N.E.C. standards so as not to oversize or undersize the building requirements.

19.17.3 Branch Circuits

Branch circuits have various capacities (15, 20, 30, 40, and 50 amperes), depending on their use. Both 240- and 120-volt circuits are furnished by the service panel. Depending on local codes, one of the following types of wiring is used in modern construction.

Nonmetallic-Sheathed Cable   This cable is inexpensive and easy to install. It is a flexible conductor that is encased with a plastic, moisture-resistant, flame-retardant, nonmetallic covering over individually insulated copper or aluminum wires. Either two- or three-wire cable is available to be used only in dry locations. For damp or underground locations, a special plastic-covered cable is used instead of the standard plastic-covered kind. Either type is available in sizes from No. 14 to No. 6 wire.

Metal-Armored Cable (AC) This cable is covered with a spiral, flexible metal sheathing that gives the wiring more protection from physical damage, yet it is unsuitable for damp locations.

Metal Conduit Metal conduit is more expensive and offers still more protection for the wiring. The conduit is installed with the boxes and the receptacles during construction; then the conductors are pulled through and connected. All splices must be made in junction boxes.

Types of branch circuits found in residences include the following:

- **General-purpose circuits (120 volts, 15 or 20 amperes)** These circuits serve lighting and general convenience outlets. Good circuitry layout provides several circuits in rooms or wings so that partial lighting will always be available in case of a blown fuse. Receptacle outlets are required in every kitchen, family room, sun room, bedroom, recreation room, and the like, so that no point along a wall shall be more than 6′ from an outlet. Number 12 AWG electrical conductors should be used, which will allow a maximum of 2400 watts in circuits up to 75′ in length; longer circuits require larger conductors.
- **Appliance circuits (120/240 volts, three-wire, 20 amperes)** These circuits should be provided in kitchens, dining areas, and laundries to serve appliances requiring heavier loads.
- **Individual-equipment circuits** These circuits are needed to service each of the following permanently installed equipment: electric range, washer-dryer, fuel-fired furnace, dishwasher, electric water heater, central air conditioning, bath heaters, and other similar equipment. The capacity of each circuit must be sufficient for the needs of each piece of equipment.
- **Feeder circuits** These circuits, with subdistribution panel equipment, are sometimes used for remote areas. A four-wire feeder is run from the main panel to a subpanel, and two-wire branch circuits are extended from the subpanel with a minimum of voltage drop.
- **Ground fault circuit interrupters** All 125-volt, single-phase, 15-ampere and 20-ampere receptacles installed in bathrooms, in garages with exceptions, outdoors with grade-level access, in kitchens within 6′ of a sink above countertop surfaces, and in boat houses, and one basement receptacle, must have ground fault circuit interrupter protection.

19.17.4 Low-Voltage Switching

Another variation of the conventional wiring system is the low-voltage switching system. With this system, wall switches control a low-voltage circuit (6 to 24 volts) through a relay that activates the line voltage at the outlet. Savings can be realized on the low-voltage wiring, since light bell wire can be used. There is no danger of shock at the switch, and any or all of the outlets can be controlled at a central station. Slightly different types of equipment are available for low-voltage systems.

19.17.5 Location of Outlets

For maximum convenience, the probable placement of furniture in each room should be considered before outlets are located. For instance, in a bedroom, the outlets would best be on each side of the bed. Since bedrooms and living rooms rely primarily on movable lamps for general lighting, one or two of the outlets should be connected to wall switches. Outlets should generally be placed near the corners of rooms, where they are easily accessible and there is less chance of their being obstructed by furniture

The National Electrical Code Section 210-52 states that the minimum requirement for the location of receptacles in dwelling units is such that no point along the floor line in any wall space is more than 6′ from an outlet in that space and that receptacle outlets are to be spaced an equal distance apart. *Wall space* is defined as a wall unbroken along the floor line by doorways, fireplaces, and similar openings. Each wall space 2′ or wider is to be treated individually and separately from other wall spaces within the room. The purpose of these requirements is to reduce the need for running extension cords across doorways, fireplaces, and similar openings. Duplex receptacle outlets should be no more than 12′ apart; thus, a 6′ extension cord can reach a receptacle from any point along the wall line.

The conventional height for outlets is 12″ above the floor, except in kitchens and dining areas. In kitchens they are placed just above the countertops at 4′ intervals, convenient for work centers; in dining areas they should be located near and just above table surfaces. Baths need an outlet near the lavatory or vanity and one on an opposite wall. Keep all outlets away from tubs or shower stalls, unless they are waterproof. All stationary appliances in kitchens, laundries, and utility rooms require an outlet near their place of installation. Remember to include sufficient outlets in garages, workshops, and recreation rooms, as well as waterproof outlets near terraces for outdoor living, and one near the front entrance. Ground fault receptacles are required in baths, kitchens, and all outside areas.

Figure 19–38 is an example floor plan of a small residence indicating the location of receptacle outlets. The living room outlets are split wired: the lower half of each duplex receptacle is always "hot" (energized), while the upper half can be switched on or off. It is likely that a major part of the illumination for this area will be provided by portable table lamps, and the split-wired receptacles provide a means of control for them. A duplex re-

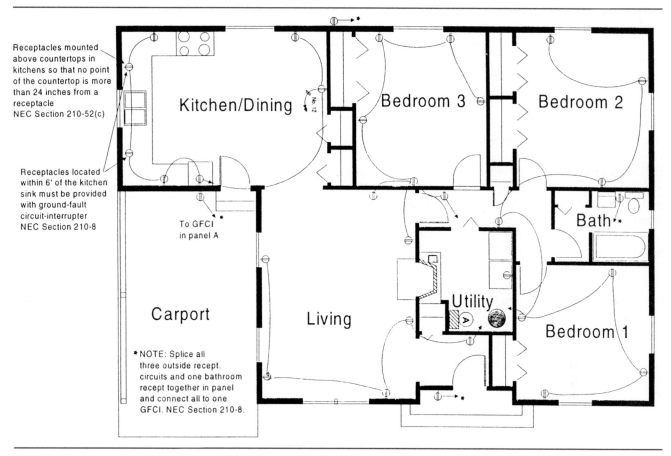

Figure 19–38 Residential receptacle plan.

Receptacles mounted above countertops in kitchens so that no point of the countertop is more than 24 inches from a receptacle
NEC Section 210-52(c)

Receptacles located within 6' of the kitchen sink must be provided with ground-fault circuit-interrupter
NEC Section 210-8

To GFCI
in panel A

Kitchen/Dining

Bedroom 3

Bedroom 2

Bath

Carport

Living

Utility

Bedroom 1

*NOTE: Splice all three outside recept. circuits and one bathroom recept together in panel and connect all to one GFCI. NEC Section 210-8.

ceptacle is located in the vestibule for cleaning purposes and is connected to the living room circuit.

The utility room has two receptacles: one for the washer on a separate circuit and a second for ironing. The kitchen receptacles are arranged so that in addition to the number of branch circuits determined by calculation, two or more 20-ampere, small-appliance, branch circuits are provided. These circuits serve all receptacle outlets, including refrigeration equipment in the kitchen, pantry, breakfast room, dining room, or similar area of the house.

The duplex receptacle installed in the bathroom is connected to the same circuit as the three outside receptacles. They are connected to a ground fault circuit interrupter (GFCI). As mentioned previously, ground fault circuit interrupters must be installed in a variety of areas, including all outside receptacles, and receptacles used in bathrooms, residential garages, unfinished basements, and crawl spaces, and receptacles within 6' of a kitchen sink or a bar sink. The bathroom receptacles can be connected to either a GFCI circuit or a GFCI receptacle. Typically a residence will require at least two GFCI cir-

cuits: one for smaller appliances and one for outdoor receptacles.

The electric range, clothes dryer, and water heater all operate at 240 volts AC, and each will be fed with a separate circuit.

19.17.6 Location of Lighting Fixtures

Interior lighting must be adequate for the activities of each area, yet subtle variations in the amount of general light and the careful use of supplementary or accent lighting give the overall lighting a less monotonous quality. Unusual architectural features must be given special consideration. Well-lighted interiors are often decidedly more interesting in the evening than during the daylight. Artificial light must be regarded as a definite part of the total design of the building.

Although fluorescent lighting is more efficient than incandescent, interiors of living areas will usually be more flattering in the warmer light of incandescent lamps. In bedrooms, also, such lamps are frequently used with or without the addition of a ceiling fixture. In work areas where abundant lighting is necessary for specific tasks, fluorescent-type light may be found more acceptable. However, glare and harsh light levels should be avoided. For example, a luminous suspended ceiling with uniformly spaced fluorescent tubes above produces a very pleasing method of lighting a kitchen. Fluorescent indirect lighting is also suitable for cove installation (Fig. 19–39). Work counters in kitchens can often be provided with indirect lighting placed beneath wall cabinets. Other than the general lighting of each room, provision must be made for supplementary activity lighting with outlets and/or fixtures properly placed. Unusual architectural features can be given added interest by the use of recessed spots or other types of isolated lighting; be sure that any special lighting has an outlet symbol indication.

Usually, small rooms have one centrally located ceiling fixture. Larger rooms and long hallways have several fixtures arranged so that uniform lighting is provided throughout. In bathrooms, light should fall on the face of the person using the mirrors and vanities (Fig. 19–40). In dining rooms, the ceiling fixture should be located directly above the table placement. Closets can be lighted

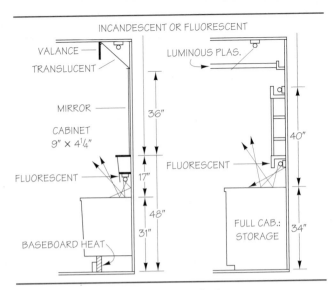

Figure 19–40 Indirect lighting in bathrooms.

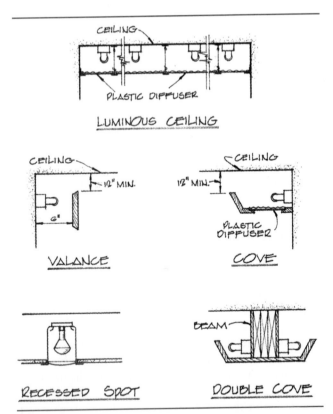

Figure 19–39 Methods for providing indirect lighting in living areas.

with door-operated or chain-pull outlets, or they may receive sufficient light from adjacent fixtures placed in front of their door openings. Surface-mounted light fixtures must meet the minimum clearance requirements as specified in the N.E.C.

All entrances and outdoor living areas should have provision for lighting, with outlets operated from inside wall switches.

Students should acquaint themselves with the many types of lighting fixtures now available and with the energy codes that specify the maximum allowable wattage per square foot in each room. Each lighting situation has a variety of workable solutions; only experience can produce the most dramatic results.

19.17.7 Location of Light Switches

Wall switches are generally placed 4' above the floor and should be located near the latch side of door openings for easy access. Switch placement must conform to the traffic patterns within the house. Switches should be located so that, as people go from one room to another, they can switch lights on before entering or upon leaving. Rooms with several exits and stairways should be provided with a three-way switch at each exit so that the lighting can be turned on or off from all exits, or in the case of stairs, from each level. Lighting can be operated from two stations by the use of *two three-way switches* (Fig. 19–41). Usually, low-voltage switching is used when it becomes necessary to control the lighting from three or more stations. Low-voltage switches are indicated on the

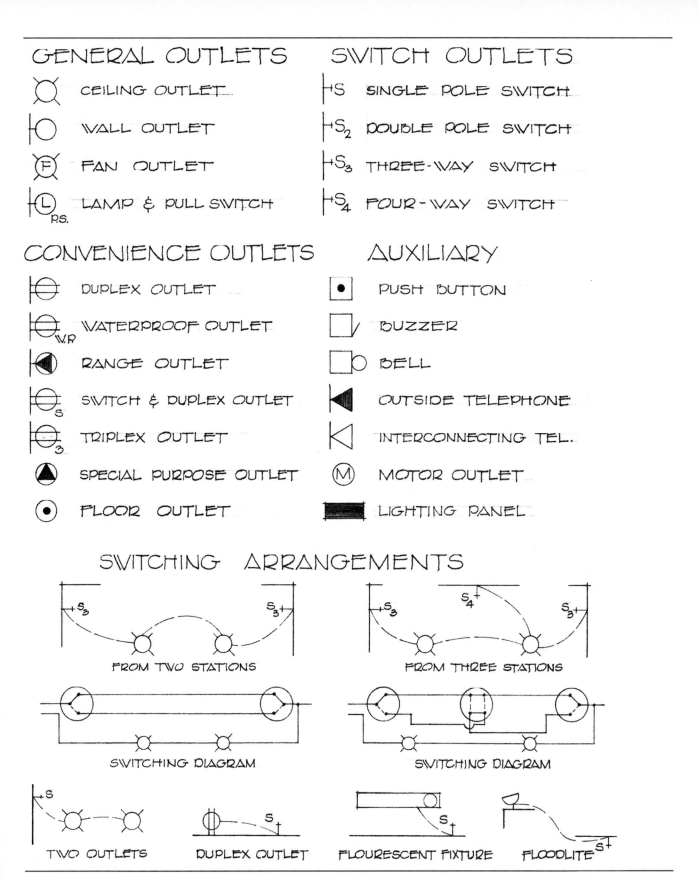

Figure 19–41 Electrical symbols for working drawings.

drawing with the same symbols as other switches. Silent mercury switches are recommended in and near bedrooms. In living rooms where variations of light intensity may be useful in creating different moods, dimmer switches can be indicated. Other types of switches are available for providing convenient lighting outlet control.

19.17.8 Special Outlets

The following special outlets are usually indicated on the electrical plan with appropriate symbols:

1. *Entrance signals* (bell or chimes) are best located in central locations such as hallways. Low-voltage current is provided by a transformer located near the service panel. Circuits to entrance door push buttons utilize No. 18 AWG bell wire.
2. *Telephone outlets* should be placed in convenient locations.
3. *Intercom systems* become practical in rambling-type homes; panels and stations are indicated on the plan.
4. *Television antenna outlets* can be provided in new home construction for connection to C.A.T.V. service.
5. *Automatic fire- and/or burglar-alarm systems* are available for home protection. Locate alarm bells in the master bedroom.
6. *Built-in entertainment or stereo systems* can be

shown with built-in main and remote speakers located in various living areas.

DRAWING THE ELECTRICAL PLAN

The electrical information on residential working drawings is usually shown on the regular floor plan. However, if excessive information requires a separate drawing, as is often needed with complex homes and commercial buildings, use a separate floor plan. Follow these general steps in drawing the electrical plan:

1. Locate the service panel, show the symbol, and indicate its amperage information.
2. Show all receptacle outlet symbols; use three-wire symbols for all outlets requiring 240 voltage, such as range, dryer, and air conditioner.
3. Show all ceiling and wall lighting outlets.
4. Show all special outlets such as telephone and doorbell.
5. Show all switches and their outlet connections (use an irregular curve).
6. If necessary, complete an electrical fixture schedule and symbol legend.
7. Complete titles and necessary notes.

REVIEW QUESTIONS

1. In planning a satisfactory residential heating system, what basic factors must you consider?
2. How can the architecture contribute to the efficiency of the heating and/or cooling system?
3. What is meant by *degree of comfort* in reference to heating-cooling systems?
4. What are the advantages of forced-warm-air heating systems? Hydronic systems? Radiant systems?
5. Would a heat-pump system be efficient for heating a home in your locality? Why or why not?
6. Why are filters placed in the return duct side of a warm-air heating system?
7. What type of warm-air furnace must be used with heat ducts embedded in a concrete slab-on-ground floor?

8. How is the air that is removed by kitchen and bath exhaust fans usually replaced in a well-designed system?
9. What three controls are necessary in making a warm-air system fully automatic?
10. Why has electricity become popular as a home heating energy source?
11. Why is the two-pipe hot-water system more satisfactory than the one-pipe system in home heating?
12. What general type of insulation is the most efficient with radiant heating systems?
13. What two methods of home heating commonly employ baseboard registers?
14. Which type of heating system is most easily combined with conventional cooling equipment?

15. If a room has a calculated heat loss of 3200 Btu's per hour, what wattage of resistance wire would be needed to satisfy the heat?

16. In heat-pump installations, what is the primary source of heat supplied to the rooms?

17. What mechanism reverses the cycles from heating to cooling in the heat pump?

18. What is the purpose of an outside thermostat?

19. What are the disadvantages of water chillers in residential cooling?

20. Are partitions between heated rooms considered in the calculation of heat losses?

21. Which part of a residential plumbing system operates by gravity? Which part by pressure?

22. How can you reduce initial plumbing costs when designing a home?

23. What is the purpose of a soil trap in plumbing fixtures?

24. What size hot-water heater will be sufficient for a family of three if the home contains automatic laundry equipment?

25. What factors determine the correct size of septic tank and the length of tile drainage lines?

Southface Energy and Environmental Resource Center

"Money is a good slave but a bad master."
—HOMER

Large-scale commercial building design and construction are topics beyond the scope of this text. The building materials and construction systems used in these types of buildings tend to be very different from those used in residential construction. Smaller commercial buildings, on the other hand, are often designed and built using materials and construction systems very similar to those used today in residential construction. A major area of difference between the two types of buildings is in the building code requirements. Commercial building codes tend to be more demanding than residential codes.

In practice, the designer of a small commercial building works closely with a client, who most likely would furnish definite requirements for a project. Designers must thoroughly understand the activities for which a building is intended before they can do a creditable job of planning the project and completing the set of drawings. The following general factors are often essential in planning a small commercial building.

1. Zoning regulations
2. Code requirements
3. Site and contextual analysis
4. Land utilization plan
5. Vehicular and pedestrian traffic
6. Parking facilities
7. Erosion control
8. Landscaping
9. Building systems
10. Structural systems
11. Enclosure systems
12. Mass and form

13. Interior space requirements
14. Public space
15. Private space
16. Community space
17. Restrooms
18. Facilities for handicapped persons
19. Horizontal and vertical circulation (corridors, stairs, and the like)
20. Reception and receiving space
21. Administrative and clerical space
22. Sales space
23. Production and manufacturing space
24. Customer and client services space
25. Meeting and training space
26. Warehouse and storage space
27. Custodial space
28. Heating, ventilating, and air-conditioning (HVAC) space
29. Acoustical and noise control
30. Interior design and furnishings

The Southface Energy and Environmental Resource Center shown on the chapter opening (page 622) and on page 623 was developed as a state-of-the-art demonstration center for saving energy, water, and natural resources. The 6000-sq-ft, 3-story facility is residential style in order to demonstrate the applicability of a wide variety of energy-efficient features to both residential and small commercial construction. The upper floor houses office space for the Southface staff. The lower floor houses classroom facilities and demonstration space. The architects for this project were Pimsler Hoss Architects of Atlanta. The project is described in detail in the boxed material on pages 626–675.

20.1
PLANNING THE DRAWING LAYOUT

Because of the size of the Southface Resource Center, it was convenient to draw the floor plans to the $1/4'' = 1'-0''$ scale; larger buildings, naturally, may necessitate the use of the $1/8'' = 1'-0''$ scale, provided that the corresponding size of paper is used. Each plan view became suitable for placing on a single sheet. The plans for the Resource Center, reduced to fit the size of this book, appear in Figs. 20–1 through 20–41 on pages 635–675.

20.2
DRAWING THE PLAN
(Figs. 20–1 through 20–5)

The first step in drawing a plan of this type is to lay out the exterior walls of the building. After the exact widths of the walls are determined, the outline of the plan is blocked in, and all partition walls are lightly drawn without regard for window and door openings. Then the door and window openings are located according to the scheme of the exterior and the traffic flow of the interior. Only construction lines are used up to this point.

The walls are then darkened and refined, window and door symbols are drawn, and minor wall outlines are completed. Manufacturers' literature is consulted before the doors and windows are drawn. Plumbing fixtures are drawn with the use of a template. Minor details are added throughout the plan. Exterior platform ramps and steps are drawn with the help of information from the site plan contour levels. Stair layouts are developed with the proper number of risers and treads between floors. (Codes are consulted for acceptable riser heights and tread widths.) Then door identification is shown, room titles are lettered, and all general callouts are added.

Next, dimensions are inserted in positions where they do not interfere with linework. Notice that, along with important exterior dimensions placed around the periphery of the plan, a continuous line of dimensions is shown through the interior in each direction to locate all partitions accurately.

In some cases, it is advisable to begin a project with a typical wall section rather than the plan. Much of the wall information must be known before the plan can be drawn.

20.3
DRAWING THE ELEVATIONS
(Figs. 20–7 and 20–8)

In starting the elevations, the designer first draws the finish floor levels. Then, after establishing the desired ceiling heights, he or she draws the major horizontal lines from information taken from the preliminary section view. Lengths of the elevation view features, however, are projected directly from the plan, as in residential drawings. Windows, offsets, doors, and other features are drawn from preliminary sketches with the help of manufacturers' literature. Where indicated, the trim members are completed, and the finish floor and the roof level are added. To complete the drawing, the grade line, footings, stairs, ramps, railings, material symbols, notes, and titles are added as shown. For convenience in construction, the footing surfaces are labeled for their correct elevation below grade.

20.4
DRAWING THE FULL
SECTIONS (Fig. 20–9)

The longitudinal and transverse sections are drawn after all detail information has been completed on both the plan and the elevations. These section views reveal all visible interior information on the cutting planes completely through the building. To conserve time, the longitudinal section is blocked in on the sheet placed directly over the left-side elevation drawing and the transverse section over the front elevation drawing. In both cases, the floor plan is oriented and attached to the board above the section view so that interior features can be easily projected. In completing the sections, the designer must occasionally duplicate information found on elevation views.

20.5
DRAWING THE DETAILS
(Figs. 20–10 through 20–12)

Although the drawing of details is a time-consuming part of the architectural routine, the drafter is required to draw all those necessary for each project. Details should be made consistent with each other, as well as with the general concept of the building. Major details must be accurately drawn before portions of the plan or elevations can be completed; inaccuracies may result in costly building errors.

As stated previously, in drawing details, the designer must continually refer to *Sweet's File* or manufacturers' catalogs for actual sizes and other information about pertinent materials. The relationship of members in details and especially the joining of different materials must be well thought out before the detail is drawn.

A detail may be typical in that it represents similar construction throughout the building, or it may be the construction of an individual feature; in either case, the title should correctly describe the application.

In dimensioning detail drawings, the design must show important sizes and must take care to include dimensions that relate the detail to major dimensions shown on the plan or the elevation. Titles and scales are shown on each detail drawing. Efficient organization of small details on a large sheet comes with drafting experience; neatness is the result of careful planning.

20.6
INTEGRATING DRAWINGS

Working drawings for most commercial buildings include drawings prepared by registered engineers. The engineering disciplines frequently involved in a commercial building project are civil, structural, mechanical, and electrical. It is important to coordinate the drawings prepared by the engineers with drawings prepared by the architect. The engineers should be consulted during the development of the schematic drawings so that provisions for engineering concerns can be addressed at an early stage. The following sections of this chapter highlight activities performed by the various engineers and stress the importance of integrating engineering drawings with the architectural drawings.

20.7
CIVIL DRAWINGS

A civil engineer usually prepares the property line survey and the topographic map, gathers soil data, and locates existing utilities on a drawing before preliminary studies begin. In many instances, the civil engineer is engaged to draw the site plan showing the building location, roads, pavements, utilities, topography, and site drainage. It is important that the building plan fit the site, complement the topography, and accommodate utility connections.

20.8
STRUCTURAL DRAWINGS
(Figs. 20–14 through 20–20)

The foundation, floor, and roof framing plans are developed by the structural engineer. Data concerning the soil conditions are important to the structural engineer to size the footings correctly and to determine the proper foundation system for the building. The structural engineer gives valuable guidance to the designer regarding the location of columns, beams, bearing walls, and roof framing systems. The civil engineer, the designer, and the structural engineer must meet frequently to exchange information concerning the structural integrity of the building.

20.9
MECHANICAL DRAWINGS
(Figs. 20–21 through 20–32)

Plumbing and HVAC drawings are prepared by the mechanical engineer. Plumbing plans show plumbing fixtures, hot- and cold-water supply pipe, and sanitary drainage pipe. Existing utility information gathered by the civil engineer, and the type, number, and location of plumbing fixtures determined by the designer, are important to the mechanical engineer.

The mechanical engineer is frequently asked to prepare an analysis of the space-conditioning needs for the building. Numerous types of heating and cooling systems are available. It is imperative that the proper type of system be selected. The designer, the structural engineer, and the mechanical engineer must carefully integrate the HVAC system into the design of the building. Adequate space for the mechanical equipment and sufficient structural supports for the system are essential.

Some buildings are required to have fire protection systems. The mechanical engineer usually develops a sprinkler system plan for fire protection for such buildings.

ELECTRICAL DRAWINGS
(Figs. 20–33 through 20–40)

The electrical engineer prepares the lighting plan and the power wiring plan for commercial buildings. In addition to the basic electrical plans, the electrical engineer is usually responsible for developing communications plans, which consist of telephone, intercom, fire detection, and alarm systems. Even some small buildings have massive amounts of electronic equipment. The designer and the electrical engineer must stay abreast of the users' needs and must carefully coordinate the location of all of the electrical features of the building.

DRAWING THE REFLECTED CEILING PLAN

Ceilings in many commercial buildings are quite complex. It is desirable to draw a reflected ceiling plan that shows the ceiling surface, including the suspended ceiling grid system, lighting fixtures, HVAC diffusers, fire protection equipment (smoke detectors, sprinkler heads, and alarm devices), exit signs, sound and communications speakers, security equipment, walls, columns, and skylights and other openings in the ceiling. The reflected ceiling plan is drawn after the architectural, structural, mechanical, and electrical plans are complete. Trace the floor plan walls. Do not show door or window openings. Lay the tracing over the mechanical and electrical plans, and draw the equipment that is located on the ceiling.

DRAWING THE ROOF PLAN
(Fig. 20–6)

Roofs on most commercial buildings vary in shape, height, and materials. Using the floor plan and the structural roof framing plan, draw the outline of the roof edge. Draw ridges, valleys, stair towers, elevator penthouses, and skylights and all other roof penetrations. Using mechanical plans as a reference, locate and draw HVAC units, exhaust fans, plumbing vents, roof drains, and the like, on the roof plan. Use arrows to indicate the direction of water flow on the roof surface.

✔ THE SOUTHFACE ENERGY AND ENVIRONMENTAL RESOURCE CENTER

The Southface Energy and Environmental Resource Center was developed by the Southface Energy Institute, in cooperation with the U.S. Department of Energy, Lawrence Berkeley Laboratory, National Renewable Energy Laboratory, Oak Ridge National Laboratory, and corporate and individual sponsors. The home and office areas showcase innovative ideas for saving energy, water, and other natural resources; for reducing waste and using recycled materials; and for maintaining a healthy indoor environment. There are also smart ideas for designing accessible buildings for people with physical disabilities.

How much would it cost to build using these ideas? This question is difficult to answer. Some of the products presented here may cost more to buy, but they will save money by lowering energy and water bills, lasting longer, and requiring less maintenance. Many technologies also improve comfort and protect your health and the health of our planet.

A better question might be, "How much will it cost NOT to build using these ideas?" Standard buildings waste hundreds of dollars each year in needless energy and water use; they produce tons of construction debris and atmospheric pollutants; they suffer from moisture problems; and they have areas that never seem to be comfortable. In addition, they ignore indoor air-quality problems which can endanger health, especially the health of children, the elderly, and anyone suffering illness.

THE FIRST LEVEL (FIG. 20–3)

The 2300-sq-ft first level contains many energy and environmental products and design ideas that make sense for any building.

Energy-Efficient Lighting

Lighting accounts for 25% of all electricity consumed in the United States. The center's overall lighting design and use of efficient fixtures reduces energy costs for lighting by over 50% and increases comfort, esthetics, and productivity. The overhead light shelf scatters light across the ceiling. The shelf houses T-8 fluorescent lamps and electronic ballasts that save energy and have excellent color rendition.

The Great Room

- **High-performance windows** The Pella Designer Series Classic windows feature argon-filled, double low-emissivity coated, insulated glass. A wood thermal break between the two panes of glass reduces heat flow through the frame. The total window R value is 2.96 (U value 0.34). The low-E coating (LBL rating 0.32) screens out ultraviolet radiation, which helps reduce fading of materials.

- **Passive solar windows** The major glass area faces south for maximum solar gain in winter. The south-facing windows equal approximately 7% of the floor area of the main level. A 2′ overhang shades the direct overhead summer sun yet still allows the lower winter sun to enter the building.

- **Thermal mass floor** During a winter day, sunlight passes through the south-facing windows and warms the tile floor in the sunroom, the great room, the kitchen, and the laundry room. The floor stores the solar heat for use at night or during cloudy weather. The 1/2″ thick floor tile is set in a 2.25″ thick mortar bed and provides about 170 cu ft (20,000 lb) of thermal mass. The tile floor is in direct sunlight and has a surface area 5.5 times greater than the area of the south-facing windows.

- **Floor tile** The glazed porcelain floor tile is manufactured by Summitville Tile of greater than 70% postindustrial recycled materials. The recycled body content and the glaze content give the tile superior strength, increased temperature resistance, and reduced absorption. The results are lower maintenance and greater wear.

- **Safe and efficient fireplace** Standard fireplaces are less than 10% efficient, send room air up the chimney, and can leak dangerous flue gases into the home, endangering health. The Heat-N-Glo natural gas, direct-vent fireplace has an American Gas Association–certified, sealed combustion chamber which prevents spillage of flue gases into the home. A duct brings outside air to the combustion chamber and vents all flue gases directly to the outdoors. The unit delivers 20,000 Btu's of heat per hour at a combustion efficiency exceeding 65%.

- **Vaulted ceiling** Vaulted ceilings often allow outside air to leak into the building through gaps in the framing. Weatherstripping gaskets were stapled to the framing of the ceiling before the drywall was installed to form an airtight seal. The ceiling is insulated with Owens Corning Fiberglass® batts (R-30C).

- **Video center** The glass used in the 27″ Panasonic color television is darker than conventional tubes to combat the degrading effect of ambient light. The power consumption is a thrifty 100 watts. The programmable on/off timer and sleep timer can reduce energy waste. The unit can also serve as a crime deterrent by providing sound at designated times.

The East Deck

- **Exterior door** The door leading to the outdoor deck is provided by Andersen and features double glazing, with low-emissivity coating and inert gas fill (R-3.3, U-0.30). The low-profile threshold enhances safety and accessibility.

- **Deck flooring** The deck flooring, manufactured by Choice Deck, is made of 51% recycled cedar and 49% recycled plastic. The material is cut and nailed much as wood decking is.

- **Fireplace vent** The inner circle of the direct-vent flue exhausts gases from the fireplace. The outer circle carries outdoor air to the sealed chamber for combustion.

- **Minimum east and west glass** East and west windows receive three times more sunlight in summer than those facing south. Minimizing windows on the east and the west, or keeping them well shaded, prevents the house from overheating.

The Main Hall

- **Water fountain** The fountain is designed to be wheelchair accessible.

The Half-Bath

- **Water-saving toilet** Flushing toilets can account for almost 40% of the water use in a home. The Eco-Lite toilet is manufactured by Kohler and features a dual-flush trip lever. Pressing the shorter lever delivers an efficient 1.1-gal flush for liquid or light waste. Pressing the longer lever gives a 1.6-gal flush.

- **Exhaust fans** Standard bath fans often exhaust fewer than 20 cu ft of air per minute (cfm), are noisy (3–4 sones), waste energy, and are not designed for continuous operation. The Panasonic fan exhausts 90 cfm through a straight metal duct to the outside, is extremely quiet (about 1 sone), uses only 17 watts, and has a motor designed for continuous use with little maintenance. Ultra-quiet (0.5 sone), 15-watt units provide 70 cfm of spot ventilation in the upstairs bathroom and copy room.

The Master Bath

- **Accessible design guidelines** The design of every home should allow for ease of movement and operation by persons with physical limitations.
- **Roll-in shower** The 3'-0" opening and barrier-free threshold for the shower enables easy wheelchair access. The hand-held showerhead further increases accessibility.
- **Accessible plumbing fixtures** The counter top provides underneath wheelchair access. Lever handles on all plumbing fixtures, grab bars, and barrier-free areas around toilets enhance safety and accessibility in all restrooms.
- **Water-saving plumbing fixtures** The 1.6-gal toilet from Kohler saves almost 2 gal per flush compared with standard toilets. Water-saving toilets and faucets (2.5-gal/min flow) are required in new construction and are an excellent money-saving retrofit for existing homes.
- **Exhaust fan** Removing moisture from bathing areas is important in controlling interior relative humidity. The Panasonic fan exhausts 110 cu ft of air per minute of operation, uses a thrifty 20 watts, and produces a quiet 1.5 sones. A timer allows the fan to exhaust the moisture that lingers in the air after people have finished their showers.
- **East window** The translucent glass in the east window allows daylight while maintaining privacy. Due to its small size, it does not contribute significantly to solar gains in summer.

The Master Bedroom

- **Healthy homes** The key to creating a healthy indoor environment is to keep pollutants out of the building, isolate unavoidable pollutants from people, test to see whether there are unsafe pollutant levels, and ventilate to dilute concentrations.
- **Low-emission paints** Formaldehyde and other volatile organic compounds (VOCs) can pose a health risk to some people. The 100% acrylic Benjamin Moore Pristine paints used throughout the interior have zero VOCs. Pristine products require less drying time than other paints and have no lingering odor.
- **Natural fiber carpet** The carpeting was manufactured by Shaw Industries of wool and has no petrochemical feedstocks. Wool carpeting has excellent wear and is treated to repel moths.

- **Airtight and controlled ventilation** Airtight construction prevents outdoor pollutants, such as radon, mold, pollen, and dust, from entering the home. Reducing uncontrolled air leakage also enables the controlled ventilation system to provide fresh air in the right amounts to the right places. The insulated foundation blocks and the structural insulated wall and roof panels minimize air leakage. Penetrations in these components and around rough openings for windows and doors were sealed with spray urethane foam from Insta-foam.
- **Ventilation and dehumidification system** Outdoor air is not always "fresh" and may need to be cleaned and dehumidified, especially in humid climates. The DEC/Thermastor Ultra-Aire Air Purifying Dehumidifier provides fresh-air ventilation, air filtration, and high-capacity humidity control.
- **Air filtration and cleaning** A standard panel filter for heating and cooling equipment has a dust-spot efficiency of less than 5% and can remove only large dirt particles. Many electrostatic filters do little better, with dust-spot efficiencies of less than 15%. Each of the three mechanical systems for the building has 30% dust-spot efficiency prefilters. A Honeywell electronic air cleaner (95%) and Durafil Series cartridge filter (85%) provide high-level filtration.
- **Central vacuum** Common portable household vacuums do not effectively trap dust mites and small particulates. The Panasonic central vacuum cleaner removes many of these contaminants.
- **Radon-mitigation system** Radon is a colorless, odorless, radioactive gas that can occur in high concentrations in some soils. Many energy-efficiency features, such as balanced house pressures and airtight construction, minimize radon entry into the home. Beneath the lower-level concrete floor slab is a gravel bed that allows radon to be collected and passively vented via a solid plastic pipe through an interior wall to above the roof. If tests detect elevated interior radon levels, a fan can be easily added to the pipe to exhaust radon from beneath the slab.
- **Carbon dioxide monitor** High concentrations of carbon dioxide (CO_2) in the air can indicate poor ventilation. A CO_2 sensor in the classroom on the lower level monitors air quality in that area and operates a fresh-air control.

The Library

- **Insulated form foundation (Fig. 20–10)** The 10' high basement walls were constructed by pouring concrete into Diamond Snap-Form, a stay-in-place form made of Perform Guard expanded polystyrene (EPS). This EPS does not contain CFCs, HCFCs, or HFCs and is treated with a borate chemical to deter termite and other insect infestations. The 14″ thick wall is made by pouring 10″ of concrete between two, 2″ layers of EPS. The Portland Cement Association, Georgia Concrete Products Association, and Blue Circle Williams provided the concrete.

- **Insulated foundation waterproofing system** The below-grade foundation walls are coated with Tuff-N-Dri, a polymer-modified, asphalt emulsion provided by Koch Materials. One-inch thick Warm-N-Dri fiberglass sheathing (R-5) covers the coating and directs water to the French drain at the footer.
- **Structural insulated panels** The above-grade walls and major roof areas were constructed with structural insulated panels provided by the Structural Insulated Panel Association. The R-control panels used were manufactured by AFM with an expanded polystyrene (EPS) core and outer skins of oriented strand board. The EPS is treated with a borate chemical to resist insect infestation. The panels are connected every 4′ with a $2 \times$ spline. They are approximately twice as strong as conventional stud framing and use about 30% less wood. The 5.5″ wall panels are R-24, and the 8″ roof panels are R-30. The panels provide for tighter construction and more continuous insulation than standard framing.
- **Air sealing and insulating band joists** The band joist on multistory houses can be difficult to air-seal and insulate properly. The cavity between the floor joists on the lower level was insulated with scrap blocks of EPS and then sealed with Insta-foam, a one-part expanding urethane foam. The band for the upper level was sealed and insulated using GreenStone blown cellulose insulation that was mixed with an adhesive.
- **Blown-in-place insulation (Fig. 20–12)** A section of the roof is framed with dimensional lumber and is insulated by dry Greenstone cellulose blown behind a mesh fabric. Cellulose, which is treated with a flame-retardant chemical, completely fills the framing cavity and reduces air leakage.
- **Airtight drywall and fiberglass batt insulation** Framing details, such as those for tray and vaulted ceilings, can allow outdoor air to leak into interior building cavities. Standard density insulation does not block this air leakage. The vaulted ceiling in the great room was sealed by stapling weatherstripping gaskets at key junctions in the framing before installing the drywall (a technique known as the *airtight drywall approach*). The ceiling was insulated with Owens Corning high-density fiberglass batts (R-30C).

The Home Office

- **Telecommuting** The home office equipment enables telecommuting—working at home to avoid the need for commuting, which can save energy and reduce air pollution.
- **Energy Star equipment** The computer, printer, and other office equipment are Energy Star–rated. This joint program of the U.S. Department of Energy and the Environmental Protection Agency helps consumers choose products that save money while reducing energy use and pollution.
- **On-line services** The computer can be used to access energy and environmental information sources through the World Wide Web access provided by MindSpring Enterprises.

The Dining Room

- **Natural light and ventilation** The concentration of windows on the south and north provides natural light and cross-ventilation throughout the house. East and west windows and overhead glass can cause overheating in summer. There is minimal glass on the east, and the entire west wall is well shaded by the porch and the garage. The cupola is an energy-efficient alternative to skylights. It allows for daylight but has an insulated roof to prevent heat loss and to shade the overhead summer sun.
- **Recycled wood flooring** Old wood has character and can be stronger than wood milled from today's fast-growing species. The recycled heart-pine wood flooring distributed by Home Depot was milled from old timbers that were salvaged from demolished buildings by the Wood Cellar.
- **Sustainable wood products** The birch shelving in the library is made from wood certified to be harvested in an environmentally sustainable manner by Scientific Certifications Systems, an independent certification company. The wood is harvested from the Seven Islands Land Company and distributed by A. E. Sampson & Son.

The Laundry Room

- **Clothes washer** Roughly 98% of the clothes washers in U.S. homes use vertical-axis technology, where clothes are agitated in a tub full of water. This process is more damaging to clothes and uses an average of 44 gal of water per load. The Frigidaire Gallery is a front-loading, horizontal-axis washer, where clothes are gently tumbled in and out of water. The water level is automatically adjusted to the size of the load, saving up to 8000 gal of water.

- **Clothes dryer** The companion Frigidaire Gallery gas clothes dryer is stacked on top of the clothes washer to save space. The dryer is vented to the outside with metal duct and has a vent cap to reduce air leakage into the house when the unit is not in use.
- **Solar water heater** The American Energy Technologies solar water heater is designed to provide at least 50 gal per day of 122°F hot water for a family of three people. A $2' \times 10'$ solar collector on the roof of the breezeway to the garage provides approximately 20,000 Btu's per day of solar water heating The solar-heated water transfers heat via a side-arm heat exchanger into a 40-gal backup electric water heater.
- **Geothermal water heating** The Addison geothermal heat pump has a de-superheating heat exchanger which captures waste heat from air conditioning in summer to provide domestic water heating.
- **Solar electric shingle controls** The integrated photovoltaic shingles produce direct electrical current from sunlight. The electricity flows to an inverter which converts it to alternating current (AC). The alternating current is used here at the building or is fed back to the utility. The solar shingles can produce 5 to 6 watts of AC per square foot of peak power in full-sun conditions and approximately 25 watt-hours per square foot daily.

The Kitchen

- **Refrigerator** The Kitchenaid Superba side-by-side refrigerator is 38% more efficient than federal energy standards require. The unit uses R-134, a refrigerant that poses a reduced risk of ozone depletion and global warming. The vacuum panel insulation is made without ozone-depleting chemicals.
- **Dishwasher** Dishwashers use 5.8 gal of water less on average than washing dishes by hand, which adds up to a savings of 1800 gal of water annually, along with reduced energy use from heating less water. The Maytag IntelliSense dishwasher automatically adjusts water heating and cycle time, yielding savings of up to 21% on energy and 27% on water, and a 19% reduction in wash-time length, versus standard appliances.
- **Kitchen range** The Amana downdraft modular electric range has a smooth-top ceramic cooking surface, quartz halogen heating elements, and an indoor grill with embedded heating elements.

- **Small appliances** The Panasonic rice cooker has a microprocessor which saves energy by monitoring the heat, the cooking levels, and the type of cooking stage being used. This "fuzzy" technology promises reduced energy use while providing more consistent results for better-tasting foods. The toaster uses fewer watts than standard models; yet it makes toast more evenly, and the sides stay cool to the touch.
- **Cabinets** The top of the sink and a counter surface are set at 34″ above the floor, rather than the standard 36″, to enhance accessibility.
- **Water circulation** It can be irritating as well as wasteful to let water flow down the drain waiting for it to heat up. The Metlund System shunts cooled water from the hot-water line into the cold-water supply. A push button activates a pump which sends cooled water in the hot-water supply lines back to the water heater.

THE SECOND LEVEL (Fig. 20–4)

- **Solar electric shingles** The integrated photovoltaic (PV) shingles resemble conventional fiberglass roofing shingles and produce 5 to 6 watts AC per square foot of peak power in full-sun conditions. United Solar Systems and Energy Conversion Devices manufacture the $1' \times 10'$ PV modules of amorphous silicon cells deposited on a thin, flexible, lightweight stainless steel substrate, which is laminated in advanced polymers. The overlapping modules replace conventional shingles. There is one sealed roof penetration for each module, and wiring connections are made below the roof decking.
- **Structural insulated roof panels** The north, east, and west gables are constructed of 8″ thick, R-30, structural insulated panels. The panels are supported by the ridge and gable beams, and they do not require conventional rafters.
- **Engineered wood products** Engineered wood products, such as laminated veneer lumber (microlams), make economical use of young, small-diameter, fast-growing, lower-quality trees, rather than larger, mature trees. Engineered products are often stronger, straighter, and lighter than dimensional framing lumber. The ridge beam and all floor joists are constructed of laminated veneer lumber supplied by Furman Lumber.
- **Spray-in-place insulation** The south gable is framed with 2×10 dimensional lumber. The roof is insulated by GreenStone dry cellulose blown in behind a mesh fabric stapled to the interior face of the rafters. A water-base adhesive mixed with the cellulose was sprayed in the open band joist areas. The higher density of the cel-

lulose provides an approximately *R*-3.5 per inch of thickness and reduces air leakage.

- **Office daylighting**   The south-facing windows and overhead cupola provide for natural daylighting, which reduces the use of artificial lighting. The solid roof of the cupola prevents direct solar gain in summer.
- **Energy-efficient lighting**   The energy-efficient lighting design and products cut lighting costs by over half and give superior performance. Compact fluorescents replace incandescent bulbs throughout the building. They provide excellent color rendition, use a quarter of the electricity, produce little waste heat, and last about 10 times longer. The tubular fluorescent fixtures have energy-efficient ballasts and T-8 lamps, which also have excellent color rendition.
- **Exit signs**   The LED exit signs use only 1 watt of electricity rather than the 20 watts of a standard lamp. The units can save up to $50 annually on electricity costs, can be retrofitted for existing fixtures, and are virtually maintenance free for 25 years.
- **Recycling carpet**   Closing the loop is key to recycling. The Interface carpeting used in the office and the lower levels is designed to be returned to the manufacturing plant to be recycled into new carpet. The carpet pad is also made from recycled materials, and low-VOC adhesives are used for installation.

THE BASEMENT LEVEL (Fig. 20–2)

Restrooms

- **Water-efficient plumbing fixtures**   The Kohler toilets and faucets in each restroom are water-saving fixtures that meet federal efficiency standards. The men's restroom has a Waterless No-Flush urinal which uses a lightweight, biodegradable oil that allows urine to percolate through to the sewer, yet maintains a sanitary water trap. The urinal can save up to 3000 gal of water annually.
- **Exhaust fans**   The Panasonic bath fans are designed for commercial use and exhaust 340 cfm with a sound level of only 3 sones. Designed for continuous operation, they have a power consumption of only 95 watts.

Mechanical Room

- **Sizing heating and cooling equipment**   The size of mechanical equipment should be determined by calculating heating and cooling loads for the specific building. Standard rules of thumb, such as so much heating or cooling for a specific number of square feet of living area, lead to big errors. A unit that is too large costs more to buy, wastes energy, and provides less comfort.
- **Heating and cooling equipment efficiency**   The efficiency of mechanical systems should be determined by the estimated operating costs of the specific building. Continuous insulation, airtight construction, and high-performance windows reduce heating and cooling needs so that smaller, medium-efficiency equipment is often the best buy.
- **Duct design**   Improper duct design and installation can increase heating and cooling bills by over 30%. It can also endanger health and safety and can reduce the durability and comfort of the home. The size of each duct is calculated based on the heating and cooling needs of each room, the type of duct material used, and the length of the run. All of the ducts are located within the insulation and air barriers of the building to ensure no energy losses.

LANDSCAPE AND GROUNDS/ENERGY AND BUILDING SYSTEMS

- **Porous concrete**   The driveway and the handicap parking space are made with a special porous concrete from Blue Circle Williams that allows rainwater to pass into the gravel underneath and then percolate into the soil. The porous paving surface reduces storm-water runoff and nonpoint-source water pollution, while recharging groundwater.
- **Wildlife habitat**   Plants were selected to emphasize flower and berry production to promote wildlife (primarily birds and butterflies). The guidelines of the Georgia Wildlife Federation's Schoolyard Wildlife Habitat program help define the needs for food, shelter, water, and nesting sites of a successful wildlife garden.
- **Community gardening**   Vegetable gardens are a useful tool for providing fresh produce and building a sense of community in urban areas. The demonstration plot is a partnership with the Atlanta Gardening Alliance, led by the Atlanta Community Food Bank. Raised beds are more productive than standard rows and are more accessible to people with physical disabilities. Using organic methods in the garden eliminates the waste of energy used in the production of synthetic fertilizers and pesticides.
- **Compost bin**   Over 25% of the household waste that is dumped in landfills can be composted at home. When designed correctly, compost bins are odorless, do not encourage insects or pests, and provide a valuable soil additive.

- **Straw bale garden cottage** The cottage was constructed of straw bales and then covered with stucco. Straw is an agricultural by-product and provides a building material that is energy efficient, fire resistant, and volunteer friendly: The cottage was built by students in a hands-on workshop.
- **Grass paving** The parking space on the lower level is constructed of Grasspave. This plastic matrix provides support beneath the grass to allow for parking. The grass surface is permeable and allows rainwater to percolate into the soil, reducing runoff.
- **Photovoltaic outdoor lighting** The outdoor lights for the parking lot rely on sunlight to generate electricity. The crystalline silicon photovoltaic (PV) panel produces electricity during the day which is stored in a ground-mounted battery. The direct-current fluorescent lamp is controlled by an automatic light cycle of 8 h per night. There is battery energy storage for six days of operation without direct sun.
- **Photovoltaic motion detector** The motion detector has a photovoltaic cell that generates electricity from sunlight. The electricity is stored in a small battery inside the light fixture. The photovoltaic cell can be mounted separately from the light to ensure access to sunlight, and the fixture does not require conventional electrical service.
- **Photovoltaic landscape lights** The landscape lights use a photovoltaic panel to charge a battery during sunny periods. The stored electricity provides lighting at night.
- **Gas-fired heat pump** Although it looks much like a traditional outdoor unit for an electric air conditioner or heat pump, the York Triathlon uses natural gas to power an engine-driven heat pump. The Triathlon provides both heating and cooling for part of the building.
- **Geothermal heat-pump loops** You can't see them, but they're there! Under the driveway are eight closed-loop wells, each roughly 200′ deep, where earth temperatures stay close to 64° year-round in Atlanta. The well loops are made of polyethylene plastic pipe manufactured by Phillips Driscopipe. An Addison geothermal, or ground-source, heat pump circulates water through the piping to transfer heat to and from the deep earth.
- **Foundation moisture protection** Keeping the foundation dry discourages termite colonies, protects the building from decay, and helps maintain proper indoor relative humidity. Below grade, the basement walls have been sprayed with Tuff-N-Dri waterproof coating by Koch Materials. Over the coating is 1″ of Warm-N-Dri insulated foundation drainage board. The fiberglass board provides an R-5 per inch of thickness and shunts groundwater to a French drain at the bottom of the footer, which then drains water away from the home. (A French drain is perforated plastic pipe covered with filter fabric and gravel.)
- **Termite protection** A termite bait and monitoring system provided by DowElanco reduces risk from infestations. Bait traps are placed around the perimeter of the building. If termite colonies are detected, a special poison is introduced that termites take back to the queen and the colony is destroyed.
- **Xeriscape** The landscape design emphasizes the principles of xeriscaping (drought-tolerant landscaping) and the use of native plant species to lower maintenance and water requirements. Native plants are well adapted to the local climate and soil and are less susceptible to insects or disease. There is no automatic irrigation system. A properly designed and installed landscape should need only occasional supplemental watering after the first year.
- **Cool communities program** American Forests, in cooperation with the U.S. Department of Energy and other agencies, sponsors the Cool Communities program to reduce urban heat islands through strategic tree shade and light-colored surfaces. Many of the landscape design features provide shade and reflect sunlight, keeping the building cooler.
- **Graywater irrigation** Graywater is wastewater, usually from washing machines, showers, and lavatories, that is not contaminated with human waste. Graywater from the building drains into a subterranean gravel bed in a planted terrace.
- **Alternative sources of water** Some of the downspouts are piped to a cistern to store rainwater. A solar-powered pump moves the rainwater from the cistern through drip irrigation lines. In addition, the hydrogen fuel cell produces about 1 gal of water per hour when operating, which drains to the landscape.
- **Minimum turf** Turf requires extensive watering and maintenance. Mulch beds, ground covers, and wildflower plantings reduce turf areas. The remaining turf is mowed higher and is maintained without pesticides. These steps reduce the chemical, mechanical, and human inputs for the landscape.
- **Protection of trees during construction** Mature trees add beauty and value to the site and

ensure immediate and long-term shade and energy efficiency. The building was sited to avoid damage to existing trees, and the trees were protected by minimizing trenching, grading, and other construction activities in their root zones.

- **Recycled landscape materials** The use of stone from grading operations and mulch made from discarded framing lumber and other wood sources keeps these materials out of the landfill and reduces the direct costs of landscaping. The wood-chip mulch is the product of utility right-of-way clearing and was available free of charge. The concrete rubble used for stepping stones and stacked walls came from the sidewalk removed by a utility company.

GARAGE AND TRANSPORTATION (Fig. 20–41)

- **Garage framing** Efficient framing can reduce the use of wood by $1/3$ without compromising the strength and durability of a structure. The garage is framed with 2 × 4's on 24″ centers along the Optimum Value Engineering guidelines of the National Association of Home Builders.
- **Garage roof** Several studies indicate that roof ventilation offers little benefit on shingle life or energy performance. The roof of the garage is vented with ridge and soffit vents to offer a comparison of the longevity between the shingles on the garage and those on the unvented roof of the house. Both structures have 30-year-life, fiberglass shingles from Owens Corning, which contain copper to prevent fungal growth.
- **Electric vehicle** The electric vehicle does not have a conventional engine, gas tank, transmission, or muffler. It relies on a 137-hp electric motor powered by 26, 12-volt batteries. The vehicle can travel approximately 70 to 90 mi on a charge, with a top speed of 80 mph. Electric vehicles have no tail-pipe emissions, so they do not contribute to air pollution in cities. The electrical energy to charge the batteries can be produced by renewable energy sources or traditional power plants.
- **Electric vehicle–charging station** The charging station uses 220-volt electricity to charge the vehicle batteries. It takes approximately 3 h to charge the batteries fully through an induction transfer of electrical energy made possible by a special "paddle" connector. The electric vehicle and the charging station are provided by Georgia Power Company and the Southern Coalition for Advanced Transportation.
- **Hydrogen fuel** The outbuilding contains an electrolyzer that generates hydrogen gas (H_2)

from liquid water. The electrolyzer can produce up to 42 cu ft per hour of H_2 using 8 kilovolts of electricity. The hydrogen is used to power a fuel cell for a vehicle. The hydrogen fuel cell produces only clean water as an emission. Part of the electrical energy to generate the hydrogen gas comes from the photovoltaic roof shingles.

- **Hydrogen vehicle** The Genesis is a zero-emission vehicle which runs on a fuel cell that produces electricity by recombining hydrogen and oxygen. The vehicle has a 7.5-kilowatt fuel cell that takes 15 min to refuel. The Genesis has a maximum speed of 15 mph and a range of 45 mi.
- **Bicycle parking** Over 80% of trips in urban areas are under 3 mi. Cycling is an efficient and healthy way to commute. Many cities recognize cycling as an important strategy for reducing air pollution, and they encourage policies to make cycling safe and convenient. The Atlanta Bicycle Campaign has provided a bicycle rack for visitors.
- **MARTA mass transit** The 39 mi per day average commute for Atlantans is the nation's highest. Air pollution from vehicles endangers health and causes thousands of dollars in damage to buildings, cars, and other materials.

SOUTHFACE PARTNERS

The Southface Energy and Environmental Resource Center was developed through the resources of the U.S. Department of Energy, the Southface Energy Institute, and partners from industry. Southface thanks the following organizations for their generous support and commitment to sustainable energy and environmental technologies.

Sustaining

U.S. Department of Energy

Home Depot

Joseph B. Whitehead Foundation

Oglethorpe Power

Gold

Atlanta Gas Light Company

Georgia Power Company

Georgia Environmental Facilities Authority

Panasonic Corporation

U.S. Environmental Protection Agency

Silver

Blue Circle Williams

Dupont-Tyvek

Furman Lumber

Honeywell, Inc.
Kohler Company
Pella Corporation
Southern Coalition for Advanced Transportation
Southwire Company
Structural Insulated Panels Association
Turner Foundation
United Solar Systems Corporation

Bronze
AFM Corporation
Allied Foam Products, Inc.
American Gas Cooling Center
Benjamin Moore & Company
Carpet and Rug Institute
Energy Efficient Building Assoc.

Evans Cabinets
Flexible Products, Inc.
J.M. Huber Company
MindSpring Enterprises
Owens-Corning
Pimsler Hoss Architects
RCD Corporation
Solium, Inc.
Summitville Tile
Therma-Stor Products
York International

Special Thanks
Lawrence Berkeley Laboratory
National Renewable Energy Laboratory
Oak Ridge National Laboratory

Contributors
ACT Metlund Systems
Accel Environmental
AddisonProducts Co.
A. E. Sampson & Son, Ltd.
Air Quality Systems Inc.
Alsy Lighting
Alta Illumination
Amana Refrigeration
AERT, Inc.
Alkco
American Energy Technologies
Andersen Windows
BASF Corporation
Borie Davis, Inc.
Carolina Lumber
Certus Laboratories
Center for Rehabilitation
    Technology
Clark Pest Control
Clopay Corporation
Columbia
Cooper Lighting
Conditioned Air Assoc. of GA.
Dongia Showroom at ADAC

DowElanco
Energy Partners
Emess Lighting Inc.
Empire Heating & Air
Frigidaire Co.
GAIA Technologies
Gatley & Associates
Georgia Concrete Products
    Association
General Electric
GreenStone Industries
Heat-N-Glo
Interface Flooring Systems
Juno Lighting
KS&B Cabinets
Lightolier
Litecontrol Corp.
Lithonia Lighting
Lumatech Corp.
Lutron
Maytag Corp.
Mitor Lighting
National Fenestration Ratings
    Council

Novitas, Inc.
Osram Sylvania
Phillips Driscopipe
Portland Cement Association
Prolight
Prescolight
Presto Products/Veco Co.
RCS
Shaper Lighting
Shaw Industries
Solar Development Inc.
Solar Energy Industries
    Association
Southern GF Company
SPI Lighting Inc.
Suncatcher of Atlanta
Two Twigs
Thomas Concrete
Turner Construction Co.
Tuff-N-Dri
Waterless Co.
Whirlpool
University of Georgia School
    of Landscape Architecture

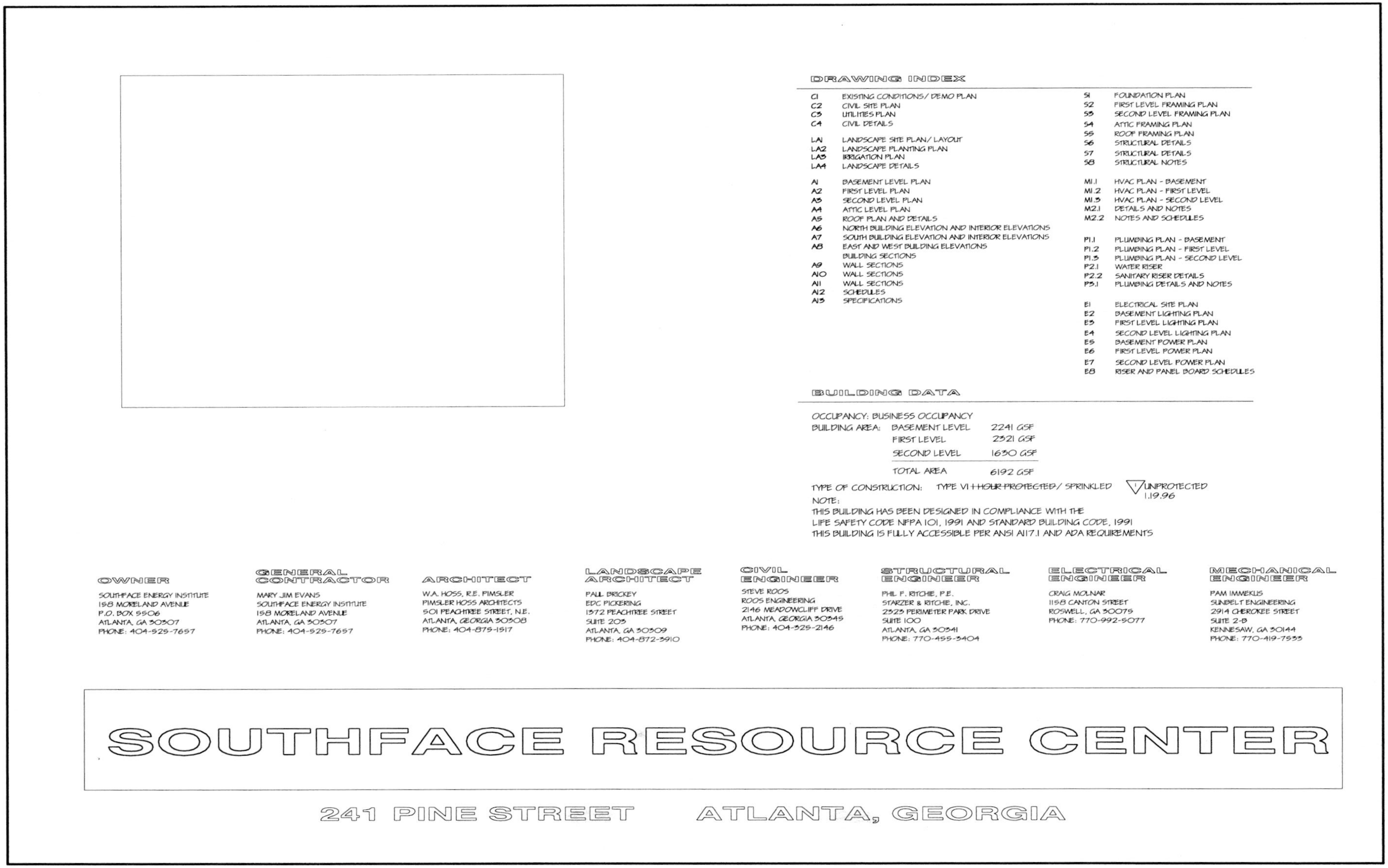

DRAWING INDEX

C1	EXISTING CONDITIONS / DEMO PLAN
C2	CIVIL SITE PLAN
C3	UTILITIES PLAN
C4	CIVIL DETAILS
LA1	LANDSCAPE SITE PLAN / LAYOUT
LA2	LANDSCAPE PLANTING PLAN
LA3	IRRIGATION PLAN
LA4	LANDSCAPE DETAILS
A1	BASEMENT LEVEL PLAN
A2	FIRST LEVEL PLAN
A3	SECOND LEVEL PLAN
A4	ATTIC LEVEL PLAN
A5	ROOF PLAN AND DETAILS
A6	NORTH BUILDING ELEVATION AND INTERIOR ELEVATIONS
A7	SOUTH BUILDING ELEVATION AND INTERIOR ELEVATIONS
A8	EAST AND WEST BUILDING ELEVATIONS
A9	BUILDING SECTIONS
A10	WALL SECTIONS
A11	WALL SECTIONS
A12	WALL SECTIONS
A13	SCHEDULES
A15	SPECIFICATIONS

S1	FOUNDATION PLAN
S2	FIRST LEVEL FRAMING PLAN
S3	SECOND LEVEL FRAMING PLAN
S4	ATTIC FRAMING PLAN
S5	ROOF FRAMING PLAN
S6	STRUCTURAL DETAILS
S7	STRUCTURAL DETAILS
S8	STRUCTURAL NOTES
M1.1	HVAC PLAN – BASEMENT
M1.2	HVAC PLAN – FIRST LEVEL
M1.3	HVAC PLAN – SECOND LEVEL
M2.1	DETAILS AND NOTES
M2.2	NOTES AND SCHEDULES
P1.1	PLUMBING PLAN – BASEMENT
P1.2	PLUMBING PLAN – FIRST LEVEL
P1.3	PLUMBING PLAN – SECOND LEVEL
P2.1	WATER RISER
P2.2	SANITARY RISER DETAILS
P3.1	PLUMBING DETAILS AND NOTES
E1	ELECTRICAL SITE PLAN
E2	BASEMENT LIGHTING PLAN
E3	FIRST LEVEL LIGHTING PLAN
E4	SECOND LEVEL LIGHTING PLAN
E5	BASEMENT POWER PLAN
E6	FIRST LEVEL POWER PLAN
E7	SECOND LEVEL POWER PLAN
E8	RISER AND PANEL BOARD SCHEDULES

BUILDING DATA

OCCUPANCY: BUSINESS OCCUPANCY

BUILDING AREA:		
	BASEMENT LEVEL	2241 GSF
	FIRST LEVEL	2321 GSF
	SECOND LEVEL	1650 GSF
	TOTAL AREA	6192 GSF

TYPE OF CONSTRUCTION: TYPE VI HOUR PROTECTED / SPRINKLED △UNPROTECTED 1.19.96

NOTE:
THIS BUILDING HAS BEEN DESIGNED IN COMPLIANCE WITH THE
LIFE SAFETY CODE NFPA 101, 1991 AND STANDARD BUILDING CODE, 1991
THIS BUILDING IS FULLY ACCESSIBLE PER ANSI A117.1 AND ADA REQUIREMENTS

OWNER

SOUTHFACE ENERGY INSTITUTE
158 MORELAND AVENUE
P.O. BOX 5506
ATLANTA, GA 30307
PHONE: 404-525-7657

GENERAL CONTRACTOR

MARY JIM EVANS
SOUTHFACE ENERGY INSTITUTE
158 MORELAND AVENUE
ATLANTA, GA 30307
PHONE: 404-525-7657

ARCHITECT

W.A. KOSS, R.E. PIMSLER
PIMSLER KOSS ARCHITECTS
901 PEACHTREE STREET, N.E.
ATLANTA, GEORGIA 30308
PHONE: 404-875-1517

CIVIL ENGINEER

STEVE ROOS
ROOS ENGINEERING
2146 MEADOWCLIFF DRIVE
ATLANTA, GEORGIA 30345
PHONE: 404-929-2146

LANDSCAPE ARCHITECT

PAUL DICKEY
EDC PICKERING
1572 PEACHTREE STREET
SUITE 205
ATLANTA, GA 30309
PHONE: 404-872-5910

STRUCTURAL ENGINEER

PHILIP F. RITCHIE, P.E.
SPANZER & RITCHIE, INC.
2525 PERIMETER PARK DRIVE
SUITE 100
ATLANTA, GA 30341
PHONE: 770-455-5404

ELECTRICAL ENGINEER

CRAIG MOLNAR
1159 CANTON STREET
ROSWELL, GA 30075
PHONE: 770-992-5077

MECHANICAL ENGINEER

PAM IMMELUS
SUNDELT ENGINEERING
2914 CHEROKEE STREET
SUITE 2-B
KENNESAW, GA 30144
PHONE: 770-419-7855

SOUTHFACE RESOURCE CENTER

241 PINE STREET   ATLANTA, GEORGIA

Figure 20-1

635

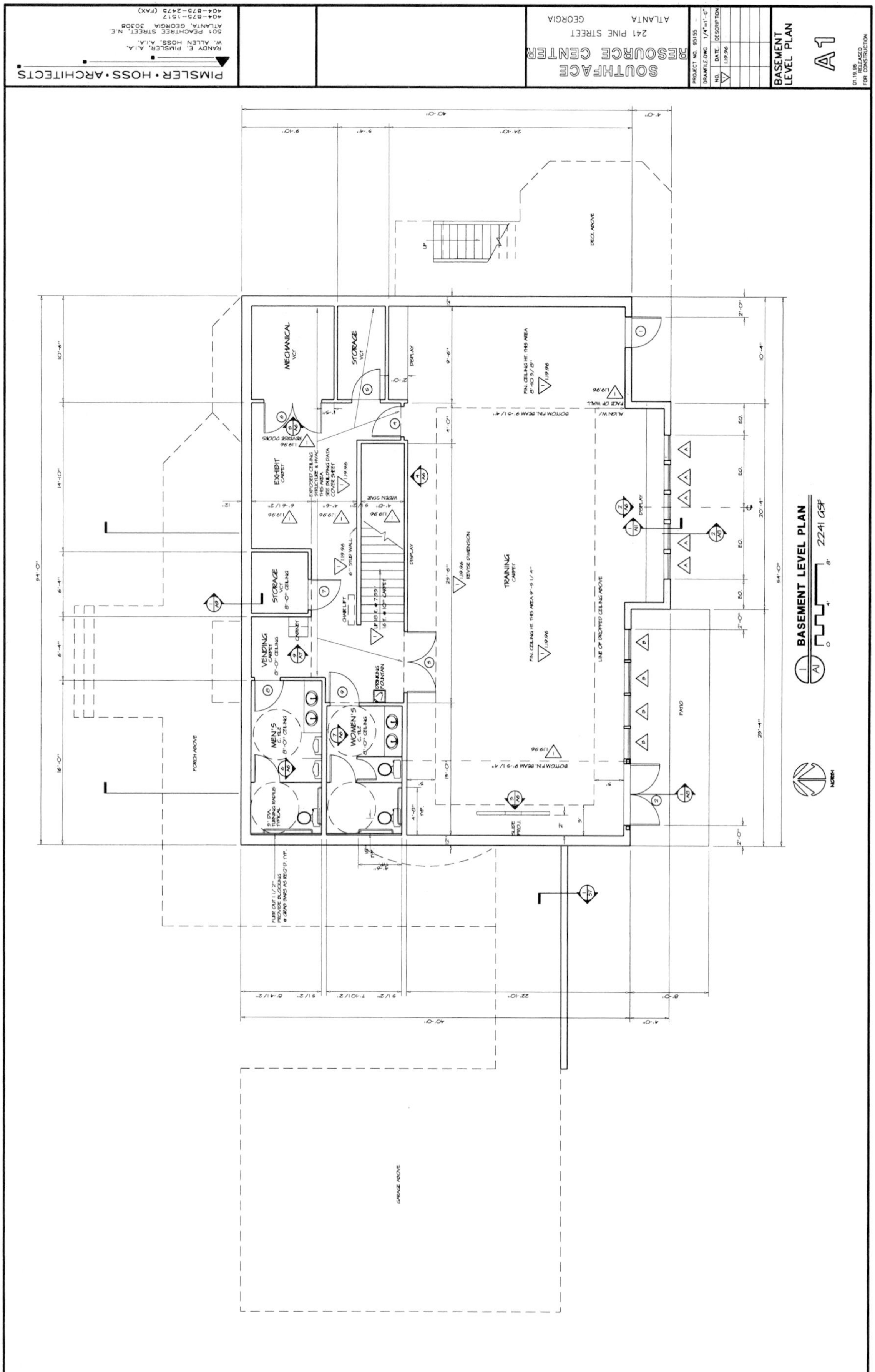

PIMSLER • HOSS • ARCHITECTS
RANDY E. PIMSLER, A.I.A.
W. ALLEN HOSS, A.I.A.
501 PEACHTREE STREET, N.E.
ATLANTA, GEORGIA 30308
404-875-1517
404-875-2475 (FAX)
SOUTHFACE
RESOURCE CENTER
241 PINE STREET
ATLANTA GEORGIA
BASEMENT LEVEL PLAN
A1
BASEMENT LEVEL PLAN
2241 GSF
MECHANICAL
STORAGE
EXHIBIT
VENDING
MEN'S
WOMEN'S
TRAINING
DISPLAY
PATIO
PORCH ABOVE
DECK ABOVE
GARAGE ABOVE
NORTH
54'-0"
40'-0"

Figure 20-2

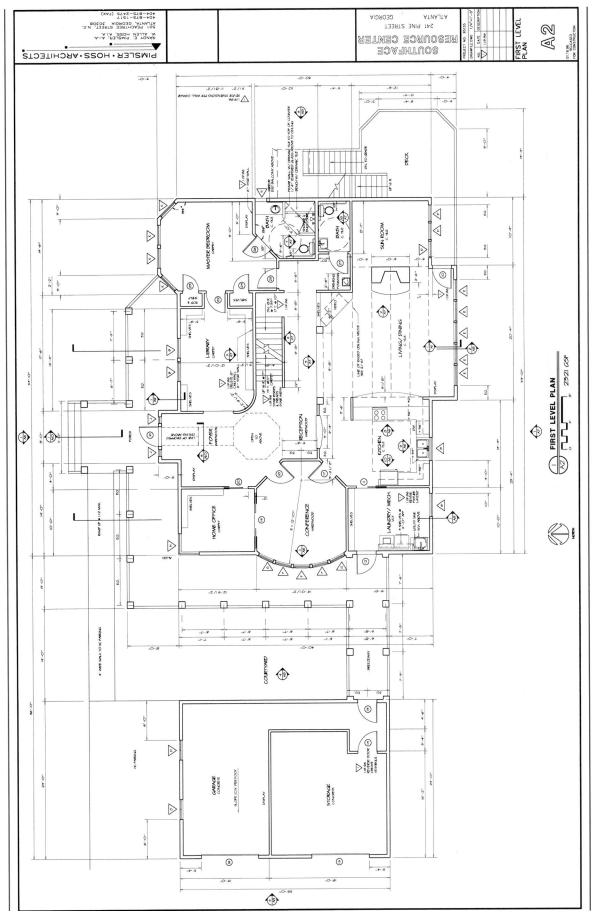

Figure 20-3

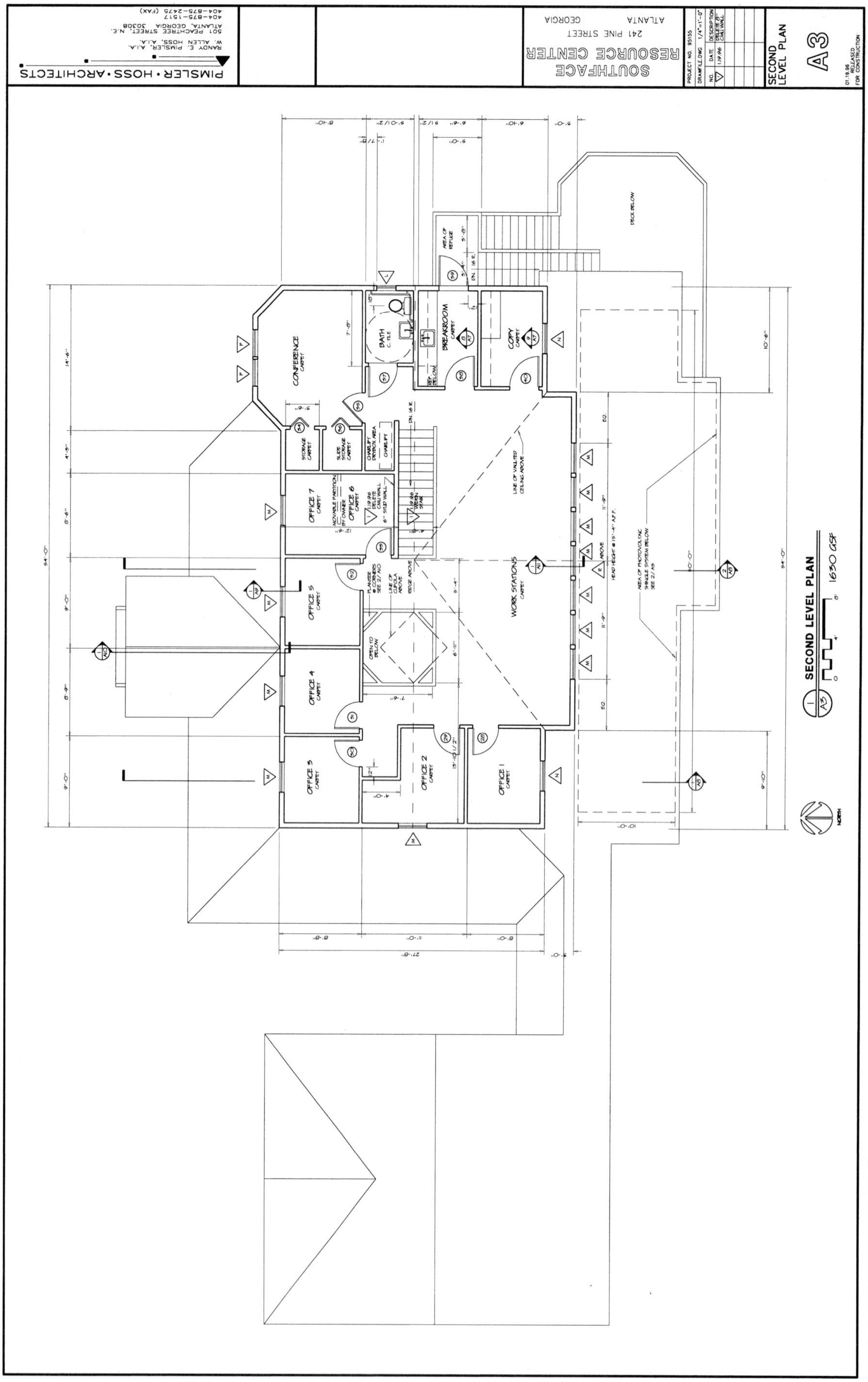
PIMSLER • HOSS • ARCHITECTS
RANDY E. PIMSLER, A.I.A.
W. ALLEN HOSS, A.I.A.
501 PEACHTREE STREET, N.E.
ATLANTA, GEORGIA 30308
404-875-1517
404-875-2475 (FAX)
SOUTHFACE
RESOURCE CENTER
241 PINE STREET
ATLANTA GEORGIA
SECOND LEVEL PLAN
A3
01.19.96 RELEASED FOR CONSTRUCTION
CONFERENCE
BATH
BREAKROOM
COPY
STORAGE
OFFICE 7
OFFICE 6
OFFICE 5
OFFICE 4
OFFICE 3
OFFICE 2
OFFICE 1
WORK STATIONS
DECK BELOW
AREA OF REFUGE
SECOND LEVEL PLAN
1630 GSF
NORTH

Figure 20–4

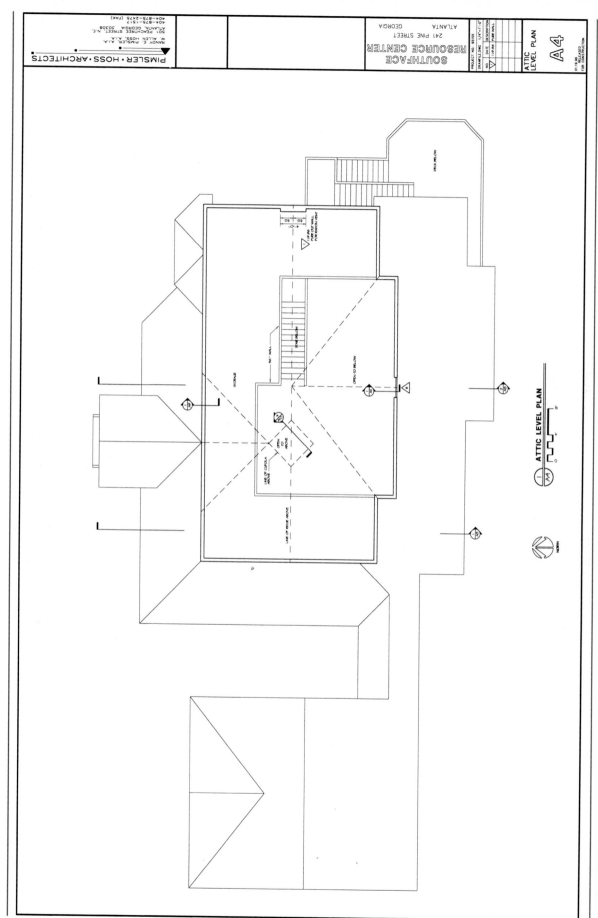

SOUTHFACE
RESOURCE CENTER

241 PINE STREET
ATLANTA   GEORGIA

ATTIC
LEVEL PLAN

A4

ATTIC LEVEL PLAN

NORTH

Figure 20-5

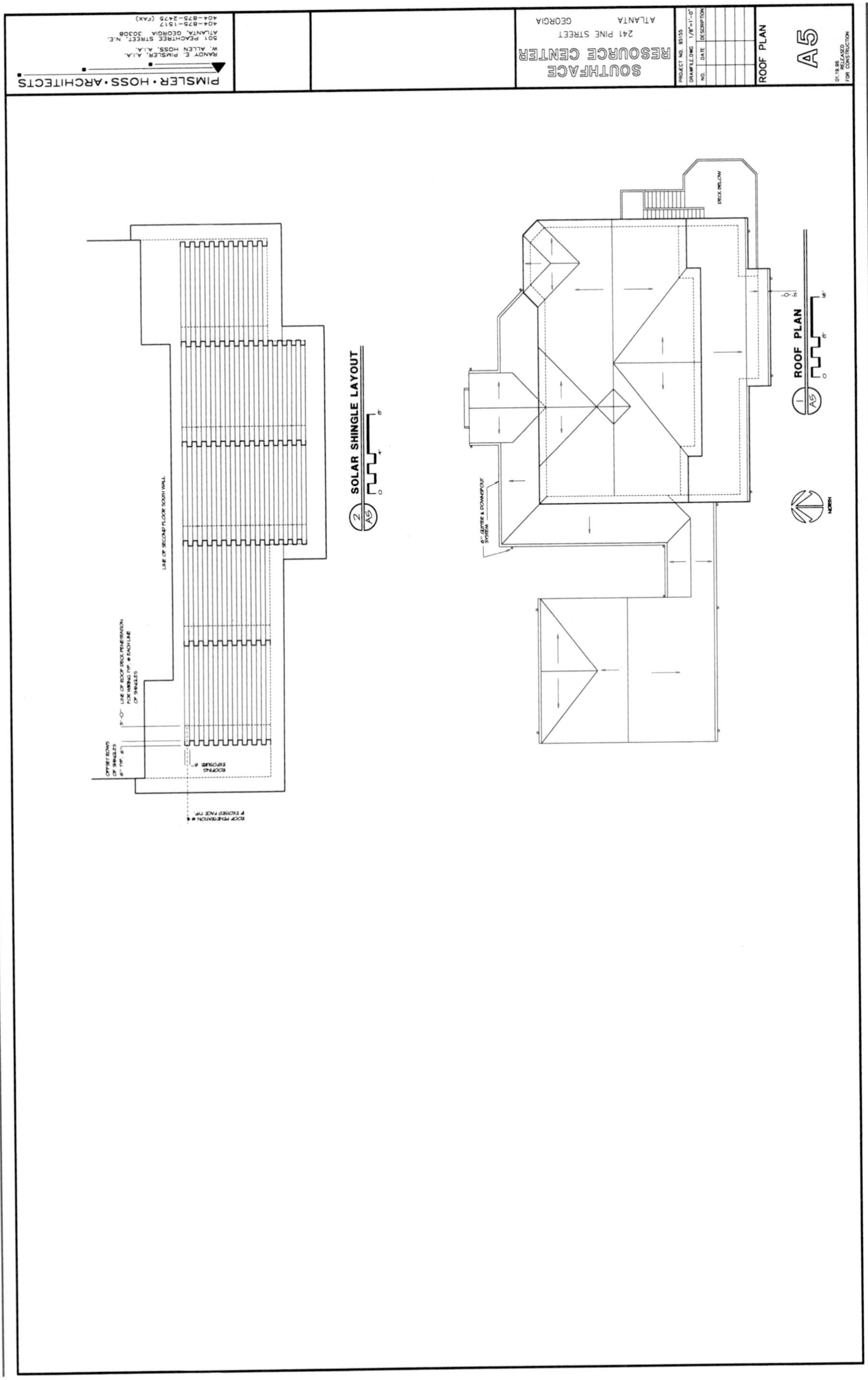

Figure 20-6

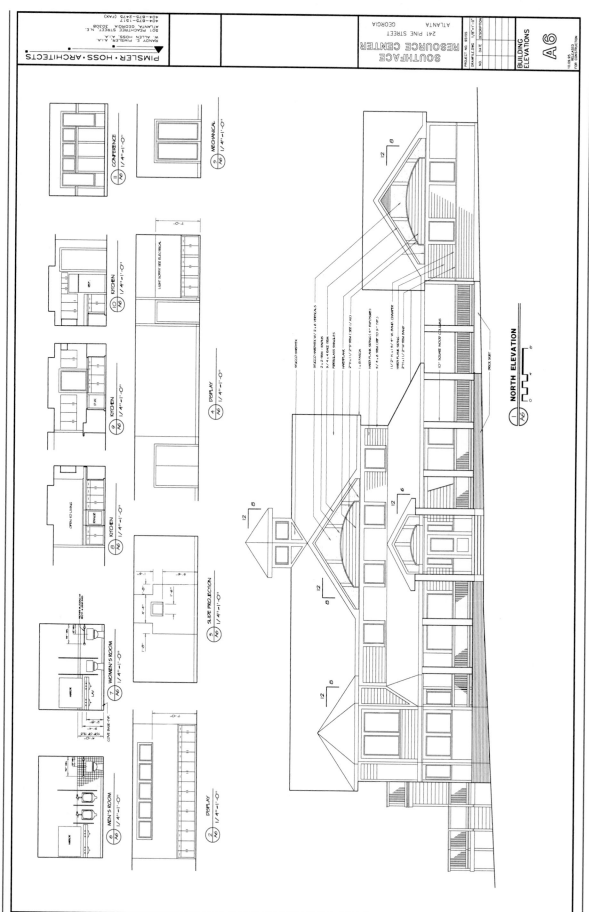

Figure 20-7

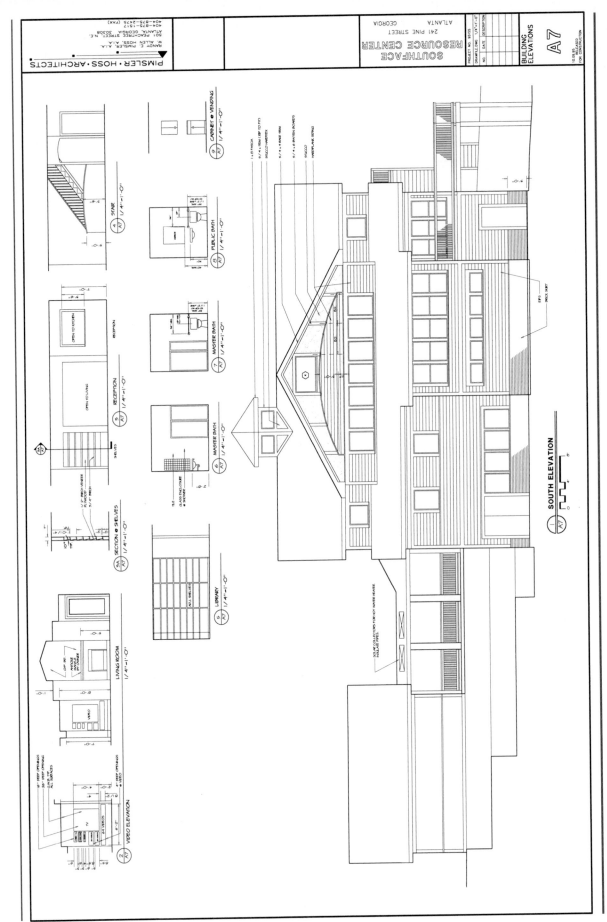

Figure 20-8

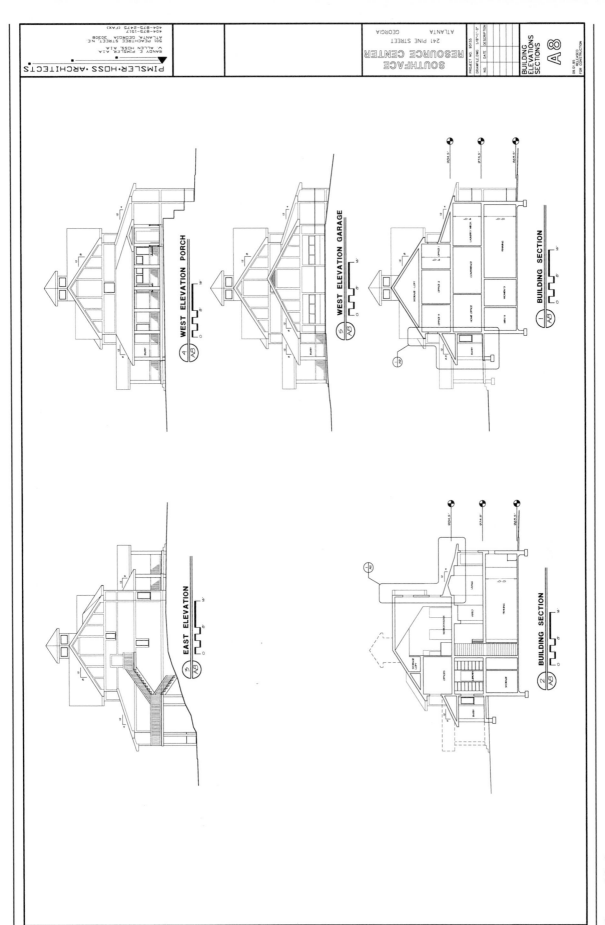

Figure 20-9

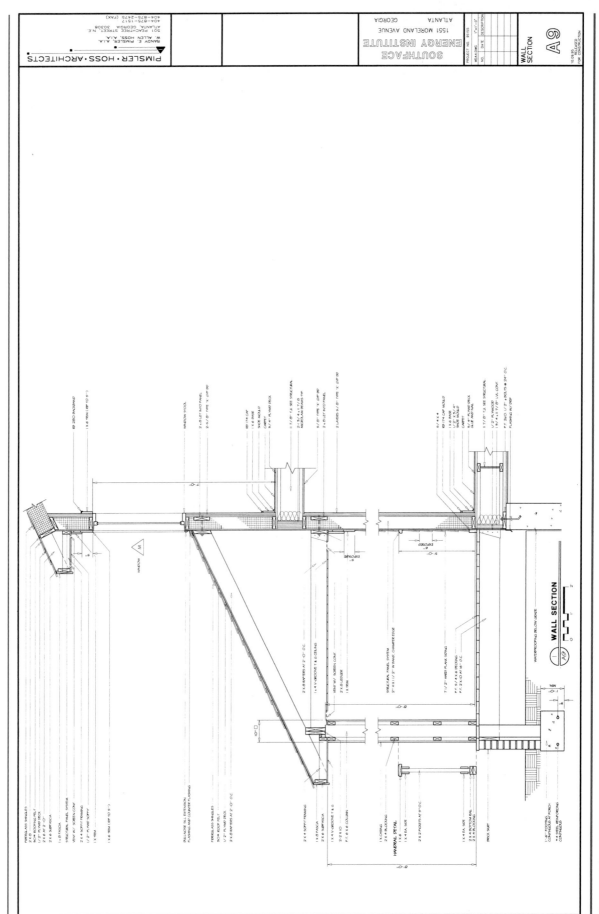

Figure 20-10

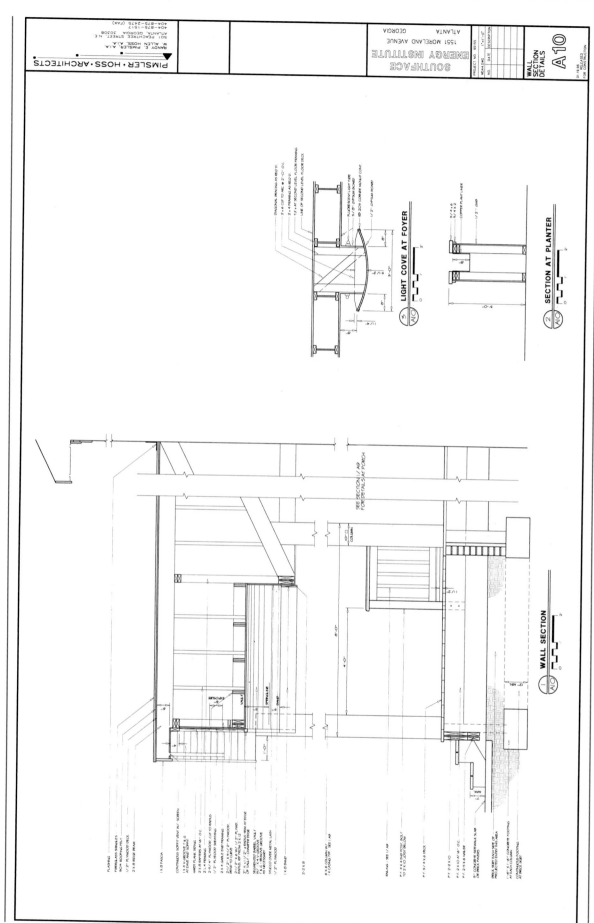

Figure 20-11

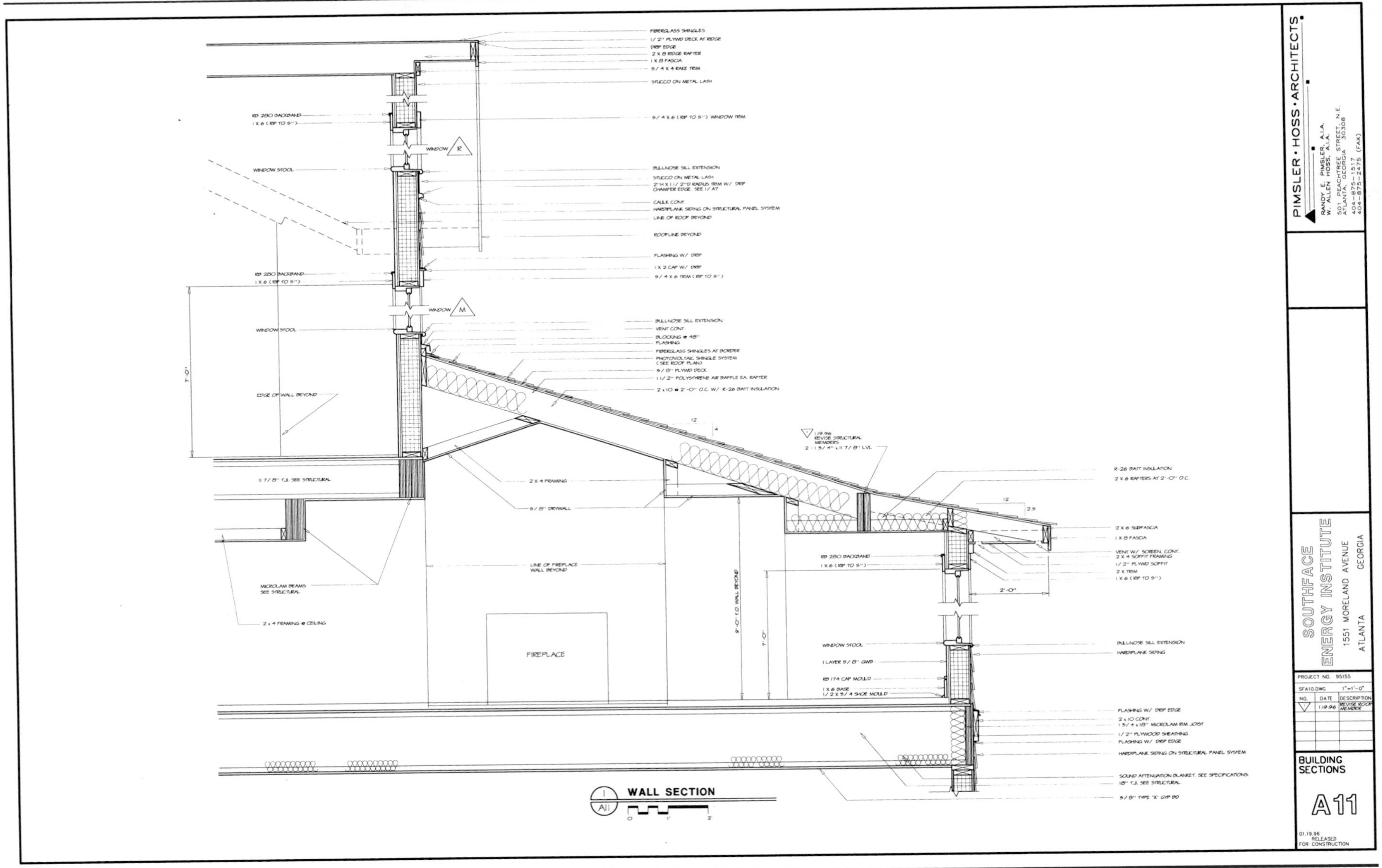
PIMSLER • HOSS • ARCHITECTS
RANDY E. PIMSLER, A.I.A.
W. ALLEN HOSS, A.I.A.
501 PEACHTREE STREET, N.E.
ATLANTA, GEORGIA 30308
404-875-1517
404-875-2475 (FAX)
SOUTHFACE
ENERGY INSTITUTE
1551 MORELAND AVENUE
ATLANTA GEORGIA
PROJECT NO. 95155
SFA10.DWG 1"=1'-0"
NO. DATE DESCRIPTION
1.19.96 REVISE ROOF MEMBER
BUILDING SECTIONS
A11
01.19.96 RELEASED FOR CONSTRUCTION
WALL SECTION
1
A11
FIREPLACE
WINDOW R
WINDOW M
2'-0"
7'-0"
12
4

Figure 20–12

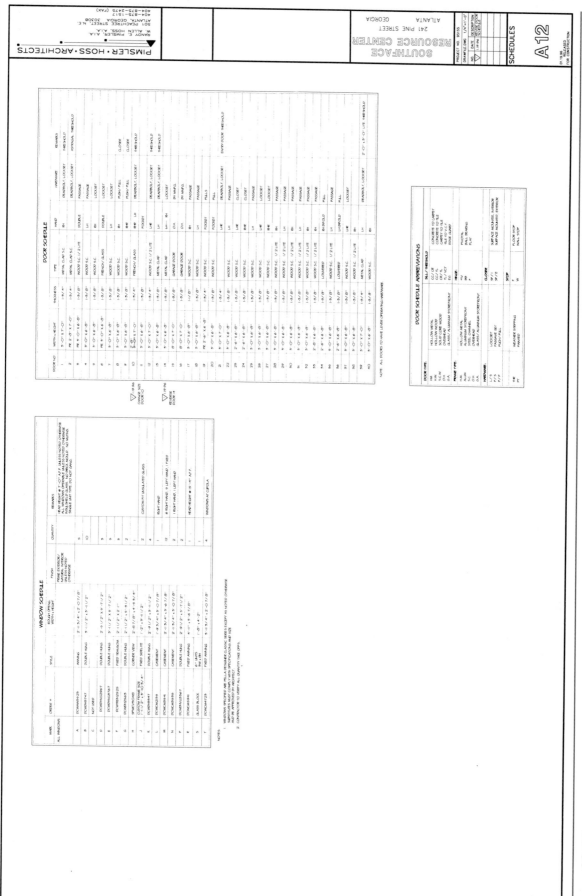

Figure 20-13

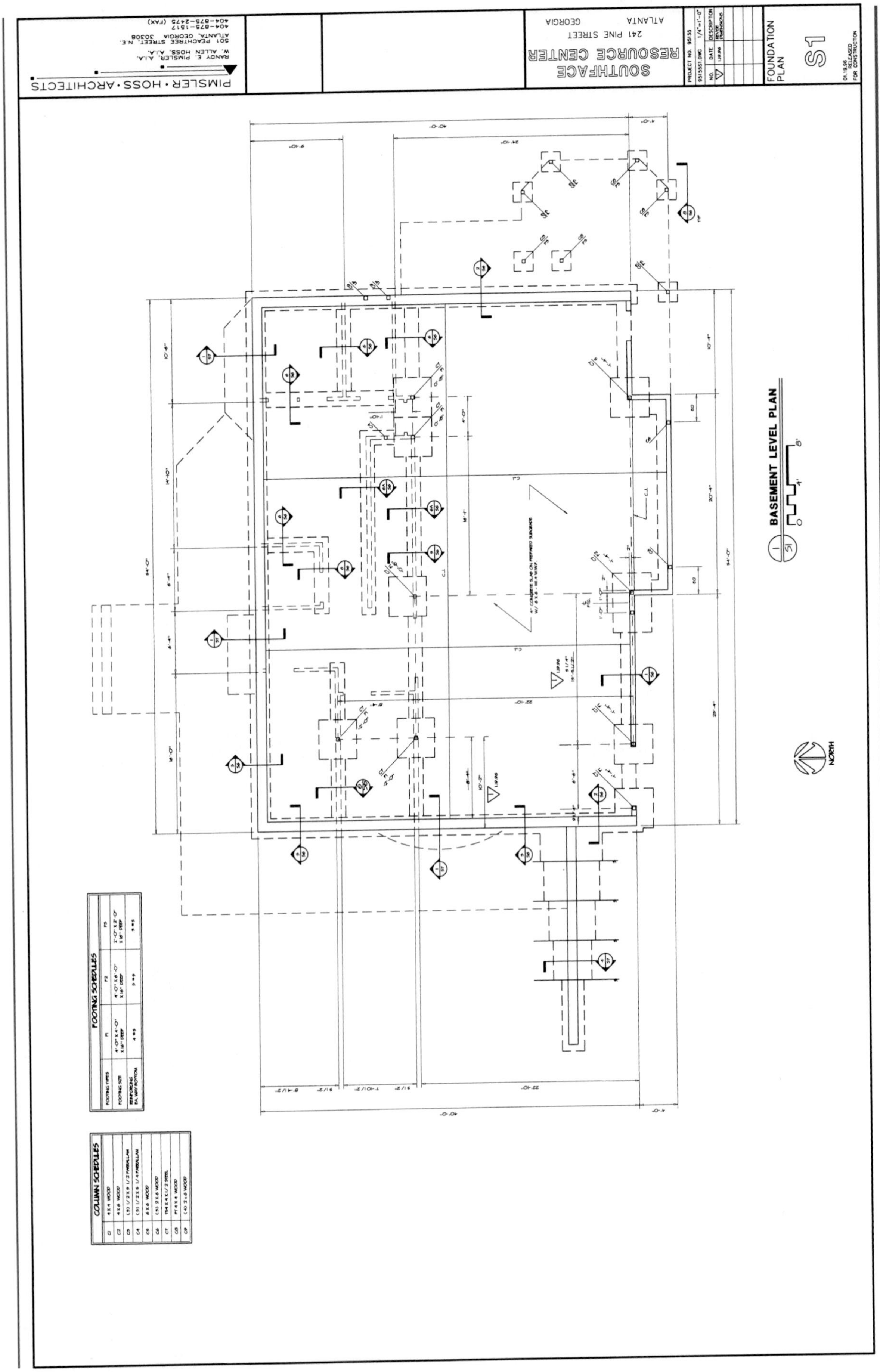
PIMSLER • HOSS • ARCHITECTS
RANDY E. PIMSLER, A.I.A.
W. ALLEN HOSS, A.I.A.
501 PEACHTREE STREET, N.E.
ATLANTA, GEORGIA 30308
404-875-1517
404-875-2475 (FAX)
SOUTHFACE RESOURCE CENTER
241 PINE STREET
ATLANTA GEORGIA
FOUNDATION PLAN
S1
BASEMENT LEVEL PLAN
NORTH
FOOTING SCHEDULES
COLUMN SCHEDULES

Figure 20–14

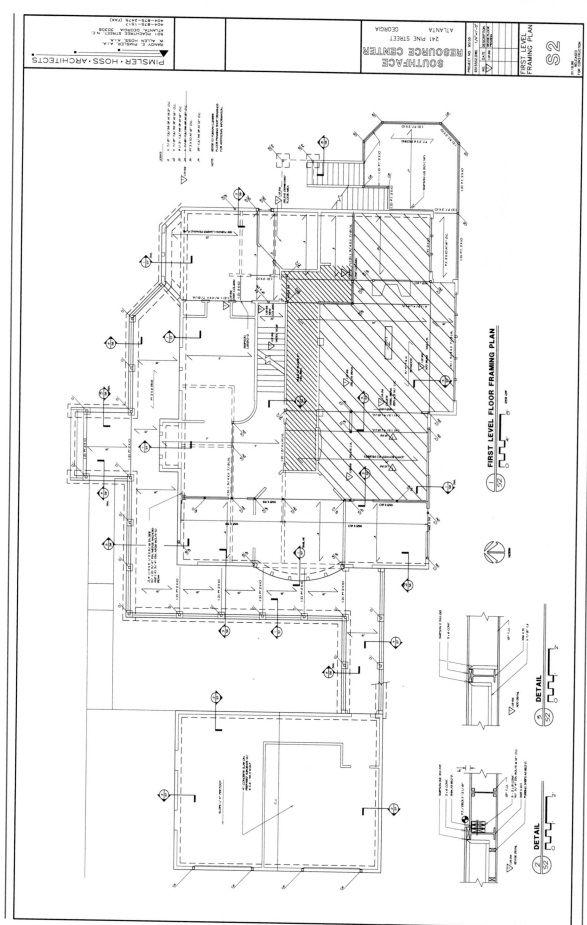

FIRST LEVEL FLOOR FRAMING PLAN

Figure 20-15

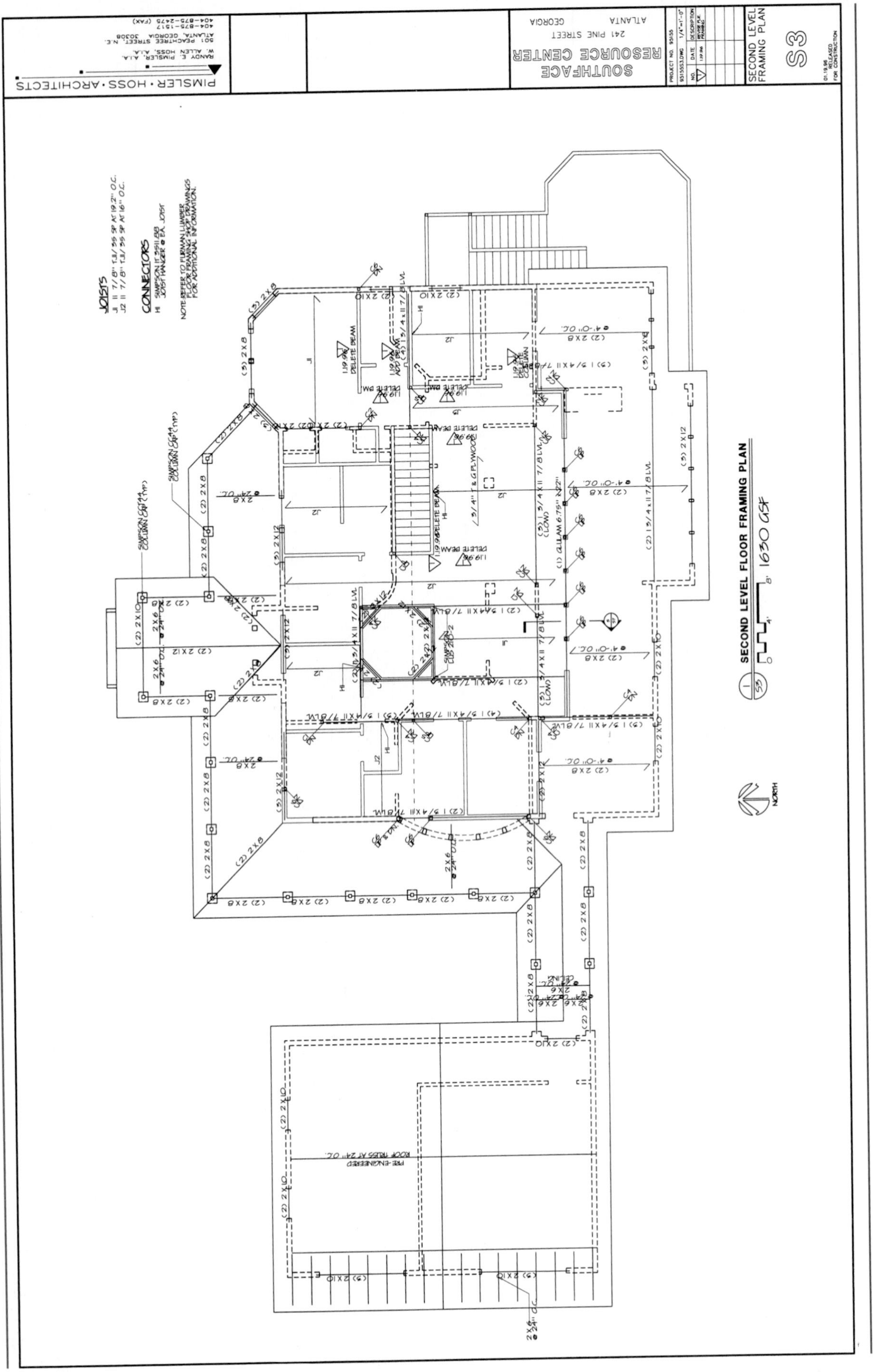

PIMSLER • HOSS • ARCHITECTS
SOUTHFACE RESOURCE CENTER
241 PINE STREET
ATLANTA GEORGIA
SECOND LEVEL FRAMING PLAN
S3
SECOND LEVEL FLOOR FRAMING PLAN
1630 GSF
NORTH

Figure 20–16

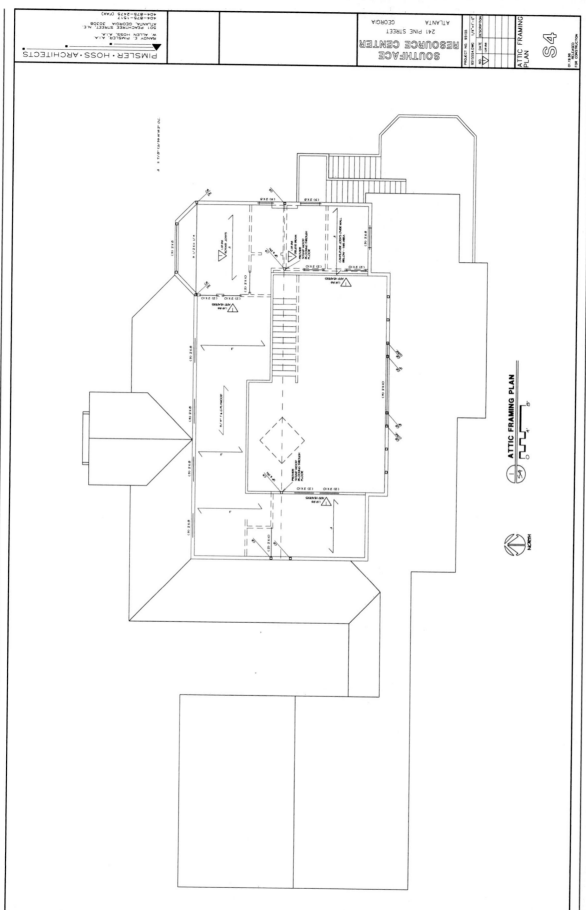

ATTIC FRAMING PLAN

NORTH

Figure 20–17

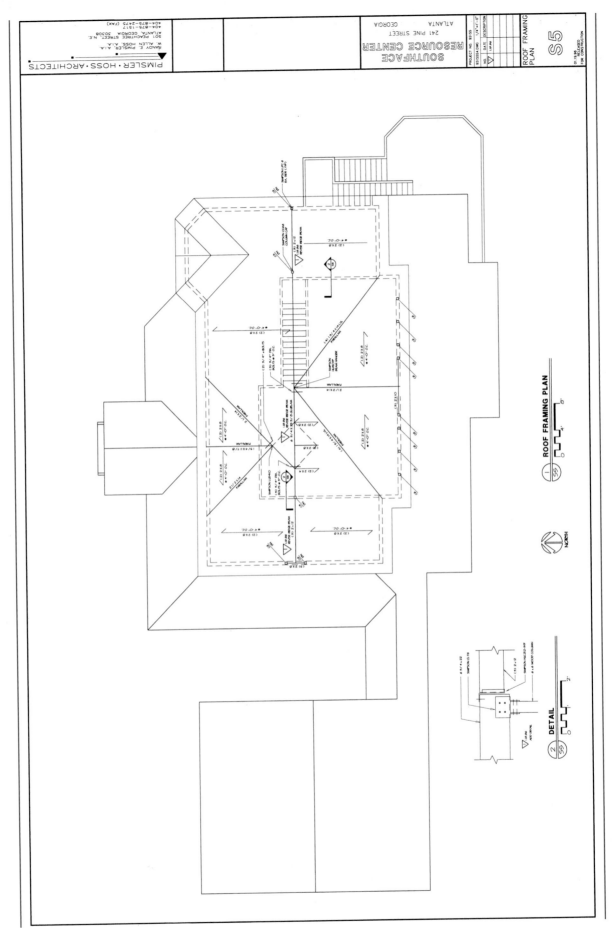

Figure 20-18

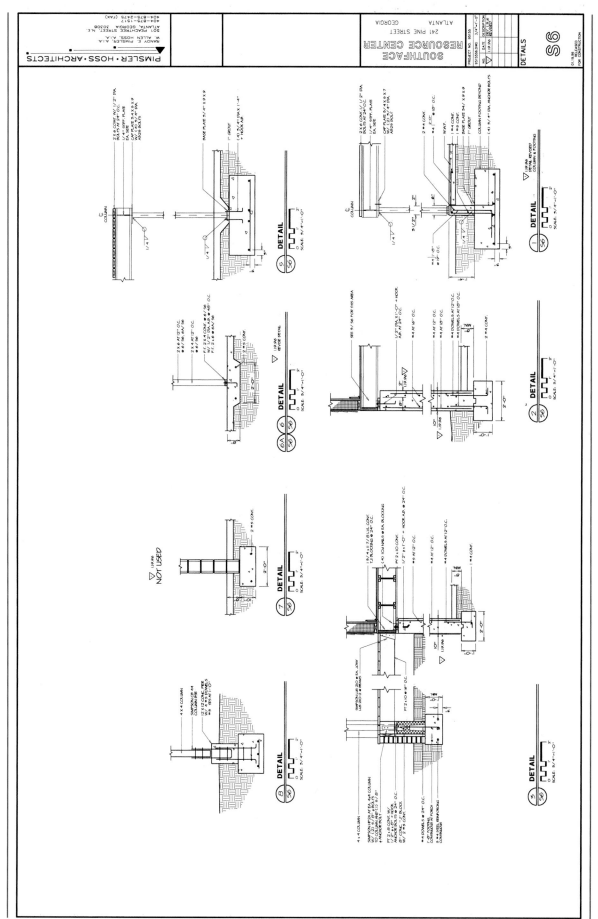

Figure 20-19

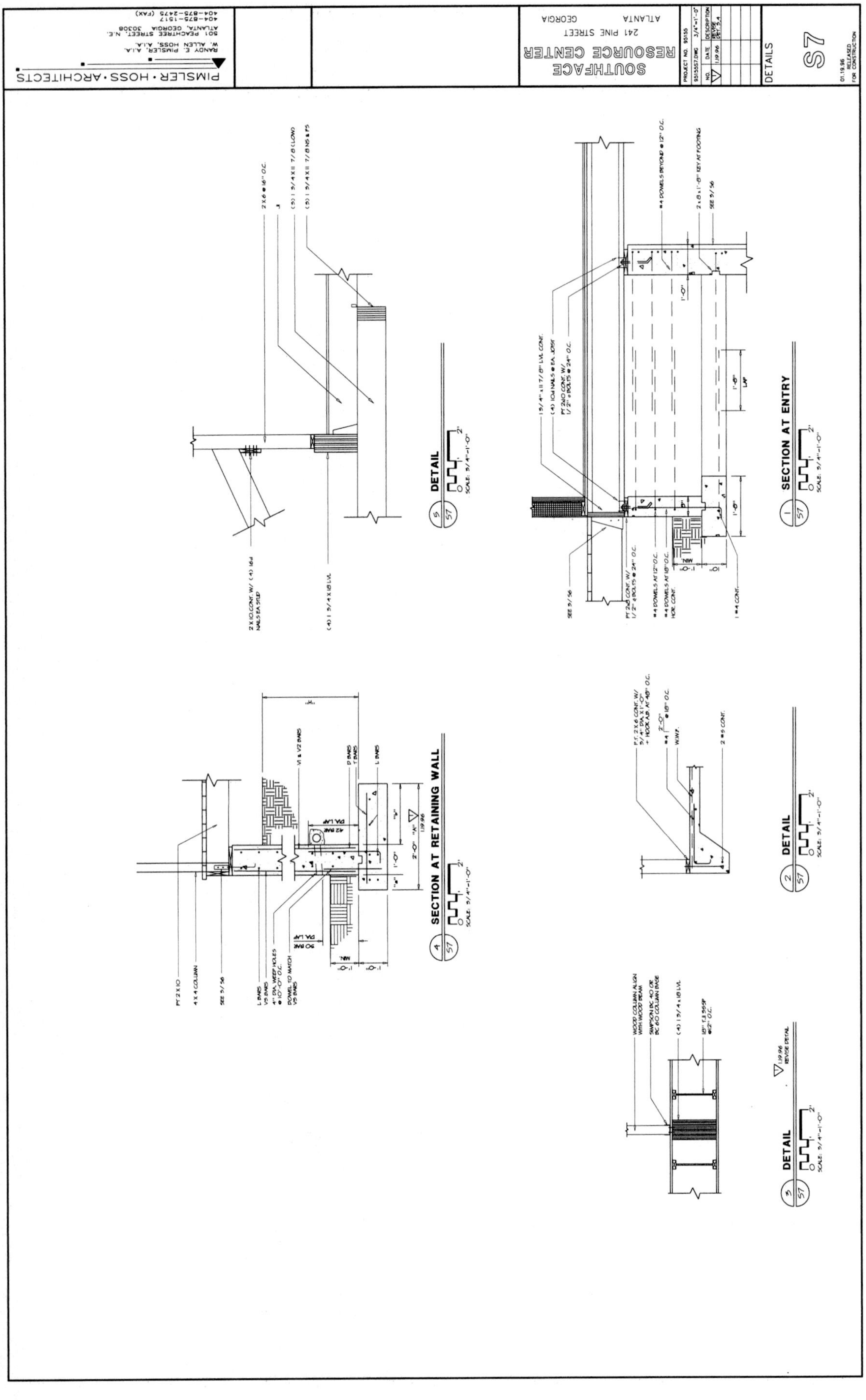

Figure 20-20

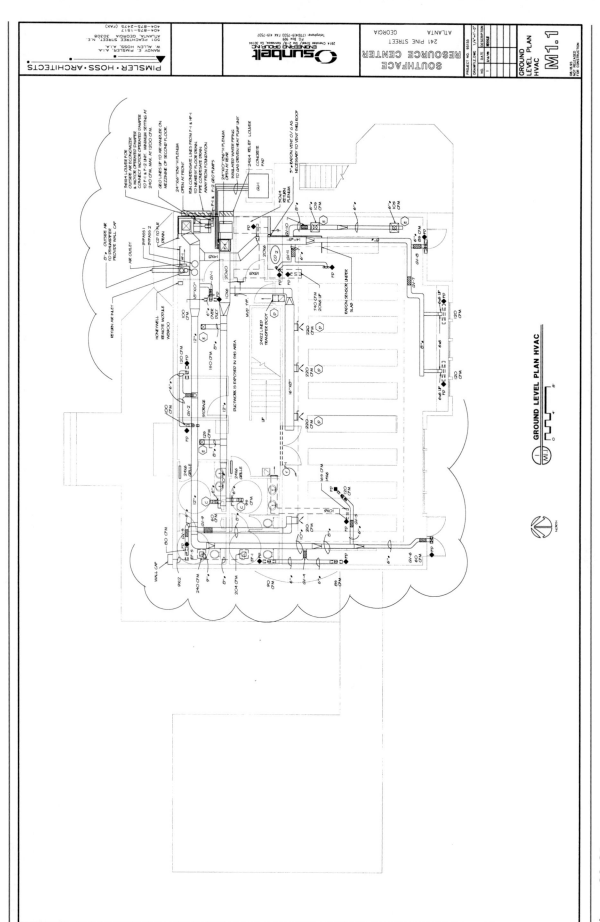

Figure 20–21

655

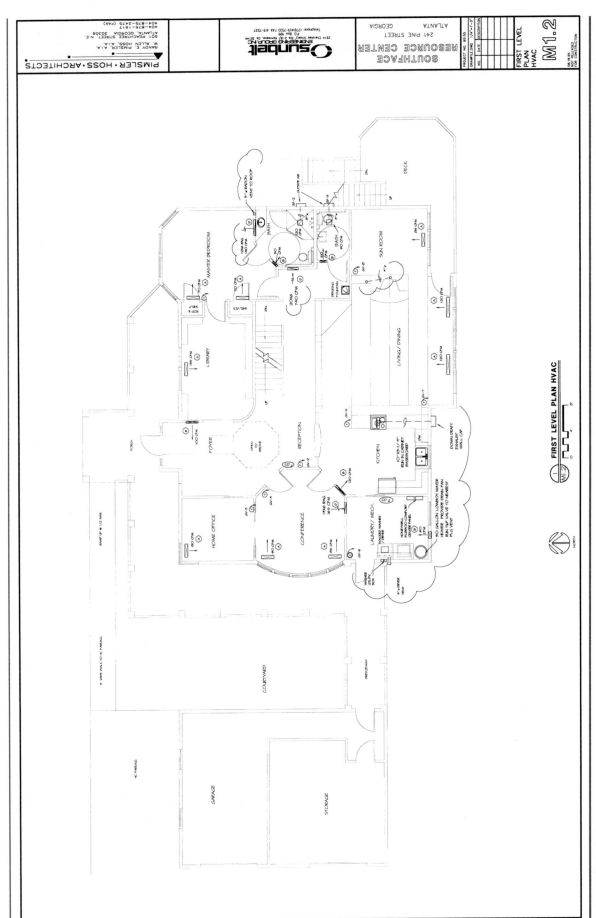

Figure 20–22

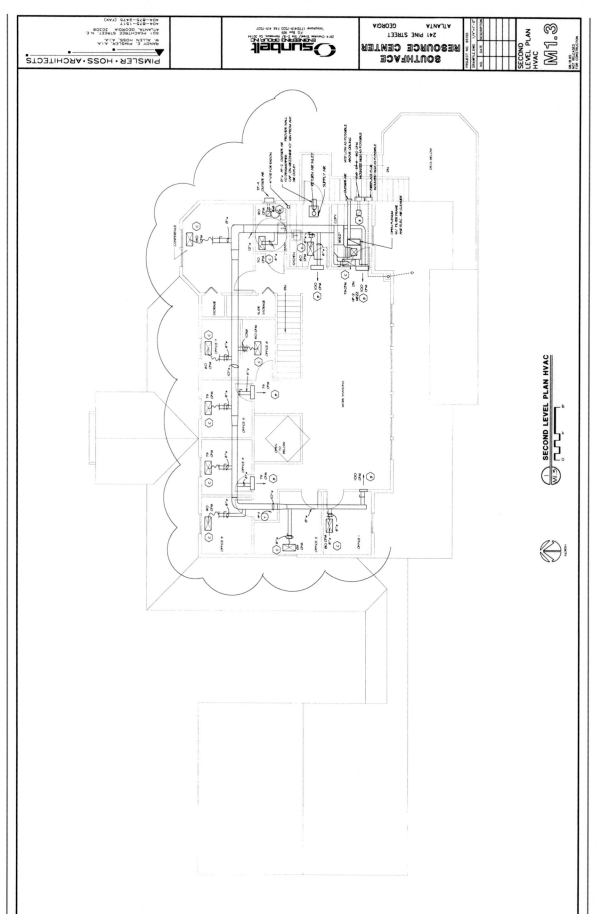

Figure 20-23

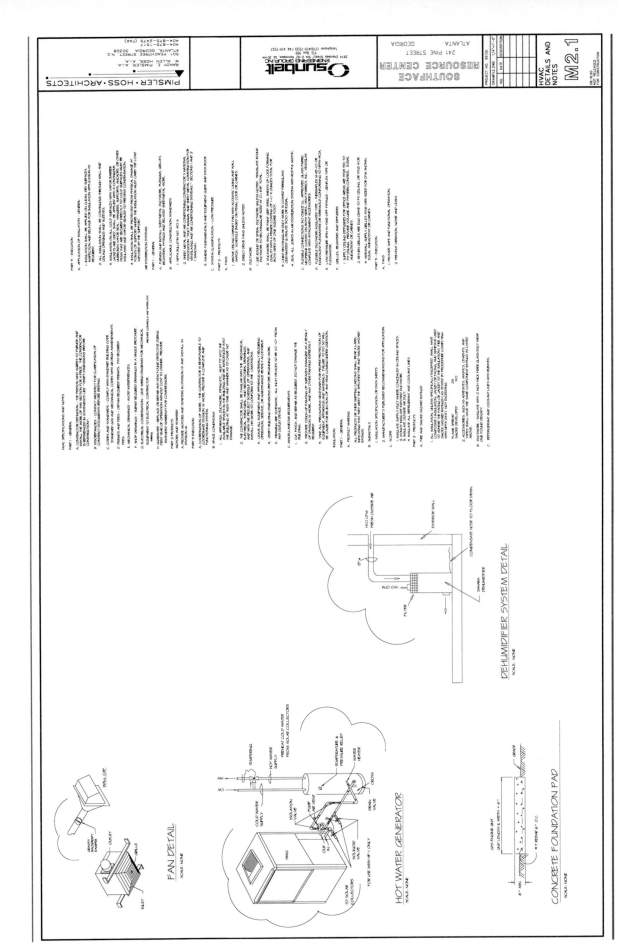

DEHUMIDIFIER SYSTEM DETAIL
SCALE: NONE

FAN DETAIL
SCALE: NONE

HOT WATER GENERATOR
SCALE: NONE

CONCRETE FOUNDATION PAD
SCALE: NONE

Figure 20-24

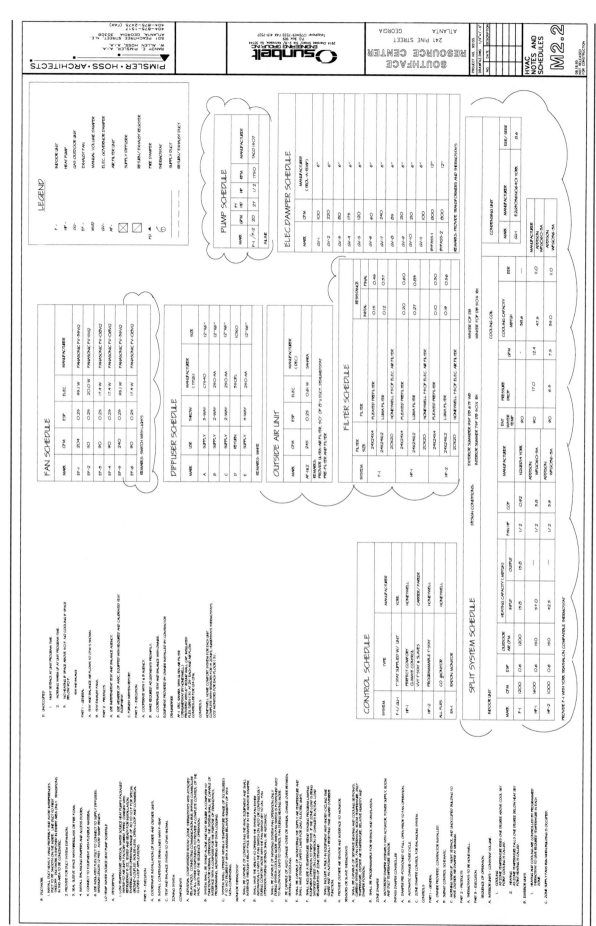

Figure 20-25

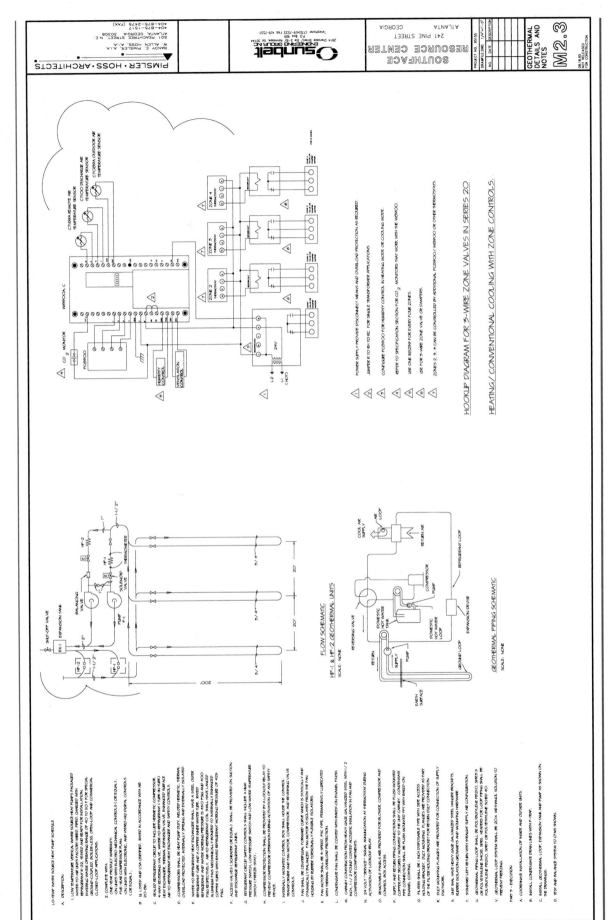

Figure 20-26

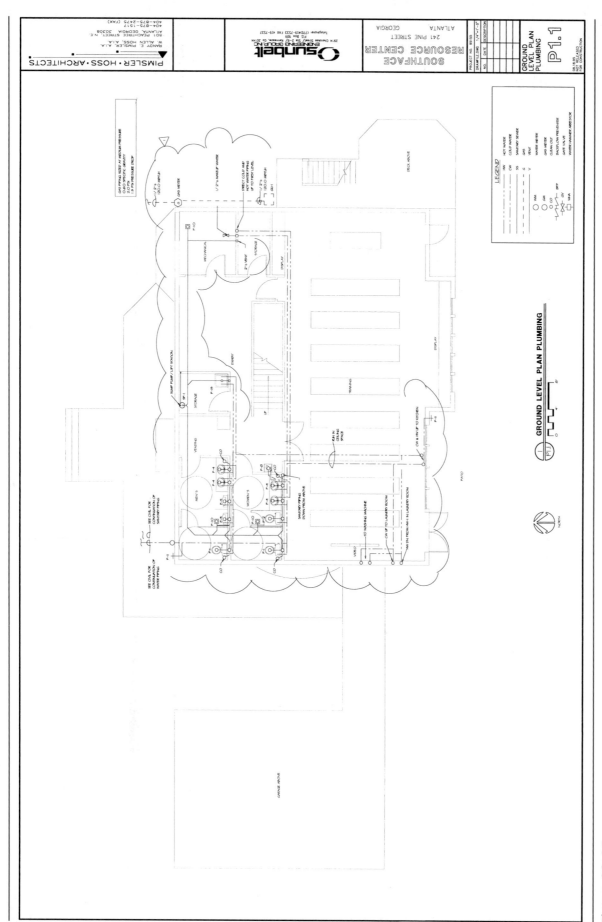

Figure 20-27

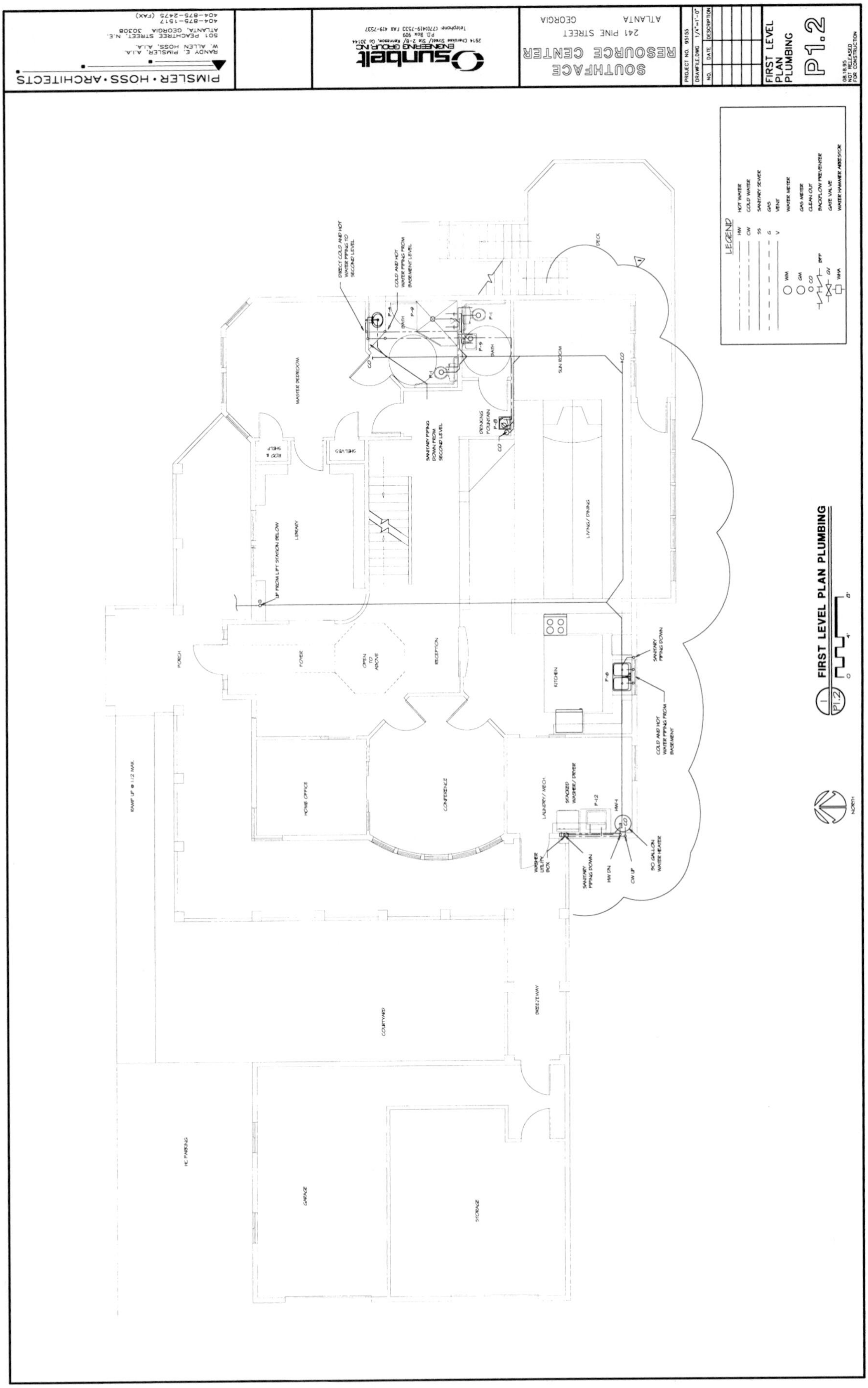
PIMSLER • HOSS • ARCHITECTS
RANDY E. PIMSLER, A.I.A.
W. ALLEN HOSS, A.I.A.
501 PEACHTREE STREET, N.E.
ATLANTA, GEORGIA 30308
404-875-1517
404-875-2475 (FAX)
sunbelt
SOUTHFACE RESOURCE CENTER
241 PINE STREET
ATLANTA GEORGIA
FIRST LEVEL PLAN PLUMBING
P1.2
LEGEND
HOT WATER
COLD WATER
SANITARY SEWER
GAS
VENT
WATER METER
GAS METER
CLEAN OUT
BACKFLOW PREVENTER
GATE VALVE
WATER HAMMER ARRESTOR
FIRST LEVEL PLAN PLUMBING
NORTH

Figure 20–28

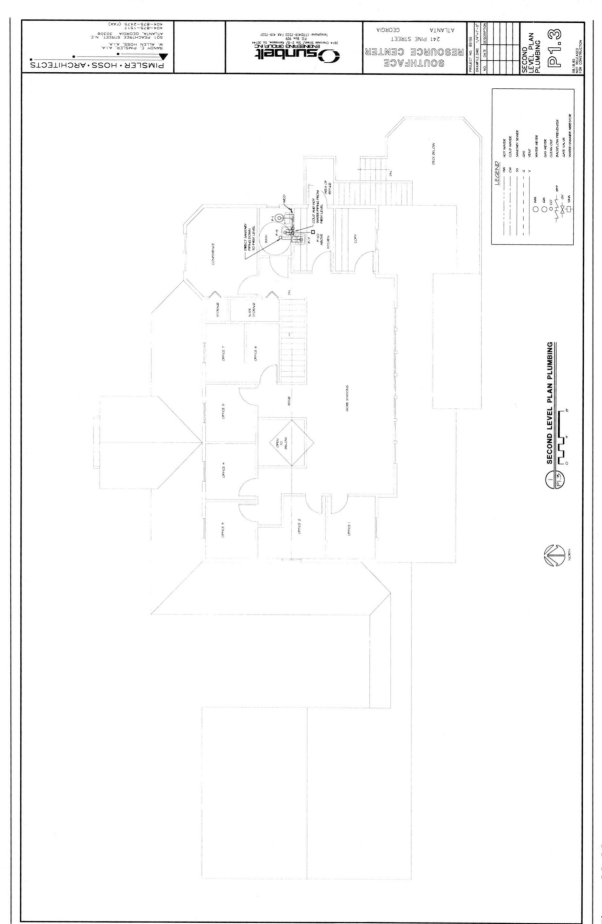

Figure 20-29

663

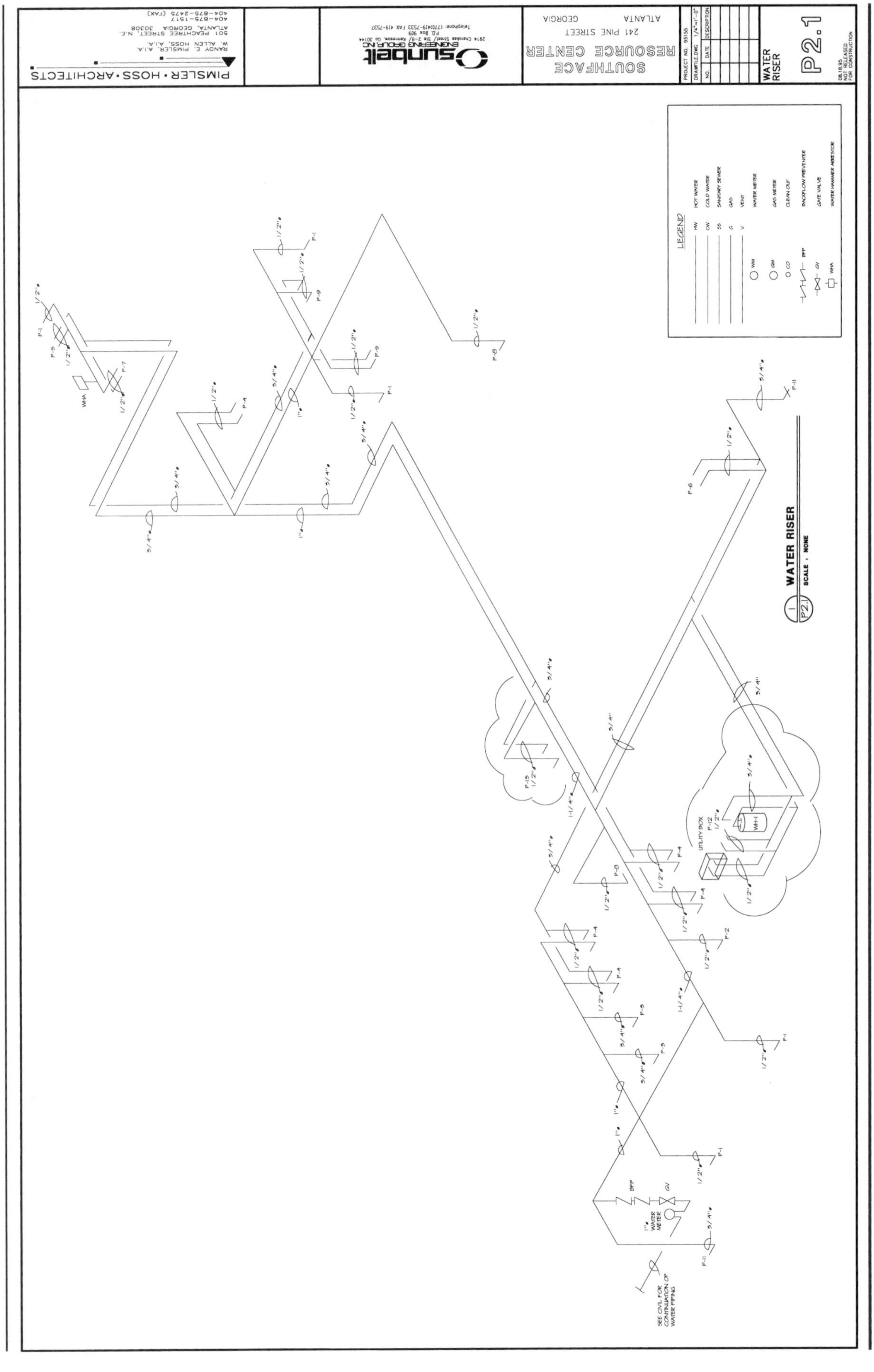

Figure 20-30

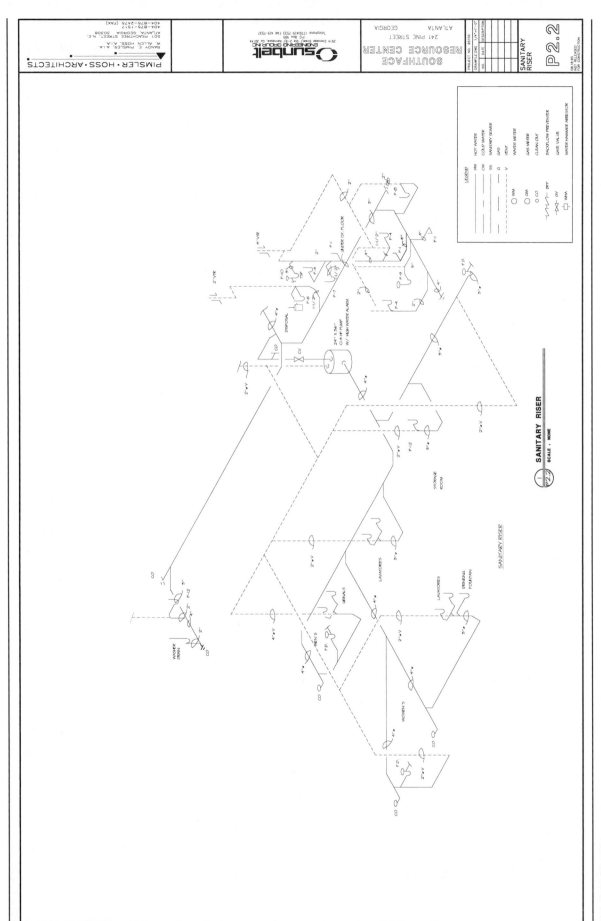

Figure 20–31

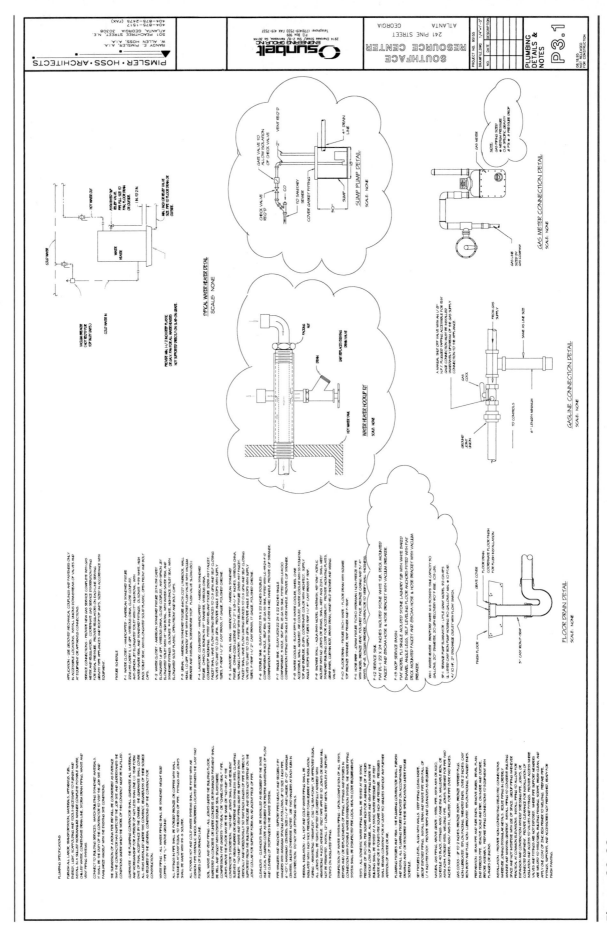

Figure 20-32

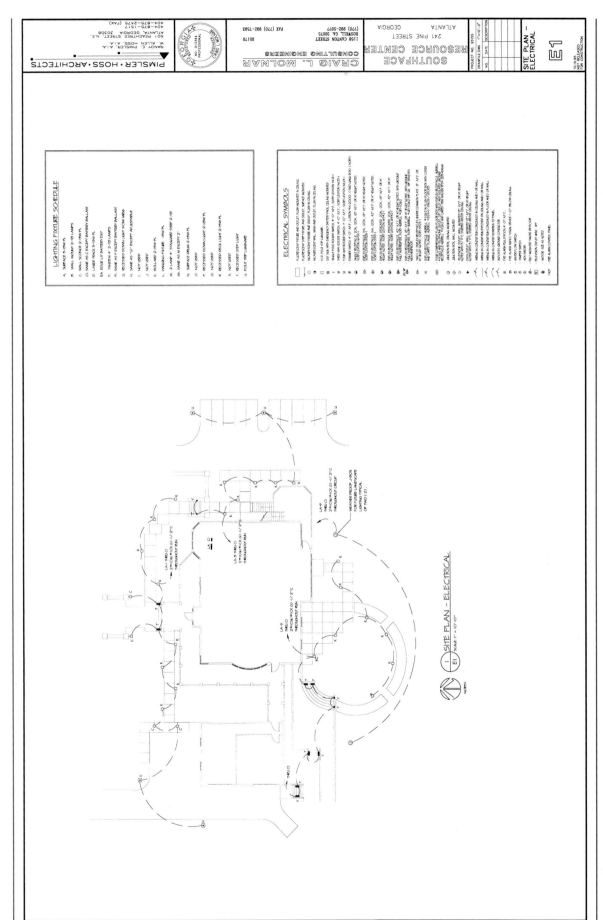

Figure 20-33

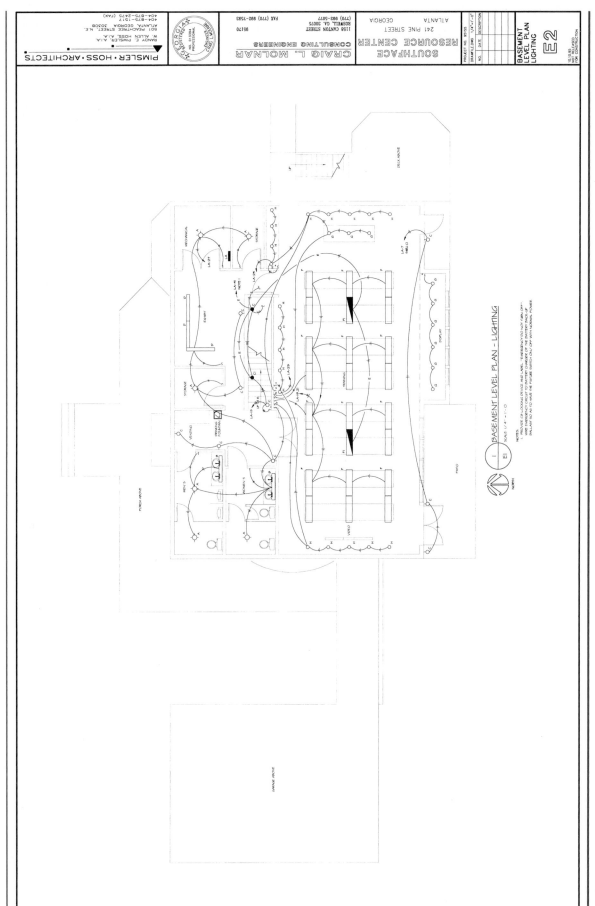

Figure 20-34

668

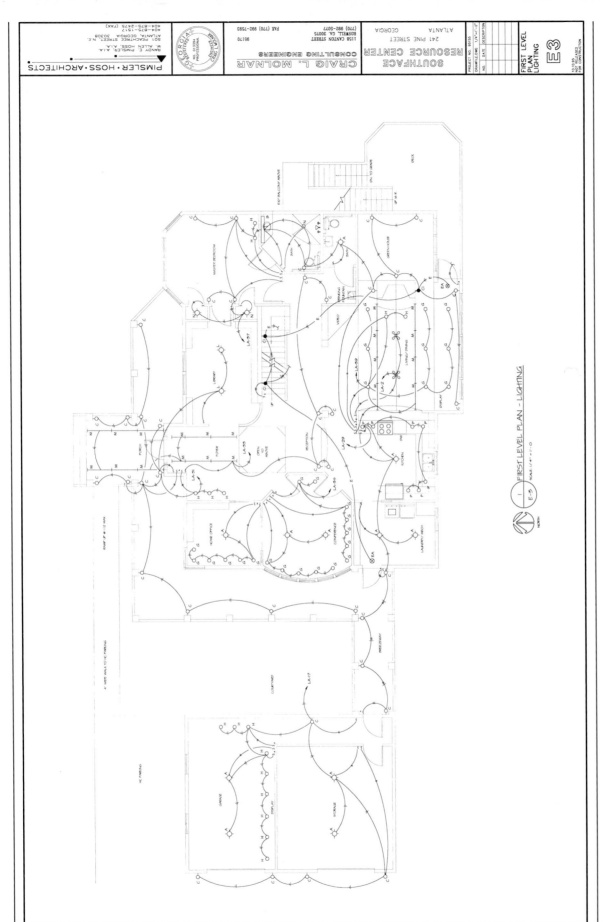

SOUTHFACE
RESOURCE CENTER

241 PINE STREET
ATLANTA
GEORGIA

FIRST LEVEL
PLAN
LIGHTING

E3

FIRST LEVEL PLAN - LIGHTING
SCALE 1/4"=1'-0

Figure 20-35

669

Figure 20-36

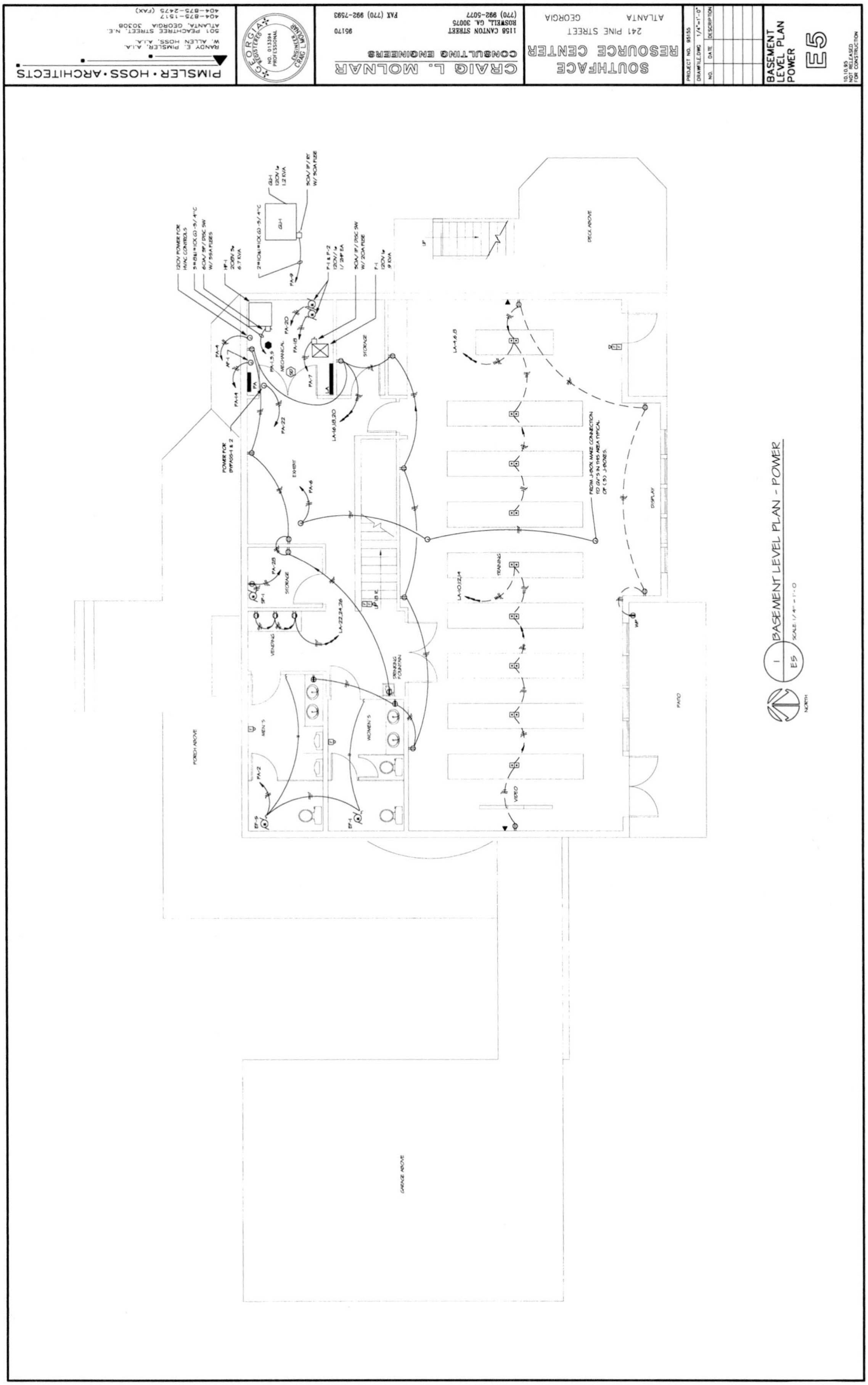

BASEMENT LEVEL PLAN - POWER
SCALE 1/4"=1'-0"

Figure 20-37

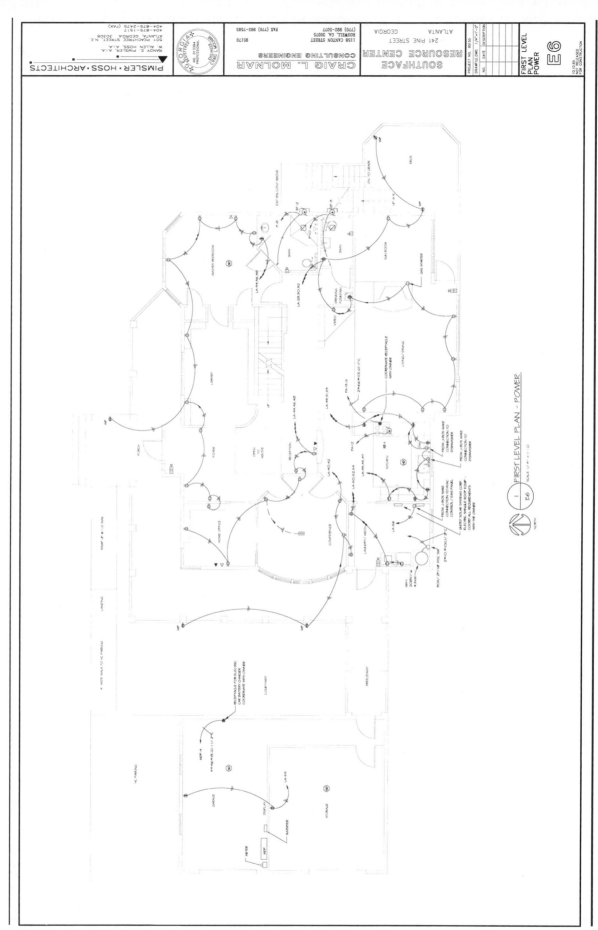

Figure 20-38

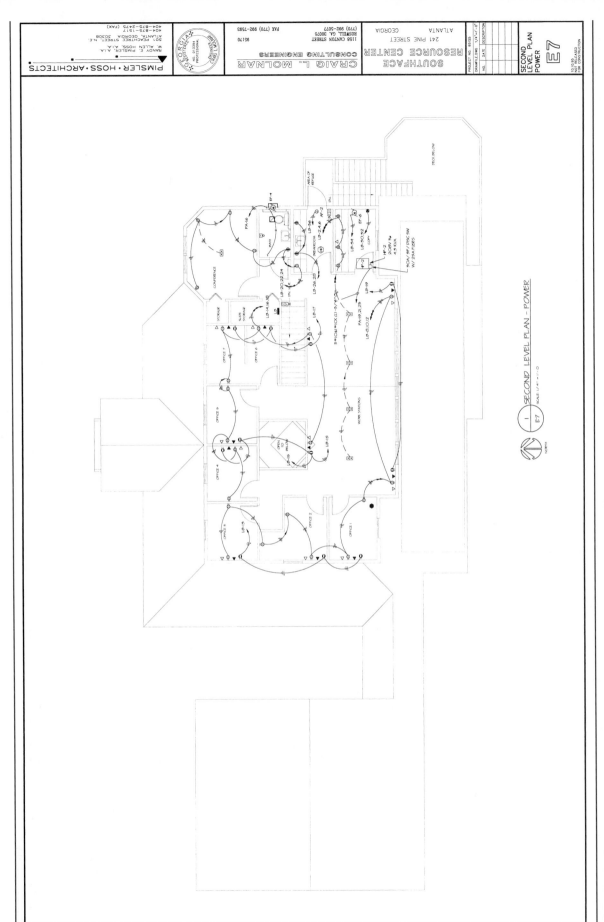

Figure 20-39

673

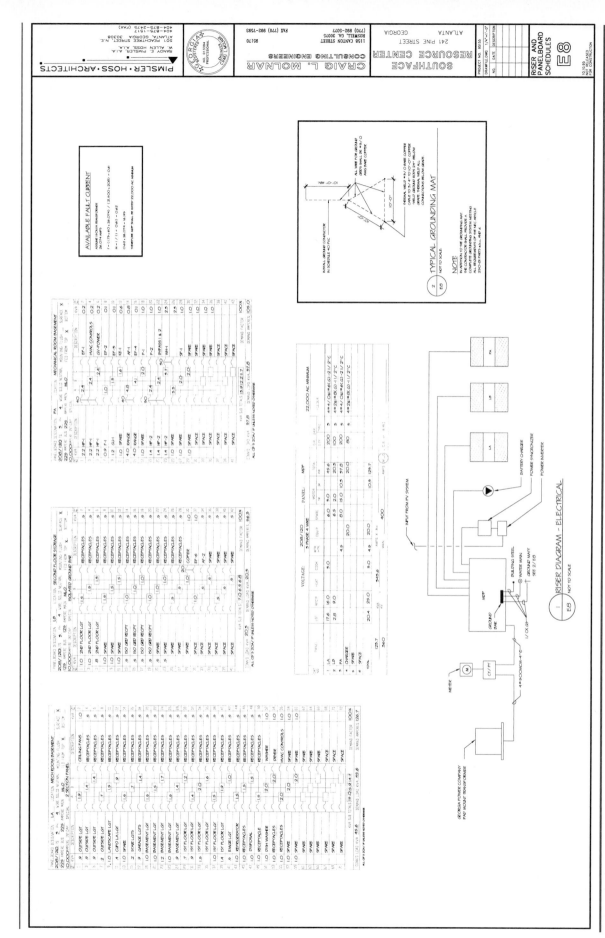

Figure 20-40

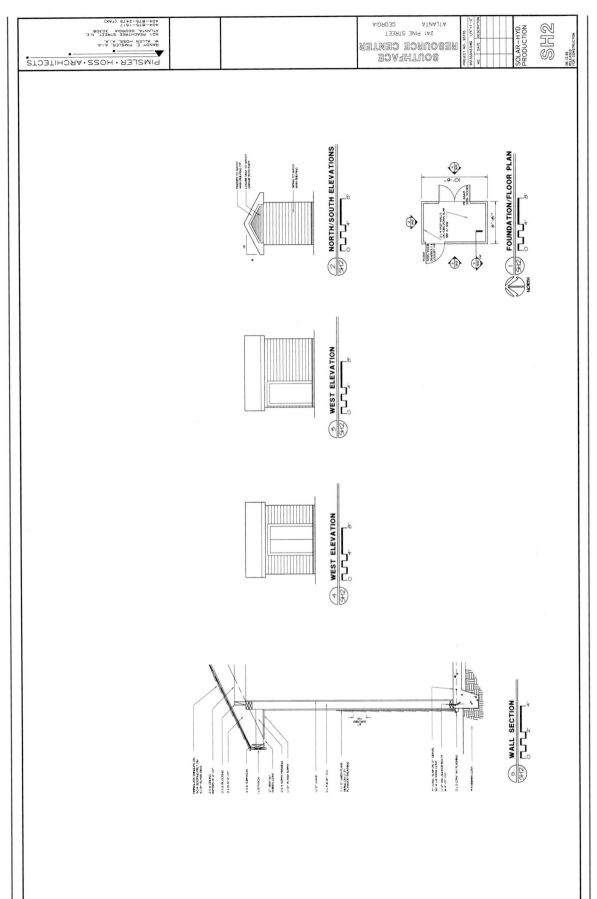

Figure 20-41

Exercises in planning and drawing commercial buildings are best undertaken under the supervision of your instructor, who can check them occasionally and offer helpful criticism during the development. Start all exercises with sketches.

1. *Wayside rest station.* Design and prepare the working drawings for a small restroom facility to be located on a level, 1-acre lot along a federal highway. Provide for the following facilities:
   (a) Parking for 10 cars, one to be handicapped accessible
   (b) Parking for 3 semitrailer trucks
   (c) Restrooms for men and for women, handicapped accessible
   (d) Trash receptacles
   (e) Water fountains
   (f) Two picnic tables and benches
   Draw the plan, the elevations, the site plan, and the necessary details.

2. *Suburban hardware store.* Design and make the working drawings for a small hardware store on a commercial lot 170' wide and 290' deep. Use C.M.U. bearing wall construction with a concrete slab-on-ground floor and a steel joist roof structure. Provide for the following facilities:
   (a) Floor area of 8000 sq ft
   (b) Display windows across front
   (c) Storage space in rear with unloading doors
   (d) Private office, bookkeeping, and lounge of 1000 sq ft on the second floor above the storage
   (e) Restrooms
   (f) Checkout counter and merchandise counters and racks
   (g) Parking spaces

3. *Small chapel.* Design and draw the working drawings for a small chapel, 6000 sq ft, intended for people of all faiths. Use wood bearing walls and trusses. The lot is 5 acres. Provide for the following facilities:
   (a) Sanctuary to seat 250 people
   (b) Two private offices of 175 sq ft each
   (c) Choir loft to seat 20 people
   (d) Restrooms for men and for women

   (e) Kitchen
   (f) Nursery
   (g) Three classrooms
   (h) Provision for mechanical equipment and janitorial care
   (i) Parking space for 70 cars

4. *Retail sales building.* Design and draw the working drawings for a laundry and retail sales building 65'-0" wide by 70'-0" deep. The lot is 130' wide by 200' deep, and it slopes to the front. Use C.M.U. and brick veneer load-bearing construction with open web roof joists. Provide the following facilities:
   (a) Laundry 1300 sq ft
   (b) Retail sales 2700 sq ft
   (c) Canopy across front of building
   (d) Toilets
   (e) Mechanical equipment room
   (f) Drive-up laundry pickup
   (g) Parking for 23 cars

5. *Library building.* Prepare the working drawings for a suburban library building with two floor levels consisting of a ground floor and a mezzanine. Use C.M.U. and brick veneer walls. The floor is to be poured concrete, and the roof is to be open web joist. The basic width is to be 54'-0", and the length 91'-0". The first-floor-to-floor height is to be 11'-4", and second-floor-to-roof height is to be 12'-8". Put columns on approximate 18'-0" centers; use one elevator, two separate stairwells, toilets, and storage. Reserve space for air-conditioning equipment. Reserve the first-floor level for entry, checkout, conference, receiving, periodicals, reference, and audiovisuals. On the mezzanine level, provide book stacks of various sizes with convenient circulation arrangements around the perimeter of the building.

   Draw the following views, using the scale 1/4" = 1'-0":
   (a) Two floor plans, first floor, and second-floor (mezzanine)
   (b) Two elevations
   (c) Site plan (convenient scale)
   (d) Transverse section

REVIEW QUESTIONS

1. Why must codes dealing with public buildings be carefully regulated?

2. List three general structural systems commonly used in commercial buildings.

3. Why are reflected ceiling plans commonly drawn for commercial buildings?

4. Why are enclosed vestibules often used in public entrances?

5. When laying out plan views of buildings with multiple floor levels, why should the drafter reserve suitable space for vertical plumbing and mechanical chases?

6. What detail is commonly drawn before the plans and the elevations are started?

7. What factors must the designer consider in deciding on the spacing of modular columns in planning a commercial building?

8. In dimensioning, how does the designer make the features in detail views relate to major features on plan and elevation views?

9. Why must masonry coursing sizes be accurately draw on details?

10. What types of professional services are usually engaged by architectural firms to complete their sets of working drawings for commercial buildings?

APPENDIX A: ABBREVIATIONS USED IN ARCHITECTURAL DRAWINGS

Alternate	ALT
Aluminum	AL or ALUM
American Institute of Architects	A.I.A.
American Institute of Architecture Students	A.I.A.S.
American Institute of Electrical Engineers	A.I.E.E.
American Institute of Steel Construction	A.I.S.C.
American National Standards Institute	A.N.S.I.
American Plywood Assoc.	A.P.A.
American Society of Heating, Refrigeration, and Air-Conditioning Engineers	A.S.H.R.A.E.
American Society of Testing and Materials	A.S.T.M.
American Standards Association, Inc.	A.S.A.
American Wire Gauge	AWG
Americans with Disabilities Act	A.D.A.
Apartment	APT
Architectural	ARCH
Association of General Contractors	A.G.C.
Avenue	AVE
Basement	BSMT
Bathroom	B
Beam	BM
Bedroom	BR
Bench mark	BM
Better	BTR
Blocking	BLKG
Board	BD
Board feet	BD FT
Board measure	BM
Boulevard	BLVD
British thermal unit	BTU
Building	BLDG
Building line	BL
Cabinet	CAB
Cast iron	CI
Ceiling	CLG
Cement	CEM
Center line	CL or ₵
Center to center	C to C
Centimeter(s)	cm
Clean out	CO
Closet	CLO
Column	COL
Combination	COMB
Composition	COMP
Concrete	CONC
Construction	CONST
Construction Specifications Institute	C.S.I.
Countersink	CSK
Cubic feet of air per minute	CFM
Cubic foot (feet)	CU FT
Cubic inch (inches)	CU IN
Cubic yard (yards)	CU YD
Detail	DET
Diagonal	DIAG
Diameter	DIA(M)
Dimension	DIM
Dining room	DR
Dishwasher	DW
Ditto	DO or ″
Divided (division)	DIV
Door	DR
Double-hung window	DHW
Double-strength grade A glass	DSA

Double-strength Grade B glass	DSB	Joint	JT
Douglas fir	DOUG FIR	Joist and plank (lumber designation)	J&P
Down	DN	Kiln-dried	KD
Downspout	DS	Kilowatt	kw
Dozen	DOZ	Kitchen	K
Drainage, waste, and vent	DWV	Laminate	LAM
Drawing	DWG	Landing	LDG
Dressed and matched (4 sides)	D4S	Laundry	LAU
Duplicate	DUP	Lavatory	LAV
Each	EA	Left hand	LH
East	E	Length	L
Elevation	ELEV	Light (pane of glass)	LT
Enclosure	ENCL	Linear feet	LIN FT
Entrance	ENT	Linen closet	L CL
Equipment	EQUIP	Living room	LR
Exterior	EXT	Louver	LVR
Exterior insulation finish system	E.I.F.S.	Lumber	LBR
Fabricate	FAB	Manufacture(r)	MFR
Fahrenheit	F	Material(s)	MTL
Federal Housing Authority	FHA	Maximum	MAX
Feet per minute	FPM	Mechanical	MECH
Finish	FIN	Medicine cabinet	MC
Finish all over	FAO	Medium	MED
Flashing	FLG	Meter(s)	m
Floor	FL	Millimeter(s)	mm
Foot board measure	FBM	Minimum	MIN
Footing	FTG	Miscellaneous	MISC
Forest Products Association	F.P.A.	Modular	MOD
Foundation	FDN	Molding	MLDG
Furring	FUR	National Bureau of Standards	N.B.S.
Gallon	GAL	National Environmental Systems Contractors Association	NESCA
Galvanized	GALV		
Galvanized iron	GI		
Gauge	GA	National Housing Authority	NHA
General contractor	GC	National Lumber Manufacturer's Association	N.L.M.A.
Glass	GL		
Gypsum	GYP		
Head	HD	National Warm Air Heating and Air Conditioning Contractors Assoc.	N.W.A.H.A.C.C.A.
Heat- and moisture-resistant rubber insulation	RHW		
Height	HGT	Nominal	NOM
Hemlock	HEM	North	N
Hexagonal	HEX	Not in contract	NIC
Home Builders Association	H.B.A.	Number	NO. or #
Horizon line	HL	Office	OFF
Horizontal	HORZ	On center	OC
Horsepower	HP	Opening	OPG
Hose bibb	HB	Ounce(s)	OZ
Hundred	C	Overhead	OH
I beam	S or I (old designation)	Paint(ed)	PNT
		Parallel	PAR or ‖
Inch(es)	IN or ″	Partition	PART
Inside diameter	ID	Penny (nail)	d
Institute of Boiler and Radiator Manufacturers	I.B.R.	Pi	π
		Picture plane	PP
Insulate (insulation)	INS	Piece	PC
Interior	INT	Plaster	PLAS or PL

Plate	PL or P$_L$	Telephone	TEL
Plumbing	PLBG	Television	TV
Plywood	PWD	Temperature	TEMP
Pounds per cubic foot	PCF	Terra-cotta	TC
Pounds per square inch	PSI	Terrazzo	TZ
Prefabricate(d)	PFB	Thousand	M
Pressure-treated	PT	Thousand board feet	MBM
Property line	PL	Threshold	THR
Quart(s)	QT	Tongue and groove	T & G
Radiator	RAD	Tread	T or TR
Radius	RA	Typical	TYP
Receptacle	RECP	Unexcavated	UNEX
Refrigerator	REFR	Unfinished	UNFIN
Revolutions per minute	RPM	U.S. standard gauge	USG
Riser	R	Vanishing point	VP
Road	RD	Ventilation	VENT
Room	RM	Vertical	VERT
Rough opening	RO	Volts	V
Round	RD	Volume	VOL
Rubber base	RB	Water closet	WC
Schedule	SCH	Water heater	WH
Second(s)	s	Waterproof(ing)	WP
Shower	SH	Weight	WT
Single-strength grade		Welded wire fabric	WWF
B glass	SSB	Welded wire mesh	WWM
Sink	SK	West Coast Lumber Inspection	
South	S	Bureau	W.C.L.I.B.
Southern Pine Inspection		Western Wood Products	
Bureau	S.P.I.B.	Assoc.	W.W.P.A.
Southern yellow pine	SYP	Wide flange (steel)	W, WF (old
Specifications	SPEC		designation)
Square foot (feet)	SQ FT	Window	WDW
Standard	STD	With	W/
Station point	SP	Wood	WD
Street	ST	Yard(s)	YD
Surface 4 sides	S4S	Yellow pine	YP

APPENDIX B:
MODULAR VERTICAL BRICK COURSING

Module	Course	Dimension	Module	Course	Dimension
	1C	$2^5/_8$″		37C	$8'\text{-}2^5/_8$″
	2C	$5^3/_8$″		38C	$8'\text{-}5^3/_8$″
1M	3C	8″	13M	39C	$8'\text{-}8$″
	4C	$10^5/_8$″		40C	$8'\text{-}10^5/_8$″
	5C	$1'\text{-}1^3/_8$″		41C	$9'\text{-}1^3/_8$″
2M	6C	$1'\text{-}4$″	14M	42C	$9'\text{-}4$″
	7C	$1'\text{-}6^5/_8$″		43C	$9'\text{-}6^5/_8$″
	8C	$1'\text{-}9^3/_8$″		44C	$9'\text{-}9^3/_8$″
3M	9C	$2'\text{-}0$″	15M	45C	$10'\text{-}0$″
	10C	$2'\text{-}2^5/_8$″		46C	$10'\text{-}2^5/_8$″
	11C	$2'\text{-}5^3/_8$″		47C	$10'\text{-}5^3/_8$″
4M	12C	$2'\text{-}8$″	16M	48C	$10'\text{-}8$″
	13C	$2'\text{-}10^5/_8$″		49C	$10'\text{-}10^3/_8$″
	14C	$3'\text{-}1^3/_8$″		50C	$11'\text{-}1^3/_8$″
5M	15C	$3'\text{-}4$″	17M	51C	$11'\text{-}4$″
	16C	$3'\text{-}6^5/_8$″		52C	$11'\text{-}6^5/_8$″
	17C	$3'\text{-}9^3/_8$″		53C	$11'\text{-}9^3/_8$″
6M	18C	$4'\text{-}0$″	18M	54C	$12'\text{-}0$″
	19C	$4'\text{-}2^5/_8$″		55C	$12'\text{-}2^5/_8$″
	20C	$4'\text{-}5^3/_8$″		56C	$12'\text{-}5^3/_8$″
7M	21C	$4'\text{-}8$″	19M	57C	$12'\text{-}8$″
	22C	$4'\text{-}10^5/_8$″		58C	$12'\text{-}10^5/_8$″
	23C	$5'\text{-}1^3/_8$″		59C	$13'\text{-}1^3/_8$″
8M	24C	$5'\text{-}4$″	20M	60C	$13'\text{-}4$″
	25C	$5'\text{-}6^5/_8$″		61C	$13'\text{-}6^5/_8$″
	26C	$5'\text{-}9^3/_8$″		62C	$13'\text{-}9^3/_8$″
9M	27C	$6'\text{-}0$″	21M	63C	$14'\text{-}0$″
	28C	$6'\text{-}2^5/_8$″		64C	$14'\text{-}2^5/_8$″
	29C	$6'\text{-}5^3/_8$″		65C	$14'\text{-}5^3/_8$″
10M	30C	$6'\text{-}8$″	22M	66C	$14'\text{-}8$″
	31C	$6'\text{-}10^5/_8$″		67C	$14'\text{-}10^5/_8$″
	32C	$7'\text{-}1^3/_8$″		68C	$15'\text{-}1^3/_8$″
11M	33C	$7'\text{-}4$″	23M	69C	$15'\text{-}4$″
	34C	$7'\text{-}6^5/_8$″		70C	$15'\text{-}6^5/_8$″
	35C	$7'\text{-}9^3/_8$″		71C	$15'\text{-}9^3/_8$″
12M	36C	$8'\text{-}0$″	24M	72C	$16'\text{-}0$″

NOTE: A *module* corresponds to 8″ concrete block course heights.

APPENDIX C:
THE METRIC SYSTEM IN CONSTRUCTION

The following guidelines should help you "see and think metric":[1]

- 1″ is just a fraction longer (1/64″) than 25 mm (1″ = 25.4 mm).
- 4″ are about 1/16″ longer than 100 mm.
- 1′ is about 3/16″ longer than 300 mm.
- 4′ are about 3/4″ longer than 1200 mm.

---

[1] "Seeing Metric," *Metric in Construction* (Sept./Oct. 1992).

- 1 millimeter (mm) is slightly less than the thickness of a dime.
- 1 meter (m) is the length of a yardstick plus about 3.3″.
- 1 gram (g) is about the mass (weight) of a large paper clip.
- 10 m equals approximately 11 yd.
- 10 m^2 equals approximately 12 sq yd.
- 10 m^3 equals approximately 13 cu yd.
- The metric equivalent of a typical 2′ × 4′ ceiling grid is 600 × 1200 mm.
- The metric equivalent of a 4′ × 8′ sheet of plywood or drywall is 1200 × 2400 mm.

Table C–1 Converting from inch and pound units to metric units.

| Measurement | From In./lb Units | To Metric | | Multiply by |
		Unit	Symbol	
Length	mile	kilometer	km	1.609 34 4
	yard	meter	m	0.914 4
	foot	meter	m	0.304 8
		millimeter	mm	304.8
	inch	millimeter	mm	25.4
Area	square mile	square kilometer	km^2	2.590 00
	acre	square meter	m^2	4.046 87
		hectare	ha	0.404 687
	square yard	square meter	m^2	0.836 127 36
	square foot	square meter	m^2	0.092 903 04
	square inch	square millimeter	mm^2	645.16
Volume	cubic yard	cubic meter	m^3	0.764 555
	cubic foot	cubic meter	m^3	0.028 316 8
		liter	L	28.316 85
	cubic inch	milliliter	mL	16.387 064
		cubic millimeter	mm^3	16 387.064

Table C–2 Metric/inch-foot length equivalents.

mm	in.	mm	ft	m	ft
1.5	1/16	300	1'-0"	1.0	3'-3"
3	1/8	400	1'-4"	1.2	4'-0"
6	1/4	450	1'-6"	2.0	6'-0"
8	3/8	600	2'-0"	2.4	8'-0"
10	3/8	750	2'-6"	3.0	10'-0"
12	1/2	900	3'-0"	4.8	16'-0"
15	5/8	1200	4'-0"	6.0	20'-0"
20	3/4	1500	5'-0"	8.0	26'-0"
25	1	1800	6'-0"	12.0	40'-0"
40	1 1/2	2400	8'-0"		
45	1 3/4	3000	10'-0"		
50	2				
65	2 1/2				
75	3				
89	3 1/2				
95	3 3/4				
100	4				
125	5				
140	5 1/2				
150	6				
200	8				
225	9				

Table C–3 Comparison between inch-foot and metric scales.

In.-Ft Scales	Ratios	Metric Scales		Remarks
		Preferred	Other	
Full Size	1:1	1:1		No change
Half full size	1:2		1:2	No change
4″ = 1′-0″	1:3			
3″ = 1′-0″	1:4			
		1:5		Close to 3″ scale
2″ = 1′-0″	1:6			
1½″ = 1′-0″	1:8			
		1:10		Between 1″ and 1½″ scales
1″ = 1′-0″	1:12			
¾″ = 1′-0″	1:16			
		1:20		Between ½″ and ¾″ scales
½″ = 1′-0″	1:24			
			1:25	Very close to ½″ scale
⅜″ = 1′-0″	1:32			
¼″ = 1′-0″	1:48			
		1:50		Close to ¼″ scale
1″ = 5′-0″	1:60			
3/16″ = 1′-0″	1:64			
⅛″ = 1′-0″	1:96			
		1:100		Very close to ⅛″ scale
1″ = 10′-0″	1:120			
3/32″ = 1′-0″	1:128			
1/16″ = 1′-0″	1:196			
		1:200		Close to 1/16″ scale
1″ = 20′-0″	1:240		1:250	Close to 1″ = 20′0″
1″ = 30′-0″	1:360			
1/32″ = 1′-0″	1:384			
1″ = 40′-0″	1:480			
		1:500		Close to 1″ = 40′0″ scale
1″ = 50′-0″	1:600			
1″ = 60′-0″	1:720			
1″ = 1 chain	1:792			
1″ = 80′-0″	1:960			
		1:1000		

SOURCE: *Metric Guide for Federal Construction.*

NOTES:

1. Metric drawing scales are expressed in nondimensional ratios.
2. Use only one unit of measure on a drawing; on large-scale site drawings, the unit should be the millimeter (mm).
3. Delete unit symbols, but provide an explanatory note ("All dimensions are shown in millimeters," for example).
4. Whole numbers always indicate millimeters; decimal numbers taken to three places always indicate meters.
5. Where basic modules are used, the recommended basic module is 100 mm, which is similar to the 4″ module (4″ = 101.6 mm).

Table C–4 Metric units used by the construction trades.

Trade	Quantity	Unit	Symbol
Surveying	length	kilometer, meter	km, m
	area	square kilometer hectare (10 000 m^2) square meter	km^2 ha (hm^2) m^2
	plane angle	degree minute second percent	° ' " %
Excavating	length volume	meter, millimeter cubic meter	m, mm m^3
Paving	length area	meter, millimeter square meter	m, mm m^2
Concrete	length area volume temperature water capacity mass cross-sectional area	meter, millimeter square meter cubic meter degree Celsius liter (cubic decimeter) megagram (metric ton) kilogram square millimeter	m, mm m^2 m^3 °C L (dm^3) Mg (t) kg mm^2
Trucking	distance volume mass	kilometer cubic meter megagram (metric ton)	km m^3 Mg (t)
Masonry	length area mortar volume	meter, millimeter square meter cubic meter	m, mm m^2 m^3
Steel	length mass mass per unit length	meter, millimeter megagram (metric ton) kilogram kilogram per meter	m, mm Mg (t) kg kg/m
Carpentry	length	meter, millimeter	m, mm
Plastering	length area water capacity	meter, millimeter square meter liter (cubic decimeter)	m, mm m^2 L (dm^3)
Glazing	length area	meter, millimeter square meter	m, mm m^2

Table C–4 (*continued*)

Trade	Quantity	Unit	Symbol
Painting	length	meter, millimeter	m, mm
	area	square meter	m^2
	capacity	liter (cubic decimeter)	L (dm^3)
		milliliter (cubic centimeter)	mL (cm^3)
Roofing	length	meter, millimeter	m, mm
	area	square meter	m^2
	slope	percent	%
		ratio of lengths	mm/mm, m/m
Plumbing	length	meter, millimeter	m, mm
	mass	kilogram, gram	kg, g
	capacity	liter (cubic decimeter)	L (dm^3)
	pressure	kilopascal	kPa
Drainage	length	meter, millimeter	m, mm
	area	hectare (10 000 m^2)	ha
		square meter	m^2
	volume	cubic meter	m^3
	slope	percent	%
		ratio of lengths	mm/mm, m/m
HVAC	length	meter, millimeter	m, mm
	volume (capacity)	cubic meter	m^3
		liter (cubic decimeter)	L (dm^3)
	air velocity	meter/second	m/s
	volume flow	cubic meter/second	m^3/s
		liter/second (cubic decimeter per second)	L/s (dm^3/s)
	temperature	degree Celsius	°C
	force	newton, kilonewton	N, kN
	pressure	pascal, kilopascal	Pa, kPa
	energy	kilojoule, megajoule	kJ, MJ
	rate of heat flow	watt, kilowatt	W, kW
Electrical	length	millimeter, meter, kilometer	mm, m, km
	frequency	hertz	Hz
	power	watt, kilowatt	W, kW
	energy	megajoule	MJ
		kilowatt hour	kWh
	electric current	ampere	A
	electric potential	volt, kilovolt	V, kV
	resistance	milliohm, ohm	mΩ, Ω

Source: *Metric in Construction* (March/April 1996)

Table C–5 Metric and U.S. inch softwood lumber sizes.

Metric Size (mm)	U.S. Nominal Size (in.)	U.S. Manufactured Size (in.)
25 × 100	1 × 4	¾ × 3½
25 × 150	1 × 6	¾ × 5½
25 × 200	1 × 8	¾ × 7¼
25 × 250	1 × 10	¾ × 9¼
25 × 300	1 × 12	¾ × 11¼
50 × 100	2 × 4	1½ × 3½
50 × 150	2 × 6	1½ × 5½
50 × 200	2 × 8	1½ × 7¼
50 × 250	2 × 10	1½ × 9¼
50 × 300	2 × 12	1½ × 11¼
100 × 100	4 × 4	3½ × 3½
100 × 150	4 × 6	3½ × 5½
100 × 200	4 × 8	3½ × 7¼
100 × 250	4 × 10	3½ × 9¼
100 × 300	4 × 12	3½ × 11¼

Table C–6 Lumber lengths.

m	ft	m	ft
1.8	6	3.9	
2.1	7	4.2	14
2.4	8	4.5	
2.7		4.8	16
3.0	10	5.1	
3.3		5.4	18
3.6	12	5.7	
		6.0	20

Table C–7 Metric steel stud lengths and their U.S. inch equivalents.

Metric (mm)	U.S. (ft and in.)	Metric (mm)	U.S. (ft and in.)
2 134	7'-0"	3 505	11'-6"
2 286	7'-6"	3 658	12'-0"
2 438	8'-0"	3 810	12'-6"
2 591	8'-6"	3 962	13'-0"
2 743	9'-0"	4 115	13'-6"
2 896	9'-6"	4 267	14'-0"
3 048	10'-0"	4 420	14'-6"
3 200	10'-6"	4 572	15'-0"
3 353	11'-0"		

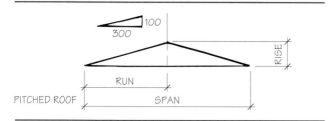

Figure C-1

Table C-8 Roof pitches.

Traditional	Metric
2/12	50/300
4/12	100/300
6/12	150/300
8/12	200/300
10/12	250/300
12/12	300/300

Table C-9 *U* factors for insulating materials.

Insulating Materials	*U* factor[a] (in.-lb)	Metric	*Thickness of Material* in.	mm
Batt blanket	0.24	1.36	1½	38
Loose wool	0.26	1.48	4	100
Vermiculite	0.24	1.36	1	25
Urethane panel	0.16	0.91	1	25
Polystyrene foam	0.20	1.14	1	25
Foam glass	0.35	1.99	1	25
Insulating board	0.36	2.04	1	25

[a]Inch-pound *U* factor × 5.678 (factor) = metric *U* factor.

Table C-10 Metric dimensions and inch equivalents of roof insulating boards.

Panels and Sheathing		Roof Boards		
Thickness:	11 mm (7/16 in.)	Thickness:	12.5 mm (0.49 in.)	
			25.5 mm (0.98 in.)	
Width:	1200 mm (47¼ in.)		38 mm (1.50 in.)	
			50 mm (1.97 in.)	
Length:	2400 mm (7 ft 10½ in.)	Width:	600 mm (23.62 in.)	
		Length:	1200 mm (47.24 in.)	

Table C-11 Gypsum board: practical metric dimensions.

Thickness	Width	Length
9.5 mm (3/8 in.)	400 mm	1200 mm (47 1/4 in.)
12.7 mm (1/2 in.)	600 mm	2400 mm (7 ft 10 1/2 in.)
		2800 mm (9 ft 2 1/4 in.)
		3000 mm (9 ft 10 in.)
15.9 mm (5/8 in.)	1200 mm	3600 mm (11 ft 9 3/4 in.)
19.0 mm (3/4 in.)		4200 mm (13 ft 9 1/3 in.)
25.4 mm (1 in.)		

Table C-12 Stair riser heights and tread widths[a].

Rise (mm)	Tread (mm)	Two Risers + One Tread (mm)
120	400	640
120	390	640
130	380	640
130	370	630
140	350	630
150	330	630
150	320	620
160	310	630
170	300	640
170	290	630
180	270	630
180	260	620
190	250	630
200	240	640
200	230	630
200	230	630

[a]Metric riser heights and tread widths are in millimeters;
2 risers + 1 tread = 610 to 640 mm.

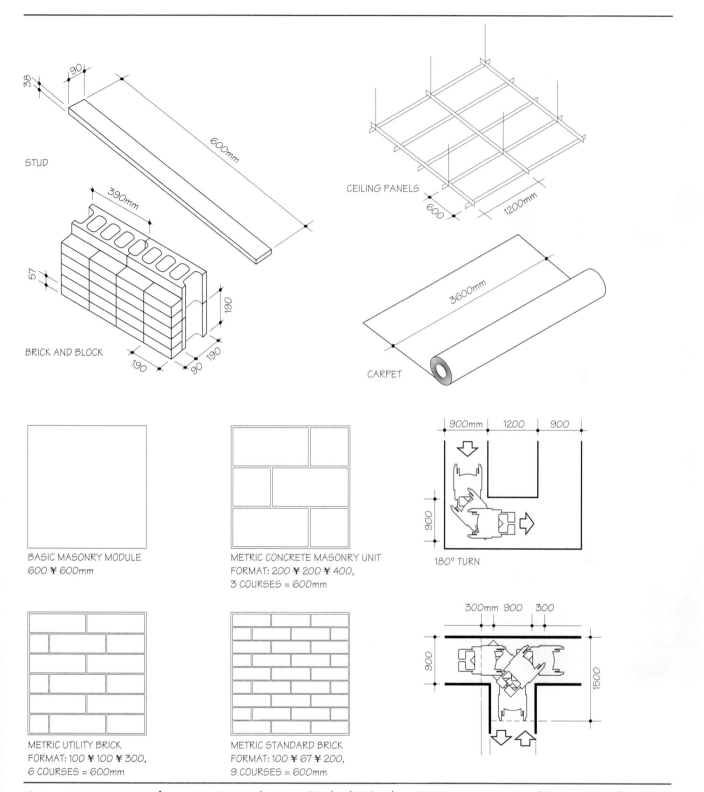

STUD

CEILING PANELS

600mm

600

1200mm

390mm

38
90

57

190

190

190

90

BRICK AND BLOCK

3600mm

CARPET

BASIC MASONRY MODULE
600 ¥ 600mm

METRIC CONCRETE MASONRY UNIT
FORMAT: 200 ¥ 200 ¥ 400,
3 COURSES = 600mm

900mm   1200   900

900

180° TURN

METRIC UTILITY BRICK
FORMAT: 100 ¥ 100 ¥ 300,
6 COURSES = 600mm

METRIC STANDARD BRICK
FORMAT: 100 ¥ 67 ¥ 200,
9 COURSES = 600mm

300mm 900   300

900

1500

Figure C–2 Metric sizes for construction applications. (Michael J. Crosbie, "Moving to Metric," *Architecture* September 1992)

Table C-13 Metric minimum and desirable room sizes.

Room	Minimum Size (mm)	m²	Desirable Size (mm)	m²
Kitchen plus dining	4000 × 4000	16		
	2500 × 5500	13.75	3500 × 6000	21
	3000 × 4000	12		
	3000 × 3000	9	3000 × 5000	15
	2500 × 4000	10		
	2500 × 3000	7.5	3000 × 3500	10.5
Kitchen plus family	4000 × 5500	22		
	4000 × 6000	24		
	4000 × 4000	16	4000 × 6000	24
	4000 × 5000	20	4000 × 7000	28
Laundry and utility	2500 × 2500	6.25		
	2000 × 4000	8		
	2500 × 3000	7.5		
	3000 × 3000	9		
	2000 × 3000	6		
Entry, rear	1000 × 1500	1.5		
	1000 × 1200	1.2		
Hall bath	1000 × 1000	1	2000 × 1500	3
	1200 × 1500	4.2		
	900 × 1200	1.08		
Dining room	3000 × 3000	9		
	3000 × 4000	12	3500 × 1500	12.25
	4000 × 4500	18	4000 × 5000	20
	3500 × 4000	14		
Living room	4000 × 5000	20	4000 × 6000	24
	4000 × 6000	24	4500 × 6000	27
Front-entrance foyer	1200 × 1800	2.16		
	2000 × 2400	4.8	2500 × 2400	6
	1600 × 1600	2.56		
Guest coat closet	720 × 1500	1.14		
	610 × 920	0.56		
	700 × 1200	0.84		
	800 × 1200	0.96		
	600 × 1200	0.72		
Den or guest room	3000 × 4000	12	4000 × 4000	16
	4000 × 4000	16	3000 × 4500	13.5

Table C–13 (*continued*).

Room	Minimum Size (mm)	m^2	Desirable Size (mm)	m^2
Master bedroom	3000 × 4000	12	4000 × 4000	16
	4000 × 4000	16	4000 × 5000	20
	4000 × 4500	18		
Closet for master bedroom	2@760 × 1500	2.28	2@730 × 1500	2.19
	2@610 × 1800	2.18	2@910 × 1800	3.27
	2@610 × 1500	1.23	2@910 × 1800	3.27
	2@700 × 1500	2.10	2@680 × 1500	2.04
Bedroom 1	3000 × 4000	12	3500 × 4000	14
	3000 × 3500	10.5	3050 × 3350	10.2
	4000 × 4000	16	4000 × 5000	20
Closet for bedroom 1	750 × 1500	1.12	2@760 × 1500	2.28
	600 × 1800	1.08	910 × 1800	1.65
	600 × 1500	0.9		
	2@750 × 1500	2.24	2@760 × 1500	2.28
Bedroom 2	2400 × 3660	8.78		
	2740 × 3350	9.18		
	3050 × 6100	18.6	3600 × 3660	13.39
	3050 × 3350	10.22	3050 × 3350	10.22
	3660 × 4300	15.74	4260 × 4880	20.78
Closet for bedroom 2	760 × 1500	1.14	2@760 × 1520	2.74
	600 × 1800	1.08	910 × 1830	1.66
	2@700 × 1500	2.10	2@670 × 1520	2.04
	600 × 1500	0.9	2@610 × 1520	1.85
Bath	1500 × 3000	4.5		
	1500 × 2500	3.75		
	2000 × 2500	5.0		
	1500 × 2000	3.0	2400 × 2400	5.76
	2000 × 2000	4.0		
	1800 × 2000	3.6	2000 × 3000	6
Garage, 2-car	7000 × 7000	49		
	5500 × 6000	33	6000 × 7000	42
Garage, 1-car	4500 × 6000	27		
	4000 × 6000	24		
	4000 × 7000	28		

Table C–14 Furniture sizes for the house.

Furniture	Length (mm)	Width (mm)	Height (mm)
Bedroom			
Single bed	2000	950	520
Night table	500	400	650
	550	350	650
King-size bed	2000	1900	520
Double bed	2000	1400–1800	520
Children's bed	1600	750	520
	1700	800	520
Kitchen and Laundry			
Washer, automatic	660–720	650–750	910
Dryer	660–690	760–790	910
Refrigerator	610–850	690–780	1420–1800
Living and Dining Rooms			
Table	1200	800	750
	1500	900	780
Round table	1100		780
Writing table	1300	700	780
	1500	800	780
Grand piano	1600	1500	1100
Sofa	1900	800	430 (seat)
	1750	800	430 (seat)
Chair	450	450	430 (seat)
Armchair	650	600	350 (seat)

C.1

METRIC: WHAT WILL CHANGE AND WHAT WILL STAY THE SAME[2]

C.1.1 Metric Module and Grid

What Will Change

- The basic building module, from 4″ to **100 mm.**
- The planning grid, from 2′ × 2′ to **600 × 600 mm.**

What Will Stay the Same

- A module and grid based on rounded, easy-to-use dimensions.

C.1.2 Drawings

What Will Change

- Units, from feet and inches to millimeters for all building dimensions and to meters for large site plans and civil engineering drawings. Unit notations are unnecessary: If there's no decimal point, it's millimeters; if there's a decimal point carried to one, two, or three places, it's meters. Centimeters are not used in construction.

- Drawing scales, from inch-fractions-to-feet to true ratios. Preferred metric scales are **1:1** (full size); **1:5** (close to 3″ = 1′-0″); **1:10** (between 1″ = 1′-0″ and 1½″ = 1′-0″); **1:20** (between ½″ = 1′-0″ and ¾″ = 1′-0″); **1:50** (close to ¼″ = 1′-0″); **1:100** (close to ⅛″ = 1′-0″); **1:200** (close to ¹⁄₁₆″ = 1′-0″); **1:500** (close to 1″ = 40′-0″); **1:1000** (close to 1″ = 80′-0″).

---

2 *Metric in Construction* (July/August 1995).

- Drawing sizes, to the ISO "A" series: **A0** (**1189 × 841 mm**, 46.8 × 33.1 in.); **A1** (**841 × 594 mm**, 33.1 × 23.4 in.); **A2** (**594 × 420 mm**, 23.4 × 16.5 in.); **A3** (**420 × 297 mm**, 16.5 × 11.7 in.); **A4** (**297 × 210 mm**, 11.7 × 8.3 in.). Of course, metric drawings can be made on any size paper.

What Will Stay the Same

- Drawing contents.

Never use dual units (both inch-pound and metric) on drawings. It increases dimensioning time, doubles the chance for errors, makes drawings more confusing, and delays the learning process.

C.1.3 Specifications
What Will Change

- Units of measure, from feet and inches to **millimeters** for linear dimensions, from square feet to **square meters** for area, from cubic yards to **cubic meters** for volume (except use **liters** for fluid volumes), and from other inch-pound units to metric units as appropriate.

What Will Stay the Same

- Everything else in the specification.

Do not use dual units in specifications except when the use of an inch-pound measure serves to clarify an otherwise unfamiliar metric measure. Then place the inch-pound unit in parentheses after the metric, for example, "7.5 kW (10 horsepower)." All unit conversions should be **checked by a professional** to ensure that rounding does not exceed allowable tolerances.

C.1.4 Floor Loads
What Will Change

- Floor load designations, from psf to kilograms per square meter (**kg/m²**) for everyday use and kilonewtons per square meter (**kN/m²**) for structural calculations.

What Will Stay the Same

- Floor load requirements.

Kilograms per square meter often are used to designate floor loads because many live and dead loads (furniture, filing cabinets, construction materials, and so on) are measured in kilograms. However, kilonewtons per square meter or kilopascals, their equivalent, are the proper measure and should be used in structural calculations.

C.1.5 Construction Products
What Will Change

- Modular products: brick, block, drywall, plywood, suspended ceiling components, and raised floors. They will undergo "hard" conversion; that is, their dimensions will change to new rounded "hard" metric numbers to fit the universal **600 × 600 mm** metric planning grid.

- A number of other products, such as concrete reinforcing bars and various kinds of fasteners. They are being converted to hard metric sizes as the result of industry initiatives.

- Products that are custom fabricated for each job (for example, cabinets, stairs, handrails, ductwork, commercial doors and windows, structural steel, and precast concrete) or poured-in-place (concrete). Such products can usually be made to any size, inch-pound or metric, with equal ease, so for metric jobs, they will simply be fabricated or formed in metric.

What Will Stay the Same

- The balance of products, since they are cut to fit at the job site (for example, framing lumber, woodwork, wiring, piping, and roofing) or are not dimensionally sensitive (for example, fasteners, hardware, electrical components, plumbing fixtures, HVAC equipment, and gravel). Such products will be just "soft" converted—that is, relabeled in metric. A 2¾″ × 4½″ wall-switch face plate will be relabeled 70 × 115 mm, and a 30-gal tank, 114 L. Eventually manufacturers may convert many of these products to new rounded "hard" metric sizes, but only when it becomes convenient to do so.

C.1.6 Studs and Other 2× Framing (Both Wood and Metal)
What Will Change

- Spacing, from 16″ to **400 mm** and from 24″ to **600 mm.**

What Will Stay the Same

- Cross sections.

2 × s are produced in fractional-inch dimensions now, so there is no need to convert them to new rounded "hard" metric dimensions; 2 × 4s may keep their traditional name, or perhaps they'll be relabeled a nominal 50 × 100 mm or a more exact size, such as 38 × 89 mm.

C.1.7 Drywall, Plywood, and Other Sheet Goods
What Will Change

- Widths, from 4′-0″ to **1200 mm.**
- Heights, from 8′-0″ to **2400 mm** and from 10′-0″ to **3000 mm.**

What Will Stay the Same

- Thicknesses, so fire, acoustic, and thermal ratings will not have to be recalculated.

Metric drywall and plywood are readily available, but with a possible cost penalty for small orders. Metric rigid insulation may not be available at this time.

C.1.8 Batt Insulation

What Will Change

- Nominal width labels, from 16″ to **16″/400 mm** and from 25″ to **24″/600 mm.**

What Will Stay the Same

- Everything else.

Batts will not change in width; they will simply have a tighter "friction fit" when installed among metric-spaced framing members.

C.1.9 Doors

What Will Change

- Height, from 6′-8″ to **2050 mm** or **2100 mm** and from 7′-0″ to **2100 mm.**
- Width, from 2′-6″ to **750 mm,** from 2′-8″ to **800 mm,** from 2′-10″ to **850 mm,** from 3′-0″ to **900 mm** or **950 mm,** and from 3′-4″ to **1000 mm.**

What Will Stay the Same

- Door thicknesses.
- Door materials and hardware.

For commercial work, doors can be ordered in any size since they normally are custom fabricated.

C.1.10 Ceiling Systems

What Will Change

- Grids and lay-in ceiling tile, air diffusers, and lighting fixtures—from 2′ × 2′ to **600 × 600 mm** and 2′ × 4′ to **600 × 1200 mm.**

What Will Stay the Same

- Grid profiles, tile thicknesses, air diffuser capacities, fluorescent tubes, and means of suspension.

C.1.11 Raised Floor Systems

What Will Change

- Grids and lay-in floor tile, from 2′ × 2′ to **600 × 600 mm.**

What Will Stay the Same

- Grid profiles, tile thicknesses, and means of support.

C.1.12 HVAC Controls

What Will Change

- Temperature units, from Fahrenheit to Celsius.

What Will Stay the Same

- All other parts of the controls.

Controls are now digital, so temperature conversions can be made with no difficulty.

C.1.13 Brick

What Will Change

- Standard brick to **90 × 57 × 190 mm.**
- Mortar joints, from ⅜″ to ½″ to **10 mm.**
- Brick module, from 2′ × 2′ to **600 × 600 mm.**

What Will Stay the Same

- Brick and mortar composition.

Of the 100 or so brick sizes currently made, 5 to 10 are within a millimeter of a metric brick, so the brick industry will have no trouble supplying metric brick.

C.1.14 Concrete Block

What Will Change

- Block sizes to **190 × 190 × 390 mm.**
- Mortar joints, from ½″ to **10 mm.**
- Block module, from 2′ × 2′ to **600 × 600 mm.**

What Will Stay the Same

- Block and mortar composition.

C.1.15 Sheet Metal

What Will Change

- Designation from "gauge" to millimeters.

What Will Stay the Same

- Thickness, which will be soft converted to tenths of a millimeter.

In specifications, use millimeters only or millimeters with the gauge in parentheses.

C.1.16 Concrete

What Will Change

- Strength designations, from psi to megapascals, rounded to the nearest 5 megapascals per ACI 318M as follows: 2500 psi to **20 MPa;** 3000 psi to **25 MPa;** 3500 psi to **25 MPa;** 4000 psi to **30 MPa;** 4500 psi to **35 MPa;** 5000 psi to **35 MPa.** (*Yes, both 3000 and 3500 psi are rounded to 25 MPa, and 4500 and 5000 psi are rounded to 35 MPa, which indicates the fairly broad allowable tolerances used in concrete strength designations.*)

What Will Stay the Same

- Everything else

C.1.17 Rebar

What Will Change

- Rebars will change in size per ASTM A615M, A616M, A617M, and A706M. New metric bar sizes are as follows: Nos. 3 and 4 to **10**; No. 5 to **15**; No. 6 to **20**; Nos. 7 and 8 to **25**; Nos. 9 and 10 to **30**; No. 11 to **35**; No. 14 to **45**; and No. 18 to **55**.

What Will Stay the Same

- Concrete.

C.1.18 Glass

What Will Change

- Cut sheet dimensions from feet and inches to millimeters.

What Will Stay the Same

- Sheet thickness, which can be rolled to any dimension and is often rolled in millimeters now. See ASTM C1036.

C.2

METRIC RESOURCES[3]

C.2.1 Design References

- **American Institute of Architects** (AIA Bookstore, 1735 New York Avenue NW, Washington, DC 20006; phone 202–626–7475. All are published by John Wiley & Sons, Professional Reference and Trade Group, 605 Third Avenue, New York, NY 10158; phone 800–225–5945).

  Architectural Graphic Standards. A metric edition is not due for several years, but current editions include a comprehensive section on metric conversion.

  The Architect's Studio Companion: Simplified Technical Guidelines for Preliminary Design, by Edward Allen and Joseph Iano. Includes dual units. 468 pp. $59.95.

  Architectural Detailing: Function, Constructability, and Aesthetics, by Edward Allen. Includes dual units. $59.95.

  Fundamentals of Building Construction: Materials and Methods, by Edward Allen. Includes dual units. $64.95.

  Neufert Architect's Data, by Ernst Neufert. Second International Edition. All units are metric. 433 pp. $55.00.

  Wiley Engineer's Desk Reference, by S. I. Heisler. Includes dual units. 566 pp. $79.95.

- **American Society of Civil Engineers** (phone 800–548–2723 for publications).

  Metric Units in Engineering—Going SI, by Cornelius Wandmacher and Ivan Johnson. $28.00.

- **Instrument Society of America** (P.O. Box 3561, Durham, NC 27702; phone 919–549–8411).

  ISA Guide to Measurement Conversions, by G. Platt. 172 pp. $50.00.

C.2.2 Cost Estimating

- **R. S. Means Company** (P.O. Box 800, Kingston, MA 02364; phone 617–585–7880).

  Means Building Construction Cost Data. 1995. Metric edition. $99.95.

- **Frank R. Walker Co.** (P.O. Box 3180, Lisle, IL 60532; phone 708–971–8989).

  Building Estimator's Reference Book. 25th Edition. $69.95.

C.2.3 Specifications

- **AIA Master Systems** (phone 703–684–9153).

  AIA Masterspec is available in either dual-unit or metric-only versions.

- **Construction Specifications Institute** (601 Madison Street, Alexandria, VA 22314–1791; phone 703–684–0300).

  CSRS Spectext contains dual units, as do all other CSI publications.

C.2.4 Building Codes

- **Building Officials and Code Administrators International** (4051 W. Flossmoor Road, Country Club Hills, IL 60477–5795; phone 312–799–2300).

  All *BOCA National Codes* are published in dual units.

- **International Conference of Building Officials** (5360 South Workman Mill Road, Whittier, CA 90601; phone 310–699–0541).

  The *1994 Uniform Codes* are published in dual units.

- **Southern Building Code Congress International, Inc.** (900 Montclair Road, Birmingham, AL 35213–1206; phone 205–591–1853).

  The *1994 Standard Codes* are published in dual units.

---

[3] *Metric in Construction* (July/August 1995).

- **National Fire Protection Association** (1 Battery-march Park, P.O. Box 9101, Quincy, MA 02269–9101; phone 800–344–3555).

  NFPA 101, *Life Safety Code*, is published in dual units, as are all NFPA standards.

C.2.5 Metric Standards

- **American Society for Testing and Materials** (1916 Race Street, Philadelphia, PA 19103; phone 215–299–5585). Call ASTM for current prices. All ASTM standards are published in metric or dual units.

  ASTM E621, *Standard Practice for the Use of Metric (SI) Units in Building Design and Construction.*

  ASTM E380, *Standard Practice for Use of the International System of Units (SI).*

  ASTM E713, *Guide for Selection of Scales for Metric Building Drawings.*

  ASTM E577, *Guide for Dimensional Coordination of Rectilinear Building Parts and Systems.*

  ASTM E835, *Guide for Dimensional Coordination of Structural Clay Units, Concrete Masonry Units, and Clay Flue Linings.*

- **American National Standards Institute, Inc.** (11 West 42nd Street, New York, NY 10036; phone 212–642–4900). Call ANSI for current prices. Many ANSI standards are available in metric units.

  ANSI/IEEE 268, *American National Standard Metric Practice.*

  ANSI/AWS A1.1, *Metric Practice Guide for the Welding Industry.*

  ANSI/IEEE 945, *Preferred Metric Units for Use in Electrical and Electronics Science and Technology.*

  ISO 1000, *SI Units and Recommendations for the Use of Their Multiples and Certain Other Units.*

C.2.6 Steel

- **American Institute of Steel Construction** (Metric Publications, 1 East Wacker Drive, Suite 3100, Chicago, IL 60601–2001; phone 800–644–2400 for publications and 312–670–5411 for software).

  Metric Properties of Structural Shapes with Dimensions According to ASTM A6M. Metric version of Part I of the *Manual of Steel Construction.* $20.00.

  Metric Conversion: Load and Resistance Factor Design Specification for Structural Steel Buildings. $20.00.

  Guide to Metric Steel Fabrication. $20.00.

AISC Database, Version 2.0, Metric Units. An ASCII data file that gives programmers electronic access to the dimensions and properties of all structural shapes—W, M, S, HP, C, MC, WT, MT, ST, L, 2L, TS, P, PX, PXX—in metric units. $60.00.

AISC for AutoCAD, Version 2.0 (for AutoCAD Releases 12 and 13), a utility that automatically draws plans, side elevations, and end elevations for the above-mentioned structural shapes. The program includes both inch-pound and metric units. $120.

- **Metrosoft** (332 Patterson Avenue, East Rutherford, NJ 07073; phone 201–438–4915). Call for prices.

  ROBOT V6, a metric-based structural analysis program.

- **American Welding Society** (550 N.W. LeJeune Road, P.O. Box 35104, Miami, FL 33135; phone 305–443–9353).

  All AWS standards include dual units.

C.2.7 Concrete

- **American Concrete Institute** (P.O. Box 19150, Detroit MI 48219; phone 313–532–2600).

  ACI 318M/318RM, *Building Code Requirements for Reinforced Concrete and Commentary.* Metric edition of ACI 318/318R. $89.25.

  ACI 318.1M/318.1RM, *Building Code Requirements for Metric Structural Plain Concrete and Commentary.* Metric edition of ACI 318.1/318.1R. $17.75.

- **Portland Cement Institute** (Order Processing, 5420 Old Orchard Road, Skokie, IL 60077; phone 800–868–6733).

  Design and Control of Concrete Mixtures—Canadian Metric Edition. $35.00.

- **Wire Reinforcing Institute** (203 Loudon Street SW, Leesburg, VA 22075; phone 703–779–2339). There is no charge for single copies of the following publications:

  Metric Welded Wire Reinforcement. TF-206, a 2-page data sheet.

  Metric Welded Wire Reinforcement for Concrete Pipe. TF-311M, a 4-page data sheet.

- **Concrete Reinforcing Steel Institute** (944 N. Plum Grove Road, Schaumburg, IL 60173; phone 708–517–1200).

  Contact CRSI for current technical information on the metrication of reinforcing steel.

C.2.8 Wood

- **American Forest and Paper Association,** formerly National Forest Products Association (1250 Connecticut Avenue, NW, Washington, DC 20036; phone 202–463–2700).

  Wood Products Metric Planning Package. 1994. 30 pp. $10.00.

- **National Particleboard Association** (18928 Premiere Court, Gaithersburg, MD 20879; phone 301–670–0604).

  Metric units have been added to the NPA/ANSI standards for particleboard and medium-density fiberboard. $6.00 each.

- **Hardwood Plywood Manufacturers Association** (P.O. Box 2789, Reston, VA 22090–2789; phone 703–435–2900).

  ANSI HPVA 1-1994, *Voluntary Standard for Hardwood and Decorative Plywood.* Includes dual units. 1992. 24 pp. $15.00.

C.2.9 Ceiling Systems

- **USG Interiors** (100 Crocker Road, Westlake, OH 44145–1089; phone 216–871–1000).

  Metric Ceiling Systems (SA905ME). No charge.

C.2.10 Metric Scales and Templates

Metric scales are available from graphic arts supply stores. Popular models are **Staedtler-Mars** 987-18-1, **Alvin** 117 PM, and **Charvoz** 30-1261.

Metric plumbing templates are available from **American Standard** (phone 703–841–9585).

C.2.11 Metric Conversion Calculators

Metric conversion calculators include the **Sharp** Model EL-344G Metric Calculator, **Texas Instruments** Model 1895II, and **Radio Shack** Model 65-828.

APPENDIX D:
TABLES FROM THE UNIFORM BUILDING CODE

Table D–1 Asphalt shingle application.

	ASPHALT SHINGLES	
	Not Permitted below 2 Units Vertical in 12 Units Horizontal (16.7% Slope)	
Roof Slope	**2 Units Vertical in 12 Units Horizontal (16.7% Slope) to Less Than 4 Units Vertical in 12 Units Horizontal (33.3% Slope)**	**4 Units Vertical in 12 Units Horizontal (33.3% Slope) and Over**
1. Deck requirement	Asphalt shingles shall be fastened to solidly sheathed roofs. Sheathing shall conform to Sections 2322.2 and 2326.12.9.	
2. Underlayment Temperate climate	Asphalt strip shingles may be installed on slopes as low as 2 inches in 12 inches (305 mm), provided the shingles are approved self-sealing or are hand sealed and are installed with an underlayment consisting of two layers of nonperforated Type 15 felt applied shingle fashion. Starting with an 18-inch-wide (457 mm) sheet and a 36-inch-wide (914 mm) sheet over it at the eaves, each subsequent sheet shall be lapped 19 inches (483 mm) horizontally.	One layer nonperforated Type 15 felt lapped 2 inches (51 mm) horizontally and 4 inches (102 mm) vertically to shed water.
Severe climate: In areas subject to wind-driven snow or roof ice buildup.	Same as for temperate climate, and additionally the two layers shall be solid cemented together with approved cementing material between the plies extending from the eave up the roof to a line 24 inches (610 mm) inside the exterior wall line of the building.	Same as for temperate climate, except that one layer No. 40 coated roofing or coated glass base shall be applied from the eaves to a line 12 inches (305 mm) inside the exterior wall line with all laps cemented together.
3. Attachment combined systems, type of fasteners	Corrosion-resistant nails, minimum 12-gage $^3/_8$-inch (9.5 mm) head, or approved corrosion-resistant staples, minimum 16-gage $^{15}/_{16}$-inch (23.8 mm) crown width. Fasteners shall comply with the requirements of Chapter 23, Division III. Fasteners shall be long enough to penetrate into the sheathing $^3/_4$ inch (19 mm) or through the thickness of the sheathing, whichever is less.	
No. of fasteners[1]	4 per 36-inch to 40-inch (914 mm to 1016 mm) strip 2 per 9-inch to 18-inch (229 mm to 457 mm) shingle	
Exposure Field of roof Hips and ridges	Per manufacturer's instructions included with packages of shingles. Hip and ridge weather exposures shall not exceed those permitted for the field of the roof.	
Method	Per manufacturer's instructions included with packages of shingles.	
4. Flashing Valleys Other flashing	Per Section 1508.2 Per Section 1509	

[1]Figures shown are for normal application. For special conditions such as mansard application and where roofs are in special wind regions, shingles shall be attached per manufacturer's instructions.

SOURCE: From the Uniform Building Code, © 1994, ICBO.

Table D–2 Wood shingle or shake application.

ROOF SLOPE	WOOD SHINGLES Not Permitted below 3 Units Vertical in 12 Units Horizontal (25% Slope) See Table 15-C	WOOD SHAKES Not Permitted below 4 Units Vertical in 12 Units Horizontal (33.3% Slope)[1] See Table 15-C
1. Deck requirement	Shingles and shakes shall be applied to roofs with solid or spaced sheathing. When spaced sheathing is used, sheathing boards shall not be less than 1 inch by 4 inches (25 mm by 102 mm) nominal dimensions and shall be spaced on centers equal to the weather exposure to coincide with the placement of fasteners. When 1-inch by 4-inch (25 mm by 102 mm) spaced sheathing is installed at 10 inches (254 mm) on center, additional 1-inch by 4-inch (25 mm by 102 mm) boards must be installed between the sheathing boards. Sheathing shall conform to Sections 2322.2 and 2326.12.9.	
2. Interlayment	No requirements.	One 18-inch-wide (457 mm) interlayment of Type 30 felt shingled between each course in such a manner that no felt is exposed to the weather below the shake butts and in the keyways (between the shakes).
3. Underlayment Temperate climate	No requirements.	No requirements.
Severe climate: In areas subject to wind-driven snow or roof ice buildup.	Two layers of nonperforated Type 15 felt applied shingle fashion shall be installed and solid cemented together with approved cementing material between the plies extending from the eave up the roof to a line 36 inches (914 mm) inside the exterior wall line of the building.	Sheathing shall be solid and, in addition to the interlayment of felt shingled between each course in such a manner that no felt is exposed to the weather below the shake butts, the shakes shall be applied over a layer of nonperforated Type 15 felt applied shingle fashion. Two layers of nonperforated Type 15 felt applied shingle fashion shall be installed and solid cemented together with approved cementing material between the plies extending from the eave up the roof to a line 36 inches (914 mm) inside the exterior wall line of the building.
4. Attachment Type of fasteners	Corrosion-resistant nails, minimum No. 14^1/$_2$-gage ^{7}/$_{32}$-inch (5.6 mm) head, or corrosion-resistant staples, when approved by the building official.	Corrosion-resistant nails, minimum No. 13-gage ^{7}/$_{32}$-inch (5.6 mm) head, or corrosion-resistant staples, when approved by the building official.
	Fasteners shall comply with the requirements of Chapter 23, Division III. Fasteners shall be long enough to penetrate into the sheathing ^{3}/$_4$ inch (19 mm) or through the thickness of the sheathing, whichever is less.	
No. of fasteners	2 per shingle	2 per shake
Exposure Field of roof Hips and ridges	Weather exposures shall not exceed those set forth in Table 15-C. Hip and ridge weather exposure shall not exceed those permitted for the field of the roof.	
Method	Shingles shall be laid with a side lap of not less than 1^1/$_2$ inches (38 mm) between joints in adjacent courses, and not in direct alignment in alternate courses. Spacing between shingles shall be approximately ^{1}/$_4$ inch (6 mm). Each shingle shall be fastened with two nails only, positioned approximately ^{3}/$_4$ inch (19 mm) from each edge and approximately 1 inch (25 mm) above the exposure line. Starter course at the eaves shall be doubled.	Shakes shall be laid with a side lap of not less than 1^1/$_2$ inches (38 mm) between joints in adjacent courses. Spacing between shakes shall not be less than ^{3}/$_8$ inch (9 mm) or more than ^{5}/$_8$ inch (16 mm) except for preservative-treated wood shakes which shall have a spacing not less than ^{1}/$_4$ inch (6 mm) or more than ^{3}/$_8$ inch (9 mm). Shakes shall be fastened to the sheathing with two nails only, positioned approximately 1 inch (25 mm) from each edge and approximately 2 inches (51 mm) above the exposure line. The starter course at the eaves shall be doubled. The bottom or first layer may be either shakes or shingles. Fifteen-inch or 18-inch (381 mm or 457 mm) shakes may be used for the starter course at the eaves and final course at the ridge.
5. Flashing Valleys Other flashing	Per Section 1508.5 Per Section 1509	

[1]When approved by the building official, wood shakes may be installed on a slope of not less than 3 units vertical in 12 units horizontal (25% slope) when an underlayment of not less than nonperforated Type 15 felt is installed.

Source: From the Uniform Building Code, © 1994, ICBO.

Table D–3 Maximum weather exposure.

GRADE LENGTH	3 UNITS VERTICAL TO LESS THAN 4 UNITS VERTICAL IN 12 UNITS HORIZONTAL (25% ≤ 33.3% SLOPE)	4 UNITS VERTICAL IN 12 UNITS HORIZONTAL (33.3% SLOPE)
	× 25.4 for mm	
Wood Shingles		
1. No. 1 16-inch	$3^3/_4$	5
2. No. 2[1] 16-inch	$3^1/_2$	4
3. No. 3[1] 16-inch	3	$3^1/_2$
4. No. 1 18-inch	$4^1/_4$	$5^1/_2$
5. No. 2[1] 18-inch	4	$4^1/_2$
6. No. 3[1] 18-inch	$3^1/_2$	4
7. No. 1 24-inch	$5^3/_4$	$7^1/_2$
8. No. 2[1] 24-inch	$5^1/_2$	$6^1/_2$
9. No. 3[1] 24-inch	5	$5^1/_2$
Wood Shakes[2]		
10. No. 1 18-inch	$7^1/_2$	$7^1/_2$
11. No. 1 24-inch	10	10
12. No. 2 18-inch tapersawn shakes	—	$5^1/_2$
13. No. 2 24-inch tapersawn shakes	—	$7^1/_2$

[1]To be used only when specifically permitted by the building official.
[2]Exposure of 24-inch (610 mm) by $^3/_8$-inch (9.5 mm) resaw handsplit shakes shall not exceed 5 inches (127 mm) regardless of the roof slope.

SOURCE: From the Uniform Building Code, © 1994, ICBO.

Table D–4 Roofing tile application for all tiles.[1]

	ROOF SLOPE $2^1/_2$ UNITS VERTICAL IN 12 UNITS HORIZONTAL (21% Slope) TO LESS THAN 3 UNITS VERTICAL IN 12 UNITS HORIZONTAL (25% Slope)	ROOF SLOPE 3 UNITS VERTICAL IN 12 UNITS HORIZONTAL (25% Slope) AND OVER
1. Deck requirements	Solid sheathing per Sections 2322.2 and 2326.12.9	
2. Underlayment In climate areas subject to wind-driven snow, roof ice damming or special wind regions as shown in Figure 16-1 of Chapter 16.	Built-up roofing membrane, three plies minimum, applied per Section 1507.6. Surfacing not required.	Same as for other climate areas, except that extending from the eaves up the roof to a line 24 inches (610 mm) inside the exterior wall line of the building, two layers of underlayment shall be applied shingle fashion and solidly cemented together with an approved cementing material.
Other climate areas		One layer heavy-duty felt or Type 30 felt side lapped 2 inches (51 mm) and end lapped 6 inches (153 mm).
3. Attachment[2] Type of fasteners	Corrosion-resistant nails not less than No. 11 gage, $^5/_{16}$-inch (7.9 mm) head. Fasteners shall comply with the requirements of Chapter 23, Division III. Fasteners shall be long enough to penetrate into the sheathing $^3/_4$ inch (19 mm) or through the thickness of the sheathing, whichever is less. Attaching wire for clay or concrete tile shall not be smaller than 0.083 inch (2.11 mm) (No. 14 B.W. gage).	
Number of fasteners[2,3]	One fastener per tile. Flat tile without vertical laps, two fasteners per tile.	Two fasteners per tile. Only one fastener on slopes of 7 units vertical in 12 units horizontal (58.3% slope) and less for tiles with installed weight exceeding 7.5 pounds per square foot (36.6 kg/m^2) having a width no greater than 16 inches (406 mm).[4]
4. Tile headlap	3 inches (76.2 mm) minimum.	
5. Flashing	Per Sections 1508.4 and 1509.	

[1]In snow areas a minimum of two fasteners per tile are required.
[2]In areas designated by the building official as being subject to repeated wind velocities to excess of 80 miles per hour (129 km/h) or where the roof height exceeds 40 feet (12 192 mm) above grade, all tiles shall be attached as follows:
   [2.1] The heads of all tiles shall be nailed.
   [2.2] The noses of all eave course tiles shall be fastened with approved clips.
   [2.3] All rake tiles shall be nailed with two nails.
   [2.4] The noses of all ridge, hip and rake tiles shall be set in a bead of approved roofer's mastic.
[3]In snow areas a minimum of two fasteners per tile are required, or battens and one fastener.
[4]On slopes over 24 units vertical in 12 units horizontal (200% slope), the nose end of all tiles shall be securely fastened.

SOURCE: From the Uniform Building Code, © 1994, ICBO.

Table D-5 Clay or concrete roofing tile application of interlocking tile with projection anchor lugs—minimum roof slope 4 units vertical in 12 units horizontal (33.3% slope).

ROOF SLOPE	4 UNITS VERTICAL IN 12 UNITS HORIZONTAL (33.3% Slope) AND OVER
1. Deck requirements	Spaced structural sheathing boards or solid roof sheathing.
2. Underlayment In climate areas subject to wind-driven snow, roof ice or special wind regions as shown in Figure 16-1.	Solid sheathing one layer of Type 30 felt lapped 2 inches (51 mm) horizontally and 6 inches (153 mm) vertically, except that extending from the eaves up the roof to line 24 inches (610 mm) inside the exterior wall line of the building, two layers of the underlayment shall be applied shingle fashion and solid cemented together with approved cementing material.
Other climates	For spaced sheathing, approved reinforced membrane. For solid sheathing, one layer heavy-duty felt or Type 30 felt lapped 2 inches (51 mm) horizontally and 6 inches (153 mm) vertically.
3. Attachment[1] Type of fasteners	Corrosion-resistant nails not less than No. 11 gage, 5/16-inch (7.9 mm) head. Fasteners shall comply with the requirements of Chapter 23, Division III. Fasteners shall be long enough to penetrate into the battens[2] or sheathing 3/4 inch (19 mm) or through the thickness of the sheathing, whichever is less. Attaching wire for clay or concrete tile shall not be smaller than 0.083 inch (2.11 mm) (No. 14 B.W. gage). Horizontal battens are required on solid sheathing for slopes 7 units vertical in 12 units horizontal (58.3% slope) and over.[1] Horizontal battens are required for slopes over 7 units vertical in 12 units horizontal (58.3% slope).[2]
No. of fasteners with: Spaced/solid sheathing with battens, or spaced sheathing[3] Solid sheathing without battens[3]	Below 5 units vertical in 12 units horizontal (41.7% slope), fasteners not required. Five units vertical in 12 units horizontal (41.7% slope) to less than 12 units vertical in 12 units horizontal (100% slope), one fastener per tile every other row. Twelve units vertical in 12 units horizontal (100% slope) to 24 units vertical in 12 units horizontal (200% slope), one fastener every tile.[4] All perimeter tiles require one fastener.[5] Tiles with installed weight less than 9 pounds per square foot (4.4 kg/m²) require a minimum of one fastener per tile regardless of roof slope. One fastener per tile.
4. Tile headlap	3-inch (76 mm) minimum.
5. Flashing	Per Sections 1508.4 and 1509.

[1]In areas designated by the building official as being subject to repeated wind velocities to excess of 80 miles per hour (129 km/h), or where the roof height exceeds 40 feet (12 192 mm) above grade, all tiles shall be attached as set forth below:
   [1.1] The heads of all tiles shall be nailed.
   [1.2] The noses of all eave course tiles shall be fastened with a special clip.
   [1.3] All rake tiles shall be nailed with two nails.
   [1.4] The noses of all ridge, hip and rake tiles shall be set in a bead of approved roofer's mastic.
[2]Battens shall not be less than 1-inch by 2-inch (25.4 mm by 51 mm) nominal. Provisions shall be made for drainage beneath battens by a minimum of 1/8-inch (3.2 mm) risers at each nail or by 4-foot-long (1219 mm) battens with at least 1/2-inch (13 mm) separation between battens. Battens shall be fastened with approved fasteners spaced at not more than 24 inches (610 mm) on center.
[3]In snow areas a minimum of two fasteners per tile are required, or battens and one fastener.
[4]Slopes over 24 units vertical in 12 units horizontal (200% slope), nose ends of all tiles must be securely fastened.
[5]Perimeter fastening areas include three tile courses but not less than 36 inches (914 mm) from either side of hips or ridges and edges of eaves and gable rakes.

SOURCE: From the Uniform Building Code, © 1994, ICBO.

Table D-6 Built-up roof-covering application.

	MECHANICALLY FASTENED SYSTEMS	ADHESIVELY FASTENED SYSTEMS
1. Deck conditions	Decks shall be firm, broom-clean, smooth and dry. Insulated decks shall have wood insulation stops at all edges of the deck, unless an alternative suitable curbing is provided. Insulated decks with slopes greater than 2 units vertical in 12 units horizontal (16.7% slope) shall have wood insulation stops at not more than 8 feet (2438 mm) face to face. Wood nailers shall be provided where nailing is required for roofing plies.	
	Solid wood sheathing shall conform to Sections 2322.2 and 2326.12.9.	Provide wood nailers where nailing is required for roofing plies (see below).
2. Underlayment	One layer of sheathing paper, Type 15 felt or other approved underlayment nailed sufficiently to hold in place, is required over board decks where openings between boards would allow bitumen to drip through. No underlayment requirements for plywood decks. Underlayment on other decks shall be in accordance with deck manufacturer's recommendations.	Not required.
3. Base ply requirements Over noninsulated decks	Over approved decks, the base ply shall be nailed using not less than one fastener for each $1^1/_3$ square feet (0.124 m²).	Decks shall be primed in accordance with the roofing manufacturer's instructions. The base ply shall be solidly cemented or spot mopped as required by the type of deck material using adhesive application rates shown in Table 15-F.
4. Mechanical fasteners	Fasteners shall be long enough to penetrate $^3/_4$ inch (19 mm) into the sheathing or through the thickness of the sheathing, whichever is less. Built-up roofing nails for wood board decks shall be minimum No. 12 gage $^7/_{16}$-inch (11.1 mm) head driven through tin caps or approved nails with integral caps. For plywood, No. 11 gage ring-shank nails driven through tin caps or approved nails with integral caps shall be used. For gypsum, insulating concrete, cementitious wood fiber and other decks, fasteners recommended by the manufacturer shall be used.	When mechanical fasteners are required for attachment of roofing plies to wood nailers or insulation stops (see below), they shall be as required for wood board decks.
5. Vapor retarder Over insulated decks	A vapor retarder shall be installed where the average January temperature is below 45°F. (7°C.), or where excessive moisture conditions are anticipated within the building. It shall be applied as for a base ply.	
6. Insulation	When no vapor retarder is required, roof insulation shall be fastened in an approved manner. When a vapor retarder is required, roof insulation is to be solidly mopped to the vapor retarder using the adhesive application rate specified in Table 15-F. See manufacturer's instructions for the attachment of insulation over steel decks.	When no vapor retarder is required, roof insulation shall be solid mopped to the deck using the adhesive application rate specified in Table 15-F. When a vapor retarder is required, roof insulation is to be solidly mopped to the vapor retarder, using the adhesive application rate specified in Table 15-F. See manufacturer's installation instructions for attachment of insulation over steel decks.
7. Roofing plies	Successive layers shall be solidly cemented together and to the base ply or the insulation using the adhesive rates shown in Table 15-F. On slopes greater than 1 unit vertical in 12 units horizontal (8.3% slope) for aggregate-surfaced, or 2 units vertical in 12 units horizontal (16.7% slope) for smooth-surfaced or cap sheet surfaced roofs, mechanical fasteners are required. Roofing plies shall be blind-nailed to the deck, wood nailers or wood insulation stops in accordance with the roofing manufacturer's recommendations. On slopes exceeding 3 units vertical in 12 units horizontal (25% slope), plies shall be laid parallel to the slope of the deck (strapping method).	
8. Cementing materials	See Table 15-G.	
9. Curbs and walls	Suitable cant strips shall be used at all vertical intersections. Adequate attachment shall be provided for both base flashing and counterflashing on all vertical surfaces. Reglets shall be provided in wall or parapets receiving metal counterflashing.	
10. Surfacing	Mineral aggregate surfaced roofs shall comply with the requirements of U.B.C. Standard 15-1 and Table 15-F. Cap sheets shall be cemented to the roofing plies as set forth in Table 15-F.	

SOURCE: From the Uniform Building Code, © 1994, ICBO.

Table D–7 Foundations for stud-bearing walls: minimum requirements.[1,2,3]

NUMBER OF FLOORS SUPPORTED BY THE FOUNDATION[4]	THICKNESS OF FOUNDATION WALL (inches) × 25.4 for mm		WIDTH OF FOOTING (inches)	THICKNESS OF FOOTING (inches)	DEPTH BELOW UNDISTURBED GROUND SURFACE (inches)
	Concrete	Unit Masonry	× 25.4 for mm		
1	6	6	12	6	12
2	8	8	15	7	18
3	10	10	18	8	24

[1]Where unusual conditions or frost conditions are found, footings and foundations shall be as required in Section 1806.1.
[2]The ground under the floor may be excavated to the elevation of the top of the footing.
[3]Interior stud bearing walls may be supported by isolated footings. The footing width and length shall be twice the width shown in this table and the footings shall be spaced not more than 6 feet (1829 mm) on center.
[4]Foundations may support a roof in addition to the stipulated number of floors. Foundations supporting roofs only shall be as required for supporting one floor.
SOURCE: From the Uniform Building Code © 1994, ICBO.

Table D–8 Wood shingle and shake side-wall exposures.

SHINGLE OR SHAKE	MAXIMUM WEATHER EXPOSURES (inches) × 25.4 for mm			
	Single-Coursing		Double-Coursing	
Length and Type	No. 1	No. 2	No. 1	No. 2
1. 16-inch (405 mm) shingles	7$\frac{1}{2}$	7$\frac{1}{2}$	12	10
2. 18-inch (455 mm) shingles	8$\frac{1}{2}$	8$\frac{1}{2}$	14	11
3. 24-inch (610 mm) shingles	11$\frac{1}{2}$	11$\frac{1}{2}$	16	14
4. 18-inch (455 mm) resawn shakes	8$\frac{1}{2}$	—	14	—
5. 18-inch (455 mm) straight-split shakes	8$\frac{1}{2}$	—	16	—
6. 24-inch (610 mm) resawn shakes	11$\frac{1}{2}$	—	20	—

SOURCE: From the Uniform Building Code, © 1994, ICBO.

Table D–9 Exposed plywood panel siding.

MINIMUM THICKNESS[1] (inch) × 25.4 for mm	MINIMUM NUMBER OF PLIES	STUD SPACING (inches) PLYWOOD SIDING APPLIED DIRECTLY TO STUDS OR OVER SHEATHING × 25.4 for mm
$\frac{3}{8}$	3	16[2]
$\frac{1}{2}$	4	24

[1]Thickness of grooved panels is measured at bottom of grooves.
[2]May be 24 inches (610 mm) if plywood siding applied with face grain perpendicular to studs or over one of the following: (1) 1-inch (25 mm) board sheathing, (2) 7/16-inch (11 mm) wood structural panel sheathing or (3) 3/8-inch (9.5 mm) wood structural panel sheathing with strength axis (which is the long direction of the panel unless otherwise marked) of sheathing perpendicular to studs.

SOURCE: From the Uniform Building Code, © 1994, ICBO.

Table D–10 Allowable spans for exposed particleboard panel siding.

GRADE	STUD SPACING (inches) ×25.4 for mm	MINIMUM THICKNESS (inches) ×25.4 for mm		
		Siding		Exterior Ceilings and Soffits
		Direct to Studs	Continuous Support	Direct to Supports
2-M-W	16	3/8	5/16	5/16
	24	1/2	5/16	3/8
2-M-1 2-M-2	16	5/8	3/8	—
2-M-3	24	3/4	3/8	—

SOURCE: From the Uniform Building Code, © 1994, ICBO.

Table D–11 Hardboard siding.

SIDING	MINIMAL NOMINAL THICKNESS (inch)	FRAMING (2" x 4") MAXIMUM SPACING	NAIL SIZE[1,2]	NAIL SPACING	
				General	Bracing Panels[3]
		×25.4 for mm			
1. LAP SIDING					
Direct to studs	3/8	16" o.c.	8d	16" o.c.	Not applicable
Over sheathing	3/8	16" o.c.	10d	16" o.c.	Not applicable
2. SQUARE EDGE PANEL SIDING					
Direct to studs	3/8	24" o.c.	6d	6" o.c. edges; 12" o.c. at intermed. supports	4" o.c. edges; 8" o.c. intermed. supports
Over sheathing	3/8	24" o.c.	8d	6" o.c. edges; 12" o.c. at intermed. supports	4" o.c. edges; 8" o.c. intermed. supports
3. SHIPLAP EDGE PANEL SIDING					
Direct to studs	3/8	16" o.c.	6d	6" o.c. edges; 12" o.c. at intermed. supports	4" o.c. edges; 8" o.c. intermed. supports
Over sheathing	3/8	16" o.c.	8d	6" o.c. edges; 12" o.c. at intermed. supports	4" o.c. edges; 8" o.c. intermed. supports

[1]Nails shall be corrosion resistant in accordance with Division III.
[2]Minimum acceptable nail dimensions (inches).

	Panel Siding (inch)	Lap Siding (inch)
	×25.4 for mm	
Shank diameter	.092	.099
Head diameter	.225	.240

[3]When used to comply with Section 2326.11.3.

SOURCE: From the Uniform Building Code, © 1994, ICBO.

Table D-12 Allowable spans for lumber floor and roof sheathing.[1,2]

SPAN (inches)	MINIMUM NET THICKNESS (inches) OF LUMBER PLACED			
	Perpendicular to Supports		Diagonally to Supports	
× 25.4 for mm	× 25.4 for mm			
	Surfaced Dry[3]	Surfaced Unseasoned	Surfaced Dry[3]	Surfaced Unseasoned
	Floors			
1. 24	$^3/_4$	$^{25}/_{32}$	$^3/_4$	$^{25}/_{32}$
2. 16	$^5/_8$	$^{11}/_{16}$	$^5/_8$	$^{11}/_{16}$
	Roofs			
3. 24	$^5/_8$	$^{11}/_{16}$	$^3/_4$	$^{25}/_{32}$

[1]Installation details shall conform to Sections 2326.9.1 and 2326.12.8 for floor and roof sheathing, respectively.
[2]Floor or roof sheathing conforming with this table shall be deemed to meet the design criteria of Section 2322.
[3]Maximum 19 percent moisture content.

SOURCE: From the Uniform Building Code, © 1994, ICBO.

Table D-13 Allowable spans for wood structural panel combination subfloor-underlayment (single-floor) continuous over two or more spans with strength axis perpendicular to supports.[1,2]

IDENTIFICATION	MAXIMUM SPACING OF JOISTS (inches)				
	× 25.4 for mm				
	16	20	24	32	48
Species Group[3]	Thickness (inches)				
	× 25.4 for mm				
1	$^1/_2$	$^5/_8$	$^3/_4$	—	—
2, 3	$^5/_8$	$^3/_4$	$^7/_8$	—	—
4	$^3/_4$	$^7/_8$	1	—	—
Span rating[4]	16 o.c.	20 o.c.	24 o.c.	32 o.c.	48 o.c.

[1]Spans limited to value shown because of possible effects of concentrated loads. Allowable uniform loads based on deflection of $^1/_{360}$ of span is 100 pounds per square foot (psf) (4.79 kN/m^2), except allowable total uniform load for $1^1/_8$-inch (29 mm) wood structural panels over joists spaced 48 inches (1219 mm) on center is 65 psf (3.11 kN/m^2). Panel edges shall have approved tongue-and-groove joints or shall be supported with blocking, unless $^1/_4$-inch (6.4 mm) minimum thickness underlayment or $1^1/_2$ inches (38 mm) of approved cellular or lightweight concrete is placed over the subfloor, or finish floor is $^3/_4$-inch (19 mm) wood strip.
[2]Floor panels conforming with this table shall be deemed to meet the design criteria of Section 2321.
[3]Applicable to all grades of sanded exterior-type plywood. See U.B.C. Standard 23-2 for plywood species groups.
[4]Applicable to underlayment grade and C-C (plugged) plywood, and single floor grade wood structural panels.

SOURCE: From the Uniform Building Code, © 1994, ICBO.

Table D–14 Classification of species.

Group 1	Group 2		Group 3	Group 4

Apitong[a] [b]
Beech, American
Birch
  Sweet
  Yellow
Douglas Fir[c]
Kapur[a]
Keruing[a] [b]
Larch, Western
Maple, Sugar
Pine
  Caribbean
  Ocote
Pine, Southern
  Loblolly
  Longleaf
  Shortleaf
  Slash
Tanoak
	Cedar, Port Orford
	Cypress
	Douglas Fir 2[c]
	Fir
	  California Red
	  Grand
	  Noble
	  Pacific Silver
	  White
	Hemlock, Western
	Lauan
	  Almon
	  Bagtikan
	  Mayapis
	  Red Lauan
	  Tangile
	  White Lauan
		Maple, Black
		Mengkulang[a]
		Meranti, Red[a] [d]
		Mersawa[a]
		Pine
		  Pond
		  Red
		  Virginia
		  Western White
		Spruce
		  Red
		  Sitka
		Sweetgum
		Tamarack
		Yellow-poplar
			Alder, Red
			Birch, Paper
			Cedar, Alaska
			Fir, Subalpine
			Hemlock, Eastern
			Maple, Bigleaf
			Pine
			  Jack
			  Lodgepole
			  Ponderosa
			  Spruce
			Redwood
			Spruce
			  Black
			  Englemann
			  White
				Aspen
				  Bigtooth
				  Quaking
				Cativo
				Cedar
				  Incense
				  Western Red
				Cottonwood
				  Eastern
				  Black (Western Poplar)
				Pine
				  Eastern White
				  Sugar

(a) Each of these names represents a trade group of woods consisting of a number of closely related species.

(b) Species from the genus Dipterocarpus are marketed collectively: Apitong if originating in the Philippines; Keruing if originating in Malaysia or Indonesia.

(c) Douglas fir from trees grown in the states of Washington, Oregon, California, Idaho, Montana, Wyoming, and the Canadian Provinces of Alberta and British Columbia shall be classed as Douglas fir No. 1. Douglas fir from trees grown in the states of Nevada, Utah, Colorado, Arizona and New Mexico shall be classed as Douglas fir No. 2.

(d) Red Meranti shall be limited to species having a specific gravity of 0.41 or more based on green volume and oven dry weight.

SOURCE: From the Uniform Building Code, © 1994, ICBO.

Table D–15 Size, height, and spacing of wood studs.

STUD SIZE (inches)	BEARING WALLS				NONBEARING WALLS	
	Laterally Unsupported Stud Height[1] (feet)	Supporting Roof and Ceiling Only	Supporting One Floor, Roof and Ceiling	Supporting Two Floors, Roof and Ceiling	Laterally Unsupported Stud Height[1] (feet)	Spacing (inches)
		Spacing (inches)				
× 25.4 for mm	× 304.8 for mm	× 25.4 for mm			× 304.8 for mm	× 25.4 for mm
1. 2 × 3[2]	—	—	—	—	10	16
2. 2 × 4	10	24	16	—	14	24
3. 3 × 4	10	24	24	16	14	24
4. 2 × 5	10	24	24	—	16	24
5. 2 × 6	10	24	24	16	20	24

[1]Listed heights are distances between points of lateral support placed perpendicular to the plane of the wall. Increases in unsupported height are permitted where justified by an analysis.
[2]Shall not be used in exterior walls.

SOURCE: From the Uniform Building Code, © 1994, ICBO.

Table D–16 Allowable spans and loads for wood structural panel sheathing and single-floor grades continuous over two or more spans with strength axis perpendicular to supports.[1,2]

SHEATHING GRADES		ROOF[3]				FLOOR[4]
		Maximum Span (inches)		Load[5] (pounds per square foot)		Maximum Span (inches)
		× 25.4 for mm		× 0.0479 for kN/m²		
Panel Span Rating	Panel Thickness (inches)	With Edge Support[6]	Without Edge Support	Total Load	Live Load	
Roof/Floor Span	× 25.4 for mm					× 25.4 for mm
12/0	5/16	12	12	40	30	0
16/0	5/16, 3/8	16	16	40	30	0
20/0	5/16, 3/8	20	20	40	30	0
24/0	3/8, 7/16, 1/2	24	20[7]	40	30	0
24/16	7/16, 1/2	24	24	50	40	16
32/16	15/32, 1/2, 5/8	32	28	40	30	16[8]
40/20	19/32, 5/8, 3/4, 7/8	40	32	40	30	20[8,9]
48/24	23/32, 3/4, 7/8	48	36	45	35	24
54/32	7/8, 1	54	40	45	35	32
60/48	7/8, 1, 1 1/8	60	48	45	35	48
SINGLE-FLOOR GRADES		ROOF[3]				FLOOR[4]
		Maximum Span (inches)		Load[5] (pounds per square foot)		Maximum Span (inches)
		× 25.4 for mm		× 0.0479 for kN/m²		
Panel Span Rating (inches)	Panel Thickness (inches)	With Edge Support[6]	Without Edge Support	Total Load	Live Load	
× 25.4 for mm						× 25.4 for mm
16 oc	1/2, 19/32, 5/8	24	24	50	40	16[8]
20 oc	19/32, 5/8, 3/4	32	32	40	30	20[8,9]
24 oc	23/32, 3/4	48	36	35	25	24
32 oc	7/8, 1	48	40	50	40	32
48 oc	1 3/32, 1 1/8	60	48	50	50	48

[1] Applies to panels 24 inches (610 mm) or wider.
[2] Floor and roof sheathing conforming with this table shall be deemed to meet the design criteria of Section 2321.
[3] Uniform load deflection limitations 1/180 of span under live load plus dead load. 1/240 under live load only.
[4] Panel edges shall have approved tongue-and-groove joints or shall be supported with blocking unless 1/4-inch (6.4 mm) minimum thickness underlayment or 1 1/2 inches (38 mm) of approved cellular or lightweight concrete is placed over the subfloor, or finish floor is 3/4-inch (19 mm) wood strip. Allowable uniform load based on deflection of 1/360 of span is 100 pounds per square foot (psf) (4.79 kN/m²) except the span rating of 48 inches on center is based on a total load of 65 psf (3.11 kN/m).
[5] Allowable load at maximum span.
[6] Tongue-and-groove edges, panel edge clips [one midway between each support, except two equally spaced between supports 48 inches (1219 mm) on center], lumber blocking, or other. Only lumber blocking shall satisfy blocked diaphgrams requirements.
[7] For 1/2-inch (13 mm) panel, maximum span shall be 24 inches (610 mm).
[8] May be 24 inches (610 mm) on center where 3/4-inch (19 mm) wood strip flooring is installed at right angles to joist.
[9] May be 24 inches (610 mm) on center for floors where 1 1/2 inches (38 mm) of cellular or lightweight concrete is applied over the panels.

SOURCE: From the Uniform Building Code, © 1994, ICBO.

Table D–17 Allowable load (lb/sq ft) for wood structural panel roof sheathing continuous over two or more spans and strength axis parallel to supports (plywood structural panels are five-ply, five-layer unless otherwise noted).[1,2]

PANEL GRADE	THICKNESS (inch)	MAXIMUM SPAN (inches)	LOAD AT MAXIMUM SPAN (psf)	
	× 25.4 for mm		× 0.0479 for kN/m²	
			Live	Total
Structural I	7/16	24	20	30
	15/32	24	35[3]	45[3]
	1/2	24	40[3]	50[3]
	19/32, 5/8	24	70	80
	23/32, 3/4	24	90	100
Other grades covered in U.B.C. Standard 23-2 or 23-3	7/16	16	40	50
	15/32	24	20	25
	1/2	24	25	30
	19/32	24	40[3]	50[3]
	5/8	24	45[3]	55[3]
	23/32, 3/4	24	60[3]	65[3]

[1] Roof sheathing conforming with this table shall be deemed to meet the design criteria of Section 2321.
[2] Uniform load deflection limitations: 1/180 of span under live load plus dead load, 1/240 under live load only. Edges shall be blocked with lumber or other approved type of edge supports.
[3] For composite and four-ply plywood structural panel, load shall be reduced by 15 pounds per square foot (0.72 kN/m2).

SOURCE: From the Uniform Building Code, © 1994, ICBO.

Table D-18 Allowable loads for particleboard roof sheathing.[1,2,3]

GRADE	THICKNESS (inch)	MAXIMUM ON-CENTER SPACING OF SUPPORTS (inches)	LIVE LOAD (pounds per square foot)	TOTAL LOAD (pounds per square foot)
		× 25.4 for mm	× 0.0479 for kN/m²	
2-M-W	$3/8$[4]	16	45	65
	$7/16$	16	105	105
	$7/16$[4]	24	30	40
	$1/2$	16	110	150
	$1/2$	24	40	55

[1]Panels are continuous over two or more spans.

[2]Uniform load deflection limitation: $1/180$ of the span under live load plus dead load and $1/240$ of the span under live load only.

[3]Roof sheathing conforming with this table shall be deemed to meet the design criteria of Section 2321.

[4]Edges shall be tongue-and-groove or supported with blocking or edge clips.

SOURCE: From the Uniform Building Code, © 1994, ICBO.

Table D-19 Braced wall panels.[1]

SEISMIC ZONE	CONDITION	CONSTRUCTION METHOD[2,3]								BRACED PANEL LOCATION AND LENGTH[4]
		1	2	3	4	5	6	7	8	
0, 1 and 2A	One story, top of two or three story	X	X	X	X	X	X	X	X	Each end and not more than 25 feet (7620 mm) on center
	First story of two story or second story of three story	X	X	X	X	X	X	X	X	
	First story of three story		X	X	X	X[5]	X	X	X	
2B, 3 and 4	One story, top of two or three story		X	X	X	X	X	X	X	Each end and not more than 25 feet (7620 mm) on center
	First story of two story or second of three story		X	X	X	X[5]	X	X	X	Each end and not more than 25 feet (7620 mm) on center but not less than 25% of building length[6]
	First story of three story		X	X	X	X[5]	X	X	X	Each end and not more than 25 feet (7620 mm) on center but not less than 40% of building length[6]

[1]This table specifies minimum requirements for braced panels which form interior or exterior braced wall lines.

[2]See Section 2326.11.3 for full description.

[3]See Section 2326.11.4 for alternate braced panel requirement.

[4]Building length is the dimension parallel to the braced wall length.

[5]Gypsum wallboard applied to supports at 16 inches (406 mm) on center.

[6]The required lengths shall be doubled for gypsum board applied to only one face of a braced wall panel.

SOURCE: From the Uniform Building Code, © 1994, ICBO.

Table D–20 Single-ply gypsum wallboard applied parallel (‖) or perpendicular (⊥) to framing members.

THICKNESS OF GYPSUM WALLBOARD (inch) × 25.4 for mm	PLANE OF FRAMING SURFACE	MAXIMUM SPACING OF FRAMING MEMBER[1] (Center to Center) (inches) × 25.4 for mm	LONG DIMENSION OF GYPSUM WALLBOARD SHEETS IN RELATION TO DIRECTION OF FRAMING MEMBERS ‖	⊥	MAXIMUM SPACING OF FASTENERS[1] (Center to Center) (inches) × 25.4 for mm Nails[3]	Screws[4]	NAILS[2]—TO WOOD × 25.4 for mm
1/2	Horizontal	16	P	P	7	12	No. 13 gage, 1 1/8″ long, 19/64″ head; 0.098″ diameter, 1 1/4″ long, annular ringed; 5d, cooler or wallboard[5] nail (0.086″ dia., 1 5/8″ long, 15/64″ head).
		24	NP	P		12	
	Vertical	16	P	P	8	16	
		24	P	P		12	
5/8	Horizontal	16	P	P	7	12	No. 13 gage, 1 5/8″ long, 19/64″ head; 0.098″ diameter, 1 3/8″ long, annular ringed; 6d, cooler or wallboard[5] nail (0.092″ dia., 1 7/8″ long, 1/4″ head).
		24	NP	P		12	
	Vertical	16	P	P	8	16	
		24	P	P		12	

Nail or Screw Fastenings with Adhesives (Maximum Center to Center in Inches)							
× 25.4 for mm							
(Column headings as above)					End	Edges	Field	
1/2 or 5/8	Horizontal	16	P	P	16	16	24
		24	NP	P	16	24	24
	Vertical	24	P	P	16	24	NR

As required for 1/2″ and 5/8″ gypsum wallboard, see above.

NOTES: Horizontal refers to applications such as ceilings. Vertical refers to applications such as walls.

‖ denotes parallel.

⊥ denotes perpendicular.   P—Permitted.   NP—Not permitted.   NR—Not required.

[1] A combination of fasteners consisting of nails along the perimeter and screws in the field of the gypsum board may be used with the spacing of the fasteners shown in the table.

For fire-resistive construction, see Tables 7-B and 7-C. For shear-resisting elements, see Table 25-I.

[2] Where the metal framing has a clinching design formed to receive the nails by two edges of metal, the nails shall not be less than 5/8 inch (16 mm) longer than the wallboard thickness, and shall have ringed shanks. Where the metal framing has a nailing groove formed to receive the nails, the nails shall have barbed shanks or be 5d, No. 13 1/2 gage, 1 5/8 inches (41 mm) long, 15/64-inch (6.0 mm) head for 1/2-inch (12.7 mm) gypsum wallboard; 6d, No. 13 gage, 1 7/8 (48 mm) inches long, 15/64-inch (6.0 mm) head for 5/8-inch (16 mm) gypsum wallboard.

[3] Two nails spaced 2 inches to 2 1/2 inches (51 mm to 64 mm) apart may be used where the pairs are spaced 12 inches (305 mm) on center except around the perimeter of the sheets.

[4] Screws shall be long enough to penetrate into wood framing not less than 5/8 inch (16 mm) and through metal framing not less than 1/4 inch (6.4 mm).

[5] For properties of cooler or wallboard nails, see Chapter 23, Division III, Table 23-III-H.

SOURCE: From the Uniform Building Code, © 1994, ICBO.

Table D–21 Application of two-ply gypsum wallboard.[1]

Thickness of Gypsum Wallboard (Each Ply) (inch) × 25.4 for mm	Plane of Framing Surface	Long Dimension of Gypsum Wallboard Sheets	Maximum Spacing of Framing Members (Center to Center) (inches) × 25.4 for mm	Maximum Spacing of Fasteners (Center to Center) (inches) × 25.4 for mm				
				Base Ply			Face Ply	
				Nails[2]	Screws[3]	Staples[4]	Nails[2]	Screws[3]
FASTENERS ONLY								
3/8	Horizontal	Perpendicular only	16	16	24	16	7	12
	Vertical	Either direction	16				8	
1/2	Horizontal	Perpendicular only	24				7	
	Vertical	Either direction	24				8	
5/8	Horizontal	Perpendicular only	24				7	
	Vertical	Either direction	24				8	
FASTENERS AND ADHESIVES								
3/8	Horizontal	Perpendicular only	16	7	12	5	Temporary nailing or shoring to comply with Section 2511.4	
Base ply	Vertical	Either direction	24	8		7		
1/2	Horizontal	Perpendicular only	24	7		5		
Base ply	Vertical	Either direction	24	8		7		
5/8	Horizontal	Perpendicular only	24	7		5		
Base ply	Vertical	Either direction	24	8		7		

[1]For fire-resistive construction, see Tables 7-B and 7-C. For shear-resisting elements, see Table 25-I.

[2]Nails for wood framing shall be long enough to penetrate into wood members not less than 3/4 inch (19 mm), and the sizes shall conform with the provisions of Table 25-G. For nails not included in Table 25-G, use the appropriate size cooler or wallboard nails as set forth in Section 2340.1.2. Nails for metal framing shall conform with the provisions of Table 25-G.

[3]Screws shall conform with the provisions of Table 25-G.

[4]Staples shall not be less than No. 16 gage by 3/4-inch (19.1 mm) crown width with leg length of 7/8 inch (22.2 mm), 1 1/8 inches (28.6 mm) and 1 3/8 inches (34.9 mm) for gypsum wallboard thicknesses of 3/8 inch (9.5 mm), 1/2 inch (12.7 mm) and 5/8 inch (15.9 mm), respectively.

SOURCE: From the Uniform Building Code, © 1994, ICBO.

E.1

FLOOR JOISTS SPAN TABLES

E.1.1 Southern Pine Floor Joists Span Tables

Maximum spans are given in feet and inches.

Table E–1 Southern pine floor joists: 30-psf live load, 10-psf dead load, L/360; sleeping rooms and attic floors.

Size	Spacing	Grade									
inches	inches on center	Dense Select Structural	Select Structural	NonDense Select Structural	No. 1 Dense	No. 1	No. 1 NonDense	No. 2 Dense	No. 2	No. 2 NonDense	No. 3
2 x 6	12	12-6	12-3	12-0	12-3	12-0	11-10	12-0	11-10	11-3	10-5
	16	11-4	11-2	10-11	11-2	10-11	10-9	10-11	10-9	10-3	9-1
	24	9-11	9-9	9-7	9-9	9-7	9-4	9-7	9-4	8-11	7-5
2 x 8	12	16-6	16-2	15-10	16-2	15-10	15-7	15-10	15-7	14-11	13-3
	16	15-0	14-8	14-5	14-8	14-5	14-2	14-5	14-2	13-6	11-6
	24	13-1	12-10	12-7	12-10	12-7	12-4	12-7	12-4	11-9	9-5
2 x 10	12	21-0	20-8	20-3	20-8	20-3	19-10	20-3	19-10	19-0	15-8
	16	19-1	18-9	18-5	18-9	18-5	18-0	18-5	18-0	17-1	13-7
	24	16-8	16-5	16-1	16-5	16-1	15-8	15-8	14-8	13-11	11-1
2 x 12	12	25-7	25-1	24-8	25-1	24-8	24-2	24-8	24-2	23-1	18-8
	16	23-3	22-10	22-5	22-10	22-5	21-11	22-5	21-1	20-3	16-2
	24	20-3	19-11	19-7	19-11	19-6	18-8	18-8	17-2	16-7	13-2

SOURCE: Southern Pine Council.
NOTE: See notes following Table E–2.

Table E–2 Southern pine floor joists: 40-psf live load, 10-psf dead load, *L*/360; all rooms except sleeping rooms and attic floors.

Size	Spacing	Grade									
inches	inches on center	Dense Select Structural	Select Structural	NonDense Select Structural	No. 1 Dense	No. 1	No. 1 NonDense	No. 2 Dense	No. 2	No. 2 NonDense	No. 3
2 x 6	12	11-4	11-2	10-11	11-2	10-11	10-9	10-11	10-9	10-3	9-4
	16	10-4	10-2	9-11	10-2	9-11	9-9	9-11	9-9	9-4	8-1
	24	9-0	8-10	8-8	8-10	8-8	8-6	8-8	8-6	8-2	6-7
2 x 8	12	15-0	14-8	14-5	14-8	14-5	14-2	14-5	14-2	13-6	11-11
	16	13-7	13-4	13-1	13-4	13-1	12-10	13-1	12-10	12-3	10-3
	24	11-11	11-8	11-5	11-8	11-5	11-3	11-5	11-0	10-6	8-5
2 x 10	12	19-1	18-9	18-5	18-9	18-5	18-0	18-5	18-0	17-3	14-0
	16	17-4	17-0	16-9	17-0	16-9	16-5	16-9	16-1	15-3	12-2
	24	15-2	14-11	14-7	14-11	14-7	14-0	14-0	13-2	12-6	9-11
2 x 12	12	23-3	22-10	22-5	22-10	22-5	21-11	22-5	21-9	20-11	16-8
	16	21-1	20-9	20-4	20-9	20-4	19-11	20-4	18-10	18-2	14-5
	24	18-5	18-1	17-9	18-1	17-5	16-8	16-8	15-4	14-10	11-10

These spans are based on the 1993 AF&PA Span Tables for Joists and Rafters and the 1994 SPIB Grading Rules. They are intended for use in covered structures or where the moisture content in use does not exceed 19 percent for an extended period of time. Loading conditions are expressed in psf (pounds per square foot). Deflection is limited to span in inches divided by 360 and is based on live load only. Check sources of supply for availability of lumber in lengths greater than 20'-0".

☐ These grades are the most commonly available.

SOURCE: Southern Pine Council.

E.1.2 Western Lumber Floor Joists Span Tables

Table E–3 Western lumber floor joists: 40-psf live load, 10-psf dead load, $L/360$. Residential occupancies include private dwelling, private apartment, and hotel guest rooms. Deck is under CABO and Standard Codes.[1]

Species or Group	Grade	Span (feet and inches)											
		2 × 6			2 × 8			2 × 10			2 × 12		
		12" oc	16" oc	24" oc	12" oc	16" oc	24" oc	12" oc	16" oc	24" oc	12" oc	16" oc	24" oc
Douglas Fir-Larch	Sel. Struc.	11-4	10-4	9-0	15-0	13-7	11-11	19-1	17-4	15-2	23-3	21-1	18-5
	No. 1 & Btr.	11-2	10-2	8-10	14-8	13-4	11-8	18-9	17-0	14-5	22-10	20-5	16-8
	No. 1	10-11	9-11	8-8	14-5	13-1	11-0	18-5	16-5	13-5	22-0	19-1	15-7
	No. 2	10-9	9-9	8-1	14-2	12-7	10-3	17-9	15-5	12-7	20-7	17-10	14-7
	No. 3	8-8	7-6	6-2	11-0	9-6	7-9	13-5	11-8	9-6	15-7	13-6	11-0
Douglas Fir-South	Sel. Struc.	10-3	9-4	8-2	13-6	12-3	10-9	17-3	15-8	13-8	21-0	19-1	16-8
	No. 1	10-0	9-1	7-11	13-2	12-0	10-5	16-10	15-3	12-9	20-6	18-1	14-9
	No. 2	9-9	8-10	7-9	12-10	11-8	10-0	16-5	14-11	12-2	19-11	17-4	14-2
	No. 3	8-6	7-4	6-0	10-9	9-3	7-7	13-1	11-4	9-3	15-2	13-2	10-9
Hem-Fir	Sel. Struc.	10-9	9-9	8-6	14-2	12-10	11-3	18-0	16-5	14-4	21-11	19-11	17-5
	No. 1 & Btr.	10-6	9-6	8-4	13-10	12-7	11-0	17-8	16-0	13-9	21-6	19-6	16-0
	No. 1	10-6	9-6	8-4	13-10	12-7	10-9	17-8	16-0	13-1	21-6	18-7	15-2
	No. 2	10-0	9-1	7-11	13-2	12-0	10-2	16-10	15-2	12-5	20-4	17-7	14-4
	No. 3	8-8	7-6	6-2	11-0	9-6	7-9	13-5	11-8	9-6	15-7	13-6	11-0
Spruce-Pine-Fir (South)	Sel. Struc.	10-0	9-1	7-11	13-2	12-0	10-6	16-10	15-3	13-4	20-6	18-7	16-3
	No. 1	9-9	8-10	7-9	12-10	11-8	10-2	16-5	14-11	12-5	19-11	17-7	14-4
	No. 2	9-6	8-7	7-6	12-6	11-4	9-6	15-11	14-3	11-8	19-1	16-6	13-6
	No. 3	8-0	6-11	5-8	10-2	8-9	7-2	12-5	10-9	8-9	14-4	12-5	10-2
Western Woods	Sel. Struc.	9-9	8-10	7-9	12-10	11-8	10-2	16-5	14-11	12-7	19-11	17-10	14-7
	No. 1	9-6	8-7	7-0	12-6	10-10	8-10	15-4	13-3	10-10	17-9	15-5	12-7
	No. 2	9-2	8-4	7-0	12-1	10-10	8-10	15-4	13-3	10-10	17-9	15-5	12-7
	No. 3	7-6	6-6	5-4	9-6	8-3	6-9	11-8	10-1	8-3	13-6	11-8	9-6

[1] Deck spans are based on normal conditions of use and assume the moisture content of lumber used in decks will not be maintained at a moisture content in excess of 19% for an extended period of time.

SOURCE: Western Wood Products Association.

Table E–4 Western lumber floor joists: 30-psf live load, 10-psf dead load, *L*/360. Residential occupancy sleeping rooms (BOCA and SBCCI only); attics with storage under the Standard Code. Does not apply in UBC areas.

Species or Group	Grade	2 × 6			2 × 8			2 × 10			2 × 12		
		12″ oc	16″ oc	24″ oc	12″ oc	16″ oc	24″ oc	12″ oc	16″ oc	24″ oc	12″ oc	16″ oc	24″ oc
Douglas Fir-Larch	Sel. Struc.	12-6	11-4	9-11	16-6	15-0	13-1	21-0	19-1	16-8	25-7	23-3	20-3
	No. 1 & Btr.	12-3	11-2	9-9	16-2	14-8	12-10	20-8	18-9	16-1	25-1	22-10	18-8
	No. 1	12-0	10-11	9-7	15-10	14-5	12-4	20-3	18-5	15-0	24-8	21-4	17-5
	No. 2	11-10	10-9	9-1	15-7	14-1	11-6	19-10	17-2	14-1	23-0	19-11	16-3
	No. 3	9-8	8-5	6-10	12-4	10-8	8-8	15-0	13-0	10-7	17-5	15-1	12-4
Douglas Fir-South	Sel. Struc.	11-3	10-3	8-11	14-11	13-6	11-10	19-0	17-3	15-1	23-1	21-0	18-4
	No. 1	11-0	10-0	8-9	14-6	13-2	11-6	18-6	16-10	14-3	22-6	20-3	16-6
	No. 2	10-9	9-9	8-6	14-2	12-10	11-2	18-0	16-5	13-8	21-11	19-4	15-10
	No. 3	9-6	8-2	6-8	12-0	10-5	8-6	14-8	12-8	10-4	17-0	14-8	12-0
Hem-Fir	Sel. Struc.	11-10	10-9	9-4	15-7	14-2	12-4	19-10	18-0	15-9	24-2	21-11	19-2
	No. 1 & Btr.	11-7	10-6	9-2	15-3	13-10	12-1	19-5	17-8	15-5	23-7	21-6	17-10
	No. 1	11-7	10-6	9-2	15-3	13-10	12-0	19-5	17-8	14-8	23-7	20-9	17-0
	No. 2	11-0	10-0	8-9	14-6	13-2	11-4	18-6	16-10	13-10	22-6	19-8	16-1
	No. 3	9-8	8-5	6-10	12-4	10-8	8-8	15-0	13-0	10-7	17-5	15-1	12-4
Spruce-Pine-Fir (South)	Sel. Struc.	11-0	10-0	8-9	14-6	13-2	11-6	18-6	16-10	14-8	22-6	20-6	17-11
	No. 1	10-9	9-9	8-6	14-2	12-10	11-3	18-0	16-5	13-10	21-11	19-8	16-1
	No. 2	10-5	9-6	8-3	13-9	12-6	10-8	17-6	15-11	13-0	21-4	18-6	15-1
	No. 3	8-11	7-9	6-4	11-4	9-10	8-0	13-10	12-0	9-9	16-1	13-11	11-4
Western Woods	Sel. Struc.	10-9	9-9	8-6	14-2	12-10	11-3	18-0	16-5	14-1	21-11	19-11	16-3
	No. 1	10-5	9-6	7-10	13-9	12-2	9-11	17-1	14-10	12-1	19-10	17-2	14-0
	No. 2	10-1	9-2	7-10	13-4	12-1	9-11	17-0	14-10	12-1	19-10	17-2	14-0
	No. 3	8-5	7-3	5-11	10-8	9-3	7-6	13-0	11-3	9-2	15-1	13-1	10-8

SOURCE: Western Wood Products Association.

CEILING JOISTS SPAN TABLES

E.2.1 Southern Pine Ceiling Joists Span Tables

Maximum spans are given in feet and inches.

Table E-5 Southern pine ceiling joists: 10-psf live load, 5-psf dead load, *L*/240; drywall ceiling, no attic storage.

Size	Spacing	Grade										
inches	inches on center	Dense Select Structural	Select Structural	NonDense Select Structural	No. 1 Dense	No. 1	No. 1 NonDense	No. 2 Dense	No. 2	No. 2 NonDense	No. 3	Standard
2 x 4	12	13-2	12-11	12-8	12-11	12-8	12-5	12-8	12-5	11-10	11-7	9-11
	16	11-11	11-9	11-6	11-9	11-6	11-3	11-6	11-3	10-9	10-0	8-7
	24	10-5	10-3	10-0	10-3	10-0	9-10	10-0	9-10	9-5	8-2	7-0
2 x 6	12	20-8	20-3	19-11	20-3	19-11	19-6	19-11	19-6	18-8	17-1	
	16	18-9	18-5	18-1	18-5	18-1	17-8	18-1	17-8	16-11	14-9	
	24	16-4	16-1	15-9	16-1	15-9	15-6	15-9	15-6	14-9	12-1	
2 x 8	12	26-0*	26-0*	26-0*	26-0*	26-0*	25-8	26-0*	25-8	24-7	21-8	
	16	24-8	24-3	23-10	24-3	23-10	23-4	23-10	23-4	22-4	18-9	
	24	21-7	21-2	20-10	21-2	20-10	20-5	20-10	20-1	19-2	15-4	
2 x 10	12	26-0*	26-0*	26-0*	26-0*	26-0*	26-0*	26-0*	26-0*	26-0*	25-7	
	16	26-0*	26-0*	26-0*	26-0*	26-0*	26-0*	26-0*	26-0*	26-0*	22-2	
	24	26-0*	26-0*	26-0*	26-0*	26-0*	25-7	25-7	24-0	22-9	18-1	

SOURCE: Southern Pine Council.
NOTE: See notes following Table E–6.

Table E–6 Southern pine ceiling joists: 20-psf live load, 10-psf dead load, L/240. Drywall ceiling. No future room development, but limited attic storage is available.

Size (inches)	Spacing (inches on center)	Dense Select Structural	Select Structural	NonDense Select Structural	No. 1 Dense	No. 1	No. 1 NonDense	No. 2 Dense	No. 2	No. 2 NonDense	No. 3	Standard
2 x 4	12	10-5	10-3	10-0	10-3	10-0	9-10	10-0	9-10	9-5	8-2	7-0
	16	9-6	9-4	9-1	9-4	9-1	8-11	9-1	8-11	8-7	7-1	6-1
	24	8-3	8-1	8-0	8-1	8-0	7-10	8-0	7-8	7-3	5-9	4-11
2 x 6	12	16-4	16-1	15-9	16-1	15-9	15-6	15-9	15-6	14-9	12-1	
	16	14-11	14-7	14-4	14-7	14-4	14-1	14-4	13-6	12-11	10-5	
	24	13-0	12-9	12-6	12-9	12-6	12-0	11-10	11-0	10-6	8-6	
2 x 8	12	21-7	21-2	20-10	21-2	20-10	20-5	20-10	20-1	19-2	15-4	
	16	19-7	19-3	18-11	19-3	18-11	18-5	18-9	17-5	16-7	13-3	
	24	17-2	16-10	16-6	16-8	15-11	15-1	15-4	14-2	13-7	10-10	
2 x 10	12	26-0*	26-0*	26-0*	26-0*	26-0*	25-7	25-7	24-0	22-9	18-1	
	16	25-0	24-7	24-1	24-5	23-2	22-2	22-2	20-9	19-9	15-8	
	24	21-10	21-6	21-1	19-11	18-11	18-1	18-1	17-0	16-1	12-10	

These spans are based on the 1993 AF&PA Span Tables for Joists and Rafters and the 1994 SPIB Grading Rules. They are intended for use in covered structures or where the moisture content in use does not exceed 19 percent for an extended period of time. Loading conditions are expressed in psf (pounds per square foot). Deflection is limited to span in inches divided by 240 and is based on live load only. Check sources of supply for availability of lumber in lengths greater than 20'-0".

*The listed maximum span has been limited to 26'-0" based on material availability.

☐ These grades are the most commonly available.

SOURCE: Southern Pine Council.

E.2.2 Western Lumber Ceiling Joists Span Tables

Table E–7 Western lumber ceiling joists: 10-psf live load, 10-lb dead load, L/360. Use these loading conditions for the following: no attic storage, ceilings where the roof slope is not steeper than 3:12, plaster ceilings.

Species or Group	Grade	2 x 4 12" oc	2 x 4 16" oc	2 x 4 24" oc	2 x 6 12" oc	2 x 6 16" oc	2 x 6 24" oc	2 x 8 12" oc	2 x 8 16" oc	2 x 8 24" oc	2 x 10 12" oc	2 x 10 16" oc	2 x 10 24" oc
Douglas Fir-Larch	Sel. Struc.	11-6	10-5	9-1	18-0	16-4	14-4	23-9	21-7	18-10	30-4	27-6	24-1
	No. 1 & Btr.	11-3	10-3	8-11	17-8	16-1	14-1	23-4	21-2	18-6	29-9	27-1	22-9
	No. 1	11-1	10-0	8-9	17-4	15-9	13-9	22-11	20-10	17-5	29-2	26-0	21-3
	No. 2	10-10	9-10	8-7	17-0	15-6	12-10	22-5	19-11	16-3	28-1	24-4	19-10
	No. 3	9-5	8-2	6-8	13-9	11-11	9-8	17-5	15-1	12-4	21-3	18-5	15-0
Douglas Fir-South	Sel. Struc.	10-4	9-5	8-3	16-3	14-9	12-11	21-5	19-6	17-0	27-5	24-10	21-9
	No. 1	10-1	9-2	8-0	15-11	14-5	12-7	20-11	19-0	16-6	26-9	24-3	20-2
	No. 2	9-10	8-11	7-10	15-6	14-1	12-3	20-5	18-6	15-9	26-0	23-7	19-3
	No. 3	9-2	7-11	6-6	13-5	11-7	9-6	16-11	14-8	12-0	20-8	17-11	14-8
Hem-Fir	Sel. Struc.	10-10	9-10	8-7	17-0	15-6	13-6	22-5	20-5	17-10	28-7	26-0	22-9
	No. 1 & Btr.	10-7	9-8	8-5	16-8	15-2	13-3	21-11	19-11	17-5	28-0	25-5	21-9
	No. 1	10-7	9-8	8-5	16-8	15-2	13-3	21-11	19-11	16-11	28-0	25-4	20-8
	No. 2	10-1	9-2	8-0	15-11	14-5	12-7	20-11	19-0	16-0	26-9	24-0	19-7
	No. 3	9-5	8-2	6-8	13-9	11-11	9-8	17-5	15-1	12-4	21-3	18-5	15-0
Spruce-Pine-Fir (South)	Sel. Struc.	10-1	9-2	8-0	15-11	14-5	12-7	20-11	19-0	16-7	26-9	24-3	21-2
	No. 1	9-10	8-11	7-10	15-6	14-1	12-3	20-5	18-6	16-0	26-0	23-8	19-7
	No. 2	9-7	8-8	7-7	15-0	13-8	11-11	19-10	18-0	15-1	25-3	22-6	18-5
	No. 3	8-8	7-6	6-1	12-8	11-0	8-11	16-0	13-11	11-4	19-7	16-11	13-10
Western Woods	Sel. Struc.	9-10	8-11	7-10	15-6	14-1	12-3	20-5	18-6	16-2	26-0	23-8	19-10
	No. 1	9-7	8-8	7-7	15-0	13-7	11-1	19-10	17-2	14-0	24-3	21-0	17-1
	No. 2	9-3	8-5	7-4	14-7	13-3	11-1	19-2	17-2	14-0	24-3	21-0	17-1
	No. 3	8-2	7-0	5-9	11-11	10-4	8-5	15-1	13-0	10-8	18-5	15-11	13-0

SOURCE: Western Wood Products Association.

Table E-8 Western lumber ceiling joists: 10-psf live load, 5-psf dead load, L/240. Use these loading conditions for the following: no attic storage, ceilings where the roof slope is not steeper than 3:12, drywall ceilings.

Species or Group	Grade	2 × 4 12" oc	2 × 4 16" oc	2 × 4 24" oc	2 × 6 12" oc	2 × 6 16" oc	2 × 6 24" oc	2 × 8 12" oc	2 × 8 16" oc	2 × 8 24" oc	2 × 10 12" oc	2 × 10 16" oc	2 × 10 24" oc
Douglas Fir-Larch	Sel. Struc.	13-2	11-11	10-5	20-8	18-9	16-4	27-2	24-8	21-7	34-8	31-6	27-6
	No. 1 & Btr.	12-11	11-9	10-3	20-3	18-5	16-1	26-9	24-3	21-2	34-1	31-0	26-4
	No. 1	12-8	11-6	10-0	19-11	18-1	15-9	26-2	23-10	20-1	33-5	30-0	24-6
	No. 2	12-5	11-3	9-10	19-6	17-8	14-10	25-8	23-0	18-9	32-5	28-1	22-11
	No. 3	10-10	9-5	7-8	15-10	13-9	11-2	20-1	17-5	14-2	24-6	21-3	17-4
Douglas Fir-South	Sel. Struc.	11-10	10-9	9-5	18-8	16-11	14-9	24-7	22-4	19-6	31-4	28-6	24-10
	No. 1	11-7	10-6	9-2	18-2	16-6	14-5	24-0	21-9	19-0	30-7	27-9	23-3
	No. 2	11-3	10-3	8-11	17-8	16-1	14-1	23-4	21-2	18-3	29-9	27-1	22-3
	No. 3	10-7	9-2	7-6	15-5	13-5	10-11	19-7	16-11	13-10	23-11	20-8	16-11
Hem-Fir	Sel. Struc.	12-5	11-3	9-10	19-6	17-8	15-6	25-8	23-4	20-5	32-9	29-9	26-0
	No. 1 & Btr.	12-2	11-0	9-8	19-1	17-4	15-2	25-2	22-10	19-11	32-1	29-2	25-2
	No. 1	12-2	11-0	9-8	19-1	17-4	15-2	25-2	22-10	19-7	32-1	29-2	23-11
	No. 2	11-7	10-6	9-2	18-2	16-6	14-5	24-0	21-9	18-6	30-7	27-8	22-7
	No. 3	10-10	9-5	7-8	15-10	13-9	11-2	20-1	17-5	14-2	24-6	21-3	17-4
Spruce-Pine-Fir (South)	Sel. Struc.	11-7	10-6	9-2	18-2	16-6	14-5	24-0	21-9	19-0	30-7	27-9	24-3
	No. 1	11-3	10-3	8-11	17-8	16-1	14-1	23-4	21-2	18-6	29-9	27-1	22-7
	No. 2	10-11	9-11	8-8	17-2	15-7	13-8	22-8	20-7	17-5	28-11	26-0	21-3
	No. 3	10-0	8-8	7-1	14-7	12-8	10-4	18-6	16-0	13-1	22-7	19-7	16-0
Western Woods	Sel. Struc.	11-3	10-3	8-11	17-8	16-1	14-1	23-4	21-2	18-6	29-9	27-1	22-11
	No. 1	10-11	9-11	8-8	17-2	15-7	12-9	22-8	19-10	16-2	28-0	24-3	19-9
	No. 2	10-7	9-8	8-5	16-8	15-2	12-9	21-11	19-10	16-2	28-0	24-3	19-9
	No. 3	9-5	8-2	6-8	13-9	11-11	9-8	17-5	15-1	12-4	21-3	18-5	15-0

SOURCE: Western Wood Products Association.

Table E-9 Western lumber ceiling joists: 20-psf live load, 10-psf dead load, L/360. Use these loading conditions for the following: limited attic storage where development is not possible; ceilings where the roof slope is steeper than 3:12; where the clear height in the attic is greater than 30"; plaster ceilings.

Species or Group	Grade	2 × 4 12" oc	2 × 4 16" oc	2 × 4 24" oc	2 × 6 12" oc	2 × 6 16" oc	2 × 6 24" oc	2 × 8 12" oc	2 × 8 16" oc	2 × 8 24" oc	2 × 10 12" oc	2 × 10 16" oc	2 × 10 24" oc
Douglas Fir-Larch	Sel. Struc.	9-1	8-3	7-3	14-4	13-0	11-4	18-10	17-2	15-0	24-1	21-10	19-1
	No. 1 & Btr.	8-11	8-1	7-1	14-1	12-9	11-2	18-6	16-10	14-8	23-8	21-6	18-7
	No. 1	8-9	8-0	7-0	13-9	12-6	10-11	18-2	16-6	14-2	23-2	21-1	17-4
	No. 2	8-7	7-10	6-10	13-6	12-3	10-6	17-10	16-2	13-3	22-9	19-10	16-3
	No. 3	7-8	6-8	5-5	11-2	9-8	7-11	14-2	12-4	10-0	17-4	15-0	12-3
Douglas Fir-South	Sel. Struc.	8-3	7-6	6-6	12-11	11-9	10-3	17-0	15-6	13-6	21-9	19-9	17-3
	No. 1	8-0	7-3	6-4	12-7	11-5	10-0	16-7	15-1	13-2	21-2	19-3	16-5
	No. 2	7-10	7-1	6-2	12-3	11-2	9-9	16-2	14-8	12-10	20-8	18-9	15-9
	No. 3	7-6	6-6	5-3	10-11	9-6	7-9	13-10	12-0	9-9	16-11	14-8	11-11
Hem-Fir	Sel. Struc.	8-7	7-10	6-10	13-6	12-3	10-9	17-10	16-2	14-2	22-9	20-8	18-0
	No. 1 & Btr.	8-5	7-8	6-8	13-3	12-0	10-6	17-5	15-10	13-10	22-3	20-2	17-8
	No. 1	8-5	7-8	6-8	13-3	12-0	10-6	17-5	15-10	13-10	22-3	20-2	16-11
	No. 2	8-0	7-3	6-4	12-7	11-5	10-0	16-7	15-1	13-1	21-2	19-3	16-0
	No. 3	7-8	6-8	5-5	11-2	9-8	7-11	14-2	12-4	10-0	17-4	15-0	12-3
Spruce-Pine-Fir (South)	Sel. Struc.	8-0	7-3	6-4	12-7	11-5	10-0	16-7	15-1	13-2	21-2	19-3	16-10
	No. 1	7-10	7-1	6-2	12-3	11-2	9-9	16-2	14-8	12-10	20-8	18-9	16-0
	No. 2	7-7	6-11	6-0	11-11	10-10	9-6	15-9	14-3	12-4	20-1	18-3	15-0
	No. 3	7-1	6-1	5-0	10-4	8-11	7-4	13-1	11-4	9-3	16-0	13-10	11-4
Western Woods	Sel. Struc.	7-10	7-1	6-2	12-3	11-2	9-9	16-2	14-8	12-10	20-8	18-9	16-3
	No. 1	7-7	6-11	6-0	11-11	10-10	9-0	15-9	14-0	11-5	19-9	17-1	14-0
	No. 2	7-4	6-8	5-10	11-7	10-6	9-0	15-3	13-10	11-5	19-5	17-1	14-0
	No. 3	6-8	5-9	4-8	9-8	8-5	6-10	12-4	10-8	8-8	15-0	13-0	10-7

SOURCE: Western Wood Products Association.

Table E–10 Western lumber ceiling joists: 20-psf live load, 10-psf dead load, L/240. Use these loading conditions for the following: limited attic storage where development of future rooms is not possible; ceilings where the roof slope is steeper than 3 : 12; where the clear height in the attic is greater than 30"; drywall ceilings.

Species or Group	Grade	2 × 4			2 × 6			2 × 8			2 × 10		
		12" oc	16" oc	24" oc	12" oc	16" oc	24" oc	12" oc	16" oc	24" oc	12" oc	16" oc	24" oc
Douglas Fir-Larch	Sel. Struc.	10-5	9-6	8-3	16-4	14-11	13-0	21-7	19-7	17-1	27-6	25-0	20-11
	No. 1 & Btr.	10-3	9-4	8-1	16-1	14-7	12-0	21-2	18-8	15-3	26-4	22-9	18-7
	No. 1	10-0	9-1	7-8	15-9	13-9	11-2	20-1	17-5	14-2	24-6	21-3	17-4
	No. 2	9-10	8-9	7-2	14-10	12-10	10-6	18-9	16-3	13-3	22-11	19-10	16-3
	No. 3	7-8	6-8	5-5	11-2	9-8	7-11	14-2	12-4	10-0	17-4	15-0	12-3
Douglas Fir-South	Sel. Struc.	9-5	8-7	7-6	14-9	13-5	11-9	19-6	17-9	15-6	24-10	22-7	19-9
	No. 1	9-2	8-4	7-3	14-5	13-0	10-8	19-0	16-6	13-6	23-3	20-2	16-5
	No. 2	8-11	8-1	7-0	14-1	12-6	10-2	18-3	15-9	12-11	22-3	19-3	15-9
	No. 3	7-6	6-6	5-3	10-11	9-6	7-9	13-10	12-0	9-9	16-11	14-8	11-11
Hem-Fir	Sel. Struc.	9-10	8-11	7-10	15-6	14-1	12-3	20-5	18-6	16-2	26-0	23-8	20-6
	No. 1 & Btr.	9-8	8-9	7-8	15-2	13-9	11-6	19-11	17-10	14-7	25-2	21-9	17-9
	No. 1	9-8	8-9	7-6	15-2	13-5	10-11	19-7	16-11	13-10	23-11	20-8	16-11
	No. 2	9-2	8-4	7-1	14-5	12-8	10-4	18-6	16-0	13-1	22-7	19-7	16-0
	No. 3	7-8	6-8	5-5	11-2	9-8	7-11	14-2	12-4	10-0	17-4	15-0	12-3
Spruce-Pine-Fir (South)	Sel. Struc.	9-2	8-4	7-3	14-5	13-1	11-5	19-0	17-3	15-1	24-3	22-1	19-3
	No. 1	8-11	8-1	7-1	14-1	12-8	10-4	18-6	16-0	13-1	22-7	19-7	16-0
	No. 2	8-8	7-11	6-8	13-8	11-11	9-8	17-5	15-1	12-4	21-3	18-5	15-0
	No. 3	7-1	6-1	5-0	10-4	8-11	7-4	13-1	11-4	9-3	16-0	13-10	11-4
Western Woods	Sel. Struc.	8-11	8-1	7-1	14-1	12-9	10-6	18-6	16-3	13-3	22-11	19-10	16-3
	No. 1	8-8	7-7	6-2	12-9	11-1	9-0	16-2	14-0	11-5	19-9	17-1	14-0
	No. 2	8-5	7-7	6-2	12-9	11-1	9-0	16-2	14-0	11-5	19-9	17-1	14-0
	No. 3	6-8	5-9	4-8	9-8	8-5	6-10	12-4	10-8	8-8	15-0	13-0	10-7

SOURCE: Western Wood Products Association.

E.3

RAFTERS SPAN TABLES

E.3.1 Southern Pine Rafters Span Tables

Maximum spans are given in feet and inches.

Table E–11 Southern pine rafters: 20-psf live load, 10-psf dead load, L/240, $C_D = 1.15$. Light roofing, drywall ceiling, snow load.

Size	Spacing	Grade									
inches	inches on center	Dense Select Structural	Select Structural	NonDense Select Structural	No. 1 Dense	No. 1	No. 1 NonDense	No. 2 Dense	No. 2	No. 2 NonDense	No. 3
2 x 6	12	16-4	16-1	15-9	16-1	15-9	15-6	15-9	15-6	14-9	12-11
	16	14-11	14-7	14-4	14-7	14-4	14-1	14-4	14-1	13-5	11-2
	24	13-0	12-9	12-6	12-9	12-6	12-3	12-6	11-9	11-4	9-1
2 x 8	12	21-7	21-2	20-10	21-2	20-10	20-5	20-10	20-5	19-6	16-5
	16	19-7	19-3	18-11	19-3	18-11	18-6	18-11	18-6	17-9	14-3
	24	17-2	16-10	16-6	16-10	16-6	16-2	16-5	15-3	14-7	11-7
2 x 10	12	26-0*	26-0*	26-0*	26-0*	26-0*	26-0	26-0*	25-8	24-6	19-5
	16	25-0	24-7	24-1	24-7	24-1	23-8	23-10	22-3	21-2	16-10
	24	21-10	21-6	21-1	21-4	20-3	19-5	19-5	18-2	17-4	13-9
2 x 12	12	26-0*	26-0*	26-0*	26-0*	26-0*	26-0*	26-0*	26-0*	26-0*	23-1
	16	26-0*	26-0*	26-0*	26-0*	26-0*	26-0*	26-0*	26-0*	25-1	20-0
	24	26-0*	26-0*	25-7	25-1	24-1	23-1	23-1	21-4	20-5	16-4

SOURCE: Southern Pine Council.
NOTE: See notes following Table E–12.

Table E–12 Southern pine rafters: 30-psf live load, 10-psf dead load, $L/240$, $C_D = 1.15$. Light roofing, drywall ceiling, snow load.

Size	Spacing	Grade									
inches	inches on center	Dense Select Structural	Select Structural	NonDense Select Structural	No. 1 Dense	No. 1	No. 1 NonDense	No. 2 Dense	No. 2	No. 2 NonDense	No. 3
2 x 6	12	14-4	14-1	13-9	14-1	13-9	13-6	13-9	13-6	12-11	11-2
	16	13-0	12-9	12-6	12-9	12-6	12-3	12-6	12-3	11-9	9-8
	24	11-4	11-2	10-11	11-2	10-11	10-9	10-11	10-2	9-9	7-11
2 x 8	12	18-10	18-6	18-2	18-6	18-2	17-10	18-2	17-10	17-0	14-3
	16	17-2	16-10	16-6	16-10	16-6	16-2	16-6	16-2	15-5	12-4
	24	15-0	14-8	14-5	14-8	14-5	14-0	14-3	13-2	12-7	10-1
2 x 10	12	24-1	23-8	23-2	23-8	23-2	22-9	23-2	22-3	21-2	16-10
	16	21-10	21-6	21-1	21-6	21-1	20-7	20-7	19-3	18-4	14-7
	24	19-1	18-9	18-5	18-6	17-6	16-10	16-10	15-9	15-0	11-11
2 x 12	12	26-0*	26-0*	26-0*	26-0*	26-0*	26-0*	26-0*	26-0*	25-1	20-0
	16	26-0*	26-0*	25-7	26-0*	25-7	24-6	24-6	22-7	21-8	17-4
	24	23-3	22-10	22-5	21-9	20-10	20-0	20-0	18-5	17-9	14-2

These spans are based on the 1993 AF&PA Span Tables for Joists and Rafters and the 1994 SPIB Grading Rules. They are intended for use in covered structures or where the moisture content in use does not exceed 19 percent for an extended period of time. Loading conditions are expressed in psf (pounds per square foot). Deflection is limited to span in inches divided by 240 and is based on live load only. The load duration factor, C_D, is 1.15 for snow loads. Check sources of supply for availability of lumber in lengths greater than 20'-0".

*The listed maximum span has been limited to 26'-0" based on material availability.

☐ These grades are the most commonly available.

SOURCE: Southern Pine Council.

E.3.2 Western Lumber Rafters Span Tables

Table E–13 Western lumber roof rafters: 20-psf live load, 30-psf dead load, $L/360$. No snow load;[1] heavy roof covering; flat roof or cathedral ceiling with plaster ceiling.

Species or Group	Grade	Span (feet and inches)											
		2 × 6			2 × 8			2 × 10			2 × 12		
		12" oc	16" oc	24" oc	12" oc	16" oc	24" oc	12" oc	16" oc	24" oc	12" oc	16" oc	24" oc
Douglas Fir-Larch	Sel. Struc.	14-4	13-0	11-4	18-10	17-2	14-10	24-1	21-10	18-1	29-3	25-8	21-0
	No. 1 & Btr.	14-1	12-9	10-5	18-6	16-2	13-2	22-9	19-9	16-1	26-5	22-10	18-8
	No. 1	13-9	11-11	9-8	17-5	15-1	12-4	21-3	18-5	15-0	24-8	21-4	17-5
	No. 2	12-10	11-1	9-1	16-3	14-1	11-6	19-10	17-2	14-1	23-0	19-11	16-3
	No. 3	9-8	8-5	6-10	12-4	10-8	8-8	15-0	13-0	10-7	17-5	15-1	12-4
Douglas Fir-South	Sel. Struc.	12-11	11-9	10-3	17-0	15-6	13-6	21-9	19-9	17-1	26-5	24-0	19-10
	No. 1	12-7	11-3	9-3	16-6	14-3	11-8	20-2	17-5	14-3	23-4	20-3	16-6
	No. 2	12-3	10-10	8-10	15-9	13-8	11-2	19-3	16-8	13-8	22-4	19-4	15-10
	No. 3	9-6	8-2	6-8	12-0	10-5	8-6	14-8	12-8	10-4	17-0	14-8	12-0
Hem-Fir	Sel. Struc.	13-6	12-3	10-9	17-10	16-2	14-2	22-9	20-8	17-9	27-8	25-1	20-7
	No. 1 & Btr.	13-3	12-0	9-11	17-5	15-5	12-7	21-9	18-10	15-5	25-3	21-10	17-10
	No. 1	13-3	11-7	9-6	16-11	14-8	12-0	20-8	17-11	14-8	24-0	20-9	17-0
	No. 2	12-7	11-0	8-11	16-0	13-11	11-4	19-7	16-11	13-10	22-8	19-8	16-1
	No. 3	9-8	8-5	6-10	12-4	10-8	8-8	15-0	13-0	10-7	17-5	15-1	12-4
Spruce-Pine-Fir (South)	Sel. Struc.	12-7	11-5	10-0	16-7	15-1	13-2	21-2	19-3	16-10	25-9	23-5	19-10
	No. 1	12-3	11-0	8-11	16-0	13-11	11-4	19-7	16-11	13-10	22-8	19-8	16-1
	No. 2	11-11	10-4	8-5	15-1	13-0	10-8	18-5	15-11	13-0	21-4	18-6	15-1
	No. 3	8-11	7-9	6-4	11-4	9-10	8-0	13-10	12-0	9-9	16-1	13-11	11-4
Western Woods	Sel. Struc.	12-3	11-1	9-1	16-2	14-1	11-6	19-10	17-2	14-1	23-0	19-11	16-3
	No. 1	11-1	9-7	7-10	14-0	12-2	9-11	17-1	14-10	12-1	19-10	17-2	14-0
	No. 2	11-1	9-7	7-10	14-0	12-2	9-11	17-1	14-10	12-1	19-10	17-2	14-0
	No. 3	8-5	7-3	5-11	10-8	9-3	7-6	13-0	11-3	9-2	15-1	13-1	10-8

[1] A 1.25 Duration of Load adjustment has been applied.

SOURCE: Western Wood Products Association.

Table E-14 Western lumber roof rafters: 20-psf snow load, 10-psf dead load, *L*/240. Roof slope is 3:12 or less; light roof covering; no ceiling finish.

Species or Group	Grade	Span (feet and inches) 2 × 6			2 × 8			2 × 10			2 × 12		
		12" oc	16" oc	24" oc	12" oc	16" oc	24" oc	12" oc	16" oc	24" oc	12" oc	16" oc	24" oc
Douglas Fir-Larch	Sel. Struc.	16-4	14-11	13-0	21-7	19-7	17-2	27-6	25-0	21-10	33-6	30-5	26-0
	No. 1 & Btr.	16-1	14-7	12-9	21-2	19-3	16-4	27-1	24-5	19-11	32-8	28-4	23-1
	No. 1	15-9	14-4	12-0	20-10	18-8	15-3	26-4	22-9	18-7	30-6	26-5	21-7
	No. 2	15-6	13-9	11-3	20-2	17-5	14-3	24-7	21-4	17-5	28-6	24-8	20-2
	No. 3	12-0	10-5	8-6	15-3	13-2	10-9	18-7	16-1	13-2	21-7	18-8	15-3
Douglas Fir-South	Sel. Struc.	14-9	13-5	11-9	19-6	17-9	15-6	24-10	22-7	19-9	30-3	27-6	24-0
	No. 1	14-5	13-1	11-5	19-0	17-3	14-5	24-3	21-7	17-8	28-11	25-1	20-5
	No. 2	14-1	12-9	10-11	18-6	16-10	13-10	23-8	20-8	16-11	27-8	24-0	19-7
	No. 3	11-9	10-2	8-3	14-10	12-10	10-6	18-1	15-8	12-10	21-0	18-2	14-10
Hem-Fir	Sel. Struc.	15-6	14-1	12-3	20-5	18-6	16-2	26-0	23-8	20-8	31-8	28-9	25-1
	No. 1 & Btr.	15-2	13-9	12-0	19-11	18-2	15-7	25-5	23-2	19-1	30-11	27-1	22-1
	No. 1	15-2	13-9	11-9	19-11	18-2	14-10	25-5	22-2	18-1	29-9	25-9	21-0
	No. 2	14-5	13-1	11-1	19-0	17-2	14-0	24-3	21-0	17-2	28-1	24-4	19-11
	No. 3	12-0	10-5	8-6	15-3	13-2	10-9	18-7	16-1	13-2	21-7	18-8	15-3
Spruce-Pine-Fir (South)	Sel. Struc.	14-5	13-1	11-5	19-0	17-3	15-1	24-3	22-1	19-3	29-6	26-10	23-5
	No. 1	14-1	12-9	11-1	18-6	16-10	14-0	23-8	21-0	17-2	28-1	24-4	19-11
	No. 2	13-8	12-5	10-5	18-0	16-2	13-2	22-9	19-9	16-1	26-5	22-10	18-8
	No. 3	11-1	9-7	7-10	14-0	12-2	9-11	17-2	14-10	12-1	19-11	17-3	14-1
Western Woods	Sel. Struc.	14-1	12-9	11-2	18-6	16-10	14-3	23-8	21-4	17-5	28-6	24-8	20-2
	No. 1	13-8	11-10	9-8	17-4	15-0	12-3	21-2	18-4	15-0	24-7	21-3	17-5
	No. 2	13-3	11-10	9-8	17-4	15-0	12-3	21-2	18-4	15-0	24-7	21-3	17-5
	No. 3	10-5	9-0	7-4	13-2	11-5	9-4	16-1	13-11	11-5	18-8	16-2	13-2

Source: Western Wood Products Association.

Table E-15 Western lumber roof rafters: 25-psf snow load, 10-psf dead load, *L*/240. Roof slope is 3:12 or less; light roof covering; no ceiling finish.

Species or Group	Grade	Span (feet and inches) 2 × 6			2 × 8			2 × 10			2 × 12		
		12" oc	16" oc	24" oc	12" oc	16" oc	24" oc	12" oc	16" oc	24" oc	12" oc	16" oc	24" oc
Douglas Fir-Larch	Sel. Struc.	15-2	13-10	12-1	20-0	18-2	15-11	25-7	23-3	20-3	31-1	28-3	24-0
	No. 1 & Btr.	14-11	13-7	11-10	19-8	17-11	15-1	25-1	22-7	18-6	30-3	26-3	21-5
	No. 1	14-8	13-4	11-2	19-4	17-3	14-1	24-4	21-1	17-3	28-3	24-5	20-0
	No. 2	14-4	12-9	10-5	18-8	16-2	13-2	22-9	19-9	16-1	26-5	22-10	18-8
	No. 3	11-2	9-8	7-10	14-1	12-2	10-0	17-3	14-11	12-2	20-0	17-3	14-1
Douglas Fir-South	Sel. Struc.	13-9	12-6	10-11	18-1	16-5	14-4	23-1	21-0	18-4	28-1	25-6	22-3
	No. 1	13-5	12-2	10-7	17-8	16-1	13-4	22-6	20-0	16-4	26-9	23-2	18-11
	No. 2	13-1	11-10	10-1	17-2	15-7	12-10	21-11	19-2	15-8	25-8	22-2	18-2
	No. 3	10-10	9-5	7-8	13-9	11-11	9-9	16-9	14-6	11-10	19-5	16-10	13-9
Hem-Fir	Sel. Struc.	14-4	13-1	11-5	18-11	17-2	15-0	24-2	21-11	19-2	29-4	26-8	23-4
	No. 1 & Btr.	14-1	12-9	11-2	18-6	16-10	14-5	23-8	21-6	17-8	28-9	25-1	20-5
	No. 1	14-1	12-9	10-10	18-6	16-10	13-9	23-8	20-7	16-9	27-6	23-10	19-5
	No. 2	13-5	12-2	10-3	17-8	15-11	13-0	22-5	19-5	15-10	26-0	22-6	18-5
	No. 3	11-2	9-8	7-10	14-1	12-2	10-0	17-3	14-11	12-2	20-0	17-3	14-1
Spruce-Pine-Fir (South)	Sel. Struc.	13-5	12-2	10-8	17-8	16-1	14-0	22-6	20-6	17-11	27-5	24-11	21-9
	No. 1	13-1	11-10	10-0	17-2	15-7	13-0	21-11	19-5	15-10	26-0	22-6	18-5
	No. 2	12-8	11-6	9-8	16-8	14-11	12-2	21-1	18-3	14-11	24-5	21-2	17-3
	No. 3	10-3	8-11	7-3	13-0	11-3	9-2	15-10	13-9	11-3	18-5	15-11	13-0
Western Woods	Sel. Struc.	13-1	11-10	10-4	17-2	15-7	13-2	21-11	19-9	16-1	26-5	22-10	18-8
	No. 1	12-8	11-0	9-0	16-1	13-11	11-4	19-8	17-0	13-11	22-9	19-9	16-1
	No. 2	12-3	11-0	9-0	16-1	13-11	11-4	19-8	17-0	13-11	22-9	19-9	16-1
	No. 3	9-8	8-4	6-10	12-2	10-7	8-8	14-11	12-11	10-7	17-3	15-0	12-3

Source: Western Wood Products Association.

Table E–16 Western lumber roof rafters: 30-psf snow load, 10-psf dead load, *L*/240. Roof slope is 3:12 or less; light roof covering; no ceiling finish.

Species or Group	Grade	2 × 6			2 × 8			2 × 10			2 × 12		
		12″ oc	16″ oc	24″ oc	12″ oc	16″ oc	24″ oc	12″ oc	16″ oc	24″ oc	12″ oc	16″ oc	24″ oc
Douglas Fir-	Sel. Struc.	14-4	13-0	11-4	18-10	17-2	15-0	24-1	21-10	19-1	29-3	26-7	22-6
Larch	No. 1 & Btr.	14-1	12-9	11-2	18-6	16-10	14-2	23-8	21-2	17-3	28-4	24-6	20-0
	No. 1	13-9	12-6	10-5	18-2	16-2	13-2	22-9	19-9	16-1	26-5	22-10	18-8
	No. 2	13-6	11-11	9-9	17-5	15-1	12-4	21-4	18-5	15-1	24-8	21-5	17-6
	No. 3	10-5	9-0	7-4	13-2	11-5	9-4	16-1	13-11	11-5	18-8	16-2	13-2
Douglas Fir-	Sel. Struc.	12-11	11-9	10-3	17-0	15-6	13-6	21-9	19-9	17-3	26-5	24-0	21-0
South	No. 1	12-7	11-5	9-10	16-7	15-1	12-6	21-2	18-9	15-3	25-1	21-8	17-9
	No. 2	12-3	11-2	9-5	16-2	14-8	12-0	20-8	17-11	14-8	24-0	20-9	17-0
	No. 3	10-2	8-9	7-2	12-10	11-2	9-1	15-8	13-7	11-1	18-2	15-9	12-10
Hem-Fir	Sel. Struc.	13-6	12-3	10-9	17-10	16-2	14-2	22-9	20-8	18-0	27-8	25-1	21-11
	No. 1 & Btr.	13-3	12-0	10-6	17-5	15-10	13-6	22-3	20-2	16-6	27-1	23-5	19-2
	No. 1	13-3	12-0	10-2	17-5	15-9	12-10	22-2	19-3	15-8	25-9	22-3	18-2
	No. 2	12-7	11-5	9-7	16-7	14-11	12-2	21-0	18-2	14-10	24-4	21-1	17-3
	No. 3	10-5	9-0	7-4	13-2	11-5	9-4	16-1	13-11	11-5	18-8	16-2	13-2
Spruce-Pine-Fir	Sel. Struc.	12-7	11-5	10-0	16-7	15-1	13-2	21-2	19-3	16-10	25-9	23-5	20-6
(South)	No. 1	12-3	11-2	9-7	16-2	14-8	12-2	20-8	18-2	14-10	24-4	21-1	17-3
	No. 2	11-11	10-10	9-0	15-9	14-0	11-5	19-9	17-1	13-11	22-10	19-10	16-2
	No. 3	9-7	8-4	6-9	12-2	10-6	8-7	14-10	12-10	10-6	17-3	14-11	12-2
Western Woods	Sel. Struc.	12-3	11-2	9-9	16-2	14-8	12-4	20-8	18-5	15-1	24-8	21-5	17-6
	No. 1	11-10	10-3	8-5	15-0	13-0	10-8	18-4	15-11	13-0	21-3	18-5	15-1
	No. 2	11-7	10-3	8-5	15-0	13-0	10-8	18-4	15-11	13-0	21-3	18-5	15-1
	No. 3	9-0	7-10	6-4	11-5	9-11	8-1	13-11	12-1	9-10	16-2	14-0	11-5

SOURCE: Western Wood Products Association.

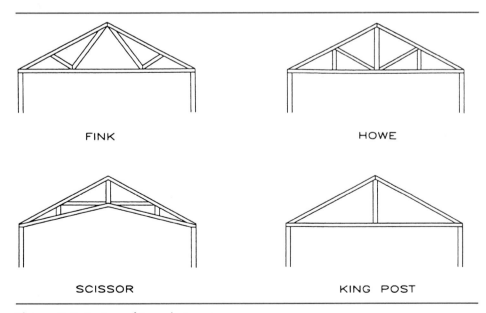

FINK

HOWE

SCISSOR

KING POST

Figure E–1 Basic roof truss designs.

E.3.3 Representative Roof Truss Spans for Selected Trusses

The following lists apply to Tables E–17 through E–20.

Design Criteria

Top chord live load	= 20 psf (958 Pa)
Top chord dead load	= 10 psf (479 Pa)
Bottom chord live load	= 0 psf (0 Pa)
Bottom chord dead load	= 10 psf (479 Pa)

Notes

Given truss spans are examples of truss spans for the given loadings and truss configurations. The tables are not intended to be used for design purposes or specific projects.

Spans have been determined in accordance with the design Specifications for Metal Plate Connected Wood Trusses, TPI-92, of the Truss Plate Institute (TPI) and the 1991 edition of the National Design Specification for Wood Construction (NDS®) of the American Forest and Paper Association (AFPA).

Tables shown are not intended to limit roof trusses to these loads, lumber species, lumber grades, and configurations. See your WTCA member fabricator for actual truss designs and solutions to custom roof profiles. Some representative spans for the configurations shown may vary with each fabricator.

Footnotes

1. Representative spans shown assume that the moisture content of the lumber does not exceed 19% at the time of fabrication and during end use.

2. The representative span for this lumber grade has been limited to the representative truss span that can be achieved by the group of lumber grades provided in these tables.

3. The representative span for this lumber grade has been limited by the representative bottom-chord panel length based on the TPI requirement of applying a 200-lb (890-N) concentrated load to represent a construction worker standing on the bottom chord.

Table E-17 Representative roof truss spans for selected Fink trusses: 24" o.c. spacing; 40-psf total design load; load duration increase of 15% (ft-in.).

			TOP CHORD		BOTTOM CHORD	
			2x4	2x6	2x4	2x6
Southern Pine	3/12	#1 Dense	29-9	42-0[3]	31-11[3]	42-0[3]
		#1	28-11	42-0[3]	30-8[3]	40-9[3]
		#2 Dense	28-7	42-0[3]	28-8	40-7[3]
		#2	27-6	40-6	27-1	38-4
	4/12	#1 Dense	31-11[3]	42-0[3]	31-11[3]	42-0[3]
		#1	31-11[3]	42-0[3]	30-8[3]	40-9[3]
		#2 Dense	31-8	42-0[3]	29-10[3]	40-7[3]
		#2	30-7	42-0[3]	27-10[3]	39-1[3]
	5/12	#1 Dense	31-11[3]	42-0[3]	31-11[3]	42-0[3]
		#1	31-11[3]	42-0[3]	30-8[3]	40-9[3]
		#2 Dense	31-11[3]	42-0[3]	29-10[3]	40-7[3]
		#2	31-10	42-0[3]	27-10[3]	39-1[3]
Douglas Fir-Larch	3/12	Sel. Str.	30-5	43-2[3]	33-2[3]	43-2[3]
		#1 & Better	28-10	42-9	30-6[3]	41-9[3]
		#1	27-10	41-3	28-3[3]	40-3[3]
		#2	26-8	39-5	25-7[3]	37-9
	4/12	Sel. Str.	33-2[3]	43-2[3]	33-2[3]	43-2[3]
		#1 & Better	32-0	43-2[3]	30-6[3]	41-9[3]
		#1	30-11	43-2[3]	28-3[3]	40-3[3]
		#2	29-7	43-2[3]	25-7[3]	38-8[3]
	5/12	Sel. Str.	33-2[3]	43-2[3]	33-2[3]	43-2[3]
		#1 & Better	33-2[3]	43-2[3]	30-6[3]	41-9[3]
		#1	32-3	43-2[3]	28-3[3]	40-3[3]
		#2	30-10	43-2[3]	25-7[3]	38-8[3]
Spruce-Pine-Fir	3/12	Sel. Str.	28-2	38-5[3]	29-4[3]	28-5[3]
		#1	25-9	38-0	23-4	32-7
		#2	25-9	38-0	23-4	32-7
	4/12	Sel. Str.	29-4[3]	38-5[3]	29-4[3]	38-5[3]
		#1	28-9	38-5[3]	25-0[3]	36-5[3]
		#2	28-9	38-5[3]	25-0[3]	36-5[3]
	5/12	Sel. Str.	29-4[3]	38-5[3]	29-4[3]	38-5[3]
		#1	29-4[3]	38-5[3]	25-0[3]	36-5[3]
		#2	29-4[3]	38-5[3]	25-0[3]	36-5[3]
Hem-Fir	3/12	Sel. Str.	29-1	39-9[3]	30-9[3]	39-9[3]
		#1	26-10	39-7	26-10[3]	37-11[3]
		#2	25-8	37-9	24-5[3]	35-2[3]
	4/12	Sel. Str.	30-9[3]	39-9[3]	30-9[3]	39-9[3]
		#1	29-10	39-9[3]	26-10[3]	37-11[3]
		#2	28-6	39-9[3]	24-5[3]	35-2[3]
	5/12	Sel. Str.	30-9[3]	39-9[3]	30-9[3]	39-9[3]
		#1	30-9[3]	39-9[3]	26-10[3]	37-11[3]
		#2	29-8	39-9[3]	24-5[3]	35-2[3]

SOURCE: *Metal Plate Connected Wood Truss Handbook*, Wood Truss Council of America, Madison, WI, 1993. Used by permission of WTCA.

Table E–18 Representative roof truss spans for selected Howe trusses: 24″ o.c. spacing; 40-psf total design load; load duration increase of 15% (ft-in.).

			TOP CHORD		BOTTOM CHORD	
			2x4	2x6	2x4	2x6
Southern Pine	3/12	#1 Dense	29-6	43-9	37-10	43-9[2]
		#1	28-9	42-9	36-3	43-9[2]
		#2 Dense	28-4	41-11	32-3	43-9[2]
		#2	27-4	40-3	30-5	42-10
	4/12	#1 Dense	32-10	48-7	42-5	48-7[2]
		#1	32-0	47-5	40-9[3]	48-7[2]
		#2 Dense	31-6	46-5	37-3	48-7[2]
		#2	30-4	44-7	35-1	48-7[2]
	5/12	#1 Dense	34-3	50-7	42-5[3]	50-7[2]
		#1	33-4	49-5	40-9[3]	50-7[2]
		#2 Dense	32-10	48-4	39-8[3]	50-7[2]
		#2	31-8	46-4	36-11[3]	50-7[2]
Douglas Fir-Larch	3/12	Sel. Str.	30-2	44-9	44-0[3]	44-9[2]
		#1 & Better	28-7	42-6	37-3	44-9[2]
		#1	27-8	40-11	33-8	44-9[2]
		#2	26-6	39-2	30-5	42-8
	4/12	Sel. Str.	33-7	49-8	44-0[3]	49-8[2]
		#1 & Better	31-10	47-1	40-6[3]	49-8[2]
		#1	30-8	45-1	37-5[3]	49-8[2]
		#2	29-5	43-5	33-10[3]	48-11
	5/12	Sel. Str.	35-1	51-9	44-0[3]	51-9[2]
		#1 & Better	33-2	49-1	40-6[3]	51-9[2]
		#1	32-0	47-2	37-5[3]	51-9[2]
		#2	30-8	45-1	33-10[3]	51-4[3]
Spruce-Pine-Fir	3/12	Sel. Str.	27-11	41-4	35-6	41-4[2]
		#1	25-7	37-9	25-9	35-6
		#2	25-7	37-9	25-9	35-6
	4/12	Sel. Str.	31-1	45-11	38-11[3]	45-11[2]
		#1	28-6	42-0	30-2	42-0
		#2	28-6	42-0	30-2	42-0
	5/12	Sel. Str.	32-6	47-11	38-11[3]	47-11[2]
		#1	29-10	43-9	33-2[3]	46-11
		#2	29-10	43-9	33-2[3]	46-11
Hem-Fir	3/12	Sel. Str.	28-10	42-9	40-10[3]	42-9[2]
		#1	26-8	39-4	31-7	42-9[2]
		#2	25-6	37-6	27-11	39-2
	4/12	Sel. Str.	32-2	47-6	40-10[3]	47-6[2]
		#1	29-8	43-8	35-6[3]	47-6[2]
		#2	28-4	41-7	32-3	45-7
	5/12	Sel. Str.	33-8	49-6	40-10[3]	49-6[2]
		#1	30-11	45-5	35-6[3]	49-6[2]
		#2	29-6	43-3	32-5[3]	46-9[3]

SOURCE: *Metal Plate Connected Wood Truss Handbook*, Wood Truss Council of America, Madison, WI, 1993. Used by permission of WTCA.

Table E–19 Representative roof truss spans for selected scissors trusses: 24″ o.c. spacing; 40-psf total design load; load duration increase of 15% (ft-in.).

			TOP CHORD		BOTTOM CHORD	
			2x4	2x6	2x4	2x6
Southern Pine	T.C. 3/12 B.C. 1.5/12	#1 Dense	21-9	32-5	24-6	32-5[2]
		#1	21-2	31-7	23-6	32-5[2]
		#2 Dense	20-11	31-2	20-2	28-6
		#2	20-1	29-11	19-0	26-8
	T.C. 4/12 B.C. 2/12	#1 Dense	24-2	35-11	29-9	35-11[2]
		#1	23-6	35-1	28-5	35-11[2]
		#2 Dense	23-3	34-6	24-9	35-2
		#2	22-5	33-2	23-5	32-10
	T.C. 5/12 B.C. 2.5/12	#1 Dense	25-11	38-7	33-10	38-7[2]
		#1	25-3	37-8	32-5	38-7[2]
		#2 Dense	25-0	37-0	28-6	38-7[2]
		#2	24-1	35-7	26-11	37-11
Douglas Fir-Larch	T.C. 3/12 B.C. 1.5/12	Sel. Str.	22-1	33-1	30-11	33-1[2]
		#1 & Better	21-0	31-5	24-10	33-1[2]
		#1	20-4	30-5	22-0	30-10
		#2	19-6	29-0	19-7	27-1
	T.C. 4/12 B.C. 2/12	Sel. Str.	24-7	36-8	36-6	36-8[2]
		#1 & Better	23-5	34-11	29-9	36-8[2]
		#1	22-8	33-9	26-8	36-8[2]
		#2	21-9	32-3	23-10	33-2
	T.C. 5/12 B.C. 2.5/12	Sel. Str.	26-6	39-5	39-3	39-5[2]
		#1 & Better	25-2	37-6	33-7	39-5[2]
		#1	24-5	36-3	30-2	39-5[2]
		#2	23-5	34-8	27-2	37-11
Spruce-Pine-Fir	T.C. 3/12 B.C. 1.5/12	Sel. Str.	20-4	30-5	22-8	30-5[2]
		#1	18-7	27-7	15-6	21-3
		#2	18-7	27-7	15-6	21-3
	T.C. 4/12 B.C. 2/12	Sel. Str.	22-9	33-10	27-7	33-10[2]
		#1	20-10	30-10	19-4	26-4
		#2	20-10	30-10	19-4	26-4
	T.C. 5/12 B.C. 2.5/12	Sel. Str.	24-6	36-4	31-7	36-4[2]
		#1	22-6	33-3	22-5	30-10
		#2	22-6	33-3	22-5	30-10
Hem-Fir	T.C. 3/12 B.C. 1.5/12	Sel. Str.	21-1	31-5	28-8	31-5[2]
		#1	19-6	29-1	20-3	28-0
		#2	18-8	27-9	17-5	24-2
	T.C. 4/12 B.C. 2/12	Sel. Str.	23-6	34-11	34-1	34-11[2]
		#1	21-10	32-4	24-8	34-5
		#2	20-10	30-10	21-5	29-10
	T.C. 5/12 B.C. 2.5/12	Sel. Str.	25-4	37-6	37-4	37-6[2]
		#1	23-6	34-9	28-2	37-6[2]
		#2	22-5	33-1	24-7	34-6

SOURCE: *Metal Plate Connected Wood Truss Handbook*, Wood Truss Council of America, Madison, WI, 1993. Used by permission of WTCA.

Table E–20 Representative roof truss spans for selected king post trusses: 24″ o.c. spacing; 40-psf total design load; load duration increase of 15% (ft-in.).

			TOP CHORD		BOTTOM CHORD	
			2x4	2x6	2x4	2x6
Southern Pine	3/12	#1 Dense	17-2	25-1	21-6[3]	25-1[2]
		#1	16-9	24-7	20-8[3]	25-1[2]
		#2 Dense	16-4	23-9	20-1[3]	25-1[2]
		#2	15-8	22-7	18-9[3]	25-1[2]
	4/12	#1 Dense	17-10	26-0	21-6[3]	26-0[2]
		#1	17-4	25-5	20-8[3]	26-0[2]
		#2 Dense	16-11	24-6	20-1[3]	26-0[2]
		#2	16-2	23-3	18-9[3]	26-0[2]
	5/12	#1 Dense	18-2	26-7	21-6[3]	26-7[2]
		#1	17-8	25-11	20-8[3]	26-7[2]
		#2 Dense	17-3	25-0	20-1[3]	26-7[2]
		#2	16-6	23-8	18-9[3]	26-3[3]
Douglas Fir-Larch	3/12	Sel. Str.	17-8	25-10	22-4[3]	25-10[2]
		#1 & Better	16-6	24-2	20-7[3]	25-10[2]
		#1	15-9	23-0	19-0[3]	25-10[2]
		#2	15-0	21-10	17-3[3]	25-10[2]
	4/12	Sel. Str.	18-4	26-9	22-4[3]	26-9[2]
		#1 & Better	17-1	24-11	20-7[3]	26-9[2]
		#1	16-3	23-8	19-0[3]	26-9[2]
		#2	15-5	22-5	17-3[3]	26-0[3]
	5/12	Sel. Str.	18-9	27-4	22-4[3]	27-4[2]
		#1 & Better	17-5	25-4	20-7[3]	27-4[2]
		#1	16-7	24-1	19-0[3]	27-0[3]
		#2	15-8	22-10	17-3[3]	26-0[3]
Spruce-Pine-Fir	3/12	Sel. Str.	16-5	24-0	19-9[3]	24-0[2]
		#1	14-9	21-5	16-10[3]	24-0[2]
		#2	14-9	21-5	16-10[3]	24-0[2]
	4/12	Sel. Str.	17-1	24-10	19-9[3]	24-10[2]
		#1	15-3	22-1	16-10[3]	24-6[3]
		#2	15-3	22-1	16-10[3]	24-6[3]
	5/12	Sel. Str.	17-6	25-5	19-9[3]	25-5[2]
		#1	15-6	22-6	16-10[3]	24-6[3]
		#2	15-6	22-6	16-10[3]	24-6[3]
Hem-Fir	3/12	Sel. Str.	17-1	24-11	20-8[3]	24-11[2]
		#1	15-4	22-3	18-1[3]	24-11[2]
		#2	14-7	21-2	16-6[3]	23-8[3]
	4/12	Sel. Str.	17-10	25-11	20-8[3]	25-11[2]
		#1	15-10	22-11	18-1[3]	25-5[3]
		#2	15-1	21-10	16-6[3]	23-8[3]
	5/12	Sel. Str.	18-3	26-6	20-8[3]	26-6[2]
		#1	16-1	23-4	18-1[3]	25-5[3]
		#2	15-4	22-2	16-6[3]	23-8[3]

SOURCE: *Metal Plate Connected Wood Truss Handbook*, Wood Truss Council of America, Madison, WI, 1993. Used by permission of WTCA.

DESIGN DATA FOR BEAMS

Design data for beams are included in Tables E–21 through E–33. Computations for bending are based on the live load indicated plus 10 psf of dead load. Computations for deflection are based on the live load only. All beams in the table were designed to extend over a single span, and the following formulas were used:

For Type A

$$M = \frac{wL^2}{8} \quad \text{and} \quad D = \frac{5wL^4(12)^3}{384EI}$$

To use the tables, first determine the span, the live load to be supported, and the deflection limitation. Then select from the tables the proper size of beam with the corresponding required values for fiber stress in bending (f) and modulus of elasticity (E). The beam used should be of a grade and a species that meet these minimum values. You can determine the maximum span for a beam of specific size, grade, and species by reversing these steps.

Table E-21 Floor and roof beams.

Required values for fiber stress in bending (f) and modulus of elasticity (E) for the sizes shown to support safely a live load of 20 pounds per square foot within a deflection limitation of $l/240$.

SPAN OF BEAM	NOMINAL SIZE OF BEAM	6' - 0" f	6' - 0" E	7' - 0" f	7' - 0" E	8' - 0" f	8' - 0" E
10'	2-3x6	1070	780000	1250	910000	1430	1040000
	1-3x8	1235	680000	1440	794000	1645	906000
	2-2x8	1030	570000	1200	665000	1370	760000
	1-4x8	880	485000	1030	566000	1175	646000
	3-2x8	685	380000	800	443000	915	506000
	2-3x8	615	340000	720	397000	820	453000
	2-2x10	630	273000	735	219000	840	364000
11'	2-3x6	1295	1037000	1510	1210000	1730	1382000
	1-3x8	1490	905000	1740	1056000	1990	1206000
	2-2x8	1245	754000	1450	880000	1660	1005000
	1-4x8	1065	647000	1245	755000	1420	862000
	3-2x8	830	503000	970	587000	1105	670000
	2-3x8	745	453000	870	529000	995	604000
	2-2x10	765	363000	890	424000	1020	484000
12'	2-3x6	1545	1346000	1800	1571000	2060	1794000
	1-3x8	1775	1175000	2070	1371000	2370	1566000
	2-2x8	1480	980000	1725	1144000	1970	1306000
	1-4x8	1270	840000	1480	980000	1690	1120000
	3-2x8	985	653000	1150	762000	1315	870000
	2-3x8	890	588000	1035	686000	1185	784000
	1-6x8	755	483000	880	564000	1005	644000
	2-2x10	910	472000	1060	551000	1210	629000
	1-3x10	1090	566000	1275	660000	1455	754000
13'	2-3x6	1815	1711000	2110	1997000	2415	2281000
	1-3x8	2085	1494000	2430	1743000	2780	1991000
	2-2x8	1740	1245000	2025	1453000	2315	1660000
	1-4x8	1490	1067000	1735	1245000	1985	1422000
	3-2x8	1160	830000	1350	969000	1545	1106000
	2-3x8	1045	747000	1215	872000	1390	996000
	1-6x8	885	614000	1040	716000	1185	818000
	2-2x10	1070	600000	1245	700000	1420	800000
	1-3x10	1280	719000	1495	839000	1710	958000
14'	2-2x8	2015	1555000	2350	1815000	2685	2073000
	3-2x8	1340	1037000	1570	1210000	1790	1382000
	2-3x8	1210	933000	1410	1089000	1610	1244000
	1-6x8	1025	766000	1200	894000	1370	1021000
	1-3x10	1485	899000	1730	1049000	1980	1198000
	2-2x10	1235	749000	1445	874000	1650	998000
	1-4x10	1060	642000	1240	749000	1415	856000
	3-2x10	825	499000	965	582000	1100	665000
	2-3x10	740	449000	865	524000	990	598000
15'	3-2x8	1540	1275000	1800	1488000	2055	1699000
	2-3x8	1390	1148000	1620	1340000	1850	1530000
	1-6x8	1180	943000	1375	1100000	1570	1257000
	1-3x10	1705	1105000	1990	1289000	2270	1473000
	2-2x10	1420	921000	1660	1075000	1895	1228000
	1-4x10	1220	789000	1420	921000	1625	1052000
	3-2x10	950	614000	1105	717000	1265	818000
	2-3x10	850	553000	995	645000	1135	737000
	1-6x10	735	464000	855	541000	980	618000
	4-2x10	710	461000	830	538000	945	614000
	2-2x12	960	512000	1120	597000	1280	682000
16'	3-2x8	1755	1548000	2045	1806000	2340	2063000
	2-3x8	1580	1393000	1840	1626000	2105	1857000
	2-2x10	1615	1118000	1890	1305000	2155	1490000
	1-4x10	1385	958000	1615	1118000	1845	1277000
	3-2x10	1075	745000	1260	869000	1435	993000
	2-3x10	970	671000	1130	783000	1290	894000
	1-6x10	835	563000	975	657000	1130	750000
	4-2x10	810	559000	945	652000	1080	745000
	1-8x10	615	413000	715	482000	815	550000
	1-3x12	1310	746000	1530	871000	1750	994000
	2-2x12	1090	621000	1275	725000	1455	828000
17'	2-2x10	1825	1341000	2125	1565000	2430	1787000
	1-4x10	1565	1149000	1825	1341000	2085	1532000
	3-2x10	1215	894000	1420	1043000	1620	1192000

SPAN OF BEAM	NOMINAL SIZE OF BEAM	6' - 0" f	6' - 0" E	7' - 0" f	7' - 0" E	8' - 0" f	8' - 0" E
17'	2-3x10	1095	804000	1280	938000	1460	1072000
	1-6x10	945	675000	1100	788000	1260	900000
	4-2x10	910	670000	1065	782000	1215	894000
	1-8x10	690	495000	805	578000	910	660000
	1-3x12	1480	894000	1725	1043000	1975	1192000
	2-2x12	1235	745000	1440	869000	1645	993000
	1-4x12	1060	639000	1230	746000	1410	852000
	3-2x12	820	497000	960	580000	1095	663000
18'	2-2x10	2045	1592000	2385	1858000	2725	2123000
	1-4x10	1755	1364000	2045	1592000	2340	1819000
	3-2x10	1365	1061000	1590	1238000	1815	1415000
	2-3x10	1270	955000	1480	1114000	1695	1273000
	1-6x10	1060	801000	1235	935000	1415	1068000
	4-2x10	1020	796000	1195	929000	1365	1062000
	1-8x10	780	588000	910	686000	1040	784000
	1-3x12	1660	1062000	1935	1239000	2210	1416000
	2-2x12	1380	885000	1615	1033000	1845	1180000
	1-4x12	1185	758000	1385	885000	1580	1011000
	3-2x12	920	590000	1075	688000	1230	786000
	2-3x12	830	531000	970	620000	1105	708000
19'	3-2x10	1520	1248000	1775	1456000	2025	1664000
	2-3x10	1365	1123000	1595	1310000	1825	1497000
	1-6x10	1170	943000	1365	1100000	1560	1257000
	4-2x10	1140	936000	1330	1092000	1520	1248000
	2-4x10	975	802000	1140	936000	1300	1070000
	1-8x10	860	691000	1005	806000	1145	921000
	1-3x12	1850	1249000	2155	1457000	2465	1665000
	2-2x12	1540	1041000	1800	1215000	2055	1388000
	1-4x12	1320	892000	1540	1041000	1760	1190000
	3-2x12	1025	694000	1200	810000	1370	926000
	2-3x12	925	624000	1080	728000	1230	832000
	1-6x12	805	531000	940	620000	1070	708000
20'	3-2x10	1685	1456000	1965	1699000	2245	1942000
	2-3x10	1515	1310000	1770	1529000	2020	1747000
	1-6x10	1300	1099000	1515	1282000	1735	1465000
	4-2x10	1260	1092000	1475	1274000	1685	1456000
	2-4x10	1080	936000	1265	1092000	1445	1248000
	1-8x10	960	806000	1120	941000	1280	1075000
	2-2x12	1705	1214000	1990	1417000	2275	1619000
	1-4x12	1465	1040000	1710	1214000	1950	1387000
	3-2x12	1140	809000	1330	944000	1520	1079000
	2-3x12	1025	728000	1195	850000	1365	971000
	1-6x12	970	620000	1130	723000	1295	826000
	2-4x12	730	520000	855	607000	975	694000
21'	3-2x10	1855	1685000	2165	1966000	2475	2247000
	2-3x10	1670	1516000	1950	1827000	2225	2088000
	1-6x10	1430	1273000	1670	1485000	1905	1697000
	4-2x10	1390	1264000	1625	1475000	1855	1686000
	2-4x10	1195	1083000	1390	1264000	1590	1444000
	1-8x10	1050	933000	1225	1089000	1400	1244000
	2-2x12	1880	1405000	2195	1640000	2510	1874000
	1-4x12	1615	1204000	1880	1405000	2150	1606000
	3-2x12	1255	937000	1465	1093000	1670	1249000
	2-3x12	1130	843000	1320	984000	1505	1124000
	1-6x12	970	717000	1130	837000	1295	956000
	2-4x12	805	602000	940	702000	1075	802000
22'	1-6x10	1580	1463000	1845	1707000	2105	1951000
	4-2x10	1525	1453000	1780	1696000	2035	1938000
	2-4x10	1310	1245000	1530	1453000	1745	1660000
	1-8x10	1160	1073000	1355	1252000	1545	1431000
	1-4x12	1770	1384000	2065	1615000	2360	1846000
	3-2x12	1375	1077000	1605	1257000	1835	1436000
	2-3x12	1240	969000	1445	1130000	1655	1291000
	1-6x12	1080	825000	1260	963000	1440	1100000
	2-4x12	885	692000	1035	807000	1180	922000
	4-2x12	1035	808000	1205	943000	1375	1078000
	5-2x12	825	646000	965	754000	1105	862000
	3-3x12	825	639000	965	746000	1105	852000

Table E-22 Floor and roof beams.

Required values for fiber stress in bending (*f*) and modulus of elasticity (*E*) for the sizes shown to support safely a live load of 20 pounds per square foot within a deflection limitation of *l*/300.

SPAN OF BEAM	NOMINAL SIZE OF BEAM	6'-0" f	6'-0" E	7'-0" f	7'-0" E	8'-0" f	8'-0" E
10'	2-3x6	1070	975000	1250	1138000	1430	1300000
	1-3x8	1235	850000	1440	992000	1645	1133000
	2-2x8	1030	712000	1200	831000	1370	949000
	1-4x8	880	606000	1030	707000	1175	808000
	3-2x8	685	475000	800	554000	915	633000
	2-3x8	615	425000	720	496000	820	566000
	2-2x10	630	341000	735	398000	840	455000
11'	2-3x6	1295	1296000	1510	1512000	1730	1727000
	1-3x8	1490	1131000	1740	1320000	1990	1508000
	2-2x8	1245	942000	1450	1099000	1660	1256000
	1-4x8	1065	809000	1245	944000	1420	1078000
	3-2x8	830	629000	970	734000	1105	838000
	2-3x8	745	566000	870	660000	995	754000
	2-2x10	765	454000	890	530000	1020	605000
12'	2-3x6	1545	1682000	1800	1963000	2060	2242000
	1-3x8	1775	1469000	2070	1714000	2370	1958000
	2-2x8	1480	1225000	1725	1429000	1970	1633000
	1-4x8	1270	1050000	1480	1225000	1690	1400000
	3-2x8	985	816000	1150	952000	1315	1088000
	2-3x8	890	735000	1035	858000	1185	980000
	1-6x8	755	604000	880	705000	1005	805000
	2-2x10	910	590000	1060	688000	1210	786000
13'	1-3x8	2085	1867000	2430	2179000	2780	2439000
	2-2x8	1740	1556000	2025	1816000	2315	2074000
	1-4x8	1490	1334000	1735	1557000	1985	1778000
	3-2x8	1160	1037000	1350	1210000	1545	1382000
	2-3x8	1045	934000	1215	1090000	1390	1245000
	1-6x8	885	767000	1040	895000	1185	1022000
	2-2x10	1070	750000	1245	875000	1420	1000000
	1-3x10	1280	899000	1495	1049000	1710	1198000
	1-4x10	915	642000	1070	749000	1220	856000
14'	3-2x8	1340	1296000	1570	1512000	1790	1727000
	2-3x8	1210	1166000	1410	1361000	1610	1554000
	1-6x8	1025	957000	1200	1117000	1370	1276000
	1-3x10	1485	1124000	1730	1312000	1980	1498000
	2-2x10	1235	936000	1445	1092000	1650	1248000
	1-4x10	1060	802000	1240	936000	1415	1069000
	3-2x10	825	624000	965	728000	1100	832000
	2-3x10	740	561000	865	655000	990	748000
	1-6x10	640	471000	745	550000	850	628000
	4-2x10	620	468000	720	546000	825	624000
	2-2x12	835	520000	975	607000	1115	693000
15'	3-2x8	1540	1594000	1800	1860000	2055	2125000
	2-3x8	1390	1435000	1620	1675000	1850	1913000
	1-6x8	1180	1179000	1375	1376000	1570	1572000
	1-3x10	1705	1381000	1990	1612000	2270	1841000
	2-2x10	1420	1151000	1660	1343000	1895	1534000
	1-4x10	1220	986000	1420	1151000	1625	1314000
	3-2x10	950	767000	1105	895000	1265	1022000
	2-3x10	850	691000	995	806000	1135	921000
	1-6x10	735	580000	855	677000	980	773000
	4-2x10	710	576000	830	672000	945	768000
	2-2x12	960	640000	1120	747000	1280	853000
	1-4x12	825	549000	960	641000	1100	732000
16'	2-3x8	1580	1741000	1840	2032000	2105	2321000
	2-2x10	1615	1397000	1890	1630000	2155	1862000
	1-4x10	1385	1197000	1615	1397000	1845	1596000
	3-2x10	1075	931000	1260	1086000	1435	1241000
	2-3x10	970	839000	1130	979000	1290	1118000
	1-6x10	835	704000	975	821000	1130	938000
	4-2x10	810	699000	945	816000	1080	932000
	1-8x10	615	516000	715	602000	815	688000
	1-3x12	1310	932000	1530	1087000	1750	1242000
	2-2x12	1090	776000	1275	905000	1455	1034000
	1-4x12	935	666000	1090	777000	1250	888000
	3-2x12	730	518000	850	604000	970	690000

SPAN OF BEAM	NOMINAL SIZE OF BEAM	6'-0" f	6'-0" E	7'-0" f	7'-0" E	8'-0" f	8'-0" E
17'	2-2x10	1825	1676000	2130	1956000	2435	2234000
	1-4x10	1565	1437000	1825	1677000	2085	1915000
	3-2x10	1215	1117000	1420	1303000	1625	1489000
	2-3x10	1095	1005000	1280	1173000	1460	1340000
	1-6x10	945	844000	1100	985000	1260	1125000
	4-2x10	910	837000	1065	977000	1215	1116000
	1-8x10	690	619000	805	722000	910	825000
	1-3x12	1480	1117000	1725	1303000	1975	1489000
	2-2x12	1235	931000	1440	1086000	1645	1241000
	1-4x12	1060	799000	1230	932000	1410	1065000
	3-2x12	820	621000	960	725000	1095	828000
	2-3x12	740	559000	865	652000	990	745000
18'	1-4x10	1755	1705000	2045	1990000	2340	2273000
	3-2x10	1365	1326000	1590	1547000	1815	1767000
	2-3x10	1270	1194000	1480	1393000	1695	1592000
	1-6x10	1060	1001000	1235	1168000	1415	1334000
	4-2x10	1020	995000	1195	1161000	1365	1326000
	1-8x10	780	735000	910	858000	1040	980000
	1-3x12	1660	1327000	1935	1549000	2210	1769000
	2-2x12	1380	1106000	1615	1291000	1845	1474000
	1-4x12	1185	947000	1385	1105000	1580	1262000
	3-2x12	920	737000	1075	860000	1230	982000
	2-3x12	830	664000	970	775000	1105	885000
	1-6x12	720	565000	840	659000	960	753000
19'	3-2x10	1520	1560000	1775	1820000	2025	2079000
	2-3x10	1365	1404000	1595	1638000	1825	1871000
	1-6x10	1170	1179000	1365	1376000	1560	1572000
	4-2x10	1140	1170000	1330	1365000	1520	1560000
	2-4x10	975	1002000	1140	1169000	1300	1336000
	1-8x10	860	864000	1005	1008000	1145	1152000
	1-3x12	1850	1561000	2155	1822000	2465	2081000
	2-2x12	1540	1301000	1800	1518000	2055	1734000
	1-4x12	1320	1115000	1540	1301000	1760	1486000
	3-2x12	1025	867000	1200	1012000	1370	1156000
	2-3x12	925	780000	1080	910000	1230	1040000
	1-6x12	805	664000	940	775000	1070	885000
20'	3-2x10	1685	1820000	1965	2124000	2245	2426000
	2-3x10	1515	1637000	1770	1910000	2020	2182000
	1-6x10	1300	1374000	1515	1603000	1735	1831000
	4-2x10	1260	1365000	1475	1593000	1685	1819000
	2-4x10	1080	1170000	1265	1365000	1445	1560000
	1-8x10	960	1007000	1120	1175000	1280	1342000
	2-2x12	1705	1517000	1990	1770000	2275	2022000
	1-4x12	1465	1300000	1710	1517000	1950	1733000
	3-2x12	1140	1011000	1330	1180000	1520	1348000
	2-3x12	1025	910000	1195	1062000	1365	1213000
	1-6x12	970	775000	1130	904000	1295	1003000
	4-2x12	855	759000	995	886000	1135	1012000
21'	2-3x10	1670	1895000	1950	2211000	2225	2526000
	1-6x10	1430	1591000	1670	1857000	1905	2121000
	4-2x10	1390	1580000	1625	1844000	1855	2106000
	2-4x10	1195	1354000	1390	1580000	1590	1805000
	1-8x10	1050	1166000	1225	1361000	1400	1554000
	2-2x12	1880	1756000	2195	2049000	2510	2341000
	1-4x12	1615	1505000	1880	1756000	2150	2006000
	3-2x12	1255	1171000	1465	1366000	1670	1561000
	2-3x12	1130	1054000	1320	1230000	1505	1405000
	1-6x12	970	896000	1130	1046000	1295	1194000
	4-2x12	940	878000	1100	1025000	1255	1170000
	2-4x12	805	752000	940	877000	1075	1002000
22'	4-2x10	1525	1816000	1780	2119000	2035	2421000
	2-4x10	1310	1556000	1530	1816000	1745	2074000
	1-8x10	1160	1341000	1355	1565000	1545	1787000
	1-4x12	1770	1730000	2065	2019000	2360	2306000
	3-2x12	1375	1346000	1605	1571000	1835	1794000
	2-3x12	1240	1211000	1445	1413000	1655	1614000
	1-6x12	1080	1031000	1260	1203000	1440	1374000
	4-2x12	1030	1010000	1205	1179000	1375	1346000
	2-4x12	885	865000	1035	1009000	1180	1153000
	5-2x12	825	807000	965	942000	1105	1076000
	3-3x12	825	799000	965	932000	1105	1065000

Table E–23 Floor and roof beams.

Required values for fiber stress in bending (*f*) and modulus of elasticity (*E*) for the sizes shown to support safely a live load of 20 pounds per square foot within a deflection limitation of *l*/360.

SPAN OF BEAM	NOMINAL SIZE OF BEAM	f (6'-0")	E (6'-0")	f (7'-0")	E (7'-0")	f (8'-0")	E (8'-0")
10'	2-3x6	1070	1170000	1250	1365000	1430	1560000
	1-3x8	1235	1020000	1440	1192000	1645	1359000
	2-2x8	1030	855000	1200	997000	1370	1140000
	1-4x8	880	727000	1030	847000	1175	969000
	3-2x8	685	570000	800	667000	915	759000
	2-3x8	615	510000	720	600000	820	679000
	1-6x8	525	419000	615	489000	700	558000
11'	2-3x6	1295	1555000	1510	1815000	1730	2073000
	1-3x8	1490	1357000	1740	1584000	1990	1809000
	2-2x8	1245	1131000	1450	1320000	1660	1507000
	1-4x8	1065	970000	1245	1132000	1420	1293000
	3-2x8	830	754000	970	880000	1105	1005000
	2-3x8	745	679000	870	793000	995	906000
	1-6x8	635	558000	740	651000	845	744000
12'	1-3x8	1775	1762000	2070	2056000	2370	2349000
	2-2x8	1480	1470000	1725	1716000	1970	1959000
	1-4x8	1270	1260000	1480	1470000	1690	1680000
	3-2x8	985	979000	1150	1143000	1315	1305000
	2-3x8	890	882000	1035	1029000	1185	1176000
	1-6x8	775	724000	880	846000	1005	966000
	2-2x10	910	708000	1060	826000	1210	943000
	1-3x10	1090	849000	1275	991000	1455	1132000
13'	2-2x8	1740	1867000	2025	2179000	2315	2490000
	1-4x8	1490	1600000	1735	1867000	1985	2133000
	3-2x8	1160	1245000	1350	1453000	1545	1659000
	2-3x8	1045	1120000	1215	1308000	1390	1494000
	1-6x8	885	921000	1040	1074000	1185	1227000
	2-2x10	1070	900000	1245	1050000	1420	1200000
	1-3x10	1280	1078000	1495	1258000	1710	1437000
	1-4x10	915	771000	1070	900000	1220	1028000
14'	3-2x8	1340	1555000	1570	1815000	1790	2073000
	2-3x8	1210	1399000	1410	1633000	1610	1866000
	1-6x8	1025	1149000	1200	1341000	1370	1531000
	1-3x10	1485	1348000	1730	1573000	1980	1797000
	2-2x10	1235	1123000	1445	1311000	1650	1497000
	1-4x10	1060	963000	1240	1123000	1415	1284000
	3-2x10	825	748000	965	873000	1100	997000
	2-3x10	740	673000	865	786000	990	897000
	1-6x10	640	565000	745	660000	850	753000
	4-2x10	620	561000	720	655000	825	749000
	2-2x12	835	624000	975	728000	1115	832000
15'	3-2x8	1540	1912000	1800	2232000	2055	2548000
	2-3x8	1390	1722000	1620	2010000	1850	2295000
	1-6x8	1180	1414000	1375	1650000	1570	1885000
	1-3x10	1705	1657000	1990	1933000	2270	2209000
	2-2x10	1420	1381000	1660	1612000	1895	1842000
	1-4x10	1220	1183000	1420	1381000	1625	1578000
	3-2x10	950	921000	1105	1075000	1265	1227000
	2-3x10	850	829000	995	967000	1135	1105000
	1-6x10	735	696000	855	811000	980	927000
	4-2x10	710	691000	830	807000	945	921000
	2-2x12	960	768000	1120	895000	1280	1023000
	1-4x12	825	659000	960	769000	1100	878000
16'	2-2x10	1615	1677000	1890	1957000	2155	2235000
	1-4x10	1385	1437000	1615	1677000	1845	1915000
	3-2x10	1075	1117000	1260	1303000	1435	1489000
	2-3x10	970	1006000	1130	1174000	1290	1341000
	1-6x10	835	844000	975	985000	1130	1125000
	4-2x10	810	838000	945	978000	1080	1117000
	1-8x10	615	619000	715	723000	815	825000
	1-3x12	1310	1119000	1530	1306000	1750	1491000
	2-2x12	1090	931000	1275	1087000	1455	1242000
	1-4x12	935	799000	1090	932000	1250	1066000
	3-2x12	730	622000	850	725000	970	828000
	2-3x12	655	559000	765	652000	875	745000
17'	2-2x10	1825	2011000	2125	2347000	2430	2680000
	1-4x10	1565	1723000	1825	2011000	2085	2298000
	3-2x10	1215	1341000	1420	1564000	1620	1788000
	2-3x10	1095	1206000	1280	1407000	1460	1608000
	1-6x10	945	1012000	1100	1182000	1260	1350000
	4-2x10	910	1005000	1065	1173000	1215	1341000
	1-8x10	690	742000	805	867000	910	990000
	1-3x12	1480	1341000	1725	1564000	1975	1788000
	2-2x12	1235	1117000	1440	1303000	1645	1489000
	1-4x12	1060	958000	1230	1119000	1410	1278000
	3-2x12	820	745000	960	870000	1095	994000
	2-3x12	740	671000	865	782000	990	894000
18'	3-2x10	1365	1591000	1590	1857000	1815	2122000
	2-3x10	1270	1492000	1480	1671000	1695	1909000
	1-6x10	1060	1201000	1235	1402000	1415	1602000
	4-2x10	1020	1194000	1195	1393000	1365	1593000
	1-8x10	780	882000	910	1029000	1040	1176000
	1-3x12	1660	1593000	1935	1858000	2210	2124000
	2-2x12	1380	1327000	1615	1549000	1845	1770000
	1-4x12	1185	1137000	1385	1327000	1580	1516000
	3-2x12	920	885000	1075	1032000	1230	1179000
	2-3x12	830	796000	970	930000	1105	1062000
	1-6x12	720	678000	840	790000	960	904000
	4-2x12	690	663000	805	774000	920	884000
19'	3-2x10	1520	1872000	1775	2184000	2025	2496000
	2-3x10	1365	1684000	1595	1965000	1825	2245000
	1-6x10	1170	1414000	1365	1650000	1560	1885000
	4-2x10	1140	1404000	1330	1638000	1520	1872000
	2-4x10	975	1203000	1140	1404000	1300	1605000
	1-8x10	860	1036000	1005	1209000	1145	1381000
	2-2x12	1540	1561000	1800	1822000	2055	2082000
	1-4x12	1320	1338000	1540	1561000	1760	1785000
	3-2x12	1025	1041000	1200	1215000	1370	1385000
	2-3x12	925	936000	1080	1092000	1230	1248000
	1-6x12	805	796000	940	930000	1070	1062000
	4-2x12	770	780000	900	910000	1025	1040000
20'	2-3x10	1515	1965000	1770	2293000	2020	2620000
	1-6x10	1300	1648000	1515	1923000	1735	2197000
	4-2x10	1260	1638000	1475	1911000	1685	2184000
	2-4x10	1080	1404000	1265	1638000	1445	1872000
	1-8x10	960	1209000	1120	1411000	1280	1612000
	2-2x12	1705	1821000	1990	2125000	2275	2428000
	1-4x12	1465	1560000	1710	1821000	1950	2080000
	3-2x12	1140	1213000	1330	1416000	1520	1618000
	2-3x12	1025	1092000	1195	1275000	1365	1456000
	1-6x12	970	930000	1130	1085000	1295	1239000
	4-2x12	855	911000	995	1063000	1135	1214000
	2-4x12	730	780000	850	910000	975	1040000
21'	4-2x10	1390	1896000	1625	2212000	1855	2529000
	2-4x10	1195	1624000	1390	1896000	1590	2166000
	1-8x10	1050	1399000	1225	1633000	1400	1866000
	1-4x12	1615	1806000	1880	2107000	2150	2409000
	3-2x12	1255	1405000	1465	1639000	1670	1873000
	2-3x12	1130	1264000	1320	1476000	1505	1686000
	1-6x12	970	1075000	1130	1255000	1295	1434000
	4-2x12	940	1054000	1100	1230000	1255	1404000
	2-4x12	805	903000	940	1053000	1075	1203000
	3-3x12	750	834000	875	973000	1000	1112000
	1-8x12	720	789000	840	921000	960	1052000
	1-10x12	570	623000	665	727000	760	830000
22'	2-4x10	1310	1867000	1530	2179000	1745	2490000
	1-8x10	1160	1609000	1355	1878000	1545	2146000
	3-2x12	1375	1615000	1605	1885000	1835	2154000
	2-3x12	1240	1453000	1445	1695000	1655	1936000
	1-6x12	1080	1237000	1260	1444000	1440	1650000
	4-2x12	1035	1212000	1205	1414000	1375	1617000
	2-4x12	885	1038000	1035	1210000	1180	1383000
	5-2x12	825	969000	965	1131000	1105	1293000
	3-3x12	825	958000	965	1119000	1105	1278000
	1-8x12	790	907000	920	1058000	1055	1209000
	1-10x12	625	716000	730	835000	835	954000

Table E-24 Floor and roof beams.

Required values for fiber stress in bending (f) and modulus of elasticity (E) for the sizes shown to support safely a live load of 30 pounds per square foot within a deflection limitation of l/240.

SPAN OF BEAM	NOMINAL SIZE OF BEAM	6'-0" f	6'-0" E	7'-0" f	7'-0" E	8'-0" f	8'-0" E
10'	2-3x6	1430	1170000	1670	1365000	1905	1560000
	1-3x8	1645	1020000	1920	1190000	2195	1360000
	1-4x8	1175	727000	1370	848000	1565	969000
	3-2x8	915	570000	1070	665000	1220	760000
	2-3x8	820	510000	955	595000	1095	680000
	2-4x8	590	364000	690	425000	785	485000
	2-2x10	840	409000	980	477000	1120	545000
11'	2-3x6	1725	1555000	2015	1815000	2300	2073000
	1-3x8	1990	1357000	2320	1584000	2655	1809000
	1-4x8	1420	970000	1660	1132000	1895	1293000
	3-2x8	1105	754000	1290	880000	1475	1005000
	2-3x8	995	679000	1160	792000	1325	905000
	2-4x8	710	485000	830	566000	945	646000
	2-2x10	1020	544000	1190	635000	1360	725000
12'	1-4x8	1690	1260000	1970	1470000	2255	1679000
	3-2x8	1315	979000	1535	1142000	1755	1305000
	2-3x8	1185	882000	1385	1029000	1580	1176000
	2-4x8	845	630000	985	735000	1125	840000
	1-6x8	1005	724000	1175	845000	1340	965000
	2-2x10	1210	708000	1410	826000	1615	944000
	3-2x10	810	472000	945	551000	1080	629000
	2-3x10	725	424000	845	495000	965	565000
13'	1-4x8	1985	1600000	2315	1867000	2645	2133000
	3-2x8	1545	1245000	1805	1453000	2060	1659000
	2-3x8	1390	1120000	1620	1307000	1855	1493000
	2-4x8	990	801000	1155	935000	1320	1068000
	1-6x8	1180	921000	1375	1075000	1575	1228000
	2-x10	1425	900000	1665	1050000	1900	1200000
	3-2x10	950	600000	1110	700000	1265	800000
	2-3x10	855	540000	1000	630000	1140	720000
	1-4x10	1220	923000	1425	1079000	1625	1230000
14'	3-2x8	1790	1555000	2090	1815000	2385	2073000
	2-3x8	1610	1400000	1880	1634000	2145	1866000
	2-4x8	1150	1000000	1340	1167000	1535	1333000
	1-6x8	1370	1149000	1600	1341000	1825	1532000
	2-2x10	1650	1123000	1925	1310000	2200	1497000
	3-2x10	1100	748000	1285	873000	1465	997000
	2-3x10	990	673000	1155	785000	1320	897000
	1-4x10	1415	963000	1650	1124000	1885	1283000
	1-6x10	915	943000	1070	1100000	1220	1257000
	2-4x10	705	481000	825	561000	940	641000
15'	2-4x8	1320	1230000	1540	1435000	1760	1640000
	1-6x8	1570	1414000	1830	1650000	2095	1885000
	2-2x10	1895	1381000	2210	1612000	2525	1841000
	3-2x10	1260	921000	1470	1075000	1680	1228000
	2-3x10	1135	829000	1325	967000	1515	1105000
	1-4x10	1620	1183000	1890	1380000	2160	1577000
	1-6x10	980	696000	1145	812000	1305	928000
	2-4x10	810	592000	945	691000	1080	789000
	4-2x10	945	691000	1105	806000	1260	921000
	1-8x10	720	510000	840	595000	960	680000
	2-2x12	1280	768000	1495	896000	1705	1024000
	1-4x12	1095	658000	1280	768000	1460	877000
16'	2-2x10	2155	1677000	2515	1957000	2875	2235000
	3-2x10	1435	1117000	1675	1303000	1915	1489000
	2-3x10	1290	1004000	1505	1174000	1720	1341000
	1-4x10	1845	1437000	2155	1677000	2460	1915000
	1-6x10	1115	844000	1300	985000	1485	1125000
	2-4x10	925	719000	1080	839000	1235	958000
	4-2x10	1075	838000	1255	978000	1435	1117000
	1-8x10	815	619000	950	722000	1085	825000
	2-2x12	1455	931000	1700	1086000	1940	1241000
	1-4x12	1250	799000	1460	932000	1665	1065000
	3-2x12	970	621000	1130	725000	1295	828000
	2-3x12	875	559000	1020	652000	1165	745000

SPAN OF BEAM	NOMINAL SIZE OF BEAM	6'-0" f	6'-0" E	7'-0" f	7'-0" E	8'-0" f	8'-0" E
17'	3-2x10	1620	1341000	1890	1565000	2160	1787000
	2-3x10	1460	1206000	1705	1407000	1945	1607000
	1-4x10	2085	1723000	2435	2011000	2780	2297000
	1-6x10	1255	1012000	1465	1181000	1675	1349000
	2-4x10	1040	862000	1215	1006000	1385	1149000
	4-2x10	1215	1005000	1420	1173000	1620	1340000
	1-8x10	920	742000	1075	866000	1225	989000
	2-2x12	1645	1117000	1920	1303000	2195	1489000
	1-4x12	1410	958000	1645	1118000	1880	1277000
	3-2x12	1095	745000	1280	869000	1460	993000
	2-3x12	985	671000	1150	783000	1313	894000
	4-2x12	820	559000	955	652000	1095	869000
18'	2-3x10	1695	1432000	1980	1671000	2260	1909000
	1-6x10	1415	1201000	1650	1401000	1885	1601000
	2-4x10	1170	1023000	1365	1194000	1560	1364000
	4-2x10	1360	1194000	1590	1393000	1815	1592000
	1-8x10	1040	882000	1215	1029000	1385	1176000
	2-2x12	1840	1327000	2150	1549000	2455	1769000
	1-4x12	1580	1137000	1845	1327000	2105	1516000
	3-2x12	1230	885000	1435	1033000	1640	1180000
	2-3x12	1105	796000	1290	929000	1475	1061000
	4-2x12	920	663000	1075	774000	1225	884000
	2-4x12	790	569000	920	664000	1055	758000
	5-2x12	735	531000	860	620000	980	708000
19'	2-4x10	1300	1203000	1515	1404000	1735	1604000
	4-2x10	1520	1404000	1775	1638000	2025	1871000
	1-8x10	1145	1036000	1335	1209000	1525	1381000
	1-4x12	1760	1338000	2055	1561000	2345	1783000
	3-2x12	1370	1041000	1600	1215000	1825	1388000
	2-3x12	1230	936000	1435	1092000	1640	1248000
	4-2x12	1025	780000	1195	910000	1365	1040000
	2-4x12	880	669000	1025	781000	1175	892000
	5-2x12	820	624000	955	728000	1095	832000
	1-6x12	1070	796000	1250	929000	1425	1061000
	3-3x12	820	617000	955	720000	1095	822000
	1-8x12	785	584000	915	681000	1045	778000
20'	4-2x12	1680	1638000	1960	1911000	2240	2183000
	1-8x10	1280	1209000	1495	1411000	1705	1611000
	3-2x12	1520	1213000	1775	1415000	2025	1617000
	2-3x12	1365	1092000	1595	1274000	1820	1456000
	4-2x12	1025	910000	1195	1062000	1365	1213000
	2-4x12	975	780000	1140	910000	1300	1040000
	5-2x12	910	728000	1060	849000	1215	970000
	1-6x12	1295	930000	1510	1084000	1725	1239000
	3-3x12	910	720000	1060	840000	1215	960000
	1-8x12	870	682000	1015	795000	1160	909000
	1-10x12	690	538000	805	628000	920	717000
	4-3x12	680	546000	795	637000	905	728000
21'	1-8x10	1400	1399000	1635	1633000	1865	1865000
	3-2x12	1670	1405000	1950	1640000	2225	1873000
	2-3x12	1505	1264000	1755	1475000	2005	1685000
	4-2x12	1255	1054000	1465	1230000	1675	1405000
	2-4x12	1075	903000	1255	1054000	1435	1204000
	5-2x12	1005	843000	1175	984000	1340	1124000
	1-6x12	1295	1075000	1510	1255000	1725	1434000
	3-3x12	1005	833000	1175	972000	1340	1110000
	1-8x12	960	789000	1120	921000	1280	1052000
	1-10x12	760	623000	885	727000	1015	830000
	4-3x12	750	632000	875	737000	1000	842000
	2-3x14	1085	774000	1265	903000	1445	1032000
22'	2-3x12	1655	1453000	1930	1696000	2205	1940000
	4-2x12	1375	1212000	1605	1414000	1835	1615000
	2-4x12	1180	1038000	1380	1211000	1575	1384000
	5-2x12	1100	969000	1285	1131000	1465	1292000
	1-6x12	1440	1237000	1680	1443000	1920	1649000
	3-3x12	1100	958000	1285	1118000	1465	1277000
	1-8x12	1055	907000	1230	1058000	1405	1209000
	1-10x12	830	716000	970	835000	1105	954000
	4-3x12	825	727000	965	848000	1100	969000
	2-3x14	1190	890000	1390	1039000	1585	1186000
	1-6x14	1045	765000	1220	893000	1395	1020000
	3-3x14	795	589000	930	687000	1060	785000
	2-4x14	820	601000	955	701000	1095	801000

Table E-25 Floor and roof beams.

Required values for fiber stress in bending (f) and modulus of elasticity (E) for the sizes shown to support safely a live load of 30 pounds per square foot within a deflection limitation of $l/300$.

SPAN OF BEAM	NOMINAL SIZE OF BEAM	6'-0" f	6'-0" E	7'-0" f	7'-0" E	8'-0" f	8'-0" E
10'	2-3x6	1430	1462000	1670	1706000	1905	1948000
	1-3x8	1645	1275000	1920	1488000	2195	1699000
	1-4x8	1175	909000	1370	1061000	1565	1212000
	3-2x8	915	712000	1070	831000	1220	949000
	2-3x8	820	637000	955	743000	1095	849000
	2-4x8	590	455000	690	531000	785	606000
	2-2x10	840	511000	980	596000	1120	681000
11'	1-3x8	1990	1696000	2320	1979000	2655	2261000
	1-4x8	1420	1212000	1660	1414000	1895	1615000
	3-2x8	1105	942000	1290	1099000	1475	1255000
	2-3x8	995	849000	1160	991000	1325	1132000
	2-4x8	710	606000	830	707000	945	808000
	2-2x10	1020	680000	1190	793000	1360	906000
	1-3x10	1220	817000	1425	953000	1625	1089000
12'	1-4x8	1690	1575000	1970	1838000	2255	2099000
	3-2x8	1315	1224000	1535	1428000	1755	1631000
	2-3x8	1185	1102000	1385	1286000	1580	1469000
	2-4x8	845	787000	985	918000	1125	1049000
	1-6x8	1005	905000	1175	1056000	1340	1206000
	2-2x10	1210	885000	1410	1033000	1615	1180000
	3-2x10	810	590000	945	688000	1080	786000
	2-3x10	725	530000	845	618000	965	706000
13'	1-4x8	1985	2000000	2315	2334000	2645	2666000
	3-2x8	1545	1556000	1805	1816000	2060	2074000
	2-3x8	1390	1400000	1620	1634000	1855	1866000
	2-4x8	990	1001000	1155	1168000	1320	1334000
	1-6x8	1180	1151000	1375	1343000	1575	1534000
	2-2x10	1425	1125000	1665	1313000	1900	1500000
	3-2x10	950	750000	1110	875000	1265	1000000
	2-3x10	855	675000	1000	788000	1140	900000
	1-4x10	1220	1154000	1425	1347000	1625	1538000
14'	3-2x8	1790	1944000	2090	2268000	2385	2591000
	2-3x8	1610	1750000	1880	2042000	2145	2333000
	2-4x8	1150	1250000	1340	1459000	1535	1666000
	1-6x8	1370	1436000	1600	1676000	1825	1914000
	2-2x10	1650	1404000	1925	1638000	2200	1871000
	3-2x10	1100	935000	1285	1091000	1465	1246000
	2-3x10	990	841000	1155	981000	1320	1121000
	1-4x10	1415	1204000	1650	1405000	1885	1605000
	1-6x10	915	1179000	1070	1376000	1220	1572000
	2-4x10	705	601000	825	701000	940	801000
15'	2-4x8	1320	1537000	1540	1794000	1760	2049000
	1-6x8	1570	1767000	1830	2062000	2095	2355000
	2-2x10	1895	1726000	2210	2014000	2525	2301000
	3-2x10	1260	1151000	1470	1343000	1680	1534000
	2-3x10	1135	1036000	1325	1209000	1515	1381000
	1-4x10	1620	1479000	1890	1726000	2160	1971000
	1-6x10	980	870000	1145	1015000	1305	1160000
	2-4x10	810	740000	945	863000	1080	986000
	4-2x10	945	864000	1105	1008000	1260	1152000
	1-8x10	720	637000	840	743000	960	849000
	2-2x12	1280	960000	1495	1120000	1705	1280000
	1-4x12	1095	822000	1280	959000	1460	1096000
16'	2-2x10	2155	2096000	2515	2446000	2875	2794000
	3-2x10	1435	1396000	1675	1629000	1915	1861000
	2-3x10	1290	1257000	1505	1467000	1720	1675000
	1-4x10	1845	1796000	2155	2096000	2460	2394000
	1-6x10	1115	1055000	1300	1231000	1485	1406000
	2-4x10	925	899000	1080	1049000	1235	1198000
	4-2x10	1075	1047000	1255	1222000	1435	1395000
	1-8x10	815	774000	950	903000	1085	1032000
	2-2x12	1455	1164000	1700	1358000	1940	1552000
	1-4x12	1250	999000	1460	1166000	1665	1332000
	3-2x12	970	776000	1130	905000	1295	1034000
	2-3x12	875	699000	1020	816000	1165	932000

SPAN OF BEAM	NOMINAL SIZE OF BEAM	6'-0" f	6'-0" E	7'-0" f	7'-0" E	8'-0" f	8'-0" E
17'	3-2x10	1620	1676000	1890	1956000	2160	2234000
	2-3x10	1460	1507000	1705	1759000	1945	2009000
	1-6x10	1255	1265000	1465	1476000	1675	1686000
	2-4x10	1040	1077000	1215	1257000	1385	1435000
	4-2x10	1215	1256000	1420	1466000	1620	1674000
	1-8x10	920	927000	1075	1082000	1225	1236000
	2-2x12	1645	1396000	1920	1629000	2195	1861000
	1-4x12	1410	1197000	1645	1397000	1880	1596000
	3-2x12	1095	931000	1280	1086000	1460	1241000
	2-3x12	985	839000	1150	979000	1315	1118000
	4-2x12	820	699000	955	816000	1095	932000
	2-4x12	705	599000	820	699000	940	799000
18'	2-3x10	1695	1790000	1980	2089000	2260	2386000
	1-6x10	1415	1501000	1650	1752000	1885	2000000
	2-4x10	1170	1279000	1365	1492000	1560	1705000
	4-2x10	1360	1492000	1590	1741000	1815	1989000
	1-8x10	1040	1102000	1215	1286000	1385	1469000
	2-2x12	1840	1659000	2150	1936000	2455	2211000
	1-4x12	1580	1421000	1845	1658000	2105	1894000
	3-2x12	1230	1106000	1435	1291000	1640	1474000
	2-3x12	1105	995000	1290	1161000	1475	1326000
	4-2x12	920	829000	1075	967000	1225	1105000
	2-4x12	790	711000	920	830000	1055	948000
	5-2x12	735	664000	860	775000	980	885000
19'	1-6x10	1570	1767000	1830	2062000	2095	2355000
	2-4x10	1300	1504000	1515	1755000	1735	2005000
	4-2x10	1520	1755000	1775	2048000	2025	2339000
	1-8x10	1145	1295000	1335	1511000	1525	1726000
	1-4x12	1760	1672000	2055	1951000	2345	2229000
	3-2x12	1370	1301000	1600	1518000	1825	1734000
	2-3x12	1230	1170000	1435	1365000	1640	1560000
	4-2x12	1025	975000	1195	1138000	1365	1300000
	2-4x12	880	836000	1025	976000	1175	1114000
	5-2x12	820	780000	955	910000	1095	1040000
	1-6x12	1070	995000	1250	1161000	1425	1326000
	3-3x12	820	771000	955	900000	1095	1028000
20'	1-8x10	1280	1511000	1495	1763000	1705	2014000
	3-2x12	1520	1516000	1775	1769000	2025	2021000
	2-3x12	1365	1365000	1595	1593000	1820	1819000
	4-2x12	1025	1137000	1195	1327000	1365	1516000
	2-4x12	975	975000	1140	1138000	1300	1300000
	5-2x12	910	910000	1060	1062000	1215	1213000
	1-6x12	1295	1162000	1510	1356000	1725	1549000
	3-3x12	910	900000	1060	1050000	1215	1200000
	1-8x12	870	852000	1015	994000	1160	1136000
	1-10x12	690	672000	805	784000	920	896000
	4-3x12	680	682000	795	796000	905	909000
	2-3x14	985	836000	1150	976000	1315	1114000
21'	3-2x12	1670	1756000	1950	2049000	2225	2341000
	2-3x12	1505	1580000	1755	1844000	2005	2106000
	4-2x12	1255	1317000	1465	1537000	1675	1755000
	2-4x12	1075	1129000	1255	1317000	1435	1505000
	5-2x12	1005	1054000	1175	1230000	1340	1405000
	1-6x12	1295	1344000	1510	1568000	1725	1791000
	3-3x12	1005	1041000	1175	1215000	1340	1388000
	1-8x12	960	986000	1120	1151000	1280	1314000
	1-10x12	760	779000	885	909000	1015	1038000
	4-3x12	750	790000	875	922000	1000	1053000
	2-3x14	1085	967000	1265	1128000	1445	1289000
	1-6x14	950	832000	1110	971000	1265	1109000
22'	4-2x12	1375	1515000	1605	1768000	1835	2019000
	2-4x12	1180	1297000	1380	1513000	1575	1729000
	5-2x12	1100	1211000	1285	1413000	1465	1614000
	1-6x12	1440	1546000	1680	1804000	1920	2061000
	3-3x12	1100	1197000	1285	1397000	1465	1596000
	1-8x12	1055	1134000	1230	1323000	1405	1511000
	1-10x12	830	895000	970	1044000	1105	1193000
	4-3x12	825	909000	965	1061000	1100	1212000
	2-3x14	1190	1112000	1390	1298000	1585	1482000
	1-6x14	1045	956000	1220	1116000	1395	1274000
	3-3x14	795	736000	930	859000	1060	981000
	2-4x14	820	751000	955	1114000	1095	1001000

Required values for fiber stress in bending (f) and modulus of elasticity (E) for the sizes shown to support safely a live load of 30 pounds per square foot within a deflection limitation of I/360.

SPAN OF BEAM	NOMINAL SIZE OF BEAM	6'-0" f	6'-0" E	7'-0" f	7'-0" E	8'-0" f	8'-0" E
10'	2-3x6	1430	1754000	1670	2047000	1905	2338000
	1-3x8	1645	1530000	1920	1785000	2195	2039000
	1-4x8	1175	1091000	1370	1273000	1565	1454000
	3-2x8	915	854000	1070	997000	1220	1138000
	2-3x8	820	76400C	955	891000	1095	1018000
	2-4x8	590	546000	690	637000	785	728000
	2-2x10	840	613000	980	715000	1120	817000
11'	1-3x8	1990	2035000	2320	2375000	2655	2713000
	1-4x8	1420	1454000	1660	1697000	1895	1938000
	3-2x8	1105	1130000	1290	1319000	1475	1506000
	2-3x8	995	1019000	1160	1189000	1325	1358000
	2-4x8	710	727000	830	848000	945	969000
	2-2x10	1020	816000	1190	952000	1360	1088000
	1-3x10	1220	980000	1425	1144000	1625	1306000
12'	1-4x8	1690	1890000	1970	2206000	2255	2519000
	3-2x8	1315	1469000	1535	1714000	1755	1958000
	2-3x8	1185	1322000	1385	1543000	1580	1762000
	2-4x8	845	944000	985	1102000	1125	1258000
	1-6x8	1005	1086000	1175	1267000	1340	1448000
	2-2x10	1210	1062000	1410	1239000	1615	1416000
	3-2x10	810	708000	945	826000	1080	944000
	2-3x10	725	636000	845	742000	965	848000
	1-4x10	1040	909000	1215	1061000	1385	1212000
13'	3-2x8	1545	1867000	1805	2179000	2060	2489000
	2-3x8	1390	1680000	1620	1960000	1855	2239000
	2-4x8	990	1201000	1155	1401000	1320	1600000
	1-6x8	1180	1381000	1375	1612000	1575	1841000
	2-2x10	1425	1350000	1665	1575000	1900	1799000
	3-2x10	950	900000	1110	1050000	1265	1200000
	2-3x10	855	810000	1000	945000	1140	1080000
	1-4x10	1220	1385000	1425	1616000	1625	1846000
	1-6x10	735	679000	855	792000	980	905000
14'	2-4x8	1150	1500000	1340	1750000	1535	1999000
	1-6x8	1370	1723000	1600	2011000	1825	2297000
	2-2x10	1650	1684000	1925	1965000	2200	2245000
	3-2x10	1100	1122000	1285	1309000	1465	1496000
	2-3x10	990	1009000	1155	1177000	1320	1345000
	1-4x10	1415	1445000	1650	1686000	1885	1926000
	1-6x10	915	1415000	1070	1651000	1220	1886000
	2-4x10	705	721000	825	841000	940	961000
	4-2x10	825	842000	960	983000	1100	1122000
	1-8x10	625	622000	730	726000	835	829000
15'	2-2x10	1895	2071000	2210	2417000	2525	2760000
	3-2x10	1260	1381000	1470	1612000	1680	1841000
	2-3x10	1135	1243000	1325	1450000	1515	1657000
	1-4x10	1620	1775000	1890	2071000	2160	2366000
	1-6x10	980	1044000	1145	1218000	1305	1392000
	2-4x10	810	888000	945	1036000	1080	1184000
	4-2x10	945	1037000	1105	1210000	1260	1382000
	1-8x10	720	764000	840	891000	960	1018000
	2-2x12	1280	1152000	1495	1344000	1705	1536000
	1-4x12	1095	986000	1280	1151000	1460	1314000
	3-2x12	855	768000	1000	896000	1140	1024000
	2-3x12	770	691000	900	806000	1025	921000
16'	3-2x10	1435	1675000	1675	1955000	1915	2233000
	2-3x10	1290	1508000	1505	1760000	1720	2010000
	1-6x10	11.5	1266000	1300	1477000	1485	1687000
	2-4x10	925	1079000	1080	1259000	1235	1438000
	4-2x10	1075	1256000	1255	1466000	1435	1674000
	1-8x10	815	929000	950	1084000	1085	1238000
	2-2x12	1455	1397000	1700	1630000	1940	1862000
	1-4x12	1250	1199000	1460	1399000	1665	1598000
	3-2x12	970	931000	1130	1086000	1295	1241000
	2-3x12	875	839000	1020	979000	1165	1118000
	4-2x12	730	699000	850	816000	975	932000
	2-4x12	625	599000	730	699000	835	798000

SPAN OF BEAM	NOMINAL SIZE OF BEAM	6'-0" f	6'-0" E	7'-0" f	7'-0" E	8'-0" f	8'-0" E
17'	2-3x10	1460	1808000	1705	2110000	1945	2410000
	1-6x10	1255	1518000	1465	1771000	1675	2023000
	2-4x10	1040	1292000	1215	1508000	1385	1722000
	4-2x10	1215	1507000	1420	1758000	1620	2009000
	1-8x10	920	1112000	1075	1298000	1225	1482000
	2-2x12	1645	1675000	1920	1955000	2195	2233000
	1-4x12	1410	1436000	1645	1676000	1880	1914000
	3-2x12	1095	1117000	1280	1303000	1460	1498000
	2-3x12	985	1007000	1150	1175000	1315	1342000
	4-2x12	820	839000	955	979000	1095	1118000
	2-4x12	705	719000	820	839000	940	958000
	5-2x12	655	671000	765	783000	875	894000
18'	2-4x10	1170	1535000	1365	1791000	1560	2046000
	4-2x10	1360	1790000	1590	2089000	1815	2386000
	1-8x10	1040	1322000	1215	1543000	1385	1762000
	2-2x12	1840	1991000	2150	2323000	2455	2654000
	1-4x12	1580	1705000	1845	1990000	2105	2273000
	3-2x12	1230	1327000	1435	1549000	1640	1769000
	2-3x12	1105	1194000	1290	1393000	1475	1591000
	4-2x12	920	995000	1075	1161000	1225	1326000
	2-4x12	790	853000	920	995000	1055	1137000
	5-2x12	735	797000	860	930000	980	1062000
	1-6x12	960	1016000	1120	1186000	1280	1354000
	3-3x12	740	787000	865	918000	985	1049000
19'	2-4x10	1300	1805000	1515	2106000	1735	2406000
	1-8x10	1145	1554000	1335	1813000	1525	2071000
	3-2x12	1370	1561000	1600	1822000	1825	2081000
	2-3x12	1230	1404000	1435	1638000	1640	1871000
	4-2x12	1025	1170000	1195	1365000	1365	1560000
	2-4x12	880	1003000	1025	1170000	1175	1337000
	5-2x12	820	936000	955	1092000	1095	1248000
	1-6x12	1070	1194000	1250	1393000	1425	1592000
	3-3x12	820	925000	955	1079000	1095	1233000
	1-8x12	785	877000	915	1023000	1045	1169000
	1-10x12	620	692000	725	807000	825	922000
	4-3x12	615	702000	715	819000	820	936000
20'	3-2x12	1520	1819000	1775	2123000	2025	2425000
	2-3x12	1365	1638000	1595	1911000	1820	2183000
	4-2x12	1025	1364000	1195	1592000	1365	1818000
	2-4x12	975	1170000	1140	1365000	1300	1560000
	5-2x12	910	1092000	1060	1274000	1215	1456000
	1-6x12	1295	1394000	1510	1627000	1725	1858000
	3-3x12	910	1080000	1060	1260000	1215	1440000
	1-8x12	870	1022000	1015	1193000	1160	1362000
	1-10x12	690	806000	805	941000	920	1074000
	4-3x12	680	818000	795	955000	905	1090000
	2-3x14	985	1003000	1150	1170000	1315	1337000
	1-6x14	860	862000	1005	1006000	1145	1149000
21'	2-3x12	1505	1896000	1755	2213000	2005	2527000
	4-2x12	1255	1580000	1465	1844000	1675	2106000
	2-4x12	1075	1355000	1255	1581000	1435	1806000
	5-2x12	1005	1265000	1175	1476000	1340	1686000
	1-6x12	1295	1613000	1510	1882000	1725	2150000
	3-3x12	1005	1249000	1175	1457000	1340	1665000
	1-8x12	960	1183000	1120	1380000	1280	1577000
	1-10x12	760	935000	885	1091000	1015	1246000
	4-3x12	750	948000	875	1106000	1000	1264000
	2-3x14	1085	1160000	1265	1354000	1445	1546000
	1-6x14	950	998000	1110	1165000	1265	1330000
	3-3x14	725	769000	845	897000	965	1025000
22'	4-2x12	1375	1818000	1605	2122000	1835	2423000
	2-4x12	1180	1556000	1380	1816000	1575	2074000
	5-2x12	1100	1453000	1285	1696000	1465	1937000
	3-3x12	1100	1436000	1285	1676000	1465	1914000
	1-8x12	1055	1361000	1230	1588000	1405	1814000
	1-10x12	830	1074000	970	1253000	1105	1432000
	4-3x12	825	1091000	965	1273000	1100	1454000
	2-3x14	1190	1334000	1390	1557000	1585	1778000
	1-6x14	1045	1147000	1220	1338000	1395	1529000
	3-3x14	795	883000	930	1030000	1060	1177000
	2-4x14	820	901000	955	1051000	1095	1201000
	4-3x14	595	667000	695	778000	795	889000

Table E-27 Floor and roof beams.

Required values for fiber stress in bending (f) and modulus of elasticity (E) for the sizes shown to support safely a live load of 40 pounds per square foot within a deflection limitation of l/240.

SPAN OF BEAM	NOMINAL SIZE OF BEAM	6'-0" f	6'-0" E	7'-0" f	7'-0" E	8'-0" f	8'-0" E
10'	2-3x6	1785	1560000	2085	1820000	2380	2079000
	1-3x8	2055	1360000	2400	1587000	2740	1813000
	2-2x8	1710	1134000	1995	1323000	2280	1512000
	1-4x8	1470	969000	1715	1131000	1960	1291000
	1-6x8	875	558000	1020	651000	1165	744000
	2-2x10	1050	545000	1225	636000	1400	726000
	1-3x10	1260	655000	1470	764000	1680	873000
11'	2-2x8	2070	1509000	2415	1761000	2760	2011000
	1-4x8	1775	1293000	2070	1509000	2365	1723000
	1-6x8	1055	743000	1230	867000	1405	990000
	2-2x10	1275	725000	1490	846000	1700	966000
	1-3x10	1525	872000	1780	1017000	2030	1162000
	1-4x10	1090	623000	1270	727000	1455	830000
	3-2x10	850	484000	990	565000	1135	645000
12'	1-4x8	2110	1680000	2460	1960000	2810	2239000
	1-6x8	1255	965000	1465	1126000	1670	1286000
	3-2x8	1645	1305000	1920	1523000	2190	1739000
	2-2x10	1510	944000	1760	1101000	2010	1258000
	1-3x10	1820	1132000	2125	1321000	2425	1509000
	1-4x10	1300	808000	1515	943000	1735	1077000
	3-2x10	1010	629000	1180	734000	1345	838000
	2-3x10	905	565000	1055	659000	1205	753000
13'	1-6x8	1475	1228000	1720	1433000	1965	1637000
	2-3x8	1735	1493000	2025	1742000	2315	1990000
	2-4x8	1235	1068000	1440	1246000	1645	1423000
	2-2x10	1780	1200000	2075	1400000	2370	1600000
	3-2x10	1185	800000	1380	934000	1580	1066000
	1-3x10	2130	1439000	2485	1679000	2840	1918000
	2-3x10	1070	720000	1250	840000	1425	960000
	1-4x10	1525	1230000	1780	1435000	2035	1640000
	2-4x10	760	514000	890	600000	1015	685000
14'	2-4x8	1435	1333000	1675	1555000	1915	1777000
	3-2x10	1375	997000	1605	1163000	1830	1329000
	2-3x10	1235	897000	1440	1047000	1645	1196000
	1-4x10	1770	1284000	2065	1498000	2360	1711000
	2-4x10	830	641000	1025	748000	1175	854000
	3-3x10	825	599000	960	699000	1100	798000
	1-6x10	1145	1257000	1335	1467000	1525	1675000
	1-8x10	780	553000	910	645000	1040	737000
	4-2x10	1030	749000	1200	874000	1375	998000
	2-2x12	1395	832000	1630	971000	1860	1109000
15'	3-2x10	1575	1228000	1840	1433000	2100	1637000
	2-3x10	1420	1105000	1655	1289000	1890	1473000
	2-4x10	1010	789000	1175	921000	1345	1052000
	3-3x10	945	737000	1100	860000	1260	982000
	1-6x10	1225	928000	1430	1083000	1635	1237000
	1-8x10	900	680000	1050	793000	1200	906000
	4-2x10	1180	921000	1375	1075000	1575	1228000
	2-2x12	1600	1024000	1865	1195000	2130	1365000
	3-2x12	1065	683000	1240	797000	1420	910000
	1-3x12	1920	1229000	2240	1434000	2560	1638000
	4-2x12	800	512000	935	597000	1065	682000
	2-3x12	960	614000	1120	716000	1280	818000
16'	3-2x10	1795	1489000	2095	1738000	2395	1985000
	2-3x10	1610	1341000	1880	1565000	2145	1787000
	2-4x10	1155	959000	1350	1119000	1540	1278000
	3-3x10	1075	894000	1255	1043000	1435	1192000
	1-6x10	1395	1125000	1625	1313000	1860	1500000
	1-8x10	1020	825000	1190	962000	1360	1100000
	4-2x10	1345	1117000	1570	1303000	1790	1489000
	2-2x12	1820	1241000	2120	1448000	2425	1654000
	3-2x12	1210	828000	1410	966000	1610	1104000
	4-2x12	910	621000	1060	724000	1215	828000
	5-2x12	730	497000	850	580000	975	662000
	2-3x12	1095	745000	1280	869000	1460	993000

SPAN OF BEAM	NOMINAL SIZE OF BEAM	6'-0" f	6'-0" E	7'-0" f	7'-0" E	8'-0" f	8'-0" E
17'	2-3x10	1825	1608000	2130	1876000	2430	2143000
	2-4x10	1300	1149000	1520	1341000	1735	1532000
	3-3x10	1215	1073000	1420	1252000	1620	1430000
	1-8x10	1150	989000	1340	1154000	1535	1318000
	3-2x12	1370	993000	1600	1159000	1825	1323000
	4-2x12	1025	745000	1195	869000	1365	993000
	5-2x12	820	596000	955	695000	1095	794000
	2-3x12	1230	895000	1435	1044000	1640	1193000
	3-3x12	820	590000	955	688000	1095	786000
	2-4x12	880	639000	1025	746000	1175	852000
	1-6x12	1070	761000	1250	888000	1425	1014000
	1-8x12	785	558000	915	651000	1045	744000
18'	2-3x10	2120	1909000	2475	2228000	2825	2545000
	2-4x10	1460	1364000	1705	1592000	1945	1818000
	3-3x10	1365	1273000	1595	1485000	1820	1697000
	1-8x10	1300	1176000	1515	1372000	1730	1568000
	3-2x12	1540	1180000	1800	1377000	2050	1573000
	4-2x12	1150	884000	1340	1032000	1530	1178000
	5-2x12	920	708000	1075	826000	1225	944000
	2-3x12	1380	1061000	1610	1238000	1840	1414000
	3-3x12	920	700000	1075	817000	1225	933000
	2-4x12	990	759000	1155	886000	1320	1012000
	1-6x12	1200	903000	1400	1054000	1600	1204000
	1-8x12	880	663000	1025	774000	1175	884000
19'	1-8x10	1430	1381000	1670	1612000	1905	1841000
	3-2x12	1710	1388000	1995	1620000	2280	1850000
	4-2x12	1280	1040000	1495	1214000	1705	1386000
	5-2x12	1025	832000	1195	971000	1365	1109000
	2-3x12	1540	1248000	1795	1456000	2050	1663000
	3-3x12	1025	823000	1195	960000	1365	1097000
	2-4x12	1100	892000	1280	1041000	1465	1189000
	1-6x12	1335	1061000	1560	1238000	1780	1414000
	1-8x12	980	779000	1145	909000	1305	1038000
	3-4x12	735	595000	860	694000	980	793000
	4-3x12	770	624000	900	728000	1025	832000
	2-6x12	670	1063000	780	1240000	895	1417000
20'	3-2x12	1900	1617000	2220	1887000	2530	2155000
	4-2x12	1280	1213000	1495	1415000	1705	1617000
	5-2x12	1135	971000	1325	1133000	1515	1294000
	3-3x12	1135	960000	1325	1120000	1515	1279000
	2-4x12	1220	1040000	1425	1214000	1625	1386000
	1-6x12	1620	1240000	1890	1447000	2160	1653000
	1-8x12	1085	909000	1265	1061000	1445	1212000
	3-4x12	810	693000	945	809000	1080	924000
	4-3x12	850	728000	990	849000	1135	970000
	2-6x12	740	620000	865	723000	985	826000
	1-10x12	860	717000	1005	837000	1145	956000
	2-3x14	1230	891000	1435	1040000	1640	1188000
21'	3-2x12	2090	1873000	2440	2186000	2785	2497000
	4-2x12	1570	1405000	1830	1640000	2095	1873000
	5-2x12	1255	1124000	1465	1312000	1675	1498000
	3-3x12	1255	1111000	1465	1296000	1675	1481000
	2-4x12	1345	1204000	1570	1405000	1795	1605000
	1-8x12	1200	1052000	1400	1228000	1600	1402000
	3-4x12	895	803000	1045	937000	1195	1070000
	4-3x12	935	843000	1090	984000	1245	1124000
	2-6x12	820	717000	955	837000	1095	956000
	1-10x12	950	831000	1110	970000	1265	1108000
	2-3x14	1355	1032000	1580	1204000	1805	1375000
	1-6x14	1190	887000	1390	1035000	1585	1182000
22'	4-2x12	1720	1616000	2005	1886000	2295	2154000
	5-2x12	1375	1292000	1605	1508000	1830	1722000
	3-3x12	1375	1277000	1605	1490000	1830	1702000
	3-4x12	985	923000	1150	1077000	1315	1230000
	4-3x12	1030	969000	1200	1131000	1375	1292000
	2-6x12	900	825000	1050	963000	1200	1100000
	1-10x12	1035	955000	1205	1114000	1380	1273000
	2-3x14	1485	1187000	1730	1385000	1980	1582000
	1-6x14	1305	1020000	1525	1190000	1740	1360000
	2-4x14	1025	801000	1195	935000	1365	1068000
	3-3x14	995	785000	1160	916000	1325	1046000
	3-4x14	680	534000	795	623000	905	712000

Required values for fiber stress in bending (f) and modulus of elasticity (E) for the sizes shown to support safely a live load of 40 pounds per square foot within a deflection limitation of l/300.

SPAN OF BEAM	NOMINAL SIZE OF BEAM	6'-0" f	6'-0" E	7'-0" f	7'-0" E	8'-0" f	8'-0" E
10'	1-3x8	2055	1700000	2400	1984000	2740	2266000
	2-2x8	1710	1417000	1995	1654000	2280	1889000
	1-4x8	1470	1211000	1715	1413000	1960	1614000
	1-6x8	875	697000	1020	813000	1165	929000
	2-2x10	1050	681000	1225	795000	1400	908000
	1-3x10	1260	819000	1470	956000	1680	1092000
	1-4x10	900	585000	1050	683000	1200	780000
11'	2-2x8	2070	1886000	2415	2201000	2760	2514000
	1-4x8	1775	1616000	2070	1886000	2365	2154000
	1-6x8	1055	929000	1230	1084000	1405	1238000
	2-2x10	1275	906000	1490	1057000	1700	1208000
	1-3x10	1525	1090000	1780	1272000	2030	1453000
	1-4x10	1090	779000	1270	909000	1455	1038000
	3-2x10	850	605000	990	706000	1135	806000
12'	1-6x8	1255	1206000	1465	1407000	1670	1607000
	3-2x8	1645	1631000	1920	1903000	2190	2174000
	2-2x10	1510	1180000	1760	1377000	2010	1573000
	1-3x10	1820	1415000	2125	1651000	2425	1886000
	1-4x10	1300	1010000	1515	1179000	1735	1346000
	3-2x10	1010	786000	1180	917000	1345	1048000
	2-3x10	905	706000	1055	824000	1205	941000
	1-6x10	785	594000	915	693000	1045	792000
	2-4x10	650	505000	760	589000	865	673000
13'	1-6x8	1475	1535000	1720	1791000	1965	2046000
	2-3x8	1735	1866000	2025	2178000	2315	2487000
	2-4x8	1235	1335000	1440	1558000	1645	1779000
	3-2x10	1185	1000000	1380	1167000	1580	1333000
	2-2x10	1780	1500000	2075	1750000	2370	2000000
	1-3x10	2130	1799000	2485	2099000	2840	2398000
	2-3x10	1070	900000	1250	1050000	1425	1200000
	1-4x10	1525	1537000	1780	1794000	2035	2049000
	2-4x10	760	642000	890	749000	1015	856000
14'	2-4x8	1435	1666000	1675	1944000	1915	2221000
	3-2x10	1375	1246000	1605	1454000	1830	1661000
	2-3x10	1235	1121000	1440	1308000	1645	1494000
	1-4x10	1770	1605000	2065	1873000	2360	2139000
	2-4x10	880	801000	1025	935000	1175	1068000
	3-3x10	825	749000	960	874000	1100	998000
	1-6x10	1145	1571000	1335	1833000	1525	2094000
	1-8x10	780	691000	910	806000	1040	921000
	4-2x10	1030	936000	1200	1092000	1375	1248000
	2-2x12	1395	1040000	1630	1214000	1860	1386000
15'	3-2x10	1575	1535000	1840	1791000	2100	2046000
	2-3x10	1420	1381000	1655	1612000	1890	1841000
	2-4x10	1010	986000	1175	1151000	1345	1314000
	3-3x10	945	921000	1100	1075000	1260	1228000
	1-6x10	1225	1160000	1430	1354000	1635	1546000
	1-8x10	900	850000	1050	992000	1200	1133000
	4-2x10	1180	1151000	1375	1343000	1575	1534000
	2-2x12	1600	1280000	1865	1494000	2130	1706000
	3-2x12	1065	854000	1240	997000	1420	1138000
	1-3x12	1920	1536000	2240	1792000	2560	2047000
	4-2x12	800	640000	935	747000	1065	853000
	2-3x12	960	767000	1120	895000	1280	1022000
16'	3-2x10	1795	1861000	2095	2172000	2395	2481000
	2-3x10	1610	1676000	1880	1956000	2145	2234000
	2-4x10	1155	1199000	1350	1399000	1540	1598000
	3-3x10	1075	1117000	1255	1303000	1435	1489000
	1-6x10	1395	1406000	1625	1641000	1860	1874000
	1-8x10	1020	1031000	1190	1203000	1360	1374000
	4-2x10	1345	1396000	1570	1629000	1790	1861000
	2-2x12	1820	1551000	2120	1810000	2425	2067000
	3-2x12	1210	1035000	1410	1208000	1610	1380000
	4-2x12	910	776000	1060	905000	1215	1034000
	5-2x12	730	621000	850	725000	975	828000
	2-3x12	1095	931000	1280	1086000	1460	1241000
17'	2-3x10	1825	2010000	2310	2345000	2430	2679000
	2-4x10	1300	1436000	1520	1676000	1735	1914000
	3-3x10	1215	1341000	1420	1565000	1620	1787000
	1-8x10	1150	1236000	1340	1442000	1535	1647000
	3-2x12	1370	1241000	1600	1448000	1825	1654000
	4-2x12	1025	931000	1195	1086000	1365	1241000
	5-2x12	820	745000	955	869000	1095	993000
	2-3x12	1230	1119000	1435	1306000	1640	1492000
	3-3x12	820	737000	955	860000	1095	982000
	2-4x12	880	799000	1025	932000	1175	1065000
	1-6x12	1070	951000	1250	1110000	1425	1268000
	1-8x12	785	697000	915	813000	1045	929000
18'	2-4x10	1460	1705000	1705	1990000	1945	2273000
	3-3x10	1365	1591000	1595	1857000	1820	2121000
	1-8x10	1300	1470000	1515	1715000	1730	1959000
	3-2x12	1540	1475000	1800	1721000	2050	1966000
	4-2x12	1150	1105000	1340	1289000	1530	1473000
	5-2x12	920	885000	1075	1033000	1225	1180000
	2-3x12	1380	1326000	1610	1547000	1840	1767000
	3-3x12	920	875000	1075	1021000	1225	1166000
	2-4x12	990	949000	1155	1107000	1320	1265000
	1-6x12	1200	1129000	1400	1317000	1600	1505000
	1-8x12	880	829000	1025	967000	1175	1105000
	3-4x12	660	632000	770	737000	880	842000
19'	3-3x10	1520	1872000	1775	2184000	2025	2495000
	3-2x12	1710	1735000	1995	2025000	2280	2313000
	4-2x12	1280	1300000	1495	1517000	1705	1733000
	5-2x12	1025	1040000	1195	1214000	1365	1386000
	2-3x12	1540	1560000	1795	1820000	2050	2079000
	3-3x12	1025	1029000	1195	1201000	1365	1372000
	2-4x12	1100	1115000	1280	1301000	1465	1486000
	1-6x12	1335	1326000	1560	1547000	1780	1767000
	1-8x12	980	973000	1145	1135000	1305	1297000
	3-4x12	735	744000	860	868000	980	992000
	4-3x12	770	780000	900	910000	1025	1040000
	2-6x12	670	1329000	780	1551000	895	1771000
20'	3-2x12	1900	2021000	2220	2358000	2530	2694000
	4-2x12	1280	1516000	1495	1769000	1705	2021000
	5-2x12	1135	1214000	1325	1417000	1515	1618000
	3-3x12	1135	1200000	1325	1400000	1515	1600000
	2-4x12	1220	1300000	1425	1517000	1625	1733000
	1-6x12	1620	1550000	1890	1809000	2160	2066000
	1-8x12	1085	1136000	1265	1326000	1445	1514000
	3-4x12	810	866000	945	1011000	1080	1154000
	4-3x12	850	910000	990	1062000	1135	1213000
	2-6x12	740	775000	865	904000	985	1033000
	1-10x12	860	896000	1005	1046000	1145	1194000
	2-3x14	1230	1114000	1435	1300000	1640	1485000
21'	4-2x12	1570	1756000	1830	2049000	2095	2341000
	5-2x12	1255	1405000	1465	1640000	1675	1873000
	3-3x12	1255	1389000	1465	1621000	1675	1851000
	2-4x12	1345	1505000	1570	1756000	1795	2006000
	1-8x12	1200	1315000	1400	1535000	1600	1753000
	3-4x12	895	1004000	1045	1172000	1195	1338000
	4-3x12	935	1054000	1090	1230000	1245	1405000
	2-6x12	820	896000	955	1046000	1095	1194000
	1-10x12	950	1039000	1110	1212000	1265	1385000
	2-3x14	1355	1290000	1580	1505000	1805	1719000
	1-6x14	1190	1109000	1390	1294000	1585	1478000
	2-4x14	930	871000	1085	1016000	1240	1161000
22'	4-2x12	1720	2020000	2005	2357000	2295	2693000
	5-2x12	1375	1615000	1605	1885000	1830	2153000
	3-3x12	1375	1596000	1605	1862000	1830	2127000
	3-4x12	985	1154000	1150	1347000	1315	1538000
	4-3x12	1030	1211000	1200	1413000	1375	1614000
	2-6x12	900	1031000	1050	1203000	1200	1374000
	1-10x12	1035	1194000	1205	1393000	1380	1592000
	2-3x14	1485	1484000	1730	1732000	1980	1978000
	1-6x14	1305	1275000	1525	1488000	1740	1700000
	2-4x14	1025	1001000	1195	1168000	1365	1334000
	3-3x14	995	981000	1160	1145000	1325	1308000
	3-4x14	680	667000	795	778000	905	889000

Table E-29 Floor and roof beams.

Required values for fiber stress in bending (f) and modulus of elasticity (E) for the sizes shown to support safely a live load of 40 pounds per square foot within a deflection limitation of l/360.

SPAN OF BEAM	NOMINAL SIZE OF BEAM	6'-0" f	6'-0" E	7'-0" f	7'-0" E	8'-0" f	8'-0" E
10'	1-3x8	2055	2040000	2400	2381000	2740	2719000
	2-2x8	1710	1701000	1995	1985000	2280	2267000
	1-4x8	1470	1453000	1715	1696000	1960	1937000
	1-6x8	875	837000	1020	977000	1165	1116000
	2-2x10	1050	817000	1225	953000	1400	1089000
	1-3x10	1260	982000	1470	1146000	1680	1309000
	1-4x10	900	702000	1050	819000	1200	936000
11'	1-4x8	1775	1939000	2070	2263000	2365	2585000
	1-6x8	1055	1114000	1230	1300000	1405	1485000
	2-2x10	1275	1087000	1490	1268000	1700	1449000
	1-3x10	1525	1308000	1780	1526000	2030	1743000
	1-4x10	1090	934000	1270	1090000	1455	1245000
	3-2x10	850	726000	990	847000	1135	968000
	2-3x10	765	654000	890	763000	1020	872000
12'	1-6x8	1255	1447000	1465	1689000	1670	1929000
	3-2x8	1645	1957000	1920	2284000	2190	2609000
	2-2x10	1510	1416000	1760	1652000	2010	1887000
	1-3x10	1820	1698000	2125	1981000	2425	2263000
	1-4x10	1300	1212000	1515	1414000	1735	1615000
	2-3x10	905	847000	1055	988000	1205	1129000
	3-2x10	1010	943000	1180	1100000	1345	1257000
	1-6x10	785	713000	915	832000	1045	950000
	2-4x10	650	606000	760	707000	865	808000
13'	2-4x8	1235	1602000	1440	1869000	1645	2135000
	2-2x10	1780	1800000	2075	2100000	2370	2400000
	3-2x10	1185	1200000	1380	1400000	1580	1600000
	2-3x10	1070	1080000	1250	1260000	1425	1440000
	1-4x10	1525	1845000	1780	2153000	2035	2459000
	2-4x10	760	771000	890	900000	1015	1028000
	3-3x10	710	719000	830	839000	945	958000
	1-6x10	920	906000	1075	1057000	1225	1208000
	4-2x10	890	899000	1040	1049000	1185	1198000
14'	2-4x8	1435	1999000	1675	2333000	1915	2665000
	3-2x10	1375	1495000	1605	1745000	1830	1993000
	2-3x10	1235	1345000	1440	1570000	1645	1793000
	2-4x10	880	961000	1025	1121000	1175	1281000
	3-3x10	825	898000	960	1048000	1100	1197000
	1-6x10	1145	1885000	1335	2200000	1525	2513000
	1-8x10	780	829000	910	967000	1040	1105000
	4-2x10	1030	1123000	1200	1310000	1375	1497000
	2-2x12	1395	1248000	1630	1456000	1860	1663000
	3-2x12	930	832000	1085	971000	1240	1109000
15'	3-2x10	1575	1842000	1840	2150000	2100	2455000
	2-3x10	1420	1657000	1655	1934000	1890	2209000
	2-4x10	1010	1183000	1175	1380000	1345	1577000
	3-3x10	945	1105000	1100	1289000	1260	1473000
	1-6x10	1225	1392000	1430	1624000	1635	1855000
	1-8x10	900	1020000	1050	1190000	1200	1360000
	4-2x10	1180	1381000	1375	1612000	1575	1841000
	2-2x12	1600	1536000	1865	1792000	2130	2047000
	3-2x12	1065	1024000	1240	1195000	1420	1365000
	1-3x12	1920	1843000	2240	2151000	2560	2457000
	4-2x12	800	768000	935	896000	1065	1024000
	2-3x12	960	921000	1120	1075000	1280	1228000
16'	2-3x10	1610	2011000	1880	2347000	2145	2681000
	2-4x10	1155	1438000	1350	1678000	1540	1917000
	3-3x10	1075	1341000	1255	1565000	1435	1787000
	1-6x10	1395	1687000	1625	1969000	1860	2249000
	1-8x10	1020	1237000	1190	1443000	1360	1649000
	4-2x10	1345	1675000	1570	1955000	1790	2233000
	2-2x12	1820	1861000	2120	2172000	2425	2481000
	3-2x12	1210	1242000	1410	1449000	1610	1655000
	4-2x12	910	931000	1060	1086000	1215	1241000
	5-2x12	730	745000	850	869000	975	993000
	2-3x12	1095	1117000	1280	1303000	1460	1489000
	3-3x12	730	737000	850	860000	975	982000

SPAN OF BEAM	NOMINAL SIZE OF BEAM	6'-0" f	6'-0" E	7'-0" f	7'-0" E	8'-0" f	8'-0" E
17'	2-4x10	1300	1723000	1520	2011000	1735	2297000
	3-3x10	1215	1609000	1420	1878000	1620	2145000
	1-8x10	1150	1483000	1340	1731000	1535	1977000
	3-2x12	1370	1489000	1600	1738000	1825	1985000
	4-2x12	1025	1117000	1195	1303000	1365	1489000
	5-2x12	820	894000	955	1043000	1095	1192000
	2-3x12	1230	1342000	1435	1566000	1640	1789000
	3-3x12	820	885000	955	1033000	1095	1180000
	2-4x12	880	958000	1025	1118000	1175	1277000
	1-6x12	1070	1141000	1250	1331000	1425	1521000
	1-8x12	785	837000	915	977000	1045	1116000
	3-4x12	585	639000	680	746000	780	852000
18'	3-3x10	1365	1909000	1595	2228000	1820	2545000
	1-8x10	1300	1764000	1515	2058000	1730	2351000
	3-2x12	1540	1770000	1800	2065000	2050	2359000
	4-2x12	1150	1326000	1340	1547000	1530	1767000
	5-2x12	920	1062000	1075	1239000	1225	1416000
	2-3x12	1380	1591000	1610	1857000	1840	2121000
	3-3x12	920	1050000	1075	1225000	1225	1400000
	2-4x12	990	1138000	1155	1328000	1320	1517000
	1-6x12	1200	1354000	1400	1580000	1600	1805000
	1-8x12	880	994000	1025	1160000	1175	1325000
	3-4x12	660	758000	770	884000	880	1010000
	4-3x12	690	796000	805	929000	920	1061000
19'	4-2x12	1280	1560000	1495	1820000	1705	2079000
	5-2x12	1025	1248000	1195	1456000	1365	1663000
	2-3x12	1540	1872000	1795	2185000	2050	2495000
	3-3x12	1025	1234000	1195	1440000	1365	1645000
	2-4x12	1100	1338000	1280	1561000	1465	1783000
	1-6x12	1335	1591000	1560	1857000	1780	2121000
	1-8x12	980	1168000	1145	1363000	1305	1557000
	3-4x12	735	892000	860	1041000	980	1189000
	4-3x12	770	936000	900	1092000	1025	1248000
	2-6x12	670	1594000	780	186000	895	2125000
	1-10x12	775	923000	905	1077000	1035	1230000
	2-3x14	1110	1146000	1295	1337000	1480	1528000
20'	4-2x12	1280	1819000	1495	2123000	1705	2425000
	5-2x12	1135	1456000	1325	1699000	1515	1941000
	3-3x12	1135	1440000	1325	1680000	1515	1919000
	2-4x12	1220	1560000	1425	1820000	1625	2079000
	1-8x12	1085	1363000	1265	1591000	1445	1817000
	3-4x12	810	1039000	945	1212000	1080	1385000
	4-3x12	850	1092000	990	1274000	1135	1456000
	2-6x12	740	930000	865	1085000	985	1240000
	1-10x12	860	1075000	1005	1254000	1145	1433000
	2-3x14	1230	1336000	1435	1559000	1640	1781000
	1-6x14	1075	1149000	1255	1341000	1430	1532000
	2-4x14	845	903000	985	1054000	1125	1204000
21'	5-2x12	1255	1686000	1465	1967000	1675	2247000
	3-3x12	1255	1666000	1465	1944000	1675	2221000
	2-4x12	1345	1806000	1570	2107000	1795	2407000
	1-8x12	1200	1578000	1400	1841000	1600	2103000
	3-4x12	895	1204000	1045	1405000	1195	1605000
	4-3x12	935	1264000	1090	1475000	1245	1685000
	2-6x12	820	1075000	955	1254000	1095	1433000
	1-10x12	950	1246000	1110	1454000	1265	1661000
	2-3x14	1355	1548000	1580	1806000	1805	2063000
	1-6x14	1190	1330000	1390	1552000	1585	1773000
	2-4x14	930	1045000	1085	1219000	1240	1393000
	3-3x14	905	1032000	1055	1204000	1205	1375000
22'	5-2x12	1375	1938000	1605	2262000	1830	2583000
	3-3x12	1375	1915000	1605	2235000	1830	2553000
	3-4x12	985	1384000	1150	1615000	1315	1845000
	4-3x12	1030	1453000	1200	1695000	1375	1937000
	2-6x12	900	1237000	1050	1443000	1200	1649000
	1-10x12	1035	1432000	1205	1671000	1380	1909000
	2-3x14	1485	1780000	1730	2077000	1980	2373000
	1-6x14	1305	1530000	1525	1785000	1740	2039000
	2-4x14	1025	1201000	1195	1401000	1365	1600000
	3-3x14	995	1177000	1160	1373000	1325	1569000
	3-4x14	680	801000	795	935000	905	1068000
	1-8x14	955	1122000	1115	1309000	1275	1495000

Table E–30 Adjustment factors.

SIZE FACTOR, C_F

Tabulated bending, tension, and compression parallel to grain design values for dimension lumber 2" to 4" thick shall be multiplied by the following size factors:

SIZE FACTORS, C_F

Grades	Width	F_b Thickness 2" & 3"	F_b Thickness 4"	F_t	F_c
Select Structural, No. 1 & Btr. No. 1, No. 2, No. 3	2", 3", & 4"	1.5	1.5	1.5	1.15
	5"	1.4	1.4	1.4	1.1
	6"	1.3	1.3	1.3	1.1
	8"	1.2	1.3	1.2	1.05
	10"	1.1	1.2	1.1	1.0
	12"	1.0	1.1	1.0	1.0
	14" & wider	0.9	1.0	0.9	0.9
Stud	2", 3", & 4"	1.1	1.1	1.1	1.05
	5" & 6"	1.0	1.0	1.0	1.0
Construction & Standard	2", 3", & 4"	1.0	1.0	1.0	1.0
Utility	4"	1.0	1.0	1.0	1.0
	2" & 3"	0.4	—	0.4	0.6

REPETITIVE MEMBER FACTOR, C_r

Bending design values, F_b, for dimension lumber 2" to 4" thick shall be multiplied by the repetitive member factor, $C_r = 1.15$, when such members are used as joists, truss chords, rafters, studs, planks, decking, or similar members which are in contact or spaced not more than 24" on centers; are not less than 3 in number; and are joined by floor, roof, or other load distributing elements adequate to support the design load.

FLAT USE FACTOR, C_{fu}

Bending design values adjusted by size factors are based on edgewise use (load applied to narrow face). When dimension lumber is used flatwise (load applied to wide face), the bending design value, F_b, shall also be multiplied by the following flat use factors:

FLAT USE FACTORS, C_{fu}

Width	Thickness 2" & 3"	Thickness 4"
2" & 3"	1.0	—
4"	1.1	1.0
5"	1.1	1.05
6"	1.15	1.05
8"	1.15	1.05
10" & wider	1.2	1.1

WET SERVICE FACTOR, C_M

When dimension lumber is used where moisture content will exceed 19% for an extended time period, design values shall be multiplied by the appropriate wet service factors from the following table:

WET SERVICE FACTORS, C_M

F_b	F_t	F_v	$F_{c\perp}$	F_c	E
0.85*	1.0	0.97	0.67	0.8**	0.9

* when $(F_b)(C_F) \leq 1150$ psi, $C_M = 1.0$
** when $(F_c)(C_F) \leq 750$ psi, $C_M = 1.0$

Table E–30 *(continued)*

SHEAR STRESS FACTOR, C_H

Tabulated shear design values parallel to grain have been reduced to allow for the occurrence of splits, checks, and shakes. Tabulated shear design values parallel to grain, F_b, shall be permitted to be multiplied by the shear stress factors specified in the following table when length of split, or size of check or shake is known and no increase in them is anticipated. When shear stress factors are used for Redwood, a tabulated design value of $F_v = 80$ psi shall be assigned for all grades of Redwood dimension lumber. Shear stress factors shall be permitted to be linearly interpolated.

SHEAR STRESS FACTORS, C_H

Length of split on wide face of 2″ (nominal) lumber	C_H	Length of split on wide face of 3″ (nominal) and thicker lumber	C_H	Size of shake* in 2″ (nominal) and thicker lumber	C_H
no split	2.00	no split	2.00	no shake	2.00
1/2 × wide face	1.67	1/2 × narrow face	1.67	1/6 × narrow face.........	1.67
3/4 × wide face	1.50	3/4 × narrow face	1.50	1/4 narrow face	1.50
1 × wide face	1.33	1 × narrow face	1.33	1/3 × narrow face	1.33
1-1/2 × wide face or more....	1.00	1-1/2 × narrow face or more	1.00	1/2 × narrow face or more ...	1.00
				*Shake is measured at the end between lines enclosing the shake and perpendicular to the loaded face.	

SOURCE: *National Design Specification for Wood Construction Supplement,* American Forest and Paper Association, 1993.

Table E–31 Base design values for visually graded dimension lumber, all species except southern pine (see Table E–32). Tabulated design values are for normal load duration and dry service conditions. See NDS 2.3 for a comprehensive description of design value adjustment factors. Use with Table E–30.

Species and commercial grade	Size classification	Design values in pounds per square inch (psi)						Grading Rules Agency
		Bending F_b	Tension parallel to grain F_t	Shear parallel to grain F_v	Compression perpendicular to grain $F_{c\perp}$	Compression parallel to grain F_c	Modulus of Elasticity E	
ASPEN								
Select Structural		875	500	60	265	725	1,100,000	
No. 1	2″-4″ thick	625	375	60	265	600	1,100,000	
No. 2		600	350	60	265	450	1,000,000	NELMA
No. 3	2″ & wider	350	200	60	265	275	900,000	NSLB
Stud		475	275	60	265	300	900,000	WWPA
Construction	2″-4″ thick	700	400	60	265	625	900,000	
Standard		375	225	60	265	475	900,000	
Utility	2″-4″ wide	175	100	60	265	300	800,000	
BEECH-BIRCH-HICKORY								
Select Structural		1450	850	100	715	1200	1,700,000	
No. 1	2″-4″ thick	1050	600	100	715	950	1,600,000	
No. 2		1000	600	100	715	750	1,500,000	
No. 3	2″ & wider	575	350	100	715	425	1,300,000	NELMA
Stud		775	450	100	715	475	1,300,000	
Construction	2″-4″ thick	1150	675	100	715	1000	1,400,000	
Standard		650	375	100	715	775	1,300,000	
Utility	2″-4″ wide	300	175	100	715	500	1,200,000	

Species and commercial grade	Size classification	Design values in pounds per square inch (psi)						Grading Rules Agency
		Bending F_b	Tension parallel to grain F_t	Shear parallel to grain F_v	Compression perpendicular to grain $F_{c\perp}$	Compression parallel to grain F_c	Modulus of Elasticity E	
COTTONWOOD								
Select Structural	2"-4" thick	875	525	65	320	775	1,200,000	NSLB
No. 1		625	375	65	320	625	1,200,000	
No. 2		625	350	65	320	475	1,100,000	
No. 3	2" & wider	350	200	65	320	275	1,000,000	
Stud		475	275	65	320	300	1,000,000	
Construction	2"-4" thick	700	400	65	320	650	1,000,000	
Standard		400	225	65	320	500	900,000	
Utility	2"-4" wide	175	100	65	320	325	900,000	
DOUGLAS FIR-LARCH								
Select Structural		1450	1000	95	625	1700	1,900,000	WCLIB WWPA
No. 1 & Btr	2"-4" thick	1150	775	95	625	1500	1,800,000	
No. 1		1000	675	95	625	1450	1,700,000	
No. 2	2" & wider	875	575	95	625	1300	1,600,000	
No. 3		500	325	95	625	750	1,400,000	
Stud		675	450	95	625	825	1,400,000	
Construction	2"-4" thick	1000	650	95	625	1600	1,500,000	
Standard		550	375	95	625	1350	1,400,000	
Utility	2"-4" wide	275	175	95	625	875	1,300,000	
DOUGLAS FIR-LARCH (NORTH)								
Select Structural	2"-4" thick	1300	800	95	625	1900	1,900,000	NLGA
No. 1/No. 2		825	500	95	625	1350	1,600,000	
No. 3	2" & wider	475	300	95	625	775	1,400,000	
Stud		650	375	95	625	850	1,400,000	
Construction	2"-4" thick	950	575	95	625	1750	1,500,000	
Standard		525	325	95	625	1400	1,400,000	
Utility	2"-4" wide	250	150	95	625	925	1,300,000	
DOUGLAS FIR-SOUTH								
Select Structural		1300	875	90	520	1550	1,400,000	WWPA
No. 1	2"-4" thick	900	600	90	520	1400	1,300,000	
No. 2		825	525	90	520	1300	1,200,000	
No. 3	2" & wider	475	300	90	520	750	1,100,000	
Stud		650	425	90	520	825	1,100,000	
Construction	2"-4" thick	925	600	90	520	1550	1,200,000	
Standard		525	350	90	520	1300	1,100,000	
Utility	2"-4" wide	250	150	90	520	875	1,000,000	
EASTERN HEMLOCK-TAMARACK								
Select Structural		1250	575	85	555	1200	1,200,000	NELMA NSLB
No. 1	2"-4" thick	775	350	85	555	1000	1,100,000	
No. 2		575	275	85	555	825	1,100,000	
No. 3	2" & wider	350	150	85	555	475	900,000	
Stud		450	200	85	555	525	900,000	
Construction	2"-4" thick	675	300	85	555	1050	1,000,000	
Standard		375	175	85	555	850	900,000	
Utility	2"-4" wide	175	75	85	555	550	800,000	
EASTERN SOFTWOODS								
Select Structural		1250	575	70	335	1200	1,200,000	NELMA NSLB
No. 1	2"-4" thick	775	350	70	335	1000	1,100,000	
No. 2		575	275	70	335	825	1,100,000	
No. 3	2" & wider	350	150	70	335	475	900,000	
Stud		450	200	70	335	525	900,000	
Construction	2"-4" thick	675	300	70	335	1050	1,000,000	
Standard		375	175	70	335	850	900,000	
Utility	2"-4" wide	175	75	70	335	550	800,000	

Table E–31 (continued)

Species and commercial grade	Size classification	Design values in pounds per square inch (psi)						Grading Rules Agency
		Bending F_b	Tension parallel to grain F_t	Shear parallel to grain F_v	Compression perpendicular to grain $F_{c\perp}$	Compression parallel to grain F_c	Modulus of Elasticity E	
EASTERN WHITE PINE								
Select Structural	2"-4" thick	1250	575	70	350	1200	1,200,000	
No. 1		775	350	70	350	1000	1,100,000	
No. 2		575	275	70	350	825	1,100,000	
No. 3	2" & wider	350	150	70	350	475	900,000	NELMA
Stud		450	200	70	350	525	900,000	NSLB
Construction	2"-4" thick	675	300	70	350	1050	1,000,000	
Standard		375	175	70	350	850	900,000	
Utility	2"-4" wide	175	75	70	350	550	800,000	
HEM-FIR								
Select Structural	2"-4" thick	1400	900	75	405	1500	1,600,000	
No. 1 & Btr.		1050	700	75	405	1350	1,500,000	
No. 1	2" & wider	950	600	75	405	1300	1,500,000	
No. 2		850	500	75	405	1250	1,300,000	
No. 3		500	300	75	405	725	1,200,000	WCLIB
Stud		675	400	75	405	800	1,200,000	WWPA
Construction	2"-4" thick	975	575	75	405	1500	1,300,000	
Standard		550	325	75	405	1300	1,200,000	
Utility	2"-4" wide	250	150	75	405	850	1,100,000	
HEM-FIR (NORTH)								
Select Structural	2"-4" thick	1300	775	75	370	1650	1,700,000	
No. 1/No. 2		1000	550	75	370	1450	1,600,000	
No. 3	2" & wider	575	325	75	370	850	1,400,000	
Stud		775	425	75	370	925	1,400,000	NLGA
Construction	2"-4" thick	1150	625	75	370	1750	1,500,000	
Standard		625	350	75	370	1500	1,400,000	
Utility	2"-4" wide	300	175	75	370	975	1,300,000	
MIXED MAPLE								
Select Structural	2"-4" thick	1000	600	100	620	875	1,300,000	
No. 1		725	425	100	620	700	1,200,000	
No. 2		700	425	100	620	550	1,100,000	
No. 3	2" & wider	400	250	100	620	325	1,000,000	
Stud		550	325	100	620	350	1,000,000	NELMA
Construction	2"-4" thick	800	475	100	620	725	1,100,000	
Standard		450	275	100	620	575	1,000,000	
Utility	2"-4" wide	225	125	100	620	375	900,000	
MIXED OAK								
Select Structural	2"-4" thick	1150	675	85	800	1000	1,100,000	
No. 1		825	500	85	800	825	1,000,000	
No. 2		800	475	85	800	625	900,000	
No. 3	2" & wider	475	275	85	800	375	800,000	
Stud		625	375	85	800	400	800,000	NELMA
Construction	2"-4" thick	925	550	85	800	850	900,000	
Standard		525	300	85	800	650	800,000	
Utility	2"-4" wide	250	150	85	800	425	800,000	
NORTHERN RED OAK								
Select Structural	2"-4" thick	1400	800	110	885	1150	1,400,000	
No. 1		1000	575	110	885	925	1,400,000	
No. 2		975	575	110	885	725	1,300,000	
No. 3	2" & wider	550	325	110	885	425	1,200,000	
Stud		750	450	110	885	450	1,200,000	NELMA
Construction	2"-4" thick	1100	650	110	885	975	1,200,000	
Standard		625	350	110	885	750	1,100,000	
Utility	2"-4" wide	300	175	110	885	500	1,000,000	

Table E–31 (continued)

Species and commercial grade	Size classification	Design values in pounds per square inch (psi)						Grading Rules Agency
		Bending F_b	Tension parallel to grain F_t	Shear parallel to grain F_v	Compression perpendicular to grain $F_{c\perp}$	Compression parallel to grain F_c	Modulus of Elasticity E	
NORTHERN SPECIES								
Select Structural	2"-4" thick	950	450	65	350	1100	1,100,000	NLGA
No. 1/No. 2		575	275	65	350	825	1,100,000	
No. 3	2" & wider	350	150	65	350	475	1,000,000	
Stud		450	200	65	350	525	1,000,000	
Construction	2"-4" thick	675	300	65	350	1050	1,000,000	
Standard		375	175	65	350	850	900,000	
Utility	2"-4" wide	175	75	65	350	550	900,000	
NORTHERN WHITE CEDAR								
Select Structural		775	450	60	370	750	800,000	NELMA
No. 1	2"-4" thick	575	325	60	370	600	700,000	
No. 2		550	325	60	370	475	700,000	
No. 3	2" & wider	325	175	60	370	275	600,000	
Stud		425	250	60	370	300	600,000	
Construction	2"-4" thick	625	375	60	370	625	700,000	
Standard		350	200	60	370	475	600,000	
Utility	2"-4" wide	175	100	60	370	325	600,000	
RED MAPLE								
Select Structural		1300	750	105	615	1100	1,700,000	NELMA
No. 1	2"-4" thick	925	550	105	615	900	1,600,000	
No. 2		900	525	105	615	700	1,500,000	
No. 3	2" & wider	525	300	105	615	400	1,300,000	
Stud		700	425	105	615	450	1,300,000	
Construction	2"-4" thick	1050	600	105	615	925	1,400,000	
Standard		575	325	105	615	725	1,300,000	
Utility	2"-4" wide	275	150	105	615	475	1,200,000	
RED OAK								
Select Structural		1150	675	85	820	1000	1,400,000	NELMA
No. 1	2"-4" thick	825	500	85	820	825	1,300,000	
No. 2		800	475	85	820	625	1,200,000	
No. 3	2" & wider	475	275	85	820	375	1,100,000	
Stud		625	375	85	820	400	1,100,000	
Construction	2"-4" thick	925	550	85	820	850	1,200,000	
Standard		525	300	85	820	650	1,100,000	
Utility	2"-4" wide	250	150	85	820	425	1,000,000	
REDWOOD								
Clear Structural		1750	1000	145	650	1850	1,400,000	RIS
Select Structural		1350	800	80	650	1500	1,400,000	
Select Structural, open grain		1100	625	80	425	1100	1,100,000	
No. 1	2"-4" thick	975	575	80	650	1200	1,300,000	
No. 1, open grain		775	450	80	425	900	1,100,000	
No. 2	2" & wider	925	525	80	650	950	1,200,000	
No. 2, open grain		725	425	80	425	700	1,000,000	
No. 3		525	300	80	650	550	1,100,000	
No. 3, open grain		425	250	80	425	400	900,000	
Stud		575	325	80	425	450	900,000	
Construction	2"-4" thick	825	475	80	425	925	900,000	
Standard		450	275	80	425	725	900,000	
Utility	2"-4" wide	225	125	80	425	475	800,000	

Species and commercial grade	Size classification	Design values in pounds per square inch (psi)						Grading Rules Agency
		Bending F_b	Tension parallel to grain F_t	Shear parallel to grain F_v	Compression perpendicular to grain $F_{c\perp}$	Compression parallel to grain F_c	Modulus of Elasticity E	
SPRUCE-PINE-FIR								
Select Structural	2″-4″ thick	1250	675	70	425	1400	1,500,000	NLGA
No. 1/No. 2		875	425	70	425	1100	1,400,000	
No. 3	2″ & wider	500	250	70	425	625	1,200,000	
Stud		675	325	70	425	675	1,200,000	
Construction	2″-4″ thick	975	475	70	425	1350	1,300,000	
Standard		550	275	70	425	1100	1,200,000	
Utility	2″-4″ wide	250	125	70	425	725	1,100,000	
SPRUCE-PINE-FIR (SOUTH)								
Select Structural		1300	575	70	335	1200	1,300,000	NELMA
No. 1	2″-4″ thick	850	400	70	335	1050	1,200,000	NSLB
No. 2		750	325	70	335	975	1,100,000	WCLIB
No. 3	2″ & wider	425	200	70	335	550	1,000,000	WWPA
Stud		575	250	70	335	600	1,000,000	
Construction	2″-4″ thick	850	375	70	335	1200	1,000,000	
Standard		475	225	70	335	1000	900,000	
Utility	2″-4″ wide	225	100	70	335	650	900,000	
WESTERN CEDARS								
Select Structural		1000	600	75	425	1000	1,100,000	WCLIB
No. 1	2″-4″ thick	725	425	75	425	825	1,000,000	WWPA
No. 2		700	425	75	425	650	1,000,000	
No. 3	2″ & wider	400	250	75	425	375	900,000	
Stud		550	325	75	425	400	900,000	
Construction	2″-4″ thick	800	475	75	425	850	900,000	
Standard		450	275	75	425	650	800,000	
Utility	2″-4″ wide	225	125	75	425	425	800,000	
WESTERN WOODS								
Select Structural		875	400	70	335	1050	1,200,000	WCLIB
No. 1	2″-4″ thick	650	300	70	335	925	1,100,000	WWPA
No. 2		650	275	70	335	875	1,000,000	
No. 3	2″ & wider	375	175	70	335	500	900,000	
Stud		500	225	70	335	550	900,000	
Construction	2″-4″ thick	725	325	70	335	1050	1,000,000	
Standard		400	175	70	335	900	900,000	
Utility	2″-4″ wide	200	75	70	335	600	800,000	
WHITE OAK								
Select Structural		1200	700	110	800	1100	1,100,000	NELMA
No. 1	2″-4″ thick	875	500	110	800	900	1,000,000	
No. 2		850	500	110	800	700	900,000	
No. 3	2″ & wider	475	275	110	800	400	800,000	
Stud		650	375	110	800	450	800,000	
Construction	2″-4″ thick	950	550	110	800	925	900,000	
Standard		525	325	110	800	725	800,000	
Utility	2″-4″ wide	250	150	110	800	475	800,000	

Table E–31 *(continued)*

Species and commercial grade	Size classification	Design values in pounds per square inch (psi)						Grading Rules Agency
		Bending F_b	Tension parallel to grain F_t	Shear parallel to grain F_v	Compression perpendicular to grain $F_{c\perp}$	Compression parallel to grain F_c	Modulus of Elasticity E	
YELLOW POPLAR								
Select Structural	2"-4" thick	1000	575	75	420	900	1,500,000	
No. 1		725	425	75	420	725	1,400,000	
No. 2		700	400	75	420	575	1,300,000	NSLB
No. 3	2" & wider	400	225	75	420	325	1,200,000	
Stud		550	325	75	420	350	1,200,000	
Construction	2"-4" thick	800	475	75	420	750	1,300,000	
Standard		450	250	75	420	575	1,100,000	
Utility	2"-4" wide	200	125	75	420	375	1,100,000	

SOURCE: *National Design Specification for Wood Construction Supplement,* American Forest and Paper Association, 1993.

NOTES:

1. **Lumber dimensions.** Tabulated design values are applicable to lumber that will be used under dry conditions such as in most covered structures. For 2" to 4" thick lumber, the DRY dressed sizes shall be used regardless of the moisture content at the time of manufacture or use. In the calculation of design values, the natural gain in strength and stiffness that occurs as lumber dries has been taken into consideration, as well as the reduction in size that occurs when unseasoned lumber shrinks. The gain in load-carrying capacity due to increased strength and stiffness resulting from drying more than offsets the design effect of size reductions due to shrinkage.

2. **Stress-rated boards.** Stress-rated boards of nominal 1", $1^1/4$", and $1^1/2$" thickness, 2" and wider, of most species are permitted the design values shown for Select Structural, No. 1 & Btr, No. 1, No. 2, No. 3, Stud, Construction, Standard, Utility, Clear Heart Structural, and Clear Structural grades as shown in the 2" to 4" thick categories herein, when graded in accordance with the stress-rated board provisions in the applicable grading rules. Information on stress-rated board grades applicable to the various species is available from the respective grading rules agencies. Information on additional design values may also be available from the respective grading agencies.

Table E-32 Adjustment factors.

SIZE FACTOR, C_F ───────────────────────

Appropriate size adjustment factors have already been incorporated in the tabulated design values for most thicknesses of Southern Pine and Mixed Southern Pine dimension lumber. For dimension lumber 4" thick, 8" and wider (all grades except Dense Structural 86, Dense Structural 72, and Dense Structural 65), tabulated bending design values, F_b, shall be permitted to be multiplied by the size factor, $C_F = 1.1$. For dimension lumber wider than 12" (all grades except Dense Structural 86, Dense Structural 72, and Dense Structural 65), tabulated bending, tension, and compression parallel to grain design values for 12" wide lumber shall be multiplied by the size factor, $C_F = 0.9$. When the depth, d, of Dense Structural 86, Dense Structural 72, or Dense Structural 65 dimension lumber exceeds 12", the tabulated bending design value, F_b, shall be multiplied by the following size factor:

$$C_F = (12/d)^{1/9}$$

REPETITIVE MEMBER FACTOR, C_r ───────────────────────

Bending design values, F_b, for dimension lumber 2" to 4" thick shall be multiplied by the repetitive member factor, $C_r = 1.15$, when such members are used as joists, truss chords, rafters, studs, planks, decking, or similar members which are in contact or spaced not more than 24" on centers; are not less than 3 in number; and are joined by floor, roof, or other load distributing elements adequate to support the design load.

FLAT USE FACTOR, C_{fu} ───────────────────────

Bending design values adjusted by size factors are based on edgewise use (load applied to narrow face). When dimension lumber is used flatwise (load applied to wide face), the bending design value, F_b, shall also be multiplied by the following flat use factors:

FLAT USE FACTORS, C_{fu}

	Thickness	
Width	2" & 3"	4"
2" & 3"	1.0	—
4"	1.1	1.0
5"	1.1	1.05
6"	1.15	1.05
8"	1.15	1.05
10" & wider	1.2	1.1

WET SERVICE FACTOR, C_M ───────────────────────

When dimension lumber is used where moisture content will exceed 19% for an extended time period, design values shall be multiplied by the appropriate wet service factors from the following table (for Dense Structural 86, Dense Structural 72, and Dense Structural 65, use tabulated design values for wet service conditions without further adjustment):

WET SERVICE FACTORS, C_M

F_b	F_t	F_v	$F_{c\perp}$	F_c	E
0.85*	1.0	0.97	0.67	0.8**	0.9

* when $(F_b)(C_F) \le 1150$ psi, $C_M = 1.0$
** when $F_c \le 750$ psi, $C_M = 1.0$

SHEAR STRESS FACTOR, C_H

Tabulated shear design values parallel to grain have been reduced to allow for the occurrence of splits, checks, and shakes. Tabulated shear design values parallel to grain, F_b, shall be permitted to be multiplied by the shear stress factors specified in the following table when length of split, or size of check or shake is known and no increase in them is anticipated. When shear stress factors are used for Southern Pine and Mixed Southern Pine, a tabulated design value of $F_v = 90$ psi shall be assigned for all grades of Southern Pine and Mixed Southern Pine dimension lumber. Shear stress factors shall be permitted to be linearly interpolated.

SHEAR STRESS FACTORS, C_H

Length of split on wide face of 2″ (nominal) lumber	C_H	Length of split on wide face of 3″ (nominal) and thicker lumber	C_H	Size of shake* in 2″ (nominal) and thicker lumber	C_H
no split	2.00	no split	2.00	no shake	2.00
1/2 × wide face	1.67	1/2 × narrow face	1.67	1/6 × narrow face	1.67
3/4 × wide face	1.50	3/4 × narrow face	1.50	1/4 narrow face	1.50
1 × wide face	1.33	1 × narrow face	1.33	1/3 × narrow face	1.33
1-1/2 × wide face or more	1.00	1-1/2 × narrow face or more	1.00	1/2 × narrow face or more	1.00
				*Shake is measured at the end between lines enclosing the shake and perpendicular to the loaded face.	

SOURCE: *National Design Specification for Wood Construction Supplement,* American Forest and Paper Association, 1993.

Table E–33 Design values for visually graded southern pine dimension lumber. Tabulated design values are for normal load duration and dry service conditions, unless specified otherwise. See NDS 2.3 for a comprehensive description of design value adjustment factors. Use with Table E–32.

| Species and commercial grade | Size classification | Design values in pounds per square inch (psi) | | | | | | Grading Rules Agency |
		Bending F_b	Tension parallel to grain F_t	Shear parallel to grain F_v	Compression perpendicular to grain $F_{c\perp}$	Compression parallel to grain F_c	Modulus of Elasticity E	
MIXED SOUTHERN PINE								
Select Structural		2050	1200	100	565	1800	1,600,000	
No. 1	2"-4" thick	1450	875	100	565	1650	1,500,000	
No. 2		1300	775	90	565	1650	1,400,000	
No. 3	2"-4" wide	750	450	90	565	950	1,200,000	
Stud		775	450	90	565	950	1,200,000	
Construction	2"-4" thick	1000	600	100	565	1700	1,300,000	
Standard		550	325	90	565	1450	1,200,000	
Utility	4" wide	275	150	90	565	950	1,100,000	
Select Structural		1850	1100	90	565	1700	1,600,000	
No. 1	2"-4" thick	1300	750	90	565	1550	1,500,000	
No. 2		1150	675	90	565	1550	1,400,000	
No. 3	5"-6" wide	675	400	90	565	875	1,200,000	
Stud		675	400	90	565	875	1,200,000	
Select Structural		1750	1000	90	565	1600	1,600,000	SPIB
No. 1	2"-4" thick	1200	700	90	565	1450	1,500,000	
No. 2	8" wide	1050	625	90	565	1450	1,400,000	
No. 3		625	375	90	565	850	1,200,000	
Select Structural		1500	875	90	565	1600	1,600,000	
No. 1	2"-4" thick	1050	600	90	565	1450	1,500,000	
No. 2	10" wide	925	550	90	565	1450	1,400,000	
No. 3		525	325	90	565	825	1,200,000	
Select Structural		1400	825	90	565	1550	1,600,000	
No. 1	2"-4" thick	975	575	90	565	1400	1,500,000	
No. 2	12" wide	875	525	90	565	1400	1,400,000	
No. 3		500	300	90	565	800	1,200,000	
SOUTHERN PINE								
Dense Select Structural		3050	1650	100	660	2250	1,900,000	
Select Structural		2850	1600	100	565	2100	1,800,000	
Non-Dense Select Structural		2650	1350	100	480	1950	1,700,000	
No. 1 Dense		2000	1100	100	660	2000	1,800,000	
No. 1	2"-4" thick	1850	1050	100	565	1850	1,700,000	
No. 1 Non-Dense		1700	900	100	480	1700	1,600,000	
No. 2 Dense	2"-4" wide	1700	875	90	660	1850	1,700,000	
No. 2		1500	825	90	565	1650	1,600,000	
No. 2 Non-Dense		1350	775	90	480	1600	1,400,000	
No. 3		850	475	90	565	975	1,400,000	
Stud		875	500	90	565	975	1,400,000	
Construction	2"-4" thick	1100	625	100	565	1800	1,500,000	
Standard		625	350	90	565	1500	1,300,000	
Utility	4" wide	300	175	90	565	975	1,300,000	

Table E–33 *(continued)*

Species and commercial grade	Size classification	Design values in pounds per square inch (psi)						Grading Rules Agency
		Bending F_b	Tension parallel to grain F_t	Shear parallel to grain F_v	Compression perpendicular to grain $F_{c\perp}$	Compression parallel to grain F_c	Modulus of Elasticity E	
Dense Select Structural		2700	1500	90	660	2150	1,900,000	
Select Structural		2550	1400	90	565	2000	1,800,000	
Non-Dense Select Structural		2350	1200	90	480	1850	1,700,000	
No. 1 Dense	2"-4" thick	1750	950	90	660	1900	1,800,000	
No. 1		1650	900	90	565	1750	1,700,000	
No. 1 Non-Dense	5"-6" wide	1500	800	90	480	1600	1,600,000	
No. 2 Dense		1450	775	90	660	1750	1,700,000	
No. 2		1250	725	90	565	1600	1,600,000	
No. 2 Non-Dense		1150	675	90	480	1500	1,400,000	
No. 3		750	425	90	565	925	1,400,000	
Stud		775	425	90	565	925	1,400,000	
Dense Select Structural		2450	1350	90	660	2050	1,900,000	SPIB
Select Structural		2300	1300	90	565	1900	1,800,000	
Non-Dense Select Structural		2100	1100	90	480	1750	1,700,000	
No. 1 Dense	2"-4" thick	1650	875	90	660	1800	1,800,000	
No. 1		1500	825	90	565	1650	1,700,000	
No. 1 Non-Dense	8" wide	1350	725	90	480	1550	1,600,000	
No. 2 Dense		1400	675	90	660	1700	1,700,000	
No. 2		1200	650	90	565	1550	1,600,000	
No. 2 Non-Dense		1100	600	90	480	1450	1,400,000	
No. 3		700	400	90	565	875	1,400,000	
Dense Select Structural		2150	1200	90	660	2000	1,900,000	
Select Structural		2050	1100	90	565	1850	1,800,000	
Non-Dense Select Structural		1850	950	90	480	1750	1,700,000	
No. 1 Dense	2"-4" thick	1450	775	90	660	1750	1,800,000	
No. 1		1300	725	90	565	1600	1,700,000	
No. 1 Non-Dense	10" wide	1200	650	90	480	1500	1,600,000	
No. 2 Dense		1200	625	90	660	1650	1,700,000	
No. 2		1050	575	90	565	1500	1,600,000	
No. 2 Non-Dense		950	550	90	480	1400	1,400,000	
No. 3		600	325	90	565	850	1,400,000	
Dense Select Structural		2050	1100	90	660	1950	1,900,000	
Select Structural		1900	1050	90	565	1800	1,800,000	
Non-Dense Select Structural		1750	900	90	480	1700	1,700,000	
No. 1 Dense	2"-4" thick	1350	725	90	660	1700	1,800,000	
No. 1		1250	675	90	565	1600	1,700,000	
No. 1 Non-Dense	12" wide	1150	600	90	480	1500	1,600,000	
No. 2 Dense		1150	575	90	660	1600	1,700,000	
No. 2		975	550	90	565	1450	1,600,000	
No. 2 Non-Dense		900	525	90	480	1350	1,400,000	
No. 3		575	325	90	565	825	1,400,000	

Species and commercial grade	Size classification	Design values in pounds per square inch (psi)						Grading Rules Agency
		Bending F_b	Tension parallel to grain F_t	Shear parallel to grain F_v	Compression perpendicular to grain $F_{c\perp}$	Compression parallel to grain F_c	Modulus of Elasticity E	
SOUTHERN PINE		(Dry service conditions—19% or less moisture content)						
Dense Structural 86	2"-4" thick	2600	1750	155	660	2000	1,800,000	
Dense Structural 72		2200	1450	130	660	1650	1,800,000	SPIB
Dense Structural 65	2" & wider	2000	1300	115	660	1500	1,800,000	
SOUTHERN PINE		(Wet service conditions)						
Dense Structural 86	2-1/2"-4" thick	2100	1400	145	440	1300	1,600,000	
Dense Structural 72		1750	1200	120	440	1100	1,600,000	SPIB
Dense Structural 65	2-1/2" & wider	1600	1050	110	440	1000	1,600,000	

SOURCE: *National Design Specification for Wood Construction Supplement,* American Forest and Paper Association, 1993.

NOTES:

1. **Lumber dimensions.** Tabulated design values are applicable to lumber that will be used under dry conditions such as in most covered structures. For 2" to 4" thick lumber, the DRY dressed sizes shall be used regardless of the moisture content at the time of manufacture or use. In the calculation of design values, the natural gain in strength and stiffness that occurs as lumber dries has been taken into consideration, as well as the reduction in size that occurs when unseasoned lumber shrinks. The gain in load-carrying capacity due to increased strength and stiffness resulting from drying more than offsets the design effect of size reductions due to shrinkage.

2. **Stress-rated boards.** Information for various grades of southern pine stress-rated boards of nominal 1", 1¼", and 1½" thickness, 2" and wider, is available from the southern pine Inspection Bureau (SPIB) in the "Standard Grading Rules for Southern Pine Lumber."

3. **Spruce Pine.** To obtain recommended design values for spruce pine graded to SPIB rules, multiply the appropriate design values for mixed southern pine by the corresponding conversion factor shown below, and round to the nearest 100,000 psi for E; to the next lower multiple of 5 psi for F_v and $F_{c\perp}$; to the next lower multiple of 50 psi for F_b, F_t, and F_c if 1000 psi or greater, 25 psi otherwise.

4. **Size factor.** For sizes wider than 12", use size factors for F_b, F_t, and F_c specified for the 12" width. Use 100% of the F_v, $F_{c\perp}$, and E specified for the 12" width.

Conversion factors for determining design values for spruce pine

	Bending, F_b	Tension Parallel to Grain, F_t	Shear Parallel to Grain, F_v	Compression Perpendicular to Grain, $F_{c\perp}$	Compression Parallel to Grain, F_c	Modulus of Elasticity, E
Conversion factor	0.78	0.78	0.98	0.73	0.78	0.82

In the following examples of heating and cooling calculations using the *Manual J Worksheet*, there are numerous references to tables contained in *Manual J*. These tables are quite extensive, and it would not be appropriate to include them here. You should obtain a copy of *Manual J* and review the calculation procedures using values from the tables appropriate for their location. The footnote below lists the source for *Manual J*.

F.1

EXAMPLE PROBLEM: HEAT LOSS CALCULATION

Data must be obtained from drawings, or by a field inspection, before load calculations can be made. The data required includes the following:

A. **Measurements to determine areas:**
   1. Overall area of windows and doors
   2. Gross areas of walls exposed to outside conditions
   3. Gross area of partitions
   4. Gross areas of walls below grade
   5. Area of ceilings or floors adjacent to unconditioned space
   6. Floor area for each room

Closets and halls are usually included with adjoining rooms. Large closets or entrance halls should be considered separately. Wall, floor, or ceiling dimensions can be rounded to the nearest foot. Window dimensions are recorded to the nearest inch. Measure the size of the window or door opening. Do not include the frame.

B. **Construction details:**
   1. Window type and construction
   2. Door type and construction
   3. Wall construction
   4. Ceiling construction
   5. Roof construction
   6. Floor construction

C. **Temperature differences:**
   1. Temperature differences across components exposed to outside conditions
   2. Temperature differences across partitions, floors, and ceilings adjacent to unoccupied spaces

Table 1, located in back of *Manual J*, lists the outside de-

sign temperatures for various locations. Room temperatures are determined by the owner or the builder based on recommendations by the heating and cooling contractor, or they are prescribed by applicable codes. ACCA recommends 70°F.

F.1.1 Calculation Procedure

Once areas, construction details, and temperature differences are determined, the tables in the back of *Manual J* and the *Worksheet for Manual J* (Fig. F–3) can be used to calculate heat loss. The total heat loss for a room is the sum of the heat lost through each structural component of the room. To calculate the heat lost through any component, multiply the HTM found in the tables by the area of the component.

HTM values for temperature differences that fall between those listed in the tables can be interpolated as follows: For example, a brick wall above grade, construction number 12-F, is subjected to a design temperature difference of 63°F. Table 2 in *Manual J* indicates an HTM of 4.2 Btu/h per square foot at 60°F and 4.6 Btu/h per square foot at 65°F. Select HTM = 4.4 Btu/h per square foot, which is approximately equal to 4.2 + 3/5 (4.6 − 4.2) Btu/h per square foot.

> EXAMPLE 1: Figure F–1 represents a house located in Cedar Rapids, Iowa. Figure F–2 lists the construction details. Assume that the inside design temperature is 70°F. From Table 1 in *Manual J*, the outside design temperature is −5°F.

Figure F–2 illustrates the completed worksheet. The following is a line-by-line explanation of the procedure:

Line 1. Identify each area that is heated.

Lines 2 and 3. Enter the room dimensions. The dimensions shown were from Figure F–1.

Line 4. Enter the ceiling height for reference. The direction the room faces is not a concern when you are making the heat loss calculation, but it is used in the heat gain calculations.

Lines 5A through 5D. Enter the gross area for the walls. For rooms with more than one exposure, use one line for each exposure. For rooms with more than one type of wall construction, use one line for each type of construction. Find the construction number in the tables in the back of *Manual J*. Enter this number on the appropriate line.

Example: The gross area of the west living room wall is 168 sq ft. This wall is listed in Table 2 in *Manual J*. The construction number is 12-D.

Lines 6A through 6C. Enter the area and the orientation of windows and glass doors for each room. Determine construction numbers from the tables, and enter them. Determine the temperature difference across the glass, and read the HTM for heating from

---

[1] The heat loss and heat gain calculation examples contained in this appendix were taken from *Manual J*, 7th ed., *Residential Load Calculation*, Air Conditioning Contractors of America (ACCA), 1712 New Hampshire Avenue NW, Washington, DC 20009.

Table F–1 Assumed design conditions and construction (heating).

	Const. No.	HTM
A. Determine outside design temperature, 5° db, (Table 1 in *Manual J*)		
B. Select inside design temperature, 70° db		
C. Design temperature difference, 75°		
D. Windows		
Living room and dining room—Clear, fixed glass, double glazed, wood frame (Table 2 in *Manual J*)	3A	41.3
Basement—Clear glass, metal casement windows, with storm (Table 2)	2C	48.8
Others—Double-hung, clear, single glass and storm, wood frame (Table 2)	2A	35.6
E. Doors: Metal, urethane-core, no storm (Table 2)	11E	14.3
F. First-floor walls: Basic frame construction with ½" asphalt board (R-11) (Table 2)	12d	6.0
Basement wall: 8" concrete block (Table 2)		
Above-grade height: 3' (R = 5)	14b	10.8
Below-grade height: 5' (R = 5)	15b	5.5
G. Ceiling: Basic construction under vented attic with insulation (R-19) (Table 2)	16d	4.0
H. Floor: Basement floor, 4" concrete (Table 2)	21a	1.8
I. All movable windows and doors have certified leakage of 0.5 cfm per running foot of crack (without storm); envelope has plastic vapor barrier; and major cracks and penetrations have been sealed with caulking material. No fireplace; all exhausts and vents are dampered; all ducts are taped.		

the tables. Enter these values. Multiply the window area by its HTM to determine the heat loss through that window. Enter the heat loss in the column marked "Btuh Htg."

Example: The living room has 40 sq ft of wood frame, fixed double-glass windows. The construction number of the window is 3A. The temperature difference across the window will be based on winter design conditions. The design temperature difference is $70° - (-5°) = 75°F$. The HTM listed for 75°F on line 1–C of the tables is 41.3 Btu/h per square foot. The heat loss through the window is 40 sq ft × 41.3 Btu/h per square foot = 1652.

Example: The workshop has 4 sq ft of metal frame awning glass (plus storm) windows. The construction number is 2C. If the shop temperature is 70°F, the design temperature difference is $70 - (-5) = 75°F$. The HTM is 48.8 Btuh/h per square foot. Heat loss through the window is 4 sq ft × 48.8 Btu/h per square foot = 195 Btuh.

Line 7. Not required for the heating calculation.

Line 8. Enter the area, the construction number, and the HTM for wood or metal doors. Multiply the HTM by the door area, and enter the heat loss through the door.

Example: The main entrance (hall A) has a 20-sq-ft, metal, urethane-core door. The construction number of the door is 11-E. The design temperature difference is $70° - (-5) = 75°F$. The HTM is 14.3 Btu/h per square foot. The heat loss through the door is 14.3 Btu/h per square foot × 20 sq ft = 286 Btuh.

Lines 9A through 9D. For each room, subtract the window and door areas from the corresponding gross wall, and enter the net wall areas and the corresponding construction number. Determine the temperature difference across each wall, and enter the HTM. Multiply the HTM by the wall area, and enter the heat loss through the wall.

Example: The west wall in the living room has a net area of 128 sq ft (168 sq ft − 40 sq ft). The wall

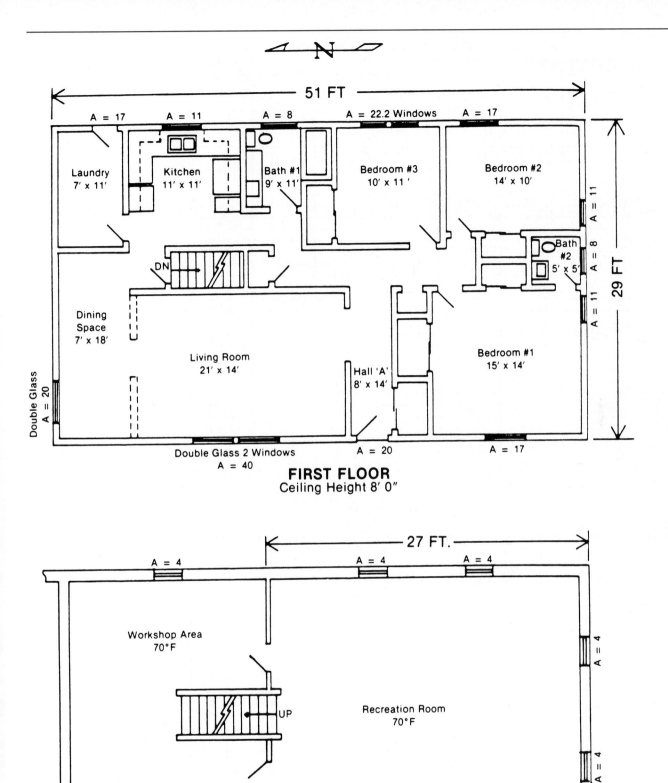

Figure F–1 Heat loss example problem: basement and first-floor plans.

				Entire House			1 Living			2 Dining			3 Laundry			4 Kitchen			5 Bath 1		
1	Name of Room			Entire House			1 Living			2 Dining			3 Laundry			4 Kitchen			5 Bath 1		
2	Running Ft. Exposed Wall			160			21			25			18			11			9		
3	Room Dimensions Ft.			51 x 29			21 x 14			7 x 18			7 x 11			11 x 11			9 x 11		
4	Ceiling Ht. Ft Directions Room Faces			8			8 West			8 North			8			8 East			8 East		
TYPE OF EXPOSURE	Const No.	HTM Htg	HTM Clg	Area or Length	Btuh Htg	Btuh Clg	Area or Length	Btuh Htg	Clg	Area or Length	Btuh Htg	Clg	Area or Length	Btuh Htg	Clg	Area or Length	Btuh Htg	Clg	Area or Length	Btuh Htg	Clg
5 Gross a	12-d			1280			168			200			144			88			72		
Exposed b	14-b			480																	
Walls & c	15-b			800																	
Partitions d																					
6 Windows a	3-A	41.3		60	2478		40	1652		20	826										
& Glass b	2-C	48.8		20	976																
Doors Htg. c	2-A	35.6		105	3738											11	392		8	285	
d																					
7 Windows North																					
& Glass E&W																					
Doors Clg. South																					
8 Other Doors	11-E	14.3		37	529								17	243							
9 Net a	12-d	6.0		1078	6468		128	768		180	1080		127	762		77	462		64	384	
Exposed b	14-b	10.8		460	4968																
Walls & c	15-b	5.5		800	4400																
Partitions d																					
10 Ceilings a	16-d	4.0		1479	5916		294	1176		126	504		77	308		121	484		99	396	
b																					
11 Floors a	21-a	1.8		1479	2662																
b																					
12 Infiltration HTM		70.6		222	15673		40	2824		20	1412		17	1200		11	777		8	565	
13 Sub Total Btuh Loss = 6+8+9+10+11+12					47808			6420			3822			2513			2115			1630	
14 Duct Btuh Loss		0%			—			—			—			—			—				
15 Total Btuh Loss = 13 + 14					47808			6420			3822			2513			2115			1630	
16 People @ 300 & Appliances 1200																					
17 Sensible Btuh Gain = 7+8+9+10+11+12+16																					
18 Duct Btuh Gain		%																			
19 Total Sensible Gain = 17 + 18																					

DO NOT WRITE IN SHADED BLOCKS

6 Bedroom 3			7 Bedroom 2			8 Bath 2			9 Bedroom 1			10 Hall			11 Rec. Room			12 Shop & Utility			
10			24			5			29			8			83			77			2
10 x 11			14 x 10			5 x 5			15 x 14			8 x 14			27 x 29			24 x 29			3
8 East			8 E & S			8 South			8 S & W			8 West			8 E & S			8 East			4
Area or Length	Btuh Htg	Clg	Area or Length	Btuh Htg	Clg	Area or Length	Btuh Htg	Clg	Area or Length	Btuh Htg	Clg	Area or Length	Btuh Htg	Clg	Area or Length	Btuh Htg	Clg	Area or Length	Btuh Htg	Clg	
80			192			40			232			64			249			231			5
															415			385			
															16	781		4	195		6
22	783		28	997		8	285		28	997												
																					7
												20	286								8
58	348		164	984		32	192		204	1224		44	264								9
															233	2516		227	2452		
															415	2283		385	2118		
110	440		140	560		25	100		210	840		112	448									10
															783	1409		696	1253		11
22	1553		28	1977		8	565		28	1977		20	1412		16	1130		4	282		12
	3124			4518			1142			5038			2410			8119			6300		13
	—			—			—			—			—			—			—		14
	3124			4518			1142			5038			2410			8119			6300		15
																					16
																					17
																					18
																					19

Figure F–2 Example heat gain calculation. (Do not write in the shaded blocks.)

construction number is 12-D. The temperature difference is 75°F, and the HTM is 6.0 Btu/h per square foot. The heat loss through the wall is 128 sq ft × 6.0 Btu/h per square foot = 786 Btuh.

Example: The basement wall surrounding the recreation room has a net area of 233 sq ft (above grade), and the construction number is 14-B. The HTM for a 75°F temperature difference is 10.8 Btu/h per square foot. The heat loss above grade is 10.8 × 233 = 2516 Btuh. The net area below grade is 415 sq ft. The construction number is 15-B. The HTM is 5.5 Btu/h per square foot. The heat loss below grade is 415 sq ft × 5.5 Btu/h per square foot = 2283 Btuh.

Lines 10A and 10B. Enter the ceiling area and the construction number for ceilings exposed to a temperature difference. Determine the temperature difference across the ceiling, and enter the HTM. Multiply the HTM by the ceiling area, and enter the heat loss through the ceiling.

Example: An *R*-19 insulated living room ceiling has an area of 294 sq ft. The construction number is 16-D. Since the attic is vented, the temperature difference is $70° - (-5°) = 75°F$. The HTM is 4.0 Btu/h per square foot and the heat loss through the ceiling is 294 sq ft × 4.0 Btu/h per square foot = 1176 Btuh.

Example: The recreation room ceiling will have no heat loss since the temperature difference is zero.

Lines 11A and 11B. Enter the floor area and the floor construction number for floots subject to a temperature difference. Determine the HTM, and calculate the heat loss through the floor.

Example: The recreation-room slab floor has an area of 783 sq ft. The construction number is 21. The design temperature difference is 75°F, and the HTM is 1.8 Btu/h per square foot. The heat loss through the floor is 783 sq ft × 1.8 Btu/h per square foot = 1409 Btuh.

Line 12. Use Table 5 and Calculation Procedure A in *Manual J* to calculate the winter infiltration HTM, and enter this value on line 12. For each room, enter the total sq ft of the window and door openings on line 12. Finally, compute the infiltration heat loss due to infiltration for each room by multiplying the winter infiltration HTM value by the appropriate sq ft of window and door openings. Enter these results on line 12. Refer to Figure F–5 for an example of the infiltration HTM calculation.

Line 13. Calculate the subtotal heat loss for each room and for the entire house. (Add lines 6, 8, 9, 10, 11, and 12.)

Line 14. Calculate and enter the duct heat loss for each room. In this problem, the ducts are in the heated space, so duct losses can be ignored.

Line 15. Add the duct losses to the room losses. This sum is the total heat required for each room and for

the structure. If ventilation air is not introduced through the equipment, you can use the sum for the "Entire House" column to size the heating equipment.

If ventilation air is used, the heat required to temper this air must be added to the total heat required by the structure. This calculation can be made on the front of the *Worksheet for Manual J* form (Fig. F–3) in the panel titled "Heating Summary." In either case, the output capacity of the heating equipment shall not be less than the calculated loss.

Since ventilation was not included in the example problem, the design heating load is calculated as follows:

Line 15 heat loss (Btuh) = 47,808 (Entire house)
Ventilation cfm   = 0
Ventilation heat (Btuh)   = 0
Design heating load   = 47,808 (House) + 0 (Vent)
                           = 47,808 Btuh

If this problem included 100 cfm of outside air for ventilation, the design heating load would be calculated as follows:

Line 15 heat loss (Btuh) = 47,808 (Entire house)
Ventilation cfm   = 100
Design temperature
  difference   = 75°F
Ventilation heat (Btuh)   = 1.1 × 100 cfm × 75°F
                           = 8250 Btuh
Design heating load   = 47,808 + 8250
                           = 56,058 Btuh

F.2

EXAMPLE PROBLEM: HEAT GAIN CALCULATION

The heat gain calculation will be made for the same structure used for the heat loss calculation. In both examples, the measurements and the construction details are the same.

F.2.1 Design Temperature Differences

The design temperature difference is the air temperature difference across a structural component. Table 1 in *Manual J* lists the outside design temperatures for various locations. Assume the inside design temperature to be 75°F, 55% relative humidity. You should estimate unconditioned space temperatures as closely as possible by considering the location and the use of the space in question. Use the design temperature difference to select the appropriate HTM for the structural components.

F.2.2 Calculation Procedure

Once the area, construction details, and temperature differences are determined, you can use the data in the back of *Manual J* and the *Worksheet for Manual J* (Fig. F–3) to

FORM J—1
Including Calculation Procedures A, B, C, D
Copyright by the
Air Conditioning
Contractors of America
1513 16th Street N.W.
Washington, D.C. 20036
Printed in U.S.A.
1986

Plan No. _____
Date _____
Calculated by _____

WORKSHEET FOR MANUAL J

LOAD CALCULATIONS FOR RESIDENTIAL AIR CONDITIONING

For: Name Example Problem _____

Address _____

City and State or Province _____

By: Contractor_____

Address _____

City _____

Design Conditions

Winter	Summer

Outside db ____-5____ °F Inside db ___70___ °F   Outside db ___88___ °F Inside db ___75___ °F

Winter Design Temperature Difference ___75___ °F   Summer Design Temperature Difference ___15___ °F

Room RH___55%___ Daily Range__ M __

Heating Summary

Total Heat Loss for Entire House (Line 15) = ____47,808____ Btuh

Ventilation CFM = ____none____ Winter Design Temperature Difference = ____ °F

Heat Required for Ventilation Air = 1.1 X _____ CFM X _____ °F = ___0___ Btuh

Design Heating Load Requirement = ___47,808___ (house) ___0___ (Vent) = ___47,808___ Btuh

Cooling Summary

Total Sensible Gain ___16,669___ Btuh (Calculation Procedure D) Design Temperature Swings

Total Latent Gain + ___3,835___ Btuh (Calculation Procedure D) Normal 3° (X) 4.5° ( )

Total = Sens. + Lat. = ___20,504___ Btuh   Ventilation CFM = ___none___

Equipment Summary

Make _____ Model _____ Type _____

Heating Input (Btuh) _____ Heating Output (Btuh) _____ Efficiency _____

Sensible Cooling (Btuh _____ Latent Cooling (Btuh) _____ Total (Btuh) _____

COP/EER/SEER/HSPF _____ Cooling CFM _____ Heating CFM _____

Space Thermostat   Heat (   )   Cool (   )   Heat/Cool (   )   Night Setback (   )

Construction Data

Windows _____   Floor _____

_____   _____

_____   Partitions _____

Doors _____   _____

_____   _____

Walls _____   Basement Walls _____

_____   _____

_____   _____

Roof _____   Ground Slab_____

_____   _____

Ceiling _____   _____

_____   _____

Figure F-3 Example load calculation summary.

determine the heat gain. The total heat gain is the sum of the heat gains through the building envelope (solar gain, transmission, infiltration, and internal loads). The HTMs for various structural components are in Tables 3 and 4. The HTM values for temperature differences that fall between those listed in the table should be interpolated as discussed. Caculate internal gains and infiltration gains by using the procedures outlined in Section V in *Manual J*.

Data must be obtained from drawings or by a field inspection before load calculations can be made. The data required include the following:

A. Measurements to determine areas:
1. Running feet of exposed wall
2. Length and width of rooms and house
3. Ceiling heights
4. Dimensions of windows and doors

B. Area calculations:
1. Gross area of walls exposed to outdoor conditions
2. Gross area of partitions
3. Gross area of walls below grade
4. Areas of windows and doors

Table F–2 Assumed design conditions and construction (cooling).

	Const. No.	HTM
A. Outside design temperature: rounded to 90° db, 38 grains (Table 1 in *Manual J*)		
B. Daily temperature range: Medium (Table 1) 88° db		
C. Inside design conditions: 75°F, 55% RH; design temperature difference = (90 − 75 = 15)		
D. Types of shading: Venetian blinds on all first-floor windows, no shading, basement		
E. Windows: All clear, double-glass on first floor (Table 3A in *Manual J*)		
North		14
East or west		44
South		23
All clear, single-glass (plus storm) in basement, Table 3A; use double glass		
East		70
South		36
F. Doors: Metal, urethane-core, no storm, 0.50 cfm/ft	11e	3.5
G. First-floor walls: Basic frame construction with ½″ asphalt board (R-11) (Table 4 in *Manual J*)	12d	1.5
Basement wall: 8″ concrete block, above grade: 3′ (R-5) (Table 4)	14b	1.6
8″ concrete block, below grade: 5′ (R-5) (Table 4)	15b	0
H. Partition: 8″ concrete block furred, with insulation (R-5), ΔT approx. 0°F (Table 4)	13n	0
I. Ceiling: Basic construction under vented attic with insulation (R-19), dark roof (Table 4)	16d	2.1
J. Occupants: 6 (Assumed 2 per bedroom, but distributed 3 in living, 3 in dining)		
K. Appliances: Add 1200 Btuh to kitchen		
L. Ducts: Located in conditioned space (Table 7B in *Manual J*)		
M. Wood and carpet floor over unconditioned basement, ΔT approx. 0°F	19	0
N. The envelope was evaluated as having average tightness. (Refer to the construction details at the bottom of Figure 3–3 in *Manual J*.)		
O. Equipment to be selected from manufacturers' performance data.		

5. Areas of ceilings under an attic or unconditioned space and/or roof ceiling combinations
6. Areas of floors exposed to the outdoors or floors over an unconditioned space or over a crawl space and/or basement floors
7. Running feet of exposed perimeter for slab-on-grade floors

Closets and halls are usually included with adjoining rooms. Entrance halls should be considered separately. Wall, floor, or ceiling dimensions can be rounded to the nearest foot. Window dimensions are recorded to the nearest inch. Measure the size of window or door openings; do not include the frame.

C. Construction details:
1. Exposed walls and partitions
2. Windows and glass doors
3. Panel doors
4. Ceilings and roof ceilings
5. Floors and ground slabs

D. Temperature differences:
1. Temperature differences across all components exposed to outdoor conditions
2. Temperature differences across partitions and all floors and ceilings that are adjacent to unconditioned spaces

EXAMPLE 2: Figure F–4 represents a house located in Cedar Rapids, Iowa. Figure F–2 lists the construction details. Assume that the inside design temperature is 75°F. From Table 1 in *Manual J*, the summer design temperature is 88°db, 75°wb, with 38-grain moisture difference and a medium (M) daily range. Figure F–2 shows the completed worksheet. Note that the outside design temperature is rounded from 88°db to 90°db to expedite the calculations. Rounding the design temperature difference up by 3° or down by 1° will not produce any serious errors in the calculations.

The following is a line-by-line explanation of the procedure:

Line 1. Identify each area.

Lines 2 and 3. Enter the pertinent dimensions from Figure F–4.

Line 4. For reference, enter the ceiling height and the direction the glass faces.

Lines 5A through 5D. Enter the gross wall area for the various walls. For rooms with more than one exposure, use one line for each exposure. For rooms with more than one type of wall construction, use one line for each type of construction. Find the construction number in the tables in the back of *Manual J*. Enter this number on the appropriate line.

Example: The gross area of the west living room wall is 168 sq ft. This wall is listed in Table 4, number 12, line D, in *Manual J*. The construction number is 12-d.

Line 6. Not required for cooling calculations.

Line 7. Enter the areas of windows and glass doors for the various rooms and exposures. Use the drawings and construction details, or determine by inspection the types of windows used in each room. Also note the shading and the exposure. Refer to the tables in the back of *Manual J*, and select the HTM for each combination of window, shading, and exposure. Enter the HTM values in the column designated "Btuh Clg." Multiply each window area by its corresponding HTM to determine the heat gain through the window. Enter this value in the column "Btuh Clg."

Example: The living room has 40 sq ft of west-facing glass. The window is double pane with drapes or blinds. The design temperature difference is rounded to 15°F. The HTM listed in Table 3A in *Manual J* (double glass, drapes or venetian blinds, design temperature difference of 13°F), is Btu/h per square foot. The heat gain is

44 Btu/h per square foot × 40 sq ft = 1760 Btuh.

Example: The dining area has 20 sq ft of north-facing glass. The HTM listed in Table 3A (double glass, draperies, 15°F temperature difference) is 14 Btu/h per square foot. The heat gain is

14 Btu/h per square foot × 20 sq ft = 280 Btuh.

Example: The glass in the basement recreation room is single pane with storm. The design temperature difference is rounded to 15°F. The room has 8 sq ft of east-facing glass and 8 sq ft of south-facing glass. From Table 3A (double pane, clear glass, 15°F design temperature difference), the HTMs are 70 Btu/h per square foot for the east and 36 Btu/h per square foot for the south. The heat gain for the east window is

70 Btu/h per square foot × 8 sq ft = 560 Btuh

The heat gain for the south window is

36 Btu/h per square foot × 8 sq ft = 288 Btuh

Line 8. For each room, enter the area of any doors that are not glass. From Table 4 of *Manual J*, No. 11-E, select the HTM, and enter this value on the worksheet. To calculate the heat gain through the door, multiply the HTM by the area of the door. Enter the heat gain in the appropriate column:

Laundry door 17 sq ft × 3.5 Btu/h per square foot = 60 Btuh

Lines 9A through 9D. For each room, subtract the window and door areas from the corresponding gross wall area, and enter the net wall areas and corresponding construction numbers. From Table 4, select the HTM for each wall. Multiply the HTM by the appropriate net wall area, and enter the heat gain through the wall.

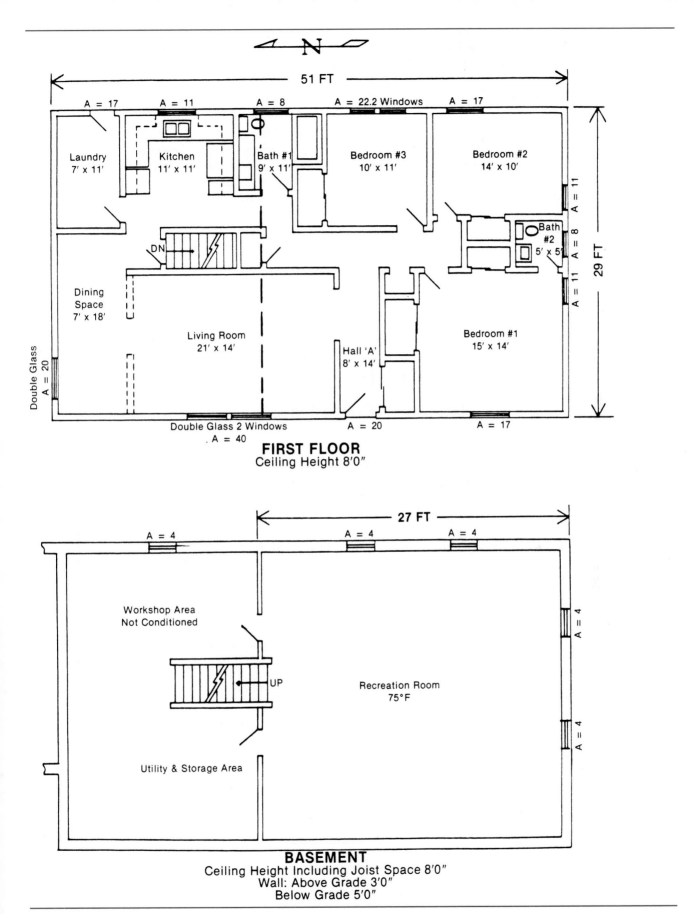

FIRST FLOOR
Ceiling Height 8'0"

BASEMENT
Ceiling Height Including Joist Space 8'0"
Wall: Above Grade 3'0"
Below Grade 5'0"

Figure F-4 Heat gain example problem: basement and first-floor plans.

Example: The west wall in the living room has a net area of 168 sq ft − 40 sq ft = 128 sq ft. The design temperature difference is rounded to 15°F, and the daily range is M. From Table 4, No. 12-D, the HTM is 1.5 Btu/h per square foot. The heat gain through the living room wall is

1.5 Btu/h per square foot × 128 sq ft = 192 Btuh

Example: The basement wall in the recreation room has a net area of 233 sq ft above grade and 410 sq ft below grade. From Table 4, No. 14-B, the HTM is

1.6 Btu/h per square foot. The heat gain through the above-grade wall is

1.6 Btu/h per square foot × 233 sq ft = 373 Btuh

(Below-grade walls need not be included in the heat gain calculation.)

Lines 10A and 10B. Enter the ceiling area for the various rooms and the construction number for the corresponding ceiling. Determine the HTM from Table 4, and enter it. Multiply the HTM by the ceiling area, and enter the heat gain.

Table 5

Infiltration Evaluation

Winter air changes per hour

Floor Area	900 or less	900 - 1500	1500 - 2100	over 2100
Best	0.4	0.4	0.3	0.3
Average	1.2	1.0	0.8	0.7 *
Poor	2.2	1.6	1.2	1.0

For each fire place add:		Best	Average	Poor
		0.1	0.2	0.6

Average - Plastic vapor barrier, major cracks and penetrations sealed, tested leakage of windows and doors between 0.25 and 0.50 CFM per running foot of crack, electrical fixtures which penetrate the envelope not taped or gasketed, vents and exhaust fans dampered, combustion air from indoors, intermittent ignition and flue damper, some duct leakage to unconditioned space.

Procedure A - Winter Infiltration HTM Calculation

1. Winter Infiltration CFM
   0.70 AC/HR x 16269 Cu. Ft. x 0.0167 = 190 CFM
   Volume

2. Winter Infiltration Btuh
   1.1 x 190 CFM x 75 Winter TD = 15675 Btuh

3. Winter Infiltration HTM
   15675 Btuh ÷ 222 Total Window = 70.6 HTM
   & Door Area

* Includes Full Basement

Above Grade Volume = 51x29x(8 + 3) = 16269

Figure F–5 Infiltration HTM calculation.

Example: The living room ceiling has an area of 294 sq ft. The (dark roof) construction number is (16-D), the design temperature difference is 15, and the daily range is M. From Table 4, No. 16-D, the HTM is 2.1 Btu/h per square foot. The heat gain through the living room ceiling is

2.1 Btu/h per square foot × 294 sq ft = 617 Btuh

Lines 11A and 11B. For a room that will experience a gain through the floor, enter the floor area and the corresponding construction number. Determine the HTM from Table 4, and enter it. Multiply the HTM by the appropriate area, and enter the heat gain.

Line 12. Use Table 5 and Calculation Procedure B in *Manual J* to calculate the summer infiltration HTM, and enter this value on line 12. For each room, enter the total sq ft of the window and door openings on line 12. Finally, compute the infiltration heat gain due to infiltration for each room by multiplying the summer infiltration HTM value by the appropriate sq ft of window and door openings. Enter these results on line 12. Refer to Figure 6–4 in *Manual J* for an example of this calculation.

Lines 13 through 15. These lines are not used for the cooling calculation.

Line 16. Enter sensible internal loads due to appliances and occupants for the rooms.

Line 17. For each room, add all of the cooling loads (lines 7, 8, 9, 10, 11, 12, and 16), and enter the totals on line 17.

Line 18. If the duct system is installed in an unconditioned space, enter an allowance for the duct gain for each room on line 18. Refer to Table 7-B in *Manual J* for the duct gain multipliers.

Line 19. For the entire house and for each room, add line 17 (structure gain) to line 18 (duct gain), and enter the space-sensible gain on the form.

Note: Line 19 on the form provides information only on the space-sensible loads. Equipment selection requires calculation of the space-latent loads, and if ventilation is used, the sensible- and latent-ventilation loads must also be calculated.

Calculation Procedure C. Use Calculation Procedure C in *Manual J* to calculate the latent infiltration load for the entire house. Refer to Figure 6–5 in *Manual J* for an example of the Procedure C calculation.

Calculation Procedure D. Use Calculation Procedure D in *Manual J* to estimate the sensible- and latent-ventilation loads and to estimate the latent loads that are produced by internal sources. Also use Calculation Procedure D to estimate the total sensible and the total latent loads that must be satisfied by the cooling equipment. Refer to Figure 6–5 for an example of these calculations. For equipment to be selected from manufacturers' performance data, use RSM = 1.0 (refer to Table 6).

GLOSSARY OF CONSTRUCTION TERMS

A

Abut To join the end of a construction member.

Acre A unit of land measurement having 43,560 sq ft.

Adhesive A natural or synthetic material, usually in liquid form, used to fasten or adhere materials together.

Adobe construction Construction using sun-dried units of adobe soil for walls; usually found in the southwestern United States.

Aggregate Gravel (coarse) or sand (fine) used in concrete mixes.

Air-dried lumber Lumber that has been dried by unheated air to a moisture content of approximately 15%.

Anchors Devices, usually metal, used in building construction to secure one material to another.

Angle A piece of structural steel having an L-shaped cross section and equal or unequal legs.

Apron Inside window trim placed under the stool and against the wall.

Arcade An open passageway usually surrounded by a series of arches.

Arch A curved structure designed to support itself and the weight above.

Areaway Recessed area below grade around the foundation to allow light and ventilation into a basement window.

Arris The sharp edge formed by two surfaces; usually on moldings.

Asbestos board A fire-resistant sheet made from asbestos fiber and portland cement.

Ash pit An enclosed opening below a fireplace to collect ashes.

Asphalt shingles Composition roof shingles made from asphalt-impregnated felt covered with mineral granules.

ASTM American Society for Testing and Materials.

Astragal T-profiled molding usually used between meeting doors or casement windows.

Atrium An open court within a building.

Attic The space between the roof and the ceiling in a gable house.

Awning window An outswinging window hinged at the top of the sash.

Axis A line around which something rotates or is symmetrically arranged.

B

Backfill Earth used to fill in areas around foundation walls.

Backsplash A protective strip attached to the wall at the back edge of a counter top.

Balcony A deck projecting from the wall of a building above ground level.

Balloon frame A type of wood framing in which the studs extend from sill to eaves without interruption.

Balusters Small, vertical supports for the railing of a stairs.

Balustrade A series of balusters supporting the railing of a stairs or balcony.

Bannister A handrail with supporting posts on a stairway.

Bargeboard The finish board covering the projecting portion of a gable roof.

Bar joist A light steel structural member fabricated with a top chord, a bottom chord, and web members.

Baseboard The finish trim board covering the interior wall where the wall and the floor meet.

Batt A type of insulation designed to be installed between framing members.

Batten The narrow strips of wood nailed vertically over the joints of boards to form board-and-batten siding.

Batter boards Horizontal boards at exact elevations nailed to posts just outside the corners of a proposed building. Strings are stretched across the boards to locate the outline of the foundation for workers.

Bays Uniform compartments within a structure, usually within a series of beams, columns, and so on.

Bay windows A group of windows projecting from the

wall of a building. The center is parallel to the wall, and the sides are angular. A bow window is circular.

Beam Horizontal structural member, usually heavier than a joist.

Bearing plate A metal plate that provides support for a structural member.

Bearing wall A wall that supports a weight above in addition to its own weight.

Bench mark A mark on some permanent object fixed to the ground from which land measurements and elevations are taken.

Bidet Low plumbing fixture in luxury bathrooms for bathing one's private parts.

Blind nailing Method of nailing to conceal nails.

Blocking Small wood pieces in wood framing to anchor or support other major members.

Board measure The system of lumber measurement. A unit is 1 board ft, which is 1′ sq by approximately 1″ thick.

Bond beam Continuous, reinforced concrete block course around the top of masonry walls.

Bonds The arrangement of masonry units in a wall.

Brick Small masonry units made from clay and baked in a kiln.

Brick veneer A facing of brick on the outer side of wood frame or masonry.

Bridging Thin wood or metal pieces fastened diagonally at midspan between floor joists to act as both tension and compression members for the purposes of stiffening and spreading concentrated loads.

Buck Frame for a door, usually made of metal.

Building line An imaginary line on a plot beyond which the building may not extend.

Built-up roof A roofing composed of layers of felt impregnated with pitch, coal tar, or asphalt. The top is finished with crushed stone or minerals. It is used on flat or low-pitched roofs.

Bullnose Rounded edge units.

Butt Type of joint having the pieces edge to edge or end to end. Also a type of door hinge allowing the edge of a door to butt into the jamb.

Buttress Vertical masonry or concrete support, usually larger at the base, which projects from and strengthens a wall.

C

Call out A note on a drawing with a leader to the feature.

Cantilever A projecting beam or structural member anchored at only one end.

Cant strip An angular board used to eliminate a sharp, right angle, usually on roof decks.

Cap Covering for a wall or post.

Carport A garage not fully enclosed.

Casement window A window with one or two sashes that hinge on their sides. They may open either in or out.

Casing The trim around a window or door opening.

Caulking A soft, waterproof material used to fill open joints or cracks.

Cavity wall A masonry wall having a 2″ air space between brick wythes.

Cement A fine, gray powder made from lime, silica, iron oxide, and alumina that when mixed with water and aggregate produces concrete.

Chamfer The beveled edge formed by removing the sharp corner of a material.

Channel A piece of structural steel having a C-shaped cross section.

Chase A vertical space within a building for ducts, pipes, or wires.

Chord The lower horizontal member of a truss.

Cleanout Accessible fitting on plumbing pipe which can be removed to clean sanitary drainage pipe.

Cleat A small board fastened to another member to serve as a brace or support.

Clerestory A portion of an interior rising above adjacent roof tops and having windows.

Collar beam A horizontal member tying opposing rafters below the ridge in roof framing.

Column A vertical supporting member.

Concrete A building material made from cement, aggregate, and water.

Concrete masonry unit (C.M.U.) A concrete block extruded from cement, aggregate, and water.

Conduit, electrical A metal pipe in which wiring is installed.

Contour A line on a map connecting all points with the same elevation.

Coping A cap or top course of masonry on a wall to prevent moisture penetration.

Corbel A projection of masonry from the face of a wall, or a bracket used for support of weight above.

Core The inner layer of plywood. It may be veneer, solid lumber, or fiberboard.

Corner board A vertical board forming the corner of a building.

Corner brace A diagonal brace at the corner of a wood frame wall to stiffen and prevent cracking.

Cornice The molded projection of the roof overhang at the top of a wall.

Cornice return The short portion of a molded cornice that returns on the gable end of a house.

Counter flashing Flashing used under cap flashing.

Cove A concave molding usually used on horizontal inside corners.

Crawl space The shallow space below the floor of a house built above the ground. Generally, it is surrounded with the foundation wall.

Cricket A device used at roof intersections to divert rain water.

Cripple A structural member that is cut less than full length, such as a studding piece above a window or door opening.

Crown molding A molding used above eye level, usually the cornice molding under the roof overhang.

Cul-de-sac A court or street that has no outlet and that provides a turnaround for vehicles.

Cupola A small, decorative structure that is placed on a roof, usually a garage roof, and that can be used as a ventilator.

Curtain wall An exterior wall that provides no structural support.

D

Dado joint A recessed joint on the face of a board to receive the end of a perpendicular board.

Damper A movable plate that regulates the draft through a flue or duct.

Dampproofing Material used to prevent passage of moisture.

Dead load The weight of the structure itself and the permanent components fastened to it.

Deck Exterior floor, usually extended from the outside wall.

Deflection The deviation of the central axis of a beam from normal when loaded.

Dimension lumber Framing lumber that is 2″ thick and 4″–12″ wide.

Dome A roof in the shape of a hemisphere used on a structure.

Doorjamb Two vertical pieces held together by a head jamb forming the inside lining of a door opening.

Doorstop The strips on the doorjambs against which the door closes.

Dormer A projection on a sloping roof framing a vertical window or vent.

Double glazing Two pieces of glass with air between to provide insulation.

Double-hung A type of window having two sashes which can be operated vertically.

Downspout A pipe for carrying rainwater from the roof to the ground or sewer connection.

Dressed size The actual finish size of lumber after surfacing.

Drip A projecting construction member or groove below the member to throw off rainwater.

Drywall construction Interior wall covering with sheets of gypsum rather than traditional plaster.

Ducts Sheet-metal conductors for air distribution throughout a building.

Duplex outlet Electrical wall outlet having two plug receptacles.

E

Earth berm An area of raised earth.

Easement A right or privilege to a piece of property held by someone other than the owner. Usually the right to run utility lines, underground pipe, or passageways on property.

Eaves The lower portion of the roof that overhangs the wall.

Efflorescence The forming of white stains on masonry walls from moisture within the walls.

Ell An extension or wing of a building at right angles to the main section.

Escutcheon The decorative metal plate used around the keyhole on doors or around a pipe extending through the wall.

Excavation A cavity or pit produced by digging the earth in preparation for construction.

Expansion joint A flexible joint used to prevent cracking or breaking because of expansion and contraction due to temperature changes.

Exterior insulation finish system (E.I.F.S.) An exterior wall finish made from styrofoam, cement, fiberglass, and an acrylic coating.

F

Façade The face or front elevation of a building.

Face brick Brick of better quality used on the face of a wall.

Fascia The outside horizontal member on the edge of a roof or overhang.

Fasteners General term for metal devices, such as nails, bolts, screws, and so on, used to secure structural members within a building.

Fenestration The arrangement of window and door openings in a wall.

Fiberboard Fabricated structural sheets made from wood fiber and adhesive under pressure.

Fill Sand, gravel, or loose earth used to bring a subgrade up to a desired level around a building.

Firecut An angular cut at the end of a floor joist resting on a masonry wall.

Fire rating A fire-resistance classification assigned to a building material or assembly.

Firestop A tight closure of a concealed space with incombustible material to prevent the spreading of fire.

Fire wall A fire-resistant masonry wall between sections of a building for the purpose of containing a fire.

Flagstone Thin, flat stones used for floors, steps, walks, and so on.

Flange The top or bottom pieces that project from a web of a structural steel member.

Flashing Sheet metal or other material used in roof or wall construction to prevent water from seeping into the building.

Flat-plate collector A solar energy collector made from metal piping with glass over.

Flitch beam A built-up beam formed by a metal plate sandwiched between two wood members and bolted together for additional length.

Floor joist Structural member of a floor.

Flue The passage in a chimney through which smoke, gases, and fumes escape to the outer air.

Footing Poured concrete base upon which the foundation wall rests.

Frieze The flat board of cornice trim that is fastened to the wall.

Frost line The deepest level of frost penetration in soil. This depth varies in different climates. Footings must be placed below the frost line to prevent a rupturing of the foundation.

Furring strips Thin strips of wood fastened to walls or ceilings for leveling and for receiving the finish surface material.

G

Gable The triangular end of a gable-roofed house.

Gambrel roof A roof with two pitches, the lower slope steeper than the upper.

Girder A heavy structural member supporting lighter structural members of a floor or a roof.

Glazing Placing of glass in windows and doors.

Grade beam A horizontal member that is between two supporting piers at or below grade and that supports a wall or a structure above.

Gradient The inclination of a road, piping, or the ground, expressed in percent.

Gravel stop A strip of metal with a vertical lip used to retain the gravel around the edge of a build-up roof.

Grounds Wood strips fastened to walls before plastering to serve as screeds and nailing base for trim.

Grout Thin cement mortar used for leveling and filling masonry cavities.

Gusset A plywood or metal plate used to strengthen the joints of a truss.

Gutter A metal or wood trough for carrying water from a roof.

Gyp board Gypsum sheets covered with paper that are fastened to walls and ceilings with nails or screws.

H

Hanger A metal strap used to support piping or the ends of joists.

Header In framing, the joists placed at the ends of a floor opening and attached to the trimmers. In masonry work, the small end of a masonry unit.

Hearth The incombustible floor in front of and within the fireplace.

Heartwood The central portion of wood within the tree, which is stronger and more decay-resistant than the surrounding sapwood.

Hip rafter The diagonal rafter that extends from the plate to the ridge to form the hip.

Hip roof A roof that rises by equally inclined planes from all four sides of a building.

Hose bibb A water faucet made for the threaded attachment of a hose.

House drain Horizontal sewer piping within a building which carries the waste from the soil stacks.

House sewer The watertight soil pipe extending from the exterior of the foundation wall to the public sewer.

Humidifier A device, generally attached to the furnace, which supplies or maintains correct humidity levels in a building.

I

I beam A structural steel shape with a web and flange components, having an I-shaped cross section.

Incandescent lamp A lamp within which a filament gives off light when sufficiently heated by an electric current.

Insulating concrete Concrete with vermiculite added to produce lightweight, insulating concrete for subfloors and roofs.

Interior trim The general term for all of the finish molding, casing, baseboard, and so on, applied within a building by finish carpenters.

J

Jack rafter A rafter shorter than a common rafter; especially used in hip-roof framing.

Jalousie A type of window having a number of small, unframed yet movable pieces of glass.

Jamb The vertical members of a finished door opening.

Joinery A general woodworking term used for all better-class wood-joint construction.

Joist A horizontal structural member supported by bearing walls, beams, or girders in floor or ceiling framing.

Joist hanger A metal strap to carry the ends of floor joists.

K

Keystone The wedged center stone at the crown of an arch.

Kiln-dried lumber Lumber that has been properly dried and cured to produce a higher grade of lumber than that which has been air dried.

King post The center upright strut in a truss.

Kip A unit of 1000-lb load.

Knee wall A low wall resulting from 1½-story construction.

Knocked down Unassembled; refers to construction units requiring assembly after being delivered to the job.

L

Laitance Undesirable surface water that forms on curing concrete.

Lally column A steel column used in light construction.

Laminated beam A beam made of superimposed layers of similar materials by uniting them with glue and pressure.

Landing A platform between flights of stairs or at the termination of stairs.

Lap joint A joint produced by lapping two similar pieces of material.

Lath A metal, wood, or gypsum base for plastering.

Lattice A framework of crossed or interlaced wood or metal strips.

Leader A vertical pipe or downspout that carries rainwater from the gutter to the ground or storm sewer.

Ledger strip A strip of lumber fastened to the lower part of a beam or girder on which notched joists are attached.

Light A single pane of glass in a window or door.

Lineal foot A 1′ measurement along a straight line.

Lintel A horizontal support member across the head of a door or window opening.

Live load Loads other than *dead loads* on a building such as wind, snow, and people.

Load-bearing wall A wall designed to support the weight imposed upon it from above.

Lookout A short wooden framing member used to support an overhanging portion of a roof. It extends from the wall to the underside surfacing of the overhang.

Lot line The line forming the legal boundary of a piece of property.

Louver An opening or slatted grill allowing ventilation while providing protection from rain.

Luminaire An electric lighting fixture used within a room.

M

Mansard roof A hip-type roof having two slopes on each of the four sides.

Masonry A general term for construction of brick, stone, concrete block, or similar materials.

Mastic A flexible adhesive for adhering building materials.

Matte finish A finish free from gloss or highlights.

Membrane A thin layer of material used to prevent moisture penetration.

Metal wall ties Corrugated metal strips used to tie masonry veneer to wood walls.

Millwork A general term that includes all dressed lumber that has been molded, shaped, or preassembled at the mill.

Miter joint A joint made with ends or edges of two pieces of lumber cut at a 45° angle and fitted together.

Modular construction Construction in which the size of all components has been based upon a standardized unit of measure.

Moisture barrier A sheet material that retards moisture penetration into walls, floors, ceilings, and so on.

Monolithic Term used for concrete construction poured and cast in one piece without joints.

Mortar A mixture of cement, sand, lime, and water used to bond masonry units.

Mortice A hole, slot, or recess cut into a piece of wood to receive a projecting part (tenon) made to fit.

Mosaic Small, colored tile, glass, stone, or similar material arranged to produce a decorative surface.

Mullion The structural member between a series of windows.

Muntin A small bar separating the glass lights in a window.

N

Newel post The main post supporting a handrail at the bottom or top of a stairs.

Nominal size The size of lumber before dressing, rather than its actual size.

Nonbearing wall A wall supporting no load other than its own weight.

Nonferrous metal A metal containing no iron, such as copper, brass, or aluminum.

Nosing The rounded edge of a stair tread.

O

On-center A method of indicating the spacing between framing members by stating the measurement from the center of one member to the center of the succeeding one.

Open web joist *See* Bar joist.

Outlet Any type of electrical box allowing current to be drawn from the electrical system for lighting or appliances.

Overhang The projecting area of a roof or upper story beyond the wall of the lower part.

P

Pallet A rugged wood skid used to stack and mechanically handle units of masonry.

Panel A flat, rectangular surface framed with a thicker material.

Parapet A low wall or railing, usually around the edge of a roof.

Parge coat A thin coat of cement plaster applied to a masonry wall for refinement of the surface or for damp-proofing.

Parquet flooring Flooring, usually of wood, laid in an alternating or inlaid pattern to form various designs.

Parting stop Thin strips set into the vertical jambs of a double-hung window to separate the sash.

Partition A wall that divides areas within a building.

Party wall A wall between two adjoining buildings in which both owners share, such as a common wall between row houses.

Passive solar system An integral energy system using only natural and architectural components to utilize solar energy.

Penny A term used to indicate the size of nails, abbreviated "d." Originally, it specified the price per hundred nails (for example, 6-penny nails cost 6¢ per hundred nails).

Pergola An open, structural framework over an outdoor area, usually covered with climbing shrubs or vines to form an arbor.

Periphery The entire outside edge of an object.

Pier A masonry pillar usually below a building to support the door framing.

Pilaster A rectangular pier attached to a wall for the purpose of strengthening the wall. Also a decorative column attached to a wall.

Pile A long shaft of wood, steel, or concrete driven into the earth to support a building.

Pitch The slope of a roof, usually expressed as a ratio.

Plastic laminate A thin, melamine-surfaced sheet used to cover counter tops.

Plat A graphic description of a surveyed piece of land, indicating the boundaries, location, and dimensions. The plat, recorded in the appropriate county official's office, also contains information as to easements, restrictions, and lot numbers, if any.

Plate The top horizontal member of a row of studs in a frame wall.

Plumb Said of an object when it is in true vertical position as determined by a plumb bob or a vertical level.

Plywood A relatively thin building material made by gluing layers of wood together.

Poché The darkening of areas on a drawing to aid in readability.

Post-and-beam construction A type of building frame in which roof and floor beams rest directly over wall posts.

Precast Concrete units that are cast and finished at the plant rather than at the site of construction.

Prime coat The first coat of paint that serves as a filler and sealer in preparation for finish coats.

Purlins Horizontal roof members laid over trusses to support rafters.

Q

Quarry tile Unglazed, machine-made tile used for floors.

Quarter round Small molding presenting the profile of a quarter circle.

Quarter-sawed oak Oak lumber, usually flooring, that has been sawed so that the medullary rays showing on end-grain are nearly perpendicular to the face of the lumber.

Quoins Large, squared stones set in the corners of a masonry building for appearance's sake.

R

Rabbet (or Rebate) A groove cut along the edge or end of a board to receive another board, producing a rabbet joint.

Radiant heating A method of heating with the use of radiating heat rays.

Rafter A roof structural member running from the wall plate to the ridge. There are jack, hip, valley, and common rafters. The structural members of a flat roof are usually called *roof joists*.

Rake joint A mortar joint that has been recessed by tooling before it sets up.

Rake molding Gable molding attached on the incline of the gable. The molding must be a different profile to match similar molding along the remaining horizontal portions of the roof.

Random rubble Stonework having irregularly shaped units and coursing.

Register Opening in air duct, usually covered with a grill.

Reinforced concrete Concrete containing steel bars or wire mesh to increase its structural qualities.

Retaining wall A heavy wall that supports an earth embankment.

Reveal The side of an opening for a window or door, between the frame and the outer surface of a wall.

Ribbon A wood strip set into studs to support floor joists in balloon framing.

Ridge The top edge of a roof where two slopes meet.

Ridgeboard The highest horizontal member in a gable roof; it is supported by the upper ends of the rafters.

Riprap Irregular stones thrown together loosely to form a wall or soil cover.

Rise The vertical height of a roof or stairs.

Rocklath Paper-covered gypsum sheets used as a plaster base.

Rough hardware All of the concealed fasteners in a building such as nails, bolts, hangers, and so on.

Rough opening As unfinished opening in the framing into which doors, windows, and other units are placed.

Rowlock A special brick coursing placed at the exterior windowsill.

Rubble Irregular, broken stone.

Run The horizontal distance of a flight of stairs, or the horizontal distance from the outer wall to the ridge of a roof.

R value The unit of thermal resistance in rating insulating materials; higher values indicate better insulators.

S

Saddle A small gable roof placed in back of a chimney on a sloping roof to shed water and debris.

Sanitary sewer Drainage pipe that transports sewage from buildings.

Sash An individual frame into which glass is set.

Scab A short piece of lumber fastened to a butt joint for strength.

Schedule A list of similar items and information about them, such as a window schedule.

Scribing Marking and fitting a piece of lumber to an irregular surface such as masonry.

Scuttle A small opening in a ceiling to provide access to an attic or roof.

Section A unit of land measurement, usually 1 mile square. A section contains approximately 640 acres, and there are 36 sections to a township. Also a drawing showing the cut-open view of an object.

Septic tank A concrete or steel underground tank used to reduce sewage by bacterial action.

Shake A handsplit wood shingle.

Sheathing The rough boarding or covering over the framing of a house.

Shim A thin piece of material used to true up or fill a space between two members.

Shoe mold The small molding covering the joint between the flooring and the baseboard on the inside of a room.

Shoring Planks or posts used to support walls or ceilings during construction.

Siding The outside finish covering on a frame wall.

Sill The horizontal exterior member below a window or door opening. The wood member placed directly onto the foundation wall in wood frame construction.

Skylight A window in a flat roof.

Sleepers Wood strips placed over or in a concrete slab to receive a finish wood floor.

Smoke chamber The enlarged portion of a chimney flue directly above the fireplace.

Soffit The underside of an overhang such as a cornice or stairs.

Soil stack The vertical pipe in a plumbing system that carries the sewage.

Solar collector A device used to collect the sun's heat.

Soleplate The horizontal member of a frame wall resting on the rough floor, to which the studs are nailed.

Span The horizontal distance between supports for joists, beams, or trusses.

Specifications The written instructions that accompany a set of working drawings.

Square A unit of measure—100 sq ft. Commonly used in reference to the amount of roofing material to cover 100 sq ft.

Stile The vertical member on a door or panel.

Stirrup A metal, U-shaped strap used to support the end of a framing member.

Stool The horizontal interior member of trim below a window.

Story A complete horizontal portion of a building having a continuous floor.

Stretcher course A row of masonry in wall with the long side of the units exposed to the exterior.

Stringer The inclined structural member supporting the treads and risers of a stairs; sometimes it is visible next to the profile of the stairs.

Stucco A cement plaster finish applied to exterior walls.

Studs The vertical framing members of a wall.

Subflooring Any material nailed directly to floor joists. The finish floor is attached over the subflooring.

Sunspace A glassed-in area for the collection of solar heat.

Suspended ceiling A finish ceiling hung below the underside of the building structure, either the floor or the roof.

T

Tail joist A relatively shorter joist that joins against a header or trimmer in floor framing.

Tensile strength The greatest longitudinal stress that a structural member can bear without adverse effects (breaking or cracking).

Termite shield Sheet metal placed over masonry to prevent the passage of termites into wood.

Terra-cotta Baked clay and sand formed into masonry units.

Terrazzo flooring Wear-resistant flooring that is made of marble chips or small stones embedded in cement and that has been polished smooth.

Thermal conductor A substance capable of transmitting heat.

Thermostat An automatic device for controlling interior temperatures.

Threshold The beveled metal, stone, or wood member directly under a door.

Title Legal evidence of the ownership of property.

Toe nail Nailing at an angle to the wood fiber.

Tongue The narrower extension on the edge of a board that is received by the groove of an adjacent board.

T post A post built up of studs and blocking to form the intersection of the framing of perpendicular walls.

Transom A hinged window over a door.

Trap A U-shaped pipe below plumbing fixtures to create a water seal and to prevent sewer odors and gases from being released into the habitable areas.

Tread The horizontal surface member of a stairs upon which the foot is placed.

Treated wood Wood that has been chemically treated to prevent decay and insect infestation.

Trim A general term given to the moldings and finish members on a building. Its installation is called *finish carpentry*.

Trimmer The longer floor framing member around a rectangular opening into which a header is joined.

Trombe wall A passive-heating concept consisting of a south-facing masonry wall with glazing in front. Solar radiation is absorbed by the wall, converted to heat, and conducted and radiated into the building.

Truss Structural members arranged and fastened in triangular units to form a rigid framework for support of loads over a long span.

Truss joists A structural framing member fabricated with a thin wood web and wood flanges.

V

Valley rafter The diagonal rafter at the intersection of two intersecting, sloping roofs.

Vapor barrier A watertight material used to prevent the passage of moisture or water vapor into and through walls.

Veneered construction A type of wall construction in

which frame or masonry walls are faced with other exterior surfacing materials.

Vent stack A vertical soil pipe connected to the drainage system to allow ventilation and pressure equalization.

Vestibule A small entrance room.

W

Wainscot The surfacing on the lower part of an interior wall that is finished differently from the remainder of the wall.

Wallboard Large sheets of gypsum or fiberboard that are usually nailed to framing to form interior walls.

Wall tie A small metal strip or steel wire used to bind tiers of masonry in cavity-wall construction, or to bind brick veneer to the wood frame wall in veneer construction.

Water closet A toilet.

Water table A horizontal member extending from the surface of an exterior wall so as to throw off rainwater from the wall. Water level below ground.

Weatherstripping A strip of fabric or metal fastened around the edges of windows and doors to prevent air infiltration.

Web The member between the flanges of a steel beam, or the vertical and diagonal members between the top and bottom chords of a truss or bar joist.

Weep hole Small holes in masonry cavity walls to release water accumulation to the exterior.

Wide flange A structural steel beam with a web and top and bottom flanges.

Winder The radiating or wedge-shaped treads at the turns of some stairs.

Wythe Pertaining to a single-width masonry wall.